SCHISTOSOMA
Biology, Pathology and Control

SCHISTOSOMA
Biology, Pathology and Control

Editor

Barrie GM Jamieson
Department of Zoology and Entomology
School of Biological Sciences
University of Queensland
Brisbane, QLD
Australia

CRC Press
Taylor & Francis Group
Boca Raton London New York

CRC Press is an imprint of the
Taylor & Francis Group, an **informa** business

A SCIENCE PUBLISHERS BOOK

Cover Acknowledgement

· The illustration of The zoonotic life cycle of *Schistosoma japonicum*: Reproduced with permission from Madeleine Flynn, Queensland Institute of Medical Research Berghofer, Brisbane, Australia

· The illustration of A female *Schistosoma japonicum* within the gynaecophoric canal of a male, SEM: Courtesy of Breanna Dunn and Malcolm Jones.

CRC Press
Taylor & Francis Group
6000 Broken Sound Parkway NW, Suite 300
Boca Raton, FL 33487-2742

First issued in paperback 2020

© 2017 by Taylor & Francis Group, LLC
CRC Press is an imprint of Taylor & Francis Group, an Informa business

No claim to original U.S. Government works

ISBN-13: 978-1-4987-4425-6 (hbk)
ISBN-13: 978-0-367-78269-6 (pbk)

Library of Congress Cataloging-in-Publication Data

Names: Jamieson, Barrie G. M. (Barrie Gillean Molyneux), editor.
Title: Schistosoma : biology, pathology, and control / editor, Barrie G.M. Jamieson.
Description: Boca Raton, FL : CRC Press, [2016] | "A Science Publishers book." | Includes bibliographical references and index.
Identifiers: LCCN 2016029024| ISBN 9781498744256 (hardback : alk. paper) | ISBN 9781498744263 (e-book)
Subjects: | MESH: Schistosoma | Schistosomiasis
Classification: LCC RC182.S24 | NLM QX 355 | DDC 616.9/63--dc23
LC record available at https://lccn.loc.gov/2016029024

Visit the Taylor & Francis Web site at
http://www.taylorandfrancis.com

and the CRC Press Web site at
http://www.crcpress.com

Preface

This book is dedicated to the memory of Theodor Maximillian Bilharz 1825-1862.

Commemoration of Theodor Maximillian Bilharz. A stamp issued in 1962 by the United Arab Republic (then Egypt and Libya) on the centenary of his death and the Bilharz Lunar crater, named in 1976 by the International Astronomical Union, photographed by the Lunar Reconaissance Orbiter.

In 1851 Theodor Maximillian Bilharz, working at the Kasr-el-Aini Hospital in Cairo under the supervision of Wilhelm Griesinger, discovered, in blood from the portal venules of Egyptians, long white worms that with the naked eye appeared to be nematodes but which under the microscope proved to be a "Distomum". He showed the trematode to have two sexes, in coitus and he named it *Distomum haematobium*, as published in 1853. In 1856, seven years before his untimely death from typhus, the genus *Bilharzia*, was named in honor of the young worker, though later termed

Schistosoma. These investigations, which are further detailed in Chapter 2, marked the starting point of the vast proliferation of researches on *Schistosoma*, a major human scourge rivalling malaria, over more than 150 subsequent years. They form the basis for the chapters of the present book.

Posthumous recognition of Bilharz has taken many forms. Among the commemorations were the tribute paid by the Egyptian government in issuing, in 1962, a portrait stamp to mark the centenary of his death and an appropriate recognition was the laying of the corner stone of the Theodor Bilharz Research Institute in Cairo in that year. A cosmic tribute was the naming of the Bilharz lunar crater after him.

The present volume attempts to cover the biology, pathology and control of *Schistosoma* species which infect humans. The scope and depth of treatment are considered to exceed those of other, albeit valuable, treatments. An unparalleled team of experts deals with Origins and Evolutionary Radiation of *Schistosoma* (*Tim Littlewood, Bonnie Webster*); Schistosomiasis: Paleopathological Perspectives and Historical Notes (*Robert Bergquist, Helmut Kloos, Aynalem Adugna*); Life Cycles of Schistosomiasis (*Allen Ross, Li Yuesheng*); *Schistosoma* Intermediate Host Snails (*Henry Madsen*); *Schistosoma* Egg (*Malcolm K Jones, Renata Russo Frasca Candido, Tim St Pierre, Robert Woodward, John Kusel, Carlos Graeff Teixeira*); Miracidium of *Schistosoma* (*Barrie Jamieson, Wilfried Haas*); *Schistosoma* Sporocysts (*Timothy Yoshino, Benjamin Gourbal, André Théron*); Cercaria of *Schistosoma* (*Martin Kašný, Wilfried Haas, Barrie Jamieson, Petr Horák*); Schistosomula (*Geoffrey Gobert, Sujeevi Nawaratna*); Tegument and External Features of *Schistosoma* (with Particular Reference to Ultrastructure) (*Geoffrey Gobert, Leigh Schulte, Malcolm Jones*); Alimentary Tract of *Schistosoma* (*Xiao Hong Li, Alan Wilson*); Nervous and Sensory System of *Schistosoma* (*Barrie Jamieson*); Reproductive System of *Schistosoma* (*Malcolm Jones, Barrie Jamieson, Jean-Lou Justine*); Spermatozoa, Spermatogenesis and Fertilization in *Schistosoma* (*Barrie Jamieson, Jean-Lou Justine*); Ova and Oogenesis in *Schistosoma* (*Christoph Grevelding, Steffen Hahnel, Zhigang Lu*); Excretory System of Schistosomes (*John Kusel*); Acute Schistosomiasis (*Li Yuesheng, Allen Ross*); Chronic Schistosomiasis (*David Olveda, Allen Ross*); Neuroschistosomiasis: Pathogenesis and Clinical Manifestations (*Allen Ross, Richard Huntsman*); Subtle Morbidity in Schistosomiasis (*Richard Olds, Jennifer Friedman*); Diagnostic Tests for Schistosomiasis (*Robert Bergquist, Govert van Dam, Jing Xu*); Control of Schistosomiasis (*Alan Fenwick OBE, Fiona Fleming, Lynsey Blair*); Chemotherapy Against Schistosomiasis (*Remigio Olveda, Allen Ross*); Schistosomiasis Vaccine Development: The Missing Link (*Robert Bergquist, Donald McManus*); Geospatial Surveillance and Response Systems for Schistosomiasis (*John Malone, Robert Bergquist, Laura Rinaldi*); Future Directions: The Road to Elimination (*Allen Ross, Remigio Olveda, Li Yuesheng*).

The editor is deeply grateful to these distinguished researchers for agreeing to give their knowledge in countering the scourge of schistosomiasis and thanks the publishers (Science Publishers and CRC Press); the library services of the University of Queensland and authors and publishers who have allowed materials to be reproduced.

Bilharz TM. 1853. Fernere Beobachtungen über das die Pfortader des Menschen bewohnende *Distomum Haematobium* und sein Verhältnis zu gewissen pathologischen Bildungen, von Dr. Th. Bilharz in Cairo (aus brieflichen Mittheilungen an Professor v. Siebold vom 29 März 1852). Zeitschrift für Wissenschaftliche Zoologie 4: 72–76.

Barrie GM Jamieson
School of Biological Sciences
University of Queensland

Contents

CHAPTER

Origins and Evolutionary Radiation of *Schistosoma*

DTJ Littlewood* and Bonnie L Webster

1.1 INTRODUCTION

Schistosomes are a diverse group of digenean blood flukes transmitted by intermediate fresh-water snail hosts. These obligate parasites are members of the family Schistosomatidae and all have two-host life cycles. The cercarial stage penetrates the epithelial surface of the definitive vertebrate host and then develops into dioecious flatworms, with males and females living within the vascular systems of their hosts. Over half the diversity of schistosomes is found in birds (67 species) and fewer in mammals (30 species). Genera infecting mammals include *Schistosoma*, *Bivitellobilharzia*, *Heterobilharzia* and *Schistosomatium*. The genus *Schistosoma*, is comprised of 23 recognized species, many of which are of medical and veterinary importance. Recent molecular studies are revealing some cryptic diversity and the likelihood that more species are yet to be recognized and discovered. At least 7 of the described *Schistosoma* species contribute to the neglected tropical disease (NTD) schistosomiasis, with an estimated 240 million people infected and >750 million at risk across 78 countries or territories; 90% of reported cases are from sub-Saharan Africa (SSA) (Colley *et al.* 2014). Animal schistosomiasis caused by worms of the genus *Schistosoma* may also be a problem causing mortality and morbidity in livestock and wildlife, but the extent of disease in non-humans is poorly reported and largely unknown in terms of economic loss. They also have implications for human health acting as reservoirs of infection or causing zoonotic transmission (Standley *et al.* 2012).

Unraveling the origins and evolutionary radiation of *Schistosoma* has generated much interest, particularly with regard to their intermediate and definitive host associations (see also Bergquist, Chapter 2). Understanding the diversity of species, the relationships amongst those infecting humans and other member of the genus and the relationship with snail host and geography, are all fundamental components in explaining their natural history, but also underpin any efforts to control the diseases they inflict. Several theories have been hypothesized concerning the origins and diversification of *Schistosoma*. This chapter briefly reviews these competing scenarios, the diversity within the genus and gaps in our understanding of the evolutionary origins.

The advent of molecular systematics has provided a means by which robust phylogenies for schistosomes can be pursued; something not readily achievable with morphology alone, even across the family (Carmichael 1984; Morand and Müller-Graf 2000). Although not fully resolved,

Department of Life Sciences, Natural History Museum, London SW7 5BD, UK.
* Corresponding author

the multi-gene phylogenies so far provide an opportunity to infer patterns of evolutionary radiation, host use and biogeography through time. However, in the absence of direct evidence such as fossils, inferences from phylogenies are necessarily limited by the quality of the data used to infer them and the various assumptions invoked in developing a plausible narrative. Concerns are numerous and include: (i) an expectation or at least desire, that gene-trees accurately reflect species trees, (ii) an acknowledgement that our understanding of intermediate and definitive host use is incomplete, (iii) poor understanding of host-specificity and (iv) our uncertainty as to what extent, if any, schistosomes and their snail hosts share a cophylogenetic history. As we have no understanding of extinct species of *Schistosoma*, we must rely entirely on inferences from extant taxa, their current distribution patterns and their documented occurrence in various host and snail vector species.

As agents of a major disease where considerable efforts are being undertaken to control and possibly eliminate infections, species history, phylogeography, radiation, propensity towards hybridisation and host use are all features requiring consideration and knowledge. Accurate diagnostics, an appreciation of species diversity and range are all key for monitoring, treating and controlling schistosomiasis. In spite of our considerable understanding of key schistosome species, our knowledge of *Schistosoma* diversity is demonstrably incomplete, given the discoveries of new species and new lineages recently made from sampling small mammals (Hanelt *et al.* 2009) and snail hosts in Africa (Morgan *et al.* 2003a) and Asia (Devkota *et al.* 2015). These new lineages are being revealed through molecular analyses adding new data to a developing phylogenetic framework of *Schistosoma* species and isolates. Environmental sampling (e.g., water) and greater attention to sampling intermediate (snail) and wildlife (definitive) hosts across suitable habitats (especially where freshwater and snails are present), will undoubtedly reveal greater species diversity. *Schistosoma* species tend to be restricted to a taxonomically narrow group of snail host genera. However, the snail genera themselves are relatively species rich and can be difficult to differentiate (e.g., Kane *et al.* 2008) and it is widely recognized that traditional morphological methods of identification of both hosts and snail vectors can be challenging and unreliable for the non-expert (Adema *et al.* 2012; see also Madsen, Chapter 3). Recent consideration of snail-parasite interactions are revealing the importance of this interplay in future control methods (Giannelli *et al.* 2016).

1.2   A PHYLOGENY FOR *SCHISTOSOMA*

Early estimates of schistosome phylogenies relied on sequencing the complete small subunit and partial large subunit fragments of ribosomal RNA genes, *ssRNA* and *lsRNA* (Lockyer *et al.* 2003; Snyder 2004). The addition of short fragments of mitochondrial cytochrome oxidase subunit I genes (*cox1*) provided additional resolution (Webster *et al.* 2006) and this suite of genes has provided a framework for incorporating new previously unsampled taxa (Hanelt *et al.* 2009), as well as recognizing new species and lineages (Devkota *et al.* 2015). Figure 1.1 shows a consensus phylogeny indicating species relationships as best determined to date. This is a synthesis of phylogenies, rather than one reflecting a particular analysis, matched against known snail and vertebrate host use, per species of *Schistosoma*.

1.3   SPECIES DIVERSITY AND DISTRIBUTION

The origins of dioecy in blood flukes and the colonization of the venous system, with its consequent requirements of expelling eggs to the external environment has been considered in detail elsewhere (e.g., Platt and Brooks 1997; Loker and Brant 2006), but has clearly played a role in the successful radiation of these trematodes. The distribution of *Schistosoma* species strongly correlates to the distribution of their species specific intermediate snail hosts, which each species has adapted to, enabling transmission. Radiation through time and space has required successful completion of a life cycle linking freshwater snails and terrestrial mammalian hosts. Through speciation, likely driven by allopatry and maintained by differences in host use since their origins, the members of *Schistosoma* genus have successfully radiated throughout the tropical and semi-tropical regions of the world, remaining particularly prevalent in developing parts of the world with high disease burden now primarily observed in Asia, Africa and South America. Multiple infections in hosts are known (Rollinson *et al.* 1990; Knowles *et al.* 2015). The fragmentary nature of freshwater habitats, the seasonality of infection, site-specificity of infection, specific host-schistosome compatibilities (both intermediate snail and definitive vertebrate) and phylogenetic distinctness have all helped maintain species and geographical boundaries.

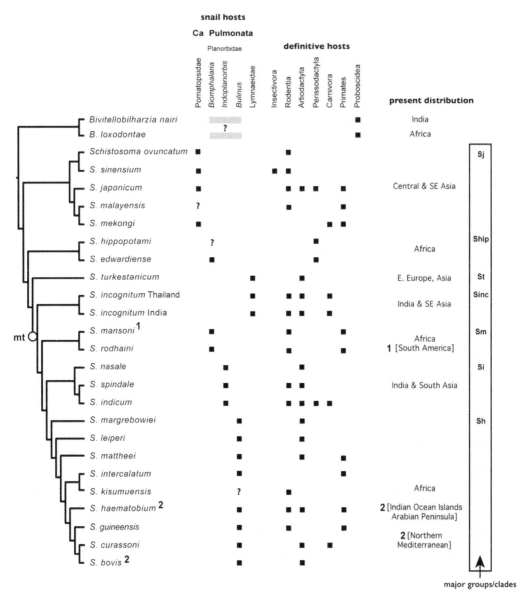

FIG. 1.1 Summary phylogeny of *Schistosoma* species in relation to its sister genus *Bivitellobilharzia*. Snail hosts are indicated by genus or family; Ca – Caenogastropoda, ? – unknown but likely. Definitive host use is indicated by larger vertebrate group. Present distribution of major groups/clades indicated, additionally with notes: **1** – *S. mansoni* also present in South America, as a result of the slave trade, **2** – *S. haematobium* and *S. bovis* each distributed more broadly outside Africa and moving northwards recently into Mediterranean islands. A major change in the mitochondrial (mt) genome order is indicated on the phylogeny by **O**.

In earlier studies, the *Schistosoma* genus was divided into 4 major groups based on geography, snail-host specificity and egg morphology. This somewhat classical grouping of taxa remains useful and viable since current estimates of phylogeny largely reflect the groupings (= clades; see Fig. 1.1).

The *S. japonicum* group (Fig. 1.1 **Sj**) is found throughout eastern and south eastern Asia including China, The Philippines and Malaysia with transmission restricted to freshwater snails within the Pomatiopsidae family. There are 5 described species infecting a diverse range of definitive hosts. Among these species is *S. japonicum* a pathogen of major medical importance and also highly zoonotic being described from perissodactylids, artiodactyls (particulary water buffalo), carnivores, rodents and primates (He *et al.* 2001). *S. mekongi* is more limited in its distribution infecting humans, dogs and possibly pigs (Crosby and Garnham 2009). *S. malayensis*, *S. sinesium* and *S. ovuncatum* are all primarily parasites of rats.

The *S. mansoni* group (Fig. 1.1 **Sm**) consists of 2 species which utilize snails of the genus *Biomphalaria*. *S. mansoni* is a major pathogen of humans that persists throughout sub-Saharan Africa and is also the only *Schistosoma* species that is found in South America, restricted to areas of Brazil, Venezuela, Surinam and parts of the Caribbean (Morgan *et al.* 2005). It can also infect rodents and this has caused problems with control efforts in regions of Latin America where the parasite has adapted its chronobiology to ensure transmission through rodents (Alarcon de Noya *et al.* 2012). *S. rodhaini* is primarily a parasite of rodents and is restricted to specific areas of Africa. These taxa are know to hybridize in the wild (Morgan *et al.* 2003b).

The *S. indicum* group (Fig. 1.1 **Si**) inhabits the western and southern Asian regions, including India, Sri Lanka and Thailand and is transmitted through *Indoplanorbis* snails. There are primarily 3 species including *S. indicum*, *S. spindale* and *S. nasale*, all of which are commonly found in animals such as ungulates, horses, pigs and possibly dogs. However, more recent molecular work has revealed greater diversity and more genetically distinct clades (Devkota *et al.* 2015). *S. indicum* is associated predominantly with artiodactylids infecting a wide range of species, but perissodactyls have also been found naturally infected (Loker 1983). *S. nasale* is restricted to cattle, causing the veterinary condition known as 'snoring disease' (Bont *et al.* 1989), as they infect nasal sinuses, a habit also found in some bird schistosomes. *S. spindale* is mainly found in ungulates but appears to have a relatively low host specificity. The differences in site-specificity may well reflect that members of this group often co-infect the same host, e.g., goats (Agrawal and Rao 2011).

The *S. haematobium* group (Fig. 1.1 **Sh**) is the largest and most diverse 'group' of schistosome species including 9 described species that are the most widespread (Africa, Indian Ocean Islands, Arabian Peninsula and Mediteranian regions) and prevalent, accounting for the majority of human and domestic livestock infections. Central to this group is *S. haematobium*, responsible for urogenital schistosomiasis with the other species of this group all causing intestinal schistosomiasis. *S. intercalatum* and *S. guineensis*, infect humans in isolated foci in central Africa, but how many people are infected and where transmission occurs remains unknown. *S. margrebowiei*, *S. leiperi*, *S. mattheei*, *S. curassoni* and *S. bovis* are parasites of domestic livestock and wild ungulates, mainly in Africa, with major veterinary and economic impact, but not widely researched. The exception is the more recently described *S. kisumuensis* which has only been found in rodents in Kenya.

Natural interactions and introgressive hybridisation between members of the *S. haematobium* group species is now known to be common. This raises questions about their true species boundaries and host associations (Webster *et al.* 2006; Rollinson *et al.* 2009). The interactions between *S. haematobium* and *S. guineensis* have been linked to the disappearance of *S. guineensis* in Loum, Cameroon through introgressive hybridisation (Webster *et al.* 2003; Tchuem Tchuenté *et al.* 2003; Webster *et al.* 2006). Hybridisation between *S. haematobium* and two related ruminant-infecting species (*S. bovis* and *S. curassoni*) in West Africa, raises concerns about increased transmission potential, spread/introduction, host range/switching and creation of reservoir hosts/zoonotic transmission (Huyse *et al.* 2009; Webster *et al.* 2013). These concerns have been highlighted most recently where a *S. haematobium*/*S. bovis* hybrid from West Africa, has been introduced into Corsica, resulting in an outbreak of human urogenital schistosomiasis in locals and tourists (Boissier *et al.* 2015; see also Jamieson and Justine, Chapter 14). There is also evidence that observations of the cattle schistosome *S. mattheei* in humans in southern Africa are in fact hybrids between *S. haematobium* and *S. mattheei*, creating another potential zoonotic threat (Wright and Ross 1980; Kruger 1990).

The *S. hippopotami* group (Fig. 1.1 **Ship**). Molecular data has resolved these 2 distinct clades. One contains the species that have been associated with hippopotami (*S. edwardiense* and *S. hippopotami*) and another as yet undescribed taxon (from cercariae recovered from *Ceratophallus* snails). Little is known about the biology of these three taxa – two species known from hippos with unidentified snail vectors and another lineage described from cercariae recovered from snail hosts but the definitive host(s) remains unknown. Although not shown in the phylogeny, the unnamed taxon is sister to the 2 species found in hippos (Morgan *et al.* 2003a).

Other groups (Fig. 1.1 **St** and **Sinc**). Not a clade as referred to by Lawton *et al.* (2011), these paraphyletic lineages are also not necessarily best described as a "proto- *S. mansoni* clade [sic]". *S. turkestanicum* has been recorded by Kumar and de Burbure (1986) as present in China, India, Mongolia, Pakistan, Iran, Iraq, Russia and Turkey although its biology is relatively poorly known. The clade **Sinc** is comprised of two taxa, each of which is described as *S. incognitum*, but different geographic isolates from Thailand and India are as different from one another phylogenetically as any other well recognised pairs of sister taxa in the genus (Webster and Littlewood 2012).

1.4   ORIGINS AND BIOGEOGRAPHY

Both an Asian origin and an African origin for the *Schistosoma* genus have been postulated previously, each based on snail distribution patterns and fossil records and different interpretations of which schistosome-snail (Asian-pomatiopsid or African-pulmonate) were the first infections; see Webster *et al.* (2006) and Lawton *et al.* (2011) for reviews. The African origin scenario first came to prominence based on snail host distribution patterns across the globe, involving a Gondwanan origin and subsequent tectonic plate shifts dictating patterns and directions of movement (Davis 1993). Although compelling the narrative has its problems as times are largely speculative as to when schistosomes arose. From molecular phylogenies of schistosomes, favored scenarios of radiation have focused on an Asian origin of *Schistosoma* (e.g., Snyder and Loker 2000; Lockyer *et al.* 2003; Morgan *et al.* 2003a), with subsequent radiations following emergence into new geographic areas or the use of new host groups, as reflected by the phylogeny (Fig. 1.1). However, scenarios built from early phylogenetic reconstructions had not incorporated data from what is now considered to be the sister group of *Schistosoma* or considered the implications.

The sister group to *Schistosoma* is resolved as *Bivitellobilharzia* (Brant *et al.* 2006), a genus known only from elephants, but with unknown snail hosts. Only *B. nairi*, from the Indian elephant, has been sequenced but we postulate that, when sequenced, *B. loxodontae* from the African elephant will be resolved as its sister species in line with its morphological similarity (Farley 1971) and taxonomic assignment. That both *Schistosoma* and *Bivitellobilharzia* each split into Asian and African lineages (Fig. 1.1) makes it difficult to infer the plesiomorphic (ancestral) geographic origin of *Schistosoma* based on the phylogenetic placement of schistosomes alone (Webster and Littlewood 2012). The derived nature of a major mitochondrial genome rearrangement within *Schistosoma* (Littlewood *et al.* 2006; Webster and Littlewood 2012), at the point where *S. incognitum* diverges, confirms only that there was continued movement between Asia and Africa throughout the radiation of the group. Lawton *et al.* (2011) explored these scenarios further and with reference to chromosomal (cytogenetic) evolution favored an 'out of Asia' scenario, in line with previous molecular phylogenies.

A review of definitive host use by *Schistosoma* reveals their propensity for host switching (Fig. 1.1). Given the incomplete and complex historical biogeography of key vertebrate host lineages such as artiodactyls (including hippos, goats, pigs, cattle, antelope, etc.), elephants and rodents, it seems unlikely that the origins of *Schistosoma* can be resolved with only a cursory consideration of host data alone. Patterns of snail host use seem clearer but only at a higher taxonomic scale. A deeper knowledge relies on a wider consideration of (extinct and extant) mammalian host diversity and diversification, snail diversification and a more comprehensive assessment of geographic range of schistosome species. Some species are considerably widespread and may have had a pivotal role in diversifications into both Africa and Asia, e.g., *S. turkestanicum.* The hosts of *S. turkestanicum* include many livestock species used for meat and milk, as well as horses and camels (see review in Wang *et al.* 2009), which would likely have been widespread and moving considerable distances with humans.

Loss of freshwater habitats, elimination of disease from both wild, domestic and human populations and the rapidly changing environments between Asia and Africa may well have made it more difficult, if not impossible, to better estimate the origins and radiation of *Schistosoma.* However, efforts to better understand the importance and distribution of non-human hosts, is worth pursuing. A small but important study by Hanelt *et al.* (2010) indicated that amongst a relatively small sample of small mammals (9 rodent and 1 insectivore species) 6% were infected with *Schistosoma.* Even with low infection prevalence (1.5%) the sheer population size of these small mammal communities suggests that their impact in terms of spread and their role as reservoir hosts for human infections, could be enormous whilst also holding the key to better understanding historical and present day biogeography. Clearly, a broader sampling of snail and non-human hosts together with the molecular characterization of their schistosome species is required.

1.5 LITERATURE CITED

Adema CM, Bayne CJ, Bridger JM, Knight M, Loker ES, Yoshino TP, Zhang SM. 2012. Will all scientists working on snails and the diseases they transmit please stand up? PLoS Neglected Tropical Diseases 6 (12): e1835.

Agrawal MC, Rao VG. 2011. Indian schistosomes: a need for further investigations. Journal of Parasitology Research 2011: 250868.

Alarcon de Noya B, Pointier JP, Colmenares C, Théron A, Balzan C, Cesari IM, Gonzalez S, Noya O. 1997. Natural *Schistosoma mansoni* infection in wild rats from Guadeloupe: parasitological and immunological aspects. Acta Tropica 68: 11–21.

Boissier J, Mone H, Mitta G, Bargues MD, Molyneux D, Mas-Coma S. 2015. Schistosomiasis reaches Europe. Lancet Infectious Diseases 15: 757–758.

Brant SV, Loker ES. 2005. Can specialized pathogens colonize distantly related hosts? Schistosome evolution as a case study. PLoS Pathogens 1: 16716–16719.

Brant SV, Morgan JA, Mkoji GM, Snyder SD, Rajapakse RP, Loker ES. 2006. An approach to revealing blood fluke life cycles, taxonomy and diversity: provision of key reference data including DNA sequence from single life cycle stages. Journal of Parasitology 92: 77–88.

Carmichael AC. 1984. *Phylogeny and Historical Biogeography of the Schistosomatidae.* PhD Thesis. Michigan: Michigan State University.

Colley DG, Bustinduy AL, Secor WE, King CH. 2014. Human schistosomiasis. Lancet 383: 2253–2264.

Crosby A, Garnham BB. 2009. Experimental models of pulmonary vascular disease in schistosomiasis. Experimental Models in Pulmonary Vascular Diseases 1: 39–41.

Davis GM. 1993. Evolution of prosobranch snails transmitting Asian *Schistosoma*; coevolution with *Schistosoma*: a review. Progress in Clinical Parasitology 3: 145–204.

Devkota R, Brant SV, Loker ES. 2015. The *Schistosoma indicum* species group in Nepal: presence of a new lineage of schistosome and use of the *Indoplanorbis exustus* species complex of snail hosts. International Journal for Parasitology 45: 857–870.

Farley J. 1971. A review of the family Schistosomatidae: excluding the genus *Schistosoma* from mammals. Journal of Helminthology 45: 289–320.

Giannelli A, Cantacessi C, Colella V, Dantas-Torres F, Otranto D. 2016. Gastropod-borne helminths: a look at the snail-parasite interplay. Trends in Parasitology 32: 255–264.

Hanelt B, Brant SV, Steinauer ML, Maina GM, Kinuthia JM, Agola LE, Mwangi IN, Mungai BN, Mutuku MW, Mkoji GM, Loker ES. 2009. *Schistosoma kisumuensis* n. sp. (Digenea: Schistosomatidae) from murid rodents in the Lake Victoria Basin, Kenya and its phylogenetic position within the *S. haematobium* species group. Parasitology 136: 987–1001.

Hanelt B, Mwangi IN, Kinuthia JM, Maina GM, Agola LE, Mutuku MW, Steinauer ML, Agwanda BR, Kigo L, Mungai BN, Loker ES, Mkoji GM. 2010. Schistosomes of small mammals from the Lake Victoria Basin, Kenya: new species, familiar species and implications for schistosomiasis control. Parasitology 137: 1109–1118.

He YX, Salafsky B, Ramaswamy K. 2001. Host–parasite relationships of *Schistosoma japonicum* in mammalian hosts. Trends in Parasitology 17: 320–324.

Huyse T, Webster BL, Geldof S, Stothard JR, Diaw OT, Polman K, Rollinson D. 2009. Bidirectional introgressive hybridization between a cattle and human schistosome species. PLoS Pathogens 5(9): e1000571.

Kane RA, Stothard JR, Emery AM, Rollinson D. 2008. Molecular characterization of freshwater snails in the genus *Bulinus*: a role for barcodes? Parasites & Vectors 1(1): 15.

Knowles SC, Webster BL, Garba A, Sacko M, Diaw OT, Fenwick A, Rollinson D, Webster JP. 2015. Epidemiological interactions between urogenital and intestinal human schistosomiasis in the context of praziquantel treatment across three West African countries. PLoS Neglected Tropical Diseases 9(10): e0004019.

Kruger FJ. 1990. Frequency and possible consequences of hybridization between *Schistosoma haematobium* and *S. mattheei* in the Eastern Transvaal Lowveld. Journal of Helminthology 64: 333–336.

Kumar V, de Burbure G. 1986. Schistosomes of animals and man in Asia. Helminthological Abstracts (series A) 55: 469–480.

Lawton SP, Hirai H, Ironside JE, Johnston DA, Rollinson D. 2011. Genomes and geography: genomic insights into the evolution and phylogeography of the genus *Schistosoma*. Parasites & Vectors 4: 131.

Lockyer AE, Olson PD, Ostergaard P, Rollinson D, Johnston DA, Attwood SW, Southgate VR, Horak P, Snyder SD, Le TH, Agatsuma T, McManus DP, Carmichael AC, Naem S, Littlewood DT. 2003. The phylogeny of the Schistosomatidae based on three genes with emphasis on the interrelationships of *Schistosoma* Weinland, 1858. Parasitology 126: 203–224.

Loker ES. 1983. A comparative study of the life-histories of mammalian schistosomes. Parasitology 87: 343–369.

Loker ES, Brant SV. 2006. Diversification, dioecy and dimorphism in schistosomes. Trends in Parasitology 22: 521–528.

Morand S, Müller-Graf CD. 2000. Muscles or testes? Comparative evidence for sexual competition among dioecious blood parasites (Schistosomatidae) of vertebrates. Parasitology 120: 45–56.

Morgan JA, Dejong RJ, Adeoye GO, Ansa ED, Barbosa CS, Bremond P, Cesari IM, Charbonnel N, Correa LR, Coulibaly G, D'Andrea PS, De Souza CP, Doenhoff MJ, File S, Idris MA, Incani RN, Jarne P, Karanja DM, Kazibwe F, Kpikpi J, Lwambo NJ, Mabaye A, Magalhaes LA, Makundi A, Mone H, Mouahid G, Muchemi GM, Mungai BN, Sene M, Southgate V, Tchuenté LA, Théron A, Yousif F, Zanotti-Magalhaes EM, Mkoji GM, Loker ES. 2005. Origin and diversification of the human parasite *Schistosoma mansoni*. Molecular Ecology 14: 3889–3902.

Morgan JA, DeJong RJ, Kazibwe F, Mkoji GM, Loker ES. 2003a. A newly-identified lineage of *Schistosoma*. International Journal for Parasitology 33: 977–985.

Morgan JA, DeJong RJ, Lwambo NJ, Mungai BN, Mkoji GM, Loker ES. 2003b. First report of a natural hybrid between *Schistosoma mansoni* and *S. rodhaini*. Journal of Parasitology 89: 416–418.

Platt TR, Brooks DR. 1997. Evolution of the schistosomes (Digenea: Schistosomatoidea): the origin of dioecy and colonization of the venous system. Journal of Parasitology Research 83: 1035–1044.

Rollinson D, Southgate VR, Vercruysse J, Moore PJ. 1990. Observations on natural and experimental interactions between *Schistosoma bovis* and *S. curassoni* from West Africa. Acta Tropica 47: 101–114.

Rollinson D, Webster JP, Webster B, Nyakaana S, Jorgensen A, Stothard JR. 2009. Genetic diversity of schistosomes and snails: implications for control. Parasitology 136: 1801–1811.

Snyder SD. 2004. Phylogeny and paraphyly among tetrapod blood flukes (Digenea: Schistosomatidae and Spirorchiidae). International Journal for Parasitology 34: 1385–1392.

Snyder SD, Loker ES. 2000. Evolutionary relationships among the Schistosomatidae (Platyhelminthes: Digenea) and an Asian origin for *Schistosoma*. Journal of Parasitology Research 86: 283–288.

Standley CJ, Dobson AP, Stothard JR. 2012. Out of Animals and Back Again: Schistosomiasis as a Zoonosis in Africa. *In*: Bagher Rokni M, editor. *Schistosomiasis.* http://www.intechopen.com/books/schistosomiasis/out-of-animals-and-back-again-schistosomiasis-as-a-zoonosis-in-africa: InTech.

Steinauer ML, Hanelt B, Agola LE, Mkoji GM, Loker ES. 2009. Genetic structure of *Schistosoma mansoni* in western Kenya: the effects of geography and host sharing. International Journal for Parasitology 39: 1353–1362.

Tchuem Tchuenté LA, Southgate VR, Jourdane J, Webster BL, Vercruysse J. 2003. *Schistosoma intercalatum*: an endangered species in Cameroon? Trends in Parasitology 19: 389–393.

Wang CR, Chen J, Zhao JP, Chen AH, Zhai YQ, Li L, Zhu XQ. 2009. *Orientobilharzia* species: neglected parasitic zoonotic agents. Acta Tropica 109: 171–175.

Webster BL, Littlewood DTJ. 2012. Mitochondrial gene order change in *Schistosoma* (Platyhelminthes: Digenea: Schistosomatidae). International Journal for Parasitology 42: 313–321.

Webster BL, Southgate VR. 2003. Compatibility of *Schistosoma haematobium*, *S. intercalatum* and their hybrids with *Bulinus truncatus* and *B. forskalii*. Parasitology 127: 231–242.

Webster BL, Southgate VR, Littlewood DT. 2006. A revision of the interrelationships of *Schistosoma* including the recently described *Schistosoma guineensis*. International Journal for Parasitology 36: 947–955.

Webster BL, Tchuem Tchuenté LA, Southgate VR. 2007. A single-strand conformation polymorphism (SSCP) approach for investigating genetic interactions of *Schistosoma haematobium* and *Schistosoma guineensis* in Loum, Cameroon. Parasitological Research 100: 739–745.

Webster BL, Diaw OT, Seye MM, Webster JP, Rollinson D. 2013. Introgressive hybridization of *Schistosoma haematobium* group species in Senegal: species barrier break down between ruminant and human schistosomes. PLoS Neglected Tropical Diseases 7(4): e2110.

Wright CA, Ross GC. 1980. Hybrids between *Schistosoma haematobium* and *S. mattheei* and their identification by isoelectric focusing of enzymes. Transactions of the Royal Society of Tropical Medicine and Hygiene 74: 326–332.

Schistosomiasis: Paleopathological Perspectives and Historical Notes

Robert Bergquist[1],*, *Helmut Kloos*[2] and *Aynalem Adugna*[3]

2.1 INTRODUCTION

The many different species of the digenetic flatworms of the genus *Schistosoma* in existence are generally adapted to a specific, mammalian definitive host and a snail intermediate host producing parasite/snail/mammal life cycles that only occasionally interact across the various species involved. *Schistosoma japonicum*, however, is uncharacteristically zoonotic with a large number of definitive hosts, though this trait is not followed through with respect to its snail host, where it depends completely on one genus, generally even on one species, *Oncomelania hupensis*. The great majority of human infections are caused by *S. mansoni* and *S. haematobium* in sub-Saharan Africa and *S. japonicum* in China, The Philippines and Indonesia. *S. mekongi* and *S. intercalatum* are not only in minority, they are also geographically limited, the former being exclusively found in foci near the border between Cambodia and Laos and the latter existing only in the vicinity of the Congo River and in Lower Guinea on the African continent.

Including the mechanism developed by this parasite to evade the definitive host's immune response, first described by Smithers and Terry (1969), the evolution of the schistosome 'double life', with the adult worm stage radiating into a terrestrial vertebrate, must have required thousands of years. Even if the worm's adaptation to a dioecious lifestyle in a homoeotherm mammal from a probable, previous, hermaphroditic situation in a poikilotherm mollusc seems improbable, an understanding of the intermediate steps is emerging (Platt and Brooks 1997). For example, the reproductive steps in the definitive host required a series of compensatory changes to offset progeny limitations, such as forming dioecious permanent worm pairs with pronounced longevity and strong fecundity. These worm-pairs further needed to colonize areas near conduits to the outside (intestines or bladder) to allow the eggs produced by the female worm to reach water. This is not only necessary for hatching and subsequently reaching the intermediate snail host, but also to ensure that cercariae, when released into the water from infected snails, seek and penetrate the skin of humans. Although an enigmatic chain of events, the result is continuation of the parasite's

[1] UNICEF/UNDP/World Bank/WHO Special Programme for Research and Training in Tropical Diseases (TDR), World Health Organization, Geneva, Switzerland.

[2] Department of Epidemiology and Biostatistics, University of California, San Francisco, USA.

[3] Department of Geography, Sonoma State University, Sonoma, California, USA.

* Corresponding author

life cycle and is as much the key to schistosomiasis epidemiology now as it was in Antiquity and prehistory.

There is little doubt that schistosomiasis was a constant menace for people in the tropics, and humankind must have been associated with this infection long before civilizations first appeared. Water, the common link for humans, snails and schistosomes, satisfies mundane human needs for drinking, washing and fishing and has done so since times immemorial. With the development of fixed human settlements and agriculture, especially when combined with irrigation, the need for water expanded exponentially and so did its role in the transmission of schistosomiasis (Kloos and Thompson 1979). Although we know little about the intensity of the disease in ancient times, it must have been considerably lower before fixed human settlements emerged. Recent estimates suggest that over 250 million people are infected, which translates into a global burden of 3.3 million disability-adjusted life years (DALYs) (Murray *et al.* 2012; Hotez *et al.* 2014), figures even higher than the 200 million infected believed to have existed by the end of last century (Chitsulo *et al.* 2000).

It was not until the early 1900s that the discovery of the parasite's life cycle made it possible to entertain realistic ideas of counteractive measures. This discovery translated into prophylaxis and disease prevention, but it took another 70 years before a safe and reliable drug was introduced. Not long after extensive drug trials covering endemic areas in all continents had proved the efficacy and safety of praziquantel (Davis and Wegner 1979), mass drug administration (MDA) started to make strong inroads with respect to morbidity in previously highly endemic areas. Although reinfection remained commonplace, pathology due to chronic disease could now be managed through scheduled, repeated drug treatment.

The despair of people with schistosomiasis in the endemic areas, as well as that of scientists and clinicians just 40 years ago, is difficult to visualize today, now that praziquantel has enabled improved control of large areas. The focus of this chapter, however, is neither on control nor on chemotherapy but on the history of the disease and its ravages since Antiquity. Even though we cannot (yet) reach back to its very origin, we shall visit surviving written records, learn about parasite eggs found in human remains, envisaging the disease at the dawn of history and make an effort to tease out the loose ends of prehistoric times.

2.2 PARASITE DIVERSIFICATION

We lack hard evidence of the presence of schistosomiasis in prehistoric times. The parasite could first have adapted to various monkey species and baboons and then crossed over to humans, as advocated by Adamson (1976). It is more probable, though, that the parasite's life cycle was simultaneously established in primordial primates and hominids, giving credence to the idea that the origin of human schistosomiasis substantially predates recorded history. Deprès *et al.* (1992) have investigated development from the genetic point of view and propose that the schistosome might have captured the human host from other animal hosts in Africa 1–10 million years ago, coinciding with the time the first hominids invaded the savannas.

Geographical separation rather than choice of host appears to be the most important factor in the diversification of the parasite, as pointed out by Morgan *et al.* (2005). Beer *et al.* (2010) have suggested that the *Schistosoma* genus evolved 70–120 million years ago in Gondwanaland, i.e., the supercontinent consisting of today's Africa, Antarctica, Australia, India and South America. Davis (1992) developed this idea further based on the snail record, maintaining that *S. mansoni* and *S. haematobium* appeared in Africa from a common ancestor more than 120 million years ago and that the ancestral schistosome moved to Asia on the Indian plate 70–148 million years ago to produce the *S. indicum* group, which eventually diversified into the *S. japonicum* of the Far East.

More recent ideas based on genetic analysis of the parasite suggest that the original parasite instead appeared in what is today northern India (see also Littlewood and Webster, Chapter 1). According to this hypothesis, it spread eastward first developing into the *S. japonicum* ancestor (Snyder and Loker 2000; Lawton *et al.* 2011), then transporting itself in the opposite direction by the widespread mammal migration in the late Miocene epoch, eventually reaching Africa between 1 and 4 million years ago where it diversified into *S. mansoni* and *S. haematobium* (Lawton *et al.* 2011). An even later time for the *mansoni/haematobium* split is suggested by Morgan *et al.* (2005) based on the relatively recent, strong variation of East African schistosome specimens, indicating an East African origin for *S. mansoni* of only 0.30–0.43 million years ago. Interestingly, this timeframe not only follows the arrival of the snail host but also coincides with early human evolution.

Since the *S. indicum* group has a closer affinity with the African group of schistosomes than with the other Asian ones (Attwood *et al.* 2002), it would not be too farfetched to think that the Indian Plate ferried the first schistosomes to Asia from Africa as suggested by Davis (1992). This plate moved north after it disconnected from Africa as early as the Mesozoic Era 150–160 million years ago. However, it stayed together with Madagascar for another 90 million years before finally breaking off and coming into close contact with southern Asia some 25 million years ago (van Hinsbergen *et al.* 2012). This scenario produces an unbroken Madagascar-India-Asia connection but to satisfy the 'ferry hypothesis' there must also have been a mainland Africa-Madagascar connection within a suitable timeframe, demanding the origin of *Schistosoma* (or a Schistosomatidae ancestor) to be at least 30 million years older in order to fit this scenario. The snail record may provide this connection. Phylogenetic studies indicate that the three *Bulinus* species endemic in Madagascar, *Bu. obtusispira*, *Bu. bavayii* and *Bu. liratus*, are related to the sub-Saharan *Bu. africanus* group, *Bu. forskalii* group and the *Bu. truncatus/tropicus* group, respectively (Stothard *et al.* 2001). Further studies are required to evaluate the divergence of these species within the timeframe of the 'ferry hypothesis' Fascinating as this is, we will leave this discussion here in the absence of further evidence and instead move to the first hominids.

2.3   EVOLUTIONARY TRENDS IN HOST SELECTION

Australopithecus afarensis, better known as 'Lucy', may have provided the emergence of segmental DNA duplication, a mutation type that increased strongly in the primate genomes, most remarkably in the human one (Sassa 2013), leading to bipedalism, increased height and brain development. This hominid evolved in early Pliocene becoming extinct about three million years ago, with our own genus (*Homo*) thought to have split off from this evolutionary chain at least 1 million years before its demise (Kimbel and Delezene 2009). The australopithecines survived for about 1 million years in East Africa living close to freshwater bodies providing accessible food in the form of shellfish. Such areas were not only suitable for the intermediate snail hosts, whose prehistoric shells remain as evidence, but were most probably also frequented by baboons, which constitute an important *S. mansoni* reservoir to this day (Fenwick 1969; Wright 1970). In this case, there should have been ample possibilities for the transfer of schistosomes between definitive and reservoir host species, which is also indicated by evidence of host-parasite co-evolution in the snail-schistosome system (Webster *et al.* 2004). However, lacking any shred of paleoepidemiological evidence from the 10^5 to 10^6-year level, we may be left in a perpetual conundrum as to the fundamentals of human schistosomiasis ecology, in particular how the first chapter of the story unfolded.

Most experts on human evolution and early migration agree that early humans split into *Homo neanderthalensis* and *H. sapiens* about 1.8 million years ago and that the former dispersed into Eurasia at an early time. However, there is debate about how and when the latter gained a foothold in the Middle East, eventually replacing *H. neanderthalensis* and peopling the rest of the world. The weight of current evidence favors the hypothesis that all present-day human populations descended from a common ancestor tribe in East Africa that first appeared 150,000 to 200,000 years ago and subsequently produced the small bands of humans who reached the Arabian Peninsula across Bab-el Mandeb at the southern end of the Red Sea (Finlayson 2005). The first move may have occurred as early as around 125,000 years ago, but there were probably several retrenchments with the earliest settlements being overrun by *H. neanderthalensis* from the North. It is now believed that the first *H. sapiens* established permanent communities in present-day Yemen, around 75,000 years ago (Armitage *et al.* 2011). If this hypothesis holds, it seems unlikely that human schistosomiasis originated in the Middle East, as argued by some, as this would set the upper time limit for the schistosome/human association at less than 75,000 years to develop. The longer period of 1.8 million years would be more plausible, but would have required a long-term, close relationship between *H. sapiens* and *H. neanderthalensis* in the Middle East, which is also unlikely.

An alternative hypothesis, based on biogeographic, paleohydrographic, archaeological and phylogenic evidence, has *H. sapiens* and schistosomiasis moving together from sub-Saharan Africa via a humid corridor that dispersed into both Egypt and the Middle East. Climatic conditions during the Holocene were associated with a "green Sahara" characterized by humid climatic conditions, which are increasingly considered to have permitted early modern man to move out of Africa via the Sahara and the Nile Valley during different pluvial periods around 120,000 years ago (Osborne *et al.* 2008; Castaneda *et al.* 2009; Drake *et al.* 2011).

2.4   THE SNAIL RECORD

The taxonomy and distribution of extant snail hosts is covered by Madsen (Chapter 4) while this overview deals exclusively with evidence of snails from previous times. Subfossil shells of *Bu. truncatus, Biomphalaria pfeifferi and Bi. alexandrina* have been recovered from at least 16 sites between Israel and the Saharan Desert as far west as Mali and as far south as Chad, Sudan and Kenya and of *Bu. truncatus* at two sites in the Upper Egyptian Nile Valley (Wendorf *et al.* 1976; Beadle 1981, p. 184; Mienis 1992, 2011; Bocxlaer and Verschuren 2011; Abou-El-Naga 2013). This, in combination with the known, heavy reliance by prehistoric populations on fishing, suggests that schistosomiasis transmission could even have occurred in those and other Neolithic Saharan settlements as recently as within the last 5,000-years (see Fig. 2.1 and Table 2.1). Molecular studies of the phylogeny of planorbid snails and fossil evidence strongly suggest that *Bulinus* originated in Africa and that *Biomphalaria* evolved from a neotropical *Bi. glabrata*-like ancestor (Morgan *et al.* 2002). Although earlier hypotheses stated that *Biomphalaria* existed in Africa prior to the breakup of Gondwanaland, recent phylogenetic studies indicate that African species of this genus colonized Africa from South America in the past 1–5 million years after the split of the continents (DeJong *et al.* 2001). The view that *Biomphalaria* colonized Africa from the Americas, attributed to birds and floating materials from those continents taking place long after the breakup of Gondwanaland, is supported by close links to the ancestral Planorbidae in the New World 200–300 million years ago, the significantly lower diversity of mitochondrial sequences of *Bi. pfeifferi* than *Bi. glabrata* and the fossil record of *Biomphalaria* in Africa from the last 1–2 million years (Morgan *et al.* 2002). The oldest *Bi. pfeifferi* haplotypes occurred in southern Africa and they appear to have expanded their range in East Africa less than 100,000 years ago (DeJong *et al.* 2003). Molecular studies also indicate that both *Bi. pfeifferi* and *Bu. truncatus* colonized North Africa and Egypt from sub-Saharan Africa via the humid Saharan corridor (DeJong *et al.* 2003; Kane *et al.* 2008; Zein-Eddine *et al.* 2014). *Bu. truncatus, Bi. pfeifferi* and *Bi. alexandrina* still live in oases, ponds and springs around many of their prehistoric sites (Brown 2005; Van Damme 1984; Van Bocxlaer and Verschurren 2011). Several putative *Biomphalaria* species, including *Bi. gaudi, Bi. rhodesiensis, Bi. stanleyi, Bi. sudanica* and the now extinct *Bi. barthi*, most of them found in East African lakes and the Sahel regions (Fig. 2.1), are closely related to other *Bi. pfeifferi* populations (DeJong *et al.* 2003). The geographic range of these species and the *S. haematobium* intermediate host *Bu. globosus* was probably much more extensive in Central and East Africa in prehistoric times than the distribution of their fossils suggests (Fig. 2.1). This view is supported by the wide distribution of most of these species today and the confinement of their subfossil shells to geological strata from dry phases of the late Pleistocene, which provided good conditions for the preservation of their shells (Van Damme and Gautier 2013). Although we might get sufficient information to state whether human schistosomiasis came from the South or the East, we will never really know the details, so let us now focus closer to our own era where there is real evidence to be found, even if it is scattered and somewhat limited.

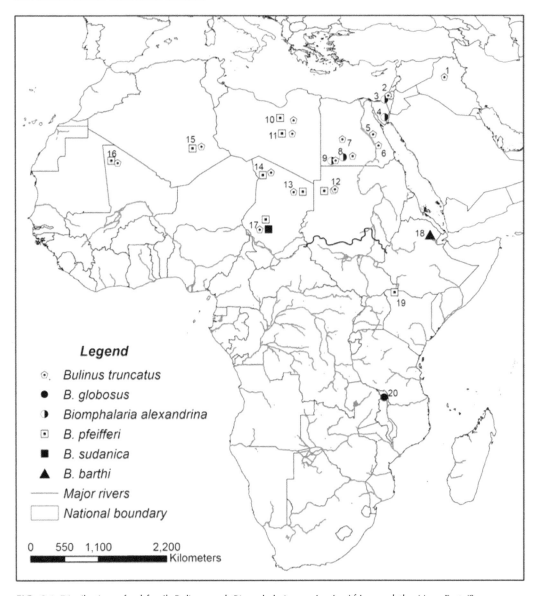

FIG. 2.1 Distribution of subfossil *Bulinus* and *Biomphalaria* species in Africa and the Near East (Sources: see Table 2.1).

TABLE 2.1 Chronology of subfossil *Bulinus* and *Biomphalaria* species in Africa and the Near East.

Site*	Locality	Species	Chronology	Source
1	Irrigated area, Iraq	*Bu. truncatus*	4,000 BC-300 AD	Zakaria 1959
2	Jericho, Israel	*Bu. truncatus*	1,650 BC	Biggs 1960
3	Ta'alat Soreq Stream, Israel	*Bu. truncatus, Bi. alexandrina*	Late Pleistocene-Holocene	Mienis 2011
4	Wadi Gibba, Sinai	*Bi. alexandrina*	Neolithic	Mienis 1992
5	Edfu, Egypt	*Bu. truncatus*	Upper Paleolithic	Gautier 1976
6	Isna, Egypt	*Bu. truncatus*	Upper Paleolithic	Gautier 1976
7	Kharga Oasis, Egypt	*Bu. truncatus*	Holocene	Wendorf *et al.* 1976
8	Bir Tarfawi, Egypt	*Bu. truncatus, Bi. alexandrina*	Paleolithic-Neolithic	Gautier 1993; Wendorf *et al.* 1976
9	Bir Sahara, Egypt	*Bu. truncatus, Bi. pfeifferi*	Paleolithic- Neolithic	Gautier 1976; Wendorf *et al.* 1976
10	Wadi Behar Belama, Libya	*Bu. truncatus, Bi. pfeifferi*	Paleolithic	Osborne *et al.* 2008
11	Wadi Quoduin, Libya	*Bu. truncatus, Bi. pfeifferi*	130,000-117,000 yrs B.P.	Osborne *et al.* 2008
12	Wadi Howar, Sudan	*Bu. truncatus, Bi. pfeifferi*	Late Quaternary	Kröpelin 1993
13	Ounianga, Chad	*Bu. truncatus, Bi. pfeifferi*	Early Holocene	Bocxlaer and Verschurren 2011
14	Tibesti, Chad	*Bu. truncatus, Bi. pfeifferi*	Late Quaternary	Van Damme 1984
15	Hoggar, Algeria	*Bu. truncatus, Bi. pfeifferi*	Late Quaternary	Van Damme 1984
16	Bassin de Taoudenni, Mali	*Bu. truncatus, Bi. pfeifferi*	Holocene	Rosso 1983
17	Lake Chad Basin	*Bu. truncatus, Bi. pfeifferi, Bi. sudanica*	Late Pleistocene-Early Holocene	Lévêque 1967; Malek 1958; Van Damme 1980
18	Assaita, Ethiopia	*Bi. barthi*	Pleistocene	Brown 2005 (p. 271)
19	Lake Turkana, Kenya	*Bi. pfeifferi*	No information	Beadle 1981 (p. 184)
20	Lake Malawi, Malawi	*Bu. globosus*	Early Pleistocene	Van Damme and Gautier 2013

*Codes are the same as in Fig. 2.1.

2.5 SCHISTOSOMIASIS IN ANTIQUITY

Not surprisingly, the oldest available records of schistosomiasis emanate from the Middle East, one of the first areas in the world to sustain larger numbers of people, starting to develop agriculture and live together in more or less permanent communities. An archaeological excavation of the ancient cemetery Tell Zeidan in the valley of the Euphrates River in modern-day northern Syria recently reported the finding of a schistosome egg in the mummified remains of a child's corpse, buried between 6,000 and 6,500 years ago (Anastasiou *et al.* 2014). This represents the oldest case of the disease so far, but additional, useful information may appear once all the material has been sifted through. It is clear that the egg emanates from a schistosome, but the authors leave the exact species open. Although they seem sure that the egg belongs to the *S. haematobium* clade, they have not been more precise, suggesting only that it is either *S. haematobium* or *S. intercalatum* (Anastasiou *et al.* 2014). Apart from the fact that the latter species exists exclusively in the western part of the African continent, at least nowadays, there are also several other reasons that speak in favor of *S. haematobium*, e.g., the fact that a disease causing blood-stained urine is mentioned in ancient medical texts from this part of the world (Adamson 1976) and the recovery across the Middle East of ancient *Bulinus* shells (Kloos and David 2002; Mienis 2011), i.e., the remains of the intermediate host of modern-day *S. haematobium*. Acid-fast staining (Ziehl-Neelsen) might provide supporting, biological evidence since *S. intercalatum* eggs are the only kind of schistosome egg stained red by this technique (Southgate 1976).

There is still no hard evidence of *S. mansoni* infection in humans during Antiquity in the Middle East. However, fossilized (*Bi. alexandrina* Mienis 2011) and *Bi. pfeifferi* and *Bi. arabica* transmitted *S. mansoni* in most countries on the Arabian Peninsula and in the Levant until very recently, with transmission continuing in Oman, Yemen and Saudi Arabia (Abdel Azim and Gismann 1956; IAMAT 2015). Molecular and ecological studies indicate that *Bi. pfeifferi* has been endemic in the Arabian Peninsula for a long time and that this snail is well adapted to the harsh desert environment (Mintsa-Nguema *et al.* 2013), further arguing for its potential role as a host of *S. mansoni* in Antiquity.

2.5.1 Egypt

Available data support the presence of schistosomiasis in the Nile Valley as early as 5,000 years ago. After Ruffer's first report (1910) of the disease in 3,000 years old mummies, other authors demonstrated similar results in mummies of variable time periods (Deelder *et al.* 1990; Miller *et al.* 1992; Contis and David 1996; Kloos and David 2002; Coon 2005; Hibbs *et al.* 2011). However, the disease must have had a much wider distribution in Antiquity, as intermediate snail hosts were then widely spread both south and east of the Mediterranean. Although the disease has been a scourge for millennia and the symptoms must have been well-known, it was not perceived as a specific, medical entity until a French army surgeon under Napoleon reported the symptoms of urinary schistosomiasis in soldiers during the French Egyptian campaign 1798–1801 (Renault 1808). His pathognomonic description only lacked the cause of the problem, whose uncovering would take another half century and include many arguments put forward by famous scientists of the period.

Theodor Maximilian Bilharz's observations in 1851 in Egypt ensure the perpetual connection of this German pathologist's name with schistosomiasis. The disease is also associated with Bilharz's mentor Carl Theodor von Siebold and his supervisor Wilhelm Griesinger at the University of Tübingen, Germany. When Griesinger was named Director of the Egyptian Department of Hygiene in 1850, he accepted on the condition that his assistant Bilharz accompany him. Thus, the then 25-year old Bilharz (Fig. 2.2) started his work at the medical school in Cairo connected with the prestigious Kasr el Aïn Hospital that should lead to his great discovery only because autopsies could be carried out there despite general religious resistance in the country at the time.

During the first year and a half of their stay in Egypt, Bilharz and Grisinger performed over 400 *post mortems*. Already within one year, Bilharz was able to report back to von Siebold about the discovery of three new flatworm species capable of infecting humans as well as noting the presence of other parasites, including hookworm and guinea-worm. He wrote to Siebold:

"As helminths in general and those who attack humans in particular are concerned, I think Egypt is the best country to study them. Nematodes in particular populate the intestines of the indigenous population in unimaginable quantities. It is not unusual to encounter 100 individuals of Strongylus duodenalis, 20–40 Ascaris, 10–20 Trichocephalus and close to 1000 Oxyuris. My attention soon turned to the liver and associated structures; in the blood from v. portae I found a number of long, white worms that with the naked eye appeared to be nematodes. A look in the microscope revealed an excellent Distomum with flat body and a twisted tail. These are a few leaves of a saga as wonderful as the best of Thousand and One Nights – if I succeeded in putting it all together".

Realizing the difference from nematodes and that the 'twisted tails' were in fact two tails of paired worms, he wrote slightly later:

"I have not told you yet about the new phases into which my worm of the portal vein has entered. This has not developed into a fairy tale, as I had assumed, but something more miraculous—a trematode, with separate sexes. The worm which I had described to you in my last letter was the male. When I examined the intestinal veins more carefully...., I soon found samples of the worm which harboured a grey thread in the canal of their tails. You can picture my surprise when I saw that a trematode projected out of the anterior opening of the canal".

FIG. 2.2 Portrait of Theodor Bilharz and his grave in the Cairo German Cemetery.

The letters received from Bilharz were published by von Siebold, together with extensive editorial comments, in the journal for which he was the editor (Bilharz 1853). Bilharz produced drawings of paired flatworms as well as eggs and attributed them to '*Distoma*', a currently obsolete genus that was used for various digenetic flukes, whose members are now placed in different genera.

When Griesinger returned to Germany in 1852, Bilharz became head physician at the Department of Internal Medicine at Kasr el Aïn Hospital. He was appointed Professor of Clinical Medicine three years later, but the death in 1856 of Egypt's ruler Abbas I, who supported the German side in the ongoing scientific competition between Germany and France, resulted in a change from

German to French hegemony in the Egyptian academic sphere. Bilharz was more or less forced to leave his position and move to the 'calmer' chair of Descriptive Anatomy, which tenure he held until his death (from typhus) in 1862 at the age of 37. The genus *Bilharzia* was created in his honor in 1856, but only two years later the name *Schistosoma* (meaning split body) was officially adopted by the International Commission on Zoological Nomenclature based on the morphology of the parasite. The parasite discovered by Bilharz thus became known as *Schistosoma haematobium* and remained the supposedly only species in Africa for half a century.

The idea of there being only one schistosome species in Africa that was present in both the mesenteric veins and those around the bladder, held sway until Louis Westenra Sambon described *S. mansoni* (Sambon 1907) and named it after his teacher Sir Patrick Manson of the London School of Tropical Medicine. Manson had already noted that the eggs passed in the urine from schistosomiasis patients invariably had a terminal spine, whilst those found in the feces had a lateral one. Interestingly however, Bilharz had already 50 years earlier reported on worms that looked different from the ones he first had reported on. Although he just took them for abnormal worms, he was bothered by the many different types of eggs seen in these cases. As both worms and eggs, as well as the pathology produced, were difficult to decipher and as he did not encounter this picture very often, neither Bilharz nor anyone else discussed this further. In retrospect one understands that the bewildering picture Bilharz reported must have been due to double-infection by both species.

There can be no doubt that what Bilharz discovered was *S. haematobium*, not only because of his description of the adult worm and the eggs, but also because this species was predominant in the Nile at that time. However, his discovery of *S. mansoni* must also be accepted even though considerably less detail was given and it is unclear whether or not he realized that it was a separate species. This should not be as surprising as it might at first seem since *S. mansoni* was uncommon in Egypt in the 18th century and did not become more frequent there until the 20th century (Jobin 1999). *Bu. truncatus*, the sole transmitter of *S. haematobium* in Egypt, appears to have existed in the Nile since ancient times, as evidenced by its recovery from palaeolithic sites in Upper Egypt (Fig. 2.1) (Abou-El-Naga 2013). It was widespread in the river until the construction of barrages north of Cairo, the Low Dam at Aswan in 1900 and the Nasser Dam in 1964, the associated extension of perennial irrigation and increasing urban-based water pollution. The changing ecology of the Nile facilitated the spread of *Bi. alexandrina* from the Nile delta to Upper Egypt and caused the near disappearance of *Bu. truncatus* from the Delta (Watts and El Katsha 1995; Kloos and David 2010; Abou-El-Naga 2013).

Importantly, Bilharz not only described the worms he discovered in generous detail, but he also associated them with the 'white exuberances of cancerous aspect' in the mucous membranes of the bladder, intestines, ureters and seminal glands that he saw during his numerous dissections of the human hosts. The idea of infection thus dawned on him much ahead of its time. The parallel to Snow (1855) stopping the cholera epidemic in London 1854 is interesting in this connection, as the bacterial origin of the epidemic was not demonstrated until 1883. In contrast to Snow, Bilharz was fortunate that he could actually see the agent that he linked to the pathology at the *post mortem* investigations. By realizing the possibility of a connection between the parasites invading the body and the pathology observed, Bilharz should correctly be described as the father of the discipline we now call Infectious Diseases. It took a rather long time for this to become mainstream knowledge, but the idea of the nature of infection was quickly accepted after Pasteur's and Koch's work in the 1870s.

With regard to schistosomiasis, investigations centered on how the parasite entered the human body. However, this would remain an open question for more than 60 years after Bilharz's discovery and his early thoughts about parasitic infections. Indeed, the question was not be settled until Robert Leiper (1916) at the London School of Tropical Medicine and Hygiene not only demonstrated the parasite's life cycle, but also confirmed Bilharz's and Sambon's idea of two different species. In his ground-breaking paper, Leiper further showed that the two species had different snail intermediate hosts. Naturally, all these conclusions did not arise out of the blue in a short span of time, but previous contributions by others, notably Cobbold (1872) and Looss (1908), played a significant role in the direction that Leiper's research was to take. The uncovering of the life cycle of a related parasite, the liver fluke *Fasciola hepatica* (Thomas 1883), should also be mentioned as this was an important catalyst for Leiper. The proposed theories by and controversies between, these researchers as well as other eminent scientists of the day that proceeded the unravelling of the schistosome life cycle are vividly described by Jordan (2000) in his account of the scientific work on the various aspects of schistosomiasis that followed.

Another towering figure in the history of diseases in the Egypt of old is Sir Armand Ruffer who worked there in the early 1900s. The discipline of paleopathology was ascribed to this British physician after he had subjected mummies from the Egyptian Twentieth Dynasty to autopsy and found calcified *S. haematobium* eggs in the kidneys of two of them (Ruffer 1910). Not only was this the very first example of retrospective diagnosis of individuals who had lived many thousands years ago, but it also indicated that the disease was already common in ancient Egypt. Paleopathology, supported by the advent of modern diagnostic methods, grew in fits and starts and eventually this new discipline began to provide information on social and ecological aspects of ancient infections (Brown *et al.* 1996). Results presented by Hibbs *et al.* (2011), based on a test for the presence of schistosome antigens in the remains of 237 bodies from three cemeteries from the Ballana period (350–550 AD) in present-day Sudanese Nubia, confirmed the positive results of an earlier study of 23 human remains from the same period in the same area (Miller *et al.* 1992). Although Hibbs *et al.* (2011) found varying prevalence rates according to the area (cemetery) investigated, the overall prevalence was similar in both studies. Together with the references above on the presence of schistosomiasis in the Nile Valley in Antiquity, the conclusion must be that schistosomiasis has been common there for at least 5,000 years, and continued to be so until the advent of praziquantel.

Jordan (2002) brings up the speculation about the meaning of the Egyptian word '*aaa*', depicted in Egyptian hieroglyphs as a urinating phallus and commonly referred to in surviving ancient texts of medical interest, e.g., the Hearst Papyrus (9 times), the Berlin Papyrus (11 times) and the Ebers Papyrus (20 times), as mentioned by Colley (1996). Some authors argue that '*aaa*' should be taken to mean haematuria and specifically haematuria due to schistosomiasis (Pfister 1913; Ebell 1927; Jonckheere 1944) but this contention is now deemed unfounded (Westendorf 1992; Nunn 1996; Nunn and Tapp 2000). Today's preferred interpretation of '*aaa*' is that it refers to the way a disease gains entry to the body, i.e., it was believed that evil spirits impregnated victims with poison during the night (von Deines 1961; Nunn 1996). The correct interpretation of '*aaa*' remains an open question and even if it is obvious that haematuria must have been observed in the endemic areas along the Nile, it must have been so common that it was not regarded as a sign of disease. Ebbell (1927) also mentions '*hrrw*', that could possibly mean worms but it must be regarded as highly unlikely that this word would implicate schistosomes, as they could only have been observed *post mortem* and even then been difficult to recognize by the naked eye.

2.5.2 Sub-Saharan Africa

Although the earliest evidence of African schistosomiasis comes from Egypt, sub-Saharan Africa is, and has probably always been, the most heavily infected part of the world with respect to schistosomiasis. This situation has worsened in modern times due to dam constructions, widespread irrigation schemes and strong population growth. It makes no sense to try to pin down the first publications of schistosomiasis country by country, but it is of interest to note that *S. intercalatum*, one of the five schistosome strains capable of causing human infection, is found only in this region. The first description of this species was described by Fisher (1934) and relates to findings in the former Belgian Congo. We now know that there are two strains of this species in West Africa, one in Zaire and one in Cameroon (Bjorneboe 1978) with most of the relevant research performed in the latter country (Tchuem Tchuenté *et al.* 2003). Two species of freshwater snails act as the intermediate host, *Bu. forskalii* in Cameroon and *Bu. africanus* in Zaire (Tchuem Tchuenté *et al.* 2003).

2.5.3   The Far East

The disease we now recognize as schistosomiasis japonica was first known as Asiatic schistosomiasis and had been found in several countries in Southeast Asia. Mortality was high and morbidity was generally more pronounced than that due to other schistosome species: patients presented with marked liver enlargement, often accompanied by bloody diarrhea, itching skin and occasionally fever. These early, serious symptoms are due to the higher egg output from the *S. japonicum* worms, which often induce intensive tissue reactions since the eggs commonly are released in aggregates (Chen and Mott 1988).

The endemic areas today include China and The Philippines with three minor foci in Sulawesi, Indonesia (*S. japonicum*) and along the Mekong River near the border between Laos and Cambodia (*S. mekongi*). The snail hosts transmitting *S. japonicum* and *S. mekongi* belong to two sub-families of the family Pomatiopsidae, which appears to have originated in Gondwanaland and radiated via Australasia to The Philippines, Japan, China and finally to Southeast Asia (Liu *et al.* 2014). According to estimates based on the molecular clock, *S. japonicum* and *S. mekongi* were already adapted to the pre-human host about 3.8 million years ago, more than 2.8 million years prior to *S. mansoni* and *S. haematobium* (Standley *et al.* 2012). We have no records of *S. mekongi* before 1957 but ample documentation of *S. japonicum* covering more than 2,000 years is available from China and there is interesting information on how the causative organism was discovered in Japan, the only country where the disease has been officially eradicated. The Japanese schistosomiasis control programme has been described in great detail by Kajihara and Hirayama (2011), demonstrating that once the epidemiology had been determined, elimination followed by eradication was a straightforward affair. Chinese control specialists, on the other hand, having a more complicated situation with both large stretches of flooded lowlands and mountainous areas, some of which difficult to reach, are just coming to the elimination stage based on a very well managed control programme that has had an uninterrupted run since the mid 1950s (Utzinger *et al.* 2005).

Very recent phylogenetic reconstruction based on complete mitochondrial genome sequences has shown that *S. japonicum* originated in the lake area of China and radiated (together with its human host) to the mountainous areas about 5,000 years ago, to Japan around 7,000 years ago and to The Philippines and Indonesia about 4 kya (Yin *et al.* 2015).

2.5.3.1   China

Symptoms resembling those of *Katayama disease* in Japan (see below) can be found in old books of traditional Chinese medicine referring to times more than 2,400 years ago (Mao and Shao 1982). In addition, splenomegaly and ascites (possibly due to chronic schistosomiasis) are mentioned in *Ling-Su*, a treatise claimed to be written by Huang-Di, a mythical Emperor said to have lived about 4,700 years ago (Mao 1986). More tangible evidence appeared when two corpses from around 2,100 years ago were discovered in the Chinese provinces of Hunan (1971) and Hubei (1975). The *S. japonicum* eggs identified in both these human remains, presumably from well-to-do persons, indicate that the disease cannot have been uncommon at that time (Wei *et al.* 1980; Chen and Feng 1999; Chen 2014). For the time being, this represents the oldest hard evidence of the disease in ancient China.

The first confirmed Chinese schistosomiasis patient was diagnosed in 1905 by Oliver Tracy Logan, an American physician working in a missionary hospital in Hunan Province (Logan 1905). However, there seems not to have been any immediate follow-up, at least not according to available literature, which is somewhat curious since there is no reason to believe that the high prevalence (60%) reported by Totell (1924) 18 years later should have been unexpected. The time lag before Totell's follow-up of Logan's original finding is, however, surprising when one learns that both Logan and Totell worked at the same hospital. Totell perhaps thought that Logan's patient was atypical and the findings unconvincing until the first large systemic study on the disease appeared (Faust and Meleney 1924). This study, covering the three provinces Jiangsu, Zhejiang and Guangdong, also provided evidence that the intermediate snail host was *Oncomelania hupensis*, the same genus (but different species) of snail that played this role in Japan. The work by Faust and Meleney was published as a monograph and obviously had a large readership, since many papers related to the prevalence, morbidity and control of schistosomiasis japonica were published soon afterwards, both in Chinese and in international medical journals (Chen 2014). Hsu and Wu (1941) also reviewed the literature and summed up the distribution of schistosomiasis based on their wide experience of field surveys. At this time, schistosomiasis was shown to be common in 12 provinces in central and southern China with the number of people infected estimated at around 10 million.

One of the first actions of the new Government that took over in 1949 was to create a programme tasked with responsibility for schistosomiasis control in all endemic areas. The Chinese national schistosomiasis control programme was initiated in 1955 and not long afterwards, Maegraith (1958) pointed out the huge public health impact and economic significance of this disease underlining the importance of recognizing the problem. He noted the strong political will to do something about the situation and witnessed the founding of a working control strategy based on local resources (community participation). The national control programme established offices at all administrative levels, each linked to an expert advisory committee for schistosomiasis. At the provincial level, there was a 'Leading Group of Schistosomiasis (Endemic Diseases) Control', while special institutions were created at the prefecture, county and township levels. The latter offices had the responsibility to carry out local day-to-day control measures according to the higher-level plans of action. Provincial institutes for parasitic diseases still exist in all endemic provinces, with the Shanghai National Institute for Parasitic Diseases, now part of the Chinese Center for Disease Control and Prevention (China CDC) coordinating all field activities (Zhou *et al.* 2012).

We understand from later publications (Mao and Shao 1982) that the schistosomiasis control programme worked well from the start and has continued to do so by successfully reducing the endemic areas as well as the number of new cases year by year. Effective delivery and intersectoral collaboration between all governmental ministries that had connection with the disease, i.e., those dealing with health, agriculture, education, forestry and water conservancy, constituted an important part of this success story. Crucially, as in Japan, there was a strong emphasis on environmental management for the control of the intermediate snail host. In addition, interventions were usually implemented in an integrated fashion and readily adapted to local eco-epidemiological settings. However, there was a long pause in the written literature until the update by Mao and Shao, referred to above, that provided information regarding the human and the zoological distribution of the disease as well as an attempt to estimate its incidence. Although schistosomiasis must have played a big role in the panorama of Chinese ailments since time immemorial, one wonders why the medical literature lacks continuity until the latest 30 years. Possible answers must include the obvious probability that much of the literature was not translated. However, it might also be a reflection of the fact that scientific reporting in scientific journals developed quite late in China and that many of the records were personal notes, which are difficult to retrieve or have in fact not survived.

In contrast to the other schistosome species, *S. japonicum* is a zoonosis infecting a wide spectrum of wild and domestic animals and the reservoir in domestic animals keeps large swathes of land endemic, which might be more to blame for human infection than anything else. In the early 1950s, there were an estimated 14,300 km^2 of *Oncomelania* snail habitats in China and about 1.2 million infected cattle, which translates into more than 100 million people at risk (Chen and

Feng 1999). The historic peak of human prevalence of the disease that Mao Zedong gave the name 'God of Plague' was between 10.5 million (Mao and Shao 1982) and 11.8 million (Chen and Feng 1999). We will never know which estimate is closest, but it is safe to say that both prevalence and intensity of disease in the endemic areas were unusually high in those days. Numerous deaths, broken families and destroyed villages followed in the wake. There were many tales of the toll and names like "big belly village", "no man's village", "widows' village" or "village without villagers" were once common in the endemic areas (Chen 2014). Indeed, schistosomiasis was the major cause of death in the endemic areas, accounting for as much as 90% of the mortality and according to a local survey in Jiangxi Province from the 1950s, a total of 1,362 villages were destroyed, 26,000 families passed away and 310,000 residents left their homes (Chen 2014).

In the early 1990s, the World Bank committed a US$ 71 million loan for schistosomiasis control in China with the Chinese Government providing USD 82 million as counterpart funds (Yuan *et al.* 2000). This World Bank Loan Project (WBLP), at the time the largest amount of money ever allocated for one disease, ran for 10 years (1992–2001) with the stated goal of enhancing morbidity control through large-scale praziquantel administration to humans and bovines. This key strategy was complemented with health education and limited environmental management to control the snail population. The specific objectives were to reduce the *S. japonicum* prevalence in both humans and bovines by at least 40% and to lower the snail density by 50%. An important feature of the WBLP was standardized implementation, monitoring of control measures and careful documentation of the achievements made over time plus economic evaluation (Yuan *et al.* 2000; Utzinger *et al.* 2005). The progress achieved included reduction of the human prevalence from 10% in 1989 to 5% six years later, while the average bovine infection rate decreased from 13% to 9%. In the remaining time of the project, the number of those infected continued down from an estimated 865,000 people in 1995 (MOH 1998) to about 700,000 in 2001 (Utzinger *et al.* 2005). In addition, transmission had been controlled in many counties of the seven endemic provinces. The final evaluation, carried out in 2002 comparing outcome measures with baseline ones, revealed that most of the specific project objectives had been met except that the diminishing trend of snail infections stalled and rates kept fluctuating at a low level rather than disappearing completely (Chen *et al.* 2005).

After the WBLP, the snail-infested areas started to expand again and the number of people infected with *S. japonicum* increased somewhat with higher numbers of acute cases. As these trends were picked up by the surveillance system, schistosomiasis control was upgraded to the highest priority level for communicable diseases in 2004 (Engels 2005). The State Council issued new plans for prevention and control marking the shift from morbidity control to an integrated strategy. For example, farmers were given mechanized farm equipment in exchange for the removal of cattle from the snail-infested areas in a move to reduce contamination of the fields (Wang *et al.* 2009). These activities involved a number of governmental sectors such as those dealing with agriculture, forestry, water conservancy, environment, education and others as needed, with the combined actions resulting in a reduction of the number of infected cases to an estimated 325,800 in 2010 and the total area of snail habitats estimated at 3,700 km^2, i.e., less than a quarter of the 1955 estimate (Lei *et al.* 2010). The most recent data published estimate that there were only 250,000 infected people in 171 counties in 2012 (Li *et al.* 2013), which supports the current aim to achieve elimination of schistosomiasis by the early 2020s (Zhou *et al.* 2012).

2.5.3.2 Japan

Although the historical roots of schistosomiasis in Japan cannot be followed as far back as in China and Egypt, several Japanese warlords of the 16th and 17th century apparently had symptoms typical of schistosomiasis and the Kofu basin was almost certainly endemic in the Edo period (1600–1867) (Kajihara and Hirayama 2011). From the discovery point of view, the Japanese situation parallels

Bilharz's work, but the enigma caused by the infection was unraveled from a direction opposite to that in Egypt. Long before the cause was known, schistosomiasis was clinically well characterized, mainly due to the higher egg output from the *S. japonicum* species. Maki *et al.* (2001) remind us that just a few years before Bilharz arrived in Egypt, Dr. Yoshinao Fujii had reported (in 1847) signs of a 'new' disease in a relatively limited area in Hiroshima Prefecture along the south-eastern coast of Honshu, the main Japanese island:

> *"During the past 2 or 3 years, farmers have had small eruptions on their legs when they entered the water to cultivate the rice field. The eruptions are unendurably painful and itchy. Cows and horses also show the same symptoms. Most of the residents suffer from this disease and they consider that the symptoms are due to the lacquer spread out in this area in ancient times".*

Fujii had not seen these pathological signs before but realized that it was a serious disease that often led to death. He also noted that it was not exclusively infecting humans but that various animals were also affected and frequently died. He came to understand that the disease was geographically focal and existed in various parts of southern Japan, often along rivers or in low-lying areas that were frequently flooded. Several counties in the Chūgoku region south of Honshu, among them Katayama (that gave the disease its first name), were particularly hard hit. In Japan, in contrast to the situation in Egypt, disease was limited to only three main endemic areas: 1) the Hiroshima and Okayama Prefectures in the Chūgoku region; 2) the northern part of Kyushu island, situated south of Honshu and only separated from Chūgoku by a narrow strait of the sea; 3) the Kofu Basin in the centre of Honshu (Tanaka and Tsuji 1997). In addition, there were a few smaller endemic areas northeast of Tokyo, mainly scattered around what is present day Narita Airport (Tanaka and Tsuji 1997).

While it took more than 60 years from Bilharz's discovery until the life cycle and route of infection was determined, it took almost as long until the disease in Japan was understood. Fujii's presumption that the disease he had discovered was due to a poison reflects the universal lack of the concept of infection in his day. The time for a paradigm shift was not ripe until the end of the 19th century when scientists finally started to hypothesize that the cause of *Katayama disease* could be a parasite. Some imagined an agent similar to that causing malaria (Nakahama 1884), while others' thoughts went in the direction of the liver fluke (Oka 1886), the latter idea being quite close to the mark. However, the correct explanation was not reached until 1904 – not long before *S. mansoni* was discovered in the West. Fijiro Katsurada, Professor of Medicine at the Okayama Medical School, had been investigating the *Suishuchoman disease* in patients from Yamanashi in the Kofu Basin when he noted eggs in the feces of five of his twelve patients, the symptoms of whom were reminiscent of those described by Fujii 40 years earlier. Katsurada thought that the eggs looked similar to those seen by Bilharz in Egypt, making him suspect that the disease was caused by a related parasite. Although he had at that time no opportunity to carry out human *post-mortem* examinations, he was able to investigate the disease in household animals:

> *"Since I had previously ascertained that trematodes (e.g., P. westermani) which most often invade humans are also not infrequently found in cats and dogs, I therefore believed that a trematode causing our disease could be found in these animals. I therefore autopsied two dogs and a cat and in the latter I found a part of a male trematode. I later received a second cat from the county of Yamanashi and found in the portal vein as well as in its tributaries numerous trematodes which were in the exact form as that discovered in the first cat".*

Katsurada soon realized that *Suishuchoman disease* was identical to *Katayama disease* and described the morphological features of the worms he found, including the eggs and pointing out differences from those described in Egypt (Katsurada 1904). He later described the eggs in the liver and intestinal walls in deceased humans, making a number of observations on the pathology and noting that the new schistosome species never caused the bladder pathology seen in *S. haematobium* infections. Based on these findings, he named the 'new' trematode *Schistosoma japonicum*.

In Europe, speculation abounded on how the schistosome entered the human host. Looss, for example, did not accept the need for intermediate hosts, believing instead that the miracidium was the infective agent. In contrast, Manson correctly felt that this was a two-step procedure with the miracidium passing into an intermediate snail host to eventually transform into cercariae, infective for humans. However, this was purely guesswork based on the similarity to *Fasciola*, whose life cycle had recently been determined by Thomas (1883). Although Katsurada originally agreed with Looss, he changed his mind when Kan Fujinami and his assistant Hachitaro Nakamura of Kyoto University heard that local people in Katayama attributed their disease to wading in muddy rice paddies. Fujinami and Nakamura carried out a series of experiments exposing calves taken from disease-free areas of Japan to local water making it clear that the infection had to be percutaneous (Fujinami and Nakamura 1909). Finally, Miyairi and Suzuki (1913) found the small, intermediate snail host *O. nosophora*, which brought the discussions to a close.

The complete schistosome life cycle remained unknown and this question was as high on the agenda in Europeas in Egypt and Japan. Manson, at the London School of Tropical Medicine and Hygiene, proposed that investigations should focus on *S. japonicum* as the parasites could be transmitted to experimental animals. He suggested that Leiper (who would later elucidate the *S. mansoni* and *S. haematobium* life cycles) travel to Shanghai in China to find out more about the oriental schistosome. This was also of some concern to the Royal Navy, which had been informed that some of its officers stationed on the River Yangtze in China had been affected by a mysterious disease that might have something to do with this parasite. The Advisory Committee of the Tropical Disease Research Fund and the British Admiralty agreed to send a joint team consisting of Robert Leiper and Edward Atkinson, a Royal Navy surgeon famous for his part in Scott's 1910–1913 Antarctic 'Terra Nova' expedition, to Shanghai to investigate the validity of the disease. The two men arrived in Shanghai in late 1913 and immediately began to set up a laboratory in a native house-boat. However, before they could start their investigation, World War I broke out and Atkinson was ordered back to England with immediate effect. Leiper, however, had time to travel to Japan, where he carried out some animal experiments before he too had to return home.

It fell to two Japanese investigators, Keinosuka Miyairi and his assistant Masatsuga Suzuki, to elucidate the first complete life cycle of a species of schistosome (Miyairi and Suzuki 1913). Their initial report describes how they had mixed cow dung containing eggs of *S. japonicum* with water to stimulate egg hatching and noted activation of the miracidia within the egg shells followed by wriggling of the larval worm and finally hatching as the egg shell broke open. They then went on to repeat these observations using human feces containing eggs, noting the same picture.

Epidemiological surveys began in 1910 in Kofu. The total number of cases of schistosomiasis japonica detected in Japan in 1920 was about 8,000, a figure reduced year-by-year down to 438 cases in 1970 (Tanaka and Tsuji 1997). Reflecting the strong focality of the infection in Japan, the number of infected people was rather modest compared to the situation in Egypt and China. The low number in Japan is peculiar when one considers the parasite's zoonotic character and it is difficult not to speculate that the topography must play an important role. Indeed, very similar conditions to Japan's endemic areas can be seen in China's mountainous endemic areas in the Sichuan and Yunnan provinces.

Various control activities were instituted, e.g., storing 'night soil', i.e., fecal matter intended for use as fertilizer in the fields, for at least two weeks to make sure that no live eggs remained. Cows and buffaloes were replaced with horses, which are more resistant to this infection. Sodium pentachlorophenate (NaPCP) was extensively sprayed in the fields in an effort to control the snail intermediate host and when sodium tartar emetic (Stibnal) became available in 1921, drug treatment of infected people started. However, environmental management was probably the most effective approach as the *Oncomelania* snail host is amphibious and cannot survive long without water. Snail habitats were destroyed by wetlands being drained to reclaim land for agriculture and ditches around the rice fields were cemented to keep snails out of the paddies. Taken together, these activities were successful. Although positive results did not occur immediately, the political will did not waver in demanding long-term, rigorous adherence to the approach chosen. Naturally, some areas were freed of the infection more rapidly than others but it took more than 50 years before final success. However, once the goal of elimination (defined as reducing the disease prevalence to a point where it is no longer a public health threat), had been achieved the control programme swiftly moved to eradication, which refers to the permanent reduction to zero of new cases in a defined geographical area. The last new human infection in Japan was in Kofu in 1977 and although snails were eradicated in most areas by 1983, a limited number of uninfected snails still remained at Kofu and Obitsu at the end of the last century (Tanaka and Tsuji 1997).

2.5.3.3 The Philippines

Schistosomiasis was first reported in The Philippines as a supplementary observation of schistosome eggs in the large intestinal wall and the liver at the autopsy of a man, who had died of other reasons (Wooley 1906). Eggs were now actively looked for with many positive findings, both at autopsy and stool examination. The disease was later shown to be endemic in many of the Philippine islands, with 25,000 to 30,000 people estimated to be infected in 1921, a number that had increased to 300,000 in 1948 (Garcia 1976). The Second World War disrupted epidemiological surveying initially, a situation that was reversed when an outbreak of schistosomiasis occurred among soldiers of the Allied Armed Forces under the command of General McArthur on the island of Leyte in 1945. Twenty years later, 24 endemic provinces had been identified with about one million people infected and five million at risk (Garcia 1976). Since the early 1980s, MDA with praziquantel has become the mainstay of control. According to recent studies the prevalence ranges from 1% to 50% in different endemic zones and the situation is complicated by sustained disease transmission due to water buffaloes and cattle contaminating the fields (Olveda *et al.* 2014).

2.5.3.4 Indonesia

Schistosomiasis had not been reported from Indonesia until Muller and Tesch (1937) described the disease in a 35-year old male from a village near Lake Lindu, situated in an isolated valley with the same name in Central Sulawesi. The patient died shortly afterwards and tissue sections taken at autopsy subsequently revealed eggs identified belonging to *S. japonicum* (Brug and Tesch 1937). Soon after the discovery of this first case, the disease was also found in Napu, another Sulawesian valley but the strain of the intermediate snail species (*O. lindoensis*) was not demonstrated until 35 years later (Davis and Carney 1973). Satrija *et al.* (2015) have recently reviewed the schistosomiasis situation in Sulawesi, including the Bada Valley, a third endemic area that was not discovered until 2008. The first prevalence survey around Lake Lindu indicated that as many as 56% of the population were infected by *S. japonicum*, while a survey in 1973 in Napu Valley showed a prevalence as high as 72% in some places (Garjito *et al.* 2008). In contrast, the prevalence in Bada Valley was only 0.8% at the time when it was first detected (2008). The disease in these three valleys is believed to be sustained by *Rattus* spp. reservoir hosts whose prevalence rates range from 0 to 20% (Garjito *et al.* 2008). Although an integrated control approach reduced the human prevalence to an average of 0.5% in Lindu and 1% in Napu by 2006, more recent data indicate that it is increasing again (Satrija *et al.* 2015).

2.5.3.5 Continental Southeast Asia

The first human case of *S. mekongi* infection was reported in Laos (Vic-Dupont *et al.* 1957) and later also in Cambodia (Audebaud *et al.* 1968). This 'new' schistosome strain was, however, not designated as a separate species until 1978 (Ohmae *et al.* 2004). *S. mekongi* causes severe intestinal and hepatosplenic disease with high mortality rates (Muth *et al.* 2010). Infections in the provinces of northern Cambodia and southern Laos were common in the early 1970s and 1990s but the disease has since been partly controlled (Muth *et al.* 2010).

Neotricula aperta is the only known intermediate host of *S. mekongi* (Attwood *et al.* 2001), which is exclusively found in the Mekong River Basin at the border between these two countries, where an estimated 1.5 million people live (Campbell 2004). Two separate clades of *N. aperta* have been found: a spring-dwelling form in northern Laos and another that prefers the ecology produced by the Mekong River in southern Laos and Cambodia; this divergence is dated at 9.3 million years ago with further radiation into sub-clades across Cambodia and Laos around 5 million years ago (Attwood *et al.* 2008). Historical events, rather than ecology, might best explain the absence of *S. mekongi* from most of Laos (Attwood *et al.* 2008).

2.5.4   Latin America

The relatively recent colonization of Brazil, Surinam, Venezuela and the Caribbean islands by *S. mansoni* from West Africa is supported by strong genetic similarities between schistosomes from both continents (Desprès *et al.* 1993; DeJong *et al.* 2001; Morgan *et al.* 2005). The infection is supposed to have been introduced by the African slave trade which in Brazilian ports started in the 16th century (Guimarães *et al.* 2012). However, the parasite could only get a foothold where the imported parasites could be supported by indigenous intermediate snail hosts, e.g., *Bi. glabrata*. The Northeast of Brazil became endemic first, since the ports of Salvador and Recife received the highest number of African slaves, but as there were also shipping routes to other parts of the New World from Africa, it is unclear whether the infection reached all of today's endemic areas directly or if there were also transfers within the New World. In Brazil, schistosomiasis soon spread to the South with large-scale population movements in the early 18th century, mainly due to the discovery of gold and diamonds in the Minas Gerais State in southeastern Brazil (Guimarães *et al.* 2012).

The first case description of schistosomiasis in Brazil was unequivocally demonstrated in 1904 in a patient in the State of Bahia in the Northeast, but this finding did not appear in the literature until four years later (Pirajá da Silva 1908). In this paper, the author reported that using stool microscopy he found eggs with lateral spines; this roused his curiosity but he was at a loss to explain what they were. Only after reading about the discovery of *S. mansoni* by Sambon (1907) did he realize what he had found, making it possible to publish the paper with full explanations. Pirajá da Silva performed three autopsies on suspected cases, finding schistosomes in all of them. He described their characteristics concluding that they were not *S. haematobium*. All this is described in great detail by Katz in a recent paper (2008).

2.6   CHANGE OVER TIME

The number of people infected with schistosomiasis increased naturally with the numbers of humans as we move from Antiquity and pre-Antiquity reaching a plateau by around 1980, after which morbidity declined sharply thanks to large-scale use of praziquantel. Although overall prevalence has not yet declined correspondingly, some countries have made impressive progress, e.g., China, Brazil and Egypt. In addition, the disease has disappeared from Japan and seemingly also from Puerto Rico and most Caribbean islands. In the Middle East there are only a few foci left, which is also the case for Africa north of the Sahara. The set of three pictures shown here (Fig. 2.3) summarizes this situation although, admittedly, the situation in Antiquity cannot be better than speculation, mainly based on Figure 2.1.

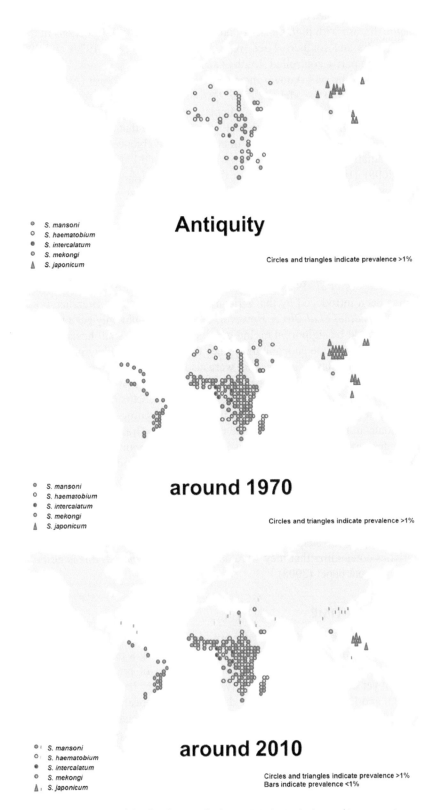

FIG. 2.3 Estimates of the distribution of schistomiasis in Antiquity, and in recent times.

2.7   CLOSING COMMENTS

All we know about schistosomiasis in ancient times is based on the finding of a limited number of schistosome eggs in mummies from Egypt, Middle East and China in combination with fossilized shells of snails that could have acted as intermediate hosts. It might still be safe to say that the general geographical distribution, with exception of the New World and Madagascar (that was still without human settlements at the time of Antiquity), has changed only little over the succeeding 5,000 years. Exceptional progress in countering this parasite has been achieved in Japan and China in modern times and the total number of schistosomes (as an expression of human prevalence and intensity of infection) probably passed its peak in the 1980s when large-scale MDA with the then new broad-spectrum anthelmintic drug praziquantel got under way. Surprisingly, however, the total number of those infected in the world is still increasing, a fact that must be ascribed to the ongoing population explosion and the creation of new transmission sites in areas characterized by land colonization and expanding irrigation in the tropical areas.

TABLE 2.2 *Schistosoma*: overview of ancient and historical findings.

Year	Indications/species	Country/place	Reference
Prehistoric		Unspecified conjecture	
6,000 BC	*S. haematobium*?	Mesopotamia (Euphrates)	Anastasiou *et al.* 2014
4,700 BC	Symptoms	China (written records)*	Mao 1986
2,100 BC	*S. japonicum*	China (Hubei Province)	Chen 2014
350–550	*S. mansoni*	Egypt (Ballana cemetery)	Miller *et al.* 1992; Hibbs *et al.* 2011
15–1800s	Symptoms	Japan (written records)**	Kajihara and Hirayama 2011
1500s	*S. mansoni*	Latin America	Guimarães *et al.* 2012
14–1500s	*S. mansoni*	France (medieval latrine)	Bouchet *et al.* 2002
1808	Symptoms	Egypt	Renault 1808***
1851	*S. haematobium*	Egypt	Bilharz 1853
1847	Symptoms	Japan (Fujii)	Tanaka and Tsuji 1997
1904	*S. japonicum*	Japan	Katsurada 1904
1905	*S. japonicum*	China	Logan 1905
1906	*S. japonicum*	The Philippines	Wooley 1906
1907	*S. mansoni*	England	Sambon 1907
1908	*S. mansoni*	Brazil	Pirajá da Silva 1908
1913	*S. japonicum* life cycle	Japan	Miyairi and Suzuki 1913
1916	*S. mansoni* life cycle	England	Leiper 1916
1916	*S. haematobium* life cycle	England	Leiper 1916
1934		Zaire (Belgian Congo)	Fisher 1934
1937	*S. japonicum*	Sulawesi, Indonesia	Müller and Tesch 1937; Brug and Tesch 1937
1957	*S. mekongi*	Laos	Vic-Dupont *et al.* 1957
1971	Snail host in Sulawesi	Sulawesi, Indonesia	Davis and Carney 1973

*Physicians at the Court of Emperor Hung-Di
**Contemporary physicians
***French Army physician

Schistosomiasis can be ascertained to have existed for at least 6,000 years, but the discovery of the causative infectious agent is recent; dating back less than 165 years (Table 2.2). The early 1980s marked a turning point as it was then that the large-scale effect of praziquantel turned out to be so profound and it now makes sense to think of the elimination of schistosomiasis as a public health threat.

Strangely, in Europe and the Middle East, there is a gap of about 1,300 years between the youngest Egyptian mummies shown to have been infected and the earliest description of schistosomiasis as a distinct disease by Renault (1808). As there was a constant exchange between Europe and the Near and Middle East, it is surprising that we have virtually no records covering this long period. The single exception is a *S. mansoni* egg found in the remains from a French latrine dating from AD 1450–1550, supposedly emanating from a European traveler or an early African immigrant (Bouchet *et al.* 2002). However, the existence of written, medieval records in China and Japan, although few, give hope that similar, yet unrecognized medical records from Greco-Romans times, might exist.

2.8   LITERATURE CITED

Abdel Azim M, Gismann A. 1956. Bilharziasis survey in southwest Asia-covering Iraq, Israel, Jordan, Libanon, Sa'udi Arabia and Syria: 1950–1951. Bulletin of the World Health Organization 14: 403–456.

Abou-El-Naga IF. 2013. *Biomphalaria alexandrina* in Egypt: past, present and future. Journal of Bioscience 38: 665–672.

Adamson PB. 1976. Schistosomiasis in antiquity. Medical History 20: 176–188.

Anastasiou E, Lorentz KO, Stein GJ, Mitchell PD. 2014. Prehistoric schistosomiasis parasite found in the Middle East. The Lancet Infectious Diseases 14: 553–554.

Attwood SW, Upatham ES, Southgate VR. 2001. The detection of *Schistosoma mekongi* infections in a natural population of *Neotricula aperta* at Khong Island, Laos and the control of Mekong schistosomiasis. Journal of Molluscan Studies 67(3): 400–405.

Attwood SW, Upatham ES, Meng XH, Qiu DC, Southgate VR. 2002. The phylogeography of Asian *Schistosoma* (Trematoda: Schistosomatidae). Parasitology 125(Pt 2): 99–112.

Attwood SW, Fatih FA, Campbell I, Upatham ES. 2008. The distribution of Mekong schistosomiasis, past and future: preliminary indications from an analysis of genetic variation in the intermediate host. Parasitology International 57(3): 256–70. doi: 10.1016/j.parint.2008.04.003.

Armitage SJ, Jasim SA, Marks AE, Parker AG, Usik VI, Uerpmann H-P. 2011. The Southern Route "Out of Africa": evidence for an early expansion of modern humans into Arabia. Science 331(6016): 453–456.

Audebaud G, Tournier-Lasserve C, Brumpt V, Jolly M, Mazaud R, Imbert X, Bazillio R. 1968. 1st case of human schistosomiasis observed in Cambodia (Kratie area). Bulletin de la Société de Pathologie Exotique et de Ses Filiales 61: 778–784.

Beadle LC. 1981. *The Inland Waters of Tropical Africa: An Introduction.* New York: Longman. 475 p.

Beer SA, Voronin MV, Zazornova OP, Khrisanfova GG, Semenova SK. 2010. Phylogenetic relationships among schistosomatidae. Med Parazitol (Mosk) Meditsinskaia Parazitologiia i Parazitarnye Bolezni (Moskva) 2: 53–59.

Biggs HEJ. 1960. Mollusca from prehistoric Jericho. Journal of Conchology 24: 379–387.

Bilharz TM. 1853. Fernere Beobachtungen über das die Pfortader des Menschen bewohnende *Distomum Haematobium* und sein Verhältnis zu gewissen pathologischen Bildungen, von Dr. Th. Bilharz in Cairo (aus brieflichen Mittheilungen an Professor v. Siebold vom 29 März 1852). Zeitschrift für Wissenschaftliche Zoologie 4: 72–76. (English translation in: Benjamin Harrison Kean (1912-1993), Kenneth E. Mott and Adair J. Russell: Tropical Medicine and Parasitology: Classic Investigations. 1 volume in 2. Ithaca, London: Cornell University Press, 1978. Also in Review of Infectious Diseases, Chicago 1984, 4: 727–732.

Bjorneboe A. 1978. A comparison of the characteristics of two strains of *Schistosoma intercalatum* Fisher, 1934 in mice. Journal of Helminthology 53: 195–203.

Bouchet F, Harter S, Paicheler JC, Aráujo A, Ferreira LF. 2002. First recovery of *Schistosoma mansoni* eggs from a latrine in Europe (15-16th centuries). Journal of Parasitology 88(2): 404–5.

Brown DS. 1994. *Freshwater Snails of Africa and their Medical Importance.* 2nd ed., London: Taylor Francis. 608 p.

Brug SL, Tesch JW. 1937. Parasitaire wormen aan het lindoe meer (o.a. Paloe, Celebes). Geneeskundig Tijdschrift voor Nederlands-Indië 77: 2151–2157 (In Dutch).

Campbell C. 2009. *The Mekong: Biophysical Environment of an International River Basin*. Academic Press, New York.

Castañeda IS, Mulitza S, Schefub E, Lopes dos Santos RA. 2009. Wet phases in the Sahara/Sahel region and human migration patterns in North Africa. Proceedings of the Academy of Sciences 106. doi: 10.1073/pnas.090577106.

Chen MG. 2014. Assessment of morbidity due to *Schistosoma japonicum* infection in China. Infectious Diseases of Poverty 3:6. doi: 10.1186%2F2049-9957-3-6.

Chen MG, Mott KE. 1988. Progress in assessment of morbidity due to *Schistosoma japonicum* infection: a review of recent literature. Tropical Diseases Bulletin 85(6): 1–45.

Chen MG, Feng Z. 1999. Schistosomiasis control in China. Parasitology International 48(1): 11–19.

Chen XY, Wang LY, Cai JM, Zhou XN, Zheng J, Guo JG, Wu XH, Engels D, Chen MG. 2005. Schistosomiasis control in China: the impact of a 10-year World Bank Loan Project (1992–2001). Bulletin of the World Health Organization 83: 43–48.

Chitsulo L, Engels D, Montresor A, Savioli L. 2000. The global status of schistosomiasis and its control. Acta Tropica 77(1): 41–51.

Cobbold TS. 1872. On the development of *Bilharzia haematobia*. British Medical Journal ii: 89–92.

Colley DG. 1996. Ancient Egypt and today: enough scourges to go around. Emerging Infectious Diseases 2(4): 362.

Contis G, David AR. 1996. The epidemiology of Bilharzia in ancient Egypt: 5000 years of schistosomiasis. Parasitology Today 12: 253–255.

Coon DR. 2005. Schistosomiasis: overview of the history, biology, clinicopathology and laboratory diagnosis. Clinical Microbiology Newsletter 27: 163–169.

Davis GM. 1992. Evolution of prosobranch snails transmitting Asian Schistosoma; coevolution with *Schistosoma*: a review. Progress in Clinical Parasitology 3: 145–204.

Davis GM, Carney WP. 1973. Description of *Oncomelania hupensis lindoensis*, first intermediate host of *Schistosoma japonicum* in Sulawesi (Celebes). Proceedings of the Academy of Natural Sciences of Philadelphia 125: 1–34.

Davis A, Wegner DH. 1979. Multicentre trials of praziquantel in human schistosomiasis: design and techniques. Bulletin of the World Health Organization 57(5): 767–771.

Deelder AM, Miller RL, de Jonge N, Krijger FW. 1990. Detection of schistosome antigen in mummies. The Lancet 335(8691): 724–725.

DeJong RJ, Morgan JAT, Paraense WL, Pointier J-P, Amarista M, Ayeh-Kumi PFK, Babiker A, Barbosa CS, Prémond P, Canese AP, et al. 2001. Evolutionary relationships and biogeography of *Biomphalaria* (Gastropoda: Planorbidae) with implications regarding its role as host of the human bloodfluke, *Schistosoma mansoni*. Molecular Biology and Evolution 18(12): 2225–2239.

DeJong RJ, Morgan JA, Wilson WD, Al-Jaser MH, Appleton CC, Coulibaly G, D'Andrea PS, Doenhoff DJ, Haas W, Idris MA, et al. 2003. Phylogeography of *Biomphalaria glabrata* and *B. pfeifferi*, important intermediate hosts of *Schistosoma mansoni* in the New and Old World tropics. Molecular Ecology 12(11): 3041–3056.

Desprès L, Iambert-Establet D, Combes C, Bonhomme F. 1992. Molecular evidence linking hominid evolution to recent radiation of schistosomes (Platyhelminthes: Trematoda). Molecular Phylogenetics and Evolution 1: 295–304.

Desprès L, Iambert-Establet D, Monnerot M. 1993. Molecular characterization of mitochondrial DNA provides evidence for the recent introduction of *Schistosoma mansoni* into America. Molecular and Biochemical Parasitology 60: 221–229.

Drake NA, Roger MB, Armitage SJ, Bristow CS, White KH. 2011. Ancient watercourses and biogeography of the Sahara explain the peopling of the desert. Proceedings of the National Academy of Sciences 108: 458–462.

Ebbell B. 1927. Die Ägyptischen Krankheitsnamen. Zeitschrift für Ägyptische Sprache und Altertumskunde 62: 13–20.

Faust EC, Meleney HE. 1924. Studies on schistosomiasis japonica. American Journal of Hygiene Monographic Series No. 3: 147–173.

Finlayson C. 2005. Biogeography and evolution of the genus *Homo*. Trends in Ecology and Evolution 20(8): 457–463.

Fisher AC. 1934. Study of the schistosomiasis of the Stanleyville district of the Belgian Congo. Transactions of the Royal Society of Tropical Medicine and Hygiene 28: 277–306.

Fujinami K, Nakamura H. 1909. The route of infection, the development of the parasite of Katayama disease and its infection in animals. Kyoto Medical Journal 6: 224–252 (in Japanese).

Garcia EG. 1976. Clinical studies on *Schistosomiasis japonica* in the Philippines. A Review Southeast Asian Journal of Tropical Medicine and Public Health 7: 247–256.

Garjito TA, Sudomo M, Abdullah, Dahlan M, Anis Nurwidayati A. 2008. Schistosomiasis in Indonesia: past and present. Parasitology International 57(3): 277–80.

Guimarães RJPS, Freitas CC, Dutra LV, Felgueiras CA, Drummond SC, Tibiriçá SHC, Oliveira G, Carvalho OS. 2012. Use of indicator kriging to investigate schistosomiasis in Minas Gerais State, Brazil. Journal of Tropical Medicine 837428. doi: 0.1155/2012/837428.

Hibbs AC, Secor WE, Van Gerven D, Armelagos G. 2011. Irrigation and infection: the immunoepidemiology of schistosomiasis in ancient Nubia. American Journal of Physical Anthropology 145(2): 290–298.

Hotez P, Alvarado M, Basáñez M, Bolliger I, Bourne R, Boussinesq M, Brooker S, Brown A, Buckle G, Budke C, *et al.* 2014. The global burden of disease study 2010: interpretation and implications for the neglected tropical diseases. PLoS Neglected Tropical Diseases 8: e2865.

Hsu HPH, Wu K. 1941. Schistosomiasis in China II: distribution. National Medical Journal China 27(9): 553–564.

IAMAT (International Association for Medical Assistance for Travellers). 2015. World schistosomiasis risk chart: geographic distribution of schistosomiasis and principal snail vectors. 2015 Edition. Available at: www.iamat.org.

Jobin W. 1999. *Dams and Disease: Ecological Design and Health Impacts of Large Dams, Canals and Irrigation Systems.* E & FN Spon, London. 600 p.

Jonckheere F. 1944. *Une maladie égyptienne: l'Hématurie parasitaire.* Bruxelles: Édition de la Fondation Egyptologique Reine Elisabeth, Brussels. 62p.

Jordan P. 2000. From Katayama to the Dakhla Oasis: the beginning of epidemiology and control of bilharzia. Acta Tropica 77(1): 9–40.

Kajihara N, Hirayama K. 2011. The war against a regional disease in Japan: a history of the eradication of Schistosomiasis japonica. Tropical Medical Health 39 (1 Suppl 1): 3–44. doi: 10.2149/tmh. 39-1-suppl_1-3.

Kane RA, Stothard JR, Emery AM, Rollinson D. 2008. Molecular characterization of freshwater snails in the genus *Bulinus*: a role for barcodes? Parasitology Vectors 1: 15.

Katsurada F. 1904. Determination of the cause of a newparasite disease seen in Yamanashi, Hiroshima, Saga and other prefectures. Tokyo Iji Shinshi 1371: 13–32 (in Japanese).

Katz N. 2008. The discovery of schistosomiasis mansoni in Brazil. Acta Tropica 108: 69–71.

Kimbel WH, Delezene LK. 2009. "Lucy" redux: a review of research on *Australopithecus afarensis.* Yearbook of Physical Anthropology 52: 2–48.

Kloos H, David R. 2002. The paleoepidemiology of schistosomiasis in ancient Egypt. Human Ecology Review 9: 14–25.

Kloos H, Thompson K. 1979. Schistosomiasis in Africa: an ecological perspective. The Journal of Tropical Geography 48: 31–46.

Langand J, Galinier R, Idris MA, Shaban MA, Mouahid G. 2013. Genetic diversity, fixation and differentiation of the freshwater snail *Biomphalaria pfeifferi* (Gastropoda, Planorbidae) in arid lands. Genetica. 2013 141(4-6): 171–84. doi: 10.1007/s10709-013-9715-8.

Lawton SP, Hirai H, Ironside JE, Johnston DA, Rollinson D. 2011. Genomes and geography: genomic insights into the evolution and phylogeography of the genus *Schistosoma.* Parasites and Vectors 4: 131.

Lei ZL, Zheng H, Zhang LJ, Zhu R, Guo JG, Wang LY, Chen Z, Zhou XN. 2010. Schistosomiasis endemic status in People's Republic of China in 2010. Chinese Journal of Schistosomiasis Control 2011; 23(6): 599–604. (in Chinese).

Leiper RT. 1916. On the relation between the terminal spined and lateral-spined eggs of bilharzia. British Medical Journal i: 411.

Lévêge C. 1967. Mollusques aquatiques de la zone du lac Tchad. Bulletin de 'Institut fundamental d'Afrique Noire 29: 1494–1533.

Li S, Zheng H, Gao Q, Zhang L-J, Zhu R, Xu J, Guo J, Xiao N, Zhou XN. 2013. Endemic status of schistosomiasis in People's Republic of China in 2012. Chinese Journal of Schistosomiasis Control 25(6): 557–563.

Liu L, Huo G-N, He H-B, Zhou B, Attwood SW. 2014. A phylogeny for pomatiopsidae (Gastropoda: Rissooidae): a resource for taxonomic, parasitological and biodiversity studies. BMC Evolutionary Biology 14: 29.

Logan OT. 1905. A case of dysentery in hunan province caused by the trematode *Schistosoma japonicum* 1905. China Medical Missionary Journal 19(3): 243–245.

Looss A. 1908. What is '*Schistosomum mansoni*' Sambon 1907? Annals of Tropical Medicine and Parasitology 2153–2191.

Maegraith B. 1958. Schistosomiasis in China. Lancet 271: 208–214.

Maki J, Mikami M, Maruyama S, Sakagami H, Kuwada M. 2001. Discovery of the adult *Schistosoma japonicum*, a causative agent of Schistosomiasis in the Katayama area of Hiroshima prefecture. Yakushigaku Zasshi 36(1): 32–5. (In Japanese).

Mao SP, Shao BR. 1982. Schistosomiasis control in the People's Republic of China. American Journal of Tropical Medicine and Hygiene 31(1): 92–99.

Mao SP. 1986. Recent progress in the control of schistosomiasis in China. Chinese Medical Journal 99(6): 439–443.

Mienis HK. 2011. A preliminary reconstruction of the mollusc fauna of the Lower Nahal Soreq Valley in Israel since the Late Pleistocene-Early Holocene. Archaeo+Malacology Group Newsletter No. 20: 5–8.

Mienis HK. 1992. *Biomphalaria alexandrina* from a Neolithic site in Wadi Gibba, Sinai. Soosiana 20: 25–27.

Miller RL, Armelagos GJ, Ikram S, De Jonge N, Krieger FW, Deeleder AM. 1992. Paleoepidemiology of *Schistosoma* infection in mummies. British Medical Journal 304(6826): 555–556.

Miyairi K, Suzuki M. 1913. On the development of *Schistosoma japonicum*. Tokyo Iji Shinshi 1836: 1961–1965. (Abstract in Tropical Diseases Bulletin 1914; 3, 289-290 by Kumagawa).

Morgan JAT, DeJong RJ, Jung Y, Khallaayoune K, Kock S, Mkoji GM, Loker ES. 2002. A phylogeny of planorbid snals, with implications for the evolution of *Schistosoma* parasites. Molecular Phylogenetics and Evolution 25: 477–488.

Morgan JA, DeJong RJ, Kazibwe F, Mkoji GM, Loker ES. 2003. A newly-identified lineage of Schistosoma. International Journal of Parasitology 33(9): 977–985. doi: 10.1016/S0020-7519(03)00132-2.

Morgan JA, Dejong RJ, Adeoye GO, Ansa ED, Barbosa CS, Brémond P, Cesari IM, Charbonnel N, Corrêa LR, Coulibaly G, *et al.* 2005. Origin and diversification of the human parasite *Schistosoma mansoni*. Molecular Ecology 14(12): 3889–3902.

Müller H, Tesch JW. 1937. Autochtoneinfectie met *Schistosoma japonicum* in Celebes. Geneeskundig Tijdschrift voor Nederlands-Indië 77: 2143–2150 (In Dutch).

Murray CJL, Vos T, Lozano R, *et al.* 2012. Disability-adjusted life years (DALYs) for 291 diseases and injuries in 21 regions, 1990-2010: a systematic analysis for the Global Burden of Disease Study 2010. Lancet 380: 2197–2223.

Muth S, Sayasone S, Odermatt-Biays S, Phompida S, Duong S, Odermatt P. 2010. *Schistosoma mekongi* in Cambodia and Lao People's Democratic Republic. Advances in Parasitology 72: 179–203. doi: 10.1016/S0065-308X(10)72007-8.

Nakahama, T. 1884. Morbus malarioides. Chugai Iji Shinpo 111: 1–6 (in Japanese).

Nunn JF. 1996. Chapter 3. Concepts of anatomy, physiology and pathology. Chapter 4. The pattern of disease. Chapter 5. Magic and religion in medicine. *In*: Nunn JF editor. *Ancient Egyptian Medicine*. British Museum Press, London: 42–112.

Nunn JF, Tapp E. 2000. Tropical diseases in ancient Egypt. Transactions of the Royal Society of Tropical Medicin and Hygiene 94(2): 147–53.

Ohmae H, Sinuon M, Kirinoki M, Matsumoto J, Chigusa Y, Socheat D, Matsuda H. 2004. Schistosomiasis mekongi: from discovery to control. Parasitology International 53(2): 135–42.

Oka B. 1886. A study on diagnosis and prophylaxis of an endemic disease. Chugai Iji Shinpo 161: 1–11, cont. in 162: 12–21 (in Japanese).

Olveda DU, Yuesheng Li, Olveda R, Lam AK, McManus DP, Chau TNP, Ham DA, Williams GM, Gray DJ, Ross AGP. 2014. Bilharzia in the Philippines: past, present and future. International Journal of Infectious Diseases 18: 52–56.

Osborne AH, Vance D, Rohling EJ, Barton N, Rogerson M, Fello N. 2008. A humid corridor across the Sahara for the migration of early modern humans out of Africa 120,000 years ago. Proceedings of the National Academy of Sciences 105. doi: 10.1073/pnas.0804472105.

Pfister EP. 1913. Concerning the *aaa* disease of the papyri of Ebers and Brugsch. Archiv für Geschichte der Medizin 6: 12–20.

Pirajá da Silva MA. 1908. Contribuição para o estudo da Schistosomíase na Bahia. Brazil Médico 22: 281–282.

Platt TR, Brooks DR. 1997. Evolution of the schistosomes (Digenea: Schistosomatoidea): the origin of dioecy and colonization of the venous system. Journal of Parasitology 83(6): 1035–44.

Rollinson D, Knopp S, Levitz S, Stothard JR, Tchuem Tchuenté LA, Garba A, Mohammed KA, Schur N, Person B, Colley DG, Utzinger J. 2013. Time to set the agenda for schistosomiasis elimination. Acta Tropica 128: 423–440.

Renault AJ. 1808. Notice sur l'hématurie qu'éprouvent les Européens dans la haute Egypte et la Nubie. Journal Général de Médecine, de Chirurgie et de Pharmacie 17: 366–370.

Ruffer MA. 1910. Note on the presence of "*Bilharzia haematobi*" in Egyptian mummies of the twentieth dynasty (1250-1000 B.C.) British Medical Journal 1(2557): 16.

Sambon LW. 1907. New or little known African Entozoa. Journal of Tropical Medicine and Hygiene 10: 117.

Sassa T. 2013. The role of human-specific gene duplications during brain development and evolution. The Journal of Neurogenetics 27(3): 86–96. doi: 10.3109/01677063.2013.789512.

Satrija F, Ridwan Y, Jastal, Samarang, Rauf A. 2015. Current status of schistosomiasis in Indonesia. Acta Tropica 141(Part B): 349–353.

Smithers SR, Terry RJ. 1969. The immunology of schistosomiasis. Advances in Parasitology 7: 41–93.

Snow J. 1855. *On the Mode of Communication of Cholera.* 2nd ed. Churchill, London. p. 167.

Snyder SD, Loker ES. 2000. Evolutionary relationships among the Schistosomatidae (Platyhelminthes: Digenea) and an Asian origin for Schistosoma. Journal of Parasitology 86: 283–288.

Southgate VR. 1976. Schistosomiasis at Loum, Cameroun. Parasitology Research 49: 145–159.

Standley CJ, Dobson AP, Stothard JR. 2012. Out of animals and back again: schistosomiasis as a zoonosis in Africa. *In:* Rokni MB, editor. *Schistosomiasis.* Intech. doi: 10.5772/1309, ISBN 978-953-307-852-6.

Stothard JR, Bremond P, Andriamaro L, Sellin B, Selin E, Rollinson D. 2001. *Bulinus* species on Madagascar: molecular evolution, genetic markers and compatibility with *Schistosoma haematobium*. Parasitology 123(Suppl): S261–S275.

Tanaka H, Tsuji M. 1997. From discovery to eradication of schistosomiasis in Japan: 1847-1996. International Journal for Parasitology 27(12): 1465–1480.

Tchuem Tchuenté LA, Southgate VR, Jourdane J, Webster BL, Vercruysse J. 2003. *Schistosoma intercalatum*: an endangered species in Cameroon? Trends in Parasitology 19: 141–153.

Thomas AP. 1883. Natural history of the liver fluke and the prevention of rot. Journal of the Royal Agricultural Society of England. Second Series 19: 276–305.

Totell GT. 1924. A preliminary survey of schistosomiasis infection in the region of Changteh. China Medical Journal 38: 270–274.

Utzinger J, Zhou XN, Chen MG, Bergquist R. 2005. Conquering schistosomiasis in China: the longmarch. Acta Tropica 96(2-3): 69–96.

Van Bocxlaer B, Verschuren D. 2011. Modern and early Holocene mollusc fauna of the Ounianga lakes (northern Chad): implications for the palaeohydrology of the central Sahara. Journal of Quaternary Science 16: 433–447.

van Hinsbergen D, Lippert P, Dupont-Nivet G, McQuarrie N, Doubrivine P, Spakman W, Torsvik T. 2012. Greater India basin hypothesis and a two-stage Cenozoic collision between India and Asia. Proceedings of the National Academy of Sciences 109(20): 7659–7664.

Vic-Dupont BE, Soubrane J, Halle B, Richir C. 1957. Bilharziose à *Schistosoma japonicum* à forme hépato-splénique révélée par une grande hématémèse. Bulletins et Mémoires de la Societe Médicale des Hôpitaux de Paris 73: 933–994.

von Deines H, Westendorf W. 1961. Wörterbuch der medizinischen Texte. Erste Hälfte. Berlin: Akademie-Verlag 7(Part 1): 129.

Wang LD, Chen HG, Guo JG, Zeng XJ, Hong XL, Xiong JJ, Wu XH, Wang XH, Wang LY, Xia G, Hao Y, Chin DP, Zhou XN. 2009. A strategy to control transmission of Schistosoma japonicum in China. New England Journal of Medicine 8; 360(2): 121–128.

Watts S, El Katsha S. 1995. Changing environmental conditions in the Nile Delta: Health and policy implications with special reference to schistosomiasis. International Journal of Environmental Health Research 5: 197–212.

Webster JP, Gower CM, Blair L. 2004. Do hosts and parasites coevolve? Empirical support from the *Schistosoma* system. The American Naturalist 164: S33–S51.

Wei DX, Yang WY, Huang SQ, Lu YF, Su TC, Ma JH, Hu WX, Xie NF. 1980. Parasitological studies on the ancient corpse of the Western Han Dynasty unearthed from Tomb No. 188 on Phoenix Hill at Jiangling County. Acta Wuhan Medical College 3: 1–6. (In Chinese).

Wendorf FR, Schild R, Said R, Hayes CV, Gautier A, Kobusciewicz M. 1976. The prehistory of the Egyptian Sahara. Science 193: 103–114.

Westendorf W. 1992. *Erwachen der Heilkunst: Die Medizin im Alten Ägypten*. Zürich: Artemis & Winkler (In German) 297 p.

Wooley PG. 1906. The occurrence of *Schistosoma japonicum* Vel Cattol in the Philippines. Philippine Journal of Science 1: 83–90.

Yin M, Zheng HX, Su J, Feng Z, McManus DP, Zhou XN, Jin L, Hu W. 2015. Co-dispersal of the blood fluke *Schistosoma japonicum* and Homo sapiens in the Neolithic Age. Scientific Reports 5: 18058. doi: 10.1038/srep18058.

Yuan HC, Guo JG, Bergquist R, Tanner M, Chen XY, Wang HZ. 2000. The 1992-1999 World Bank schistosomiasis research initiative in China: outcome and perspectives. Parasitology International 49: 195–207.

Zakaria H. 1959. Historical study of *Schistosoma haematobium* and its intermediate host, *Bulinus truncatus*, in central Iraq. Journal of the Faculty of Medicine Bagdad (Jami'at, Bagdad) 1: 2–10.

Zein-Eddine R. 2014. Phylogeny of seven *Bulinus* species originating from endemic areas in three African countries, in relation to the human blood fluke *Schistosoma haematobium*. Evolutionary Biology 14: 271.

Zhou XN, Jiang QW, Guo JG, Lin DD, ZhuR, Yang GJ, Yang K, LiSZ, XuJ. 2012. Road map for transmission interruption of schistosomiasis in China Zhongguo Xue Xi Chong Bing Fang Zhi Za Zhi 24 (2012), pp. 1–4 (in Chinese).

Life Cycles of Schistosomiasis

Allen GP Ross[1,]* and *Li Yuesheng*[2, 3]

3.1 INTRODUCTION AND LIFE CYCLE

The schistosome life cycle is shown in Fig. 3.1 for three species and in Fig. 3.2 for *S. japonicum* which is zoonotic. Unlike other trematodes, schistosomes are dioecious (i.e. they have separate sexes), with the adults having a cylindrical body 7 to 20 mm in length featuring two terminal suckers, the complex tegument, a blind digestive tract and reproductive organs. The worms have a complex life cycle that involves infection of fresh-water molluscs that act as intermediate hosts and the bloodstream of higher order vertebrate as definitive hosts (Shiff 2000; Ross *et al.* 2002, 2012). Schistosomes have coevolved with their molluscan and mammalian hosts resulting in a well-balanced and highly efficient means of transmission (Shiff 2000).

Adult male and female worms pair up and the latter produce numerous eggs. Mating and subsequent cohabitation occurs when the longer, more slender female schistosome is embraced within the ventral gynaecophoric canal of the male. An adult female *Schistosoma mansoni* lays 100–300 eggs per day, a *S. haematobium* female 20–200 eggs per day and a *S. japonicum* female 500–3,500 eggs per day (Pittella 1997).

The shape of the eggs and the location of their excretion differentiate the various species (Fig. 3.1) (see also Jones *et al.*, Chapter 5, Fig. 5.1). Eggs are deposited in the capillaries of the target organ that is parasitized by the breeding pair of worms (Ross *et al.* 2012). With *S. haematobium* this is the urinary system as well as the sacral and pelvic vessels whereas in intestinal schistosomiasis (*S. mansoni*, *S. japonicum*, *S. intercalatum* and *S. mekongi*) parasitization occurs in the inferior mesenteric veins (*S. mansoni*, *S. japonicum*) and the superior hemorrhoidal veins (*S. japonicum*) (Chen 1991; Pittella 1997; Waine and McManus 1997; Ross *et al.* 2002, 2012). Once deposited, a proportion of the eggs then enter the lumen of the bladder or intestines. A small amount of bleeding occurs as eggs perforate the bladder (*S. haematobium*) or intestinal mucosa en route to the urine or feces (Ross *et al.* 2002, 2012). Eggs that do not reach the mucosa are retained in the target organ tissue (specific to the species of worm) or are carried downstream in the portal circulation and in the case of an *S. mansoni* and *S. japonicum* infection, are trapped in the liver (Ross *et al.* 2012).

[1] Tropical Medicine and Global Health, Griffith University, Logan Campus, University Drive, Meadowbrook QLD 4131, Australia.

[2] Hunan Institute of Parasitic Diseases, Yueyang, Hunan Province, World Health Organisation Collaborating Centre for Research and Control on Schistosomiasis in Lake Region, Yueyang, Hunan, 414000, People's Republic of China.

[3] QIMR Berghofer Medical Research Institute, 300 Herston Rd., Brisbane QLD 4006, Australia.

* Corresponding author

Once excreted, an ovum hatches in contact with fresh water and a free-living motile form, called the miracidium, is released.

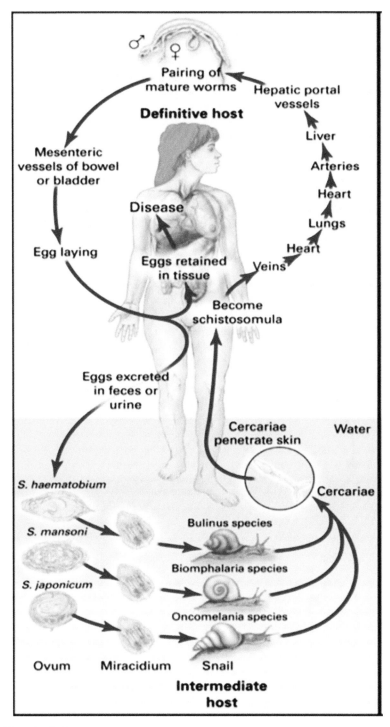

FIG. 3.1 The life cycle of schistosomiasis. From Ross AG, Hou X, Farrar J, Huntsman R, Gray DJ, McManus DP, Li YS. 2012. Journal of Neurology 12 Jan; 259(1): 22–32. Epub 2011 Jun 15. Fig. 1.

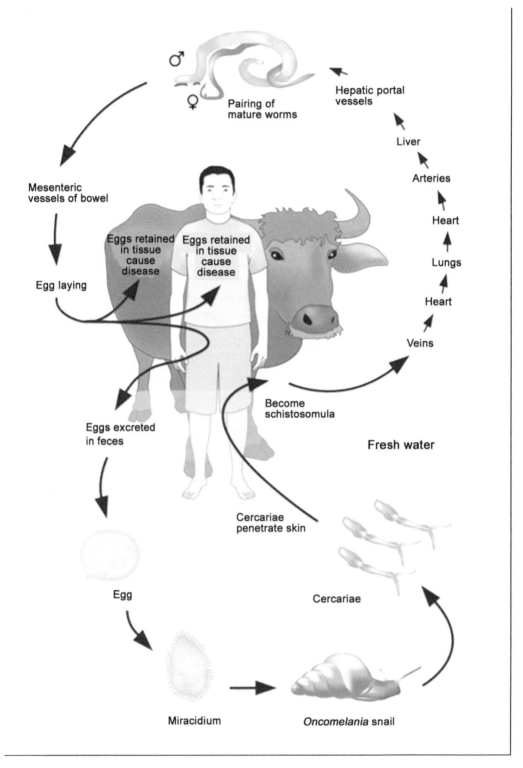

FIG. 3.2 The zoonotic life cycle of *Schistosoma japonicum.* From DU Olveda *et al.* 2013. International Journal of Infectious Diseases 18: 52–56. Fig. 1. After McManus DP *et al.* 2010. Clinical Microbiology Reviews 23(2): 442–466.

Each schistosome species infects a specific fresh-water snail intermediate host – e.g., with *S. mansoni* this is generally *Biomphalaria glabrata* (Americas) or *B. pfeifferi* (Africa); with *S. japonicum* it is *Oncomelania hupensis*. After approximately 30 days, infected snails begin to release cercariae in response to sunlight. The cercariae possess a forked tail and are free-swimming. Because cercariae are non-feeding, using their endogenous nutrient stores for energy production, they must contact with the skin of a human (or alternative mammalian host [e.g., water buffaloes, cattle, rodents, dogs, sheep, pigs and dogs] in the case of *S. japonicum*) within 12–24 hours after emerging from the snail (Ross *et al.* 2001; McManus *et al.* 2010). Once contact with the skin has occurred, percutaneous penetration is achieved by mechanical activity and by the production of proteolytic enzymes. The cercariae shed their bifurcated tails and transform their trilaminate (single-lipid-layer) tegument into an unusual heptalaminate (double-lipid-layer) form. This is exquisitely adapted to the mammalian environment, with the ability to control water and drug uptake and to modulate or subvert host immunological mediators at the schistosome surface (McManus *et al.* 2010). Now schistosomula, they leave the skin via the blood vessels and draining lymphatics and reach the lungs. After several days the male and female worms exit the lungs and arrive in the hepatic portal system, where they mature, pair up and migrate downstream to complete the life cycle.

Control includes administration of the drug praziquantel to treat the human definitive host and applications of molluscicides (e.g. niclosamide to kill the snail intermediate host).

3.2 CHAPTER SUMMARY

Schistosomiasis has a complex life cycle that involves infection of fresh-water molluscs that act as intermediate hosts and the bloodstream of higher order vertebrate as definitive hosts. Unlike other trematodes, schistosomes are dioecious, adult female worms laying up 3,000 eggs per day depending on the species. Integrated control strategies target various aspects of parasite's life cycle. For example, praziquantel is used to treat the human definitive host and molluscicides (e.g., niclosamide) are used to kill the snail intermediate host.

3.3 LITERATURE CITED

Chen MG. 1991. Relative distribution of *Schistosoma japonicum* eggs in the intestine of man: a subject of inconsistency. Acta Tropica 48: 163–171.

McManus DP, Gray DJ, Li YS, Williams GM, Feng Z, Rey-Ladino J, Stewart D, Ross AG. 2010. Schistosomiasis in the Peoples' Republic of China: the era of the Three Gorges Dam. Clinical Microbiology Reviews April 23(2): 442–466.

Olveda DU, Yuesheng Li, Olveda RN, Lam KL, McManus DP, Chau NP, Harn DA, Williams GM, Gray DJ, Ross AGP. 2014. Bilharzia in the Philippines: past, present and future. International Journal of Infectious Diseases 18: 52–56.

Pittella JE. 1997. Neuroschistosomiasis. Brain Pathology 7: 649–662.

Ross AG, Sleigh AC, Li YS, Davis GM, Williams GM, Jiang Z, Feng Z, Jingping G, McManus DP. 2001. Schistosomiasis in the People's Republic of China: prospects and challenges for the 21st century. Clinical Microbiology Reviews 14(2): 270–295.

Ross AG, Bartley PB, Sleigh AC, Olds GR, Li YS, Williams GM, McManus DP. 2002. Schistosomiasis. New England Journal of Medicine 346(16): 1212–20.

Ross AG, Hou X, Farrar J, Huntsman R, Gray DJ, McManus DP, Li YS. 2012. Neuroschistosomiasis. Journal of Neurology 12 Jan; 259(1): 22–32. Epub 2011 Jun 15.

Shiff C. 2000. Epidemiology of helminth infections. *In*: Nelson KE, Masters Williams C, Graham NM, editors. *Infectious Disease Epidemiology: Theory and Practice.* Aspen, Gaithersburg 768 p.

Waine GJ, McManus DP. 1997. Schistosomiasis vaccine development – the current picture. Bioessays 19: 435–443.

CHAPTER

Schistosoma Intermediate Host Snails

Henry Madsen

4.1 INTRODUCTION

Distribution and population dynamics of schistosome intermediate host species are important factors determining the transmission intensity given that other conditions for transmission are met, i.e., human water contact and contamination of the water bodies with schistosome eggs. As snails are an essential element for the schistosomes to complete their life cycle, control of them might be an important way to reduce transmission. During the early history of schistosomiasis control, the main focus was on snail control but this was mainly due to the lack of safe and efficient drugs for treating infected people (see also Fenwick, Chapter 22). With the release of praziquantel in the mid-1970s and the less than satisfactory results of large-scale mollusciciding campaigns, focus was changed to morbidity control through regular medical treatment of infected people (Appleton 1996). Experience has shown that reinfection occurs rapidly (Madsen *et al.* 2011; Kariuki *et al.* 2013) and morbidity due to light infections may be significant (King *et al.* 2006). In 2012, the World Health Assembly adopted a resolution (WHA 65.21) aiming to eliminate schistosomiasis in some countries. While preventive chemotherapy has been the main component for schistosomiasis control, other operational components are required to have a greater impact towards the interruption of transmission. Among these are the provision of potable water, adequate and reliable sanitation, hygiene interventions and snail control. Hence snail control could again play an important role (Rollinson *et al.* 2013).

Each of the schistosome species infecting humans has its characteristic intermediate snail host spectrum. The intermediate hosts of *S. mansoni*, *S. haematobium*, *S. intercalatum* and *S. guineensis* belong to the family Planorbidae (Brown 1994) while those of *S. japonicum*, *S. mekongi* and *S. malayensis* belong to the family Pomatiopsidae (Davis 1979; Attwood 2010; see also Littlewood and Webster, Chapter 1). Various species of *Biomphalaria* serve as intermediate hosts for *S. mansoni* (Fig. 4.1) and certain *Bulinus* species are intermediate hosts for *S. haematobium*, *S. guineensis* and *S. intercalatum*. For *S. japonicum* subspecies of *Oncomelania hupensis* are the intermediate hosts (Davis 1980) while for *S. mekongi* and *S. malayensis* it is species of the Pomatiopsidae, tribe Triculinae (Davis 1980).

Department of Veterinary Disease Biology, University of Copenhagen, Dyrlægevej 100, Frederiksberg C, Denmark.

FIG. 4.1 *Biomphalaria pfeifferi* shedding *Schistosoma mansoni* cercariae.

In order to control snails, it is essential to have data on which species are the hosts in which areas, where and when they are found, when and where they produce the infective stage of the parasite, where and when people become infected and which factors influence their distribution and infection status. Both snail distribution and schistosome transmission patterns are not stationary and show marked variation from year to year (Madsen 1996) and there may be trends in the changes due to climate change (Steensgaard *et al.* 2013). In this chapter, schistosome intermediate host species are reviewed and their biology, ecology and control is discussed.

4.2 MEDICAL MALACOLOGY

4.2.1 Introduction

The involvement of snails in the life cycle of *Schistosoma* spp. has been known since the early 20th century when Miyairi and Suzuki in 1913 showed that *Oncomelania hupensis nosophora* was the intermediate host of *S. japonicum* (Ishii 2005) and later, Leiper (1915) discovered the life cycle of *S. haematobium* and *S. mansoni* by finding *Bulinus* sp. and *Planorbis boissyi* as their respective snail hosts.

During the late 1940s and early 1950s, the World Health Organization decided to address the control of major tropical diseases, including bilharziasis (schistosomiasis) in Africa, the Middle East, Southeast Asia and South America. Because of the lack of safe and efficient drugs, early control efforts addressed control of the snails serving as intermediate hosts. With the focus on snail control in controlling schistosomiasis there was an obvious need for clarity on the species involved as intermediate hosts. At that time the nomenclature of the host snails, particularly those now belonging to the genera *Biomphalaria* and *Bulinus* in Africa, was in a state of confusion (Preface to Mandahl-Barth 1958).

Snail Identification Centers were established in Copenhagen, Paris and Salisbury to which collections could be forwarded for identification (Preface to Mandahl-Barth 1958). A monograph on African species was published by one of these centers based on the examination of a large number of specimens collected from many localities and comparison with the original descriptions and whenever possible, with type material (Mandahl-Barth 1958). Mandahl-Barth (1958) used selected anatomical characteristics of the radula, the size of reproductive organs and the shape of the shells and succeeded in organizing the snail hosts into two distinct genera, *Bulinus* and *Biomphalaria*.

In 1980, DS Brown published a book on freshwater snails of Africa and their medical importance and this was revised in 1994 (Brown 1980, 1994). This book is currently the most comprehensive overview of African freshwater snails. A number of regional field guides for snail identification were prepared by the Danish Bilharziasis Laboratory (e.g., 1986, 1998). For Neotropical *Biomphalaria* species the identification keys were published by the Pan American Health Organization (1968, Malek 1985). As for the Asian intermediate hosts of *Schistosoma*, description and revision of the Pomatiopsidae was done by GM Davis (Davis 1979).

4.2.2 Taxonomic Position

Snails belong to the phylum Mollusca and the class Gastropoda and traditionally freshwater snails were divided into two sub-classes Prosobranchia and Pulmonata (Brown 1994). These two clades, however, are not monophyletic and because major clades within the Gastropoda are difficult to define this system is convenient to maintain because most available identification keys follow this classification. A detailed description of the major clades within the Gastropoda is beyond the scope of this chapter and readers are referred to for example, Ponder and Lindberg 1997; Grande *et al.* 2004; Colgan *et al.* 2007; Ponder *et al.* 2008.

Intermediate hosts of African and Neotropical *Schistosoma* species belong to the family Planorbidae (Brown 1994) while those of Asian species belong to the family Pomatiopsidae (Davis 1979; Attwood 2010). More species can be infected experimentally but they may not transmit the parasite in nature (Davis 1980). For example, there have been reports that *S. haematobium* was transmitted by *Ferrissia* in India; at least three endemic foci of human schistosomiasis have been described in India previously and sporadic autochthonous cases and cercarial dermatitis are also not uncommon (Kali 2015). Experimental exposure of *Ferrissia tenuis* to miracidia of *S. haematobium* showed that this snail could be infected and shed cercariae (Agrawal 2012). In the north-western extremity of Africa, *Planorbarius metidjensis* is common and although it is an experimental host for *S. haematobium* (Yacoubi *et al.* 2007) there is no evidence that it is a natural host (Brown 1994).

Family Planorbidae Gray, 1840. The Planorbidae is the largest family of freshwater pulmonates comprising approximately 40 genera with a worldwide distribution (Jørgensen *et al.* 2004). Following Hubendick (1955), the Planorbidae has two subfamilies, Planorbinae and Bulininae. However, Brandt (1974) referred to 4 families, i.e., Planorbidae, Bulinidae, Physidae and Ancylidae belonging to a superfamily Ancylacea Brown 1844. Two representatives of the family Ancylidae fall within the Planorbidae (Morgan *et al.* 2003) highlighting the need for further analysis and possible reclassification of this group. Molecular analysis strongly suggested the monophyly of Planorboidea (= 'Ancyloplanorbidae') (Albrecht *et al.* 2006). The latter authors found in addition to a distinct *Burnupia* clade, two major clades that correspond to family level taxa (traditional Bulinidae and Planorbidae).

The snails are small to medium-sized with long slender tentacles and blood containing hemoglobin (Brown 1994). The shell is discoid, lens-shaped or higher ovate to turreted and the animals are sinistral, that is, the genital openings and the anus are situated on the left side, but in most of the discoid forms the shell appears to be dextral, because it is carried inverted, so that the side representing the spire (apical side) in other families is the lower side of the planorbid shell and the upper side is umbilical (Mandahl-Barth 1957a, b).

The two subfamilies, the Planorbinae and the Bulininae, are recognizable by the shape of the shells and by some anatomical features. The Planorbinae have a discoid or lens-shaped shell, a pseudobranch forming a simple lobe, a prostate gland consisting of a series of glandular tubes along the sperm a duct and a copulatory organ which is not completely introverted when not in use. The Bulininae have an ovate or turreted shell, a pseudo branch which is deeply folded and highly vascularized, a compact prostate gland and a copulatory organ ("ultrapenis") which is completely introverted when not in use (Mandahl-Barth 1958).

Genus *Biomphalaria* Preston, 1910. By the 1950s there was a total of about 35 named species in Africa, of which only ten were recognized as distinct by Mandahl-Barth (1957a, b) in his revision based on morphology (although some were divided into subspecies). Subsequent changes in this classification have been few and these are also based on morphology. Only two new taxa have been named, *B. sudanica rugosa* Mandahl-Barth 1968 and the subfossil *B. barthi* Brown 1973.

Currently, there are about 34 *Biomphalaria* species, 12 in the Old World (Africa, Madagascar and the Middle East) and 22 in the Neotropics. The origin of the genus *Biomphalaria* is American and the ancestor of *Biomphalaria glabrata* colonized Africa and speciated into all the African *Biomphalaria* species (DeJong *et al.* 2001). This probably explains why any species of the genus *Biomphalaria* found in Africa and the Middle East may serve as host for *S. mansoni* while in South America only certain *Biomphalaria* species are intermediate hosts.

African species (see examples in Fig. 4.2) are: *B. alexandrina* (Ehrenberg 1831); *B. angulosa* Mandahl-Barth 1957; *B. barthi* Brown 1973 [known only from shells]; *B. camerunensis* (Boettger 1941); *B. choanomphala* (Martens 1879); *B. pfeifferi* (Krauss 1848); *B. rhodesiensis* Mandahl-Barth 1957; *B. salinarum* (Morelet 1868); *B. smithi* Preston 1910; *B. stanleyi* (Smith 1888); *B. sudanica* (Martens 1870); and *B. tchadiensis* (Germain 1904). The most widely distributed is *B. pfeifferi* which also extends its distribution into the Arabian Peninsula. The species found in the Arabian Peninsula, *B. arabica* is a synonym of *B. pfeifferi* (DeJong *et al.* 2003). The process of delineating the species is still ongoing. Several putative species and subspecies, *B. arabica*, *B. gaudi*, *B. rhodesiensis* and *B. stanleyi*, are shown to be undifferentiated from other *B. pfeifferi* populations (DeJong *et al.* 2003).

In South America and the West Indies there are primarily three species that are important as intermediate hosts for *S. mansoni*, i.e. *B. glabrata* (Say 1818); *B. tenagophila* (Orbigny 1835); *B. straminea* (Dunker 1848). Other species maybe experimentally infected (Malek 1985), but are not important for transmission in the field. Some of the neotropical *Biomphalaria* species have extended their distribution, e.g., *B. glabrata* in Egypt (Pflüger 1982; Yousif *et al.* 1996), *B. tenagophila* in Kinshasa (Pointier *et al.* 2005) and Europe (Majoros *et al.* 2008), *B. straminea* in China (Woodruff *et al.* 1985a, b). A hybrid between *B. glabrata* and *B. alexandrina* was found in Egypt (Yousif *et al.* 1998) but Lotfy *et al.* (2005) did not find any evidence for the presence of *B. glabrata* or a hybrid between the two species.

Genus *Bulinus* Müller 1781. The shell is sinistral, invariably having a greater height than width, inversely conical, ovate or higher to almost cylindrical and turreted according to the height of the spire (Mandahl-Barth 1958).

Early in the 20th century the number of species, all defined by shell characters alone, had reached nearly one hundred (Brown 1996). The first comprehensive revision (Mandahl-Barth 1958) recognized only 20 species (some with subspecies), defined according to the radula and soft parts as well as the shell. Of these species, some have since been placed in synonymy and some then treated by Mandahl-Barth as synonyms are now regarded as distinct. With the descriptions of a few additional species there is a net gain of 17, bringing the total now recognized to about 37 (Brown 1996).

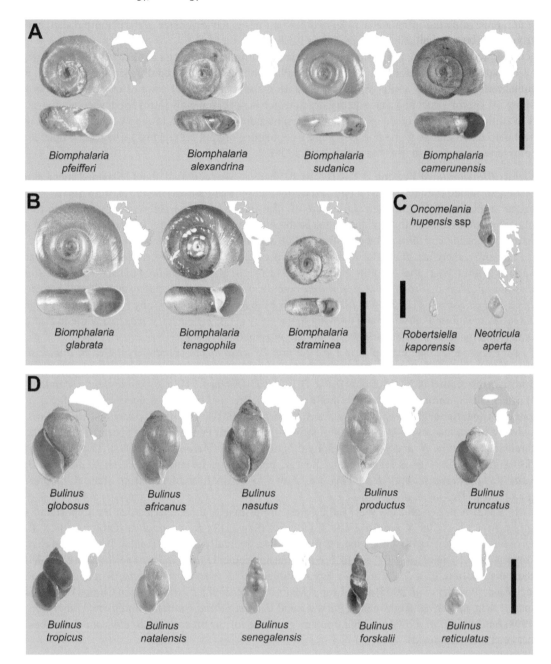

FIG. 4.2 Selected species of intermediate hosts for *Schistosoma* and their distribution. **A**. *Biomphalaria* species in Africa; **B**. Neotropics; **C**. Pomatiopsidae species in Asia; **D**. *Bulinus* species from Africa. Shell photos and maps are based on the original photographs and drawings used in preparation of slides in World Health Organization (1989).

In 1957 Mandahl-Barth united under *Bulinus* the three genera *Isidora*, *Physopsis* and *Pyrgophysa* and introduced a system of species groups, thereby allowing the name *Bulinus* to be applied to any of the species. The system has been modified since so that species are placed into four species groups, i.e., the *Bulinus africanus* group, *Bulinus forskalii* group, *Bulinus reticulatus* group and the *Bulinus truncatus/tropicus* complex (Brown 1994). These species groups appear stable throughout the majority of phylogenetic analyses with both variable and highly conserved genes (Stothard *et al.* 2001; Kane *et al.* 2008; Jørgensen *et al.* 2011). However, species identification within these groups can be unreliable when based solely on morphological data as characters are often plastic being subject to ecophenotypy or incongruence (Jørgensen *et al.* 2013).

The species of *Bulinus* (of Africa, Arabia and Indian Ocean islands) are arranged below and in alphabetical order within the species groups. The schistosome species for which they are hosts are indicated in hard bracket (*S.h, S.i.* and *S.g.* for *S. haematobium, S. intercalatum* and *S. guineensis,* respectively). For *S. haematobium* Brown (1994) provided an overview of the species involved and here we just give references to newer work. The taxa are:

B. forskalii group (= genus *Bulinus sensu stricto*): *B. barthi* Jelnes 1979; *B. bavayi* (Dautzenberg 1894); *B. beccarii* (Paladilhe 1872) [***S.h.***]; *B. browni* Jelnes 1979; *B. camerunensis* Mandahl-Barth 1957 [***S.h.***]; *B. canescens* (Morelet, 1868); *B. cernicus* (Morelet, 1867) [***S.h.***]; *B. crystallinus* (Morelet 1868; *B. forskalii* (Ehrenberg 1831) [***S.h, S.i.S.g.***]; *B. scalaris* (Dunker 1845); *B. senegalensis* Muller 1781 [***S.h.***].

Bulinus forskalii was found naturally infected with *S. haematobium* in Niger (Labbo *et al.* 2007).

B. africanus group (= genus *Physopsis*): *B. abyssinicus* (Martens 1866) [***S.h.***]; *B. africanus* (Krauss 1848) [***S.h.***]. ; *B. globosus* (Morelet 1866) [***S.h.***]; *B. hightoni* Brown and Wright 1978; *B. jousseaumei* (Dautzenberg 1890) [***S.h.***]; *B. nasutus* (Martens 1879) [***S.h.***]; *B. obtusispira* (Smith 1882) [***S.h.***]; *B. obtusus* Mandahl-Barth 1973; *B. ugandae* Mandahl-Barth 1954; *B. umbilicatus* Mandahl-Barth 1973 [***S.h.***].

B. truncatus/tropicus complex (= genus *Isidora*): *B. angolensis* (Morelet 1866); *B. depressus* Haas 1936; *B. hexaploidus* Burch 1972; *B. liratus* (Tristram 1863); *B. natalensis* (Kuster 1841); *B. nyassanus* (Smith, 1877) [***S.h.***]; *B. octoploidus* Burch 1972; *B. permembranaceus* (Preston 1912); *B. succinoides* (Smith 1877); *B. transversalis* (Martens 1897); *B. trigonus* (Martens 1892); *B. tropicus* (Krauss 1848); *B. truncatus* (Audouin 1827) [***S.h.***]; *B. yemenensis* Paggi *et al.* 1978 [***S.h.***].

Most diploid members of this group seem resistant to *S. haematobium* although some species can be experimentally infected (Brown 1994; Arouna and Remy 2005; DeKock and Wolmarens 2006). *Bulinus nyassanus,* which also is diploid (2n = 36), is, however, a very important host in Lake Malawi (Madsen *et al.* 2001).

B. reticulatus group: *B. reticulatus* Mandahl-Barth 1954; *B. wrighti* Mandahl-Barth 1965 [***S.h.***].

B. reticulatus can be infected with *S. haematobium* but would probably not play a major role in the transmission under natural conditions due to the nature of their habitats (De Kock *et al.* 2002).

Pomatiopsidae Stimpson 1865. The Pomatiopsidae belongs to the superfamily Rissooidea which is one of the largest and most diverse molluscan superfamilies, with 23 recognized recent families including marine, freshwater and terrestrial members (Criscione and Ponder 2013). Several species within this superfamily have medical and/or veterinary importance. The Pomatiopsidae comprise two subfamilies, the Pomatiopsinae Stimpson, 1865 and the Triculinae Annandale 1924 (Davis 1980; Attwood *et al.* 2003). Certain subspecies (see Fig. 4.3) of *Oncomelania hupensis* (Pomatiopsidae: Pomatiopsinae) are known to transmit *Schistosoma japonicum* (Rollinson and Southgate 1987). According to Davis (1980), these subspecies are too similar to be recognized as separate species, while others claim that some of these subspecies should be recognized as full species, i.e., *O. quadrasi* (Woodruff *et al.* 1988; Hope and McManus 1994) and *O. lindoensis* (Woodruff *et al.* 1999).

The subspecies transmitting *S. japonicum* are *Oncomelania h. chiui* (Habe and Miyazaki 1962) in Taiwan; *O. h. formosana* (Pilsbry and Hirase 1905) in Taiwan; *O. h. hupensis* Gredler 1881 in mainland China; *O. h. lindoensis* Davis and Carney 1973 in Sulawesi; *O. h. nosophora* (Robson 1915) in Japan; and *O. h. quadrasi* (Möllendorff 1895) in the Phillipines (Davis 1980). There are three subspecies on the mainland of China with discrete patterns of distribution: above the

Three Gorges of the Yangtze River (*O. h. robertsoni*) in Yunnan and Sichuan provinces; below the Three Gorges along the Yangtze River drainage (*O. h. hupensis*) with an incursion into Guangxi Province; and Fujian Province along the coast (*O. h. tangi*) (Davis *et al.* 2006). In western China, *O. h. robertsoni* is the sole intermediate host for *S. japonicum* (Attwood *et al.* 2004; Hauswald *et al.* 2011). Of these taxa, only *O. h. hupensis* has been shown to be dimorphic, with ribbed-shelled aggregates of individuals on flood plains and smooth-shelled populations in habitats elevated above the effects of floods or removed from the effects of severe annual floods by barriers (Wilke *et al.* 2000; Davis *et al.* 2006).

FIG. 4.3 *Oncomelania hupensis* subspecies **A**. *O. h. hupensis* (ribbed form). **B**. *O. h. hupensis* (smooth form). **C**. *O. h. nosophora*. **D**. *O. h. formosana*. **E**. *O. h. quadrasi*. **F**. *O. h. chiui*.

In Asia, the Triculinae is very diverse with an endemic fauna that includes over 90 species occurring along a 300 km stretch of the lower Mekong River in Thailand and Laos (Brandt 1974; Davis 1979; Attwood *et al.* 2003; Liu *et al.* 2014).

The Triculinae comprises three tribes (Davis *et al.* 1992); these are the Jullieniini, Pachydrobiini and Triculini. The Triculinae include *Neotricula aperta* (Temcharoen 1971) (Pachydrobiini), the intermediate host of *Schistosoma mekongi* Voge *et al.* 1978, primarily a parasite of humans along the Mekong River of Laos and Cambodia (see Temcharoen 1971; Attwood 2001). In addition to *N. aperta*, several other Triculinae are known to act as intermediate host for *Schistosoma* species: *Robertsiella kaporensis* Davis and Greer 1980 (Pachydrobiini) transmits *S. malayensis* Greer *et al.* 1988 in peninsular Malaysia; this is a parasite of rodents but also infects the aboriginal peoples of the region (Attwood 2001). In Malaysia, another species, *Robertsiella silvicola* also transmits *S. malayensis* (Attwood *et al.* 2005).

Three strains of *N. aperta* have been identified on the basis of body size and mantle pigmentation (Davis *et al.* 1976). All three strains are able to act as host for *S. mekongi* but in nature only the γ-strain is known to be epidemiologically significant (Attwood *et al.* 1997).

4.3 BIOLOGY AND HABITAT

Planorbidae. *Biomphalaria* and *Bulinus* species are aquatic and hermaphroditic snails which can be found in almost all types of water bodies, ranging from small temporary ponds or streams to large lakes and rivers (Brown 1994). Man-made water bodies including irrigation canals and dams are in particular excellent habitats and because of the intense human water contact, these may play a very important role in the transmission of schistosomiasis. *Biomphalaria* and *Bulinus* species are usually found in shallow water and usually in association with aquatic macrophytes, but species in some of the large lakes may be found at considerable depth, for example *Biomphalaria* species in the East African lakes (Magendantz 1972; Kazibwe *et al.* 2006; Lange *et al.* 2013) and *Bulinus nyassanus* in Lake Malawi (Wright 1967; Mandahl-Barth 1972; Madsen *et al.* 2011).

Eggs of *Biomphalaria* and *Bulinus* species are deposited in masses on plants or solid objects in the water (Fig. 4.4). Under optimal temperature conditions hatching takes place after 7–10 days. Snails may reach sexual maturity after 4–6 weeks. The average longevity varies among species and with local conditions. Their reproductive capacity is high and populations of these species may undergo marked seasonal fluctuations in density with rainfall and/or temperature being the main causative factors. Snails may be widely distributed in an area but there is a tendency for *Schistosoma* infected snails to be focally distributed due to the focality of human water contacts. Focal infection rates with schistosomes may be quite high. In large irrigation schemes, most infected *Biomphalaria* and *Bulinus* snails are found close to human settlements (Babiker *et al.* 1985; Madsen *et al.* 1987). The focality and seasonality in the transmission of schistosomes are key issues for snail control.

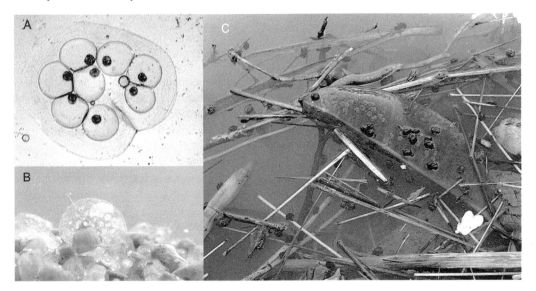

FIG. 4.4 A. Egg mass of *Bulinus truncatus*. **B.** Egg mass of *B. nyassanus*. **C.** High density of *B. globosus* and its egg masses on a leaf (turned over).

Pomatiopsidae. *Oncomelania hupensis* is a small, amphibious and dioecious snail. Females tend to be larger than males. Eggs are laid singly on solid objects (Pesigan *et al.* 1958). Hatching occurs after 10–25 days, depending on temperature and newly-hatched snails pass through an aquatic stage of 1–2 weeks. Snails reach sexual maturity after 10–16 weeks and may live for 24–35 weeks (Pesigan *et al.* 1958). Their reproductive potential is low compared to the pulmonate snails and recolonization of sites treated with molluscicide may be slow (Pesigan *et al.* 1958). *Oncomelania* ssp inhabit flood plains and especially man-made habitats, resulting from agricultural development, drainage channels, roadside ditches, rice fields and small canals and drainage canals of irrigation works, are important (Pesigan *et al.* 1958; Xu *et al.* 2005; Madsen *et al.* 2008). Snails are found primarily on the banks but some are also found in very shallow water (i.e., depth less than c. 20 cm). Habitats preferred by *O. hupensis* are shaded by vegetation and the temperature is relatively constant and cool. Current speeds above 0.14 m/s are generally unfavorable for *O. hupensis*. In eastern China, *O. hupensis* is found in smooth- and ribbed-shelled populations and this seems to be associated with the annual floods of the Yangtze River (Wilke *et al.* 2000). Also for *O. hupensis*, density of infected snails shows a focal distribution (Madsen *et al.* 2008). Davis *et al.* (2006) found that there were significantly more infected snails per area in hot spots than in non-hot spots and density of infected snails is not correlated with density of snails overall.

Neotricula aperta is found in rivers in Central Lao PDR and eastern Cambodia and there is also a spring dwelling form of the species (Attwood 2010). Population densities appear to be controlled by the annual rise and fall in the water level of the river (Upatham *et al.* 1980). During the rainy season the flow is torrential and the species seems to persist as eggs attached to the underside of stones. Female snails live less than a year and apparently lay large numbers of eggs prior to the onset of the rains.

Robertsiella kaporensis may be found in small streams attached to rocks, leaves and twigs, but are most abundant in overgrown areas where they attach to roots (Greer *et al.* 1980). The snail lay typical hydrobid eggs and the development period is about 20–30 days; the young grow to maturity in 14–15 weeks (Upatham *et al.* 1985). *Robertsiella silvicola* is found in water bodies ranging from small, spring-fed, first-order streams to tiny trickles of water draining off forested hill sides in peninsular Malaysia (Attwood *et al.* 2005).

4.4   SNAIL CONTROL

4.4.1   General

Although several countries, primarily through extensive chemotherapy campaigns, have achieved considerable progress in reducing schistosomiasis caused morbidity, there is a risk of resurgence of transmission if control efforts are reduced. In order to reduce transmission further or even eliminate it, there is need for transmission control as highlighted in the 2012 World Health Assembly resolution 65.21 on elimination of schistosomiasis (see also Fenwick *et al.*, Chapter 22). It is important that interventions are made at several levels including measures adapted to the specific local ecological settings against the intermediate host snails. The specific measures that can be implemented in a given area should be developed on the basis of a detailed knowledge about transmission and in close collaboration between stakeholders. Especially in sub-Saharan Africa, schistosomiasis is a major public health problem and transmission is often associated with water development projects such as dam and/or irrigation projects for electricity and/or food production. Clearly, it is important to provide food and energy to an increasing human population and at the same time protect it from infections related to water usage. Both water development projects and disease control efforts in should be implemented such that water resources and their associated biodiversity are affected as little as possible. Management of these systems and under some circumstances natural systems as well, is the key to ensuring minimal schistosome transmission and it might even be possible to further utilize existing schemes for food production (aquaculture) without increasing the risk of transmission.

If snail control is deemed necessary, this should be as environmentally friendly as possible. Snails can be controlled using chemicals, molluscicides, to kill snails (McCullough *et al.* 1980; McCullough 1992), biological control (McCullough 1980) and through environmental/ecological management (Madsen 1996). Implementation of snail control will depend on many factors such as infection level and morbidity in final hosts (people), freshwater habitats (ranging from large lakes or rivers to small ponds or streams/canals, rice fields or swamps, transmission pattern, target snail species and ecological concerns.

No control option used alone is likely to solve the disease problems related to trematodes. It is absolutely essential that we adopt a holistic approach and attempt to reduce the problem from as many angles as possible, i.e., treatment of people and perhaps reservoir hosts, health education, reduction of contamination with parasite eggs and control of intermediate hosts. The specific methods chosen will depend on the specific settings in a given endemic area.

4.4.2   Chemical Snail Control

A number of chemicals have been used as molluscicides in the past, but at the moment there is virtually only one compound which is commercially available, i.e., niclosamide (McCullough

1992). Most of these compounds are highly toxic to fish and that obviously would limit their use in aquaculture ponds and natural habitats where people rely on fishing. Furthermore, the costs of the chemicals are high and the operational costs for application are also high, because treatment must be repeated regularly as snail populations rebuild rapidly. There has been considerable research into the use of chemicals of plant origin to control snails (Mott 1987), but most of these compounds are also toxic to fish. Development of new formulations of existing molluscicides or development of new molluscicides is highly called for (Dai *et al.* 2008, 2010). Large-scale application of molluscicides should be avoided and if used application should be at focal transmission sites and at the right time (Rollinson *et al.* 2013).

Use of broad spectrum molluscicides to control the snail hosts of schistosomes is problematic and focus should be on compounds or formulations targeting intermediate hosts as selectively as possible. This is mainly due to environmental concerns about pesticide pollution in general (Ansara-Ross *et al.* 2012) and costs of large-scale mollusciciding operations.

In the search for new compounds, focus may be on plant molluscicides and possibly toxins from cyanobacteria. Currently there is an increasing interest in compounds from marine algae especially cyanobacteria (blue-green algae) which are known to produce potent toxins (Duke *et al.* 2010; Valério *et al.* 2010). Thus Versman (2010) presented a microbial product derived from *Pseudomonas fluorescens* that is highly selective for control of the invasive Zebra mussel, *Dreissena polymorpha*, which causes major problems in cooling systems of power plants. If systemic toxins with molluscidal effect were found in cyanobacteria and if the effect was retained in dead cells, these might be ideal for inclusion in bait formulations.

Since snails are attracted to certain chemicals such as amino acids and short-chain carboxylic acids (Attwood 1995; Thomas 1996), this might be utilized to develop highly specific controlled release formulations for selective removal of schistosome intermediate hosts (Thomas *et al.* 1980, 1983). This conceptually simple model involves using an ingestible matrix of the optimum size and composition to release specific snail attractants, arrestants and phagostimulants into the external environment and a toxicant in the gut of the target snail following ingestion (Thomas 1996). The release of the toxicant in the gut would avoid the problem of the toxicant being a repellent. In view of the ability of such third stage controlled release formulations (CRF3) to both attract snails and to induce ingestion, it should be possible to use them in small amounts and this should make them cost-effective (Thomas 1996).

4.4.3   Ecological Snail Control

Ecological measures to control pulmonate snail species include increasing water velocity beyond 0.3 m/s, regularly fluctuate the water level in reservoirs, removal of aquatic vegetation, concrete-lining of canals and others (Oomen *et al.* 1994; Laamrani *et al.* 2000; Boelee and Laamrani 2004). It is quite obvious that these measures do not necessarily work for the whole range of intermediate host species involved in schistosome transmission.

Drying of irrigation canals to control snail populations is helpful, but the critical parameter will be how fast the canal is drained, the effectiveness of the draining leaving no "dead waters" in the system and the duration of the drying-out period.

Amphibious snails are particularly sensitive to environmental control measures due to the nature of their habitats. Such measures include stream channelization, seepage control, canal lining, canal relocation with deep burial of snails, proper drainage in irrigation schemes, vegetation removal, earth filling, ponding, tree planting for drainage and improved agricultural practices. These measures may be expensive to implement but they can control *Oncomelania* effectively.

4.4.4   Biological Control

Many organisms exist that could be listed as predators, parasites, parasitoids, pathogens or competitors of snails and as such might be suggested for biological control of snails (McCullough

1980; Madsen 1996). However, most emphasis has been on other snails that could act as competitors of the intermediate hosts. Certain fish species are efficient predators of freshwater snails (Brodersen *et al.* 2002; Chimbari and Madsen 2003; Madsen and Stauffer 2011) and the inclusion of such a species in fish culture ponds in Africa might be an option to consider (polyculture) to avoid the risk schistosome transmission in such ponds. Also crustacean have received interest as snail predators for biological control (Lee *et al.* 1982; Sokolow *et al.* 2014). In choosing a competitor snail species or predator species for biological control, one should opt for a local species so as to avoid the introduction of foreign species.

4.4.5 Precautions

Although the snail fauna receives attention because of the involvement of some species as intermediate hosts for trematodes of medical or veterinary importance, it should be kept in mind that they play an important role in the functioning of freshwater ecosystems and efforts must therefore be taken to conserve their diversity. Many large lakes and rivers may have diverse snail faunas and extensive snail control operations could have serious impact on the local fauna. Therefore use of molluscicides may not be an option in large rivers or lakes, but also introduction of foreign species for biological snail control should be avoided in such habitats. Chemical measures of control are not necessarily ecologically unsound, especially if applied focally and similarly biological control measures are not necessarily sound, because the introduced agents could affect biodiversity negatively.

4.5 ISSUES RELATING TO THE INTERMEDIATE HOSTS

4.5.1 Changing Transmission Patterns

Historically, open shorelines of Lake Malawi were free from *S. haematobium* transmission, but this changed in the mid-1980s, possibly as a result of over-fishing reducing density of molluscivore fishes (Madsen *et al.* 2011). Lake Malawi, the most southerly lake in the East African Rift valley system, harbors more fish species than any other lake on Earth. In the southern part of the lake, the prevalence of *S. haematobium* was very high in some lake shore communities (Madsen *et al.* 2011) and this could be due to a recent adaptation of the parasite to utilize *Bulinus nyassanus* as host in addition to *Bulinus globosus* or the introduction of a new *Schistosoma* strain (Stauffer *et al.* 2014).

In Corsica (France) the first cases of autochtonous schistosomiasis (*Schistosoma haematobium*) were recently found (Holtfreter *et al.* 2014; Berry *et al.* 2014). It was shown that the schistosome responsible for the emergence of schistosomiasis in Corsica was due to *S. haematobium* introgressed by genes from *S. bovis* (Moné *et al.* 2015) (see also Jones *et al.*, Chapter 13). Such newly focal transmission needs rapid and focal control with molluscicides in order not to spray in other places in France and in Europe.

4.5.2 Spread of Snails

There are numerous examples of freshwater snail species, including some species of schistosome intermediate hosts, spreading beyond their natural geographic area and this is primarily a result of human intervention (Madsen and Frandsen 1989; Pointier *et al.* 2005). Snails may be spread over relatively short distances with water flow or attached to other animals, in mud on feet of or over somewhat longer distances passing alive through the digestive tract of migratory birds. However, the major mean of transport is the global trade in aquatic animals and plants.

Asian species such as *Tarebia granifera* have spread to South Africa, *Biomphalaria straminea* from South America to Asia, *Indoplanorbis exustus* to Africa and many other examples. Apple snails were introduced to Asia for food production and became invasive. Invasive species could have major impact on local biodiversity.

The many known examples of snails and parasites "travelling" over great distances, clearly points to the necessity of enforcing drastic measures which will reduce the risks considerably. In certain countries, e.g., Australia (Walker 1978) and USA (Burch 1960, 1982), a very stringent policy has been followed in order to prevent the import of undesirable animals and plants, primarily to avoid crop pests being introduced with, e.g., slugs, giant African snails and other species. Such instructions would be useful on a worldwide scale.

4.5.3   Climate Change

Climate plays an important role in the transmission of many infectious diseases; it not only determines spatial and seasonal distributions, but influences inter-annual variability, including epidemics and long-term trends (Kelly-Hope and Thomson 2008). Evidence of climate change includes the instrumental temperature record, rising sea levels and decreased snow cover in the Northern Hemisphere (Lotfy 2014). One of the most conspicuous effects of climate is an increased frequency of extreme weather conditions which can have devastating effects on the fauna in some vulnerable habitats (Kariuki *et al.* 2012).

Some studies have attempted to predict the effect of climate change on the distribution or transmission intensity of schistosomiasis using dynamic modeling (McCreesh and Booth 2014). These studies focused mainly on the effect of increasing mean temperature, with only one including changes in rainfall and none considering the effects of extreme weather events (McCreesh and Booth 2014). Global assessment of the potential impacts of anthropogenically induced climate change on vector-borne diseases suggests an increase in extent of the geographical areas susceptible to transmission of malarial *Plasmodium* parasites, dengue *Flavivirus* and *Schistosoma* worms (Martens *et al.* 1997). Future projections indicate that while the potential *S. mansoni* transmission area expands, the snail ranges are more likely to contract and/or move into cooler areas in the south and east (Stensgaard *et al.* 2013). Importantly, even though climate per se matters, the impact of humans on habitat plays a crucial role in determining the distribution of the intermediate host snails in Africa (Stensgaard *et al.* 2013). Thus, a future contraction in the geographical range size of the intermediate host snails caused by climatic changes does not necessarily translate into a decrease or zero sum change in human schistosomiasis prevalence (Stensgaard *et al.* 2013).

4.6   FUTURE RESEARCH

Detection of infected snails may not be a good predictor of transmission because prevalence of infection in snails is often low and yet infection in the final host can be high (Madsen *et al.* 2011). Application of DNA methods to detect infections in snails (Amarir *et al.* 2014) is essential to confirm absence of transmission in areas where control aims to eliminate transmission. In some situations cercariae are not necessarily produced in human water contact sites but could be carried into the sites with runoff water (Madsen *et al.* 2008) and in such situations examination of environmental DNA as used in fish surveys (Minamoto *et al.* 2012) to detect transmission sites would be useful and should be developed. Although snail control is deemed essential for elimination of schistosomiasis and mollusciciding is efficient in reducing transmission (King *et al.* 2015), we can never revert to large-scale mollusciciding programs as done in the early days of schistosomiasis control, because of the environmental impact of molluscicides. Hence there is need for more specific ways to remove the intermediate hosts from transmission sites, such as third stage controlled release formulations. Also more research should be done on environmental snail control (see also Ross *et al.*, Chapter 26).

4.7 LITERATURE CITED

Agrawal MC. 2012. Schistosomes and schistosomiasis in South Asia, doi: 10.1007/978-81-322-0539-5_2, Springer India.

Albrecht C, Kuhn K, Streit B. 2006. A molecular phylogeny of Planorboidea (Gastropoda, Pulmonata): insights from enhanced taxon sampling. Zoologica Scripta 36: 27–39.

Amarir F, Sebti F, Abbasi I, Sadak A, Fellah H, Nhammi H, Ameur B, El Idrissi AL, Rhajaoui M. 2014. *Schistosoma haematobium* detection in snails by DraI PCR and Sh110/Sm-Sl PCR: Further evidence of the interruption of schistosomiasis transmission in Morocco. Parasites and Vectors 7: 288.

Ansara-Ross TM, Wepener V, van den Brink PJ, Ross MJ. 2012. Pesticides in South African fresh waters. African Journal of Aquatic Science 37: 1–16.

Appleton CC. 1996. Molluscicides in snail control. pp. 213–228. *In*: Madsen H, Kristensen TK, Ndolovu P, editors. A Status of Research on Medical Malacology in Relation to Schistosomiasis in Africa. Proceedings from a Workshop held August 21st to 25th 1995, Harare, Zimbabwe. Blair Research Institute and Danish Bilharziasis Laboratory.

Arouna N, Remy M. 2005. Variation of the compatibility between *Schistosoma haematobium* and the *Bulinusna talensis/tropicus* complex (Gastropoda: Planorbidae) populations. African Journal of Biotechnology 4: 1010–1016.

Attwood SW, Ambu S, Meng XH, Upatham ES, Xu FS, Southgate VR. 2003. The phylogenetics of triculine snails (Rissooidea: Pomatiopsidae) from south-east Asia and southern China: historical biogeography and the transmission of human schistosomiasis. Journal of Molluscan Studies 69: 263–271.

Attwood SW, Upatham ES, Zhang YP, Yang ZQ, Southgate VR. 2004. A DNA-sequence based phylogeny for triculine snails (Gastropoda: Pomatiopsidae: Triculinae), intermediate hosts for *Schistosoma* (Trematoda: Digenea): phylogeography and the origin of *Neotricula*. Journal of Zoology 262: 47–56.

Attwood SW, Lokman HS, Ong KY. 2005. *Robertsiella silvicola*, a new species of triculine snail (Caenogastropoda: Pomatiopsidae) from Peninsular Malaysia, intermediate host of *Schistosoma malayensis* (Trematoda: Digenea). Journal of Molluscan Studies 71: 379–391.

Attwood SW. 1995. Uptake of acetate by *Neotricula aperta* (Gastropoda: Pomatiopsidae), the snail host of *Schistosoma mekongi* in the Lower Mekong Basin. Journal of Molluscan Studies 61: 109–125.

Attwood SW. 2001. Schistosomiasis in the Mekong region: epidemiology and phylogeography. Advances in Parasitology 50: 87–152.

Attwood SW. 2010. Studies on the parasitology, phylogeography and the evolution of host–parasite interactions for the snail intermediate hosts of medically important trematode genera in southeast Asia. Advances in Parasitology 73: 405–440.

Attwood SW, Kitikoon V, Southgate VR. 1997. Infectivity of a Cambodian isolate of *Schistosoma mekongi* to *Neotricula aperta* from Northeast Thailand. Journal of Helminthology 71: 183–187.

Babiker A, Fenwick A, Daffalla AA, Amin MA. 1985. Focality and seasonality of *Schistosoma mansoni* transmission in the Gezira Irrigated Area, Sudan. Journal of Tropical Medicine and Hygiene 88: 57–63.

Berry A, Moné H, Iriart X, Mouahid G, Abbo O, Boissier J, Fillaux J, Cassaing S, Debuisson C, Valentin A, Mitta G, Théron A, Magnaval JF. 2014. Schistosomiasis haematobium, Corsica, France [letter]. Emerging Infectious Diseases [Internet]. Sep [date cited]. http://dx.doi.org/10.3201/eid2009.140928.

Boelee E, Laamrani H. 2004. Environmental control of schistosomiasis through community participation in a Moroccan oasis. Tropical Medicine and International Health 9: 997–1004.

Brandt AMR. 1974. The non-marine aquatic Mollusca of Thailand. Archives für Molluskenkunde 105: 1–247.

Brodersen J, Chimbari MJ, Madsen H. 2002. An enclosure study to evaluate the effect of *Sargochromis codringtoni* on snail populations in an irrigation canal. African Zoology 37: 255–258.

Brown D. 1980. *Freshwater Snails of Africa and their Medical Importance*. London: Taylor & Francis 608 p.

Brown D. 1994. *Freshwater Snails of Africa and their Medical Importance (2nd Ed.)*. London: Taylor & Francis.

Brown DS. 1996. Systematic malacology: the foundation of snail-schistosome studies. pp. 27–45. *In*: Madsen H, Kristensen TK, Ndolovu P, editors. A Status of Research on Medical Malacology in Relation to Schistosomiasis in Africa. Proceedings from a Workshop held August 21st to 25th 1995, Harare, Zimbabwe. Blair Research Institute and Danish Bilharziasis Laboratory.

Burch JB. 1960. Some snails and slugs of quarantine significance to the United States. Sterkiana 2: 13–53.

Burch JB. 1982. Taxonomic and nomenclatural changes since 1960 in snails and slugs of quarantine significance to the United States. Malacological Review 15: 141–142.

Chimbari MJ, Madsen H. 2003. Predation on snails by an indigenous fish, *Sargochromis codringtonii*, in ponds in Zimbabwe. African Journal of Aquatic Science 28: 187–190.

Colgan DJ, Ponder WF, Beacham E, Macaranas J. 2007. Molecular phylogenetics of Caenogastropoda (Gastropoda: Mollusca). Molecular Phylogenetics and Evolution 42: 717–737.

Criscione F, Ponder WF. 2013. A phylogenetic analysis of rissooidean and cingulopsoidean families (Gastropoda: Caenogastropoda). Molecular Phylogenetics and Evolution 66: 1075–1082.

Dai JR, Coles GC, Wang W, Liang YS. 2010. Toxicity of a novel suspension concentrate of niclosamide against *Biomphalaria glabrata*. Transactions of the Royal Society of Tropical Medicine and Hygiene 104: 304–306.

Dai JR, Wang W, Liang YS, Li HJ, Guan XH, Zhu YC. 2008. A novel molluscicidal formulation of niclosamide. Parasitology Research 103: 405–412.

Danish Bilharziasis Laboratory 1986. A field guide to African freshwater snails. 3. North East African species. Second Edition. Danish Bilharziasis Laboratory 31 pp.

Danish Bilharziasis Laboratory 1998. A field guide to African freshwater snails. Introduction. Danish Bilharziasis Laboratory 28 pp.

Davis GM, Kitikoon V, Temcharoen P. 1976. Monograph on *"Lithoglyphopsis" aperta*, the snail host of Mekong river schistosomiasis. Malacologia 15: 241–287.

Davis GM, Wu WP, Williams G, Liu HY, Lu SB, Chen HG, Zheng F, McManus DP, Guo JG. 2006. Schistosomiasis japonica intervention study on Poyang lake, China: The snail's tale. Malacologia 49: 79–105.

Davis GM. 1979. The origin and evolution of the gastropod family pomatiopsidae, with emphasis on the Mekong River Triculinae. Proceedings of the Academy of Natural Sciences of Philadelphia 20: 1–120.

Davis GM. 1980. Snail hosts of Asian *Schistosoma* infecting man: evolution and coevolution. *In*: Bruce JI, Sornmani S, Asch HL, Crawford KA, editors. *The Mekong Schistosome*, Malacological Review Suppl 2: 195–238.

Davis GM. 1992. Evolution of prosobranch snails transmitting Asian Schistosoma; coevolution with Schistosoma: a review. Progress in Clinical Parasitology 3: 145–204.

De Kock KN, Wolmarans CT, Strauss HD, Killian M, Maree DC. 2002. Geographical distribution and habitats of the freshwater snail *Bulinus reticulatus* and its susceptibility to *Schistosoma haematobium* miracidia under experimental conditions. African Zoology 37: 1–6.

De Kock KN, Wolmarans CT. 2006. Distribution and habitats of *Bulinus natalensis* and its role as intermediate host of economically important helminth parasites in South Africa. African Journal of Aquatic Science 31: 63–69.

DeJong RJ, Morgan JAT, Paraense WL, Pointier JP, Amarista M, Ayeh-Kumi PFK, Babiker A, Barbosa CS, Brémond P, Canese AP, De Souza CP, Dominguez C, File S, Gutierrez A, Nino INC, Kawano T, Kazibwe F, Kpikpi J, Lwambo NJS, Mimpfoundi R, Njiokou F, Poda JN, Sene M, Velasquez LE, Yong M, Adema CM, Hofkin BV, Mkoji GM, Loker ES. 2001. Evolutionary relationships and biogeography of *Biomphalaria* (Gastropoda: Planorbidae) with implications regarding its role as host of the human bloodfluke, *Schistosoma mansoni*. Molecular Biology and Evolution 18: 2225–2239.

DeJong RJ, Morgan JAT, Wilson WD, Al-Jaser MH, Appleton CC, Coulibaly G, D'Andrea PS, Doenhoff MJ, Haas W, Idris MA, Magalhaes LA, Moné H, Mouahid G, Mubila L, Pointier JP, Webster JP, Zanotti-Magalhaes EM, Paraense WL, Mkoji GM, Loker ES. 2003. Phylogeography of *Biomphalaria glabrata* and *B. pfeifferi*, important intermediate hosts of *Schistosoma mansoni* in the New and Old World tropics. Molecular Ecology 12: 3041–3056.

Duke SO, Cantrell CL, Meepagala KM, Wedge DE, Tabanca N, Schrader KK. 2010. Natural toxins for use in pest management. Toxins 2: 1943–1962.

Grande C, Templado J, Cervera L, Zardoya R. 2004. Molecular phylogeny of Euthyneura (Mollusca: Gastropoda). Molecular Biology and Evolution 21: 303–313.

Greer GJ, Lim HK, Ow-Yang CK. 1980. Report of a freshwater hydrobiid snail from Pahang, Malaysia: a possible host for schistososomes infecting man. The Southeast Asian Journal of Tropical Medicine and Public Health 11: 146–147.

Hauswald AK, Remais JV, Xiao N, Davis GM, Lu D, Bale MJ, Wilke T. 2011. Stirred, not shaken: genetic structure of the intermediate snail host *Oncomelania hupensis robertsoni* in a historically endemic schistosomiasis area. Parasites and Vectors 4(1): 206.

Holtfreter MC, Moné H, Müller-Stöver I, Mouahid G, Richter J. 2014. *Schistosoma haematobium* infections acquired in Corsica, France, August 2013. Eurosurveillance 19(22): http://www.eurosurveillance.org/ViewArticle.aspx?ArticleId=20821.

Hope M, McManus DP. 1994. Genetic variation in geographically isolated populations and subspecies of *Oncomelania hupensis* determined by a PCR-based RFLP method. Acta Tropica 57: 75–82.

Hubendick B. 1955. Phylogeny in the Planorbidae. Transactions of the Zoological Society of London 28: 453–542.

Ishii A. 2005. Successful parasite controls in Japan. Eradication of schistosomiasis. pp. 184–193. *In*: Chen M-G, Zhou X-N, Hirayama K, editors. Schistosomiasis in Asia. Asian Parasitology series monograph, Vol. 5. The Federation of Asian Parasitologists, Chiba, Japan.

Jørgensen A, Kristensen TK, Stothard JR. 2004. An investigation of the "Ancylo-planorbidae" (Gastropoda, Pulmonata, Hygrophila): preliminary evidence from DNA sequence data. Molecular Phylogenetics and Evolution 32: 778–787.

Jørgensen A, Madsen H, Nalugwa A, Nyakaana S, Rollinson D, Stothard JR, Kristensen TK. 2011. A molecular phylogenetic analysis of *Bulinus* (Gastropoda: Planorbidae) with conserved nuclear genes. Zoologica Scripta 40: 126–136.

Jørgensen A, Stothard JR, Madsen H, Nalugwa A, Nyakaana S, Rollinson D. 2013. The ITS2 of the genus *Bulinus*: novel secondary structure among freshwater snails and potential new taxonomic markers. Acta Tropica 128: 218–225.

Kali A. 2015. Schistosome infections: an Indian perspective. Journal of Clinical and Diagnostic Research 9: DE01–DE04.

Kane RA, Stothard JR, Emery AM, Rollinson D. 2008. Molecular characterization of freshwater snails in the genus *Bulinus*: a role for barcodes? Parasites and Vectors 1: 15.

Kariuki HC, Madsen H, Sturrock RF, Ouma JH, Butterworth AEB, Dunne D, Gachuhi K, Vennervald BJ, Mwatha J, Booth M, Muchiri E. 2013. Focal mollusciciding using niclosamide (Bayluscide) in stream habitats and its effect on transmission of schistosomiasis mansoni following community-based chemotherapy at Kibwezi in Makueni District, Kenya. Parasites and Vectors 6: 107.

Kazibwe F, Makanga B, Rubaire-Akiiki C, Ouma J, Kariuki C, Kabatereine NB, Booth M, Vennervald BJ, Sturrock RF, Stothard JR. 2006. Ecology of *Biomphalaria* (Gastropoda: Planorbidae) in Lake Albert, Western Uganda: snail distributions, infection with schistosomes and temporal associations with environmental dynamics. Hydrobiologia 568: 433–444.

Kelly-Hope L, Thomson MC. 2008. Chapter 3. Climate and Infectious Diseases. *In*: Thomson MC, Garcia-Herrera R, Beniston M, editors. *Seasonal Forecasts, Climatic Change and Human Health*. Springer Science + Business Media B.V. 232 p.

King CH, Sturrock RF, Kariuki HC, Hamburger J. 2008. Transmission control for schistosomiasis – why it matters now. Trends in Parasitology 22: 575–582.

King CH, Sutherland LJ, Bertsch D. 2015. Systematic review and meta-analysis of the impact of chemical-based mollusciciding for control of *Schistosoma mansoni* and *S. haematobium* transmission. PLoS Neglected Tropical Diseases 9(12): e0004290.

Laamrani H, Khallaayoune K, Boelee E, Laghroubi MM, Madsen H, Gryseels B. 2000. Evaluation of environmental methods to control snails in an irrigation system in Central Morocco. Tropical Medicine and International Health 5: 545–552.

Labbo R, Djibrilla A, Zamanka H, Garba A, Chippaux JP. 2007. *Bulinus forskalii*: a new potential intermediate host for *Schistosoma haematobium* in Niger. Transactions of the Royal Society of Tropical Medicine and Hygiene 101: 847–848.

Lange CN, Kristensen TK, Madsen H. 2013. Gastropod diversity, distribution and abundance in habitats with and without anthropogenic disturbances in Lake Victoria, Kenya. African Journal of Aquatic Science 38: 295–304.

Lee PG, Rodrick GE, Sodeman Jr, Blake NJ. 1982. The giant Malaysian prawn, *Macrobrachium rosenbergii*, a potential predator for controlling the spread of schistosome vector snails in fish ponds. Aquaculture 28: 293–301.

Leiper RT. 1915. Observations on the mode of spread and prevention of vesical and intestinal bilharziosis in Egypt, with additions to August, 1916. Occasional lecture. The Royal Society of Medicine. p. 145–172.

Liu L, Huo GN, He HB, Zhou B, Attwood SW. 2014. A phylogeny for the pomatiopsidae (Gastropoda: Rissooidea): a resource for taxonomic, parasitological and biodiversity studies. BMC Evolutionary Biology 14(1): 29.

Lotfy WM. 2014. Climate change and epidemiology of human parasitosis in Egypt: a review. Journal of Advanced Research 5: 607–613.

Lotfy WM, DeJong RJ, Abdel-Kader A, Loker ES. 2005. A molecular survey of *Biomphalaria* in Egypt: is *B. glabrata* present? American Journal of Tropical Medicine and Hygiene 73: 131–139.

Madsen H. 1996. Ecological field studies on schistosome intermediate hosts: relevance and methodological considerations. pp. 67–87. *In*: Madsen H, Kristensen TK, Ndolovu P, editors. A Status of Research on Medical Malacology in Relation to Schistosomiasis in Africa. Proceedings from a Workshop held August 21st to 25th 1995, Harare, Zimbabwe. Blair Research Institute and Danish Bilharziasis Laboratory.

Madsen H, Bloch P, Makaula P, Phiri H, Furu P, Stauffer JR. 2011. Schistosomiasis in Lake Malawi villages. EcoHealth 8: 163–176.

Madsen H, Bloch P, Phiri H, Kristensen TK, Furu P. 2001. *Bulinus nyassanus* is an intermediate host for *Schistosoma haematobium* in Lake Malawi. Annals of Tropical Medicine and Parasitology 95: 353–360.

Madsen H, Carabin H, Balolong D, Tallo VL, Olveda R, Yuan M, MacGarvey ST. 2008. Prevalence of *Schistosoma japonicum* infection of *Oncomelania quadrasi* snail colonies in 50 irrigated and rain-fed villages of Samar province, the Philippines. Acta Tropica 105: 235–241.

Madsen H, Coulibaly G, Furu P. 1987. Distribution of freshwater snails in the river Niger basin in Mali with special reference to the intermediate hosts of schistosomes. Hydrobiologia 146: 77–88.

Madsen H, Frandsen F. 1989. The spread of freshwater snails including those of medical and veterinary importance. Acta Tropica 46: 139–146.

Madsen H, Stauffer JR. 2011. Density of *Trematocranus placodon* (Pisces: Cichlidae): a predictor of density of the schistosome intermediate host, *Bulinus nyassanus* (Gastropoda: Planorbidae), in Lake Malawi. EcoHealth 8: 177–189.

Magendantz M. 1972. The biology of *Biomphalaria choanomphala* and *B. sudanica* in relation to their role in the transmission of *Schistosoma mansoni* in Lake Victoria at Mwanza, Tanzania. Bulletin of the World Health Organization 47: 331–341.

Majoros G, Fehér Z, Deli T, Földvári G. 2008. Establishment of *Biomphalaria tenagophila* snails in Europe. Emerging Infectious Diseases 14: 1812–1814.

Malek EA. 1985. Snail hosts of schistosomiasis and other snail-transmitted diseases in Tropical America: a manual. Scientific Publication No. 478, Pan American Health Organization, World Health Organization, Washington, D.C. 20037, USA.

Mandahl-Barth G. 1972. The freshwater mollusca of Lake Malawi. Revue de Zoologie et de Botanique Africaines 86: 257–289.

Mandahl-Barth G. 1957a. Intermediate hosts of *Schistosoma*. African *Biomphalaria* and *Bulinus*. 1. *Biomphalaria*. Bulletin of the World Health Organization 16: 1103–1163.

Mandahl-Barth G. 1957b. Intermediate hosts of Schistosoma. African *Biomphalaria* and *Bulinus*. 2. *Bulinus*. Bulletin of the World Health Organization 17: 1–65.

Mandahl-Barth G. 1958. Intermediate hosts of *Schistosoma*. African *Biomphalaria* and *Bulinus*. World Health Organization Monograph Series No. 37. World Health Organization, 132 pp.

Martens WJM, Jetten TH, Focks DA. 1997. Sensitivity of malaria, schistosomiasis and dengue to global warming. Climatic Change 35: 145–156.

McCreesh N, Booth M. 2014. The effect of simulating different intermediate host snail species on the link between water temperature and schistosomiasis risk. PLoS ONE 9(7): e87892.

McCullough FS, Gayral Ph, Duncan J, Christie JD. 1980. Molluscicides in schistosomiasis control. Bulletin of the World Health Organization 58: 681–689.

McCullough FS. 1980. Biological control of the snail intermediate hosts of human *Schistosoma* spp.: a review of its present status and future propects. Acta Tropica 38: 5–13.

McCullough FS. 1992. The role of mollusciciding in schistosomiasis control. WHO/SCHISTO/92.107.

Moné H, Holtfreter MC, Allienne J-F, Mintsa-Nguéma R, Ibikounlé M, Boissier J, Berry A, Mitta G, Richter J, Mouahid G. 2015. Introgressive hybridizations of *Schistosoma haematobium* by *Schistosoma bovis* at the origin of the first case report of schistosomiasis in Corsica (France, Europe). Parasitology Research doi: 10.1007/s00436-015-4643-4.

Morgan JAT, DeJong RJ, Kazibwe F, Mkoji GM, Loker ES. 2003. A newly-identified lineage of *Schistosoma*. International Journal for Parasitology 33: 977–985.

Mott KE, editor. 1987. *Plant Molluscicides*. John Wiley & Sons, Chichester 326 p.

Oomen JMV, de Wolf J, Jobin WR. 1994. Health and irrigation. Incorporation of disease-control measures in irrigation, a multi-faceted task in design, construction, operation. Publication 45. International Institute for Land Reclamation and Improvement/ILRI. Wageningen.

Pan American Health Organization 1968. A guide for the identification of the snail intermediate hosts of schistosomiasis in the Americas. Scientific Publication No. 168, Pan American Health Organization, World Health Organization, Washington, D.C. 20037, USA. 122 p.

Pesigan TP, Hairston NG, Jauregui JJ, Garcia EG, Santos AT, Santos BC, Besa AA. 1958. Studies on *Schistosoma japonicum* infection in the Philippines. 2. The molluscan host. Bulletin of the World Health Organization 18: 481–578.

Pflüger W. 1982. Introduction of *Biomphalaria glabrata* to Egypt and other African countries. Transactions of the Royal Society of Tropical Medicine and Hygiene 76: 567.

Pointier JP, DeJong RJ, Tchuem Tchuenté LA, Kristensen TK, Loker ES. 2005. A neotropical snail host of *Schistosoma mansoni* introduced into Africa and consequences for the schistosomiasis transmission: *Biomphalaria tenagophila* in Kinshasa (Democratic Republic of Congo). Acta Tropica 93: 191–199.

Pointier JP, David P, Jarne P. 2005. Biological invasions: the case of planorbid snails. Journal of Helminthology 79: 249–256.

Ponder WF, Colgan DJ, Healy JM, Nützel A, Simone LRL, Strong EE. 2008. Caenogastropoda. pp. 331–383. *In*: Ponder WF, Lindberg DL, editors. *Phylogeny and Evolution of the Mollusca*. University of California Press, Berkeley, Los Angeles, London.

Ponder WF, Lindberg DR. 1997. Towards a phylogeny of gastropod molluscs: an analysis using morphological characters. Zoological Journal of the Linnean Society London 119: 83–265.

Rollinson D, Southgate VR. 1987. The genus *Schistosoma*: a taxonomic appraisal. *In*: Rollinson D, Simpson AJ, editors. *The Biology of Schistosomes: From Genes to Latrines*. Academic Press, London, pp. 149.

Rollinson D, Knopp S, Levitz S, Stothard JR, Tchuem Tchuenté L-A, Garba A, Mohammed KA, Schur N, Person B, Colley DG, Utzinger J. 2013. Time to set the agenda for schistosomiasis elimination. Acta Tropica 128: 423–440.

Sokolow SH, Lafferty KD, Kuris AM. 2014. Regulation of laboratory populations of snails (*Biomphalaria* and *Bulinus* spp.) by river prawns, *Macrobrachium* spp. (Decapoda, Palaemonidae): implications for control of schistosomiasis. Acta Tropica 132: 64–74.

Stauffer JR, Madsen H, Rollinson D. 2014. Introgression in Lake Malawi: increasing the threat of human urogenital schistosomiasis? EcoHealth 11: 251–254.

Stensgaard A-S, Utzinger J, Vounatsou P, Hürlimann E, Schur N, Saarnak CFL, Simoonga C, Mubita P, Kabatereine NB, Tchuenté L-AT, Rahbeka C, Kristensen TK. 2013. Large-scale determinants of intestinal schistosomiasis and intermediate host snail distribution across Africa: does climate matter? Acta Tropica 128: 378–390.

Stothard JR, Brémond P, Andriamaro L, Sellin B, Sellin E, Rollinson D. 2001. *Bulinus* species on Madagascar: molecular evolution, genetic markers and compatibility with *Schistosoma haematobium*. Parasitology 123(Suppl): S261–S275.

Temcharoen P. 1971. New aquatic molluscs from Laos. Archive für Mollusken-kunde 101: 91–109.

Thomas JD. 1996. The application of behavioural and biochemical ecology towards the development of controlled release formulations (CRFs or environmental antibodies) for the selective removal of the snail hosts of schistosome parasites. pp. 275–321. *In*: Madsen H, Kristensen TK, Ndlovu P, editors. Proceedings of '*A Status of Research on Medical Malacology in Relation to Schistosomiasis in Africa*', Zimbabwe, August 1995.

Thomas JD, Ofosu-Barko J, Patience RL. 1983. Behavioural responses to carboxylic and amino acids by *Biomphalaria glabrata* (Say), the snail host of *Schistosoma mansoni* and other freshwater molluscs. Comparative Biochemistry and Physiology 75C: 57–76.

Thomas JD, Cowley C, Ofosu-Barko J. 1980. Behavioural responses to amino and carboxylic acids by *Biomphalaria glabrata* one of the snail hosts of *Schistosoma mansoni*. 6th International Symposium on Controlled Release of Bioactive Materials. (Ed. R. Baker), pp. 433–448. Academic Press, New York & London.

Upatham ES, Kruatrachue M, Viyanant V, Khunborivan V, Kunatham L. 1985. Biology of *Robertsiellaka porensis* snails and Malaysian *Schistosoma*. Southeast Asian Journal of Tropical Medicine and Public Health 16: 1–9.

Upatham ES, Sornmani S, Thirachantra S, Sitaputra P. 1980. Field studies on the bionomics of alpha and gamma races of *Triculaaperta* in the Mekong River at Khemmarat, Ubol Ratchathani province, Thailand. *In*: Bruce JI, Sornmani S, Asch HL, Crawford KA, editors. The Mekong Schistosome, Malacological Review (Suppl 2): 239–261.

Valério E, Chaves S, Tenreiro R. 2010. Diversity and impact of prokaryotic toxins on aquatic environments: a review. Toxins 2: 2359–2410.

Versman J. 2010. ZEQUANOX. Microbial molluscicide. Marrone Bio Innovations. (Pdf-file giving presentation of the product).

Walker J. 1978. The finding of *Biomphalaria straminea* amongst fish imported into Australia. Unpublished document WHO/SCHISTO/78: 46.

Wilke T, Davis GM, Cui E, Xiao-Nung Z, Xiao Peng Z, Yi Z, Spolsky CM. 2000. *Oncomelania hupensis* (Gastropoda: Rissooidea) in eastern China: molecular phylogeny, population structure and ecology. Acta Tropica 77: 215–227.

Woodruff DS, Carpenter MP, Upatham ES, Viyanant V. 1999. Molecular phylogeography of *Oncomelania lindoensis* (Gastropoda: Pomatiopsidae), the intermediate host of *Schistosoma japonicum* in Sulawesi. Journal of Molluscan Studies 65: 21–31.

Woodruff DS, Mulvey M, Yipp MW. 1985a. Population genetics of *Biomphalaria straminea* in Hong Kong: a neotropical schistosome-transmitting snail recently introduced into China. Journal of Heredity 76: 355–360.

Woodruff DS, Mulvey M, Yipp MW. 1985b. The continued introduction of intermediate host snails of *Schistosoma mansoni* into Hong Kong. Bulletin of the World Health Organization 63: 622.

Woodruff DS, Staub KC, Upatham ES, Viyanant V, Yuan H-C. 1988. Genetic variation in *Oncomelania hupensis*: *Schistosoma japonicum* transmitting snails in China and the Philippines are distinct species. Malacologia 29: 347–361.

World Health Organization. 1989. World Health Organization slide set. Subject: the intermediate snail hosts of African schistosomiasis. Taxonomy, ecology and control. World Health Organization.

Wright CA, Klein J, Eccles DH. 1967. Endemic species of *Bulinus* in Lake Malawi. Journal of Zoology, London 151: 199–209.

Xu X-J, Zhou XN, Zhang Y. 2005. 8. Biology and control of Oncomelania snail. pp. 133–150. *In*: Chen M-G, Zhou XN, Hirayama K, editors. *Asian Parasitology, Schistosomiasis in Asia, Vol. 5*. The Federation of Asian Parasitologists, Chiba, Japan.

Yacoubi B, Zekhnini A, Moukrim A, Rondelaud D. 2007. *Bulinus truncatus*, *Planorbarius metidjensis* and endemic bilharziosis in the South-West of Morocco. Bulletin de la Société de Pathologie Exotique 100: 174–175.

Yousif F, Haroun N, Ibrahim A, El-Bardicy S. 1996. *Biomphalaria glabrata*: a new threat for schistosomiasis transmission in Egypt. Journal of the Egyptian Society of Parasitology 26: 191–205.

Yousif F, Ibrahim A, Abdel Kader A, El-Bardicy S. 1998. Invasion of the Nile Valley in Egypt by a hybrid of *Biomphalaria glabrata* and *Biomphalaria alexandrina*, snail vectors of *Schistosoma mansoni*. Journal of the Egyptian Society of Parasitology 28: 569–582.

CHAPTER

Schistosoma Egg

Malcolm K Jones[1],, Renata Russo Frasca Candido[1, 2, 3],
Tim St Pierre[3], Robert Woodward[3],
John Kusel[4] and Carlos Graeff Teixeira[2]*

5.1 INTRODUCTION

Schistosomes are parasites of the vasculature of their hosts. While the intravascular niche provides a rich source of nutrition, the location deep within tissues presents a major problem, namely, dispersal of offspring into the environment where they can infect the intermediate host. The eggs themselves orchestrate their escape, by inducing an intense cellular response that forces the egg across the endothelium and the lining tissues of organs such as the intestine or urinary bladder. The ova and oogenesis are treated by Grevelding *et al.*, Chapter 15. The eggs are central in pathogenesis and diagnosis of schistosomiasis. The females and the males, prior to reproductive maturation, will pair and then migrate to the mesenteric veins of the host, where the females will lay hundreds of eggs daily (Wilson *et al.* 2007). In the case of *S. mansoni*, adult worm pairs lay approximately 350 eggs per day and *S. japonicum* lay approximately three times more (Cheever *et al.* 1994). About a third to one half of the eggs released by the females will not reach the external environment (Cheever *et al.* 1994; Pearce and MacDonald 2002).

Schistosome eggs (Fig. 5.1) are distinctive structures, in terms of their morphology as well as their development and behavior within the host. This chapter will overview the structure of the egg, trace its development in the host tissues and examine the hatching behavior of the miracidium.

[1] School of Veterinary Sciences, The University of Queensland, Brisbane, Australia.
[2] Laboratório de Biologia Parasitária, Faculdade de Biociências e Laboratório de Parasitologia Molecular, Instituto de Pesquisas Biomédicas, Pontifícia Universidade Católica do Rio Grande do Sul, Porto Alegre, Brazil.
[3] School of Physics, The University of Western Australia, Perth, Australia.
[4] The University of Glasgow, Glasgow, Scotland.
* Corresponding author

5.2   GENERAL MORPHOLOGY OF THE EGG

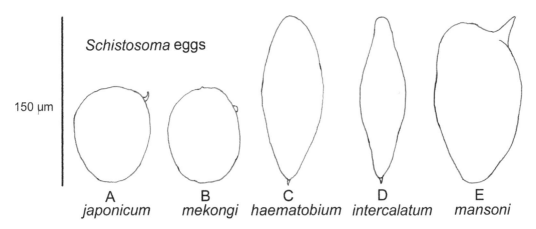

FIG. 5.1 Outlines of eggs of five species of *Schistosoma*. Redrawn from photographs in *Bustinduy AL and King CH. 2014.* In: Farrar J. editor. *Manson's Tropical Infectious Diseases* (Twenty-Third Edition). London: W.B. Saunders. P698–725. Fig. 52.3.

5.3   EMBRYOLOGY

Until the recent work of Jurberg *et al.* (2009), details of the development of the embryo in the egg have been lacking. This may be because such a study requires a range of histological and immunochemical techniques to reveal the complexity of the development of the cellular structures and organelles. Early work used a classification system in which a number of stages were identified in the development of the egg as they were deposited in the tissues of the host (Vogel 1942; Prata 1957). These stages were discrete and could be identified readily by light microscopy. The classification was used with great value by Pellegrino *et al.* (1962) to detect the effects of drugs on the ability of the female worm to produce viable eggs *in vivo* in a range of experimental animals and patients. The method is still of value (Mati and Melo 2013). More recently Hartenstein and Ehlers (2000) have described a detailed series of developmental changes in the free-living rhabdocoel *Mesostoma lingua*. This has stimulated work in the schistosomes by Jurberg *et al.* (2009) who have described the embryogenesis of *Schistosoma mansoni* using the following techniques: light microscopy alone and after fixation with several protocols and staining with dyes of a range of specificities including those for neutralgly coproteins and polysaccharides, collagen and smooth muscle and proteoglycans with a range of sulphate groups. An elegant technique of carmine staining followed by confocal microscopy (Jurberg *et al.* 2008) was also used.

These authors were able to produce a classification of the embryological development into eight stages, similar in number to that of Hartenstein and Ehlers (2000). The egg developing within the fertilised female was observed and after egg laying *in vivo* and *in vitro*, developmental processes were compared between those eggs laid in the living host (Swiss Webster mice) and after *in vitro* culture.

The stages were defined after careful observation (see Fig. 5.2, after Jurberg *et al.* 2009) and were compared with the stages found by light microscopy alone by Vogel (1942) and Prata (1957).

Stage 0: A single oocyte leaves the ovary and each oocyte is fertilised by a sperm in the seminal receptacle distal to the ovary. The fertilised egg is the zygote. This passes through the oviduct during which time it is met and surrounded by the vitelline cells from the vitelline duct. The number of vitelline cells around the zygote was found to be between 40–45 cells. This zygote plus surrounding vitellocytes is pushed into the ootype where the eggshell is formed. The egg is then moved into the uterus before being released from the female through the genital pore into the external environment, blood stream (*in vivo*) or culture medium (*in vitro*). The egg has not divided at this stage when laid either *in vivo* or *in vitro*. The vitelline cells are smaller than the zygote and are full of granules stained blue with Giemsa.

Stage 1: The first division (cleavage) of the zygote gives blastomeres of slightly different size. Their nuclei are large and round and have a densely staining nucleolus. The cytoplasm is uniformly stained with Giemsa and PAS and the authors conclude that this is indicative of protein synthesis. Vitelline cells have vacuolated cytoplasm, highly red stained in PAS-AB, which indicates the presence of neutral polysaccharides and neutral glycoproteins. The first cleavage initiated fusion of the vitelline cells into a yolk syncytium.

Stage 2: The number of embryonic cells increases to four to six cells wide and six to eight cells long. Some cells with pycnotic nuclei suggest the occurrence of apoptosis. Vitelline cells fuse completely. This stained as in Stage 1 red with PAS-AB. Some vitelline nuclei seem to adhere to the eggshell. The outer envelope begins its formation by cells detaching from each pole of the embryo and attaching to the eggshell.

Stage 3: Eggs are significantly larger than in Stage 2 and the embryo cell mass occupies two-thirds of the length of the egg. The yolk syncytium is still heavily stained red with PAS-AB. Some cell differentiation occurs with some cells concentrating in the centre of the embryo which are thought to be neuroblasts. Some signs of apoptosis continue and some peripheral cells begin the differentiation of the inner envelope. The outer envelope shows syncytium formation of the cells adhering to the shell and further syncytium formation occurs within the yolk syncytium.

Stage 4: The inner envelope detaches from the yolk syncytium and has surrounded the whole embryo. The outer envelope remains fused to the yolk syncytium and begins to reduce. The center of the embryo is occupied by the neural mass primordium (central neural mass). These neural precursors are arranged in a layer one or two cells thick around an a cellular region (neuropile). Immature cells are concentrated at the periphery of the embryo.

Stage 5: The size of the egg continues to increase. The embryo now extends into the whole volume of the egg. The outer envelope has become a thin anucleate membrane and the inner envelopeis highly stained by PAS-A Band continues to be a syncytium. The neural mass has increased in size in both its cell number and a cellular region. Beneath the inner envelope the epidermis begins to differentiate and forms a syncytium, blue staining with Giemsa, indicating active protein synthesis.

Stage 6: The lateral glands form from immature cells on each side of the neural mass primordium. These cells are highly PAS positive showing accumulations of carbohydrate rich macromolecules. Other cells, in the anterior region of the embryo form the apical gland. Also the terebratorium forms anteriorly from the epidermis. The epidermis becomes well differentiated from the inner mesenchyme. In themid-posterior region the germ cells can be identified, with large round nuclei. Muscle precursors can be seen under the epidermis.

Stage 7: The embryo now occupies the entire volume of the egg, but no movement can be seen. The unicellular lateral glands are highly red stained with PASH (periodic acid-Schiff/hematoxylin), suggesting richness in carbohydrate. The apical gland is also stained red with PASH. The neural mass undergoes nuclear condensation, so that the neural cells (neurons) have each a small nucleus. The musculature beneath the epidermis can be stained by Masson Trichrome and the longitudinal and circular muscles throughout the embryo can be revealed by monoclonal antibody and phalloidin labelling.

Stage 8: The embryos occupy the whole volume of the egg, are moving and showing flame cell activity and activity of cilia. Germ cells are found in the mid-posterior region in groups of three to five cells. The epidermal cells form epidermal plates. These are separated from the underlying musculature by a basal membrane. The outer and inner envelope retain the structure seen in stage 5.

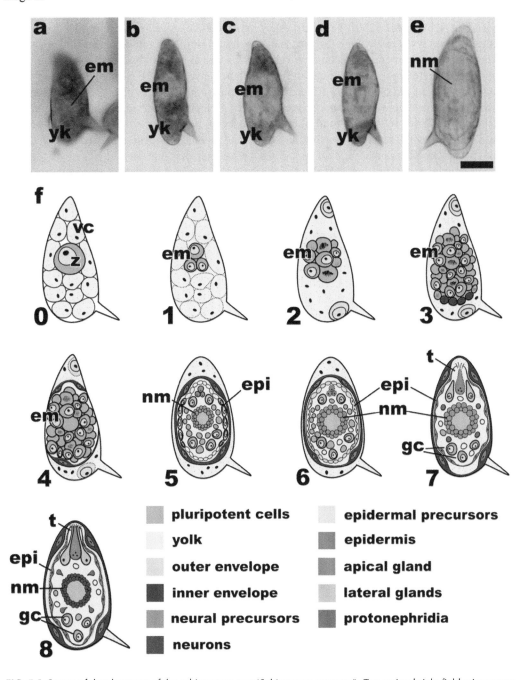

FIG. 5.2 Stages of development of the schistosome egg (*Schistosoma mansoni*). Top series: bright field microscopy stages I to V as in the method of Vogel and Prata (see text) em, embryo; epi, epidermal cells; gc, germinal cells, i.e., inner envelope; lg, lateral glands; t, terebratorium; yk, yolk, nm, neural mass, vc, vitelline cells. Diagrams 0–8 Stages described in this chapter, from Jurberg *et al.* 2009. Development Genes and Evolution 219: 219–234. Fig. 8.

The above description of the elegant work of Jurberg *et al.* (2009) makes the following points and challenges.

The schistosome mode of egg development has the organ, the vitellarium, which provides cells that closely adhere to the zygote before entering the ootype for eggshell formation. From the morphological evidence, the vitelline cells seem to contribute to eggshell formation and also in nourishment of the developing embryo, although this is uncertain without further metabolic studies. The work of Ashton *et al.* (2001) has clearly shown, using radioactive amino acids that the embryo can metabolize very actively from exogenous molecules. Most of this activity seemed to arise from protein synthesis occurring in the outer envelope. This was confirmed by immunocytochemistry, locating the secreted proteins predominantly in the outer envelope. The eggshell has pores which are used by the egg to procure host molecules for nourishment during its exposure to the host environment. These pores are also a passage way for molecules secreted by the embryo and its membranes as shown by Ashton *et al.* (2001). Such secretions from the mature egg (stage 8) are responsible for the immune response around the egg and granuloma formation. Ashton *et al.* (2001) found that six proteins, including two peptidases were secreted in this way. This is much fewer in number than the widely used soluble egg antigen (SEA). Many SEA molecules were not present in those secreted after biosynthetic labelling. Immunoreactivity shared with SEA was thought to be due to glycan epitopes carried by several of the labelled proteins. Further work on the metabolism of the egg during differentiation is awaited with great interest!

Very interesting work has been reported to inhibit the eggshell formation by altering the histone transacetylase of the parasite as it forms the egg in the ootype (Carneiro *et al.* 2014). The inhibition with herbimycin A ofenzymes involved in signaling egg development (protein tyrosine kinases) has also been effective (Knobloch *et al.* 2006).

The egg is undeveloped when laid and all the stages 1–8 occur in the host environment. The signals from the host required to complete development are not known. Swiderski (1994) has shown that before the first cleavage (stage 0) vitelline cells begin to fuse to form a syncytium. Thus the signal for this is important to determine.

The results of PAS and PAS-AB staining during egg development described by Jurberg *et al.* (2009) suggest production of neutral glycoproteins and polysaccharides and their secretion from apical and lateral glands and inner envelope. These macromolecules can pass through the eggshell pores and stimulate the complex immune response, which causes granuloma but also is used by the egg to exit the tissues (Doenhoff *et al.* 1986).

Jurberg *et al.* (2009) draw attention to the germ cells of the miracidium and suggest that they may be homologous to the neoblasts of other platyhelminths (*Schmidtea* for example). The germ cells have great proliferative ability and are rich in mitochondria, unlikeneoblasts. This suggests the need of oxidative metabolism and a possible very rapid almost explosive nature of cell division when the miracidium has entered the snail host.

There is nothing else in nature that looks like a schistosome egg. The egg of each species, formed singly in the ootype, as described above. This acts in part as an egg mould, is produced with or without, a distinctive form (Fig. 5.1), with or without a pronounced spine that is characteristic of that species. For this reason, the eggs have been a target for diagnostic approaches to schistosomiasis, as the presence of an egg in stool or urine is almost certainly diagnostic of infection with that species of schistosome. The general morphology of the eggs of schistosomes is shown in Figure 5.1. The reader is referred to Jones *et al.*, Chapter 13, Table 13.1, for details of the dimensions of eggs.

The eggs are sub-spherical to ovate, with a thin shell less than 1 µm thick (Karl *et al.* 2013). The shell bears a characteristic spine. In *Schistosoma mansoni*, the spine is a large posterolateral structure, whereas the spines are terminal and posterior in members of the *S. haematobium* group and lateral and vestigial in *S. japonicum* and other Asian species (Ford and Blankespoor 1979; Bustinduy and King 2014). The egg contains the embryo of the parasite. After its formation is finished in the ootype, the egg is ready to be released from the uterus into the host circulation, where the initial cleavages will commence (Jurberg *et al.* 2009). The eggs are in earliest stages of embryogenesis when laid and mature over a period of 3–5 days, being completely mature after 7 days (Ashton *et al.* 2001). Over this time, the egg may be forced from the host by immune phenomena, so that eggs in excreta usually contain a mature larva, the miracidium (see Jones *et al.*, Chapter 13). Most researchers engaged in schistosome biology, however, purify schistosome eggs from the liver of laboratory hosts and thus observe eggs with larvae in various stages of embryonic development.

Early studies on the structure of the schistosome egg revealed that the miracidium was enclosed with substantial extra-embryonic materials within the shell. This material has the appearance of small spheroidal cushions that became more apparent as the eggs begin to hatch (Kusel 1970). The egg has been subjected to electron microscope investigations by a range of authors so that we now have a good understanding of the structure of the eggs and how it is that they can be highly immunogenic within the host. The studies were hampered by the innate resistance of the eggs to ultrastructural analysis. The eggshell is highly resistant to fixatives commonly used in electron microscopy, so that most studies report a poor quality of tissues within the egg. This poor accessibility of fixatives has been addressed by such methods as applying a pin-prick to individual eggs (a daunting task) or more recently by the use of rapid high pressure freezing (Jones *et al.* 2008; Karl *et al.* 2013). An artefactual (but helpful) outcome of freezing under high pressure was that the method introduced micro-cracks in the shells enabling unimpeded penetration of the eggs by fixatives using a cryosubstitution apparatus. The outcomes of this method are that eggs prepared in this way have superior fixation quality, could be labelled in cytochemical studies and dynamic aspects of the hatching biology could be investigated in real time.

Descriptions of the schistosome egg, the authors honored previous champions of the schistosmiasis field by naming various internal structures of the eggs for them. Thus we see such terms as von Lichternberg's layer and Reynold's layer to describe different intraovular regions (Neill *et al.* 1988). Other authors avoided such terminology, favoring a more descriptive nomenclature that aligns with that used for other groups of platyhelminths, notably the cestodes (Swiderski 1994). The former terminology undoubtedly honors outstanding researchers but should be used advisedly as the terms are not used in the literature to describe similar structures seen within eggs of other trematodes or other groups of platyhelminths. Unravelling the homologies, divergences and convergences of platyhelminth embryogenesis is not the focus of the present study. In the following overview we will use both sets of terminology, with the descriptive term used as the major term and the honorific name for the layer or region of the egg in brackets.

5.3.1 Miracidium

The miracidium is a bilaterally symmetrical larva that lies with its long axis parallel with the long axis of the egg. Its morphology is described in detail in Jamieson and Haas, Chapter 6.

5.4 TRANSGENESIS

Transgenesis is the process in which a foreign gene is inserted into a host organism. This technique has been extensively used for a number of reasons including analysis of genes function, modelling of human diseases and development of novel therapies. Since the genome sequences of *Schistosoma mansoni*, *S. japonicum* and *S. haematobium* have been described (Berriman *et al.* 2009; Consortium 2009; Protasio *et al.* 2012; Young *et al.* 2012; Zerlotini *et al.* 2013), a new era of research in attempting to identify and elucidate the structures and functions of schistosome genes has commenced. Approaches using particle bombardment to introduce and express nucleic acids in parasites, electroporation, soaking and infection with viruses to introduce DNA have been widely tested for many years in order to advance the knowledge of the genome and transcriptome of schistosomes (Davis *et al.* 1999; Beckmann *et al.* 2007; Kines *et al.* 2006, 2010). Despite significant progress being made in the transgenesis field, there is still the need for the establishment of new methods to generate stable transgenic schistosomes. Beckmann *et al.* (2007) and Rinaldi *et al.* (2012) have targeted eggs and miracidia in transgenesis approaches aimed at transforming the germ cells of the parasites. Beckman and collaborators (2007), using particle bombardment, were able to transform miracidia with GFP reporter gene constructs and successfully reintroduce the miracidia into the schistosome cycle through infection of snails.

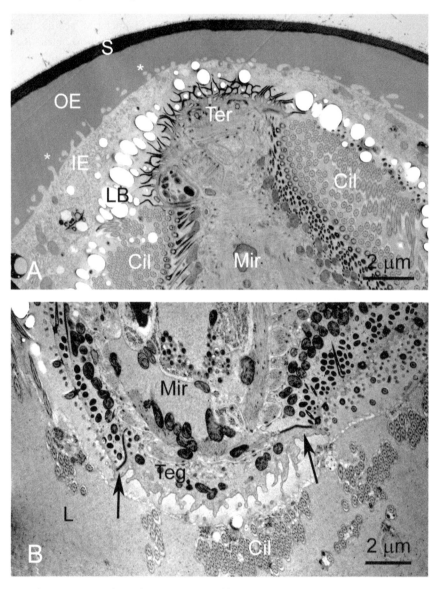

FIG. 5.3 Eggs of *S. japonicum*, shown by TEM, prepared by HPF and cryo-substitution. **A.** Anterior region of a miracidium, showing the terebratorium. The miracidium is enveloped by, in turn, the lacuna, containing lipoid bodies, the inner and outer envelopes and the shell. Inpocketings of the outer envelope (asterisk) occur at the boundary of the inner and outer envelopes. **B.** Posterior end of the miracidium in the lacuna. Note the septate desmosomes (arrows) separating the epithelial plates of the miracidial tegument. Cil, cilia; IE, inner envelope; L, lacuna; LB, lipoid bodies; Mir, miracidium; S, shell; Terebratorium. Adapted from Jones *et al.* 2008, PLoS Neglected Tropical Diseases 2(11): e334. Fig. 2.

Rinaldi and colleagues (2012) used pseudotyped murine leukemia virus (MLV) for insertional mutagenesis of schistosome chromosomes and investigated vertical transmission of transgenes through the developmental cycle of *S. mansoni* after introducing transgenes into eggs. The authors, using high-throughput sequencing revealed the genomic locations of numerous integrations of MLV and demonstrated that retroviral integration events were widely and randomly distributed along all eight schistosome chromosomes, including the Z/W sex chromosomes of *Schistosoma mansoni*, determining that the retroviral genes were transmitted through the germline. Moreover, a germline-transmitted transgene, neoR, encoding neomycin phosphotransferase permitted schistosomules to survive toxic concentrations of the aminoglycoside antibiotic G418. Those findings represent the first report of wide-scale insertional mutagenesis of schistosome chromosomes and vertical germline transmission of an integrated transgene in schistosomes, contributing to a better understanding in functional genomics of schistosomes.

5.5 STRUCTURE OF THE EGG

The egg of schistosomes has been investigated in a number of different studies for *S. mansoni* (Niell *et al.* 1988; Ashton *et al.* 2001), *S. haematobium* (Swiderski 1994) and for *S. japonicum* (Jones *et al.* 2008). The descriptions reveal a consistent morphology for the eggs of the three species. Details of the ultrastructure, from Jones *et al.* (2008) are shown in Figures 5.3 and 5.4.

5.5.1 Shell

5.5.1.1 Structure

The shell is a solid structure, about 700–1000 nm thick (Karl *et al.* 2013), composed by three layers: the peripheral microspine layer, a middle layer with intermediate electron density and an inner layer with extreme electron density (Neill *et al.* 1988). The microspines vary in length among species, measuring 200–300 nm in length and 60 nm of diameter (Karl *et al.* 2013), being much longer in *S. japonicum* eggs than those of *S. mansoni* (Ford and Blankespoor 1979; Karl *et al.* 2013). The shell has a homogenous matrix when viewed by electron microscopy. At points, the shell matrix is interrupted by pores (Neill *et al.* 1988; Karl *et al.* 2013). In *S. mansoni*, these pores appear to open externally about one third of the distance from either pole of the egg. At the base of the shell is a thin boundary region, consisting of a thin electron-lucent region sandwiched between two electron-opaque laminae. This boundary region is the first structure to disappear during hatching (Jones *et al.* 2008) and this is thought to accelerate the inflow of water into the shell.

5.5.1.2 Eggshell Chemistry

The eggshell is a tanned structure formed of cross-linked proteins. A series of studies has established that dominant proteins of the shell are tyrosine rich molecules. The egg is formed when a fertilized oocyte, produced in the ovary and released in the oviduct, is surrounded by the vitelline cells, originated by the vitelline gland (Smyth and Clegg 1959). The oocyte and the vitelline cells then move along to the ootype, which will start contractions making the vitelline cells release the proteins that will form the eggshell (Smyth and Clegg 1959; deWalick *et al.* 2011) (Fig. 5.5). The vitelline cells, in addition to producing the precursors for the shell also produce two tyrosinases that facilitate the hydroxylation of mono-phenolic tyrosine to the ortho-diphenol L-dihydroxyphenylalanine (L-DOPA). In turn, this residue is oxidized to orthoquinone. The reactive residues thus formed, rapidly adducts with nucleophilic amino acids on adjacent proteins, forming a rigid meshwork of cross-linked proteins, a process known as quinone tanning (Cordingley 1987; Fitzpatrick *et al.* 2007).

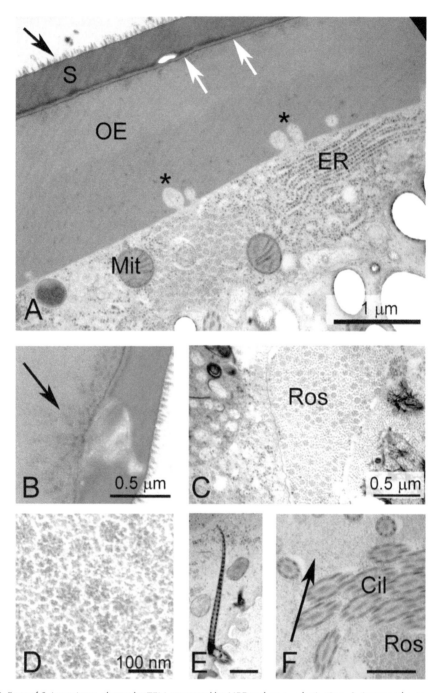

FIG. 5.4 Eggs of *S. japonicum*, shown by TEM, prepared by HPF and cryo-substitution. **A**. Inner and outer envelope. Note the boundary between the shell and outer envelope (arrows). **B**. Pore in shell. Aggregations of a granular material lie in the outer envelope just beneath the pore (arrows). **C**. Rosettes in a vacuole. **D**. High magnification view of rosettes. **E**. Ciliary rootlet of miracidial tegument. Scale bar 1 μm. **F**. High magnification of lacuna. The matrix of the lacuna is granular Scale bar 0.5 μm. Adapted from Jones *et al.* 2008, PLoS Neglected Tropical Diseases 2(11): e334. Fig. 4.

Many molecules are said to be incorporated into the eggshell, among them the major egg antigen p40, HSP 70, cytoskeletal elements and glycolytic enzymes. The significance of these molecules may lie in immune responses induced by the egg, although it is uncertain whether the proteins have functional significance (deWalick *et al.* 2011). The eggshells are also glycosylated, with F-LDN (-F) and Lewis X moieties evident on the surface (deWalick *et al.* 2011). Blocking of Lewis-X by antibodies minimizes binding of eggs to endothelia, suggesting that this glycan is important in the process of extravasation.

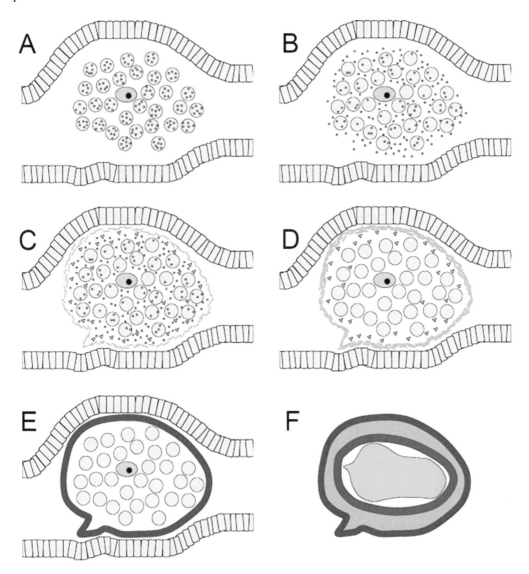

FIG. 5.5 A-F. Formation of the eggshell in the ootype. Scheme of eggshell formation in *Schistosoma mansoni.* In pink the ootype, yellow the fertilized oocyte, light blue circles are vitelline cells with vitelline droplets (orange) containing eggshell precursor proteins, green triangles are tyrosinases. In the mature egg (f) the miracidium (yellow) is surrounded by von Lichtenberg's envelope and Reynolds' layer (blue layers) and the eggshell (brown). From deWalick S, Tielens AGM, van Hellemond JJ. 2012. Experimental Parasitology 132(1): 7–13. Fig. 2.

5.5.1.3 *Physical Properties of the Eggshell*

When pores were first described in the external surface of the eggshells, it was assumed that they played an important role in the passage of metabolites and nutritional fluids, as well as the release of enzymatic secretions to help the extravasation of the eggs from the tissues to the environment (Race *et al.* 1969). Unlike most other trematodes, schistosome eggs lack an operculum and are rich in pores (Kamphoosa *et al.* 2012). Recent works have shown that *S. mansoni* and *S. japonicum* eggshell pores are partially filled with a material containing oxygen, phosphorous and iron (Fig. 5.6, after Karl *et al.* 2013). Those studies by Karl and colleagues were conducted to elucidate the processes underlying the success of a previously developed magnetic fractionation approach for the detection of parasite eggs in fecal samples using magnetic microspheres, namely the Helmintex method (Teixeira *et al.* 2007). In this latter study, Teixeira and colleagues showed that *S. mansoni* eggs bind to magnetic microspheres regardless of the adsorbed ligands on the surface of the microspheres. Also in 2007, Jones *et al.* investigated the biology of iron stores in vitelline cells of *S. japonicum* in an attempt to understand the role of the metal in early development of the parasites and discovered that the iron stored in the eggshell precursor cells is incorporated into the eggshells.

Iron is a vital element for most organisms in nature. Proteins that contain iron are associated with oxygen and electron transport and are essential for metabolism in any living organism. Schistosomes are dependent on the host iron for early development within the mammalian host and the iron seems to be absorbed through the surface tegument of the parasites (Clemens and Basch 1989). Whether the function of iron is related to the embryo development or to structures and proteins in the eggshell for stabilization during its formation is still unclear.

Despite the fact that the eggshells of *S. mansoni* and *S. japonicum* eggs contain iron and physically bind to magnetic microspheres without the application of a magnetic field, the low iron content coupled with the form of iron in the eggshells does not explain this interaction, which is unlikely to be magnetic in origin (Karl *et al.* 2013; Candido *et al.* 2015). One possibility that could explain the binding is that different electrostatic interactions between the oppositely charged surfaces of the eggs and the microspheres could be promoting this adhesion. Ford and Blankespoor (1979) noticed that *S. japonicum* eggs tend to clump together and to adhere tenaciously to glassware. The adhesion of microorganisms to surfaces is influenced by long-range, short-range and hydrodynamic forces, composed from two additive terms: electrostatic repulsion and van der Waals attraction (Ruter and Vincent 1984).

Schistosome eggs, in order to continue the life cycle, have to adhere to the vascular endothelium, cross the vessel wall and be excreted with the feces to the environment. The luminal surface of vascular endothelial cells is negatively charged (Skutelsky *et al.* 1975). File (1995), studying the interaction of schistosome eggs with venular endothelium, demonstrated that eggs deposited directly by adult worms rapidly attach to the vascular endothelium and that endothelial cells actively migrate over the eggs. Furthermore, schistosome eggshells bind platelets (Wu *et al.* 2007) the plasma proteins, VWF, fibrinogen and fibronectin (deWalick *et al.* 2014), increasing the hypothesis that all those interactions might be related to the nature of the schistosome eggshell.

5.5.2   Outer Envelope (Reynold's Layer)

The outer envelopeis an a cellular layer consisting of a network of fine extra cellular filaments and it lies immediately beneath the inner layer of the shell (Neill *et al.* 1988). A flocculent material can be seen among the filaments and may be associated with the pores. Eggs fixed during the hatching process reveal this layer to unravel rapidly, suggesting it may be bound by a water soluble component. The inner margin of this layer is pockmarked with frequent invaginations. These invaginations contain evaginations of the subjacent layer, the inner envelope. The outer envelope is formed from macromeres that detached from the embryo poles and adhered to eggshell during egg development (Jurberg *et al.* 2009) and it is very thin or absent in new laid eggs, progressively thickening from ~46 nm to 581 nm as the egg matures (Neill *et al.* 1988). In *S. mansoni* eggs, this layer fills the lateral spine before accumulating around the rest of the egg and becomes thicker at the poles of the mature eggs than in the lateral sides (Neil *et al.* 1988).

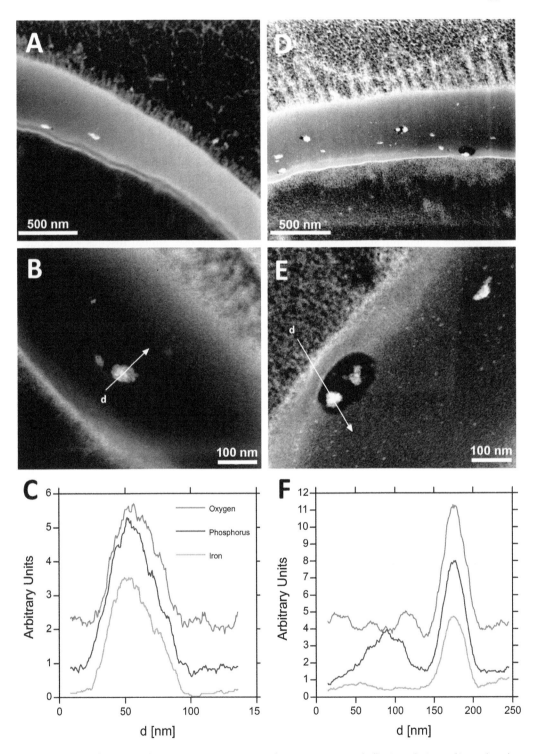

FIG. 5.6 Iron localization in the *Schistosoma mansoni* and *S. japonicum* eggshells. **A.** Inclusions of iron phosphate in the shell of *S. mansoni* at low resolution. **B.** Similar inclusions in *S. mansoni* at a higher resolution. **C.** STEM-EDS spectra for iron, phosphorus and oxygen acquired when scanning across an inclusion, along the white line (d) shown in **B.**, **D.**, **E.** and **F.** Similar observations for *S. japonicum*. From Karl *et al.* 2013. PLOS Neglected Tropical Diseases 7(5): e2219. doi: 10.1371/journal.pntd.0002219. Fig. 3.

5.5.3 Inner Envelope (von Lichtenberg's Layer)

Immature eggs lack complex structures, with some peripheral cells forming an epithelium adjacent to the eggshell. This epithelium forms a thin but complete cytoplasmic envelope surrounding the inner contents of the egg (Neill *et al.* 1988). In fully developed eggs, the inner envelope has 2–3 nuclei (Jurberg *et al.* 2009) and a thickness of 500–600 nm (Ashton *et al.* 2001). The layer has organelles indicative of a high protein synthetic capacity and is thought to be the source of many of the secreted antigens of the egg (Ashton *et al.* 2001; Schramm *et al.* 2006; Mathieson *et al.* 2011). The nuclei in this layer are usually near the poles of the egg and opposite the spine and are large, with scarce heterochromatin, prominent nucleoli and are limited by perinuclear cisterns with numerous nuclear pores. The nuclei deteriorate as the miracidium reaches maturity (Neill *et al.* 1988). Among the synthetic apparatuses of mitochondria, rough endoplasmic reticulum and Golgi bodies are vesicles (Cheever's bodies) containing rosettes, reminiscent of α-glycogen granules. These granules could be labelled with Concanavalin A lectin. ConA binds a range of molecules notably glycans containing such hexoses as mannose and glucose. Similar rosettes were evident within the lumen between the inner envelope and the embryo (see Fig. 5.7, after Jones *et al.* 2008).

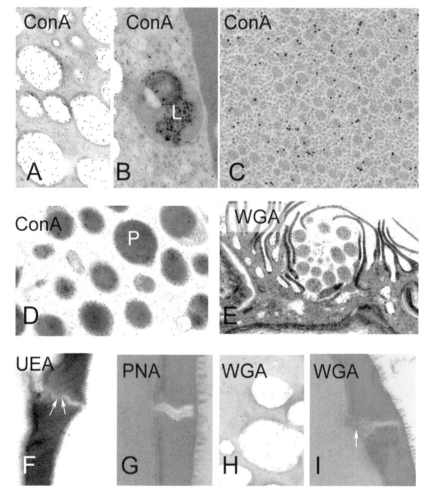

FIG. 5.7 Lectin cytochemistry shown by TEM of *S. japonicum* egg showing representative regions with positive localization with the lectin. **A–D** Con A cytochemistry. **A.** Lipoid bodies **B.** Lysosome in inner envelope. **C.** Rosettes in vacuoles. **D.** Penetration gland of miracidium. **E.** WGA labelling in lacuna surrounding the terebratorium. **F.** UEA in pore of shell. **G.** PNA in pore of shell. **H.** WGA in lipoid body. **I.** WGA in pore. From Jones MK *et al.* 2008. PLoS Neglected Tropical Diseases 2(11): e334. Fig. 7.

5.5.4 Lacuna (Lehman's Lacuna)

The miracidium lies within a space, the lacuna, which lies immediately internal to the inner envelope and it is surrounded by a dense array of materials which are packed in a ground substance that is finely granular. A range of inclusion bodies is present in the lacuna, including vacuoles filled with the rosettes described for the inner envelope (Jones *et al.* 2008). In undeveloped eggs, this cavity is filled with a fluid through which are scattered lipoid bodies, myelinoid whorls, occasional vitelline cells and fragments of residual vitelline cell cytoplasm. Small lipoid bodies are found in all stages in the lacuna and they become concentrated between Cheevers bodies at maturity (Neill *et al.* 1988). Cilia from the miracidial epidermal plates protrude into the lacuna. At the time when miracidial morphogenesis appears to be complete, there is a marked increase in α-glycogen profiles in most miracidial cells and similar profiles appear in the inner envelope and Cheever bodies (Neill *et al.* 1988). By comparison with studies of *Fasciola hepatica* and *Echinostoma caproni* and glycomic analyses of eggs, it can be postulated that the lacuna is rich in complex carbohydrates. This is borne out by cytochemical analyses that demonstrate the presence of abundant glycans in the matrix (Jones *et al.* 2008).

5.6 THE EGG: PHYSIOLOGY AND HATCHING

5.6.1 Secretions

The nature of the egg secretions has been a matter of some controversy, with estimates of different secreted proteins from scores (Cass *et al.* 2007) to only 7 (Mathieson and Wilson 2010), including the IL-4-inducing principal of *S. mansoni* eggs (IPSE), Omega-1, kappa-5 egg secreted protein 15 (ESP15), a micro-exon gene 2 (MEG-2) protein and two members of the recently described MEG-3 family (Dunne *et al.* 1991; Schramm *et al.* 2009; Mathieson and Wilson 2010). Of these IPSE/alpha-1 and Omega-1 have been intensively studied, being demonstrated to regulate immune cells inducing a strong Th2 response (Schramm *et al.* 2003; Fitzsimmons *et al.* 2005; Meevissen *et al.* 2010; Ferguson *et al.* 2015). IPSE/alpha-1 and Omega-1 were shown to have *N*-glycosylation sites occupied with glycans that carry Lewis X motifs (Wuhrer *et al.* 2006; Meevissen *et al.* 2010), a major immunogenic glycan component of schistosomes (Okano *et al.* 1999; Okano *et al.* 2001), expressed in all the stages of the *Schistosoma* life cycle, mainly found in the gut and on the tegument of adult worms, on eggshells and on the oral sucker of cercariae (Nyame *et al.* 1998; van Remoortere *et al.* 2000). These molecules are also hepatotoxic (Abdullah 2011).

Glycans. Helminths are "experts" in modulating the inflammatory responses of their hosts targeting host lectins and typically cause attenuated Th1 responses and induction of Th2 and regulatory T cell populations, favoring their survival inside the host (van Die and Cummings 2010). Non-structural proteins in the eggshell and specific *N*- and *O*-linked glycans are known to promote an immune response (deWalick *et al.* 2012). Those proteins containing a wide range of highly immunogenic complex glycans play an important role in immunomodulation once the parasite is inside the vertebrate host. The glycan Lewis X was shown to be a key factor in modulating the host immune system shift toward a Th2 response (Okano *et al.* 1999). Although the host immune system produces antibodies against antigen invasion, only after several weeks of *Schistosoma* infection a high level of antibodies production was noticed, coincidentally with the egg deposition by the adult worms, suggesting that the eggs are the major stimulus of Th2 cytokine response in *S. mansoni* infections (Grzych *et al.* 1991). This may be due to the fact that intact schistosome eggs produce and release secretions with a diverse range of products, including highly antigenic glycoproteins. These secretions induce a strong cellular response around the egg, which has the effect of expelling eggs deposited near luminal organs, such as the intestine or urinary bladder, but trapping the eggs that embolize to other solid organs, such as the liver or the central nervous system.

The carbohydrate components of glycans and glycoconjugates are crucial in the establishment of the parasite inside the host and most of the host antibody response is directed against glycan epitopes. *S. mansoni* secretes glycoproteins with glycosylation patterns that are complex and stage-specific. N-glycans from egg secretions are predominantly core-difucosylated, different from the cercarial secretions which are dominated by core-xylosylated (Jang-Lee *et al.* 2007). Schistosomes can also express glycans containing terminal α- and β-linked GalNAc residues, such as the LDN (GalNAcβ1-4GlcNAc) and its fucosylated derivative, LDNF and they are used as a backbone for further modifications, among others leading to several multifucosylated immunogenic structures that are found in egg and cercarial glycoproteins and glycolipids (van Die and Cummings 2010).

5.6.2 Escape From Host Vasculature

One of the primary postulates for the presence of the spines on the schistosome eggshell is that these structures assist the egg to extravasate. The sharp spine of *S. mansoni* and of *S. haematobium* may well assist the egg to penetrate the obstructing tissues, but as pointed out by deWalick *et al.* (2014) this physical process can hardly be the case for *S. japonicum*, in which the spine is vestigial.

The host blood system requires a number of parasite and host factors. The first is the parasite. Wood and Bacha (1983) described the adult female of *Austrobilharzia variglandis* in chickens as leaving the gynaecophoric canal of the adult male to migrate to the smallest blood vessels within the intestine to deposit the eggs. The eggs are pressed hard against the endothelium. Attempts at artificial culture of schistosome eggs laid by females *in vitro* have revealed that the eggs are highly adhesive to the polyvinyl chloride matrix of commonly used culture dishes (for example Petri Dishes). Mature eggs in feces or derived from the liver appear less sticky, suggesting that the adhesive properties are lost over time. Eggs of *S. japonicum* are laid in clumps, a consequence of the large numbers of eggs laid and of the physicochemical properties of their surfaces. Further, according to File (1995) the endothelium rapidly grows over the egg, entrapping the egg within tissues.

The adhesive property of the shell may be explained in a number of ways. First, the eggs are able to grow over time, suggesting some alterations in the bonding properties of the shell. This, in turn, may indicate that, like many polymerized compounds, the shell undergoes only partial polymerization in the ootype and the final reactions occur throughout development of the egg. This may mean that unbound reactive groups in immature eggs enhance the bonding of the shell to substrates. Alternatively or additionally, the microspines may enhance the adhesive nature of the eggs. Thirdly, early authors have hinted at the presence of secretions from the uterus and ootype coating the outside of eggs, but there have been no functional studies to determine whether this is the case.

Whatever facilitates binding, it appears that host factors come into play. There is evidence that the eggs bind to receptors on the surface of the endothelium, notably E-selectin, ICAM-1 and VCAM-1. deWalick and colleagues (2014) have shown that serum molecules, namely von Willebrand factor, fibrinogen and fibronectin bind strongly and rapidly to eggs thereby facilitating clotting. The stabilized clot in turn promotes migration of the endothelium over the egg.

At about this time, it is thought that eggs begin to secrete molecules through the pores in the shells. Pores in immature eggs have a larger size than the structures in mature eggs, suggesting that immature eggs release more secreted components.

5.6.3 Hatching

Hatching biology has been studied in *S. japonicum* and *S. mansoni*, as well as in species of *Austrobilharzia* of water fowl (Horák *et al.* 1998). The most extensive morphological analysis of hatching has been performed for *S. japonicum* and hatching biology of this species will be considered first. Considerably more data on the physiology and biochemical alterations of hatching are available for *S. mansoni* and a detailed overview of these aspects will be considered subsequently.

The egg hatching biology of *S. japonicum* is truly bizarre. Eggs are held in an isosmotic state while they are in the host or in human urine. Within this state the enclosed miracidium remains quiescent. Some minutes after reaching freshwater, the egg begins the process of hatching. The first change that can be observed in the egg is that the boundary layer between the shell and the outer envelope disappears. Jones and colleagues (2008) postulated that the inflow of water induces an internal cascade within the egg causing production and maturation of enzymes, notably leucine amino peptidases to digest this boundary layer.

Once the boundary is digested, the internal components of the egg (the miracidium, envelopes and other components) are freed to move. The first microscopical evidence of hatching is the activation of the miracidium. This can be seen initially as infrequent twitching of the cilia. The cilia become progressively more active over time. Electron microscopy of hatching eggs demonstrates a progressive dissolution of the contents of the lacuna. Jones *et al.* (2008) postulated that the inflowing water molecules bonded to the packed complex carbohydrates in this region. This caused an unraveling of the tridimensional structure of these complex macromolecules, leading to the dissolution of the contents of the lacuna and an increase in internal pressure in the egg. Throughout this process, the inner envelope also degrades, while the outer envelope remains intact.

At some point, the miracidium becomes freed from its surrounding envelopes and the contents of the lacuna and is able to become motile. The larva alters its position so that it lies with its long axis at right angles to the long axis of the egg. The increased pressure and presumably muscular activity of the larva causes the shell to rupture. Rupture for this species always occurs along the long axis of the shell.

Upon shell rupture the larva emerges through the split, but it remains enclosed within a sac formed by the outer envelope. The sac expands rapidly so that it becomes almost three times the volume of the shell. The miracidium swims "frantically" within the sac, stopping periodically to probe against the sac wall, which is formed of the fibrillar matrix of the outer envelope. By electron microscopy, the fibrils of the outer envelope appear to unravel, so that the wall of this sac forms from the outer envelope, becomes thinner. Eventually, the miracidium is able to force a hole through this wall and escapes from the sac.

Physiology of Hatching. Hatching of schistosome eggs occurs as a response to reduced osmotic strength of the freshwater environment. Conditions like temperature and light was also shown to have an effect on the rupture of the eggs. Early work has been summarized by Xu and Dresden (1986). It emerges from their review that there are uncertainties about the role both of light and temperature in the initiation of the hatching process. According to Standen (1951), temperatures like 4°C and 37°C have an inhibitory effect on the hatching of *S. mansoni* eggs purified from fecal samples, 28°C being the optimal temperature. It was also shown in the same work that bright light stimulates the emergence of the miracidium while darkness almost completely inhibits it. Interestingly, more recent investigations using eggs isolated from mice livers reported that hatching occurred equally well in light and in darkness (Kassin and Gilbertson 1976). It is important to note that the eggs used in those works were obtained from different sources and whether this aspect would change hatching conditions is still unclear.

The effect of salt concentration on hatching is more definite and hatching rate decreases with an increase in osmolarity. Xu and Dresden (1986) summarize the three theories for the mechanism of hatching. These are (1) the muscular activity of the miracidium alone; (2) the presence of a hatching enzyme and (3) the importance of changes in internal osmotic pressure and the inflow of water. As described above in the work of Jones *et al.* (2008) (Section 5.6.3. Observations), all of these factors are likely to contribute to the dramatic release of the miracidium from the eggshell.

Calcium and Hatching Enzyme. In addition, a number of factors were also thought to be involved in hatching including changes in calcium levels and the involvement of an enzyme. The drug of choice against schistosomiasis, praziquantel (Redman *et al.* 1996), induces a rapid influx of calcium into the worm tissues causing muscle contraction and paralysis (Greenberg 2005; Pica-Mattoccia *et* *al.* 2008) and also induces hatching of the eggs in a hypotonic media (Matsuda *et al.* 1983). Jones *et al.* (2008) corroborated the latter observations, suggesting that when the schistosome egg is transferred to freshwater from an isosmotic environment, an inflow of water signals calcium fluxes in the external membranes of the inner envelope of the miracidium, leading to a cascade of events that will initiate the hatching process.

The presence of an enzyme as a hatching factor of schistosome eggs was first suggested by Kusel (1970). Later on, Xu and Dresden (1986) identified a leucine aminopeptidase (LAP) in the hatching fluid of *S. mansoni* eggs and verified inhibition of the hatching process when bestatin, a specific inhibitor of LAP *in vitro*, was present. The authors suggested that the enzyme may function as an initiator of hatching, weakening the eggshell by proteolytic action, prior to rupture due to osmotic effects. The localization of the LAP, using fluorescent enzyme substrate techniques (Bogitsh 1983) revealed the enzyme to be associated with the vitelline membrane. Further work on the leucine amino peptidase has been reported by Rinaldi *et al.* (2009). They showed that two forms of LAPs could be identified. LAPs were shown to exhibit a two domain structure: a less conserved N terminal domain and a conserved C terminal domain binding Zinc atoms involved in the catalytic process. The LAPs were found in eggs but also very abundantly in free living miracidia and in the adult male and female tissues, notably the gut. These workers used bestatin to inhibit hatching but also used the RNAi interference method to reduce the synthesis of two forms of leucine amino peptidase (SmLAP1 and SmLAP2). This interference method considerably reduced the hatching process. The authors concluded that LAPs play either a direct role in the hatching or they provide essential metabolites for the completion of the process. Jones *et al.* (2008) have described (Section 5.3.3.3) the possible ways the LAPs could be altering the internal structure of the eggshell and membranes.

Shell Structure. Physical properties of the eggshell have been described above (5.5.3). The early observation of Samuelson *et al.* (1984) where the eggshell, after emergence of the miracidium, is shown by scanning electron microscopy to have a sharp but curved exit region may suggest release of tensions within the protein fabric of the shell during hatching. There may be a tension at the outside of the shell, such that perturbation within, due to osmotic, enzymatic or muscular forces may cause release of tension with bursting of the shell. This possibility could be tested by using a variety of polarising or X-ray microscopic techniques to examine the unhatched and hatched eggshell.

Another interesting work showed that immature eggs and eggs greater than 160 µm in length do not hatch (Xu and Dresden 1989). The authors proposed three possible explanations: (1) eggshells of older eggs may be difficult to break; (2) if breakage of the eggshell requires close contact between the miracidium and the shell, smaller eggs would have a better chance to hatch and (3) a necessary factor is lost as the miracidium grow older (Xu and Dresden 1990).

While the whole mechanism of hatching is still not very clear and which factors can enhance the breakage of the eggshell and release of the miracidium remains a mystery, osmolarity appears to be agreed upon as the most important condition to promote hatching of the eggs.

5.7 LITERATURE CITED

Abdulla M-H, Lim K-C, McKerrow JH, Caffrey CR. 2011. Proteomic identification of IPSE/alpha-1 as a major hepatotoxin secreted by *Schistosoma mansoni* eggs. PLoS Neglected Tropical Diseases 5(10): e1368.

Ashton PD, Harrop R, Shah B, Wilson RA. 2001. The schistosome egg: development and secretions. Parasitology 122: 329–338.

Beckmann S, Wippersteg V, El-Bahay A, Hirzmann J, Oliveira G, Grevelding CG. 2007. *Schistosoma mansoni:* germ-line transformation approaches and actin-promoter analysis. Experimental Parasitology 117(3): 292–303. doi: 10.1016/j.exppara.2007.04.007.

Berriman M, Haas BJ, LoVerde PT, Wilson RA, Dillon GP, Cerqueira GC, Mashiyama ST, Al-Lazikani B, Andrade LF, Ashton PD, *et al.* 2009. The genome of the blood fluke *Schistosoma mansoni*. Nature 460(7253): 352–358.

Bogitsh BJ. 1983. Peptidase activity in the egg-shell-enclosed embryo of *Schistosoma mansoni* using fluorescent histochemistry. Transactions of the American Microscopical Society 102: 169–172.

Bustinduy AL, King CH. 2014. Schistosomiasis. pp. 698–725. *In*: Farrar J, editor. *Manson's Tropical Infectious Diseases* (Twenty-Third Edition). London: W.B. Saunders.

Candido RR, Favero V, Duke M, Karl S, Gutiérrez L, Woodward RC, Graeff-Teixeira C, Jones MK, St Pierre TG. 2015. The affinity of magnetic microspheres for *Schistosoma* eggs. International Journal for Parasitology 45(1): 43–50.

Carneiro VC, de Abreu da Silva IC, Torres EJL, Caby S, Lancelot J, Vanderstraete M, Furdas SD, Jung M, Pierce RJ, Fantappié MR. 2014. Epigenetic changes modulate Schistosome egg formation and are a novel target for reducing transmission of Schistosomiasis. PLoS Pathogens 10(5): e1004116. doi: 10. 1371/journal.ppat.1004116.

Cass CL, Johnson JR, Califf LL, Xu T, Hernandez HJ, Stadecker MJ, Yates JR 3rd, Williams DL. 2007. Proteomic analysis of *Schistosoma mansoni* egg secretions. Molecular and Biochemical Parasitology 155(2): 84–93.

Cheever AW, Macedonia JG, Mosimann JE, Cheever EA. 1994. Kinetics of egg production and egg excretion by *Schistosoma mansoni* and *S. japonicum* in mice infected with a single pair of worms. American Journal of Tropical Medicine and Hygiene 50: 281–295.

Clemens LE, Basch PF. 1989. *Schistosoma mansoni*: effect of transferrin and growth factors on development of schistosomula *in vitro*. Journal of Parasitology 75: 417–421.

Colley DG, Bustinduy AL, Secor WE, King CH. 2014. Human schistosomiasis. The Lancet 383: 2253–64.

Consortium SjGSaFA. 2009. The *Schistosoma japonicum* genome reveals features of host-parasite interplay. Nature 460(7253): 345–351. doi: 10.1038/nature08140.

Cordingley JS. 1987. Trematode eggshells: novel protein biopolymers. Parasitology Today 3(11): 341–344.

Davis RE, Parra A, LoVerde PT, Ribeiro E, Glorioso G, Hodgson S. 1999. Transient expression of DNA and RNA in parasitic helminths by using particle bombardment. Proceedings of the National Academy of Sciences USA 96(15): 8687–8692.

deWalick S, Bexkens ML, van Balkom BW, Wu YP, Smit CH, de Groot PG, Heck AJR, Tielens AGM, van Hellemond JJ. 2011. The proteome of the insoluble *Schistosoma mansoni* eggshell skeleton. International Journal for Parasitology 41: 523–532.

deWalick S, Tielens AGM, van Hellemond JJ. 2012. *Schistosoma mansoni*: the egg, biosynthesis of the shell and interaction with the host. Experimental Parasitology 132(1): 7–13.

deWalick S, Hensbergen PJ, Bexkens ML, Grosserichter-Wagener C, Hokke CH, Deelder AM, de Groot PG, Tielens AGM, van Hellemond JJ. 2014. Binding of von Willebrand factor and plasma proteins to the eggshell of *Schistosoma mansoni*. International Journal for Parasitology 44: 263–268.

Doenhoff MJ, Hassounah O, Murare H, Bain J, Lucas S. 1986. The schistosome egg granuloma: immunopathology in the cause of host protection or parasite survival? Transactions of the Royal Society of Tropical Medicine and Hygiene 80(4): 503–514.

Dunne DW, Jones FM, Doenhoff MJ. 1991. The purification, characterization, serological activity and hepatotoxic properties of two cationic glycoproteins (α1 and w1) from *Schistosoma mansoni* eggs. Parasitology 103: 225–236.

Ferguson BJ, Newland SA, Gibbs SE, Tourlomousis P, Santos PF, Patel MN, Hall SW, Walczak H, Schramm G, Haas H, *et al.* 2015. The *Schistosoma mansoni* T2 ribonuclease omega-1 modulates inflammasome-dependent IL-1β secretion in macrophages. International Journal for Parasitology 45(13): 809–13.

File S. 1995. Interaction of schistosome eggs with vascular endothelium. Journal of Parasitology 81: 234–238.

Fitzpatrick JM, Hirai Y, Hirai H, Hoffmann KF. 2007. Schistosome egg production is dependent upon the activities of two developmentally regulated tyrosinases. FASEB Journal 21: 823–835.

Fitzsimmons CM, Schramm G, Jones FM, Chalmers W, Hoffmann KF, Grevelding CG, Wuhrer M, Hokke CH, Haas H, Doenhoff MJ, *et al.* 2005. Molecular characterization of omega-1: a hepatotoxic ribonuclease from *Schistosoma mansoni* eggs. Molecular and Biochemical Parasitology 144: 123–127.

Ford JW, Blankespoor HD. 1979. Scanning electron microscopy of the eggs of three human schistosomes. International Journal for Parasitology 9: 141–145.

Greenberg RM. 2005. Ca2+ signalling, voltage-gated Ca2+ channels and praziquantel in flatworm neuromusculature. Parasitology 131 Suppl: S97–108.

Grzych JM, Pearce E, Cheever A, Caulada ZA, Caspar P, Heiny S, Lewis F, Sher A. 1991. Egg deposition is the major stimulus for the production of Th2 cytokines in murine *Schistosomiasis mansoni*. The Journal of Immunology 146: 1322–1327.

Hartenstein V, Ehlers U. 2000. The embryonic development of the rhabdocoel flatworm *Mesostoma lingua* (Abildgaard, 1789). Development Genes and Evolution 210: 399–415.

Horák P, Kolárová L, Dvorák J. 1998. *Trichobilharzia regenti* n. sp. (Schistosomatidae, Bilharziellinae), a new nasal schistosome from Europe. Parasite 5: 349–357.

Jang-Lee J, Curwen RS, Ashton PD, Tissot B, Mathieson W, Panico M, Dell A, Wilson RA, Haslam SM. 2007. Glycomics analysis of *Schistosoma mansoni* egg and cercarial secretions. Molecular and Cellular Proteomics 6: 1485–1499.

Jones MK, McManus DP, Sivadorai P, Glanfield A, Moertel L, Belli SI, Gobert GN. 2007. Tracking the fate of iron in early development of human blood flukes. International Journal of Biochemistry and Cell Biology 39: 1646–1658.

Jones MK, Bong SH, Green KM, Holmes P, Duke M, Loukas A, McManus DP. 2008. Correlative and dynamic imaging of the hatching biology of *Schistosoma japonicum* from eggs prepared by high pressure freezing. PloS Neglected Tropical Diseases 2(11): e334.

Jurberg AD, Pascarelli BM, Pelajo-Machado M, Maldonado A Jr, Mota EM, Lenzi HL. 2008. Trematode embryology: a new method for whole-egg analysis by confocal microscopy. Development Genes and Evolution 218: 267–271.

Jurberg AD, Gonçalves T, Costa TA, de Mattos ACA, Pascarelli BM, de Manso PPA, Ribeiro-Alves M, Pelajo-Machado M, Peralta JM, Coelho PMZ, *et al.* 2009. The embryonic development of *Schistosoma mansoni* eggs: proposal for a new staging system. Development Genes and Evolution 219: 219–234.

Kamphoosa P, Jones MK, Lovas E, Srisawangwong T, Laha T, Piratae S, Thammasiri C, Suwannatrai A, Sripanidkulchai B, Eursitthichai V, *et al.* 2012. Light and electron microscopy observations of embryogenesis and egg development in the human liver fluke, Opisthorchis viverrini (Platyhelminthes, Digenea). Parasitology Research 110: 799–808.

Karl S, Gutiérrez L, Lucyk-Maurer R, Kerr R, Candido RRF, Toh SQ, Saunders M, Shaw JA, Suvorova A, Hofmann A, *et al.* 2013. The iron distribution and magnetic properties of schistosome eggshells: implications for improved diagnostics. PLoS Neglected Tropical Diseases 7(5): e2219. doi: 10.1371/journal.pntd.0002219.

Kassim O, Gilbertson DE. 1976. Hatching of *Schistosoma mansoni* eggs and observations on motility of miracidia. Journal of Parasitology 62: 715–720.

Kines KJ, Mann VH, Morales ME, Shelby BD, Kalinna BH, Gobert GN, Chirgwin SR, Brindley PJ. 2006. Transduction of *Schistosoma mansoni* by vesicular stomatitis virus glycoprotein-pseudotyped moloney murine leukemia retrovirus. Experimental Parasitology 112: 209–220. doi: 10.1016/j.exppara.2006.02.003.

Kines KJ, Rinaldi G, Okatcha TI, Morales ME, Mann VH, Tort JF, Brindley PJ. 2010. Electroporation facilitates introduction of reporter transgenes and virions into schistosome eggs. PLoS Neglected Tropical Diseases 4(2): e593. doi: 10.1371/journal.pntd.0000593.

Knobloch J, Kunz W, Grevelding CG. 2006. Herbimycin a suppresses mitotic activity and egg production of female *Schistosoma mansoni*. International Journal for Parasitology 36(12): 1261–1272.

Kusel JR. 1970. Studies on the structure and hatching of the eggs of *Schistosoma mansoni*. Parasitology 60: 79–88.

Mathieson W, Wilson RA. 2010. A comparative proteomic study of the undeveloped and developed *Schistosoma mansoni* egg and its contents: the miracidium, hatch fluid and secretions. International Journal for Parasitology 40: 617–628.

Mathieson W, Castro-Borges W, Wilson RA. 2011. The proteasome-ubiquitin pathway in the *Schistosoma mansoni* egg has development- and morphology-specific characteristics. Molecular and Biochemical Parasitology 175: 118–125.

Mati VLT, Melo AL. 2013. Current applications of oogram methodology in experimental schistosomiasis: fecundity of female *Schistosoma mansoni* and egg release in the intestine of AKR/J mice following immunomodulatory treatment with pentoxifylline. Journal of Helminthology 87(01): 115–124.

Matsuda H, Tanaka H, Nogami S, Muto M. 1983. Mechanism of action of praziquantel on the eggs of *Schistosoma japonicum*. The Japanese Journal of Experimental Medicine 53: 271–274.

Meevissen MH, Wuhrer M, Doenhoff MJ, Schramm G, Haas H, Deelder AM, Hokke CH. 2010. Structural characterization of glycans on omega-1, a major *Schistosoma mansoni* egg glycoprotein that drives Th2 responses. Journal of Proteome Research 9: 2630–2642.

Neill PJ, Smith JH, Doughty BL, Kemp M. 1988. The ultrastructure of the *Schistosoma mansoni* egg. American Journal of Tropical Medicine and Hygiene 39: 52–65.

Nyame AK, Debose-Boyd R, Long TD, Tsang VCW, Cummings RD. 1998. Expression of LeX antigen in *Schistosoma japonicum* and *S. haematobium* and immune responses to LeX in infected animals: lack of LeX expression in other trematodes and nematodes. Glycobiology 8: 615–624.

Okano M, Satoskar AR, Nishizaki K, Abe M, Harn Jr DA. 1999. Induction of Th2 responses and IgE is largely due to carbohydrates functioning as adjuvants on *Schistosoma mansoni* egg antigens. The Journal of Immunology 163: 6712–6717.

Okano M, Satoskar AR, Nishizaki K, Harn Jr DA. 2001. Lacto-N-fucopentaose III found on *Schistosoma mansoni* egg antigens functions as adjuvant for proteins by inducing Th2-type response. The Journal of Immunology 167: 442–450.

Pearce EJ, MacDonald AS. 2002. The immunobiology of schistosomiasis. National Review of Immunology 2: 499–511.

Pellegrino J, Oliveira CA, Faria J, Cunha AS. 1962. New approach to the screening of drugs in experimental schistosomiasis mansoni in mice. American Journal of Tropical Medicine and Hygiene 11: 201–215.

Pica-Mattoccia L, Orsini T, Basso A, Festucci A, Liberti P, Guidi A, Marcatto-Maggi A-L, Nobre-Santana S, Troiani A-R, Ciolo D, *et al.* 2008. *Schistosoma mansoni*: lack of correlation between praziquantel-induced intra-worm calcium influx and parasite death. Experimental Parasitology 119: 332–335.

Prata A. 1957. Biópsia retal na esquistossomose mansoni – bases e aplicações no diagnóstico e tratamento. Tese de Doutorado. Serviço Nacional de Educação Sanitária – Ministério da Saúde, Rio de Janeiro. PhD Thesis. 197 p.

Protasio AV, Tsai IJ, Babbage A, Nichol S, Hunt M, Aslett MA, De Silva N, Velarde GS, Anderson TJ, Clark RC, *et al.* 2012. A systematically improved high quality genome and transcriptome of the human blood fluke *Schistosoma mansoni*. PLoS Neglected Tropical Diseases 6(1): e1455. doi: 10.1371/journal.pntd.0001455.

Race GJ, Michaels RM, Martin JH, Larsh JE, Matthews JL. 1969. *Schistosoma mansoni* eggs: An electron microscopic study of shell pores and microbarbs. Proceedings of the Society for Experimental Biology and Medicine 130: 990–992.

Redman CA, Robertson A, Fallon PG, Modha J, Kusel JR, Doenhoff MJ, Martin RJ. 1996. Praziquantel: an urgent and exciting challenge. Parasitology Today 12: 14–20.

Rinaldi G, Morales ME, Airefaei YN, Cancela M, Castillo E, Dalton JP, Tort JF, Brindley PJ. 2009. RNA interference targeting leucine aminopeptidase blocks hatching of *Schistosoma mansoni* eggs. Molecular and Biochemical Parasitology 167: 118–126.

Rinaldi G, Eckert SE, Tsai IJ, Suttiprapa S, Kines KJ, Tort JF, Mann VH, Turner DJ, Berriman M, Brindley PJ. 2012. Germline transgenesis and insertional mutagenesis in *Schistosoma mansoni* mediated by murine leukemia virus. PLoS Pathogens 8(7): e1002820. doi: 10.1371/journal.ppat.1002820.

Ruter PR, Vincent B. 1984. Physiochemical interactions of the substratum, microorganisms and the fluid phase. pp. 21–38. *In*: Marshall KC, editor. *Microbial Adhesion and Aggregation*. Springer–Verlag AG, Berlin.

Samuelson JC, Quinn JJ, Caulfield JP. 1984. Hatching, chemokinesis and transformation of miracidia of *Schistosoma mansoni*. Journal of Parasitology 70: 321–331.

Schramm G, Falcone FH, Gronow A, Haisch K, Mamat U, Doenhoff MJ, Oliveira G, Galle J, Dahinden CA, Haas H. 2003. Molecular characterization of an interleukin-4-inducing factor from *Schistosoma mansoni* eggs. The Journal of Biological Chemistry 278: 18384–18392.

Schramm G, Gronow A, Knobloch J, Wippersteg V, Grevelding CG, Galle J, Fuller H, Stanley RG, Chiodini PL, Haas H, *et al.* 2006. IPSE/alpha-1: a major immunogenic component secreted from *Schistosoma mansoni* eggs. Molecular and Biochemical Parasitology 147: 9–19.

Schramm G, Hamilton JV, Balog CI, Wuhrer M, Gronow A, Beckmann S, Wippersteg V, Grevelding CG, Goldmann T, Weber E, *et al.* 2009. Molecular characterisation of kappa-5, a major antigenic glycoprotein from *Schistosoma mansoni* eggs. Molecular and Biochemical Parasitology 166: 4–14.

Skutelsky E, Rudich Z, Danon D. 1975. Surface charge properties of the luminal front of blood vessel walls: an electron microscopical analysis. Thrombosis Research 7: 623–634.

Smyth JD, Clegg JA. 1959. Egg-shell formation in trematodes and cestodes. Experimental Parasitology 8(3): 286–323.

Standen OD. 1951. The effects of temperature, light and salinity upon the hatching of the ova of *Schistosoma mansoni*. Transactions of the Royal Society of Tropical Medicine and Hygiene 45: 225–241.

Swiderski Z. 1994. Origin, differentiation and ultrastructure of egg envelopes surrounding the miracidia of *Schistosoma mansoni.* Acta Parasitologica 39: 64–72.

Teixeira C, Neuhauss E, Bem R, Romanzini J, Graeff-Teixeira C. 2007. Detection of *Schistosoma mansoni* eggs in feces through their interaction with paramagnetic beads in a magnetic field. PLoS Neglected Tropical Diseases 1(2): e73. doi: 10.1371/journal.pntd.0000073.

van Die I, Cummings RD. 2010. Glycan gimmickry by parasitic helminths: a strategy for modulating the host immune response? Glycobiology 20: 2–12.

van Remoortere A, Hokke CH, Van Dam GJ, Van Die I, Deelder AM, Van den Eijnden DH. 2000. Various stages of *Schistosoma* express LewisX, LacdiNAc, GalNAcb1-4 (Fuca1-3 GlcNAc and GalNAcb1-4 (Fuca1-2 Fuca1-3) GlcNAc carbohydrate epitopes: detection with monoclonal antibodies that are characterized by enzymatically synthesized neoglycoproteins. Glycobiology 10: 601–609.

Vogel H. 1942. Über Entwicklung, Lebensdauer und Tod der Eiervom *Bilharzia japonica* im Wirtsgewebe. Deutsch Tropenmedizinische Zeitschrift 46: 57–91.

Wilson MS, Mentink-Kane MM, Pesce JT, Ramalingam TR, Thompson R, Wynn TA. 2007. Immunopathology of schistosomiasis. Immunology and Cell Biology 85: 148–154.

Wood LM, Bacha Jr WJ. 1983. Distribution of eggs and the host response in chickens infected with *Austrobilharzia variglandis* (Trematoda). Journal of Parasitology 69: 682–688.

Wu YP, Lenting PJ, Tielens AGM, de Groot PG, van Hellemond JJ. 2007. Differential platelet adhesion to distinct life cycle stages of the parasitic helminth *Schistosoma mansoni*. Journal of Thrombosis and Haemostasis 5: 2146–2148.

Wuhrer M, Balog CI, Catalina MI, Jones FM, Schramm G, Haas H, Doenhoff MJ, Dunne DW, Deelder AM, Hokke CH. 2006. IPSE/alpha-1, a major secretory glycoprotein antigen from schistosome eggs, expresses the Lewis X motif on core-difucosylated Nglycans. The FEBS Journal 273: 2276–2292.

Xu YZ, Dresden MH. 1986. Leucine aminopeptidase and hatching of *Schistosoma mansoni* eggs. The Journal of Parasitology 12: 507–511.

Xu YZ, Dresden MH. 1989. *Schistosoma mansoni*: egg morphology and hatchability. The Journal of Parasitology 75: 481–483.

Xu YZ, Dresden MH. 1990. The hatching of schistosome eggs. Experimental Parasitology 70: 236–240.

Young ND, Jex AR, Li B, Liu S, Yang L, Xiong Z, Li Y, Cantacessi C, Hall RS, Xu X, *et al.* 2012. Whole-genome sequence of *Schistosoma haematobium*. Nature Genetics 44: 221–225. doi: 10.1038/ng.1065.

Zerlotini A, Aguiar ER, Yu F, Xu H, Li Y, Young ND, Gasser RB, Protasio AV, Berriman M, Roos DS, *et al.* 2013. SchistoDB: an updated genome resource for the three key schistosomes of humans. Nucleic Acids Research 41 (Database issue): D728–731. doi: 10.1093/nar/gks1087.

CHAPTER

Miracidium of *Schistosoma*

Barrie GM Jamieson[1],* and Wilfried Haas[2]

6.1 INTRODUCTION

The following account deals with the morphology and anatomy of the miracidium of *Schistosoma* species which infect humans and will chiefly be limited to electron microscopy. Host-finding and related phenomena are covered in Section 6.9. The most detailed morphological study is that of Pan (1980), for *S. mansoni*, from which this account, with that of Eklu-Natey *et al.* (1985), for four species, is largely drawn. There are other studies which refer to the morphology of the miracidium of *S. mansoni* (e.g., Faust and Hoffman 1934; Gordon *et al.* 1934, with *S. haematobium*; Watanabe 1934; Porter 1938; Olivier and Mao 1949; Ottolina 1957; Kinoti 1971; Brooker 1972; Basch and Diconza 1974; Schutte 1974; Wikel and Bogitsh 1974; Koie and Frandsen 1976; Ebrahimzadeh 1977; Meuleman *et al.* 1978; Bogitsh and Carter 1982; Dunn and Yoshino 1988; Neill *et al.* 1988; Swiderski 1994; Araque *et al.* 2003; Elkerdany 2003 and Collins *et al.* 2011). The miracidium of *S. japonicum* was briefly described by Tang (1938) and received detailed attention by Faust and Meleney (1924) and by Ozaki (1952) who described the epidermal plates (see early references therein). The miracidia of *S. japonicum* and of *S. haematobium* were examined by Loverde (1975), that of *S. haematobium* having been described earlier by Faust (1929), as figured in Chandler and Read (1961). Jones *et al.* (2008) have described the ultrastructure of the developing miracidium of *S. japonicum* before and shortly after hatching (see also Jones *et al.*, Chapter 5). Several books also refer briefly to *S. mansoni* miracidial structure (e.g., Faust 1929; Wright 1971; Rollinson and Simpson 1987; Basch 1991; Jordan *et al.* 1993; Marty and Andersen 1995; Mahmoud 2001; Secor and Colley 2005; Brehm and Lueder 2009; Bogitsh *et al.* 2013). That of *S. bovis* is treated by Lengy (1962) (Fig. 6.15). The general anatomy of the miracidium of *Schistosoma mansoni* is illustrated in Figure 6.1. The ultrastructure of the miracidium of *S. japonicum* within the egg is well illustrated by Jones *et al.*, Chapter 5.

The four species investigated by Eklu-Natey *et al.* (1985), viz., *Schistosoma mansoni, S. haematobium, S. intercalatum* and *S. japonicum*, showed differences in their respective dimensions, in the configuration of their terebratoria (apical papillae), in the shape of the epidermal plates and in the distribution of the sensory receptors. These aspects are treated below.

[1] Department of Zoology and Entomology, School of Biological Sciences, University of Queensland, Brisbane 4072, Australia.
[2] Former Section Parasitology, Institute for Zoology, University Erlangen-Nuernberg, Erlangen, Germany.
* Corresponding author

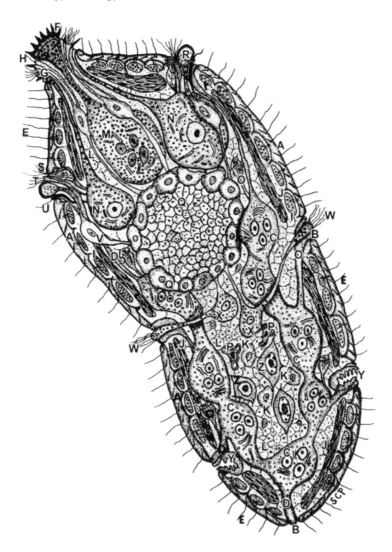

FIG. 6.1 A schematic representation of cellular architecture of the miracidium of *Schistosoma mansoni* (primarily based on electron microscopy but not drawn to scale): **A.** Epidermal plate. **B.** Epidermal ridge. **C.** Ridgecell-body. **D.** Cytoplasmic bridge. **E.** Cilium and its rootlet. **F.** Terebratorium and the profile of cytoplasmic expansions. **G.** Multiciliated deep-pit sensory papilla. **H.** Uniciliated sensory papilla. **I.** Outer circular muscle fiber. **J.** Inner longitudinal muscle fiber. **K.** Interstitial cell-body. **L.** Processes of interstitial cells. **M.** Apical gland and secretory duct. **N.** Lateral (cephalic) gland and secretory duct. **0.** Flame-cell with excretory tubule. **P.** Cell-body of common excretory tubule. **Q.** Neural mass with peripheral ganglia. **R.** Lateral papilla. **S.** Multiciliated saccular sensory organelle. **T.** Multiciliated shallow-pit sensory papilla. **U.** Uniciliated sensory papilla. **V.** Perikaryon of the neuron to lateral papilla. **W.** Multiciliated sensory papilla. **X.** Perikaryon of latter. **Y.** Excretory vesicle. **Z.** Germ cell. Modified after Pan SCT. 1980. Journal of Invertebrate Pathology 36(3): 307–372. Fig. 1.

6.2   MORPHOLOGY AND ANATOMY

6.2.1   Dimensions

The miracidium is pear-shaped, tapering posteriorly. Eklu-Natey *et al.* (1985) give detailed dimensions of the miracidia of the four chief human schistosomes. They report for miracidia of *S. haematobium* and *S. mansoni* similar mean lengths (133 ± 4 μm and 136 ± 3 μm), but widths which are significantly different (49 ± 2 μm and 54 ± 1 μm). For *S. mansoni*, Watts and Boyd

(1950) cite length 163 (156–174 μm), width 64 (63–65 μm), Porter (1938) gave 150–170 × 35–55 μm; (while Schutte (1974) gave 132 (118–153 μm) × 80 (62–96 μm). By volume, the miracidia of *S. mansoni* are bigger than those of *S. haematobium* but the miracidium of *S. intercalatum* has the greatest dimensions of all the four species (160 ± 6 μm in length and 55 ± 1 μm in width). *S. japonicum* is the smallest (averaging 77 μm in length and 48 μm in width) (Eklu-Natey *et al.* 1985), comparing with 78 to 120 μm in length and 30 to 40 μm in width, the width increasing by half as the free-swimming period approaches its end (Faust and Meleney 1924). The dimensions clearly vary with the state of contraction or extension, as shown by Lengy for *S. bovis*, which were 218 × 29 μm when extended, 92 × 55 μm contracted and 156 × 37 μm when relaxed. Ozaki (1952) gave dimensions of 79–141 μm × 26–59 μm for 100 miracidia of *S. japonicum*, depending on natural variation and the type of fixation.

6.2.2 Epidermis

The surface of the miracidium, with the exception of the apical papilla, is covered with ciliated epidermal cells containing numerous mitochondria, membranous bodies and glycogen. Nuclei are absent from these epidermal cells; mitochondria are packed along the base of each epidermal cell and between these there are numerous membrane-bound bodies. The epidermal cells are divided into a system of epidermal plates, in four tiers. Cilia appear most numerous and longest on the first tier (Pan 1980). Epidermal cells from the more posterior tiers have fewer mitochondria and cilia and often display basal cavitations (Wikel and Bogitsh 1974b). The anucleate condition of the epidermis was also noted in *S. japonicum* by Tang (1938) who observed nucleated subspherical subepidermal cells connected with it.

Intercellular ridges connecting with subepidermal cell-bodies interrupt the epidermal cells at numerous points and are joined to these cells by septate desmosomes. The ridges lack cilia (Wikel and Bogitsh 1974; Pan 1980). They connect to their respective cell-bodies by narrow cytoplasmic bridges which are lined by microtubules (Pan 1980). Each cell-body appears to contain two to three round nuclei, each with a prominent nucleolus. The nucleus is surrounded by a zone of granular endoplasmic reticulum (RER) with Golgi apparatus scattered among the RER the cisternae of which occasionally show high activity. Many mitochondria with abundant prominent cristae are usually seen at the periphery of the RER zone. Small patches of RER are also found scattered along the plasma membrane. The bulk of the cytoplasm is occupied by numerous glycogen particles (primarily β-particles) and membrane-bound vesicles. These vesicles contain stored membranes and participate in rapid formation of the tegument of the mother sporocyst during the first 24 hr after the miracidium enters the snail host. Small numbers of these vesicles are sometimes seen in the cytoplasmic bridges and the areas of ridge adjacent to the bridges. The ridges lack kinocilia and display numerous β-glycogen particles and mitochondria. Small numbers of microvilli are also present on the ridge surface (Pan 1980).

Circular and longitudinal muscle bundles underlie the epidermal cells. Sensory structures of various types are associated with the outer covering: (1) numerous "knob-like" cytoplasmic projections associated with epidermal cells, (2) bulbous, lamelloid structures with external cytoplasmic projections and (3) ciliated nerve endings with posterior epidermal tiers and ciliated nerve pits associated with apical papilla (Wikel and Bogitsh 1974b; Koie and Frandsen 1976). There are a few ciliated pits between the cells in the 1st tier and up to 12 ciliated pits with long cilia between the 2nd and 3rd tiers (Koie and Frandsen 1976). The bulbous bodies are considered by He and Ma (1981) to probably be secretory bodies of the cephalic gland cells. On each side of the miracidium, a lateral papilla of bulbous form is situated between the 1st and 2nd tiers of the ciliated epithelial plates (Koie and Frandsen 1976; Pan 1980; He and Ma 1981) (see Section 6.7.5). Lengy (1962) reports the papillae for *S. bovis*, *S. haematobium* and *S. mansoni*. In *S. japonicum* two excretory pores are found on either side in longitudinal intercellular ridges of the 3rd tier (He and

Ma 1981). The cell-bodies associated with the ridges are syncytial and are positioned along the inner side of the longitudinal muscles at the level between the neural mass and posterior end of the organism (Eklu-Natey *et al.* 1985).

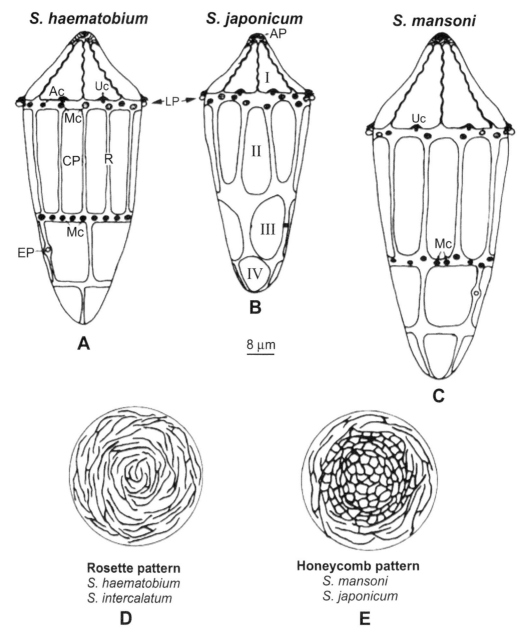

FIG. 6.2 Schematic drawing of the miracidia of **A**. *Schistosoma haematobium*. **B**. *S. japonicum*. **C**. *S. mansoni*. The relative sizes, the shape of the epidermal plates, the distribution of the sensory receptors are shown. The miracidium of *S. intercalatum,* which is not represented, resembles that of *S. haematobium,* but is larger than all the others. **D.**, **E**. Representation of the two main patterns of membrane folds observed on the terebratoria: **D**. The 'rosette' pattern in *S. haematobium* and *S. intercalatum* and **E**. The 'honeycomb' pattern in *S. mansoni* and *S. japonicum*. Abbreviations: Ac, aciliated sensory receptor. AP, apical papilla or terebratorium. CP, ciliated epidermal plate. EP, excretory pore. Mc, multiciliated sensory receptor. R, intercellular ridge. Uc, uniciliated sensory receptor. I, II, III, IV order of tiers of ciliated plates. Adapted from Eklu-Natey DT, Wuest J, Swiderski Z, Striebelt HP, Huooel H. 1985. International Journal for Parasitology 15(1): 33–42. Figs. 1–3.

6.2.3 Epidermal Plates

Eklu-Natey *et al.* (1985) have compared the miracidia of four species of *Schistosoma* by scanning electron microscopy: *S. mansoni, S. haematobium, S. intercalatum* and *S. japonicum*. The general appearance of the miracidia of the four species is similar in major respects. The body wall of the pear-shaped larva is covered by ciliated epidermal plates arranged in four tiers respectively consisting of 6 (rarely 7, Schutte 1974), 8 or 9, 4 and 3 plates, from the anterior to the posterior end (Fig. 6.2), as previously demonstrated for *S. mansoni* (Faust and Hoffman 1934; Maldonado and Acosta Matienzo 1947; Ottolina 1957; Schutte 1974; Wikel and Bogitsh 1974b and Pan 1980) and for *S. japonicum* (Tang 1938; Suzuki 1952, *fide* Ozaki 1952; He and Ma 1981). However, in *S. japonicum*, although the typical formulae occurred in 67% of the larvae, the number of plates varies from 13–29 and these may vary in shape (Ozaki 1952). The plates are separated longitudinally by the intercellular, non-ciliated ridges (described as grooves by Faust and Hoffman 1934). The first tier is said to lack ridges (Eklu-Natey *et al.* 1985) but these are reported, though illustrated in reduced form by Ozaki (1952) for *S. japonicum* (Fig. 6.3) and by Collins *et al.* (2011) for *S. mansoni* (Fig. 6.4C). Three transverse, circular ridges separate the tiers from one another and form the anterior, median and posterior rings. Various types of putative sensory receptors can be observed as small pits on the transverse rings, as illustrated by Eklu-Natey *et al.* (1985) (Fig. 6.2) and Ozaki (1952) (Fig. 6.3). In frontal view anti-phospho S/T strongly labels the plates of the first tier, revealing an hexamerous configuration with the central terebratorium (Collins *et al.* 2011) (Fig. 6.4C). According to Koie and Frandsen (1976) the number of epidermal cells in the *S. mansoni* miracidium varies from 17 to 22, corresponding roughly with the 21 epidermal plates reported by Pan (1980).

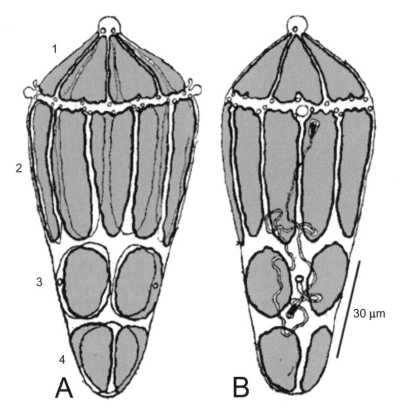

FIG. 6.3 *Schistosoma japonicum.* Silver impregnated miracidium. **A.** Ventral view of specimen. Tiers of epidermal plates are numbered. Ventral plates are shaded. Thin lines demarcate the dorsal plates. **B.** Left side, also showing part of the excretory system. Modified after Ozaki Y. 1952. Dobutsugaku Iho 25: 343–351. Fig. 1.

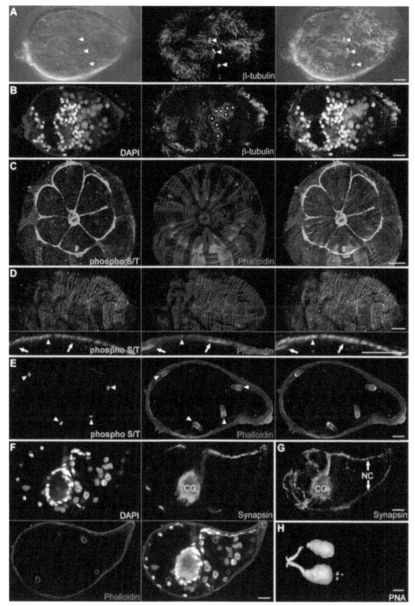

FIG. 6.4 The organ systems of miracidia of *Schistosoma mansoni*. **A**. The miracidial surface is covered by motile cilia on the epidermal plates as well as multiciliated sensory papilla (arrowheads) visualized by staining for β-tubulin and DIC optics. **B**. Mid-level confocal section showing microtubule meshwork of germ cells (asterisks). **C**. Anti-phospho S/T strongly labels the terebratorium and epidermal ridges surrounding the first tier of epidermal plates. **D**. Anti-phospho S/T also displays a weaker circumferential banding pattern similar to circumferential muscle (top). Optical cross sections indicate that the weak, superficial anti-phospho S/T labeling is at the level of circumferential muscles (arrowhead) but not longitudinal muscles (arrows). **E**. Similar to cercariae, anti-phospho S/T strongly labels the base of flame cells (arrowheads) in miracidia. **F**. and **G**. Immunofluorescence with an anti-synapsin antibody labels the cephalic ganglia and peripheral nerve structures. **F**. Mid-level confocal section showing labeling of neuropile of the cephalic ganglia with anti-synapsin (green). The neuronal cell bodies of the cephalic ganglia contain small nuclei that stain intensely with DAPI (grey) and surround the neuropile. Phalloidin staining (magenta) labels the muscle as well as flame cells. **G**. Maximum intensity projection of a miracidium stained with anti-synapsin antibody. **H**. Volume rendering of miracidium stained with PNA showing labeling of lateral (cephalic) glands and ducts. Scale Bars, 10 µm. Anterior faces on left in all panels except **C** that represents a view from the anterior surface. Abbreviations: Cephalic ganglia (CG), nerve cords (NC). From Collins JJ, King RS, Cogswell A, Williams DL, Newmark PA. 2011. PLoS Neglected Tropical Diseases 5(3): e1009. Fig. 9.

The anterior end of the miracidium consists of the terebratorium or apical papilla (*S. mansoni*, Figs. 6.6, 6.11A; *S. haematobium*, Fig. 6.7A, D; *S. japonicum*, Figs. 6.7C, 6.9), which is composed of an astomosing membrane folds and has numerous ciliated sensory receptors (see Section 6.3).

The miracidia of the four species investigated by Eklu-Natey *et al.* (1985) are clearly differentiated by their dimensions; the shape of the epidermal plates; the configuration of the apex of the terebratorium; the position of the excretory pores; and the number and distribution of the sensory receptors (Fig. 6.2).

The epidermal plates (Fig. 6.6) are all covered by locomotory cilia, which are of an increasing length from anterior to posterior (4–12 µm) in the first tier and of an equal length (12 µm) in the other tiers. The surfaces of ciliated plates and non-ciliated ridges are also covered by fine microvilli of about 50 nm (Eklu-Natey *et al.* 1985) or discordantly, 700 × 90 nm (Pan 1980) in height and diameter. The plates of the first tier are triangular in all four species studied; their longitudinal sides are sinuous. Transversely, their bases are bounded by a wavy, moderately thickened evagination, on the side of which one single uniciliated sensory receptor is located. No evident interspecific differences are found in the plates of the second and third tiers, except in *S. japonicum,* where they are elliptical in shape, where as in the other species they are oblong. The intercellular ridges consequently cover a larger area on the body of *S. japonicum* (Eklu-Natey *et al.* 1985). During the first several hours after the miracidium penetrates the snail, cisternae of the RER in the ridge cell-bodies are dilated and filled with electron-dense material (Pan 1972).

The ciliated epithelial plates (EP) in *S. mansoni* are seen by SEM to be attached to the syncytial ridges, in which they were embedded, by extensive pleated septate junctions that have 18-24 strands of intramembrane particles on the protoplasmic faces and complementary pits on the ectoplasmic faces. These junctions also appear to separate the EP plasma membrane into apical and basolateral domains with a larger number of intramembrane particles on the latter. Transformation into sporocysts (induced by placing the miracidia in salt containing medium) resulted in shedding of the plates and the syncytial ridges expanded until they formed a syncytium covering the parasite surface (Koie and Frandsen 1976; Samuelson and Caulfield 1985) but shedding does not normally occur until after the miracidium has entered the snail (see Yoshino *et al.*, Chapter 7). Basch and Diconza (1974) presumed that the new sporocyst tegument formed from material mobilized from the subepidermal region. The epidermal plates and the terebratorium rest on a thin, continuous basal lamina which in turn is bordered by the outer circular muscle fibers (Pan 1980). A developing miracidial plate in the embryo within the egg is illustrated by TEM by Neill (1988).

The glycocalyx. By transmission electron microscopy, the miracidial plates and ridges are seen to be covered by a 0.5 µm thick glycocalyx composed of a mesh of 9 to 10 nm fibrils. By indirect immunofluorescence, antibodies against cercarial glycocalyx stained miracidial plates and ridges. As the miracidia transform into sporocysts, the glycocalyx remain associated with the plates as they are sloughed. Thus miracidia possess a glycocalyx similar in structure and antigenicity to the cercarial glycocalyx (Chiang and Caulfield 1988).

Surface antigens. In *S. mansoni* monoclonal antibody reactive epitopes appear to be uniquely expressed in the intercellular ridges and submuscular, multinucleate syncytium in the miracidium/ primary sporocyst since similar molecules are absent from daughter sporocysts, cercariae, adults and snail tissues (Dunn and Yoshino 1988).

Fan *et al.* (1997) described a cDNA from miracidia and adult *S. japonicum* encoding a new member of the TM4SF family of tetraspanin antigens. Tetraspanins belong to an assemblage of surface antigens reported from mammalian and other vertebrate cells, from schistosomes, fruit flies and *Caenorhabditis elegans.* The deduced polypeptide, termed Sj25/TM4, was predicted to span the cell membrane four times, with its NH2- and COOH-termini embedded in the cytoplasm and to have two extracellular hydrophilic loops, one of which may be N-glycosylated. Sequence motifs are encoded by genes at separate loci and show interstrain variation in Chinese and Philippine isolates of *S. japonicum*. A further cDNA encoding a new tetraspanin, termed TE736 (tetraspanin 736), was isolated and characterized from *S. japonicum*. Nucleotide and deduced amino acid sequences of the cDNA revealed that TE736 was similar to the previously characterized Sm23/ Sj23/Sh23 species homologues and to Sj25/TM4 from schistosomes and to other tetraspanins. A phylogenetic comparison of the relationship of approximately 30 tetraspanins from mammals and other groups revealed that TE736, Sj25/TM4 and the two sequences from *S. mansoni* form an independent family of tetraspanins termed the Sj25 family of tetraspanins. TE736 appears to be encoded by a single gene and to be expressed in at least two life cycle stages of *S. japonicum* (Fan *et al.* 1997; Fan and Brindley 1998).

Septins. Zeraik *et al.* (2013) have shown that septins (highly conserved cytoskeletal proteins) localize in superficial structures and germ cells of the miracidium and sporocyst as revealed by application of anti-septin immunoglobulins, followed by incubation with an Alexa Fluor 633-conjugated secondary antibody. The superficial layers of the miracidium expressed septins on the ciliated epidermal plates and septins were prominent in optical sections of germs cell of miracidia. Colocalization of septin and actin was observed in the superficial optical sections of miracidia and sporocysts, though not in deeper sites (Zeraik *et al.* 2013).

6.2.4 Cilia

The structure of kinocilia (locomotory cilia) of the *S. mansoni* miracidium is similar to the conventional 9 + 2 form described for other organisms. The kinocilia arise from the epidermal plates at a right angle to the surface (Figs. 6.5, 6.6). The basal body of each cilium connects at its annulus to a long tapering rootlet (about 2 μm long) that lies at about a 20° angle to the plate surface with the tip pointing anteriorly, as also illustrated for *S. japonicum* by Jones *et al.* (2008; see also Fig. 6.10D). The basal body is embedded in the surface layer of the epidermal plate and has a characteristic centriolar arrangement of microtubules: nine peripheral triplets and no central tubules. The hollow rootlet exhibits periodic cross-banding of 65 nm with four equal subunits. Scattered among the ciliary rootlets are numerous electron-dense, membrane-bound, round to oval bodies measuring upto 500 nm in diameter. These vesicles become part of the membrane of new tegument which rapidly forms around the mother sporocyst during the early stages of transformation of the *S. mansoni* miracidium (Pan 1980). In at least *S. japonicum*, the cilia are reported to not be uniform: those between the pores of the lateral gland cells and the pair of head (cephalic) gland cells are noticeably long though decreasing in size anteriad; the cilium just anterior to the lateral glands is stiff and sickle-shaped; it is these anterior cilia which initiate swimming when the miracidium emerges from the egg (Faust and Meleney 1924). However, Lengy (1962) could not confirm the sickle shape for *S. bovis*. The differentiated crown of cilia, just anterior to the excretory pores, seen in *S. haematobium* by the former author, is not developed.

Employment here of the term 'cephalic glands' is explained below in Section 6.3.2 under Penetration Apparatus.

Numerous cristae-rich mitochondria occupy the zone between the ciliary rootlets and the basal plasma membrane of each plate. The cytoplasm of the epidermal plates has a finely granular texture and is moderately electron-dense. Small numbers of β-glycogen particles are occasionally present. The epidermal pegs are most developed in the first and last tiers. The basal lamina is a thin layer (45–90 nm thick) that follows the basal contour of the plates and epidermal ridges. It has an electron-dense fibrous middle zones sandwiched between two lighter homogeneous peripheral zones.

The basal surface of each plate is attached to both layers (outer circular and inner longitudinal) of muscle fibers, across the basal lamina, by many discontinuous intermediate (simple) tight junctions (macula adherentes) (Pan 1980).

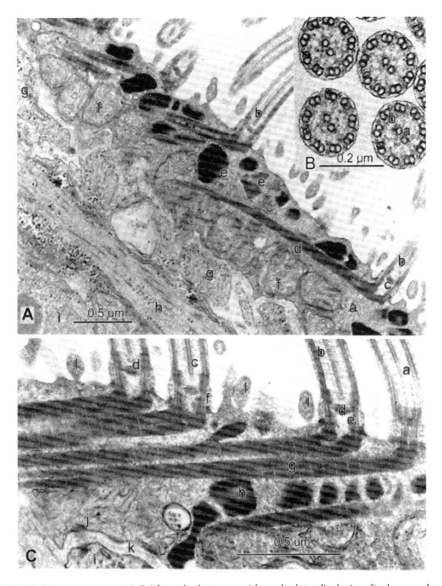

FIG. 6.5 A. *Schistosoma mansoni*. Epidermal plate. a, epidermal plate displaying finely granular ground substance; b, cilium; c, basal body (axosome); d, cross-banded rootlet of cilium; e, membrane-bound, electron-dense body; f, mitochondrion; g, outer circular muscle fiber; h, inner longitudinal muscle fiber; i. lateral (cephalic) gland. **B**. Cross sections of cilia, each with two central singlet microtubules and nine peripheral doublet microtubules. **C**. High magnification of epidermal plate. a. cilium; b, peripheral doublet microtubule; c, central single microtubule; d, basal body; f, basal body of cilium; g, cross-banded ciliary rootlet; h. membrane-bound. electron-dense body; i. outer circular muscle fiber; J, tight junction; k, basal lamina; l, microvillus. Adapted from Pan SCT. 1980. Journal of Invertebrate Pathology 36(3): 307–372. Figs. 13–15.

6.3 TEREBRATORIUM

6.3.1 General Structure

The terebratorium (*sensu* Reissinger 1923), frequently referred to as the anterior papilla, is an approximately conical anterior projection of the miracidium, containing several sensory endings as well as three openings of glands, has been described in great detail, supported by many transmission electron micrographs, by Pan (1980). It has also been described by scanning electron microscopy

(SEM) in four species, *Schistosoma haematobium, S. intercalatum, S. mansoni* and *S. japonicum* (Eklu-Natey *et al.* 1985), in *S. mansoni* (Koie and Frandsen 1976) and in *S. japonicum* (He and Ma 1981) and Kinoti (1971) provided observations on *S. matteei.* The surface covering (epithelial sheet) is regarded as a modified epidermal plate without kinocilia and presents a network of interlaced cytoplasmic expansions. The profiles of these expansions usually appear as more or less straight filopodia (ca. 1 µm long) consisting of two closely apposed membranes. The network is apparent only in tangential sections. When viewed from the surface, it appears as the septa of numerous pit-liked depressions (ca. 1 µm deep) on the terebratorium (Pan 1980) and the microvilli reported by Kinoti (1971) are not confirmed. Similar structure has been described for miracidia of *S. japonicum* and *S. haematobium* (Loverde 1975) as for *S. mansoni* (Koie and Frandsen 1976) and the rodent parasite *Schistosomatium douthitti* (Blankespoor and Van der Schalie 1976).

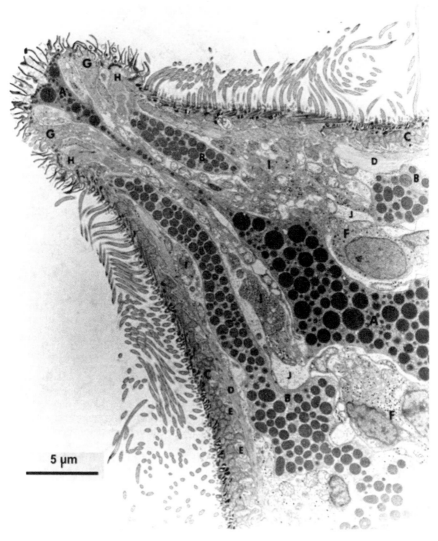

FIG. 6.6 *Schistosoma mansoni.* Sagittal section of the terebratorium and anterior area covered by the first tier of epidermal plates. **A.** Apical gland and its duct. Secretory material is being discharged at the duct opening; **B.** Lateral (cephalic) gland and its duct; **C.** Epidermal plate with cilia and rootlets; **D.** Inner longitudinal muscle fiber; **E.** Outer circular muscle fiber; **F.** extra-CNS neuron; **G.** Multiciliated deep-pit sensory papilla; **H.** uniciliated sensory papilla; **I.** Nerve containing neurosecretory vesicles; **J.** Process of interstitial cell. Modified after Pan SCT. 1980. Journal of Invertebrate Pathology 36(3): 307–372. Fig. 7.

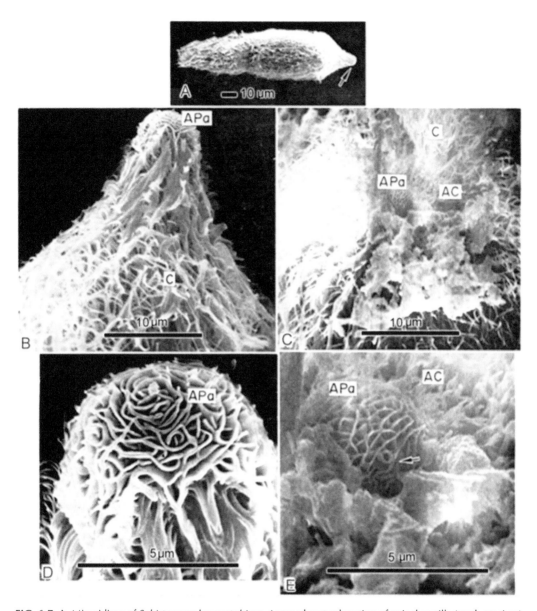

FIG. 6.7 A. Miracidium of *Schistosoma haematobium*. Arrow denotes location of apical papilla (terebratorium). **B**. Apical papilla (A pa) of *S. haematobium* miracidium surrounded by locomotor cilia (C). **C**. Apical papilla (APa) of *Schistosoma japonicum* miracidium. **D**. Apical papilla (APa) of *S. haematobium* miracidium showing putative mini-sucker pad composed of tiny sucker-like cups. **E**. Apical papilla (APa) of *S. japonicum* miracidium showing putative mini-sucker pad composed of tiny sucker-like cups, with a secretory pore (arrow) surrounded by tactile apical cilia (AC). Note that **D** shows the rosette pattern for *S. haematobium* and **E** the honeycomb pattern later defined by Eklu-Natey *et al.* (1985). Relabeled after Loverde PT. 1975. International Journal for Parasitology 5(1): 95–97. Figs. 1–5.

The septa generally consist of elongated bands arranged in interdigitating semicircles and forming closed, supposedly adhesive sucker-like cups around the ciliary sensory receptors. They show, however, interspecific variations in the pattern of their arrangement, notably as rosette like or honeycomb-like arrangements (Eklu-Natey *et al.* 1985).

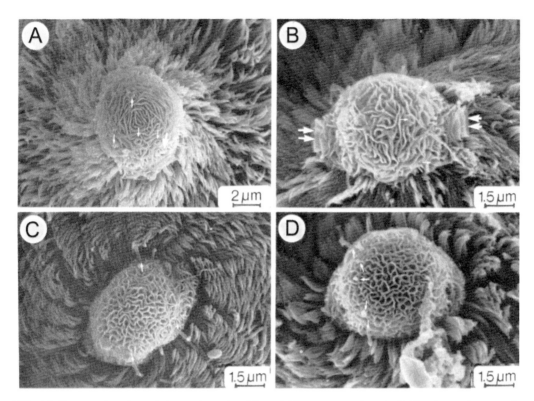

FIG. 6.8 Comparative view of the terebratoria of four *Schistosoma* species. **A, B**. The 'rosette' pattern in **A**. *S. haematobium* and **B**. *S. intercalatum*. **C, D**. The 'honeycomb' pattern in **C**. *S. japonicum* and **D**. *S. mansoni*. Note the two openings of the lateral penetration glands on the terebratorium of *S. intercalatum* (double arrow) and the ciliated receptors (small arrows). Relabeled after Eklu-Natey *et al.* (1985). International Journal for Parasitology 15(1): 33–42. Figs. 13–16.

In the miracidia of *S. haematobium* and *S. intercalatum*, the bands are concentric all over the terebratorium, with a few ramifications connecting them to adjacent bands. These terebratoria thus have the appearance of a rosette. Moreover, in *S. haematobium*, the bands are more regularly concentric than in *S. intercalatum*. In *S. mansoni* and *S. japonicum* the folds join at the apex of the terebratorium to form closed 'alveoli', causing the apex to look like a 'honeycomb' (Fig. 6.8C, D). These alveoli seem more regular in *S. mansoni*, as square or round sections (Fig. 6.8D), whereas in *S. japonicum* they are somewhat elongated (Fig. 6.8C) (Eklu-Natey *et al.* 1985). The contrasting types were well illustrated by SEM, though not so named, by Loverde (1975) for *S. haematobium* and *S. japonicum* (Fig. 6.7) though he was inclined to ascribe the conditions to different physiological states at the time of fixation. The 'rosette' and the 'honeycomb' patterns are particularly stable and specific structures and no significant modifications were observed, despite various fixation media and different strain or host origins of the material, thus supporting a true genetic difference (Eklu-Natey *et al.* 1985). These authors state that this could be used as a supplementary criterion for the differentiation of the miracidia of *S. haematobium* and *S. mansoni*, mainly in the geographic regions where they are sympatric. The similarity between the miracidia of *S. haematobium* and *S. intercalatum*, on the other hand, reflects the fact that they have been considered to belong to one species because of their resemblance at various levels. Their miracidia can be distinguished only on the basis of the irrelative dimensions. He and Ma (1981) merely describe an anastomosing network for *S. japonicum*. We may note, however, that honeycomb arrangement is also seen in a micrograph of the terebratorium of the miracidium of *Schistosomatium douthitii* by Blankespoor and Van der Schalie (1976).

In miracidia of *S. haematobium* and *S. japonicum*, the tiny sucker like cups were presumed to be used by the miracidium to facilitate attachment to the snail during penetration (Loverde 1975) (Fig. 6.7). An adhesive function of the cups was also advocated by Wright (1971) but was questioned by Pan (1980) and appears unlikely to the authors. A lateral opening (secretory pore) on the apical papilla and short stubby apical cilia (tactile or sensory) were also demonstrated (Loverde 1975).

The thin cytoplasm of the surface covering of the terebratorium is finely granular, like that of the epidermal plates and contains a few β-glycogen particles. The epithelial sheet rests on the basal lamina and is attached by septate desmosomes to the first tier of epidermal plates. The base of the epithelial sheet forms many thin "pegs" which extend into the miracidial body. Beneath the basal lamina lie several outer circular and inner longitudinal muscle fibers which are attached to the base of the epithelial sheet by simple tight junctions (macula adherentes) (Pan 1980).

The secretory duct of the apical gland (Fig. 6.6) is funnel shaped as it opens to the outside through the center of the terebratorium. The circular muscle fibers form a sphincter around the "neck" of the similarly funnel-shaped excretory duct, which is usually flattened dorsoventrally and is surrounded by four multiciliated, deep-pit nerve endings (see nervous system, Section 6.7). At least eight uniciliated nerve endings lie between the multiciliated, deep-pit nerve endings. The rims of these 12 nerve endings are attached to the terebratorium by septate desmosomes. The duct openings of the two lateral (cephalic) glands are located at the base of the terebratorium adjacent to the first tier of epidermal plates. Septate desmosomes line the rims of the openings (Pan 1980).

Lateral papillae. Small anterolateral papillae are present but the numbers reported vary greatly. Schutte (1974) gives 1–5 and for larger, less variable posterolateral papillae, 15–24, confirming numbers reported by Ottolina (1957).

The apical papilla and the lateral papillae persist for a few hours following shedding of the ciliated plates after penetration of the snail host, but the ciliated pits disappear shortly after loss of the plates (Koie and Frandsen 1976). Loss of the plates during transformation of miracidia of *S. mansoni* into sporocysts is illustrated by light microscopy by Yoshino *et al.*, Chapter 7 (Fig. 7.2).

6.3.2   Penetration Apparatus

The penetration apparatus of the terebratorium is composed of three unicellular glands. Their structure has been described by several investigators (Tang 1938; Faust and Hoffman 1934; Maldonado and Acosta Matienzo 1947; Pan 1965; Kinoti 1971; Schutte 1974; Wikel and Bogitsh 1974). Wadji (1966) termed them adhesive glands. Wikel and Bogitsh (1974) and Pan (1980) give illustrated accounts of these large glandular cells, one apical and two lateral. They occupy much of the area anterior to the neural ring. Each cell consists of a cell-body and a long secretory duct which opens through the center of the terebratorium. In the basal third of the cell-body of the apical gland there are four irregularly shaped nuclei (Tang 1938; Ottolina 1957; Schutte 1974; Pan 1980). The gland measures 41×17 μm (Schutte 1974). Only a single nucleus was reported by Wikel and Bogitsh (1974) and two nuclei by Maldonado and Acosta Matienzo (1947). Lengy (1962) reviewed supposed variation in numbers of nuclei in *S. japonicum*, *S. haematobium* and *S. mansoni* and ascribed this to difficulty of observation, concluding that the correct number is probably four. The nuclei are surrounded by a layer of free ribosomes and granular endoplasmic reticulum. The rest of the cytoplasm, including that of the duct, contains membrane-bound, round secretory droplets of various sizes (upto 2 μm in diameter), electron densities, ribosomes, RER, α- and β-glycogen particles and occasional Golgi apparatus, features suggesting active secretion (Pan 1980).

The vesicles were shown by Wikel and Bogitsh (1974b) to contain macromolecular diglycols. The cisternae of the RER are usually distended and frequently contain low to moderately electron-dense granular material which resembles the contents of secretory droplets. Many mitochondria with long cristae are scattered in the cytoplasm. The duct is lined with a parallel array of microtubules along the plasma membrane and its opening is attached to the surface layer of the terebratorium and according to Pan (1980) lacks desmosomes. However, Wikel and Bogitsh (1974b) note that at their narrowest part, near the pore, the gland cells are joined to adjacent integumetary cells by septate desmosomes. The pore is sealed by a plasma membrane with some perforations (Pan 1980).

The ducts of the pair of lateral (cephalic) glands (Fig. 6.6) are also lined with microtubules but they open at the base of the terebratorium. The rim of each pore is lined with septate desmosomes. The cell-body of each cephalic gland differs from the apical gland in having only one nucleus, as earlier reported by Lengy (1962) for *S. bovis* and its secretory droplets are generally smaller and less electron-dense (Pan 1980).

Collins *et al.* (2011) observed staining of the lateral (cephalic) glands in the *S. mansoni* miracidium with a number of lectins (Fig. 6.4H) but only a few lectins were observed staining the apical gland. This suggested that the products of the apical gland were different from those of the cephalic glands, an observation consistent with the demonstration by Pan (1980) of differences in size of secretory vesicles in these glands (Collins *et al.* 2011).

Transmission electron microscope, as distinct from SEM, images of the miracidium of *S. japonicum* appear to be limited to sections of the miracidial embryo in the egg, which show many details (Jones *et al.*, Chapter 5, Fig. 5.3). For general antomy we are largely reliant on the observations by light microscopy for this species by Faust and Meleney (1924). They considered a short solid pocket of tissue at the anterior end of the larva to be a functionless gut, as did Gordon *et al.* (1934). Faust and Meleney state that there are four nuclei buried deeply within the granular cytoplasm of this undifferentiated organ, though only a single nucleus was mentioned by Faust and Hoffman (1934) for *S. mansoni*. It is here considered that the nucleated condition of the organ, especially as there are four nuclei as seen in *S. mansoni* by Pan (1980), is consistent with recognizing it as an apical gland not withstanding the fact that a solid, nucleated parenchymal gut exists in some Platyheminthes (Acoela). These authors describe a pair of pyriform lateral glands, which they term cephalic glands, just lateral to the "gut" each of which contains a prominent vesicular nucleus and is densely packed with large granules. Again, as described by Pan (1980) for *S. mansoni*, each gland has a single nucleus. Because Faust and Meleney (1924) also describe as lateral glands a pair of mucoid glands which discharge anterolaterally well behind the terebratorium, it is here proposed to revive the term cephalic glands for those bordering the apical gland. As the apical gland is also in a cephalic position the cephalic glands might be distinguished as "lateral cephalic glands". Gordon *et al.* (1934), for *S. haematobium* and *S. mansoni*, observed evanescent refractile globules at the knob-like mucoid glands but Tang (1938) was unable to confirm any secretions from these glands, the knob-like ends of which lay between the first and second rows of epidermal plates, in *S. japonicum*. His line drawing of the miracidium is otherwise identical in almost every detail with that of Faust and Meleney.

Pan (1980) has shown that the miracidium of *S. mansoni* maintains its glandular cells for several days and discharges secretory droplets into snail tissues after entering the snail host. He draws attention to the fact that Wilson (1971) suggested that the secretory droplets of the *Fasciola hepatica* miracidium may contain zymogen. It was therefore considered probable that the apical and lateral (cephalic) glands of *S. mansoni* miracidia have the dual function of aiding the miracidium to enter the snail host and of preparing sites for intramolluscan development (Pan 1980). Significant cysteine proteinase activity has been demonstrated in miracidia of *S. mansoni* (Yoshino *et al.* 1993).

The most remarkable differentiating features observed with SEM in the four species of *Schistosoma* investigated by (Eklu-Natey *et al.* 1985) are found in the network of anastomosing membrane folds on the terebratorium (See Section 6.3.1).

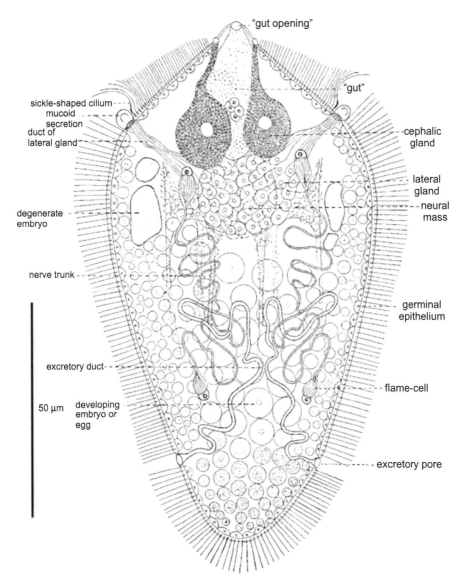

FIG. 6.9 *Schistosoma japonicum*. Anatomy of the miracidium. The "gut" and "mouth" are here considered to be the apical gland and pore, as in *S. mansoni*. A terebratorium, though present, is not illustrated. Relabeled after Faust, Meleney (1924). American Journal of Hygiene, Monograph series 3. Plate II, Fig. 11.

6.4   MUSCULATURE

Except where otherwise mentioned, the following account of the musculature is derived from Pan (1980). The musculature consists of outer circular and inner longitudinal layers, as seen by Schutte (1974). Only smooth muscle is present. Each layer of musculature is only one cell (fiber) thick and each spindle-shaped myofibril forms a muscle that runs parallel to others within the layer. In cross section, circular fibers appear almost to encircle the entire miracidium. The muscle fiber may measure 4 μm thick at the level of the nucleus. The circular muscle fibers lie close to the basal lamina and nestle between adjacent epidermal pegs. Both circular and longitudinal myofibrils are attached, across the basal lamina, to the base of the epidermal plates or epidermal ridges with discontinuous intermediate tight junctions (macula adherentes). The macula adherens for inner longitudinal fibers is usually seen attached to the epidermal pegs via the gaps between adjacent circular fibers. Occasional tight junctions between circular and longitudinal myofibrils are also present. Each muscle fiber consists of a major outer fibrillar zone containing longitudinally orientated thick and thin myofilaments and a minor inner afibrillar zone (or outpocket) containing the oval nucleus, many mitochondria, lipid droplets and abundant glycogen particles (mostly β-glycogen particles). Small numbers of β-glycogen particles are also seen scattered among the myofilaments.

The plasma membrane of the fibrillar zone occasionally forms inpocketings or invaginations that may be associated with dense bodies. The plasma membrane of the afibrillar zone frequently forms outpocketings that are associated with neuromuscular junctions. A detailed description of neuromuscular junctions is given in Section 6.7 on the nervous system. The thick myofilaments are about 25 nm and the thin about 6 nm in thickness. In cross section, each thick myofilament is surrounded by 6–10 thin myofilaments. The thick myofilaments are composed of ca. 10 subunits spaced at 500 to 100 nm. Thick and thin filaments are apparently crosslinked with oblique lines. Dense bodies (about 80 nm thick) are scattered randomly in the muscle fibers and appear to be connected to the thin myofilaments. A granular endoplasmic reticulum (sarcoplasmic reticulum) is distributed primarily along the periphery of myofibrils close to the sarcolemma (less than 30 nm from the membrane), but is also found in the center of myofibrils.

The oval nucleus (2×4 μm) contains a small nucleolus, is usually poor in heterochromatin and frequently holds a cluster of dense granules (upto 20 in number and ca. 100 nm in diameter). Each of the granules is surrounded by ribosomes. Similar dense granules are also found in the nuclei of neurons in the neural mass (see Section 6.7.1). Numerous mitochondria surrounding the nucleus contain abundant long cristae.

The simple form of miracidial muscles may reflect the fact that mobility of miracidia depends primarily on ciliary motion. The sarcoplasmic reticulum of *S. mansoni* miracidium may function similarly to the T tubules of vertebrate muscle fibers in the conduction of stimuli (Pan 1980).

Beneath the muscle layers in *Schistosoma mansoni* eggs, miracidia and primary sporocysts there are four actin-rich tubules. The tubules are in pairs, transverse to the length of the parasite and are located towards the extremities. Application of an anti-flame-cell specific antibody confirms that the tubules co-localize with flame-cells and also demonstrates that the tubule core is filled with microtubules. The additional presence of myosin in these tubules strongly suggests that they are contractile structures (Bahia *et al.* 2006).

Anti-phospho S/T staining in the miracidia was observed in a circumferential banding pattern similar to that observed for phalloidin staining of the muscle (Fig. 6.4D). It was difficult to ascertain whether this labeling was in the muscle layer itself or in the overlying epidermal plates and ridges. In addition to the superficial banding pattern, anti-phospho S/T labeling of the base of flame cells was observed (Fig. 6.4E bottom arrowheads) (Collins *et al.* 2011).

6.5 INTERSTITIAL CELLS

There are approximately 20 interstitial cells in the miracidium of *S. mansoni*. These cells fill intercellular spaces; they contain abundant complex carbohydrates and lipid droplets and are deduced to serve as reserve food stores. The small, irregularly shaped cell-bodies occupy the core area of the miracidium posterior to the neural ring and are surrounded by the cell-bodies of ridge cells. Many long oval, round or irregularly shaped processes extend from them and fill the intercellular spaces of the larva. The small irregular nucleus with a relatively large nucleolus has little heterochromatin and is surrounded by a thin layer of granular endoplasmic reticulum and ribosomes. Adjacent to the RER layer are numerous mitochondria with abundant cristae. Small patches of RER are scattered throughout the cell-bodies and their processes; they are often masked by α- and β-glycogen particles which fill much of the cytoplasm. Lipid droplets are scattered among the glycogen particles. Germinal cells, flame-cells, excretory tubules and the cell-bodies of putative chemoreceptor neurons are all surrounded by the processes of interstitial cells. These processes are also found around the neural ring and along many nerves throughout the miracidium (Pan 1980).

6.6 EXCRETORY SYSTEM

A detailed review of the adult schistosome excretory system is given by Kusel in Chapter 16. The present account is confined to the miracidium. *S. mansoni, S. haematobium* and *S.*

intercalatum have two pores of the protonephridial excretory system symmetrically located in the longitudinal ridges between the epidermal plates of the third tier (as also noted by Schutte 1974, for *S. mansoni*), usually halfway from the plate corners. In *S. japonicum* their location varies and they may even be found on the median ring (Eklu-Natey *et al.* 1985). However, Lengy (1962) described the two pores as opening at the juncture of the third and fourth series of epidermal cells. Gordon *et al.* (1934) stated that in *S. mansoni* the long axis of the flame cells lies in the anterior-posterior plane whereas in *S. haematobium* the long axis is perpendicular to this plane, being the only difference in internal structure noted between the two species.

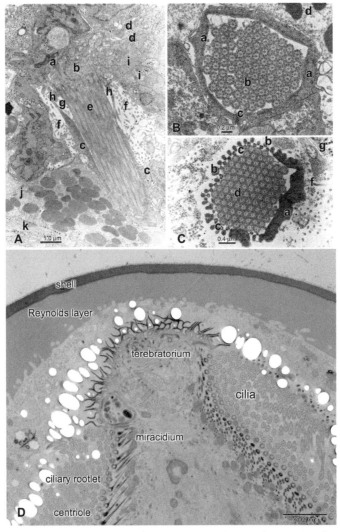

FIG. 6.10 A. *Schistosoma mansoni*. A flame-cell from a postmiracidium within 6½ hr after entering snail host, *Biomphalaria glabrata*. a, a small section of flame-cell nucleus; b, perikaryon of flame-cell; c, cylinder wall; d, cytoplasmic process of flame cell-body; e, cilia in the cylinder; f, peri-flamecell space displaying many leptotrichs; g, leptotrichs within cylinder; h, grille-work (fenestra or weir) of cylinder; i. neuropiles in the CNS; j, cell-body of lateral gland; K, longitudinal muscle fiber. **B**. Cross section of the flame-cell cylinder at midlevel. a, cylinder wall containing circular fibrils (arrow); b, cross sections of cilia within the cylinder; c. septate desmosome joining the margins of cylinder wall; d, lateral (cephalic) gland. **C**. Cross section of flame-cell cylinder at the level of the grille. a, cylinder wall; b, thick bar of the grille; c, thin bar of the grille; d, cross sections of cilia within the cylinder; e, septate desmosome joining the margins of cylinder wall; f, leptotrichs in the peri-flamecell space; g, process of interstitial cell. Relabeled after Pan SCT. 1980. Journal of Invertebrate Pathology 36(3): 307–372. Figs. 40, 43 and 44 **D**. *Schistosoma japonicum*. Nearly mature miracidium within the egg, showing ciliary rootlets arising from the centrioles. From Jones MK (unpublished).

The flame-cell (cyrtocyte in Ishii 1980, for the planarian *Bdellocephala brunnea*) of the protonephridium consists of a cell-body, a hollow cylinder (barrel or tube) and numerous cilia within the cylinder. The cell-body (cyton) is stellate and sends many slender processes into the surrounding intercellular spaces, probably as anchoring devices and for absorption. A depressed, peripherally located nucleus occupies about a third of the cell-body. The cytoplasm contains abundant free ribosomes, granular and smooth endoplasmic reticula, mitochondria, small membrane-bound vesicles and microtubules. Distal to the nucleus the cytoplasm expands into a thin sheet which is rolled into a gradually tapering hollow cylinder with its margins joined by septate desmosomes. As many as 100 cilia originating from basal bodies that are embedded in the cell-body project into the cylinder (Fig. 6.10). The flickering appearance of their motion in life is recognized in the name flame-cell. Each basal body is attached to a short, cross striated rootlet. The tapered distal end of the cylinder joins the proximal end of the excretory tubule. The proximal quarter of the barrel forms a filtration grille. For other platyhelminths the grille is alternatively termed a fenestra (e.g., Ishii 1980) or a weir (e.g., Rohde and Georgi 1983; Rohde *et al.* 1989, 1992; Xylander 1992). It contains an outer row of thick bars (filaments) and an inner row of thin filaments. In cross section, thick filaments (160 nm across) alternate with thin filaments (100 nm across) forming a zigzag line. Numerous leptotrichs (elongate projections resembling microvilli and ca. 80 nm in diameter) extend from thick and thin filaments. Leptotrichs originating from thick filaments are long and in large numbers and extend into a relatively large intercellular space around the flame-cell. Leptotrichs originating from thin filaments are short and few in number and extend into the flame-cell cylinder. The outer leptotrichs appear to be suspended in the relatively large space around the flame cell. Except for the presence of circular fibrils in the inner zone, the cytoplasm of the cylinder wall has a structure similar to that of the flame-cell body (Pan 1980). Rohde *et al.* (1992), for a mongenean, use the term 'flame bulb', which they considered to occur in all Digenea, formed by a terminal [flame] cell and a proximal canal cell but the term flame-cell is retained here as Pan did not recognize the interdigitation of the flame cell with a canal cell, though this remains to be demonstrated. Gobert and Nawaratna (Chapter 9) illustrate and briefly describe a flame-cell of the schistosomula of *S. japonicum*. The 'cap cell' has approximately 60 flagella; Tubule cells are mentioned, comprising the walls of the osmoregulatory ducts, but no distinct flame bulb is recognized.

Two excretory tubules lie on each side of the miracidium adjacent to the flame-cells and consist of elongated cells the cytoplasm of which forms a continuous tubule. The tubules on each side join a common tubule which is formed from a separate cell.

6.7 NERVOUS SYSTEM

6.7.1 The Neural Mass (CNS)

The neural mass (neural ring) in the miracidium of *S. mansoni* occupies much of the area covered by the second tier of the epidermal plates and nearly 10% of the miracidial volume (Fig. 6.1) (Pan 1980). It measures 22.3 μm in diameter (Schutte 1974). More than 20 nerves originate directly from it. It is covered with the processes of interstitial cells. These processes also extend superficially into interneuronal spaces but seldom reach into the neuropiles. The neurons are located peripherally, encircling numerous axons and dendrites (neuropiles). The neurons (3–6 μm in diameter), form a layer one to two cells thick (Fig. 6.11E). The nuclei are oval, kidney-shaped, elongated or lobulated with a small nucleolus, patches of heterochromatin and occasionally one or two clusters of electron-dense ribosomes within which are embedded several larger and denser granules (30–45 nm). The granular cytoplasm is moderately electron-dense and contains some or all of the following structures: α-(more numerous) and β-glycogen particles, lipid droplets, Golgi complex, mitochondria, granular and a granular endoplasmic reticula, free ribosomes, myelin figures and four types of membrane-bound "neurosecretory" vesicles (see Section 6.7.3, below) (Pan 1980).

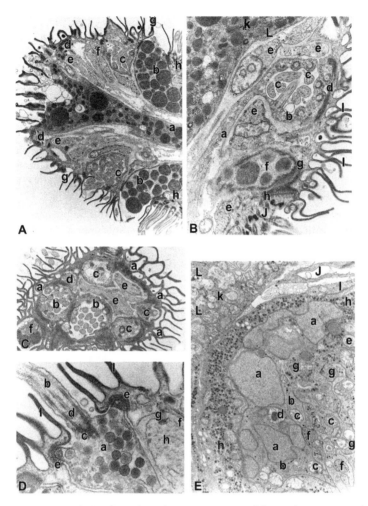

FIG. 6.11 *Schistosoma mansoni.* **A**. Terebratorium showing openings of ducts of penetration glands and ciliated sensory papillae. x 7500. **a**, duct and its opening of apical gland. The opening is covered with a membrane and the duct is lined with microtubules along the submembranous area; **b**, duct and its opening of lateral (cephalic) gland. The opening is covered with a membrane which is lined with microtubules; **c**, multiciliated, deep-pit sensory papilla; **d**, uniciliated sensory papilla; **e**, muscle fiber; **f**, nerve to uniciliated sensory papilla; **g**, profile of cytoplasmic expansions of terebratorium; **h**, epidermal plate of the first tier. **B**. Multiciliated deep-pit sensory papilla. x 14900. **a**, nerve containing Type A neurosecretory vesicles and neurotubules; **b**, wall of multiciliated deep-pit sensory papilla with Type A neurosecretory vesicles; **c**, cross section of sensory cilium lacking central single microtubules; **d**, septate desmosome joining the papilla to terebratorium; **e**, muscle fibers; **f**, duct of lateral gland; **g**, septate desmosome joining glandular duct to terebratorium; **h**, septate desmosome joining epidermal plate of the first tier to terebratorium; **i**, profile of cytoplasmic expansion of terebratorium; **j**, epidermal plate of the first tier; **k**, duct of apical gland; **l**, microtubules lining the duct membrane. **C**. Tangential section of terebratorium showing interlacing cytoplasmic expansions. x 9120. **a**, meshwork formed by cytoplasmic expansion of terebratorium; **b**, multiciliated deep-pit sensory papilla: **c**, uniciliated sensory papilla with Type C neurosecretory vesicles; **d**, nerve to sensory papilla; **e**, muscle fiber; **f**, duct opening of lateral (cephalic) gland. **D**. Uniciliated sensory papilla of terebratorium. x 29000. **a**, sensory papilla containing type C neurosecretory vesicles; **b**, sensory cilium; **c**, basal body of cilium without rootlet; **d**, axosome of cilium; **e**, septate desmosome; **f**, basal lamina; **g**, simple tight junction (macula adherens); **h**, muscle fiber; profile of cytoplasmic expansion of terebratorium. **E**. Structure of neural mass. x 7500). α-glycogen particles are seen. **a**, nucleus of neuron at the periphery; **b**, perikaryon; **c**, mitochondrion; **d**, myelin figure; **e**, axonencircled with several concentric membranes (primitive myelin sheath); **f**, axoncontaining Type C neurosecretory vesicles; **g**, axoncontaining Type A neurosecretory vesicles; **h**, process of interstitial cell containing numerous glycogen particles; **i**, nerve leaving CNS; **j**, excretory tubule; **k**, cell-body of ridge; **L**, membrane-bound vesicle in ridge cell-body. Adapted from Pan SCT. 1980. Journal of Invertebrate Pathology 36(3): 307–372. Figs. 23, 24, 27, 34 and 49.

Neuropiles are membrane-bound structures densely grouped in the central region of the CNS. The numerous neuropiles and nerves have cytoplasm and cytoplasmic inclusions similar to their perikaryons. Neurotubules are frequently present. The synapses of neuropiles form a unilateral electron-dense zone. Many synaptic vesicles are commonly present near the synapses. Some neuropiles may occasionally be circled by four or more layers of unit membrane suggesting a primitive form of myelination (Pan 1980).

The neural mass stains with the anti-synapsin antibody and is surrounded by neuronal cell bodies that are identifiable by their small ovoid nuclei which stain intensely with DAPI (Fig. 6.4F). Several nerve cords project anteriorly while only a dorsal and ventral nerve cord project posteriorly (Fig. 6.4 F, G) (Collins *et al.* 2011). The neuronal cell bodies correspond with the small nerve nuclei around the periphery of the neural mass described by Lengy (1962). Tang (1938) observed two forked lateral nerves in *S. japonicum*.

Many of the nerves terminate in the peripheral sensory organelles. Germinal cells with short processes are nestled among the cell-bodies of interstitial cells. A group of extra-CNS, bipolar neurons (about 12 in number) are positioned between the neural ring and the cell-bodies of interstitial cells. Two flame-cells are located among the cell-bodies of interstitial cells and two lie on the anterolateral sides of the neural mass (Fig. 6.1). Four of the six major types of sensory organelles are positioned in the ridges separating the epidermal plates of adjacent tiers and the rest lie in the terebratorium. The two excretory pores open through the ridges separating the plates of the third tier (Pan 1980).

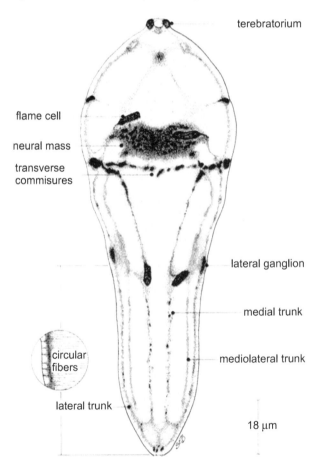

FIG. 6.12 *Schistosoma mansoni.* Diagrammatic representation of ASChE-positive areas in the miracidium. Relabeled after Bruckner DA, Voge M. 1974. The Journal of Parasitology 60(3): 437–446. Fig. 16.

The nervous system of the miracidium, was earlier demonstrated by means of acetylcholinesterase (ASChE) (Bruckner and Voge 1974) (Fig. 6.12). The thick neural mass in the anterior half of the organism and three pairs of longitudinal nerve trunks, connected to a transverse commissure, are revealed. One of these trunks, in the anterior third of the body, bears lateral thickenings (ganglia?) close to the body surface and the transverse commissure. The protonephridial flame-cells stain intensely and an intricate innervation is suggested by the staining reaction. At the anterior end of the miracidium, two intensely staining papillae appear to be part of a small circular ring [at the apex of the terebratorium] and are connected with longitudinal trunks situated close to the surface of the organism. The papillae are deduced to mark the lateral, cephalic gland ducts and the ring to be the ciliated nerve ending described by Brooker (1972). Heavy staining occurs in the area of the lateral papillae. The existence of a fine network beneath the surface of the miracidium is suggested by the presence of very delicate, darkly staining transverse lines seen in a few miracidia (Fig. 6.12). Physostigmine salicylate eliminated all AChE activity (Bruckner and Voge 1974).

Faust and Meleney (1924) describe six sub-epithelial cords, two dorsal, two lateral and two ventral, with two transverse commissures surrounding the system. The longitudinal elements extend anterior and posterior to the neural mass and are composed of fibers and cells of irregular outline and are identified with the three anterior and posterior pairs of nerve trunks in adult schistosomes. Longitudinal cords are also illustrated by Faust (1929) (Fig. 6.14A).

Confocal microscopy of miracidia has revealed phosphorylated protein kinase C associated with the neural mass, excretory vesicle, epidermis, ciliated plates, terebratorium and germinal cells, related to transformation into mother sporocysts (Ludtmann *et al.* 2009).

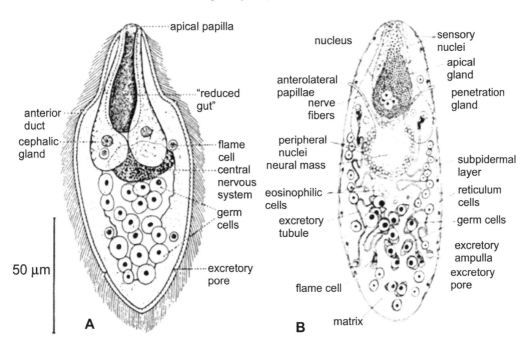

FIG. 6.13 *Schistosoma mansoni.* Miracidium. **A.** As interpreted by Cort (1919). Relabeled from Basch PF. 1991. *Schistosomes: development, reproduction and host relations.* New York: Oxford University Press. Fig. 5–2. By permission of Oxford University Press, USA. **B.** Relabeled from Schutte CHJ. 1974. South African Journal of Science 70(10): 299–302. Fig. 2a.

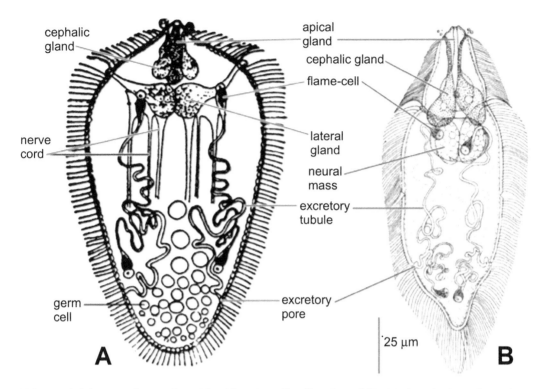

FIG. 6.14   A. *Schistosoma haematobium*. Miracidium. From Chandler AC, Read CP. 1961. *Introduction to Parasitology.* New York, London: John Wiley and Sons. 822 p. Fig. 66A. After Faust EC. (1929). *Human Helminthology.* Lea and Febiger, Philadelphia. **B**. *Schistosoma mansoni*. Miracidium. Relabeled after Faust and Hoffman (1934). Puerto Rico Journal of Public Health and Tropical Medicine 10: 1–47. Plate I, Fig. 6.

A neural mass is not illustrated by Faust and Meleney (1924) for *S. japonicum* (Fig, 6.9) nor by Faust and Hoffman (1929) for *S. haematobium* (Fig. 6.14) although all of these authors recognize its presence. Instead, a glandular structure with lateral ducts is shown. Moreover, lateral ducts extruding oil droplets were described for *S. mansoni* and *S. japonicum* by Cort (1919) who nevertheless illustrated the neural mass (Fig. 6.13). Faust and Meleney indicate that in *S. japonicum*, the viscoid glandular mass from which the ducts extend is separate from the neural mass and state that the secretion is mucoid, having observed the process of secretion. They state that this gland complex is even more conspicuous in *S. haematobium* and term it the "lateral gland". It is therefore noteworthy that a glandular mass, with lateral ducts, discrete from the neural mass is not described in the careful work of Pan (1980) for *S. mansoni* nor in the account of *S. bovis* by Lengy (1962). It is clear, however, that a neural mass and a lateral gland coexist but are closely situated. This is confirmed by the statement of Faust and Meleney that there is "a round group of true nerve ganglia with prominent cells on the side of the miracidium opposite the mucoid cell complex" and that the central nerve mass is "a round mass of cells below the mucoid glands" which latter connect with the lateral ducts. It is further confirmed by the clear statement of Faust and Hoffman (1934) that "the nerve center consists of a dense oval mass lying dorsal to the lateral secretory glands".

Schutte (1974) observed a considerable number of small cells, of which only the nuclei were visible, between the neural mass and the terebratorium, most abundant in the region of the apical and penetration gland ducts (Fig. 6.13B). They were considered to be the nuclei of sensory cells, several being nuclei of bipolar sensory cells or nerve fibers which originated in the neural mass and ended at the anterior lateral papillae. It appeared that some of the fibers passed to the terebratorium. He did not observe the strings of sensory cells mentioned by Ottolina.

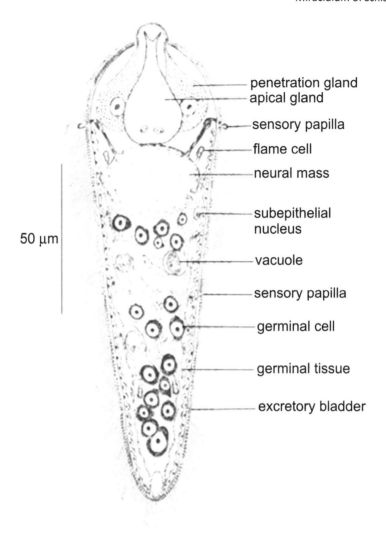

50 μm

penetration gland
apical gland
sensory papilla
flame cell
neural mass
subepithelial nucleus
vacuole
sensory papilla
germinal cell
germinal tissue
excretory bladder

FIG. 6.15 *Schistosoma bovis.* Dorsal view of miracidium. Relabeled from Lengy J. 1962. Bulletin of the Research Council of Israel Section E: Experimental Medicine 10E(I): 1–36. Fig. 3.

6.7.2   Peripheral Nerves

Peripheral nerves are essentially similar in fine structure to neuropiles but may contain more abundant neurosecretory vesicles particularly at the terminals. Usually only one type of synaptic vesicle is present in each nerve but occasionally two types may be seen. Peripheral nerves are frequently accompanied by several ganglion cells which resemble the neurons of the neural ring. Larger nerves, such as that extending to the terebratorium, contain more ganglion cells than do smaller ones such as those going to the lateral papilla (Pan 1980).

6.7.3   Neurosecretory Vesicles

At least four types of membrane-bound, round, "neurosecretory (synaptic)" vesicles are present in the nervous system of the miracidium. These vesicles are classified by Pan (1980) into Types A, B, C and D according to size and electron-dense material within.

Type A vesicles. These vesicles usually measure 70 nm in diameter and are usually electron-lucent centrally. The inner side of the membrane is coated with slightly electron-dense material which occasionally fills the entire vesicle. Type A neurosecretory vesicles are seen in some neurons and neuropiles of the CNS, in the lateral papilla and its nerves and ganglia, in the multiciliated deep-pit nerve endings of the terebratorium, in the wall of the ciliated saccular sensory organelle and in the nerves at the neuromuscular junctions.

Type B vesicles are relatively uniform in size, measuring 100 nm in diameter. The center of the vesicle is occupied by strongly electron-dense material. A narrow, distinctly electron-lucent rim separates the membrane from the central material. Type B vesicles are present in some uniciliated nerve endings in the terebratorium, in some nerve endings at neuromuscular junctions and in some neurons and neuropiles of the CNS.

Type C vesicles, ca. 120 nm in diameter, are filled with moderately dense material and are found in some uniciliated nerve endings in the terebratorium and some nerve endings at the neuromuscular junctions.

Type D vesicles are the largest (200 nm in diameter) and most conspicuous of the four types. They are uniform in size and are usually filled with moderate to strongly electron-dense material. These vesicles are seen only in a group of bipolar ganglions behind the CNS, their axons and ciliated nerve endings (Pan 1980).

6.7.4 Neuromuscular Junctions

The fine structure of neuromuscular junctions is generally similar to that of synapses between neuropiles in the CNS except that the junctions are usually unilateral. The nerve endings usually attach to "pockets" or "depressions" on the muscle. Mitochondria are frequently present in this region of the muscle. Sometimes more than three-quarters of a nerve ending may be surrounded by the plasma membrane of the invaginated muscle surface. The opposing membranes of both the muscle and nerve ending are denser than normal and are separated by a space of ca. 15 nm. The inner side of muscle membrane is coated with electron-dense material (ca. 30 nm wide) which may also be present in the gap between the membranes. The nerve ending is crowded with many small synaptic vesicles of Type A and occasionally also contains Type B (Pan 1980).

6.7.5 Peripheral Sensory Organelles

Five main types of sensory receptors are reported on the surface of the miracidia of *S. mansoni, S. haematobium, S. intercalatum* and *S. japonicum*: uniciliated, multiciliated and aciliated receptors, lateral papillae and a stellate structure (Eklu-Natey *et al.* 1985). The lateral papillae form bulbous projections on either side between the 1st and 2nd tiers of epidermal cells, as noted in Section 6.2.2. Lengy (1962) traced a pair of nerve fibers from the neural mass to the vicinity of the papillae, in *S. bovis*. Brooker (1972) observed photoreceptor-like structures near the base of a lateral papilla.

Variations (possibly artefactual) in the number and distribution of these sensory organs in the above four species are observed in individual miracidia of the same species. However, some differentiating characteristics have been observed. The uniciliated receptors are sensory endings, with a single cilium, 2–4 μm in length, which are found on the terebratorium and on the anterior ring. On the terebratorium, the cilium is surrounded by a circular, membranous fold. Two groups of five to seven receptors each, a rearranged symmetrically and are associated with a lateral gland opening. In all, 14 uniciliated receptors are seen on the terebratorium of *S. haematobium,* 12 in *S. intercalatum,* 10–12 in *S. mansoni* and 10 in *S. japonicum* (Eklu-Natey *et al.* 1985).

On the anterior ring, there are six uniciliated receptors each situated in the evagination at the base of an epidermal plate of the first tier. Two of them seem to be associated with the lateral papillae. The multiciliated receptors show more complexity in their structure, number and distribution. They are observed in three different regions of the body: on the terebratorium, on the anterior ring and on the median ring (Eklu-Natey *et al.* 1985). Silver nitrate impregnation revealed a ring of 12–14 papillae girdling the miracidium at the level of the junction of the second and third tiers of epidermal plates in *S. bovis*. Their sensory nature could not be confirmed as a nerve supply was not demonstrated (Lengy 1962). Reisinger (1923) earlier observed a ring of 12 papillae encircling the miracidium of *S. haematobium* at this level and an additional 10 encircling papillae at the junction of the first and second tiers. Similar papillary rings were reported for *S. mansoni* miracidia by Maldonado and Acosta Matienzo (1947). Lengy (1962) observed only the pair of lateral papillae at the latter junction in *S. bovis, S. haematobium* and *S. mansoni*.

Sensory structures on the surface of the miracidium can be labeled with anti-tubulin antibodies (Fig. 6.4A, arrowheads). In addition to labeling the multiciliated sensory papilla, anti-tubulin antibodies label the numerous motile cilia of the epidermal plates (Fig. 6.4B) (Collins *et al.* 2011).

Six types of putative peripheral sensory receptors were recognized by Pan (1980) in *S. mansoni*, two distributed in the terebratorium and five in the ridges between the epidermal plates.

6.7.6   Ciliated Sensory Organelles in the Terebratorium

Pan (1980) recognized seven types of ciliated sensory organs in the terebratorium: (1) multiciliated, deep-pit nerve ending (sensory papilla); (2) uniciliated nerve ending (sensory papilla); (3) lateral papilla (sensory papilla); (4) multiciliated, saccular sensory organelle; (5) multiciliated, shallow-pit nerve ending (sensory papilla); (6) uniciliated sensory nerve endings (papilla); and (7) multiciliated sensory papilla. They are briefly described below and further details may be sought in Pan (1980). Functions which have been attributed to these structures are largely speculative.

Multiciliated, Deep-Pit Nerve Ending (Sensory Papilla). In the *S. mansoni* miracidium, there are four multiciliated, deep-pit nerve endings in the terebratorium (Fig. 6.6, 6.11B, C), positioned at the level between the openings of the apical and cephalic glands. They encircle the terebratorium and are equally spaced from each other. Each sensory organelle is formed by a nerve-ending that is shaped like a pit or goblet which is attached to the epidermis of the terebratorium with a continuous septate desmosome. About 12 sensory cilia protrude slightly beyond the opening of the pit. The pit wall and its connecting nerve contain numerous Type A neurosecretory vesicles and neurotubules (Pan 1980).

Uniciliated Nerve Ending (Sensory Papilla). There are at least eight uniciliated nerve endings in the terebratorium. Four of these are distributed between and slightly anterior and the other four posterior to the four multiciliated deep-pit nerve endings. Each has a slightly enlarged, rounded tip which is attached to the epidermis with a continuous septate desmosome. There is a single cilium. The four anteriorly positioned uniciliated nerve endings contain Type B neurosecretory vesicles and the other four contain Type C vesicles.

All 12 nerves, in types 1 and 2, appear to originate from a ganglion located in the anterodorsal side of the apical gland cell-body. A thick, short nerve trunk connects the ganglion with the CNS.

Lateral Papilla (Sensory Papilla). The pair of bulbous, protruding sensory organelles is located, one on each side, in the ridge separating the first and second tiers of the epidermal plates. Each papilla is formed by spherical enlargement of a nerve ending and is enveloped by a thin layer of ridge cytoplasm. The bulb and its connecting nerve are separated from the ridge by basal lamina. Many Type A neurosecretory vesicles and neurotubules are present in the papillae and their connecting nerves. Numerous large mitochondria are present in the nerves in addition to a few lipid droplets. The mitochondria are distributed close to the base of the bulb but not within it. The surface zone of the ridge cytoplasm covering each papilla is more electron-dense than the zone adjacent to the basal lamina. The nerve is formed by a bipolar ganglion cell that connects with the CNS (Pan 1980).

Multiciliated, Saccular Sensory Organelle. This pair of sensory organelles has the most complex structure of all the peripheral sensory nerve endings. Each is located along the anterior side of the nerve to the lateral papilla. The organelle is an elongate pouch (10×2.5 μm) formed at the terminus in the CNS of an axon of a bipolar ganglion cell. The rim of the pouch opening is attached to the ridge with septate desmosomes. There is evidence that the "pouch" may open to the outside. Many Type A synaptic vesicles are present in the wall of the pouch together with mitochondria, glycogen granules and neurotubules. The basal wall of the pouch contains an extensive network of agranular endoplasmic reticulum. The lumen of the pouch contains at least four coils of concentric lamellae. Each lamella is connected to at least one cilium. The cilium originates from a basal granule embedded in the sac wall and has the "9 + 0" arrangement of microtubules (Pan 1980).

Multiciliated, Shallow-Pit Nerve Ending (Sensory Papilla). The shallow-pit nerve ending is located between the base of the lateral papilla and the attachment of the ciliated saccular sensory organelle to the ridge. There are about six relatively long sensory cilia originating from basal bodies embedded in the pit wall. The rim of the pit is attached to the ridge with septate desmosomes. The connecting nerve is one of the two axons of a bipolar neuron running between the nerves connecting the lateral papilla and the ciliated saccular sensory organelle. These three nerves along with the nerve for the uniciliated papilla (see below) run side by side and form a short nerve trunk (Pan 1980).

Uniciliated Sensory Nerve Endings (Papilla). A pair of this type of nerve ending is present, one at the posterior side of the base of each lateral papilla (6.11C, D). The fine structure of these nerve endings is essentially similar to that present in the terebratorium, except that the nerve endings here contain Type A membrane-bound vesicles. The ridge cytoplasma round the rim of the nerve ending expands to form several filopodia-like projections (nearly 3 nm long). The layer beneath the unit membrane is strongly electron-dense (Pan 1980).

Multiciliated Sensory Papilla. There are about 12 sensory nerve endings of this type that encircle the miracidium in the ridge separating the second and third tiers of epidermal plates. Each papilla is spaced at about 12 μm and is the terminus of one of the two axons of a bipolar ganglion cell. The other axon extends into the CNS at its posterior side. The perikarya of ganglia are clustered behind the neural ring and are surrounded by processes of interstitial cells (Fig. 6.6) and ridge cell-bodies. The papillae are directly exposed to the outside and are attached to the ridge by septate desmosomes. Each papilla contains about 12 sensory cilia which are separate or are bundled together and enveloped in a plasma membrane (multiciliated axoneme). The basal body of each cilium continues into a short, banded, tapered rootlet the tip of which extends into a cluster of dense particles (85 nm). Electron-dense synaptic vesicles of Type D occupy much of the individual perikaryon and its axons including the papilla. The ganglion cells contain cisternae which may be a modified agranular endoplasmic reticulum. Each perikaryon may measure 5 μm at the widest dimension and contains a chromatin-rich, irregularly shaped nucleus (Pan 1980).

6.8 GERMINAL CELLS

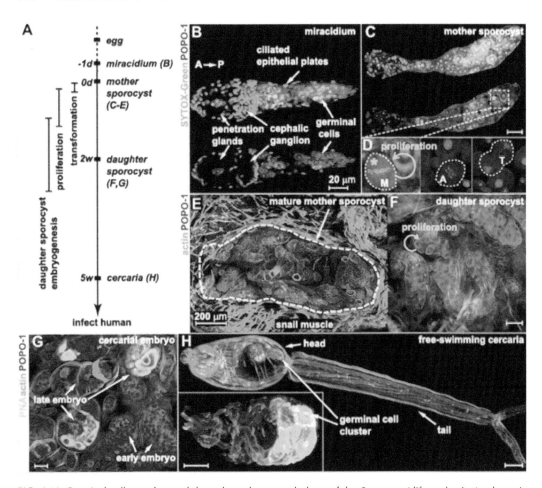

FIG. 6.16 Germinal cells are detected throughout the asexual phase of the *S. mansoni* life cycle. **A.** A schematic timeline of schistosome asexual amplification. **B–C.** Maximum intensity projections of confocal stacks (top) and single optical slices (bottom) of a POPO-1 and SYTOX-Green co-stained miracidium (B) and a sporocyst 24 hr after *in vitro* transformation (C). **D.** Representative images of cells at metaphase (M), anaphase (A) and telophase (T) (from left to right), captured in sporocysts 24 hr post-transformation. **E–G.** Cryosections of the tentacle of a *Biomphalaria glabrata* snail showing a mother sporocyst (perimeter highlighted by dashed line) with daughter sporocysts packed inside (3 weeks post infection) (E); an individual daughter sporocyst that has migrated to the digestive glands of a *B. glabrata* snail 6 weeks post infection (F); and cercarial embryos within a daughter sporocyst in the digestive glands of a *B. glabrata* snail 6 weeks post infection (G). Actin is stained with phalloidin. Peanut agglutinin (PNA) visualizes acetabular glands and ducts of the cercariae. **H.** A mature cercaria. The inset shows a magnified view of the cercarial "head" visualized with PNA and POPO-1 staining. Scale bars are 20 μm, except in (E) which is 200 μm. From Wang B, Collins JJ, Newmark PA. 2013 eLife 2: e00768. doi: 10.7554/eLife.00768: 1–15. Fig. 1.

Capron *et al.* (1965) and Pan (1980) estimated that there were about 20 germinal cells in a miracidium of *S. mansoni*. Consistent with this, Wang *et al.* (2013) found 10–20 germinal cells in the posterior half of a free swimming miracidium (see below). However, Faust and Meleney (1924) had earlier observed few to more than a hundred, Maldonado and Acosta Matienza (1947) observed 40–50, Olivier and Mao (1949) 38–61, an average of 53 and Ottolina (1957) 24 cells. Lengy (1962) observed 30–40 germinal cells in *S. bovis* miracidia, embedded in a definite germinal tissue. In *S. mansoni*, Schutte (1974) reported 27 germinal cells, with mean diameters of 10 μm and 47 smaller cells resembling the reticulum cells of the sporocyst with mean diameters of 4.6 μm. The germinal cells are embedded in a vacuolated, acellular matrix which extends as strings of material, in which reticulum cells are embedded, along the sides of the glands and neural mass. Tang (1938), for *S. japonicum*, states that the germ cells, together with the posterior pair of flame cells, are enclosed in a "germinal sac". Pan notes that the germinal cells are located behind the clustered cell-bodies of the multiciliated papillae of the presumed "chemoreceptors" and at the level of the third and fourth tiers of epidermal plates. Germinal cells are in general irregularly shaped and contain several thin processes that extend into the intercellular spaces, possibly as anchoring devices. Their cell-bodies may measure 5 μm at the widest diameter and are surrounded by processes of the interstitial cells. The granular cytoplasm of germinal cells is crowded with numerous ribosomes, many mitochondria, rare Golgi apparatus, granular endoplasmic reticulum and a few β-glycogen granules. The usually oval nucleus is large (3 μm), in relation to the cell-body, contains a prominent nucleolus (1.5 μm) and little heterochromatin (Pan 1980). Faust and Meleney (1924) stated that the germ balls did not progress beyond the one-cell stage and considered that they degenerated to form mucoid masses. However, Faust and Hoffman (1934) state that in the molluscan host these balls are destined to become the second generation of sporocysts.

Anti-β-tubulin gives strong labeling of a cytoplasmic meshwork of microtubules in the germ cells (Fig. 6.4B). It is possible that the microtubule meshwork functions to generate or maintain the cellular projections reported by Pan (Collins *et al.* 2011).

Neoblast-like Stem Cells. Collins *et al.* (2013) identified a population of neoblast-like cells in adult *S. mansoni.* Wang *et al.* (2013), using advanced techniques given in the legend to Figure 6.16, accordingly reinterpreted the germinal cells in the larval stages as neoblast-like. They demonstrated that this proliferative larval cell population (germinal cells) shares some molecular signatures with stem cells from diverse organisms, in particular neoblasts of planarians. Neoblasts are pluripotent stem cells which in planarians can regenerate injured tissue or even reconstitute a whole animal from a single cell (Newmark and Sánchez Alvarado 2002; Wagner *et al.* 2011). Wang *et al.* (2013) have identified two distinct germinal cell lineages that differ in their proliferation kinetics and expression of a nanos ortholog. They showed that a vasa/PL10 homolog is required for proliferation and maintenance of both populations, whereas argonaute 2 and a fibroblast growth factor receptor-encoding gene are required only for nanos-negative cells. They recognized the same cell types in further larval stages: mother and daughter sporocysts and cercariae. A timeline for this lineage is given in Figure 6.16 together with the expression of various tests in the larval stages. These neoblast-like cells allow multiplication of larval stages by a form of polyembryony.

6.9 TRANSMISSION TO THE SNAIL HOST

6.9.1 General Aspects

Miracidia do not feed or reproduce. Their only chance to transmit their genes is to invade an appropriate host snail. Furthermore, the *Schistosoma* species are specialized on only few snail-host species or even strains and this host-specificity enables them to manipulate the snails' internal defence mechanisms, metabolism and endocrine systems in many details. Therefore it can be expected that miracidia of each *Schistosoma* species are adapted to maximise their transmission to the appropriate host-snail spectrum. In fact, the morphological, physiological and behavioral organisation of miracidia seems to be fully dedicated to the goal of finding and invading the appropriate snail-host. This is seen considering, e.g., the hatching mechanisms, swimming behavior, responses to light, gravity, temperature, chemicals and physical boundaries.

Adaptations to Environmental Conditions. The physiological properties of miracidia correspond to the environmental condition in the habitats where the host snails occur. This is obvious, when the effect of temperature on miracidial age, activity and infectivity are considered (Prah and James 1977; Anderson *et al.* 1982). In *S. mansoni* and *S. haematobium* miracidia the life expectancy increases with decreasing temperature, from around 5 h at 35°C to approximately 16 h at 15°C. However, the activity of the miracidia and their infection success with time decline dramatically with decreasing temperatures. The interplay of these conditions enable the miracidia to infect their hosts successfully in a broad temperature range between 15 and 35°C, the maximum of infection success being at 25°C (Anderson *et al.* 1982).

Other environmental factors within the host snails' habitats are also well tolerated by miracidia. Direct sun radiation had no effect on activity and survival of *S. haematobium* miracidia within the first 30 min of exposure, but damaged the organisms severely after 1 h, presumably due to ultraviolet radiation. Artificially produced ultraviolet C radiation (UVC) impaired the larvae down to a depth of 15 cm in clear water. However, turbidity in water samples protected the larvae from ultraviolet and it was suggested that the effect of solar ultraviolet radiation is negligible in turbid waters (Prah and James 1977). On the other hand it was found that turbidity affected the ability of *S. mansoni* miracidia to infect *Biomphalaria glabrata* snails, probably due to mechanical effects of silt (Upatham 1972a). The rate of infection decreased by 31% when turbidity increased from 0 ppm (units equivalent to ppm suspended silica) to 100 ppm, but then decreased only slowly from 100 ppm to 500 ppm, an intensity which rarely occurs in natural waters. Therefore miracidia seem to be able to tolerate the degree of turbidity most frequently occurring in natural snail habitats (mean turbidity in St. Lucia snail habitats is 73 ppm). Studies applying ultraviolet B (which in contrast to UVC reaches the surface of the earth) showed that this radiation reduced the infection success of *S. mansoni* miracidia and had damaging effects on the development of the resulting sporocysts in *B. glabrata*. However, if exposed to visible light following UVB irradiation the parasites repaired the damage, even after penetrating the snails (Ruelas *et al.* 2007). Thus, the miracidia are adapted to repair UVB damage of the DNA, probably with the enzyme photolyase, which is expressed in the different *S. mansoni* stages. Under field conditions they seem able to mitigate the UV effects, as they are exposed to UVB and visible light simultaneously.

Miracidia are also adapted to the pH conditions in their snails' habitats. *S. mansoni* miracidia achieved high infection rates in *B. glabrata* snails in pH ranges between 6 and 9 and some infections occurred even at pH 5 and 10 (Upatham 1972a).

Furthermore, the salinity of host-snail habited waters is tolerated by miracidia to some extent. *S. mansoni*, *S. haematobium* and *S. mattheei* miracidia survived even longer in salinities of 1.7, 3.5 and 5.2% than in freshwater, with maximum life spans of 28, 30 and 24 h respectively at 25°C. However, in salinities of >7.0% longevity was progressively reduced and the lethal time of death of 50% of the population (LT_{50}) decreased from a maximum of 13–15 h in 1.7–3.5% to less than 2 h in 10.5%. Nevertheless, the miracidia were infective for their snails at salinities up to 3.5% in *S. mansoni* and *S. mattheei* and up to 2.5% in *S. haematobium* (Donnelly *et al.* 1984). In another study *S. mansoni* could infect *B. glabrata* at salinities of even up to 4.2% resp. 6.2% (Upatham 1972a; Chernin and Bower 1971).

Moreover the mechanisms by which schistosome miracidia hatch from the eggs reflect adaptations to optimize the transmission (Jones *et al.*, Chapter 5). For instance that hatching is mainly stimulated by a decrease in osmolarity secures that miracidia remain in the eggs in stool or urine and hatch when they arrive in freshwater.

6.9.2   Dispersal and Selection of the Hosts' Habitat

Schistosoma miracidia seem to be adapted to search for host snails in large areas of their aquatic habitat (Fig. 6.17). Within the few hours of infectiousness they swim around at high speeds. *S. mansoni* miracidia move with a speed between 2.0 and 3.0 mm/s under stress-free conditions (Prah and James 1977; Samuelson *et al.* 1984) three times as fast as the ciliate *Paramecium caudatum*, which is twice as long. A fast rotation along the longitudinal axis supports a straight movement. Theoretically a miracidium swimming 2.2 mm/s could cover a distance of 8 m in one hour. Therefore it is not surprising that *S. mansoni* miracidia were found to infect *Biomphalaria glabrata* snails at distances of up to 9 m in standing and 97 m in running waters (Upatham 1973) and in a natural pond up to 18 m (Laracuente *et al.* 1979). Many infection-studies have shown that, as expected, the infection success of miracidia in the experiments increases with increasing number of miracidia released, increasing time of exposure to the snails and with decreasing distances between miracidia and snails, due to decreasing volumes of water within the vessels used in the experiments (e.g., Chernin and Dunavan 1962; Upatham 1972 a, b 1973; James and Prah 1978; Anderson *et al.* 1982; Carter *et al.* 1982).

FIG. 6.17 A snail habitat on Lake Albert, D.R. Congo. Miracidia of *Schistosoma mansoni* infect snails and local people are exposed to cercariae released from the snails. Photo Barrie Jamieson.

Miracidia distribute in their aquatic habitat by responding mainly to the environmental cues light, gravity and temperature. The photo-orientation of at least *S. mansoni* miracidia is a phototaxis as the miracidia are able to detect the relative intensities of two separate sources of light (Mason and Fripp 1977a). The wavelengths to which the *Schistosoma* miracidia respond have been determined for *S. mansoni* and *Schistosomatium douthitti* (Wright *et al.* 1972; Wright 1974). Like two other Digenea-species they respond maximally to blue-green light between 500 and 525 nm, which penetrates deepest into clear water. *S. mansoni* miracidia prefer, in addition, red-brown light of 650 nm, which is typical of muddy waters. The responses to light, gravity and temperature conditions seem fully adapted to guiding them into the habitats preferred by their respective snail hosts. This is obvious, e.g., in three schistosome species:

S. mansoni. In choice chambers and vertical cylinders the miracidia show a positive photo-orientation in the horizontal direction and a negative geo-orientation in the vertical direction, interacting with the positive photo-orientation (Chernin and Dunavan 1962; Prah and James 1978). The preferred accumulation of the miracidia near the surface of the water fits to the distribution of most *Biomphalaria* snail species, which are found close to the water surface. However in water columns the miracidia infected snails at a depth of 2 m as intensively as at the surface (Prah and James 1978) and in a 1.20 m deep natural pond the miracidia even infected more snails at the bottom than near the surface, when they were released at the bottom (Upatham 1972b). Hence the behavior of *S. mansoni* miracidia enables them to infect snails at most water levels where host-snails occur in endemic areas.

S. haematobium. In contrast to *S. mansoni*, *S. haematobium* miracidia show a strong negative photo-orientation and a positive geo-orientation. They infected snails mainly in the darker areas of vessels and at the bottom of water columns. However they also invaded snails near the surface, particularly when they were introduced into the columns near the surface (Shiff 1968, 1969; Wajdi 1972; Prah and James 1978). Also in a natural pond the miracidia infected snails most frequently when they were presented on the bottom or at the surface in shade. However in winter, at cooler temperatures, they infected more snails near the water surface, also in the light. This was the result of a shift in their orientation: their normal negative response to light at warm temperature was reversed when the temperature dropped to 18°C or 13°C. This behavior is related to that of their host snail *Bulinus globosus* (Shiff 1974).

S. japonicum. When the miracidia were exposed to high light intensities they showed a positive photo-orientation at temperatures below 18°C, but shifted to a negative photo-orientation with increasing temperature. However at low light intensities they were photopositive. They showed a negative geo-orientation in the dark, which shifted to a positive one with increasing light intensity and increasing temperature. These behavioral patterns resembled the patterns of their *Oncomelania* host snails (Takahashi *et al.* 1961).

6.9.3 Host-directed Orientation

The host-finding mechanisms of miracidia were most intensively studied in the 1960s and 1970s. However, the results obtained by the different researches were contradictory, they have been summarized and discussed in several reviews (MacInnis 1976; Saladin 1979; Smyth and Halton 1983; Sukhdeo and Mettrick 1987; Haas *et al.* 1995; Haas and Haberl 1997; Haas 2003; Sukhdeo and Sukhdeo 2004). Meanwhile many contrasting results could be explained by differences in the methods applied and actually there is some consensus on what happens when *Schistosoma* miracidia find and invade their hosts.

6.9.3.1 Mechanisms of Orientation

Despite some initial controversy, it is now broadly accepted that miracidia approach and invade their host snails by responding to chemical cues emitted by the snails. The stimulating chemicals form an "active space" around the snails (Chernin 1970; MacInnis 1976; Hertel *et al.* 2006). The miracidia of many species, when entering the active space of a host, change their normal straight-on swimming mode and perform a vigorous increase in the rate of change of direction (RCD), leading to an accumulation around the snail (Chernin 1970; Mason Fripp 1976; Mason 1977; Samuelson *et al.* 1984; Haberl *et al.* 1995).

Once inside the active space, when miracidia approach the outer boarder of the active space, they show a sharp 180° turn that prevents them from leaving again (Roberts *et al.* 1979; Samuelson *et al.* 1984). In both orientation types, increase of RCD and turnback swimming (Fig. 6.18A) the miracidia do not adjust their swimming path according to the direction of the concentration gradient of the chemical cues. Instead they respond only to an increase or decrease of the concentration of the attractants with an undirected orientation, according to Fraenkel and Gunn (1940) a kinesis. As the response is a change of the direction of the movement and as it is released by chemicals it is a chemoklinokinesis. At least *S. mansoni* and *S. haematobium* and in other digeneans *Trichobilharzia szidati* (synonymous with *T. ocellata*), *Fasciola hepatica* and *Echinostoma caproni* show both types of chemokinesis, an increase of the RCD and turnback swimming (Haberl *et al.* 1995; Kalbe *et al.* 1997; Haberl *et al.* 2000).

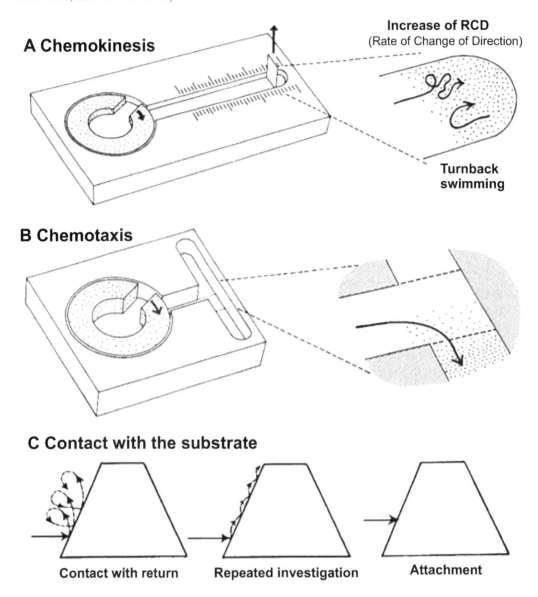

FIG. 6.18 A, B. Mechanisms of chemo-orientation of miracidia towards their snail hosts or snail conditioned water and methods to test miracidial behavior to chemical stimuli using choice chambers. Miracidia are released from the central chamber by opening a closure ring (spotted) and their swimming paths in concentration gradients of attractants added at the end of the straight or branched canal are recorded. Modified after Haas (2003). **C.** Responses of miracidia after contact with agar pyramids containing snail host signals. Direct attachment or repeated investigation are the typical responses after contact with suitable host snails. Both, contact with return and repeated investigation can also lead to attachment. Modified after MacInnis (1965). Journal of Parasitology 51: 731–746. Fig. 3.

The undirected chemokinesis orientation type seems to be an adaptation of the miracidia to their swimming mode. As they swim at high speed and rotate along their longitudinal axis it may be more economical to respond simply to changes in the concentration of the attractants instead of detecting the direction of the concentration gradients. Nevertheless miracidia of *S. japonicum* seem to approach their snail hosts with a directed chemotactical orientation (Fig. 6.18B) (Haas *et al.* 1991). They must in an unknown manner be capable to detect the direction of the attractant's concentration gradients and when the concentration gradients decrease they turn back. Chemotactic orientation may be an adaptation to the semi aquatic habitats of the oncomelanid host snails where turbulences are not present and the snail has to be located in small water trickles between physical obstacles. It is noteworthy that snail invading echinostomatid cercariae have also evolved both types of orientation, 3 species were found to use the chemokinetic strategies and one chemotaxis (reviewed by Haas 2000, 2003).

The responses of miracidia to snail-emitted chemicals greatly increase their infection success although the movements of most species are non-directional. The hosts' active space can be determined by recording the movements of the miracidia around the snail. In the system *Lymnaea stagnalis* and *Trichobilharzia szidati* it spread within 5 min to a volume 32 times of that of the snail and within 20 min to 124 times the snails' volume. The gyrating miracidia seem not to be trapped in the periphery of the odor clouds: in large water volumes the snails were much more infected when they were allowed to establish an active space before miracidia exposure (Hertel *et al.* 2006).

6.9.3.2   Chemical Host Cues

Contradictory Results. Many researchers studied the nature of the snail-emitted chemicals attracting the miracidia. This host-finding work was also spurred by the idea that miracidial attractants could be used in *Schistosoma* control technology. However, the results obtained in different former studies were contradictory (reviewed by Haas and Haberl 1997). For example for *S. mansoni* miracidia four researchers identified four very different components of snail-conditioned water (SCW) as the exclusive chemical stimuli: Amino acids (MacInnis 1965; MacInnis *et al.* 1974), magnesium ions (Stibbs *et al.* 1976), decreasing calcium concentration in relation to the magnesium concentration (Sponholtz and Short 1976) and peptides (Mason 1977). Other studies with *S. mansoni* described compounds which were supposed to attract miracidia additionally, e.g., short-chain fatty acids and sialic acid (MacInnis 1965), ammonia (Mason and Fripp 1977b; Brasio *et al.* 1985), neurotransmitters and acetylcholine antagonists (Roberts *et al.* 1978), glutathione (Disko and Weber 1979), inorganic ions (Brasio *et al.* 1985), D-glucose (Plorin and Gilbertson 1985) and β-ecdysterone (Shiff and Dossaji 1991). At least some of these disagreements have their origin in the use of inappropriate methods. Most studies did not take into account that *S. mansoni* miracidia respond highly sensitively to minimal changes in the water by the snail hosts. An incubation of water with a snail for few seconds makes the spot of inoculation attractive. However the miracidia respond also very sensitively to artificial changes in the water composition, also to chemicals which are not related to host-finding. For example, minor changes in pH stimulate an increase of random turning of the swimming organisms, but only in two of the studies cited above was the pH of the substrates controlled, which were offered to the miracidia. Caution is also necessary when aggregation of miracidia is considered in the various types of choice chambers, as the organisms aggregate also in response to toxic effects and to minor changes in the ionic composition of the water. Furthermore, variations in the preparation of SCW applied in the experiments may account for conflicting results (reviewed by MacInnis 1976; Saladin 1979). Some authors incubated the snails in distilled water and this may influence the results, especially when inorganic ions are considered as the stimulants. In addition there may be effects when SCW was prepared by snail crowding over longer time periods (reviewed by Saladin 1979).

Cues are Macromolecules. All the snail host signals described in former studies were simple small-molecular chemicals and it is unlikely that they can signal the snail-host specificity often observed in miracidial host finding. New perspectives came up as Dissous *et al.* (1986) found that an 80-kDa glycoprotein from *B. glabrata* stimulated *S. mansoni* miracidia to incorporate methionine and suggested that this glycoprotein influences snail penetration and development of the miracidia. Then it was discovered that also the host-finding behavior of miracidia is stimulated by snail-emitted macromolecules. *S. japonicum* miracidia approached *Oncomelania hupensis* snails chemotactically in response to an attractant with a molecular weight >30 kDa (Haas *et al.* 1991). Studies on the host-finding of miracidia of *S. mansoni* and *S. haematobium* miracidia, where the distinct behavioral responses of individual miracidia were investigated (methods: Fig. 6.18B, C) proved that all investigated responses, increase of the RCD, turning back and repeated investigation were released by macromolecular components of SCW with a molecular weight greater than 30 kDa or at least in *S. mansoni*, 300 kDa and the small molecular fraction of SCW had no effect (Haberl and Haas 1992; Haberl *et al.* 1995). Studies with *T. szidati*, *F. hepatica* and *E. caproni* also revealed that these species responded exclusively to macromolecular "miracidia attracting glycoproteins" (MAGs) (Kalbe *et al.* 1997, 2000; Haberl *et al.* 2000). By employing various methods to fractionate and analyse SCW the attractants could be isolated and characterized. Their saccharide chains are linked to a core protein by an *O*-glycosidic linkage, probably between threonine or serine and *N*-acetylgalactosamine. Their identification signal is encoded in a very complex carbohydrate moiety, the isolated fraction contains at least 14 different monosaccharides coupled by different types of chemical linkage. The miracidia are extremely sensitive to MAGs: An isolated SCW fraction containing MAGs and many macromolecules with molecular weights around 300 kDa stimulated the responses at a concentration of their carbohydrates as low as 0.1 ng/ml, i.e. 1 mg in 100,000 litres of water. It can be expected that the signaling molecule itself will stimulate the miracidia at even considerably lower concentration, making it interesting as a tool for schistosome control technology. However, the complex chemical nature of the MAG fraction impedes the analysis of the signaling codes.

6.9.4   Host Contact and Invasion

After contact with artificial substrates miracidia show 12 different behavioral patterns (MacInnis 1965). The patterns "repeated investigation", "attachment" (Fig. 6.18C) and "penetration" are the typical responses upon contact with suitable host snails. Experiments with agar pyramids containing SCW and its fractions revealed that in *S. mansoni* and *S. haematobium* these invasion related responses are stimulated by the same MAGs that attract the miracidia (Haberl and Haas 1992; Haberl *et al.* 1995). This would mean that the miracidia rely on a single host signal in all their invasion phases. This is in contrast to other infective stages of parasitic worms studied so far (cercariae, see Kašný *et al.*, Chapter 8; and nematodes) which respond to different host signals when they approach the host, attach to it, remain on it and penetrate. However, the host cues used by the other infective stages are small and simple molecules and the parasites achieve some host specificity by responding to sequences of different cues. That miracidia have evolved the ability to respond to uniquely complex and host specific signals may allow them to use these cues for all their host-finding phases.

However, attachment and penetration responses of miracidia are poorly studied and future research may disprove the extraordinary "one signal host finding strategy" of miracidia. For instance *S. japonicum* miracidia approached chemotactically to their snail host *Oncomelania hupensis* and to the noncompatible *B. glabrata*, but after contact with the snails they responded only to *O. hupensis* (Haas *et al.* 1991).

6.9.5   Snail-Host Specificity in Host-Finding Behavior

Each *Schistosoma* species or strain can successfully develop only in a very narrow spectrum of host snail species or strains and much is known on the immunological, biochemical, physiological and molecular biological background of the compatibility between schistosomes and their snail hosts (see Kašný *et al.* Chapter 4, Yoshino *et al.*, Chapter 7). The question arises, whether schistosome miracidia can distinguish between compatible and non-compatible snails already during their host finding behavior. This would enable them to avoid losses by unsuccessful penetration into and distraction by unsuitable snails and other aquatic organisms. In most infection experiments on the compatibility of schistosomes and snail hosts and other aquatic organisms high numbers of miracidia were exposed to the target organisms in few milliliters of water for hours. Such artificial conditions exclude normal host-finding behavior of the miracidia and they tend to attach or adhere also to incompatible host snails (Sapp and Loker 2000) and it was reported that even tadpoles were penetrated under such conditions.

However in choice chambers miracidia showed a high degree of specificity in host-finding. For instance *S. haematobium* miracidia discriminated between the host snail *Bulinus globosus* and the sympatric, non-host species *Cleopatra ferruginea* and they were even able to prefer uninfected over patent *B. globosus*. (Allan *et al.* 2009). Studies on the snail-emitted glycoproteins used by the miracidia to find and invade their snail hosts also suggest that there may be a high degree of host-specificity. Miracidia of *F. hepatica*, *T. szidati* and of an Egyptian strain of *S. mansoni* responded exclusively to a defined fraction of SCW from their compatible host snail and not to the same fraction of the other snail species stimulating the other parasites. However, two South American *S. mansoni* strains could not distinguish between their host snails and five incompatible snail species (Kalbe *et al.* 1996, 2000; Haberl *et al.* 2000; Hassan *et al.* 2003; Kalbe *et al.* 2004). This non-specific response was stimulated by another SCW fraction than that signaling the host-specificity. This shift to non-specific host-finding behavior may have occurred as the parasites had acquired their new host *B. glabrata* after being transported to South America with African slaves. The physiological background of this signal shift within a few hundred years is not understood. Comprehensive crossing experiments with an Egyptian and a Brazilian *S. mansoni* strain revealed that the non-specific host-finding mode of the Brazilian strain is genetically dominant following simple Mendelian inheritance (Kalbe *et al.* 2004). This suggests that the specificity of host recognition might be controlled by a single gene or gene complex. A low specificity in host-finding was also found in *E. caproni* miracidia (Haberl *et al.* 2000). This behavior could be an artefact due to prolonged maintenance of the parasite cycle in the laboratory in *B. glabrata* which is not the natural host of *E. caproni*.

Whether the host-specific host-finding mode is the normally prevalent behavior of digenean species is not known, as most studies on miracidium-snail compatibility excluded normal host-finding behavior by crowding miracidia and target snails. Experiments under semi-field conditions and in natural ponds revealed that various species of incompatible snails as decoys reduced the transmission to the compatible snails considerably (Upatham and Sturrock 1973; Laracuente *et al.* 1979). However, these experiments were performed using South American strains of *S. mansoni* and such decoy effects may be absent in other *Schistosoma* species or strains which employ the specific host-finding strategy. In fact, the infection success of miracidia of an Iraqi strain of *S. haematobium* could not be reduced by inserting five different incompatible snail species as decoys (Wajdi 1972).

The fact is that miracidia are able to find their snail hosts with species-specificity and it cannot be excluded that this is the normal host-finding strategy in Digenea. A question is why snails still express species-specific soluble molecules which are used as host-finding cues after probably more than 400 million years of coevolution with their parasites. One hypothesis is that the glycoproteins may serve as essential pheromones in intraspecific communication in snails and that the miracidia take advantage of these molecules for host identification.

6.10   CHAPTER SUMMARY

This chapter deals with the morphology and anatomy of the miracidium together with host finding and related phenomena for *Schistosoma* species which infect humans. Morphology is chiefly limited to electron microscopy. The structure of the miracidium has received little detailed attention in recent decades. The most detailed morphological study is that of Pan (1980), for *S. mansoni*, from which this account, with that of Eklu-Natey *et al.* (1985), for four species and the notable study by Collins *et al.* (2001) is largely drawn. Other studies which refer to the morphology of the miracidium of *S. mansoni* are listed and briefly reviewed. Transmission to the snail host is viewed in terms of adaptation to environmental conditions, dispersal and selection of the habitat of the snail, host-directed orientation, mechanisms of orientation, chemical host cues, host contact and invasion and snail host specificity in host-finding behavior.

6.11 ACKNOWLEDGMENTS

The University of Queensland Library and Document Delivery Service and staff gave excellent assistance.

6.12 LITERATURE CITED

Allan F, Rollinson D, Smith JE, Dunn AM. 2009. Host choice and penetration by *Schistosoma haematobium* miracidia. Journal of Helminthology 83(1): 33–38.

Anderson RM, Mercer JG, Wilson RA, Carter NP. 1982. Transmission of *Schistosoma mansoni* from man to snail: experimental studies of miracidial survival and infectivity in relation to larval age, water temperature, host size and host age. Parasitology 85: 339–360.

Araque W, Barrios EE, Rodriguez P, Delgado VS, Finol HJ. 2003. Ultrastructural study of the *in vitro* interaction between Biomphalaria *glabrata* hemocytes and *Schistosoma mansoni* miracidia. Memorias do Instituto Oswaldo Cruz 98(7): 905–908.

Bahia D, Avelar LGA, Vigorosi F, Cioli D, Oliveira GC, Mortara RA. 2006. The distribution of motor proteins in the muscles and flame cells of the *Schistosoma mansoni* miracidium and primary sporocyst. Parasitology 133(Part 3): 321–329.

Basch PF. 1991. *Schistosomes: Development, Reproduction and Host Relations.* New York: Oxford University Press.

Basch PF, Diconza JJ. 1974. The miracidium sporocyst transition in *Schistosoma mansoni* surface changes *in-vitro* with ultrastructural correlation. Journal of Parasitology 60(6): 935–941.

Blankespoor HD, Van der Schalie H. 1976. Attachment and penetration of miracidia observed by scanning electron microscopy. Science 191: 291–293.

Bogitsh BJ, Carter CE, Oeltmann TN. 2013. *Human Parasitology.* Elsevier. 430 p.

Bogitsh BJ, Carter OS. 1982. Developmental biology of *Schistosoma mansoni* with emphasis on the ultrastructure and enzyme histochemistry of intra molluscan stages. pp. 221–248. *In*: Harrison FW, Cowden RR, editors. *Developmental Biology of Freshwater Invertebrates.* New York: Liss.

Brasio BC, Magalhaes LA, Miller J, De Carvalho JF. 1985. Verificacao da resposta de miracidios de *Schistosoma mansoni* a substancias provenientes de moluscos planorbideos. Revista de Saude Publica 19: 154–170.

Brehm K, Lueder CGK. 2009. Introduction: Parasites. pp. 239–243. *In*: Rupp S, Sohn K, editors. *Methods in Molecular Biology.* Totowa, NJ, USA: Humana Press.

Brooker BE. 1972. The sense organs of trematode miracidia. pp. 171–180. *In*: Canning EU, Wright CW, editors. *Behavioral Aspects of Parasite Transmission.* New York: Academic Press.

Bruckner DA, Voge M. 1974. The nervous system of larval *Schistosoma mansoni* as revealed by acetycholinesterase staining. The Journal of Parasitology 60(3): 437–446.

Capron A, Deblock S, Biguet J, Clay A, Adenis L, Vernes A. 1965. Contribution à l'étude experimental de la bilharziose à *Schistosoma haematobium*. Bulletin de l'Organisation Mondiale de la Santé 32(6): 755–778.

Carter NP, Anderson RM, Wilson RA. 1982. Transmission of *Schistosoma mansoni* from man to snail: laboratory studies on the influence of snail and miracidial densities on transmission success. Parasitology 85: 361–372.

Chandler AC, Read CP. 1961. *Introduction to Parasitology.* New York, London: John Wiley and Sons. 822 p.

Chernin E. 1970. Behavioral responses of miracidia of *Schistosoma mansoni* and other trematodes to substances emitted by snails. Journal of Parasitology 56(2): 287–296.

Chernin E, Bower C. 1971. Experimental transmission of *Schistosoma mansoni* in brackish waters. Parasitology 63: 31–36.

Chernin E, Dunavan C. 1962. The influence of host-parasite dispersion upon the capacity of *Schistosoma mansoni* miracidia to infect *Australorbis glabratus*. American Journal of Tropical Medicine and Hygiene 11(4): 455–471.

Chiang CP, Caulfield JP. 1988. *Schistosoma mansoni* ultrastructural demonstration of a miracidial glycocalyx that cross-reacts with antibodies raised against the cercarial glycocalyx. Experimental Parasitology 67(1): 63–72.

Collins JJ, King RS, Cogswell A, Williams DL, Newmark PA. 2011. An atlas for *Schistosoma mansoni* organs and life cycle stages using cell type-specific markers and confocal microscopy. PLoS Neglected Tropical Diseases 5(3): e1009.

Collins JJ, Wang B, Lambrus BG, Tharp M, Iyer H, Newmark PA. 2013. Adult somatic stem cells in the human parasite, *Schistosoma mansoni*. Nature 494: 476–9. doi: 10.1038/nature11924.

Cort WW. 1919. Notes on the eggs and miracidia of the human schistosomes. University of California Publications in Zoology 18: 509–519.

Disko R, Weber L. 1979. The attraction of miracidia of *Schistosoma mansoni* to the snail *Biomphalaria glabrata*. Zentralblatt für Bakteriologie 263: 197.

Dissous C, Dissous CA, Capron A. 1986. Stimulation of *Schistosoma mansoni* miracidia by a 80 kDa glycoprotein from *Biomphalaria glabrata*. Molecular and Biochemical Parasitology 21: 203–209.

Donnelly FA, Appleton CC, Schutte CHJ. 1984. The influence of salinity on the ova and miracidia of three species of *Schistosoma*. International Journal for Parasitology 14(2): 113–120.

Dunn TS, Yoshino TP. 1988. *Schistosoma mansoni* origin and expression of a tegumental surface antigen on the miracidium and primary sporocyst. Experimental Parasitology 67(2): 167–181.

Ebrahimzadeh A. 1977. Contributions to the micromorphology of miracidia of *Schistosoma mansoni* part 1 fine structure of the tegument and its associated structures. Zeitschrift fuer Parasitenkunde 54(3): 257–268.

Eklu-Natey DT, Wüest J, Swiderski Z, Striebel HP, Huggel H. 1985. Comparative scanning electron microscope (SEM) study of miracidia of four human schistosome species. International Journal for Parasitology 15(1): 33–42.

Elkerdany ED. 2003. Ultrastructural studies of the different developmental stages of *Schistosoma mansoni* eggs and their hepatic granulomas. Journal of the Medical Research Institute 24(1): 80–103.

Fan J, Brindley PJ. 1998. Characterization of cDNAs encoding a new family of tetraspanins from schistosomes-the Sj25 family. Gene (Amsterdam) 219(1-2): 1–8.

Fan J, Hooker CW, McManus DP, Brindley PJ. 1997. A new member of the transmembrane 4 superfamily (TM4SF) of proteins from schistosomes, expressed by larval and adult *Schistosoma japonicum*. Biochimica et Biophysica Acta (BBA) – Biomembranes 1329(1): 18–25.

Faust EC. 1929. *Human Helminthology: A Manual for Clinicians, Sanitarians and Medical Zoölogists*. Philadelphia: Lea and Febiger. 616 p.

Faust EC, Hoffman WA. 1934. Studies on schistosomiasis mansoni in Puerto Rico. III. Biological Studies I. The extra mammalian phases of the life cycle. P R J Public Health and Tropical Medicine 10: 1–47.

Faust EC, Meleney HE. 1924. Studies on Schisosomiasis Japonica. American Journal of Hygiene Monographic series 3: 1–99.

Fraenkel GS, Gunn DL. 1940. *The Orientation of Animals*. Oxford: Oxford University Press.

Gordon RM, Davey TH, Peaston H. 1934. The transmission of human bilharziasis in Sierra Leone. Annals of Tropical Medicine and Parasitology 28: 323–418.

Haas W. 2000. The behavioral biology of echinostomes. pp. 175–197. *In*: Fried B, Graczyk TK, editors. *Echinostomes as Experimental Models for Biological Research*. Dordrecht: Kluwer Academic Publishers.

Haas W. 2003. Parasitic worms: strategies of host finding, recognition and invasion. Zoology 106: 349–364.

Haas W, Gui M, Haberl B, Ströbel M. 1991. Miracidia of *Schistosoma japonicum*: approach and attachment to the snail host. Journal of Parasitology 77: 509–513.

Haas W, Haberl B, Kalbe M, Körner M. 1995. Snail-host-finding by miracidia and cercariae: chemical host cues. Parasitology Today 11(12): 468–472.

Haas W, Haberl B. 1997. Host recognition by trematode miracidia and cercariae. pp. 197–227. *In*: Fried B, Graczyk TK, editors. *Advances in Trematode Biology*. Boca Raton: CRC Press.

Haberl B, Haas W. 1992. Miracidium of *Schistosoma mansoni*: a macromolecular glycoconjugate as signal for the behaviour after contact with the snail host. Comparative Biochemistry and Physiology 101A: 329–333.

Haberl B, Kalbe M, Fuchs H, Ströbel M, Schmalfuss G, Haas W. 1995. *Schistosoma mansoni* and *S. haematobium*: miracidial host-finding behaviour is stimulated by macromolecules. International Journal for Parasitology 25(5): 551–560.

Haberl B, Körner M, Spengler Y, Hertel J, Kalbe M, Haas W. 2000. Host-finding in *Echinostoma caproni*: miracidia and cercariae use different signals to identify the same snail species. Parasitology 120: 479–486.

Hassan AHM, Haberl B, Hertel J, Haas W. 2003. Miracidia of an Egyptian strain of *Schistosoma mansoni* differentiate between sympatric snail species. Journal of Parasitology 89: 1248–1250.

He Y, Ma J. 1981. Surface structures of miracidium of *Schistosoma japonicum* revealed by scanning electron microscopy. Acta Zoologica Sinica 27(4): 301–304.

Hertel J, Holweg A, Haberl B, Kalbe M, Haas W. 2006. Snail odour-clouds: spreading and contribution to the transmission success of *Trichobilharzia ocellata* (Trematoda, Digenea) miracidia. Oecologia 147: 173–180.

Ishii S. 1980. The ultrastructure of the protonephridial flame cell of the freshwater planarian *Bdellocephala brunnea.* Cell and Tissue Research 206(3): 441–449.

James C, Prah SK. 1978. The influence of physical factors on the behaviour and infectivity of miracidia of *Schistosoma mansoni* and *Schistosoma haematobium.* III. Effect of contact time and dispersion in static and flowing waters. Journal of Helminthology 52: 221–226.

Jordan P, Webbe G, Sturrock RF, editors. 1993. *Human Schistosomiasis.* Wallingford: CAB International. 465 p.

Kalbe M, Haberl B, Haas W. 1996. *Schistosoma mansoni* miracidial host-finding: species-specificity of an Egyptian strain. Parasitology Research 82: 8–13.

Kalbe M, Haberl B, Haas W. 1997. Miracidial host-finding in *Fasciola hepatica* and *Trichobilharzia ocellata* is stimulated by species-specific glycoconjugates released from the host snails. Parasitology Research 83: 806–812.

Kalbe M, Haberl B, Haas W. 2000. Finding of the snail host by *Fasciola hepatica* and *Trichobilharzia ocellata*: compound analysis of 'miracidia-attracting glycoproteins'. Experimental Parasitology 96: 231–242.

Kalbe M, Haberl B, Hertel J, Haas W. 2004. Heredity of specific host-finding behaviour in *Schistosoma mansoni* miracidia. Parasitology 128: 635–643.

Kinoti GK. 1971. The attachment and penetration apparatus of the miracidium of *Schistosoma*. Journal of Helminthology 45(2–3): 229–235.

Koie M, Frandsen F. 1976. Stereoscan observations of the miracidium and early sporocyst of *Schistosoma mansoni.* Zeitschrift fuer Parasitenkunde 50(3): 335–344.

Laracuente A, Brown RA, Jobin W. 1979. Comparison of four species of snails as potential decoys to intercept schistosome miracidia. American Journal of Tropical Medicine and Hygiene 28(1): 99–105.

Lengy J. 1962. Studies on *Schistosoma bovis* (Sonsino, 1876) in Israel. I. Larval stages from egg to cercaria. Bulletin of the Research Council of Israel Section E Experimental Medicine 10E(l): 1–36.

Loverde PT. 1975. Scanning electron microscope observations on the miracidium of *Schistosoma*. International Journal for Parasitology 5(1): 95–97.

Ludtmann MHR, Rollinson D, Emery AM, Walker AJ. 2009. Protein kinase C signalling during miracidium to mother sporocyst development in the helminth parasite, *Schistosoma mansoni*. International Journal for Parasitology 39(11): 1223–1233.

MacInnis AJ. 1965. Responses of *Schistosoma mansoni* miracidia to chemical attractants. Journal of Parasitology 51: 731–746.

MacInnis AJ. 1976. How parasites find hosts: some thoughts on the inception of host-parasite integration. pp. 3–20. *In*: Kennedy CR, editor. *Ecological Aspects of Parasitology*. Amsterdam: North-Holland Publishing Company.

MacInnis AJ, Bethel WM, Cornford EM. 1974. Identification of chemicals of snail origin that attract *Schistosoma mansoni* miracidia. Nature 248(5446): 361–363.

Mahmoud AAF, editor. 2001. *Schistosomiasis*. London: Imperial College Press. 513 p.

Maldonado JF, Acosta Matienzo J. 1947. The development of *Schistosoma mansoni* in the snail intermediate host, *Australorbis glabratus*. Puerto Rico Journal of Public Health and Tropical Medicine 22(4): 331–373.

Marty AM, Andersen EM. 1995. Helminthology. pp. 801–982. *In*: Philip ES Palmer, Reeder MM, editors. *Tropical Pathology*. Springer Berlin Heidelberg.

Mason PR, Fripp PJ. 1976. Analysis of the movements of *Schistosoma mansoni* miracidia using dark-ground photography. Journal of Parasitology 62(5): 721–727.

Mason PR, Fripp PJ. 1977a. The reactions of *Schistosoma mansoni* miracidia to light. Journal of Parasitology 63(2): 240–244.

Mason PR, Fripp PJ. 1977b. Chemical stimulation of *Schistosoma mansoni* miracidial activity. Zeitschrift für Parasitenkunde 53: 287–295.

Mason PR. 1977. Stimulation of the activity of *Schistosoma mansoni* miracidia by snail-conditioned water. Parasitology 75: 325–338.

Meuleman EA, Lyaruu DM, Khan MA, Holzmann PJ, Sminia T. 1978. Ultrastructural changes in the body wall of *Schistosoma mansoni* during the transformation of the miracidium into the mother sporocyst in the snail host *Biomphalaria pfeifferi*. Zeitschrift fuer Parasitenkunde 56(3): 227–242.

Neill PJ, Smith JH, Doughty BL, Kemp M. 1988. The ultrastructure of the *Schistosoma mansoni* egg. American Journal of Tropical Medicine and Hygiene 39(1): 52–65.

Newmark PA, Sánchez Alvarado A. 2002. Not your father's planarian: a classic model enters the era of functional genomics. Nature Reviews Genetics 3: 210–19.

Olivier L, Mao CP. 1949. The early larval stages of *Schistosoma mansoni* Sambon. 1907 in the snail host. *Australorbis glabratus* (Say. 1818). Journal of Parasitology 35(3): 267–275.

Ottolina C. 1957. Elmiracidio del *Schistosoma mansoni*. Anatomiacitologia-fisiologia 22: 1–435.

Ozaki Y. 1952. Epidermal structure of the miracidium of *Schistosoma japonicum* (Katsurada). Dobutsugaku Iho 25: 343–351.

Pan C. 1965. Studies on the host-parasite relationship between *Schistosoma mansoni* and the snail, Australorbis glabratus. American Journal of Tropical Medicine and Hygiene 14(6): 931–976.

Pan C. 1972. Formation of sporocyst tegument of *Schistosoma mansoni*. Journal of Parasitology 58(Abstract): 29–30.

Pan C. 1980. The fine structure of the miracidium of *Schistosoma mansoni*. Journal of Invertebrate Pathology 36(3): 307–372.

Plorin GG, Gilbertson DE. 1985. Behavior of *Schistosoma mansoni* miracidia in gradients and in uniform concentrations of glucose. Journal of Parasitology 71(1): 116.

Porter A. 1938. The larval Trematoda found in certain South African Mollusca with special reference to the schistosomiasis (bilharziasis). Publications of the South African Institute of Medical Research 8: 1–492.

Prah SK, James C. 1977. The influence of physical factors on the survival and infectivity of miracidia of *Schistosoma mansoni* and *Schistosoma haematobium* I. Effect of temperature and ultra-violet light. Journal of Helminthology 51: 73–85.

Prah SK, James C. 1978. The influence of physical factors on the survival and infectivity of miracidia of *Schistosoma mansoni* and *Schistosoma haematobium* II. Effect of light and depth. Journal of Helminthology 52: 115–120.

Reissinger E. 1923. Die Emunktorien des Mirazidium von *S. haematobium* Bilharz, nebstg einigen Beitragen zu dessen Anatomie und Histologie. Zoologisches Anzeiger 57: 1–20.

Roberts TM, Stibbs HH, Chernin E, Ward S. 1978. A simple quantitative technique for testing behavioral responses of *Schistosoma mansoni* miracidia to chemicals. Journal of Parasitology 64(82): 277–282.

Roberts TM, Ward S, Chernin E. 1979. Behavioral responses of *Schistosoma mansoni* miracidia in concentration gradients of snail-conditioned water. Journal of Parasitology 65(1): 41–49.

Rohde K, Watson N, Roubal F. 1989. Ultrastructure of the protonephridial system of Dactylogyrus sp. and an unidentified ancyrocephaline (Monogenea: Dactylogyridae). International Journal for Parasitology 19(8): 859–864.

Rohde K, Watson NA, Roubal FR. 1992. Ultrastructure of the protonephridial system of Anoplodiscus cirrusspiralis (Monogenea Monopisthocotylea). International Journal for Parasitology 22(4): 443–457.

Rollinson D, Simpson JG, editors. 1987. *The Biology of Schistosomes from Genes to Latrines*. London, New York: Academic Press. 472 p.

Ruelas DS, Karentz D, Sullivan JT. 2007. Sublethal effects of ultraviolet B radiation on miracidia and sporocysts of *Schistosoma mansoni*: intramolluscan development, infectivity and photoreactivation. Journal of Parasitology 93(6): 1303–1310.

Saladin KS. 1979. Behavioral parasitology and perspectives on miracidial host-finding. Zeitschrift für Parasitenkunde 60: 197–210.

Samuelson JC, Quinn JJ, Caulfield JP. 1984. Hatching, chemokinesis and transformation of miracidia of *Schistosoma mansoni*. Journal of Parasitology 70: 321–331.

Samuelson JC, Caulfield JP. 1985. Role of pleated septate junctions in the epithelium of miracidia of *Schistosoma mansoni* during transformation to sporocysts *in-vitro*. Tissue and Cell 17(5): 667–682.

Sapp KK, Loker ES. 2000. Mechanisms underlying digenean-snail specificity: role of miracidial attachment and host plasma factors. Journal of Parasitology 86(5): 1012–1019.

Schutte CHJ. 1974. Studies on the South African strain of *Schistosoma mansoni*. 1. Morphology of the miracidium. South African Journal of Science 70(10): 299–302.

Secor WE, Colley DG, editors. 2005. *Schistosomiasis*. Boston, USA: Springer. 235 p.

Shiff CJ. 1968. Location of *Bulinus* (*Physopsis*) *globosus* by miracidia of *Schistosoma haematobium*. Journal of Parasitology 54(6): 1133–1140.

Shiff CJ. 1969. Influence of light and depth on location of *Bulinus* (*Physopsis*) *globosus* by miracidia of *Schistosoma haematobium*. Journal of Parasitology 55(1): 108–110.

Shiff CJ. 1974. Seasonal factors influencing the location of *Bulinus* (*Physopsis*) *globosus* by miracidia of *Schistosoma haematobium* in nature. Journal of Parasitology 60(4): 578–583.

Shiff CJ, Dossaji SF. 1991. Ecdysteroids as regulators of host and parasite interactions: a study of interrelationships between *Schistosoma mansoni* and the host snail *Biomphalaria glabrata*. Tropical Medicine and Parasitology 42: 11–16.

Smyth JD, Halton DW. 1983. *The Physiology of Trematodes.* Cambridge: Cambridge University Press.

Sponholtz GM, Short RB. 1976. *Schistosoma mansoni* miracidia: stimulation by calcium and magnesium. Journal of Parasitology 62(1): 155–157.

Stibbs HH, Chernin E, Ward S, Karnowsky ML. 1976. Magnesium emitted by snails alters swimming behaviour of *Schistosoma mansoni* miracidia. Nature 260: 702–703.

Sukhdeo MVK, Mettrick DF. 1987. Parasite behaviour: understanding Platyhelminth responses. Advances in Parasitology 26: 73–144.

Sukhdeo MVK, Sukhdeo SC. 2004. Trematode behaviours and the perceptual worlds of parasites. Canadian Journal of Zoology 82: 292–315.

Swiderski Z. 1994. Origin, differentiation and ultrastructure of egg envelopes surrounding the miracidia of *Schistosoma mansoni*. Acta Parasitologica 39(2): 64–72.

Takahashi T, Mori K, Shigeta Y. 1961. Phototactic, thermotactic and geotactic responses of miracidia of *Schistosoma japonicum*. Japanese Journal of Parasitology 10: 686–691.

Tang CC. 1938. Some remarks on the morphology of the miracidium and cercaria of *Schistosoma japonicum*. Chinese Medical Journal, Supplement 2: 423–432.

Upatham ES. 1972a. Effects of some physico-chemical factors on the infection of *Biomphalaria glabrata* (Say) by miracidia of *Schistosoma mansoni* Sambon in St. Lucia, West Indies. Journal of Helminthology 46(4): 307–315.

Upatham ES. 1972b. Effects of water depth on the infection of *Biomphalaria glabrata* by miracidia of St. Lucian *Schistosoma mansoni* under laboratory and field conditions. Journal of Helminthology 46(4): 317–325.

Upatham ES. 1973. Location of *Biomphalaria glabrata* (Say) by miracidia of *Schistosoma mansoni* Sambon in natural standing and running waters on the West Indian Island of St. Lucia. International Journal for Parasitology 3(3): 289–297.

Upatham ES, Sturrock RF. 1973. Field investigations on the effect of other aquatic animals on the infection of *Biomphalaria glabrata* by *Schistosoma mansoni* miracidia. Journal of Parasitology 59(3): 448–453.

Wagner DE, Wang IE, Reddien PW. 2011. Clonogenic neoblasts are pluripotent adult stem cells that underlie planarian regeneration. Science 332: 811–816.

Wajdi N. 1972. Behaviour of the miracidia of an Iraqi strain of *Schistosoma haematobium*. Bulletin of the World Health Organization 46: 115–117.

Wang B, Collins JJ, Newmark PA. 2013. Functional genome characterization of neoblast-like stem cells in larval *Schistosoma mansoni*. eLife 2: e00768. doi: 10.7554/eLife.00768:1-15.

Watanabe M. 1934. Uber die Embrionalentwicklung von *Schistosoma mansoni*. Journal of the Okayama Medical Society 46: 615–664.

Watts NP, Boyd GA. 1950. *Schistosoma mansoni* miracidium under phase difference microscopy. Stain Technology 25(3): 157–160.

Wikel SK, Bogitsh BJ. 1974. *Schistosoma mansoni* penetration apparatus and epidermis of the miracidium. Experimental Parasitology 36(3): 342–354.

Wilson RA. 1971. Gland cells and secretions in the miracidiium of *Fasciola hepatica*. Parasitology 63: 225–231.

Wright CA. 1971. Flukes and snails. *In*: Carthy JD, Sutcliffe JF, editors. *The Science of Biology Series*. London: Allen and Unwin Ltd. p 168.

Wright DGS. 1974. Responses of miracidia of *Schistosoma mansoni* to an equal energy spectrum of monochromatic light. Canadian Journal of Zoology 52(7): 857–859.

Wright DGS, Lavigne DM, Ronald K. 1972. Responses of miracidia of *Schistosomatium douthitti* to monochromatic light. Canadian Journal of Zoology 50(2): 197–200.

Xylander WER. 1992. Investigations on the protonephridial system of post larval Gyrocotyle urna and Amphilina foliacea (Cestoda). International Journal for Parasitology 22(3): 287–300.

Yoshino TP, Lodes MJ, Rege AA, Chappell CL. 1993. Proteinase activity in miracidia, transformation excretory-secretory products and primary sporocysts of *Schistosoma mansoni*. Journal of Parasitology 79(1): 23–31.

Zeraik AE, Rinaldi G, Mann VH, Popratiloff A, Araujo APU, DeMarco R, Brindley PJ. 2013. Septins of Platyhelminths: identification, phylogeny, expression and localization among developmental stages of *Schistosoma mansoni*. PLOS Neglected Tropical Diseases 7(12): E2602–E2602.

CHAPTER

Schistosoma Sporocysts

Timothy P Yoshino[1],*, Benjamin Gourbal[2] and André Théron[2]

7.1 INTRODUCTION

In the complex, multi-host life cycles that typify the vast majority of digenean trematode species, the intramolluscan phase of development is critically important to successful transmission and continued maintenance of the species. This is due, in large part, to the capacity of sporocyst stages to undergo asexual reproduction, resulting in the generation of numerous cercariae, which represent either the penultimate or final transmission stage in the completion of fluke life cycles. In the case of *Schistosoma* spp., it is the snail-dwelling sporocysts that undergo asexual reproduction, greatly amplifying the population of the definitive host-infective cercariae. For human-infective schistosome species, from an epidemiological viewpoint, this amplification potential means that only a relatively small number of infected snails at water contact sites would be needed to provide a continuous source of infection to humans or other mammalian host, to maintain the life cycle (Rowel *et al.* 2015).

From the perspective of the snail host, the primary (= mother) sporocyst is of particular importance because it is at this stage of development that success or failure to establish a viable infection is determined. As discussed later in this chapter, a complex molecular interplay between parasite and host that involve both host mechanisms aimed at eliminating the sporocyst and larval counter measures serving to promote its survival within the snail host. Successful primary sporocyst infection then leads to generation of secondary (= daughter) sporocysts and eventual production of cercariae. Throughout this intramolluscan developmental process, sporocysts are not benign "houseguests" to the host, but on the contrary, cause damage/changes to tissues resulting in a myriad of alterations in host metabolism. However, larval schistosomes must maintain a delicate balance between pathology induced and sustaining basic metabolic function for host survival, at least long enough for production of the next life cycle stage. This balance has been achieved over the many millennia of interaction between snails and their schistosome parasites and continues today as a dynamic evolutionary process.

The goal of this chapter is to describe the formation and development of the schistosome sporocyst stages and summarize aspects of their physiology, molecular biology and their relationship with the snail intermediate host. Much of this review will center on *Schistosoma mansoni*, mainly due to the emphasis on this species as a research model. However, we believe that much of the information

---

[1] Department of Pathobiological Sciences, School of Veterinary Medicine, University of Wisconsin, 2115 Observatory Drive, Madison, Wisconsin 53706 USA.

[2] Host Pathogen Environment interactions, UMR 5244, University of Perpignan Via Domitia, CNRS, IFREMER, UM, F-66860 Perpignan, France.

* Corresponding author

covered will be applicable to other *Schistosoma* spp. and hopefully will serve to open up new avenues of research and inquiry in those less studied.

7.1.1   When Does a Miracidium Become a Sporocyst? Miracidium-to-Sporocyst Transformation

Soon after penetration of a suitable snail intermediate host, the free-living miracidium of *Schistosoma* spp. undergoes dramatic morphological and physiological changes in adjusting to its new host environment. Unlike fasciolid miracidia (*Fasciola hepatica* or *Fascioloides magna*), which, upon contact with the host, shed their ciliated epidermal plates outside of the snail prior to host entry (Southgate 1970; Koie *et al.* 1976; Coil 1977), schistosome miracidia enter the snail fully intact. It is within the host that ciliated plates detach as it begins to transform to the primary or mother sporocyst stage. The dramatic morphological changes miracidia undergo during the transformation process has been well documented at both the light (Meuleman 1972; Voge and Seidel 1972; Daniel *et al.* 1992) and electron (Basch and DiConza 1974; Meuleman *et al.* 1978; Loker *et al.* 1982; Gobel and Pan 1982; Samuelson and Caulfield 1985) microscopic levels and represents a critically important larval transition from the free-living to parasitic modes of existence. As illustrated in Figure 7.1, intercellular ridges separating the miracidial ciliated plates are connected by cytoplasmic bridges to underlying perikarya (nucleated subtegumentary cytons) and are firmly attached to ciliated platesvia pleated septate junctions along their adjacent plasma membranes (Meuleman *et al.* 1978; Samuelson and Caulfield 1985). Upon initiation of transformation, cytoplasmic contents from perikarya move into the intercellular ridges that separate the plates, forcing lateral expansion of ridges along the basement membrane resulting in the rounding up and pinching off of the ciliated plates. A contiguous cellular syncytium is formed by the fusion of ridge cytoplasmic membranes at the points of plate detachment and constitutes the outer tegument of the newly formed primary sporocyst.

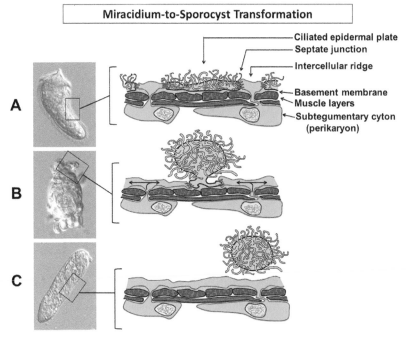

FIG. 7.1 *Schistosoma mansoni.* Schematic of miracidium-sporocyst transformation. **A.** Anatomy of the miracidial surface prior to entry into the snail host. Note the intercellular ridges (IRs) are connected to perikarya by cytoplasmic bridges and are attached to ciliated epidermal plate (CEP) membranes by septate junctions. **B.** During transformation, expansion of IRs (arrows) force CEPs to round up as the tegument is formed. **C.** Finally, fusion of IRs results in the pinching off of CEPs and formation of the continuous cellular syncytium of the tegumental (drawn by Nathan Peterson).

Early on it was discovered that miracidium-to-sporocyst transformation could be induced *in vitro* simply by exposing *S. mansoni* miracidia to snail ringers or low-tonicity salt solution (Voge and Seidel 1972; Basch and DiConza 1974) (Fig. 7.2), indicating that merely altering environmental osomlarity from ~10 mOsm/kg (freshwater) to >110 mOsm/kg (approximate molarity of snail hemolymph) was sufficient to trigger this response. More recently it has been shown that the physiological mechanism(s) responsible for regulating transformation appear to be linked to cytosolic cyclic AMP (cAMP) levels. Evidence from pharmacological studies shows that activation of membrane adenylate cyclase and corresponding increases in cAMP block *in vitro* miracidial transformation (Kawamoto *et al.* 1989; Taft *et al.* 2010) whereas lowering cAMP levels allows transformation to proceed. Larval transformation appears to utilize a protein kinase A signaling pathway in association with inhibitory dopamine D2 and/or selective serotonin reuptake receptors (Taft *et al.* 2010), although there also may be crosstalk with other pathways (e.g., PKC-p38 Erk signaling pathway; Ludtmann *et al.* 2009; Ressurreição *et al.* 2011). The finding of other biogenic amine receptors in newly-transformed primary sporocysts, including a putative histamine-like G-protein coupled receptor in the tegument (El-Shehabi *et al.* 2009) and a functionally-identified serotonin receptor involved in sporocyst motility (Boyle *et al.* 2000), provide further insights into the physiological adaptations required for the initial establishment of snail infections.

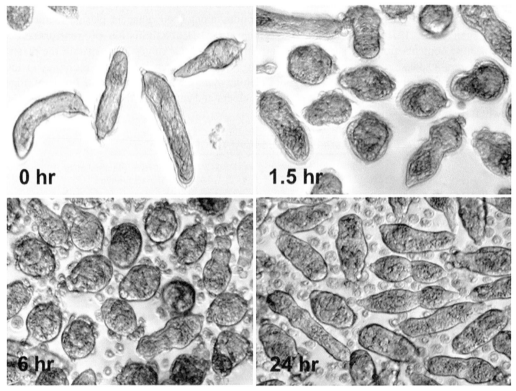

FIG. 7.2 *Schistosoma mansoni. In vitro* transformation of miracidia to primary sporocysts. **0 hr.** Newly hatched miracidia; **1.5 hr.** Miracidia placed in buffered saline (110 mOs/kg) cease swimming, while ciliated plates are starting to buckle due to intercellular rigid expansion; **6 hr.** Ciliated plates are "pinched" off and released due to fusion of intercellular ridges to form the complete tegument seen in fully transformed sporocysts (**24 hr**).

7.1.2 Intramolluscan Sporocyst and Cercarial Development

Following miracidial transformation, the primary sporocyst (PSp) undergoes further morphogenetic changes in its final location within the snail host. These locations most commonly comprise the headfoot region (including tentacles or mantle collar) not far from the point of larval entry for *S. mansoni* and *S. haematobium* (Capron *et al.* 1965; Jourdane 1982) or dispersed throughout the body for *S. japonicum* (Jourdane and Xia 1987). Within 24–48 hr post-infection (PI) many of the miracidium-associated structures, such as the terebratorium and associated sensory structures, protonephridia, penetration/lateral glands and neural mass, have degenerated or are in the process of degeneration (Schutte 1974; Meuleman *et al.* 1978), while the germinal cells and supporting reticulum cells begin multiplying in the sac-like PSp. Pan (1965) and Schutte (1974) provide detailed descriptions of secondary (daughter) sporocyst (SSp) development within *S. mansoni* PSp starting with germinal cell proliferation and formation of germinal balls (primordial SSp embryos) and proceeding through production and emergence of mature SSp from the PSp. Ultrastructural observations of SSp development within mature PSp, including formation of the primitive epithelium, generally coincide with those at the light microscope level (Gobel and Pan 1985; Yoshino and Laursen 1995). Upon maturation, SSp emerge from PSp and migrate towards the posterior end of the snail's body. Migration of these small, vermiform larvae may be passive (carried by circulating hemolymph) or active (crawling through tissues) and although some may stray to ectopic locations (headfoot or kidney), most eventually arrive at the hepatopancreas where they continue to grow and develop (Jourdane and Théron 1987).

Cercariogenesis within SSp parallels the asexual reproductive and developmental processes observed for sporocysts. Cheng and Bier (1972) and Schutte (1974) detailed the morphological progression of cercarial development for *S. mansoni* from a single germinal cell to the mature cercaria. Most of the stages described by these authors generally coincide with *in vivo* histological findings (Pan 1965). Further observations by Fournier and Théron (1985) indicate that cercarial development is compartmentalized into a specific "zone" within SSp, and exit of mature stages from sporocysts is through a defined tegumental birth pore. Ultrastructural details of cercarial development, including tegument formation (Meuleman and Holzmann 1975; Caulfield *et al.* 1988) and corresponding changes in mature SSp morphology (Fournier and Théron 1985) provide additional insights into larval development at the cellular/subcellular levels.

7.1.3 *In Vitro* Cultivation of Intramolluscan Schistosome Larvae

As previously mentioned the miracidium-sporocyst transformation can easily be accomplished under *in vitro* conditions simply by axenic isolation and culturing of schistosome miracidia in snail ringers (Fig. 7.2). This was an important finding as it set the stage for the further development of *in vitro* culture methods that would eventually be used to investigate not only early intramolluscan larval development, but also questions related to the physiologic, molecular and immunologic interactions with its snail intermediate hosts. Although initial attempts to cultivate the entire intramolluscan life cycle (miracidium to cercaria) were unsuccessful (Voge and Seidel 1972; Basch and DiConza 1974), portions of the life cycle proved to be amenable to cultivation. As examples, 7–14 day-cultured *S. mansoni* SSp from snail-derived PSp were able to establish cercaria-generating infections by transplantation into uninfected snails (Chernin 1966; DiConza and Basch 1974), while snail-derived SSp were shown to produce a second-generation of SSp (progeny SSp) when co-cultured with an *Aedes* mosquito cell line (Hansen *et al.* 1973) or a *Biomphalaria glabrata* embryonic (Bge) cell line (Hansen 1975; Hansen 1976). These latter studies employing the Bge cell line prompted a series of Bge cell-*S. mansoni* larval co-cultivation experiments that, for the first time, resulted in the continuous *in vitro* development of miracidia to first-generation SSp (Yoshino and Laursen 1995). This work was followed by the successful continuous cultivation of the complete intramolluscan cycle from miracidium to cercaria by C. J. Bayne's group (Ivenchenko *et al.* 1999) (Fig. 7.3), which represented a major breakthrough in the schistosome field. *In vitro* cercarial production, however, was very low and required long cultivation times of 6 months or more indicating that culture conditions continue to be suboptimal compared to *in vivo* infections.

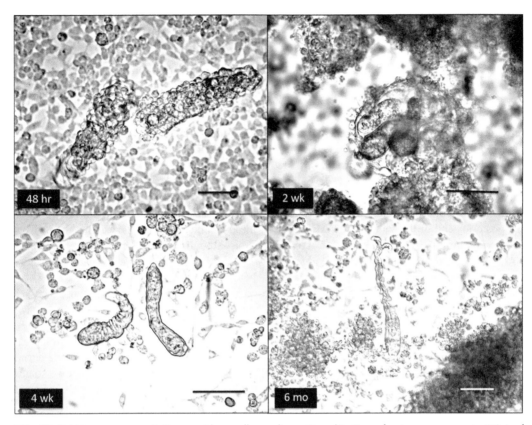

FIG. 7.3 *Schistosoma mansoni.* Sporocyst-bge cell coculture. Co-cultivation of primary sporocysts (PSp) of *Schistosoma mansoni* with cells of the *Biomphalaria glabrata* embryonic (Bge) cell line at different times of *in vitro* culture. **48 hr.** PSp are encapsulated with Bge cells that strongly adhere to the larval surface; **2 wk.** Bge cell-encapsulated PSp contain brood capsules in which motile SSp develop; **4 wk.** SSp emerge from the PSp as free motile larvae. Multiple generations of SSp may be produced under culture conditions; **6 mo.** After prolonged cultivation in co-cultures (6–9 months), cercariae emerge from sporocysts, although this is typically a rare event. Scale bars 50 µm. From Yoshino, Bickham and Bayne (2013a). Canadian Journal of Zoology 91: 391–404. Fig. 2.

Contrary to the stringent host specificity exhibited between *Schistosoma* spp. and their snail intermediate hosts (see review by Basch, 1976), the *B. glabata*-derived Bge cell line also worked well in supporting *in vitro* production of SSp of *S. japonicum* (Coustau *et al.* 1997) and secondary rediae of the fasciolid *Fascioloides magna* (Laursen and Yoshino 1999), but not sporocysts of *S. intercalatum* or *S. mattheei* (Coustau and Yoshino 2000). The differential response of divergent fluke species to Bge cells suggests that these cells are not merely supporting general larval nutritional needs, but rather, are providing chemical signals capable of regulating specific aspects of larval morphogenesis. The ability of *S. mansoni* to produce cercariae *in vitro* solely in the presence of Bge cell-conditioned media (i.e., in the physical absence of Bge cells) and *S. japonicum* sporocysts to undergo advanced *in vitro* development in the absence of encapsulating Bge cells indicate that development-promoting molecules are released from Bge cells as soluble factors. Identifying such factors should provide interesting insights into how asexual reproduction is regulated in larval flukes and perhaps, may lead to development of new control strategies by targeting these development regulatory pathways.

7.2   CELLULAR BIOLOGY: HOW SPOROCYSTS FEED AND REPRODUCE

Sporocysts of schistosomes and other trematode species, have been aptly referred to as "germinal sacs" (Wright 1971; Whitfield and Evans 1983), due to the abundance of cells/tissues devoted to reproduction contained within a single, contiguous cellular syncytium. Despite this simplistic view, sporocysts clearly possess all of the biological requirements, including biochemical and molecular processes, needed to thrive within their snail host. The most important of these requirements include (i) the ability to acquire the necessary nutrients to sustain growth and reproduction and (ii) the capacity to asexually generate multiple larval stages within an individual infected host. As described in the following sections, both of these requirements are efficiently met by the tegument (body wall) and embryonic germinal cells, respectively.

7.2.1 Tegument: At the Host-Parasite Interface

Sporocysts of schistosome, as in other fluke species, lack a mouth or gut as a means of feeding and eliminating waste. Instead the entire body of primary and secondary sporocysts is surrounded by a single, multinucleate syncytium called the tegument (see also Jones *et al.*, Chapter 10). This structure is of central importance in maintaining the viability and growth of sporocysts and physiologically, represents one of the most active of sporocyst structures. It provides the exclusive means by which nutrients are obtained, waste products and secretions are expelled and as the sole interface with the internal host environment, must provide the means for counteracting attempts by the host to eliminate ongoing infections.

In PSp the plasma membrane of the tegument is a simple lipid bilayer that is in contact with host tissue at its apical (outer) surface and with a basement membrane on its basolateral (inner) surface. The outer membrane is comprised of numerous microvilli, which greatly increases the nutrient-absorbing larval surface. It contains all of the organelles typical of eukaryotic cells including mitochondria, rough and smooth ER, ribosomes, dense and lucent vesicles, Golgi and glycogen deposits (Basch and DiConza 1974, Meuleman *et al.* 1978; Yoshino and Laursen 1995). The plasma membrane of the SSp syncytium is similar to that of PSp, but the tegument itself is much thinner and contains fewer cytoplasmic structures; mainly mitochondria, dense-core vesicles and an apical surface with microvilli and spines (Smith and Chernin 1974; DiConza and Basch 1974; Caulfield *et al.* 1988). Tegumental spines are presumed to facilitate SSp movement during their migration through host tissues.

The proteins expressed at the apical membrane surface of the PSp originate from vesicles located within subtegumentary cytons or perikarya (Dunn and Yoshino 1988). In *S. mansoni* it has been shown that during miracidia transformation (Fig. 7.1), proteins associated with these cytoplasmic vesicles are transported into expanding ridges where they are then incorporated into the tegumental membrane by vesicle-plasma membrane fusion (Fig. 7.4). A similar pattern of tegument-associated molecule distribution was observed for lectin-reactive glycoproteins or glycolipids during *S. margrebowiei* miracidial transformation (Daniel *et al.* 1992). Thus it is presumed that most, if not all, of the native proteins expressed at the sporocyst surface originate via this mechanism. Table 7.1 lists a variety of sporocyst tegumental proteins that have been identified by immunolocalization, histochemistry, protein isolation or functional experimental approaches. Based on the presence of a functional glucose transporter (Boyle *et al.* 2003), *S. mansoni* PSp, like their adult worm counterpart (Jiang *et al.* 1996), can acquire free glucose through a facilitated diffusion mechanism. Moreover, the presence of glycogen deposits in the tegument and perikarya (Meuleman *et al.* 1978) suggests that sporocysts, as in adult stages (Telens *et al.* 1990), are metabolically equipped for glucose storage and breakdown. However, unlike adult schistosomes (Krautz-Peterson *et al.* 2007), tegumental amino acid transporters have not yet been identified in sporocysts. Alternatively, snail hemoglobin, the most abundant protein found in *Biomphalaria* spp. (Lee and Cheng 1972; Granath *et al.* 1987), rapidly binds to the PSp tegument soon after *in vitro* transformation, then disappears from the surface within 24 hr post-exposure (Bayne and Hull 1988). Implied by this observation is that surface-bound hemoglobin rapidly enters the tegument, where it is presumed to under go intracellular proteolysis and distribution as amino acids for larval protein synthesis or other metabolic processes.

TABLE 7.1 Tegument-associated proteins expressed by schistosome primary (PSp) and/or secondary (SSp) sporocysts.

Tegumental protein	Stage	Species	Reference
Alkaline phosphatase	PSp, SSp	*S. mansoni*	Ivenchenko *et al.* 1999
Alkaline phosphatase	SSp	*S. mattheei, S. bovis*	Kinoti *et al.* 1971
Calmodulin	PSp	*S. mansoni*	Taft & Yoshino 2011
CD36-like scavenger receptor	PSp	*S. mansoni*	Dinguirard & Yoshino 2006
Glucose transporter	PSp	*S. mansoni*	Boyle *et al.* 2003
Histamine-related biogenic amine receptor	PSp	*S. mansoni*	El-Shehabi *et al.* 2009
p38 MAPK (activated)	PSp	*S. mansoni*	Ressurreicão *et al.* 2011
Peroxiredoxin (Prx1/2)	PSp	*S. mansoni*	Vermeire & Yoshino 2007
Protein kinase C	PSp	*S. mansoni*	Ludtmann *et al.* 2009
Septin/GTP binding protein	PSp	*S. mansoni*	Zeraik *et al.* 2013
Serotonin transport receptor	PSp	*S. mansoni*	Boyle *et al.* 2000

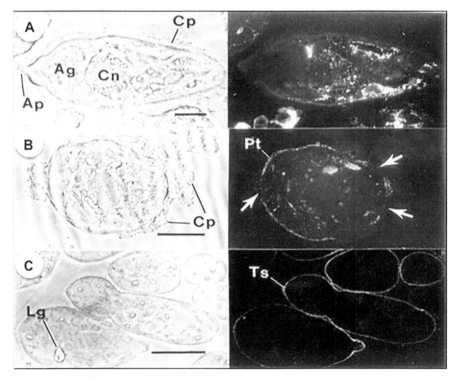

FiG. 7.4 Distribution and expression of a tegumental antigen during miracidium-to-sporocyst transformation. Immunohistochemical localization of an anti-*S. mansoni* sporocyst tegument monoclonal antibody (MAB, Boswell *et al.* 1987) during transformation. Time-course of antigen distribution from miracidium (**A**) through 6-hr transforming sporocyst (**B**) to 24-hr transformed sporocysts (**C**) demonstrates the transit of epitope-bearing vesicles from perikarya to tegument. Cp = ciliated plates, Pt = newly-formed immunoreactive tegument, Ts = tegumental surface. Arrows indicate location of attached Cp (note absence of MAB rxn). Phase contrast (left panels) and immunofluorescence (right panels), Scale bars for (**A**) and (**B**) 20 μm; (**C**) 50 μm. From Dunn and Yoshino (1988). Experimental Parasitology 67: 167–181. Fig. 1.

7.2.2 Sugar-Coated Parasites: Glycobiology of the Sporocyst Tegument

From earlier studies using plant lectins (carbohydrate (CHO)-binding proteins) as probe reagents, the presence of a layer of carbohydrates (generally referred to as the glycocalyx) was clearly demonstrated at the tegumental surface of *S. mansoni* miracidia and PSp (Yoshino *et al.* 1977; Zelck and Becker 1990; Uchikawa and Loker 1991; Johnston and Yoshino 1996). The vast majority of lectin-reactive surface glycoconjugates appear to be glycoproteins (Uchikawa and Loker 1991). Although most commercially available lectins exhibit a degree of sugar-binding specificity, they could not be used to identify the complex glycan structures exhibited at the larval surface. This situation changed dramatically with the application of mass spectrometry to the structural analysis of complex CHOs associated with schistosome glycoconjugates (Hokke *et al.* 2007; Smit *et al.* 2015). Sequencing of oligosaccharides representing N- and O-linked glycans of various stages of *S. mansoni* was followed by the creation of authentic synthetic glycans and highly specific monoclonal antibodies capable of their recognition (Van Remoortere *et al.* 2000; Robijn *et al.* 2005). By employing these antibodies, specific glycan epitopes (glycotopes) were identified at the tegumental surface of miracidia and *in vitro*-transformed PSp. Most prominent among these were lacdiNAc(LDN) [GalNAc(β1-4)GlcNAc] and fucosylated variants of LDN including F-LDN, F-LDN-F, LDN-F and a tri-mannosyl core N-glycan (Nyame *et al.* 2002; Peterson *et al.* 2009). Likewise, *in vivo* developing SSp tegumental proteins exhibited a similar pattern of glycosylation by expressing LDN, F-LDN, LDN-F and F-LDN-F glycotopes (Nyame *et al.* 2002; Lehr *et al.* 2008). Recently the enzyme systems involved in sporocyst CHO fucosylation (GDP-L-fucose synthesis, fucosyltransferases) have been investigated (Peterson *et al.* 2013a, b). Functionally, this dense sugar-coating of surface glycoproteins and lipids may be serving as a protective chemical barrier to external insults during the transition between environments. In addition, because the immune system of *Biomphalaria* spp. utilizes recognition of larval-expressed CHOs to trigger defensive reactions against the parasite, glycan expression would be critical to host responsiveness. Clearly, knowledge of the specific sugars displayed by sporocysts can provide valuable insight into the mechanisms involved in this process (see Section 7.3.2).

7.2.3 Sporocyst Reproduction: Importance to Transmission and Life Cycle Maintenance

In *S. mansoni*, after penetrating the snail, the miracidium develops near the penetration point (subepithelia of the headfoot, mantle collar, antennae) and transforms into a primary sporocyst (PSp) (Fig. 7.5). Asexual reproduction in the PSp is believed to be a continuous "budding" process in which primordial stem cells (germinal cells), attached separately or in small groups to the body wall (Wright 1973) where they undergo mitotic division resulting in the creation of multiple embryos that represent the next sporocyst generation (secondary sporocysts; SSp). Subsequently, by employing a similar mitotic process, multiple cercarial embryos are produced within mature SSp (reviewed by Basch 1991). These germinal cells possess molecular characteristics of typical stem cells (Collins and Newmark 2013; Wang *et al.* 2013), especially neoblast cells found in free living planarian flatworms (Newmark and Sanchez Alvarado 2002). This ability of neoblast-like stem cells to self-renew and differentiate accounts for the capacity of a single schistosome sporocyst to generate large numbers of SSp and cercariae.

As illustrated in Figure 7.5, fully differentiated SSp of *S. mansoni* leave the PSp by breaking through the tegument wall and migrating posteriorly towards the snail digestive gland between the 10th and 17th day after infection. Depending on the organs they pass through, two types of migration patterns are used by the SSp; (i) an active migration through host tissue and (ii) a passive migration using the circulatory system of the host. Young SSp contains 50-100 germinal cells located in the sporocyst body cavity and only start to differentiate into cercariae when they have reached their permanent location in the digestive gland of the snail. Cercarial formation within SSp appears to follow a similar asexual process ultimately leading to the production of numerous cercariae from a single infected snail. In SSp of *S. mansoni* one extremity is differentiated into a terminal cercarial exit pore that is associated with nerve cells exhibiting acetylcholinesterase activity (Théron and Fournier 1982). This pore participates in cercarial shedding and its involvement in chronobiological mechanisms has been hypothesized (Théron and Fournier 1982).

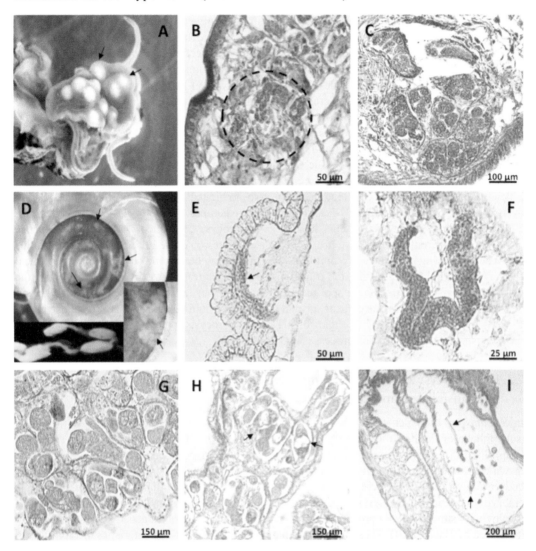

FIG. 7.5 *Schistosoma mansoni* intra-molluscan development in *Biomphalaria glabrata.* **A.** Primary sporocysts (PSp) in snail headfoot; **B.** PSp in snail foot tissue at 3 days post infection (DPI); **C.** PSp with multiple SSp at 10 DPI; **D.** SSp in snail digestive gland at 15 DPI. Appearing as white patches and harboring a vermiform shape when dissected; **E.** Secondary sporocysts (SSp) in migration in snail kidney at 10 DPI; **F.** SSp in snail digestive gland at 15 DPI; **G.** SSp full of cercariae starting their development (proliferation/differentiation) at 20 DPI; **H.** SSp full of mature cercariae at 30 DPI; **I.** Cercariae in snail mantle at 35 DPI. Parasite stages are indicated in figures by black arrows.

Because of the critical importance of progeny amplification to maintaining effective schistosome transmission and its life cycle, questions focusing on the reproductive capacity of sporocysts and strategies for maintaining chronic, long-term cercarial production have received considerable research attention. Until the 1970's and 80's, it was thought that the production of SSp from PSp led directly to the formation of the cercarial stage in SSp However, studies involving experimental infections have challenged this simple view and demonstrated in several schistosome species (*S. haematobium, S. bovis, S. rodhaini, S. mansoni*) that SSp could self-replicate, producing new generations of secondary sporocysts (SSp) and this reproductive sequence occurs as a normal mode of multiplication during intramolluscan development (Hansen 1973; Jourdane *et al.* 1987; Kechermir and Théron 1980; Touassem and Théron 1986; Théron and Touassem 1989). Different processes of sporocystogenesis have been described (Jourdane and Theron 1987) (Fig. 7.6) in which the final location within the snail of migrating SSp determines the developmental program. SSp originating from PSp migrate to the digestive gland (normal site) where they develop directly into cercariogenous sporocyst (c-SSp) (Fig. 7.6; 1-sporocystogenesis), whereas SSp that migrate to ectopic sites (headfoot or kidney), instead of producing cercariae, produce another generation of sporocysts (Fig. 7.6; 2-sporocystogenesis). These sporocystogenous sporocysts (s-SSp) were often smaller in size and contained fewer replicating young sporocyts in their cavities. These s-SSp were then shown to be capable of producing new c-SSp (Fig. 7.6; 4-sporocystogenesis). Interestingly, some c-SSp, after stopping their cercarial production due to degeneration of cercarial embryos, were able to again produce new generations of sporocysts (Fig. 7.6; 3-sporocystogenesis). SSp also may simultaneously produce new sporocysts and cercariae, although this pattern of multiplication typically was rarely observed. Finally, following their release from c-SSp, cercariae migrate anteriorly through snail tissue and reach the water environment by penetrating the snail's mantle epithelium (Fig. 7.6; 5).

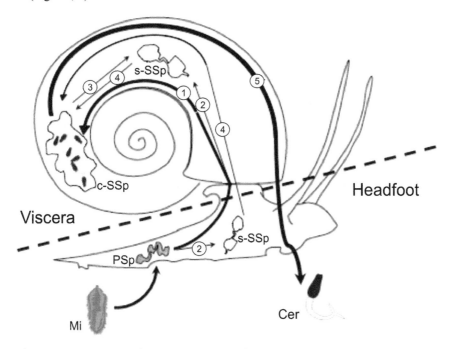

FIG. 7.6 Schematic representation of sporocystogenesis and cercariogensis. **1.** Direct cercariogenous sporocyst production. **2.** Ectopic sporocytogenous sporocysts. **3.** Sporocystogenesis after cercariogenesis. **4.** Cercariogenous sporocysts produced by sporocystogenous sporocysts. **5.** Cercaria migration and shedding. Mi, miracidium; PSp, primary sporocyst; s-SSp, sporocystogenous secondary sporocyst; c-SSp, cercariogenous sporocyst; Cer, cercariae.

In most schistosome species the production of cercariae occurs daily and stops when the host dies. Moreover, cercariae leave the snail according to daily rhythms that are unique to each species of schistosome. Analyses of the variation in cercarial productivity and release have documented the existence of chronobiological patterns of shedding. These rhythms are generally circadian (one emergence peak per day) or ultradian (two emergence peaks per day) and may be influenced by a large number of biotic and abiotic factors. However, the factors that seem to most strongly influence the chronobiology of cercarial shedding are photoperiod and the thermoperiod. Rhythmic patterns of cercarial emergence tend to maximize the probability of encountering the most permissive host. Indeed, for most species the rhythm of cercarial emission is closely correlated with the period of highest host activities. These chronobiological emission patterns have been widely demonstrated at both the interspecific and intraspecific levels (Théron 1986). In the case of schistosomiasis, knowing how cercarial rhythmicity influences cercarial densities in waters associated with transmission sites and its effect on risk of host infection are crucial in terms of public health impact and the epidemiology of disease transmission. Currently, some unresolved issues still remain. For example, the mechanisms by which external synchronizers or host internal synchronizers regulate larval emission rhythms are still unknown. Even if previous crossbreeding experiments suggest that sporocysts play an active role in the release of cercariae (Théron and Combes 1983; Théron 2015), we still do not know if photoperiod (light stimulus), temperature or other factors are acting directly on the sporocyst or the snail host to effect shedding patterns. Genomes, transcriptomes or exomes are now available for an increasing number of *Schistosoma* species and constitute excellent tools that can be applied to gain a better understanding of the genetic foundations, to identify genes (DNA/RNA seq, extreme QTL mapping) and uncover the molecular mechanisms involved in the chronobiology of cercarial shedding (Théron 2015).

7.3 SPOROCYST-SNAIL HOST INTERACTIONS

As previously described, intramolluscan development of larval digenetic trematodes is complex and involves an initial infection of the snail host by the miracidium, its subsequent transformation to the primary sporocyst stage, followed by asexual reproduction and release of secondary sporocysts and finally the formation and release of cercariae into the environment. Parasites therefore face many challenges such as obtaining enough energy to maintain, grow and multiply, while at the same time, being able to evade the internal defense system of the host (van der Knaap and Loker 1990; De Jong-Brink *et al.* 2001). To reach this goal, the parasites profoundly affect the host's energy flow, directing it to their own benefit. This is mainly accomplished by interfering with the two major regulatory systems in the snail host, the neuroendocrine system and the immune system (De Jong-Brink *et al.* 2001). How larval schistosomes manipulate these systems with the goal of establishing successful relationships with their snail hosts is discussed in the following sections.

7.3.1 Determinants of Host-Parasite Specificity: Compatibility, Susceptibility, Resistance, Suitability, Unsuitability

A basic understanding of the mechanisms underpinning snail-schistosome interactions is of great theoretical and practical interest. Indeed, effective treatment for schistosomiasis still relies on mass chemotherapy with a single drug, praziquantel. The possible emergence of praziquantel resistance in natural schistosome populations and the absence of an efficient vaccine increases the necessity of developing new approaches to control schistosomiasis (Fallon and Doenhoff 1994). An alternative control strategy could therefore be focused on tactics for blocking transmission via the aquatic snails that serve as intermediate hosts. In this context, effective control of schistosomes would depend upon a comprehension of the factors controlling their relationships with their intermediate hosts. Understanding the molecular mechanisms by which snails and schistosomes interact will be key to discovering new approaches to interrupt transmission by genetic or other means of manipulating the snail intermediate host (Tennessen *et al.* 2015).

Before describing the molecular mechanisms functioning at the core of these interactions it is important to define the terminology that has been used to describe the types of interactions observed in the various host/parasite models. Key concepts needed to understand the complexity of snail-schistosome interactions are given in Box 7.1 (modified from Bayne 2009). Schistosome infections of their snail hosts are typified by their high degree of specificity. This specificity varies between populations, strains or individuals (Basch 1976; Théron *et al.* 1997). Among the various *B. glabrata*/*S. mansoni* combinations, infection outcomes can differ widely with some parasites succeeding to infect snails whilst others fail to do so. Snails that become infected are termed susceptible or suitable hosts, while those that are not infected have traditionally been referred to as 'resistant' (Box 7.1). However, for snails incapable of being infected, the term 'resistant' now applies when parasite elimination is the result of an active host cell/tissue response, while 'unsuitable' is used when the host does not provide a physiological/physicochemical environment supportive of normal sporocyst development or growth (Sullivan and Richards 1981) (Box 7.1). Representing the other interacting half of the host-parasite relationship, the parasite brings its own genome and expressed gene repertoire into play when confronting a snail host in the form of 'infectivity' factors. Parasite-host compatibility results when there is a benign matching of resistance and infectivity factors (genotypes/phenotype) between an individual parasite and its host, whereas a mismatching of these factors by host and parasite results in incompatibility and a failed infection outcome (Box 7.1).

Box 7.1

Key Concepts of Snail-schistosome Interactions. (Modified from Bayne 2009)

- **Specificity:** ability of a miracidium to infect only some individuals of a given host species or strain.

- **Infectivity:** the capacity of individuals in a parasite population to establish viable infections in individuals within a host population.

- **Resistance:** condition in which a host prevents the development of the parasite through an active killing process.

- **Susceptibility:** condition in which a host fails to inactivate or kill a parasite, resulting in an infection that progresses to patency. This requires that the host internal environment is physiological suitable to support parasite development.

- **Unsuitability:** condition in which the physiological/physicochemical needs of the parasite are not adequately met by the host internal environment.

- **Compatibility:** a pairing of specific host and parasite genotypes/phenotypes in which an individual parasite is able to infect an individual host and completes that part of its life cycle within the host.

- **Incompatibility:** a pairing of specific host and parasite genotypes/phenotypes in which an individual parasite is incapable of infecting an individual host due to failure to complete the part of its life cycle appropriate to the host.

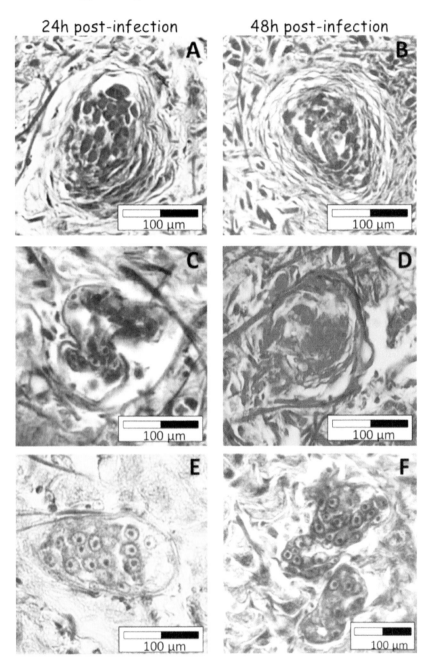

FIG. 7.7 *Schistosoma mansoni/Biomphalaria glabrata* immunobiological interactions. The outcome of infection is dependent on the nature of the host and parasite interaction. The immunobiological interactions of the parasite (*S. mansoni*) towards different strains of *B. glabrata* are either compatible/susceptible (C/S) or incompatible/resistant (I/R). Such immune interactions occur within 24 to 48 hours after infection. **A, B.** In the I/R immunobiological interactions, the transforming PSp is recognized as non-self after penetration, and is then encapsulated and killed by the hemocytes, the snail's circulating immune cells. Note the multi-layer hemocyte capsule surrounding a dead sporocyst. **C, D.** Also in I/R immune interactions, some parasites are killed by humoral factors without encapsulation. Note the degenerating parasites at 48 h, with few remaining sporocyst cells or structures. **E, F.** In the C/S imunobiological interaction, the parasite is not recognized or encapsulated and develops normally in snail tissue following penetration of the headfoot.

Unfortunately, the host-parasite interactions that result in a clear dichotomy between susceptible or resistant snails are not so simple. For example, the *B. glabrata* BS-90 strain has historically been considered as "completely resistant" to both new and old world *S. mansoni* strains (Ittiprasert and Knight 2012). However recently the *Bg* BS-90 stock was shown to be partially susceptible to the LE strain of *S. mansoni* (53%, 20 miracidium exposure) (Théron *et al.* 2014), indicating a schistosome strain-dependent shifting in host resistance phenotype. In another study, when infecting susceptible *B. glabrata* snail with 10 miracidia of *S. mansoni*, a proportion of miracidia (20–40%) were not recognized by the innate defense system of the host and developed into PSp, while the other larvae were encapsulated or degenerated within the same snail host (Théron *et al.* 1997) (Fig. 7.7). This means that within an individual snail, a specific defense response was elicited against some miracidia but not other neighboring larvae (Théron *et al.* 1997). These results support the hypothesis that the success or failure of an infection mainly depends on the compatible/incompatible status of a specific pair of host and parasite phenotypes. This phenomenon has been termed 'compatibility polymorphism', which is based on a matching phenotype model (Basch 1975; Basch 1976; Théron and Coustau 2005; Théron *et al.* 2008; Mitta *et al.* 2012). In this model, compatibility is tested independently for each miracidium and each exposed snail, in which the phenotype (susceptible vs. unsusceptible) of the host is expressed as a function of the parasite genotype it harbors and reciprocally, the phenotype (infective vs uninfective) of the parasite is expressed as a function of the genotype of the particular host that it enters (Basch 1976; Théron and Coustau 2005; Théron *et al.* 2014). Characterization of the mechanisms underlying compatibility polymorphism between *S. mansoni* and *B. glabrata* is crucial to understand which combinations of molecular determinants define the compatibility or incompatibility status of a given snail-schistosome interaction. As emphasized by Bayne (2009) the fate of a parasite is determined by several factors including recognition or its failure and immune effectors and their anti-parasite efficiency. In fact a complex interplay of the recognition capacities of the host, evasion capacities of parasites, effectiveness of killing activities of hosts and counter-defense mechanisms of parasites represent crucial elements of the system that need to be investigated and described.

7.3.2 Immune Interactions

Snail species known to serve as intermediate hosts for *Schistosoma* spp. possess an internal defense system (IDS) comprised of both cellular and humoral elements that are capable of responding to invading pathogens including larval trematodes. Phagocytic cells, termed hemocytes, found circulating within the snail's hemolymph system represent the primary immune effector cells of snails. Plasma (cell-free hemolymph) contains humoral factors including pattern recognition receptors (PRR), aggregation factors and other potential effector molecules such proteases, stress proteins, anti-bacterial peptides and the like. Most prominent among the plasma PRRs is a large and genetically-diversified family of lectin-like proteins, termed fibrinogen-related proteins (FREPs, Adema *et al.* 1997), that have been functionally linked to *B. glabrata* resistance to *S. mansoni* infection (Moné *et al.* 2010; Hanington *et al.* 2012). Although the IDS of *Oncomelania* spp. (host for *S. japonicum*) (Sasaki *et al.* 2003, 2005; Wang *et al.* 2012) and *Bulinus* spp. (host for *S. haematobium*) (Krupa *et al.* 1977; Adema *et al.* 1992) has been investigated, the vast majority of research on snail-schistosome immune interactions has focused on model systems involving various *Biomphalaria* spp. and *S. mansoni*. Because this topic has been extensively reviewed in recent years (Bayne 2009; Loker 2010; Yoshino and Coustau 2011; Mitta *et al.* 2012; Knight *et al.* 2014; Adema and Loker 2015; Coustau *et al.* 2015) the following section will overview the immune basis for parasite-host compatibility.

The immune mechanisms underlying compatibility have remained largely unknown (Bayne 2009; Yoshino and Coustau 2011; Théron *et al.* 2014). However, in order to gain a better understanding of specific genes involved in regulating snail compatibility, comparative molecular approaches have yielded important insights including the identification of some of the molecular processes involved in regulating *B. glabrata/S. mansoni* interactions. These findings support the hypothesis that resistant/incompatible snails recognize parasites and mount an appropriate encapsulation response, while susceptible/compatible snails are unable to defend themselves against infection either because they lack the capacity to recognize and react to the parasite or effector cells (hemocytes) or humoral factors are rendered ineffective by active suppression of the defense response during larval infection (Lockyer *et al.* 2012).

As discussed in Section 7.3.2.2, one mechanism by which larval schistosomes may achieve these different goals is by production and release of diverse proteins that are postulated to serve a protective function during the early miracidium-to-sporocyst transformation process by counteracting the diverse repertoire of lectin PRRs or effectors presented by the snail's IDS (Guillou *et al.* 2007; Mitta *et al.* 2012; Coustau *et al.* 2015). Thus compatibility polymorphism for a given *B. glabrata* snail invaded by a *S. mansoni* miracidium could be the result of multigenic and "multi-mechanism" processes determined by a combination of interacting factors that result in a balancing of multiple immune recognition/effector vs. anti-immune mechanisms that ultimately determine the success or failure of the infection. So what is the evidence supporting the compatibility polymorphism model?

7.3.2.1 Immune Recognition: lectin and Antigen Polymorphisms

Newly penetrated incompatible parasites are contacted by host hemocytes within 1–2 h post-infection until entirely encapsulated by 8–12 h post-infection). In contrast, compatible miracidia are not recognized or encapsulated during PSp transformation and continue to develop normally (see Fig. 7.7). These observations suggest that constitutive antigenic differences exist between compatible and incompatible parasites and this appears to be the case. Using a comparative proteomics approach, Roger *et al.* (2008a) have identified a molecularly diverse multi-gene family of schistosome antigens that share characteristics with mucin-like glycoproteins, which they have named *S. mansoni* polymorphic mucins (*Sm*PoMuc). *Sm*PoMuc proteins are: (i) highly glycosylated, (ii) only expressed by larval schistosome stages that interact with the snail intermediate host, (iii) located in the apical gland of miracidia and sporocysts, (iv) secreted and released as excretory–secretory products miracidial transformation and finally (v), exhibit a high degree of genetic polymorphism (Roger *et al.* 2008a, b).

Gene polymorphism of *Sm*PoMuc is driven by frequent genomic recombination events within this multigene family of around 10 members, as well as by frequent trans-splicing, alternative splicing and aberrant splicing event sat the transcriptional level (Roger *et al.* 2008c). Furthermore, it was determined that post-translational modifications in the form of differential glycosylation add to the structural complexity of *Sm*PoMucs. Indeed, quantitative differences between carbohydrate patterns of *Sm*PoMucs between compatible and incompatible *S. mansoni*s trains have been identified (Roger *et al.* 2008c). Of particular importance, however, was the finding that *Sm*PoMuc glycoproteins from larval *S. mansoni* are recognized by and reactive with a polymorphic, highly diversified group of immune recognition lectins of *B. glabrata* (Moné *et al.* 2010), the fibrinogen-related proteins or FREPs (Adema *et al.* 1997; Zhang *et al.* 2004; Zhang *et al.* 2008). In addition, although both *Sm*PoMuc and FREPfamilies possess extremely high molecular polymorphism, individual snails and parasites exhibit specific patterns of expression and reactivity, thus providing a basis for a phenotype-matching model of host-parasite interaction. Interestingly, in their interactome study, in addition to identifying a *Sm*PoMuc and FREP, Moné *et al.* (2010) also retrieved in the same immune molecular complex a putative opsonin, a thioester-containing protein from *B. glabrata* (*Bg* TEP) known to be involved in phagocytosis and encapsulation responses (Blandin and Levashina 2004; Blandin *et al.* 2008).

Taken together, these results suggest that larval antigenic diversity may explain the differences in compatibility between various *S. mansoni* and *B. glabrata* strains. In an incompatible combination, an immune complex composed of specific *Sm*PoMucs, FREPs and *Bg* TEP is produced, resulting in hemocyte recognition and a rapid cellular encapsulation response and killing of the parasite. In a compatible combination, *B. glabrata* PRRs (e.g., FREPs) do not recognize *Sm*PoMucs and encapsulation of *S. mansoni* does not occur. It is important to note, however, that other molecules may well be involved in the compatibility polymorphism phenomenon and the interactions of FREPs and *Sm*PoMucs alone may not completely explain the compatibility process. Recent studies have demonstrated that numerous glycoproteins from *S. mansoni* are found to interact with, and precipitate, several *B. glabrata* lectins and other proteins (e.g., galectin, C-type lectins, GREP, CREP, biomphalysin) (Yoshino *et al.* 2008; Moné *et al.* 2010; Galinier *et al.* 2013; Dheilly *et al.* 2015) indicating that the immune interactome likely is much more complex than previously expected.

7.3.2.2   Anti-Immune Responses by Sporocysts: Achieving Compatibility

During the initial first hours of infection, *S. mansoni* may employ various strategies for avoiding recognition and elimination by the host's IDS. This capability is crucial to the establishment of a compatible host-parasite relationship. Sporocyst evasion of immune reactivity may be either active (direct interference with or suppression of immune mechanisms) or passive (avoidance of immune recognition) (Yoshino and Boswell 1986). Interference of the snail immunity refers to the ability of larvae to disrupt the host immune system by counteracting the major hemocyte defense functions such as adhesion, phagocytosis and encapsulation, as well as cytotoxic activity of plasma by the release of parasite products capable of "neutralizing" or eliminating such factors. An alternative strategy by which early developing stages may be able to counteract host immunity is by passively avoiding immune recognition. Some authors have noted that miracidia and sporocysts of *S. mansoni* possess antigenic similarities with various host tissues including hemolymph (Daniel *et al.* 1992; Damian 1997; Abu-Shakra *et al.* 1999). Mimicry of host molecules (termed molecular mimicry; Damian 1989) effectively reduces parasite "antigenicity", thereby decreasing the ability of the host to recognize the parasite as foreign and mount an effective immune response.

The precise mechanisms by which parasites directly interfere with host immunity and other physiological functions are not well understood. However, the production and release of proteins during early miracidium-to-PSp and PSp-SSp transformation, variously referred to as excretory-secretory proteins (ESP; Lodes and Yoshino 1989; Guillou *et al.* 2007), secretory-excretory products (SEP; DeGaffe and Loker 1998) or larval transformation proteins (LTP; Wu *et al.* 2009), have been hypothesized to serve such a function. In *S. mansoni*, penetration of the snail intermediate host, *Biomphalaria* spp. results in the rapid shedding of ciliary epidermal plates from the larval surface during formation of the new tegumental syncytium of the developing sporocyst (Pan 1996). As illustrated in Figure 7.4, during this transformation process membrane-bound vesicles, originating from sub tegumentary perikarya, attach to and fuse with the outer tegumental membrane resulting in the release of their vesicular contents (ESP/SEP/LTP) into the surrounding snail tissue (Dunn and Yoshino 1988; Pan 1996; Guillou *et al.* 2007). Other sources of transformation proteins may also include degenerating or degraded ciliary epidermal plates released during larval development. Protection from host responses are most important during the first 24 h post-infection as the PSp is the stage that is most sensitive to immune attack by the snail host. Therefore timely release of ESP that surrounds the developing sporocyst represents a possible early immunoprotective strategy that shields the parasite during this most vulnerable stage.

Recent proteomic analyses of these released products from schistosomes or echinostomes reveal a diverse array of proteins including some with anti-immune function such as antioxidants, proteases and protease inhibitors and anti-inflammatory factors (Guillou *et al.* 2007; Wu *et al.* 2009). Earlier studies have demonstrated that schistosome larval ESPs do function in modulating the immune response through direct interference with host hemocyte effector functions or serving as a molecular deterrent of the host IDS. Functions affected include cellular adhesion, cell signaling, motility, proteinase activity, production of reactive oxygen species, phagocytosis and encapsulation (van der Knaap and Loker 1990; Bayne *et al.* 2001; Walker 2006; Bayne 2009; Yoshino and Coustau 2011; Adema and Loker 2015; Coustau *et al.* 2015). In addition, ESPs also have been shown to affect the host's energy flow. For example, proteins released by daughter sporocysts have been implicated in the modulation of snail host reproductive physiology by interfering with carbohydrate metabolism (Crews-Oyen and Yoshino 1995) or inducing inhibitory effects on host reproduction by disruption of the neuroendocrine system (De Jong-Brink *et al.* 2001). The effects of schistosome infection on host neuroendocrine function are detailed in Section 7.3.3.

7.3.2.3 *Immune Effectors and Anti-Effector Mechanisms*

As a first step to investigate the molecular determinants that play a key role in the cross-talk between snails and schistosomes, the biochemical characterization of larval proteins released during transformation was undertaken (Guillou *et al.* 2007; Wu *et al.* 2009). This has led to the identification of two main protein 'classes' with potential for disrupting snail immune effector function: (i) the proteases/protease inhibitors and (ii) the redox/antioxidant enzymes andion-binding proteins with putative anti-oxidant Fe-binding activities. These findings fit in well with a model in which parasite ESP/LTP protect sporocysts from oxidative damage and other immune effectors of the snail host.

Proteases and protease inhibitors. Hydrolytic enzymes can be crucial in host-pathogen interactions and in regulating snail compatibility. Parasite proteases and host anti-proteases likely are involved in specific interactions to control the physicochemical properties of connective tissue to facilitate or prevent migration of the parasite in snail tissue and/or hemocytes to the site of infection. In response to larval infection many host extracellular components participating in cell/cell or cell/matrix interactions, such as MFAP4 (microfibrillar associated protein), mucins, dermatopontins and matrilins, are activated to drive a dynamic remodeling of the extracellular matrix (ECM), which in turn, influences hemocyte migration and capsule formation around the parasite (Mitta *et al.* 2005; Lockyer *et al.* 2012). During this same time, transforming miracidia release many *S. mansoni* venom allergen-like (SmVAL) proteins, one of which (SmVAL9), has been shown to upregulate expression of the snail matrix metalloproteinase, MMP1, that is involved in ECM remodelling in *B. glabrata* (Yoshino *et al.* 2014). Also released are other proteases including calpain and leishmanolysin (Wu *et al.* 2009). In order to counteract these larval proteases, the snail host responds by upregulating expression of protease inhibitors like cystatin or serine protease inhibitors (Myers *et al.* 2008; Ittiprasert *et al.* 2010; Lockyer *et al.* 2012) with the 'goal' of maintaining adhesive properties of hemocytes and/or defensive factors at the parasite surface (Coustau *et al.* 2015). This illustrates the dynamic molecular interplay of host-parasite factors involved in establishing infections.

Reactive oxygen species (ROS) and antioxidant enzymes. Using the *B. glabrata/S. mansoni* model system, several recent studies have shown that ROS and ROS scavengers play an active role in the outcome of sporocyst–hemocyte interactions (Bender *et al.* 2007; Yoshino *et al.* 2008; Mourão *et al.* 2009; Ittiprasert *et al.* 2010; Moné *et al.* 2011), with hemocytes representing the primary effector cells responsible for parasite killing (Dikkeboom *et al.* 1988; Bayne *et al.* 2001; Bender *et al.* 2005). Among the ROS molecules, the most toxic form to sporocysts is hydrogen peroxide (H_2O_2), likely due to its capacity to cross cell membranes (Hahn *et al.* 2001; Bienert *et al.* 2006). This chemical is generated by Cu/Zn superoxide dismutase (Cu/Zn SOD)-catalyzed conversion of O_2^- to H_2O_2, so it is not surprising that the ability of snails to eliminate its parasite infection is directly correlated with host's capacity to generate extracellular H_2O_2 (Bender *et al.* 2005). Importantly, of three

B. glabrata Cu/Zn SOD alleles (SOD1 A, B and C), only resistant snails possessed the B allele indicating a linkage of this allele and the resistant phenotype (Goodall *et al.* 2006). Follow up experiments showed that hemocytes from snails carrying the B allele and exhibiting the resistant phenotype had an overall higher SOD1 transcript expression level than susceptible snails lacking this allele (Bender *et al.* 2007). Functionally, this is consistent with resistant snail hemocytes having a greater capacity to produce H_2O_2 and to kill encapsulated larvae more effectively. In order to counteract host ROS molecules, antioxidant activity (Connors *et al.* 1991; Vermeire and Yoshino 2007) and redox enzymes have been reported in ESPs/LTPs of *S. mansoni* sporocysts (Connors *et al.* 1991; Guillou *et al.* 2007; Wu *et al.* 2009). Sporocysts express an array of antioxidant enzymes including glutathione-S-transferase (GST), Cu/Zn and Mn SODs, glutathione peroxidase (GPx) and peroxiredoxins (Prx) that have the capability of degrading harmful extracellular oxidants and thereby protecting sporocysts from oxidative damage (Guillou *et al.* 2007; Wu *et al.* 2009; Mourão *et al.* 2009). Clearly this represents another case in which compatibility or incompatibility depend on the balancing of these host- and parasite-mediated processes.

7.3.2.4 *Immune Evasion by Stealth*

Molecular mimicry has been hypothesized as an important immune avoidance mechanism for parasites (Damian 1989). For larval schistosomes and their molluscan hosts the sharing of common molecules has been well documented. Antibodies generated against *B. glabrata* plasma or hemocytes cross react with *S. mansoni* miracidia and/or sporocyst (Yoshino and Bayne 1983; Bayne and Stephens 1983; Granath and Aspevig 1993; Agner and Granath 1995; Chacon *et al.* 2002) indicating the presence of common epitopes between host and parasite. The sharing of glycan epitopes between parasite (miracidium or PSp) and elements of the snail's immune system could be the basis of an efficient molecular mimicry strategy used by schistosomes to avoid host immune recognition (Martins-Souza *et al.* 2011; Yoshino *et al.* 2013b; Sullivan *et al.* 2014). Recent mass spectrometry analyses of CHOs and anti-CHO glycotope studies reveal an abundance of specific glycans that are naturally shared between sporocysts and their snail hosts (Lehr *et al.* 2007, 2008; Yoshino *et al.* 2012, 2013b). Moreover, inbred susceptible strains of *B. glabrata* snails possessed greater amounts of shared glycans than resistant strains, providing a correlation between molecular similarity and susceptibility (Lehr *et al.* 2010; Yoshino *et al.* 2012). Heterogeneity in glycan expression on the part of both larval schistosome and snail host is speculated to play an important role in determining parasite-host compatibility (Mitta *et al.* 2012).

7.3.3 How Parasitic Infection Influences Host Reproduction and Growth

7.3.3.1 *Effects on Reproduction: Parasitic Castration*

A direct consequence of larval schistosome infection in its snail host is the constant demand for nutrients necessary for fueling sporocyst replication and the production of cercariae. Given that the snail represents a finite energy resource, one major effect of this nutrient demand is parasitic castration (Baudoin 1975); a parasite-mediated impairment of host reproduction involving a cessation of egg production in adult snails or developmental arrest of the reproductive system in juveniles. This is believed to be a general parasite strategy for reallocating energy resources to larval growth and development, while at the same time maintaining host basal metabolic activity necessary for host survival (Théron *et al.* 1992; Lafferty and Kuris 2009).

   Although the timing of castration in adult snails following initial parasite infection can vary, reduced egg production appears to coincide with the timing of the release of first-generation SSp from PSp through early cercarial development following SSp colonization of the hepatopancreas (Pan 1965; Meuleman 1972; Looker and Etges 1979; Meier and Meier-Brook 1981; Crews and Yoshino 1989; Blair and Webster 2007). Unlike the fasciolids (Wilson and Denison 1980) and other redial-producing species, in which direct consumption of the ovotestes and other tissues by actively feeding redial stages leads to castration, sporocyst-bearing schistosome species appear to affect snail reproduction through indirect mechanisms. Direct tegumental absorption of transportable nutrients (e.g., glucose, free amino acids) by sporocysts can accounts for some loss of host nutrients (reviewed by Humphries 2011). However, De Jong-Brink (1995) argues that early in the infection cycle, when the total parasite biomass is relatively small, other indirect mechanisms, besides nutrient competition, may be responsible for the reduced reproductive capacity of the snail observed during the prepatent period. Possible mechanisms by which larval infection are indirectly interfering with host reproduction include disruption of metabolic processes affecting nutrient acquisition or conversion in reproductive organs/tissues and/or disruption of neuroendocrine (NE) control pathways involved in regulating snail reproduction.

The chemical mediators of castration are unknown, although it is likely that larval excretory-secretory proteins released during development are involved (De Jong-Brink 2001). The participation of ESP in the disruption of both snail host internal defense and reproduction suggests that these two systems may be interacting with each other. As suggested by De Jonk-Brink *et al.* (2001) many of the effects of ESPs on the neuroendocrine system would thus be mediated by factors presumably belonging to the internal defense system. This could explain why, many of the ESPs identified to date could serve a parasite protective function against the host innate immune response (Guillou *et al.* 2007; Wu *et al.* 2009). By modulating the IDS of the host, the parasite would be able to establish in host tissue and in so doing, also modify the host energy flow for its own reproductive benefit. One can also imagine that the interaction of ESP with target cells or tissues could results in alteration of physiological processes related to both host defense and reproduction. As an example, the release of cysteine proteinases from early developing *S. mansoni* PSp that degrade host hemolymph proteins (Yoshino *et al.* 1993) may have dual effects of disrupting both immune and/or reproductive functions. In terms of direct effects on reproduction ESP are capable of binding various host tissues including parenchymal cells surrounding the ovotestes (Nyame *et al.* 2002) and the albumen gland (Crews and Yoshino 1990). In the latter case, snail albumen glands incubated in *S. mansoni* SSp ES products resulted in reduced galactogen synthesis. Given that this gland represents the sole source of galactogen, the primary nutrient for developing eggs, disruption of this critical food source would have a significant impact on reproductive capacity. Similar modulations of other metabolic processes also have potential for impacting host reproduction. For example, *S. mansoni* infection has been shown to disrupt phosphate metabolism (phosphonate content, inorganic P balance) in the albumen gland and egg masses (Thompson and Lee 1985) and plasma biogenic amine (serotonin) levels (Manger *et al.* 1996) in *B. glabrata*, both of which have consequences of reducing host fecundity.

The hypothesis that schistosome-associated reproductive dysfunction is mediated by interference of the NE control of snail reproduction has grown out of extensive investigations of the pond snail *Lymnaea stagnalis* and its avian schistosome *Trichobilharzia ocellata* (see De Jong-Brink, 1995; 2001 for reviews). This work was prompted by early findings that various neuropeptide hormones synthesized and released by the dorsal bodies (dorsal body hormone, DBH) (Geraerts and Joosse 1975) and caudodorsal cells (caudodorsal cell hormone, CDCH; calfluxin) (reviewed by Roubos *et al.* 2010) of the snail's central nervous system (= cerebral ganglia) were involved in regulating development of the reproductive system and its functions. In an extensive series of investigations it was demonstrated that *T. ocellata* infection in *L. stagnalis* exerted a down-regulating effect on these hormonal systems with decreased reproductive capacity (castration) being the outcome (De Jong-Brink 1995). It was subsequently shown that the mechanism by which the avian schistosome interferred with the NE system of *Lymnaea* was through a parasite-induced host peptide factor, termed schistosomin. This peptide was isolated from infected *Lymnaea* plasma and shown to exert anti-gonadotrophic activity by inhibiting CDCH, calfluxin and DBH production (Hordijk *et al.* 1991a, 1991b; Schallig *et al.* 1991). However, these findings appear to be in contrast to schistosomin's effect on *B. glabrata* during *S. mansoni* infection. While *Biomphalaria* snails possess a similar neuroanatomy (cerebral ganglia with CDCs and DBs) as *L. stagnalis* (Roubos and van de Ven 1987) and produce CDCH (van Minnen *et al.* 1992), expression of *B. glabrata* schistosomin was restricted mainly to snail embryos and juveniles and was not induced upon infection or during larval development (Zhang *et al.* 2009). Although parasite manipulation of the NE system of *Biomphalaria* may still be involved in host castration, it is unlikely that schistosomin plays a role in this process. Compared to *Trichobilarzia* and *Lymnaea*, prepatent *S. mansoni* infection seems to exert greater effect on host metabolic processes and nutrient availability, which may be contributing more significantly to reproductive dysfunction than NE effects (Humphries 2011).

7.3.3.2 Effects on Growth: Gigantism

In *Lymnaea stagnalis* infected with the avian schistosome *Trichobilharzia*, juvenile snails grow larger due to accelerated shell growth, while organ/tissue growth is not affected. The outcome of this differential enhanced growth is an increase in hemocoel volume, resulting in more space for an expanding larval population. Termed "giant growth" (De Jong-Brink 1995), schistosome-induced growth in *L. stagnalis* has been hypothesized to be associated with parasite effects on host metabolism, possibly through modulation of the NE system (Joosse and van Elk 1986). In studies of *S. mansoni* infections in *B. glabrata*, gigantism is less pronounced and is usually manifested only in early infections and/or in snails infected as juveniles (Meuleman 1972; Meier and Meier-Brook 1981; Crews and Yoshino 1989). Also, unlike *Lymnaea* snails, *S. mansoni*-infected *B. glabrata* exhibited enhanced growth of the digestive gland early in infection, followed by a decrease in organ mass 2–4 weeks post-infection (Théron *et al.* 1992). Despite differences in the ways schistosomes influence snail growth, the fact remains that the sporocysts are highly capable of manipulating the host internal environment to their advantage. Understanding the underlying mechanisms responsible for these host-modulating effects remains an important future challenge for researchers.

7.4 BEYOND PHENOTYPE: THE PROMISE OF GENOMICS AND TRANSCRIPTOMICS IN THE POST-GENOMIC ERA

With the completion of sequencing and ongoing efforts to assemble and annotate the genomes of the *S. mansoni* (Berriman *et al.* 2009), *S. japonicum* (Zhou *et al.* 2009) and *S. haematobium* (Young *et al.* 2012), a wealth of new and valuable information on specific genes, their expression, encoded proteins and functions is now becoming available. However, the vast majority of investigations at the molecular level have focused on the mammalian stages of the parasite, from the infective cercaria to adult worms, with less attention paid to intramolluscan stages. Due to the minute size of the parasite and their localization in host tissue, it is extremely difficult to recover these parasite larval stages under *in vivo* conditions. As a result, most of the research conducted on intramolluscan stages to date has been performed using miracidia and *in vitro* transformed primary sporocysts (PSp). Although this body of research has provided insights into the molecular mechanisms of development, signaling, host interactions and immune evasion, at present, there is still very little is known regarding the specific genes regulating these processes within the snail intermediate host. Therefore this represents a critically important gap in our knowledge of schistosome biology.

To begin filling this information gap, previous approaches utilized quantitative PCR or DNA microarray analyses to determine gene expression profiles in larval schistosomes (Verjovski-Almeida *et al.* 2003; Vermeire *et al.* 2004, 2006; Jolly *et al.* 2007; Vermeire and Yoshino 2007; Gobert *et al.* 2009). These approaches provided important information on expression of *known* genes, but did not permit the discovery of new genes or molecules associated with different developmental stages of larval schistosome or their host interactions. Subsequent application of serial analysis of gene expression (SAGE) focusing on miracidia and PSp of *S. mansoni* has provided a more global view of expressed genes without requiring prior knowledge of genes being analyzed (Williams *et al.* 2007; Taft *et al.* 2009). For example, differential expression of genes encoding stress proteins (HSP), Ca-binding proteins, anti-oxidant enzymes and secreted egg proteins were found between miracidia, 6-day cultured sporocysts and 20-day cultured sporocysts (Taft *et al.* 2009). New gene discovery approaches are now available to enhance our view of the specific genes regulating intramolluscan schistosome development including high throughput pyro-sequencing technologies (RNAseq). Using the schistosome genome data bases and ever-evolving bioinformatics tools, gene discovery and transcriptomic analyses are being developed to provide easy, inexpensive ways to analyze specifically gene expression of the different larval stages in specific host snail tissues throughout its intramolluscan development. These kinds of efforts are ongoing in different laboratories currently working on *S. mansoni*/snail interactions.

A major challenge currently facing researchers mining genomic information of *Schistosoma* spp. is the lack of suitable tools to effectively evaluate and characterize the function of larval-expressed genes and their products. Demonstrating transcription and translation of an identified gene is only the first step in addressing the actual function it serves within an organism. In this regard, having methods to functionally validate a gene and its encoded protein is critically important to assess how potential candidates are involved in snail/schistosome interactions. This challenge is being met, at least in part, through the use of RNA interference involving adult and intramollucan stages (Skelly 2006; Yoshino *et al.* 2010; Da'dara and Skelly 2015), in which introduction of dsRNA of genes of interest into parasites results in a specific expression knockdown of target transcripts. Also promising has been the use of transgenic approaches for introducing and expressing genes of interest in order to test their function within larval schistosomes. Successful introduction and transient expression of transgene constructs for HSP70 and cysteine proteinase ER60 were demonstrated for *S. mansoni* miracidia and sporocysts (Wippersteg *et al.* 2002, 2003; Heyers *et al.* 2003), while Beckmann *et al.* (2007) accomplished multi-stage expression (F_0 miracidia to F_2 adults) of a GFP-actin transgene construct. Functionally active transgenes also have been successfully introduced and expressed in schistosome sporocysts using a VSV glycoprotein-pseudotyped murine retroviral vector (Kines *et al.* 2006). Viral infection and larval transduction of marker genes have been shown to be chromosomally integrated and germline transmitted between stages in the *S. mansoni* life cycle (Rinaldi *et al.* 2012). With the rapidly developing field of genome editing, functional manipulation of parasitic flatworm genomes using the CRISPR/Cas or similar system may soon become commonplace providing much deeper understanding how individual gene participates in the complexities of schistosome biology (Schwartz and Sternberg 2014). For now, however, the integration of the genomic, transcriptomic and proteomic information, together with advances in the genetic manipulation of individual genes, will continue to provide valuable insights into the molecular architecture shaping the biology, pathogenesis and host-parasite interactions of the human blood flukes (Han *et al.* 2009). Importantly, it is anticipated that functional genomics analyses will provide an invaluable resource for the research community to develop much needed new control tools for the eradication of this important and neglected tropical disease.

7.5 CONCLUSIONS

The sporocyst stages occupy a centrally important position in the life cycle of schistsome and other digenetic trematode, species. Their capacity to reproduce asexually within the snail intermediate host by creating multiple secondary sporocyst generations and in turn, large numbers of cercariae, is essential to maintaining sufficient populations of infective larvae for transmission to the mammalian definitive host, including humans. However, in order to successfully complete its development within the snail host, the sporocyst must adapt to an internal host environment that provides for all of its physico-chemical and physiological needs, while at the same time, defending itself against aggressive host responses aimed at parasite elimination. Therefore, the eventual outcome (infection vs. no infection) of each encounter between a miracidium and snail depends on a balance between interacting larval infectivity factors and host susceptibility/resistance factors that will determine the compatibility/incompatibility of the relationship. We are only beginning to understand the basic mechanisms underlying compatibility between *S. mansoni* and *B. glabrata*, although it is clearly multifactorial and probably varies among populations. Further comparative studies should investigate these different factors in different populations of *Biomphalaria* snails and their corresponding schistosome parasites, since each combination exhibits different co-evolutionary histories. This type of analysis would be facilitated by the use of discreet host phenotypes selected for their spectrum of compatibility/incompatibility towards several strains of *S. mansoni* (Théron *et al.* 2014). These population level studies will be crucial to understanding the complexities of compatibility polymorphism as it applies to schistosome-snail relationships on a global scale and should lead to the development of novel approaches to controling larval infections within the snail host.

7.6 ACKNOWLEDGMENTS

Published research cited in this review for TPY was supported by the NIH, NIAID (RO1AI015503). BG is supported by ANR JCJC INVIMORY (ANR-13-JSV7-0009). Thanks to Nathalie Dinguirard for assistance with figures and chapter editing.

7.7 LITERATURE CITED

Abu-Shakra M, Buskila D, Shoenfeld Y. 1999. Molecular mimicry between host and pathogen: examples from parasites and implication. Immunology Letters 67(2): 147–152.

Adema CM, Harris RA, van Deutekom-Mulder EC. 1992. A comparative study of hemocytes from six different snails: morphology and functional aspects. Journal of Invertebrebate Pathology 59: 24–32.

Adema CM, Hertel LA, Miller RD, Loker ES. 1997. A family of fibrinogen-related proteins that precipitates parasite-derived molecules is produced by an invertebrate after infection. Proceedings of the National Academy of Sciences USA 94: 8691–5696.

Adema CM, Loker ES. 2015. Digenean-gastropod host associations inform on aspects of specific immunity in snails. Developmental and Comparative Immunology 48: 275–283.

Agner AE, Granath WO Jr. 1995. Hemocytes of schistosome-resistant and susceptible *Biomphalaria glabrata* recognize different antigens on the surface of *Schistosoma mansoni* sporocysts. Journal of Parasitology 81: 179–186.

Basch PF. 1975. An interpretation of snail-trematode infection rates: specificity based on concordance of compatible phenotypes. International Journal for Parasitology 5: 449–452.

Basch PF. 1976. Intermediate host specificity in *Schistosoma mansoni*. Experimental Parasitology 36:150–169.

Basch PF. 1991. *Schistosomes – Development, Reproduction and Host Relations.* Oxford University Press, Inc., New York, Oxford. 248 p.

Basch PF, DiConza JJ. 1974. The miracidium-sporocyst transition in *Schistosoma mansoni*: surface changes *in vitro* with ultrastructural correlation. Journal of Parasitololgy 60: 935–941.

Baudoin M. 1975. Host castration as a parasitic strategy. Evolution 29: 335–352.

Bayne CJ. 2009. Successful parasitism of vector snail *Biomphalaria glabrata* by the human blood fluke (trematode) *Schistosoma mansoni*: a 2009 assessment. Molecular and Biochemical Parasitology 165: 8–18.

Bayne CJ, Hahn UK, Bender RC. 2001. Mechanisms of molluscan host resistance and of parasite strategies for survival. Parasitology 123(Suppl): S159–S167.

Bayne CJ, Hull CJ. 1988. The host-parasite interface in molluscan schistosomiasis: biotin as a probe for sporocyst and hemocyte surface peptides. Veterinary Parasitology 29: 131–142.

Bayne CJ, Stephens JA. 1983. *Schistosoma mansoni* and *Biomphalaria glabrata* share epitopes: antibodies to sporocysts bind host snail hemocytes. Journal of Invertebrate Pathology 42: 221–223.

Beckmann S, Wippersteg V, El-Bahay A, Hirzmann J, Oliveira G, Grevelding CG. 2007. *Schistosoma mansoni*: germ-line transformation approaches and actin-promoter analysis. Experimental Parasitology 117: 292–303.

Bender RC, Broderick EJ, Goodall CP, Bayne CJ. 2005. Respiratory burst of *Biomphalaria glabrata* hemocytes: *Schistosoma mansoni*-resistant snails produce more extracellular H_2O_2 than susceptible snails. Journal of Parasitology 91: 275–279.

Bender RC, Goodall CP, Blouin MS, Bayne CJ. 2007. Variation in expression of *Biomphalaria glabrata* SOD1: a potential controlling factor in susceptibility/resistance to *Schistosoma mansoni*. Developmental and Comparative Immunology 31: 874–8.

Bienert GP, Schjoerring JK, Jahn TP. 2006. Membrane transport of hydrogen peroxide. Biochimica Biophysica Acta 1758: 994–1003.

Berriman M, Haas BJ, LoVerde PT, Wilson RA, Dillon GP, Cerqueira GC, Mashiyama ST, Al-Lazikani B, Andrade LF, Ashton PD, *et al.* 2009. The genome of the blood fluke *Schistosoma mansoni*. Nature 460: 352–358.

Blair L, Webster JP. 2007. Dose-dependent schistosome-induced mortality and morbidity risk elevates host reproductive effort. Journal of Evolutionary Biology 20: 54–61.

Blandin S, Levashina EA. 2004. Thioester-containing proteins and insect immunity. Molecular Immunology 40: 903–908.

Blandin SA, Marois E, Levashina EA. 2008. Antimalarial responses in *Anopheles gambiae*: from a complement-like protein to a complement-like pathway. Cell Host Microbe 3: 364–374.

Boswell CA, Yoshino TP, Dunn TS. 1987. Analysis of tegumental surface proteins of *Schistosoma mansoni* primary sporocysts. Journal of Parasitology 73: 778–786.

Boyle JP, Wu XJ, Shoemaker CB, Yoshino TP. 2003. Using RNA interference to manipulate endogenous gene expression in *Schistosoma mansoni* sporocysts. Molecular and Biochemical Parasitology 128: 205–215.

Boyle JP, Zaide JV, Yoshino TP. 2000. *Schistosoma mansoni*: effects of serotonin and serotonin receptor antagonisits on motility and length of primary sporocysts *in vitro*. Experimental Parasitology 94: 217–226.

Capron A, Deblock S, Biguet J, Clay A, Adenis L, Vernes A. 1965. Contribution a l'etude experimentale de la bilharziose a *Schistosoma haematobium*. Bulletin of the World Health Organization 32: 755–778.

Caulfield JP, Yuan HC, Cianci CML, Hein A. 1988. *Schistosoma mansoni*: development of the cercarial glycocalyx. Experimental Parasitology 65: 10–19.

Chacon N, Losada S, Noya B, Alarcon de Noya B, Noya O. 2002. Antigenic community between *Schistosoma mansoni* and *Biomphalaria glabrata*: on the search of candidate antigens for vaccines. Memórias do Instituto Oswaldo Cruz 97: 99–104.

Cheng TC, Bier JW. 1972. Studies on molluscan schistosomiasis: an analysis of the development of the cercaria of *Schistosoma mansoni*. Parasitology 64: 129–141.

Chernin E. 1966. Transplantation of larval *Schistosoma mansoni* from infected to uninfected snails. Journal of Parasitology 52: 473–482.

Coil WH. 1977. The penetration of *Fascioloides magna* miracidia into the snail host *Fossaria bulimoides*. Zeitschrift für Parasitenkunde 52: 53–59.

Collins JJ, Newmark PA. 2013. It's no fluke: the planarian as a model for understanding schistosomes. PLoS Pathogens 9(7): e1003396.

Connors VA, Lodes MJ, Yoshino TP. 1991. Identification of *Schistosoma mansoni* sporocyst excretory-secretory anti-oxidant molecule and its effect on superoxide production by *Biomphalaria glabrata* hemocytes. Journal of Invertebrate. Pathology 58: 387–395.

Coustau C, Ataev G, Jourdane J, Yoshino TP. 1997. *Schistosoma japonicum: in vitro* cultivation of miracidium to daughter sporocyst using a *Biomphalaria glabrata* embryonic cell line. Experimental Parasitology 87: 77–87.

Coustau C, Gourbal B, Duval D, Yoshino TP, Adema CM, Mitta G. 2015. Advances in gastropod immunity from the study of the interaction between the snail *Biomphalaria glabrata* and its parasites: a review of research progress over the last decade. Fish and Shellfish Immunology 46: 5–16.

Coustau C, Yoshino TP. 2000. Flukes without snails: advances in the *in vitro* cultivation of intramolluscan stages of trematodes. Experimental Parasitology 94: 62–66.

Crews AE, Yoshino TP. 1989. *Schistosoma mansoni*: effect of infection on reproduction and gonadal growth in *Biomphalaria glabrata*. Experimental Parasitology 68: 326–334.

Crews AE, Yoshino TP. 1990. Influence of larval schistosomes on polysaccharide synthesis in albumen glands of *Biomphalaria glabrata*. Parasitology 101: 351–359.

Crews-Oyen AE, Yoshino TP. 1995. *Schistosoma mansoni*: characterization of excretory-secretory polypeptides synthesized *in vitro* by daughter sporocysts. Experimental Parasitology 80: 27–35.

Damian RT. 1989. Molecular mimicry: parasite evasion and host defense. Current Topics in Microbiology and Immunology 145: 101–115.

Damian RT. 1997. Parasite immune evasion and exploitation: reflections and projections. Parasitology 115(Suppl): S169–S175.

Daniel BE, Preston TM, Southgate VR. 1992. The *in vitro* transformation of the miracidium to the mother sporocyst of *Schistosoma margrebowiei*; changes in the parasite surface and implications for interactions with snail plasma factors. Parasitology 104: 41–49.

Da'dara AA, Skelly PJ. 2015. Gene suppression in schistosomes using RNAi. Methods in Molecular Biology 1201: 143–164.

DeGaffe G, Loker ES. 1998. Susceptibility of *Biomphalaria glabrata* to infection with *Echinostoma paraensei*: correlation with the effect of parasite secretory-excretory products on host hemocyte spreading. Journal of Invertebrate Pathology 71: 64–72.

De Jong-Brink M, Elsaadany M, Soto MS. 1991. The occurrence of schistosomin, an antagonist of female gonadotropic hormones, is a general phenomenon in haemolymph of schistosome-infected freshwater snails. Parasitology 103: 371–378.

De Jong-Brink M. 1995. How schistosomes profit from the stress responses they elicit in their hosts. Advances in Parasitology 35: 177–256.

De Jong-Brink M, Bergamin-Sassen M, Soto MS. 2001. Multiple strategies of schistosomes to meet their requirements in the intermediate snail host. Parasitology 123(Suppl): S129–S141.

Dheilly NM, Duval D, Mouahid G, Emans R, Allienne JF, Galinier R, Genthon C, Dubois E, Du Pasquier L, Adema CM, Grunau C, Mitta G, Gourbal B. 2015. A family of variable immunoglobulin and lectin domain containing molecules in the snail *Biomphalaria glabrata*. Developmental and Comparative Immunology 48: 234–243.

Dikkeboom R, Bayne CJ, van der Knaap WP, Tijnagel JM. 1988. Possible role of reactive forms of oxygen in *in vitro* killing of *Schistosoma mansoni* sporocysts by hemocytes of *Lymnaea stagnalis*. Parasitology Research 75: 148–154.

DiConza JJ, Basch PF. 1974. Axenic cultivation of *Schistosoma mansoni* daughter sporocysts. Journal of Parasitology 50: 757–763.

Dictus WJ, De Jong-Brink M, Boer, HH. 1987. A neuropeptide (calfluxin) is involved in the influx of calcium into mitochondria of the albumen gland of the freshwater snail *Lymnaea stagnalis*. General and Comparative Endocrinology 65: 439–450.

Dinguirard N, Yoshino TP. 2006. Potential role of a CD36-like class B scavenger receptor in the binding of modified low-density lipoproteins (acLDL) to the tegumental surface of *Schistosoma mansoni* sporocysts. Molecular and Biochemical Parasitology 146: 37–46.

Dunn TS, Yoshino TP. 1988. *Schistosoma mansoni*: origin and expression of a tegumental surface antigen on the miracidium and primary sporocyst. Experimental Parasitology 67: 167–181.

El-Shehabi F, Vermeire JJ, Yoshino TP, Riberiro P. 2009. Developmental expression analysis and immunolocalization of a biogenic amine receptor in *Schistosoma mansoni*. Experimental Parasitology 122: 17–27.

Fallon PG, Doenhoff MJ. 1994. Drug-resistant schistosomiasis: resistance to praziquantel and oxamniquine induced in *Schistosoma mansoni* in mice is drug specific. American Journal of Tropical Medicine and Hygiene 51: 83–88.

Fournier A, Théron A. 1985. Sectorisation morpho-anatomique et fonctionelle du sporocyste-fils de *Schistosoma mansoni*. Zeitschrift für Parasitenkunde 71: 325–336.

Galinier R, Portela J, Moné Y, Allienne JF, Henri H, Delbecq S, Mitta G, Gourbal B, Duval D. 2013. Biomphalysin, a new beta pore-forming toxin involved in *Biomphalaria glabrata* immune defense against *Schistosoma mansoni*. PLOS-Pathogen 9(3): e1003216.

Geraerts WP, Joosse J. 1975. Control of vitellogenesis and of growth of female accessory sex organs by the dorsal body hormone (DBH) in the hermaphroditic freshwater snail *Lymnaea stagnalis*. General and Comparative Endocrinology 27: 450–464.

Gobel E, Pan JP. 1985. Ultrastructure of the daughter sporocyst and developing cercaria of *Schistosoma japonicum* in experimentally infected snails, *Oncomelania hupensishupensis*. Zeitschrift für Parasitenkunde 71: 227–240.

Gobert GN, Moertel L, Brindley PJ, McManus DP. 2009. Developmental gene expression profiles of the human pathogen *Schistosoma japonicum*. BMC Genomics 10: 128.

Goodall CP, Bender RC, Brooks JK, Bayne CJ. 2006. *Biomphalaria glabrata* cytosolic copper/zinc superoxide dismutase (SOD1) gene: association of SOD1 alleles with resistance/susceptibility to *Schistosoma mansoni*. Molecular and Biochemical Parasitology 147: 207–210.

Granath WO Jr, Aspevig JE. 1993. Comparison of hemocyte components from *Schistosoma mansoni* (Trematoda)-susceptible and resistant *Biomphalaria glabrata* (Gastropoda) that cross-react with larval schistosome surface proteins. Comparative Biochemistry and Physiology B 104: 675–680.

Granath WO Jr, Spray FJ, Judd RC. 1987. Analysis of *Biomphalaria glabrata* (Gastropoda) hemolymph by sodium dodecyl sulfate-polyacrylamide gel electrophoresis, high-performance liquid chromatogrphy, and immunoblotting. Journal of Invertebrate Pathology 49: 198–208.

Guillou F, Roger E, Moné Y, Rognon A, Grunau C, Théron A, Mitta G, Coustau C, Gourbal BE. 2007. Excretory–secretory proteome of larval *Schistosoma mansoni* and *Echinostoma caproni*, two parasites of *Biomphalaria glabrata*. Molecular and Biochemical Parasitology 155: 45–56.

Hahn UK, Bender RC, Bayne CJ. 2001. Killing of *Schistosoma mansoni* sporocysts by hemocytes from resistant *Biomphalaria glabrata*. Journal of Parasitology 87: 292–299.

Han ZG, Brindley PJ, Wang SY, Chen Z. 2009. *Schistosoma* genomics: new perspectives on schistosome biology and host-parasite interaction. Annual Review of Genomics and Human Genetics 10: 211–40.

Hanington PC, Forys MA, Loker ES. 2012. A somatically diversified defense factor, FREP3, is a determinant of snail resistance to schistosome infection. PLOS-Neglected Tropical Diseases 6(3): e1591.

Hansen EL, Perez-Mendez G, Long S, Yarwood E. 1973. *Schistosoma mansoni*: emergence of progeny-daughter sporocysts in monxenic culture. Experimental Parasitology 21: 373–379.

Hansen EL. 1975. Secondary daughter sporocysts of *Schistosoma mansoni*: their occurrence and cultivation. Annals of the New York Academy of Sciences 266: 426–436.

Hansen EL. 1976. A cell line from embryos of *Biomphalaria glabrata* (Pulmonata): establishment and characteristics. pp. 75–97. *In*: Maramorosch K, editor. *Invertebrate Tissue Culture: Research Applications*. Academic Press, New York.

Heyers O, Walduck AK, Brindley PJ, Bleiss W, Lucius R, Dorbic T, Wittig B, Kalinna BH. 2003. *Schistosoma mansoni* miracidia transformed by particle bombardment infect *Biomphalaria glabrata* snails and develop into transgenic sporocysts. Experimental Parasitology 105: 174–178.

Hokke CH, Deelder AM, Hoffmann KF, Wuhrer M. 2007. Glycomics-driven discoveries in schistosome research. Experimental Parasitology 117: 275–283.

Hordijk PL, Ebberink RHM, De Jong-Brink M, Joosse J. 1991a. Isolation of schistosomin, a neuropeptide which antagonizes gonadotropic hormones in the freshwater snail. European Journal of Biochemistry 195: 131–136.

Hordijk PL, Ebberink RHM, De Jong-Brink M, Joosse J. 1991b. Purification and characterization of the neuropeptide schistosomin which inhibits the bioactivity of gonadotropic hormones in trematode-infected *Lymnaea stagnalis*. pp. 340–345. *In*: Kits KS, Boer HH, Joosse J, editors. *Molluscan Neurobiology*. North-Holland Publishing, Amsterdam.

Humphries J. 2011. Effects of larval schistosomes on *Biomphalaria* snails. pp. 103–125. *In*: Toledo R, Fried B, editors. *Biomphalaria Snails and Larval Trematodes*. Springer, New York.

Ittiprasert W, Knight M. 2012. Reversing the resistance phenotype of the *Biomphalaria glabrata* snail host *Schistosoma mansoni* infection by temperature modulation. PLoS-Pathogen 8(4): e1002677.

Ittiprasert W, Miller A, Myers J, Nene V, El-Sayed NM, Knight M. 2010. Identification of immediate response genes dominantly expressed in juvenile resistant and susceptible *Biomphalaria glabrata* snails upon exposure to *Schistosoma mansoni*. Molecular and Biochemical Parasitology 169: 27–39.

Ivanchenko MG, Lerner JP, McCormick RS, Toumadje A, Allen B, Fischer K, Hedstrom O, Helmrich A, Barnes DW, Bayne CJ. 1999. Continuous *in vitro* propagation and differentiation of cultures of the intramolluscan stages of the human parasite *Schistosoma mansoni*. Proceedings of the National Academy of Sciences USA 96: 4965–4970.

Jiang J, Skelly PJ, Shoemaker CB, Caulfield JP. 1996. *Schistosoma mansoni*: the glucose transport protein SGTP4 is present in tegumental multilamellar bodies, discoid bodies, and the surface lipid bilayers. Experimental Parasitology 82: 201–210.

Johnston LA, Yoshino TP. 1996. Analysis of lectin- and snail plasma-binding glycopeptides associated with the tegumental surface of the primary sporocysts of *Schistosoma mansoni*. Parasitology 112: 469–479.

Jolly ER, Chin CS, Miller S, Bahgat MM, Lim KC, DeRisi J, McKerrow JH. 2007. Gene expression patterns during adaptation of a helminth parasite to different environmental niches. Genome Biology 8(4): R65.

Joosse J, van Elk R. 1986. *Trichobilharzia ocellata*: physiological characterization of giant growth glycogen depletion and absence of reproductive activity in the intermediate snail host, *Lymnaea stagnalis*. Experimental Parasitology 62: 1–13.

Jourdane J. 1982. Etude des mecanismes de rejet dans les couples mollusque-schistosme incompatibles a partir d' infestations par voie naturelle et par transplantations microchirurgicales de stades parasitaires. Acta Tropica 39: 325–335.

Jourdane J, Théron A, Combes C. 1980. Demonstration of several sporocysts generations as a normal pattern of reproduction of *Schistosoma mansoni*. Acta Tropica 37: 177–182.

Jourdane J, Xia MY. 1987. The primary sporocyst stage in the life cycle of *Schistosoma japonicum* (Trematoda, Digenea). Transaction of the American Microscopical Society 106: 364–372.

Jourdane J, Théron A. 1987. Larval development: eggs to cercariae. pp. 83–113. *In*: Erasmus DA, editor. *The Biology of Schistosomes*. Academic Press Ltd., London, New York.

Kawamoto F, Shozawa A, Kumada N, Kojima. 1989. Possible roles of cAMP and Ca2+ in the regulation of miracidial transformation in *Schistosoma mansoni*. Parasitology Research 75: 368–374.

Kechemir N, Théron A. 1980. Existence of replicating sporocysts in the development cycle of *Schistosoma haematobium*. Journal of Parasitology 66: 1068–1070.

Kines KJ, Mann VH, Morales ME, Shelby BD, Kalinna BH, Gobert GN, Chirgwin SR, Brindley PJ. 2006. Transduction of *Schistosoma mansoni* by vesicular stomatitis virus glycoprotein-pseudotyped moloney murine leukemia retrovirus. Experimental Parasitology 112: 209–220.

Kinoti GK, Bird RG, Barker M. 1971. Electron microscope and histochemical observations on the daughter sporocyst of *Schistosoma mattheei* and *Schistosoma bovis*. Journal of Helminthology 45: 237–244.

Knight M, Arican-Goktas HD, Ittiprasert W, Odoemelam EC, Miller AN, Bridger JM. 2014. Schistosomes and snails: a molecular encounter. Frontiers in Genetics 5: 230.

Koie M, Christensen N, Nansen P. 1976. Stereoscan studies of eggs, free-swimming and penetrating miracidia and early sporocysts of *Fasciola hepatica*. Zeitschrift für Parasitenkunde 51: 79–90.

Krautz-Peterson G, Radwanska M, Ndegwa D, Shoemaker CB, Skelly PJ. 2007. Optimizing gene suppression in schistosomes using RNA interference. Molecular and Biochemical Parasitology 153: 194–202.

Krupa PL, Lewis LM, Del Vecchio P. 1977. *Schistosoma haematobium* in *Bulinus guernei*: electron microscopy of hemocyte-sporocyst interactions. Journal of Invertebrate Pathology 30: 35–45.

Lafferty KD, Kuris AM. 2009. Parasitic castration: the evolution and ecology of body snatchers. Trends in Parasitology 25: 564–572.

Laursen JR, Yoshino TP. 1999. *Biomphalaria glabrata* embryonic (Bge) cell line supports *in vitro* miracidial transformation and early larval development of the deer liver fluke, *Fascioloides magna*. Parasitology 118: 187–194.

Lee FO, Cheng TC. 1972. *Schistosoma mansoni:* alterations in total protein and hemoglobin in the hemolymph of infected *Biomphalaria glabrata*. Experimental Parasitology 31: 203–216.

Lehr T, Beuerlein K, Doenhoff MJ, Grevelding CG, Geyer R. 2008. Localization of carbohydrate determinants common to *Biomphalaria glabrata* as well as to sporocysts and miracidia of *Schistosoma mansoni*. Parasitology 135: 931–942.

Lehr T, Frank S, Natsuka S, Geyer H, Beuerlein K, Doenhoff MJ, Hase S, Geyer R. 2010. N-glycosylation patterns of hemolymph glycoproteins from *Biomphalaria glabrata* strains expressing different susceptibility to *Schistosoma mansoni* infection. Experimental Parasitology 126: 592–602.

Lehr T, Geyer H, Maass K, Doenhoff MJ, Geyer R. 2007. Structural characterization of N-glycans from the freshwater snail *Biomphalaria glabrata* cross-reacting with *Schistosoma mansoni* glycoconjugates. Glycobiology 17: 82–103.

Lockyer AE, Emery AM, Kane RA, Walker AJ, Mayer CD, Mitta G, Coustau C, Adema CM, Hanelt B, Rollinson D, Noble LR, Jones CS. 2012. Early differential gene expression in haemocytes from resistant and susceptible *Biomphalaria glabrata* strains in response to *Schistosoma mansoni*. PLOS One 7(12): e51102.

Lodes MJ, Yoshino TP. 1989. Characterization of excretory-secretory proteins synthesized *in vitro* by *Schistosoma mansoni* primary sporocysts. Journal of Parasitology 75: 853–862.

Loker ES. 2010. Gastropod immunobiology. Advances in Experimental Medicine and Biology 708: 17–43.

Loker ES, Bayne CJ, Buckley PM, Kruse KT. 1982. Ultrastructure of encapsulation of *Schistosoma mansoni* mother sporocysts by hemocytes of juveniles of the 10-R2 strain of *Biomphalaria glabrata*. Journal of Parasitology 68: 84–94.

Looker DL, Etges FJ. 1979. Effect of *Schistosoma mansoni* infection on fecundity and perivitelline fluid composition in *Biomphalaria glabrata*. Journal of Parasitology 65: 880–885.

Ludtmann MH, Rollinson D, Emery AM, Walker AJ. 2009. Protein kinase C signalling during miracidium to mother sporocyst development in the helminth parasite, *Schistosoma mansoni*. International Journal for Parasitology 39: 1223–1233.

Manger P, Li J, Christensen BM, Yoshino TP. 1996. Biogenic monoamines in the freshwater snail, *Biomphalaria glabrata*: influence of infection by the human blood fluke, *Schistosoma mansoni*. Comparative Biochemistry and Physiology 114: 227–234.

Martins-Souza RL, Pereira CA, Rodrigues L, Araújo ES, Coelho PM, Corrêa Jr A, Negrão-Corrêa D. 2011. Participation of N-acetyl-D-glucosamine carbohydrate moieties in the recognition of *Schistosoma mansoni* sporocysts by haemocytes of *Biomphalaria tenagophila*. Memórias do Instituto Oswaldo Cruz 106: 884–891.

Meier M, Meier-Brook C. 1981. *Schistosoma mansoni*: effect on growth, fertility and development of distal male organs in *Biomphalaria glabrata* exposed to miracidia at different ages. Zeitschrift für Parasitenkunde 66: 121–131.

Meuleman EA. 1972. Host-parasite interrelationships between the freshwater pulmonate *Biomphalaria pfeifferi* and the trematode *Schistosoma mansoni*. Netherlands Journal of Zoology 22: 355–427.

Meuleman EA, Holzmann PJ. 1975. The development of the primitive epithelium and true tegument in the cercaria of *Schistosoma mansoni*. Zeitschrift für Parasitenkunde 45: 307–318.

Meuleman EA, Lyaruu DM, Khan MA, Holzmann PJ, Sminia T. 1978. Ultrastructural changes in the body wall of *Schistosoma mansoni* during the transformation of the miracidium into the mother sporocyst in the snail host *Biomphalaria pfeifferi*. Zeitschrift für Parasitenkunde 56: 227–242.

Mitta G, Adema CM, Gourbal B, Loker ES, Theron A. 2012. Compatibility polymorphism in snail/schistosome interactions: From field to theory to molecular mechanisms. Developmental and Comparative Immunology 37: 1–8.

Mitta G, Galinier R, Tisseyre P, Allienne JF, Girerd-Chambaz Y, Guillou F, Bouchut A, Coustau C. 2005. Gene discovery and expression analysis of immune-relevant genes from *Biomphalaria glabrata* hemocytes. Developmental and Comparative Immunology 29: 393–407.

Moné Y, Gourbal B, Duval D, Du Pasquier L, Kieffer-Jaquinod S, Mitta G. 2010. A large repertoire of parasite epitopes matched by a large repertoire of host immune receptors in an invertebrate host/parasite model. PLOS Neglected Tropical Diseases 4(9). pii: e813.

Moné Y, Ribou AC, Cosseau C, Duval D, Théron A, Mitta G, Gourbal B. 2011. An example of molecular co-evolution: reactive oxygen species (ROS) and ROS scavenger levels in *Schistosoma mansoni/ Biomphalaria glabrata* interactions. International Journal for Parasitology 41: 721–30.

Mourão MM, Dinguirard N, Franco GR, Yoshino TP. 2009. Role of the endogenous antioxidant system in the protection of *Schistosoma mansoni* primary sporocysts against exogenous oxidative stress. PLOS Neglected Tropical Diseases 3(11): e550.

Myers J, Ittiprasert W, Raghavan N, Miller A, Knight M. 2008. Differences in cysteine protease activity in *Schistosoma mansoni*-resistant and -susceptible *Biomphalaria glabrata* and characterization of the hepatopancreas cathepsin B full-length cDNA. Journal of Parasitology 94: 659–668.

Newmark PA, Sánchez Alvarado A. 2002. Not your father's planarian: a classic model enters the era of functional genomics. Nature Review Genetics 3: 210–219.

Nyame AK, Yoshino TP, Cummings RD. 2002. Differential expression of LacdiNAc, fucosylated LacdiNAc, and Lewis x glycan antigens in intramolluscan stages of *Schistosoma mansoni*. Journal of Parasitology 88: 890–897.

Pan CT. 1965. Studies on the host-parasite relationship between *Schistosoma mansoni* and the snail *Australorbis glabratus*. American Journal of Tropical Medicine and Hygiene 14: 931–976.

Pan CT. 1996. *Schistosoma mansoni*: the ultrastructure of larval morphogenesis in *Biomphalaria glabrata* and of associated hot-parasite interactions. Japanese Journal of Medical Science and Biology 49: 129–149.

Peterson NA, Hokke CH, Deelder AM, Yoshino, TP. 2009. Glycotope analysis in miracidia and primary sporocysts of *Schistosoma mansoni*: differential expression during the miracidium-to-sporocyst transformation. International Journal for Parasitology 39: 1331–1344.

Peterson NA, Anderson TK, Wu XJ, Yoshino TP. 2013a. *In silico* analysis of the fucosylation-associated genome of the human blood fluke *Schistosoma mansoni*: cloning and characterization of the enzymes involved in GDP-L-fucose synthesis and Golgi import. Parasites and Vectors 6: 201.

Peterson NA, Anderson TK, Yoshino TP. 2013b. *In silico* analysis of the fucosylation-associated genome of the human blood fluke *Schistosoma mansoni*: cloning and characterization of the fucosyltransferase multigene family. PLOS One 8(5): e63299.

Ressurreição M, Rollinson D, Emery AM, Walker AJ. 2011. A role for p38 mitogen-activated protein kinase in early post-embryonic development of *Schistosoma mansoni*. Molecular and Biochemical Parasitology 180: 51–55.

Rinaldi G, Eckert SE, Tsai IJ, Suttiprapa S, Kines KJ, Tort JF, Mann VH, Turner DJ, Berriman M, Brindley PJ. 2012. Germline transgenesis and insertional mutagenesis in *Schistosoma mansoni* mediated by murine leukemia virus. PLOS Pathogen 8(7): e1002820.

Robijn ML, Wuhrer M, Kornelis D, Deelder AM, Geyer R, Hokke CH. 2005. Mapping fucosylated epitopes on glycoproteins and glycolipids of *Schistosoma mansoni* cercariae, adult worms and eggs. Parasitology 130: 67–77.

Roger E, Mitta G, Moné Y, Bouchut A, Rognon A, Grunau C, Boissier J, Théron A, Gourbal BE. 2008a. Molecular determinants of compatibility polymorphism in the *Biomphalaria glabrata/Schistosoma mansoni* model: new candidates identified by a global comparative proteomics approach. Molecular and Biochemical Parasitology 157: 205–216.

Roger E, Gourbal B, Grunau C, Pierce RJ, Galinier R, Mitta G. 2008b. Expression analysis of highly polymorphic mucin proteins (Sm PoMuc) from the parasite *Schistosoma mansoni*. Molecular and Biochemical Parasitology 157: 217–227.

Roger E, Grunau C, Pierce RJ, Hirai H, Gourbal B, Galinier R, Emans R, Cesari IM, Cosseau C, Mitta G. 2008c. Controlled chaos of polymorphic mucins in a metazoan parasite (*Schistosoma mansoni*) interacting with its invertebrate host (*Biomphalaria glabrata*). PLOS Neglected Tropical Diseases 2(11): e330.

Roubos EW, van de Ven AMH. 1987. Morphology of neurosecretory cells in basommatophoran snails: homologous with egg-laying and growth hormone-producing cells of *Lymnaea stagnalis*. General and Comparative Endocrinology 67: 7–23.

Roubos EW, Jenks BG, Xu L, Kuribara M, Scheenen WJJM, Kozicz T. 2010. About a snail, a toad and rodents: animal models for adaptation research. Frontiers in Endocrinology 1: 1–18.

Rowel C, Fred B, Betson M, Sousa-Figueiredo JC, Kabatereine NB, Stothard JR. 2015. Environmental epidemiology of intestinal schistosomiasis in Uganda: population dynamics of *Biomphalaria* (Gastropoda: Planorbidae) in Lake Albert and Lake Victoria with observations on natural infections with digenetic trematodes. BioMed Research International, Art ID 717261.

Samuelson JC, Caulfield JP. 1985. Role of pleated septate junctions in the epithelium of miracidia of *Schistosoma mansoni* during transformation to sporocysts *in vitro*. Tissue and Cell 17: 667–682.

Sapp KK, Loker ES. 2000. Mechanisms underlying digenean-snail specificity: role of miracidial attachment and host plasma factors. Journal of Parasitology 86: 1012–1019.

Sasaki Y, Furuta E, Kirinoki M, Seo N, Matsuda H. 2003. Comparative studies on the internal defense system of schistosome-resistant and -susceptible amphibious snail *Oncomelania nosophora* 1. Comparative morphological and functional studies on hemocytes from both snails. Zoological Science 20: 1215–1222.

Sasaki Y, Kirinoki M, Chigusa Y. 2005. Comparative studies of the defense mechanism against *Schistosoma japonicum* of schistosome-susceptible and -resistant *Oncomelania nosophora*. Parasitology International 54: 157–165.

Schallig HD, Sassen MJ, Hordijk PL, De Jong-Brink M. 1991. *Trichobilharzia ocellata*: influence of infection on the fecundity of its intermediate snail host *Lymnaea stagnalis* and cercarial induction of the release of schistosomin, a snail neuropeptide antagonizing female gonadotropic hormones. Parasitology 102: 85–91.

Schutte CHJ. 1974. Studies on the South African strain of *Schistosoma mansoni*-Part 2: the intra-molluscan larval stages. South African Journal of Science 79: 327–346.

Schwartz HT, Sternberg PW. 2014. Transgene-free genome editing by germline injection of CRISPR/Cas RNA. Methods in Enzymology 546: 441–457.

Skelly PJ. 2006. Gene silencing in flatworms using RNA interference. pp. 423–434. *In*: Maule AG, Marks NJ, editors. *Parasitic Flatworms: Molecular Biology, Immunology and Physiology*. CABI, Oxford shire, UK.

Smit CH, van Diepen A, Nguyen DL, Wuhrer M, Hoffmann KF, Deelder AM, Hokke CH. 2015. Glycomic analysis of life stages of the human parasite *Schistosoma mansoni* reveals developmental expression profiles of functional and antigenic glycan motifs. Molecular and Cellular Proteomics 14: 1750–1769.

Smith JH, Chernin E. 1974. Ultrastructure of young mother and daughter sporocysts of *Schistosoma mansoni.* Journal of Parasitology 60: 85–89.

Southgate VR. 1970. Observations on the epidermis of the miracidium and on the formation of the tegument of the sporocyst of *Fasciola hepatica.* Parasitology 61: 177–190.

Sullivan JT, Belloir JA, Beltran RV, Grivakis A, Ransone KA. 2014. Fucoidan stimulates cell division in the amebocyte-producing organ of the schistosome-transmitting snail *Biomphalaria glabrata.* Journal of Invertebrate Pathology 123: 13–16.

Sullivan JT, Richards CS. 1981. *Schistosoma mansoni,* NIH-SM-PR-2 strain, in susceptible and nonsusceptible stocks of *Biomphalaria glabrata:* comparative histology. Journal of Parasitology 67: 702–708.

Taft AS, Vermeire JJ, Bernier J, Birkeland SR, Cipriano MJ, Papa AR, McArthur AG, Yoshino T. 2009. Transcriptome analysis of *Schistosoma mansoni* larval development using serial analysis of gene expression (SAGE). Parasitology 136: 469–485.

Taft AS, Yoshino TP. 2011. Cloning and functional characaterization of two calmodulin genes during larval development in the parasitic flatworm *Schistosoma mansoni.* Journal of Parasitology 97: 72–81.

Telens AG, van den Heuvel JM, Man den Bergh SG. 1990. Substrate cycling between glucose 6-phosphate and glycogen occurs in *Schistosoma mansoni.* Molecular and Biochemical Parasitology 39: 109–116.

Tennessen JA, Théron A, Marine M, Yeh JY, Rognon A, Blouin MS. 2015. Hyper diverse gene cluster in snail host conveys resistance to human schistosome parasites. PLOS Genetics 11(3): e1005067.

Théron A. 1986. Chronobiology of schistosome development in the snail host. Parasitology Today 2: 192–194.

Théron A. 2015. Chronobiology of trematode cercarial emergence: from data recovery to epidemiological, ecological and evolutionary implications. Advances in Parasitology 88: 123–164.

Théron A, Combes C. 1983. Genetic analysis of the shedding pattern of *Schistosoma mansoni* cercariae by hybridization of species with early and late emission peaks. Comp Rendu del'Academiedes Sciences Paris 297(Serie III): 571–574.

Théron A, Coustau C. 2005. Are *Biomphalaria* snails resistant to *Schistosoma mansoni?* Journal of Helminthology 79: 187–191.

Théron A, Coustau C, Rognon A, Gourbière S, Blouin MS. 2008. Effects of laboratory culture on compatibility between snails and schistosomes. Parasitology 135: 1179–1188.

Théron A, Pages JR, Rognon A. 1997. *Schistosoma mansoni:* distribution patterns of miracidia among *Biomphalaria glabrata* snail as related to host susceptibility and sporocyst regulatory processes. Experimental Parasitology 85: 1–9.

Théron A, Rognon A, Gourbal B, Mitta G. 2014. Multi-parasite host susceptibility and multi-host parasite infectivity: a new approach of the *Biomphalaria glabrata/Schistosoma mansoni* compatibility polymorphism. Infection, Genetic and Evolution 26: 80–88.

Théron A, Touassem R. 1989. *Schistosoma rodhaini:* intramolluscan larval development, migration and replication processes of daughter sporocysts. Acta Tropica 46: 39–45.

Théron A, Gérard C, Moné H. 1992. Early enhanced growth of the digestive gland of *Biomphalaria glabrata* infected with *Schistosoma mansoni:* side effect or parasite manipulation? Parasitology Research 78(5): 445–450.

Théron A, Fournier A. 1982. Mise en evidence de structures nerveuses dans le sporocyste-fils du Trématode *Schistosoma mansoni.* Compte Rendu de l'Académie des Sciences Paris 294 (Serie III): 365–369.

Thompson SN, Lee RW. 1985. ^{31}P NMR studies on adenylates and other phosphorus metabolites in the schistosome vector *Biomphalaria glabrata.* Journal of Parasitology 71: 652–661.

Touassem R, Théron A. 1986. Study on the intramolluscal development of *Schistosoma bovis:* demonstration of three patterns of sporocystogenesis by daughter sporocysts. Parasitology 92: 337–341.

Uchikawa R, Loker ES. 1991. Lectin-binding properties of the surfaces of *in vitro*-transformed *Schistosoma mansoni* and *Echinostoma paraensei* sporocysts. Journal of Parasitology 77: 742–748.

van der Knaap WP, Loker ES. 1990. Immune mechanisms in trematode-snail interactions. Parasitology Today 6: 175–182.

van Minnen J, Schallig HD, Ramkema MD. 1992. Identification of putative egg-laying hormone containing neuronal systems in gastropod molluscs. General and Comparative Endocrinology 86: 96–102.

van Remoortere A, Hokke CH, van Dam GJ, van Die I, Deelder AM, van den Eijnden DH. 2000. Various stages of *Schistosoma* express Lewis(x), LacdiNAc, GalNAc beta1-4 (Fuc alpha1-3) GlcNAc and GalNAc beta1-4 (Fuc alpha1-2Fuc alpha1-3) GlcNAc carbohydrate epitopes: detection with monoclonal antibodies that are characterized by enzymatically synthesized neoglycoproteins. Glycobiology 10: 601–609.

Vasquez RE, Sullivan JT. 2001. Further characterization of passively transferred resistance to *Schistosoma mansoni* in the snail intermediate host *Biomphalaria glabrata*. Journal of Parasitology 87: 1360–1365.

Verjovski-Almeida S, DeMarco R, Martins EA, Guimarães PE, Ojopi EP, Paquola AC, Piazza JP, Nishiyama MY Jr, Kitajima JP, Adamson RE, *et al.* 2003. Transcriptome analysis of the acoelomate human parasite *Schistosoma mansoni*. Nature Genetics 35: 148–157.

Vermeire JJ, Boyle JP, Yoshino TP. 2004. Differential gene expression and the effects of *Biomphalaria glabrata* embryonic (Bge) cell factors during larval *Schistosoma mansoni* development. Molecular and Biochemical Parasitology 135: 153–157.

Vermeire JJ, Taft AS, Hoffmann KF, Fitzpatrick JM, Yoshino TP. 2006. *Schistosoma mansoni*: DNA microarray gene expression profiling during the miracidium-to-mother sporocyst transformation. Molecular and Biochemical Parasitology 147: 39–47.

Vermeire JJ, Yoshino TP. 2007. Antioxidant gene expression and function in *in vitro*-developing *Schistosoma mansoni* mother sporocysts: possible role in self-protection. Parasitology 134: 1369–1378.

Voge M, Seidel JS. 1972. Transformation *in vitro* of miracidia of *Schistosoma mansoni* and *S. japonicum* into young sporocysts. Journal of Parasitology 58: 699–704.

Wang B, Collins JJ III, Newmark PA. 2013. Functional genomic characterization of neoblast-like stem cells in larval *Schistosoma mansoni*. eLife 2: e00768.

Wang H, Zhao QP, Nie P, Jiang MS, Song J. 2013. Identification of differentially expressed genes in *Oncomelania hupensis* chronically infected with *Schistosoma japonicum*. Expimental Parasitology 130: 374–383.

Walker AJ. 2006. Do trematode parasites disrupt defence-cell signaling in the snail hosts? Trends in Parasitology 22: 154–159.

Whitfield PJ, Evans NA. 1983. Parthenogenesis and asexual multiplication among parasitic platyhelminths. Parasitology 86: 121–160.

Williams DL, Sayed AA, Bernier J, Birkeland SR, Cipriano MJ, Papa AR, McArthur AG, Taft A, Vermeire JJ, Yoshino TP. 2007. Profiling *Schistosoma mansoni* development using serial analysis of gene expression (SAGE). Experimental Parasitology 117: 246–258.

Wilson RA, Denison J. 1980. The parasitic castration and gigantism of *Lymnaea truncatula* infected with the larval stages of *Fasciola hepatica*. Zeitschrift für Parasitenkunde 61: 109–119.

Wippersteg V, Kapp K, Kunz W, Jackstadt WP, Zahner H, Grevelding CG. 2002. HSP70-controlled GFP expression in transiently transformed schistosomes. Molecular and Biochemical Parasitology 120: 141–150.

Wippersteg V, Ribeiro F, Liedtke S, Kusel JR, Grevelding CG. 2003. The uptake of texas red-BSA in the excretory system of schistosomes and its colocalisation with ER60 promoter-induced GFP in transiently transformed adult males. International Journal for Parasitology 33: 1139–1143.

Wright CA. 1971. *Flukes and Snails*. Allen and Unwin, Ltd., London, 168 p.

Wu X-J, Sabat G, Brown JF, Zhang M, Taft A, Peterson NA, Harms A, Yoshino TP. 2009. Proteomic analysis of *Schistosoma mansoni* proteins released during *in vitro* miracidium-to-sporocyst transformation. Molecular and Biochemical Parasitology 164: 32–44.

Yoshino TP, Bayne CJ. 1983. Mimicry of snail host antigens by miracidia and primary sporocysts of *Schistosoma mansoni*. Parasite Immunology 5: 317–28.

Yoshino TP, Bickham U, Bayne CJ. 2013a. Molluscan cells in culture: primary cell cultures and cell lines. Canadian Journal of Zoology 91(1): 391–404.

Yoshino TP, Boswell CA. 1986. Antigen sharing between larval trematodes and their snail hosts: how real a phenomenon in immune evasion? pp. 221–238. *In*: Lackie AM, editor. *Immune Mechanisms in Invertebrate Vectors*. Clarendon Press, Oxford.

Yoshino TP, Boyle JP, Humphries JE. 2001. Receptor–ligand interactions and cellular signalling at the host–parasite interface. Parasitology 123(Suppl): S143–S157.

Yoshino TP, Brown M, Wu XJ, Jackson CJ, Ocadiz-Ruiz R, Chalmers IW, Kolb M, Hokke CH, Hoffmann KF. 2014. Excreted/secreted *Schistosoma mansoni* venom allergen-like 9 (SmVAL9) modulates host extracellular matrix remodelling gene expression. International Journal for Parasitology 44: 551–563.

Yoshino TP, Cheng TC, Renwrantz LR. 1977. Lectin and human blood group determinants of *Schistosoma mansoni*: alteration following *in vitro* transformation of miracidium to mother sporocyst. Journal of Parasitology 63: 818–824.

Yoshino TP, Coustau C. 2011. Immunobiology of *Biomphalaria*-trematode interactions. pp. 159–189. *In*: Toledo R, Fried B, editors. *Biomphalaria Snails and Larval Trematodes*. Springer, New York.

Yoshino TP, Dinguirard N, Kunert J, Hokke CH. 2008. Molecular and functional characterization of a tandem-repeat galectin from the freshwater snail *Biomphalaria glabrata*, intermediate host of the human blood fluke *Schistosoma mansoni*. Gene 411: 46–58.

Yoshino TP, Laursen JR. 1995. Production of *Schistosoma mansoni* daughter sporocysts from mother sporocysts maintained in synxenic culture with *Biomphalaria glabrata* embryonic (Bge) cells. Journal of Parasitology 81: 714–722.

Yoshino TP, Lodes MJ, Rege AA, Chappell CL. 1993. Proteinase activity in transformation excretory-secretory products and primary sporocysts of *Schistosoma mansoni*. Journal of Parasitology 79: 23–31.

Yoshino TP, Wu XJ, Liu HD, Gonzalez LA, Deelder AM, Hokke CH. 2012. Glycotope sharing between snail hemolymph and larval schistosomes: larval transformation products alter shared glycan patterns of plasma proteins. PLOS Neglected Tropical Diseases 6(3): e1569.

Yoshino TP, Wu XJ, Gonzalez LA, Hokke CH. 2013b. Circulating *Biomphalaria glabrata* hemocyte subpopulations possess shared schistosome glycans and receptors capable of binding larval glycoconjugates. Experimental Parasitology 133: 28–36.

Young ND, Jex AR, Li B, Liu S, Yang L, Xiong Z, Li Y, Cantacessi C, Hall RS, Xu X, *et al.* 2012. Whole-genome sequencing of *Schistosoma haematobium*. Nature Genetics 44: 221–225.

Zahoor Z, Davies AJ, Kirk RS, Rollinson D, Walker AJ. 2010. Larval excretory-secretory products from the parasite *Schistosoma mansoni* modulate HSP70 protein expression in defense cells of its snail host, *Biomphalaria glabrata*. Cell Stress and Chaperones 15: 639–650.

Zelck U, Becker W. 1990. Lectin binding to cells of *Schistosoma mansoni* sporocysts and surrounding *Biomphalaria glabrata* tissue. Journal of Invertebrate Pathology 55: 93–99.

Zelck UE, Von Janowsky B. 2004. Antioxidant enzymes in intramolluscan *Schistosoma mansoni* and ROS-induced changes in expression. Parasitology 128: 493–501.

Zelck UE, Gege BE, Schmid S. 2007. Specific inhibitors of mitogen-activated protein kinase and PI3-K pathways impair immune responses by hemocytes of trematode intermediate host snails. Developmental and Comparative Immunology 31: 321–331.

Zeraik AE, Rinaldi G, Mann VH, Popratiloff A, Araujo AP, Demarco R, Brindley PJ. 2013. Septins of Platyhelminthes: identification, phylogeny, expression and localization among developmental stages of *Schistosoma mansoni*. PLOS Neglected Tropical Diseases 7: e2602.

Zhang SM, Adema CM, Kepler TB, Loker ES. 2004. Diversification of Ig superfamily genes in an invertebrate. Science 305: 251–254.

Zhang SM, Nian H, Wang B, Loker ES, Adema CM. 2009. Schistosomin from the snail *Biomphalaria glabrata*: expression studies suggest no involvement in trematode-mediated castration. Molecular and Biochemical Parasitology 165: 79–86.

Zhang SM, Zeng Y, Loker ES. 2008. Expression profiling and binding properties of fibrinogen-related proteins (FREPs), plasma proteins from the schistosome snail host *Biomphalaria glabrata*. Innate Immunity 14: 175–189.

Zhou Y, Zheng H, Chen Y, Zhang L, Wang K, Guo J, Huang Z, Zhang B, Huang W, Jin K, *et al.* 2009. The *Schistosoma japonicum* genome reveals features of host-parasite interplay. Nature 460: 345–351.

CHAPTER

Cercaria of *Schistosoma*

Martin Kašný[1,2,], Wilfried Haas[3], Barrie GM Jamieson[4] and Petr Horák[1]*

8.1 INTRODUCTION

Cercaria of the genus *Schistosoma* represents a free-swimming developmental stage that to complete its life cycle must contact the mammalian skin. Thereafter, it penetrates the skin and transforms into an intravertebrate stage – the schistosomulum. Therefore, the cercaria represents an infective stage, a link between the intermediate (snail) and the definitive (mammal) hosts. The cercaria lives for only a brief time in fresh water; it apparently does not feed and if it fails to rapidly penetrate the host and continue development its stored energy resources become exhausted and it dies (Coles 1972; Dorsey *et al.* 2002; McKerrow *et al.* 2006).

The following account of the morphology and ultrastructure of the cercaria of *Schistosoma* is based on a high number of fundamental historical works, such as those written by, e.g., Morris (1971), Yamaguti (1971), Stirewalt (1974), Samuelson and Caulfield (1985). Among recent contributions in the field it is necessary to mention comprehensive works of Dorsey *et al.* (2002) and Collins *et al.* (2011). In addition to outlining the morphology and anatomy, this chapter investigates dispersal of cercariae and selection of the definitive host, temporal correlation with the host activities, shedding (emergence) from the snail intermediate host, longevity and infectivity of cercariae, behavior after contact with the host and the penetration process as the key point in host invasion. Many of the processes mentioned above are dependent on the action of molecular factors produced by cercariae; therefore, the data on important proteins, saccharides and other substances and their expression in the cercarial body will also be presented.

Origin of the cercaria from the sporocyst is briefly discussed by Yoshino *et al.* (Chapter 7) and its transition into the schistosomulum by Gobert and Nawaratna (Chapter 9).

8.2 MORPHOLOGY OF MATURE CERCARIA

8.2.1 General Morphology

Cercariae of various trematodes can differ in their gross morphology (body size and shape, presence/absence of particular body parts and organs or their formation); as a consequence they can

[1] Department of Parasitology, Faculty of Science, Charles University in Prague, Czech Republic, Viničná 7, Prague 128 00, Czech Republic.

[2] Department of Botany and Zoology, Faculty of Science, Masaryk University, Kamenice 753/5 (A31), Brno 625 00, Czech Republic.

[3] Former Section Parasitology, Institute for Zoology, University Erlangen-Nürnberg, Staudtstrasse 5, Erlangen D-91058, Germany.

[4] Department of Zoology and Entomology, School of Biological Sciences, University of Queensland, Brisbane 4072, Australia.

* Corresponding author

be classified into several morphological groups, such as, e.g., gymnocephalous, amphistome, echinostome, monostome, pleurolophocercous, xiphidio and furcocercous. Cercariae of schistosomes belong to the group of furcocercous apharyngeate cercariae. They consist of approximately 1000 cells (Dorsey *et al.* 2002).

It should be mentioned that, before recognizing cercariae as developmental stages of digeneans (flukes), these larvae were described as independent components of the water plankton. Later on, it became clear that they represent larvae of digeneans, but the corresponding adult worms were unknown. As a consequence, the term (genus name) *Cercaria* was introduced at that time; such generic names are gradually replaced by valid names of digeneans for which relation of particular larvae and adults has been disclosed.

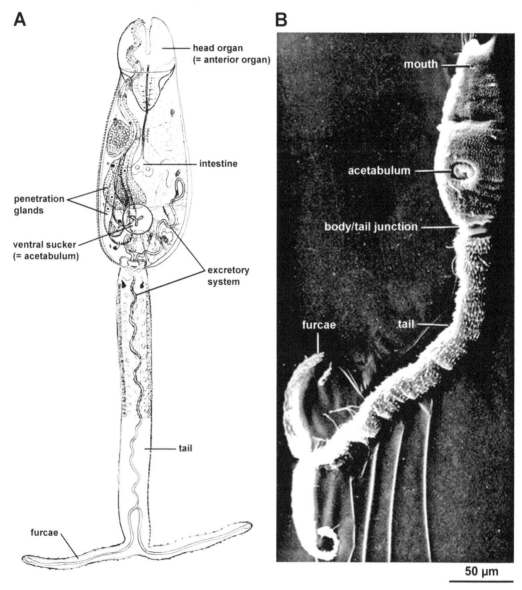

FIG. 8.1 A. Cercarial morphology in *Schistosoma haematobium*. From Capron A. 1965. Bulletin of the World Health Organization 32(6): 755–778. Figs. 23, 25, 26. **B**. A scanning electron micrograph of a *Schistosoma mansoni* cercaria. From Dorsey CH, Cousin CE, Lewis FA, Stirewalt MA. 2002. Micron 33(3): 279–323. Fig. 1.

The *Schistosoma* spp. cercariae possess a body (after penetrating mammalian skin, this part gives rise to schistosomulum) and a tail with bifurcation (see *S. haematobium*, Fig. 8.1A; *S. mansoni*, Fig. 8.1B). The tail serves for swimming. To attach to the host skin, the body contains an anterior organ (muscular conus) with outlets of penetration glands and ventrally located sucker (acetabulum). The supposedly non-functional oral opening is located subterminally. Spines cover the body segment and tail, being most prominent on the tail. The length of the cercaria of *Schistosoma* varies with the state of contraction, but in *S. mansoni* it is reported to average at least 125–154 μm (Cheng and Bier 1972) or as much as 500 μm (Dorsey *et al.* 2002) (see below).

8.2.2 Surface

The surface of schistosome cercariae provides protection against the outer hypo-osmotic environment. Therefore, it comprises not only a neodermis, but also a thick glycocalyx. The term neodermis (= tegument) reflects the general organization of all neodermatans: it is formed by tegumental syncytium with an anucleated cytoplasmic region delimited by cytoplasmic membranes (outer and inner) and continuous with nucleated cell bodies (cytons containing various vesicles) deeper in the parenchyma; the latter cells are linked with the anucleated cytoplasmic region by cytoplasmic connections (Fig. 8.2) (see adult tegument in Gobert *et al.*, Chapter 10).

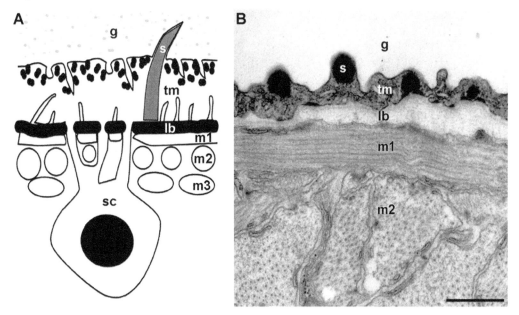

FIG. 8.2 Tegument. **A.** Scheme of tegument. tm; tegument, s; tegumental spine, lb; lamina basalis, m1; circular muscle fibers, m2; longitudinal muscle fibers, m3; diagonal muscle fibers. From Tyler S, Hooge M. 2004. Canadian Journal of Zoology 82: 194–210. Fig. 1B. **B.** *Schistosoma mansoni*. A cross section of a cercaria through the tegument of the anterior organ. tm, tegument; s, tegumental spine; lb, lamina basalis; m1, circular muscle fibers; m2, longitudinal muscle fibers. Scale bar 0.5 μm. From Dorsey CH, Cousin CE, Lewis FA, Stirewalt MA. 2002. Micron 33(3): 279–323. Scale bar 0.5 μm. Fig. 3A.

The definitive tegument is preceded by the so-called primitive epithelium which is derived from the tegument of the daughter sporocyst inside the intermediate snail host. The final tegument is formed from peripherally located somatic cells of the cercarial embryo which expand and coalesce beneath the primitive epithelium. The latter degenerates and disappears (Meuleman and Holzmann 1975). Later on, a carbohydrate-rich glycocalyx is formed. The definitive tegument is similar to that in adult schistosomes, but it is thinner, measuring 0.5 μm at the anterior body part and 0.2 μm at the tail part (Morris 1971; Hockley 1973). The surface area of the cercarial body was estimated at 20,000 μm² (Samuelson and Caulfield 1985), containing also posteriorly facing tegumental spines (Fig. 8.3A, B) and sensory receptors (Samuelson and Caulfield 1985; Dorsey *et al.* 2002). The spines can be labeled with phalloidin (Fig. 8.3A), because they are rich in filamentous actin (Mair *et al.* 2003). The tegument turns inward to line the mouth, esophagus, excretory pores and distal areas of multiciliated sensory pits. As for the anucleated region, the granular cytoplasm contains scattered small mitochondria, inclusion bodies and other components (Kemp 1970; Morris 1971; Smyth and Halton 1983); the inclusion bodies are spheroidal, elongated or discoid (Dorsey *et al.* 2002). A basal plasma membrane separates the tegument from the underlying basal lamina. There are four types of cells, perikarya of which are associated with the tegument: subtegumental cells I, II, III and IV; they occur within the parenchyma of the body, but not the tail (Fig. 8.4). Each subtegumental cell consists of a nucleus surrounded by cytoplasm containing inclusion bodies specific to the cell type (Dorsey *et al.* 2002).

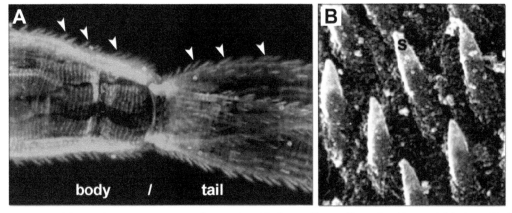

FIG. 8.3 A. Spines (white arrows). From Collins JJ, King RS, Cogswell A, Williams DL, Newmark PA. 2011. PLoS Neglected Tropical Diseases 5(3): e1009. Fig. 5A; **B**. Spines (s). From Samuelson JC and Caulfield JP. 1985. The Journal of Cell Biology 100(5): 1423–1434. Fig. 5.

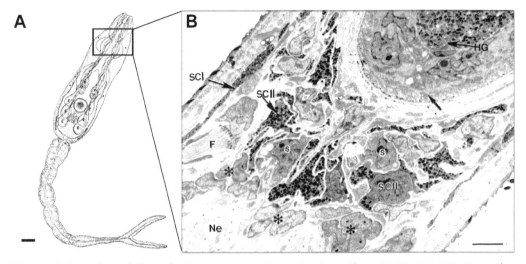

FIG. 8.4 A. General morphology of *Schistosoma mansoni* cercariae. From Cheng TC, Bier JW. 1972. Parasitology 64: 129–141. Fig. 16. **B**. Subtegumental cells, low magnification electron micrograph showing a section of a cercaria through the upper body segment and lower anterior organ regions. HG; the posterior end of the head gland with the muscular cone of the anterior organ that separates it from the body segment region (arrow), F; flame cell, SCI; subtegumental cell I, SCII; subtegumental cell II, s; body segment support cells, Ne; neuropile, *; nuclei of neurons. Scale bar 3 μm. From; Dorsey CH, Cousin CE, Lewis FA, Stirewalt MA. 2002. Micron 33(3): 279–323. Fig. 4.

The outer tegumental surface is covered by a thick glycocalyx, a sugar-rich and highly immunogenic layer (thickness of about 1–2 µm) which represents a negatively charged "coat" over the entire body (composition – 82% carbohydrates and 18% proteins); the glycocalyx of the body is less anionic than that of the tail (Nanduri *et al.* 1991). In SEM studies, *S. mansoni* cercarial glycocalyx is composed of 15–40 nm particles and a fibrous material (Samuelson and Caulfield 1985; Nanduri *et al.* 1991). The carbohydrate chains are linked to proteins of the outer tegumental membrane by O-glycosidic bonds via N-acetyl-D-galactosamine linked to serine or threonine of proteins, but some N-glycans linked to asparagine can also be detected (Hynman 1951; Nyame *et al.* 1988; Samuelson and Caulfield 1985; Basch 1991; Nanduri *et al.* 1991; Xu *et al.* 1994; Dorsey *et al.* 2002). The major monosaccharide constituents of body glycocalyx are L-fucose (51%) forming diverse multifucosylated glycans/structures, D-galactose (30%) and a low amount of D-glucosamine, D-galactosamine and D-mannose (Nanduri *et al.* 1991; Cummings and Nyame 1996; Khoo *et al.* 2001). D-glucose is the dominant carbohydrate of the tail glycocalyx (about 80%), while the amount of L-fucose is significantly reduced (less than ca 5%) (Nanduri *et al.* 1991). Among L-fucose-containing saccharides, Lewis-X antigen has been recognized (e.g., Robijn *et al.* 2005). In the later phases of the life cycle, the Lewis-X is apparently used by schistosomes as a mimicry component and probably helps them to evade the immune system of the host (Wuhrer *et al.* 2000).

In addition to spines, several types of receptors are present in surface layers; they are concentrated along the longitudinal nerve trunks (Richard 1971). There are tens of receptors (Nuttman 1971; Short and Cartrett 1973) serving for photo, mechano- and/or chemo-reception and they are involved in host-seeking and -invasion (many receptors are located at the anterior organ). Frequently, these receptors are associated with sensory cilia; the shape of sensory papilla and the number of cilia determine the morphological type of receptors (Dorsey *et al.* 2002 and references therein, Fig. 8.5). Currently, six different types of sensory papillae have been described, such as bulbous sensory nerve ending I, multiciliated pits, bulbous projecting sensory nerve ending II, uniciliated sheathed papillae, uniciliated unsheathed papillae and multiciliated unsheathed papillae (see also Gobert *et al.*, Chapter 9; Gobert *et al.*, Chapter 10, for the adult).

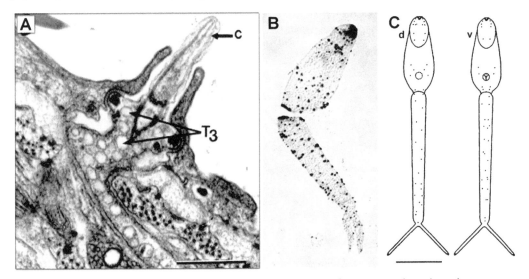

FIG. 8.5 A. Examples of receptors of *Schistosoma mansoni* cercariae. c; a cilium projects above the surface tegument. T3; type III neurosecretory vesicles. Scale bars 0.5 µm. From Dorsey CH, Cousin CE, Lewis FA, Stirewalt MA. 2002. Micron 33(3): 279–323. Fig. 21. **B.** Cercaria of *S. mansoni*. Whole organism stained with acetylthiocholine iodide, showing reaction with sensory papillae. (180 x) From Bruckner DA and Vage M. 1974. Journal of Parasitology 60(3): 437–446. Fig. 6. **C.** Cercaria of *S. mansoni*, papillae stained with silver nitrate. d, dorsal papillae; v, ventral papillae. Scale bar 100 µm. From Wagner A. 1961. Journal of Parasitology 47: 614–618. Figs. 3, 4.

These receptors are linked to nerve endings/processes that are putatively responsible for transmission of the signal (excitation). Surface distribution of receptors can be mapped by several methods, e.g., SEM, silver staining (the most traditional method), anti-tubulin antibody (for cilia-containing receptors) or phalloidin (for "discs" encircling cilia projecting through the tegument) (Mair *et al.* 2003).

8.2.3 Cercarial Digestive System

The digestive system is primordial and contains mouth, esophagus and intestine/cecum that bifurcates to form short pouch-like appendices (blind sacs). Although the cecal lumen often contains several bodies of varying shapes and electron-density (Dorsey *et al.* 2002), the cercaria is widely believed to be non-feeding stage. The mouth is an invagination of the tegument and leads to the buccal cavity. The esophagus joins the buccal cavity, passes through the muscular cone (anterior/head organ) and connects to the bifid cecum; the esophagus is associated with subesophageal cells. The bifid cecum is a blind pouch connected to the caudal end of the esophagus by a complex septate desmosome. It is uncertain whether the cecum is syncytial. Its luminal surface is elaborated into many microvilli while the outer surface is irregular and folded (Fig. 8.6) (Dorsey *et al.* 2002). Particular parts of the digestive system are surrounded by inner circular and outer longitudinal musculature (Dorsey *et al.* 2002).

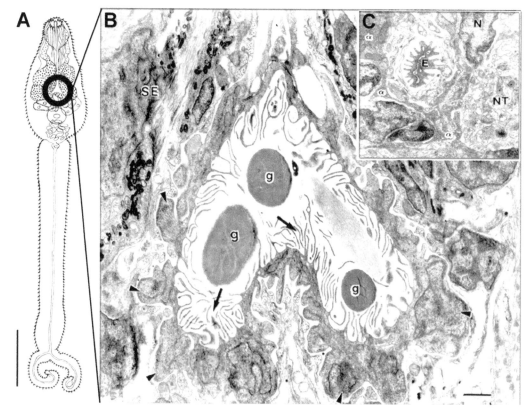

FIG. 8.6 A. *Schistosoma mansoni* cercariae (the length of the entire organism is ca 500 μm). **B.** Longitudinal section through the bifid cecum. The walls of the bifid cecum are irregular in thickness with protruding nucleated processes (arrowheads). Microvilli (arrows) project into the lumen of the bifid cecum which contains large electron-dense granules (g). Scale bar 0.5 μm. **C.** A cross section of the anterior part of body with esophagous. α; support cell with anastomosing processes, encircling the esophagus (E), NT; two nerve trunks, N; the nucleus of a neuron. Scale bars 1 μm. Dorsey CH, Cousin CE, Lewis FA, Stirewalt MA. 2002. Micron 33(3): 279–323. Figs. 2, 27B, 35A.

8.2.4   Muscle Tissue

Muscle tissue of the cercariae can be visualized e.g., by phalloidin (Collins *et al.* 2011). It is represented by two types of musculature: subtegumental muscle layers and organ-associated muscles. As for the subtegumental muscles, there are three layers of muscles separated from the overlying tegument by lamina basalis: an external layer of circular muscles, below which are situated longitudinal muscles followed by interweaved diagonal fibers in the anterior portion of the body (Mair *et al.* 2003). The arrangement of subtegumental muscles resembles that in free-living flatworms or monogeneans (Hyman 1951). The organ-associated muscles can be found elsewhere in the body and they constitute the anterior (head) organ and ventral sucker (acetabulum). The acetabulum is a cup-shaped muscular organ covered with tegument that is continuous with that of the general body surface (Cousin *et al.* 1981). It consists mainly of an extensive complex of circular and longitudinal muscles. In life it is actively protruded and retracted and has an important role in skin exploration and anchoring before the penetration (Stirewalt and Kruidenier 1961). The anterior (head) organ is a complex muscular cone at the anterior end of cercaria; again, it is covered with tegument that is continuous with that of the general body surface. The processes of penetration glands pass through the cone to open terminally where numerous sensory papillae/receptors are located. In addition, the outlets of some other (head) glands are present and the oral opening occurs subterminally.

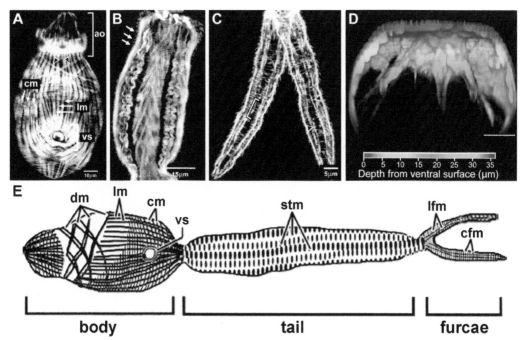

FIG. 8.7 Confocal scanning laser micrographs of *Schistosoma mansoni* cercaria stained with phalloidin–FITC showing the different muscle fiber systems in the main body, tail and furcae **A.** general view of the ventral body wall musculature, comprising largely circular (cm) and longitudinal (lm) muscle fibers. ao; anterior cone, vs; ventral sucker or acetabulum. **B.** a braided arrangement of the central striated tail musculature was evident in a single specimen. The surface shows numerous surface spines (arrows). **C.** overview of the paired terminal furcae, which contain few longitudinal fibers (arrowheads) and multiple circular or transverse fibers (bracket). **D.** ventral sucker. Scale shown below indicates the color-coding of distances from the ventral surface (i.e. the colors transition from red (ventral) to blue (dorsal) moving deeper into the animal). From Collins JJ, King RS, Cogswell A, Williams DL, Newmark PA. 2011. PLoS Neglected Tropical Diseases 5(3): e1009. Fig. 1H. **E.** A schematic representation of gross muscle fiber organization. Cfm, circular muscle fibers in the furcae; cm, circular muscle fibers; dm, diagonal muscle fibers; lfm, longitudinal muscle fibers in the furcae; lm, longitudinal muscle fibers; stm, striated tail musculature; vs, ventral sucker. From Mair GR, Maule AG, Fried B, Day TA, Halton DW. 2003. The Journal of Parasitology 89(3): 623–625. Fig. 1B; Fig. 2A, I, J.

Muscle cells (myocytes) enclosed in sarcolemma are either embedded completely in the granular matrix of the parenchyma or surrounded by the parenchyma plus the basal lamina of the tegument. Myocytes extend in long processes (myofibers) that may either contain thick and/or thin myofilaments or are afibrillar/granular with only few myofilaments; myocytes may be unipolar or multipolar (Dorsey *et al.* 2002) (Fig. 8.7). Large myofibers are characteristic of the subtegumental musculature, the muscle cone, the ventral sucker and the radial muscles. Small myofibers with sparse myofilaments accompany the mouth, buccal cavity, esophagus, acetabular glands, bifid cecum, flame-cells, nerve trunks, excretory tubules and bladder (Dorsey *et al.* 2002; Mair *et al.* 2003; Cavalcanti *et al.* 2009).

The cercarial tail is a deciduous locomotor organ consisting of two major parts, the stem and the furcae (Fig. 8.7B, C). The anterior fourth of the stem has longitudinal and circular muscles and a gradual formation of vertical striations more posteriorly. The remainder of the tail stem contains muscles with a striated appearance. The muscles of the furcae are morphologically similar to the peripheral muscles of the body, but their arrangement is different (Dorsey *et al.* 2002).

8.2.5 Nervous System

The nervous system of the cercaria consists of two central ganglia connected by a commissure. They have a diameter of 1.5 μm and are the largest component of the nervous system (Dorsey *et al.* 2002 and references therein). The central ganglia surround the esophagus anterior to the bifid cecum (Cousin *et al.* 1981). Neuronal processes form the neuropile of the central ganglia and neuronal perikarya form the periphery. A pair of nerve trunks projects from the central ganglia anteriorly and posteriorly. In addition, six nerve trunks project longitudinally at regular intervals around the periphery of the central ganglia anteriad and posteriad and supply the anterior organ and the tail musculature (details, including much cytology, in Dorsey *et al.* 2002). The nervous system of the cercarial tail has particularly been visualized by synapsin immunostaining, revealing extensive neural projections within the tail (Fig. 8.8) (Collins *et al.* 2011).

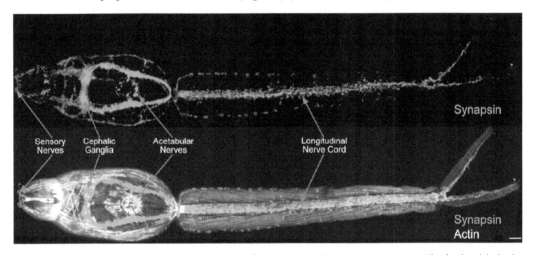

FIG. 8.8 The nervous system of cercariae. Immunofluorescence with an anti-synapsin antibody that labels the cephalic ganglia and many peripheral neural structures. Below, synapsin labeling is shown together with phalloidin staining. Scale bars 10 μm. From Collins JJ, King RS, Cogswell A, Williams DL, Newmark PA. 2011. PLoS Neglected Tropical Diseases 5(3): e1009. Fig. 4A.

8.2.6 Excretory System

In their brief aquatic phase, cercariae may regulate water balance by controlling permeability of the tegument and/or glycocalyx and by excretion of excess fluid by the osmoregulatory/protonephridial system (Wilson and Webster 1974; Faghiri and Skelly 2009; Collins *et al.* 2011). The system consists of protonephridia (flame cells), collecting tubules, bladder and excretory pores (Fig. 8.9). Each flame cell is equipped with a prominent bundle of long cilia which project into the lumen of the first tubule cell (Dorsey *et al.* 2002); the ciliary tuft of a flame-cell sits in a barrel or basket-like structure formed by interdigitation between the flame-cell and first tubule cell (Dorsey *et al.* 2002). The distribution of protonephridia in the cercarial body has been suggested as a differential marker for certain groups of digeneans in general and schistosomes in particular; however, species cannot usually be recognized by this criterion (Agrawal 2012). Contrary to the previous report by Dorsey *et al.* (2002) that cercariae have six pairs of flame cells, Collins *et al.* (2011) found that they typically have five pairs of flame cells: four pairs in the body and a pair in the anterior tail (Fig. 8.9A).

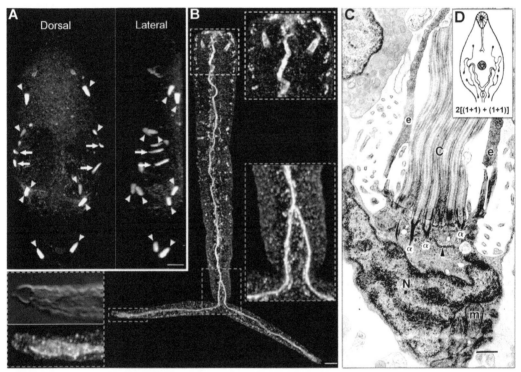

FIG. 8.9 The protonephridial system of cercariae (*Schistosoma mansoni*). **A.** Immunofluorescence with an anti-beta-tubulin antibody to label the ciliated tufts of the flame cells (arrowheads) and ciliated regions of the protonephridial tubules (white arrows) within the body and tail. Dorsal view is shown to the left and a lateral view is shown to the right. **B.** Immunofluorescence with an anti-phospho Y antibody that labels the barrel of the flame cells and the excretory duct. Inserts are magnified views of the indicated regions. Upper box, flame cells of the tail. Lower box, the protonephridial tubule splits anterior to the point of tail bifurcation. Box around the tip of the furcae, the protonephridial tubule extends to the nephridiopore at the tip of the furcae. Mid-level DIC optical section showing nephridiopore and tegumental structure at tip of tail. Scale bars 10 µm. From Collins JJ, King RS, Cogswell A, Williams DL, Newmark PA. 2011. PLoS Neglected Tropical Diseases 5(3): e1009. Fig. 3A, B. **C.** A longitudinal section through a flame cell located in the body segment. N; the cell body with its large nucleus and scant cytoplasm containing granular endoplasmic reticulum (arrowhead), α; basal bodies, m; mitochondria, C; cilia. The relationship of the flame cell entrance to the osmoregulatory tubule (e) is shown. Scale bar 0.5 µm. From Dorsey CH, Cousin CE, Lewis FA, Stirewalt MA. 2002. Micron 33(3): 279–323. Fig. 41A. **D.** Formula of schistosome cercarial protonephridia. From Smyth JD 1994. *Introduction to Animal Parasitology*. 3rd Ed. Cambridge University Press, Cambridge, 549 pp. Fig. 52.

The lumen of the collecting tubules is encircled by a series of elongated tubule cells, the opposing membranes of which are joined by long septate desmosomes. In the cercarial body, the tubules extend from the anterior pair of flame-cells at the level of the muscle cone of the anterior organ. Small tubules from the flame cells connect with secondary tubules to form a pair of main tubules with thick walls. Each main tubule divides into dorsal and ventral tubules. The two ventral branches curve inward and enter the bladder, parallel to one another. The two dorsal branches do not enter the bladder, but extend posteriorly and enter the tail on opposite sides of the bladder, ending with a pair of flame cells (Dorsey *et al.* 2002).

Two small ducts exit the bladder and become one tubule, extending the length of the tail (Dorsey *et al.* 2002 and references therein); the tubule splits anterior to the bifurcation of the tail (Fig. 8.9B; blue inset). The nephridiopores are located at the tips of the tail furcae (Fig. 8.9B).

8.2.7   Genital Primordium

The genital primordium that gives rise via the schistosomula to the sexual organs of the adults emerges as a group of undifferentiated cells localized in cercarial parenchyma. It is located in the ventral area of the body. The cells are arranged in a rod-shaped mass that extends from the posterior end of the postacetabular glands anteriorly to the ventral sucker (Fig. 8.10). Cells within this aggregation have heterochromatic nuclei and small amount of cytoplasm (Dorsey *et al.* 2002; Galaktionov and Dobrovolski 2003).

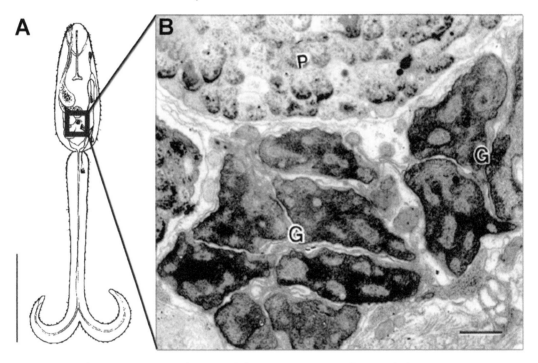

FIG. 8.10 A. *Schistosoma* cercariae (scheme). Scale bar 0.1 mm. From Frandsen F and Christensen NO. 1984. Acta Tropica 41(2): 181–202. Fig 4A. **B.** A group of germinal cells (G) located below a postacetabular gland fundus (P). This group of cells extends from the distal end of the fundus to the ventral sucker. Scale bar 0.5 μm. From Dorsey CH, Cousin CE, Lewis FA, Stirewalt MA. 2002. Micron 33(3): 279–323. Fig. 25A.

8.2.8   Penetration Glands

The penetration (acetabular) glands (Fig. 8.11A–G) are located close to the ventral sucker (acetabulum). Two pairs of penetration glands, circumacetabular (in many publications preacetabular) are situated antero-dorso-laterally to the acetabulum and three pairs, postacetabular posteriorly to the acetabulum. This arrangement seems to be conserved among schistosomes. The above mentioned five pairs of unicellular gland cells contain substances involved in skin adhesion and penetration (e.g., proteolytic enzymes, see chapter 8.4.6). They occupy a large portion (1–2 thirds) of the cercarial body. For example, in cercariae of *Trichobilharzia regenti*, a bird schistosome, the glands occupy ca one third of cercarial body volume (circumacetabular 12%, postacetabular glands ca 15% and head gland 6%) (Fig. 8.11) and their inner pH ranges between 7.00–7.50 (Ligasová *et al.* 2011).

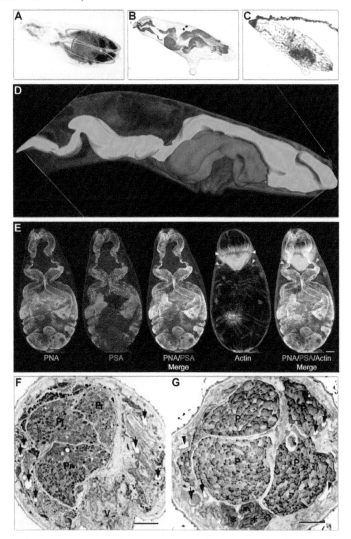

FIG. 8.11 Penetration glands. **A.** Cercariae of *S. mansoni* (ventral view) stained with oxidized apomorfine, red circle circumacetabular glands, green circles postacetabular glands. From Bruckner DA. 1974. Journal of Parasitology 60 (5): 752–756. Fig. 2. **B.** Cercariae of bird schistosome *Trichobilharzia regenti* (lateral view) stained with lithium carmine – postacetabular glands and **C.** Alizarin – circumacetabular glands. **D.** Three-dimensional model of *T. regenti* acetabular glands; Cy3-azide and Alexa Fluor® 488 stained cercariae were employed for the reconstruction; postacetabular glands are in green, circumacetabular glands in red. From Ligasová A. Bulantová J, Šebesta O, Kašný M, Koberna K., Mikeš L. 2011. Parasites and Vectors 4: 162. Fig. 1C. **E.** The secretory glands of *S. mansoni* cercariae. (A) Maximum confocal projections showing labeling of the entire circum- and post-acetabular gland system with lectins PNA and PSA. Yellow arrowheads indicate musculature surrounding the anterior ducts of the acetabular glands and white arrowheads indicate the muscle cone. Scale bar 10 μm. From Collins JJ, King RS, Cogswell A, Williams DL, Newmark PA. 2011. PLoS Neglected Tropical Diseases 5(3): e1009. Fig. 2A. **F.** A cross section through the posterior body segment showing pre-acetabular gland fundi (Pr), osmoregulatory tubules (arrows) and the extruded ventral sucker (V). **G.** A cross section through the posterior body segment showing sections of postacetabular gland fundi (P), a flame cell, two main osmoregulatory tubules (arrowheads), osmoregulatory tubules (arrows) and germinal cells (asterisks). Scale bar 3 μm. Dorsey CH, Cousin CE, Lewis FA, Stirewalt MA. 2002. Micron 33(3): 279–323. Fig. 24A, B; (Figs. B, C; photo M. Kašný).

Chemical composition of the acetabular glands was investigated by adoption of histochemical methods. In *S. mansoni* cercariae the circumacetabular glands are acidophilic and eosinophilic and can be stained by dyes with affinity for calcium such as alizarin and purpurin (Stirewalt and Kruidenier 1961). *S. mansoni* is known to possess an extremely high concentration of calcium in the circumacetabular glands (Dresden and Edlin 1974) present in the form of carbonate (Dresden and Asch 1977) or an ionic form (Dorsey and Stirewalt 1977). It probably plays a role in regulation of gland peptidase activity (Dresden and Edlin 1974; McKerrow *et al.* 1985), disaggregation of host skin proteoglycans (Landsperger *et al.* 1982), prevention of cercarial surface damage (Modha *et al.* 1998), etc. The presence of calcium was also disclosed in the circumacetabular glands of two schistosomes parasitizing birds, *T. regenti* and *T. szidati* (Mikeš *et al.* 2005). The postacetabular glands consist of six gland cells possessing long fundi and cell processes (Fig. 8.11); they are basophilic and after fixation, they can be stained with lithium carmine, aniline blue, thionine or toluidine blue (Stirewalt and Kruidenier 1961). They react with periodic acid-Schiff (PAS) which suggests the presence of reducing saccharides (possibly polysaccharides, glycoproteins or glycolipids) (e.g., Linder 1985). Apomorphine can be used to differentiate both types of glands in schistosome cercariae (Bruckner and Vage 1974; Mikeš *et al.* 2005).

Probably two unicellular head glands represent another gland type that occurs within the muscular head organ of cercariae (see also Fig. 8.4). Fundi of these cells taper into a system of multiple processes that open into the tegument at the anterior end of cercarial body (Morris 1971; Dorsey 1976; Dorsey *et al.* 2002). Although the exact role of the head gland is unknown, it is hypothesized that it provides material (phospholipids) for cercarial tegument injured during penetration (Dorsey 1976) or produces secretions that may be involved in the process of penetration (Curwen and Wilson 2003).

8.3 SOURCES OF CERCARIAL ENERGY

Swimming *S. mansoni* cercariae maintain largely aerobic energy metabolism, the Krebs cycle being the main terminal of carbohydrate breakdown (Bruce *et al.* 1971; Coles and Hill 1972; Coles 1972); glycogen is considered to be the main energy store. If the cercaria fails to quickly find and penetrate its natural definitive host, its energy stores become exhausted and it dies (Lewert and Para 1966; Bruce *et al.* 1971; Coles 1972; Lawson and Wilson 1980). Free-swimming cercariae 18 h old have largely depleted their endogenous glycogen and after the first 24 h of swimming activity, cercariae lose up to 79% of labeled glucose (Para *et al.* 1970). Experimentally, metabolic processes of cercariae can be stimulated by exogenous sources of energy, e.g., cercariae extensively metabolize exogenous glucose (Para *et al.* 1970; Bruce *et al.* 1971).

Availability of energy stores is also directly linked with cercarial infectivity – the decreasing infectivity in aging cercariae is associated with the depletion of glycogen stores. In fact, the cercariae utilize about 20–30% of the glycogen reserves of their bodies just for penetration of mouse upper epidermis layers (reviewed by Lawson and Wilson 1980). The cercariae are well adapted to these conditions; when swimming, they utilize more glycogen of the tail and save glycogen stores of the body (Lawson and Wilson 1980). Furthermore, an increase of osmolarity, as it occurs during skin penetration in combination with glucose and warmth (and not fatty acids that stimulate penetration behavior) trigger spreading of the glucose transporter protein SGTP4 on the surface of the tegument (Skelly and Shoemaker 2000). This will enable the newly formed schistosomula to utilize glucose of the host epidermis and to survive with depleted glycogen stores.

In earlier works, the presence of cercarial metabolic enzymes has been confirmed biochemically in cercarial samples (Conde-del Pino *et al.* 1968; Coles and Hill 1972; Coles 1973). In recent studies working with *schistosoma* cercarial DNA/RNA and extracts, a number of metabolic enzymes was identified at nucleic and amino acid sequential level (e.g., Jolly *et al.* 2007; Gobert *et al.* 2009; Liu *et al.* 2015).

8.4 BEHAVIOR AND TRANSMISSION OF CERCARIA

Schistosoma life cycles are fully dependent on the infection success of the cercariae. Cercariae do not feed or reproduce and therefore it can be expected that their morphological, physiological, biochemical, molecular and behavioral organization is fully adapted to maximize their infection success. This is particularly obvious considering their shedding patterns, with subsequent host-finding behavior.

Schistosoma cercariae are very small when compared with their hosts and the physical and chemical cues emanating from the hosts are effective at very short distances only. Therefore the transmission success of cercariae depends mainly on probability that they come in contact with their mobile hosts. It can be expected that the behavior of cercariae is adapted to maximize chances for such an encounter. In fact, cercariae which actively invade their hosts show behavioral patterns, which enable them to be present in the space frequently visited by the host and in the time when the host is present (Combes *et al.* 1994). The accumulation of cercariae in the space frequented by the host is based on adaptations of their swimming behavior, leading to dispersal and selection of the habitats frequented by the hosts (reviewed by Haas 1992; Haas 1994).

8.4.1 Shedding Patterns of Cercariae

Schistosoma cercariae emerge from their host snails in defined circadian rhythms and these shedding periods of different *Schistosoma* species fit to the times in which an encounter with the respective hosts is most probable (reviews: Combes *et al.* 1994; Théron 2015) (Fig. 8.12).

The spectrum ranges from shedding in the morning when the vertebrate hosts visit water reservoirs, to the middle of the day when humans have contact with the water and to the evening and night, when nocturnal rodents are active. The shedding rhythm of *S. margrebowiei* cercariae even has two distinct emergence peaks at dawn and dusk, coinciding with the times when the antelope hosts water (Raymond and Probert 1991).

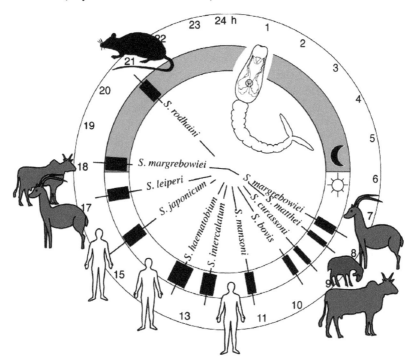

FIG. 8.12 Cercarial shedding patterns of *Schistosoma* species infecting different hosts. From Théron A. 2015. *Advances in Parasitology* 88: 123–164. Fig. 13.

Circadian shedding patterns can also vary in different populations of the same *Schistosoma* species. For example in *S. mansoni* on the island of Guadeloupe the cercariae emerge in the late morning (peak emergence at 11:00 h) in habitats where humans are the hosts and in the late afternoon (peak at 16:00 h) in foci where rats are infected. In habitats where humans and rats were the hosts, both, the early and late shedding pattern occurred, with an additional, intermediate pattern (a single peak at 13:00 h). Experiments showed, that the intermediate shedding pattern resulted after crossing parasites with early and late patterns; the heredity followed simple Mendelian rules (Théron and Combes 1988; Théron and Pointier 1995). In Oman, a *S. mansoni* strain isolated from rats showed a strictly nocturnal emergence pattern with a mean peak at 20:00 h, whereas in other areas, where humans are the hosts, diurnal shedding patterns occurred with peaks at 11:00 and 16:00 h (Mouahid *et al.* 2012). Similar diversities of circadian shedding patterns were found also in *S. japonicum* and *S. haematobium* cercariae (reviewed by Théron 2015).

The predominant synchronizer of the rhythmic emergence of *Schistosoma* cercariae is the photoperiod. In *S. mansoni* cercariae emergence is stimulated at light intensities as low as 100–200 ergs/cm^2/sec and at different visible portions of the light spectrum (Asch 1972). When the snails are maintained in constant conditions of light and dark the circadian shedding rhythm disappears (Williams *et al.* 1984) and an inversion of the photoperiod provokes an inversion of the cercarial emergence rhythms (e.g., Asch 1972; Valle *et al.* 1973; Glaudel and Etges 1973). Only when there is no photoperiod can the rhythmic shedding also be maintained by oscillation in temperature (Valle *et al.* 1973).

It was suggested that the cercarial shedding patterns may be stimulated by circadian activities of the snail hosts. However, many findings support the view that the rhythm is intrinsic to the parasite stages. In *S. mansoni* no correlation was found between the locomotor activity and heartbeat rate of the snail host and the number of cercariae emerging or between the rhythms of snail activity and cercarial shedding (Williams and Gilbertson 1983; Williams *et al.* 1984). The circadian shedding patterns of defined *Schistosoma* strains did not differ when the cercariae were produced in different strains or even different species of snail hosts (reviewed by Théron 2015), even when the snail species belonged to different genera (Mouahid and Théron 1986). Moreover, when snails were infected with two *Schistosoma* strains or species, with differing shedding patterns, each parasite shed cercariae in its distinct rhythm, independent of the emerging rhythm of the other parasite. That the rhythm of cercarial release emanates from the parasite stages is further supported by findings that the rhythms of the different *Schistosoma* species/strains are genetically controlled (reviewed by Théron 2015). This has been demonstrated by cross-breeding of *S. mansoni* strains (Théron and Combes 1988), hybridization between *S. mansoni* and *S. rodhaini* (Théron 1989), *S. bovis* and *S. intercalatum* (Pages and Théron 1990) and between *S. haematobium* and *S. intercalatum* (Pages and Théron 1990).

An unresolved question is to what extent cercariae or/and daughter sporocysts participate in the cercarial release. Sporocysts of *S. mansoni* possess a birth pore with acetylcholinesterase activity and nervous and muscular structures which could enable release of the cercariae by actively opening and closing the pore (see also Yoshino *et al.*, Chapter 7). However, sporocysts of other species such as *S. haematobium, S. bovis* and *S. intercalatum* have no birth pore and the cercariae leave the sporocyst by actively rupturing the body wall of the sporocyst. This indicates an active role of the cercariae in the emerging rhythm. The active role is particularly obvious in *S. haematobium* cercariae; they show a typical diurnal shedding pattern, but they leave the snail, in addition, by responding to dark stimuli (Nojima and Sato 1978, 1982). In this connection the *S. haematobium* cercariae seem to use their specific sensitivity for shadow stimuli to increase their abundance around a potential host.

8.4.2 Swimming Behavior: Movement Patterns

Spontaneously swimming *Schistosoma* cercariae perform an intermittent swimming mode consisting of an active phase in which the cercariae show fast, upward directed, backward (tail first) swimming and a passive phase, in which the cercariae sink down using the forked tail as a drag anchor. In the active phase, the cercariae generate the backward thrust by body and tail oscillation and using the spread furcae as oar blades (Graefe *et al.* 1967) (Fig. 8.13A); the estimated speed of this movement is 1.2 mm/s. In response to environmental and host cues the cercariae can also swim in a forward (body first) direction by snake-like movements and using the folded fork branches like a tail fin (Graefe *et al.* 1967; see also Fig. 8.13B).

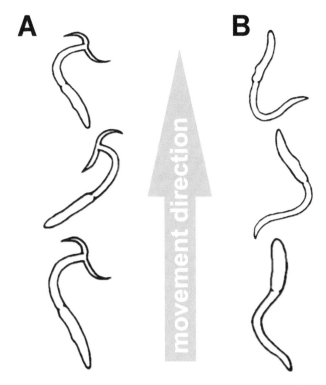

FIG. 8.13 Swimming mode of *S. mansoni* cercariae. **A.** Forward directed swimming – tail first. **B.** Forward swimming – body first. Drawing based on slow motion pictures of Haas 1976.

Although the interplay of active swimming and passive sinking determines the presence of the parasites in particular water levels, little is known on this behavior mechanism in *Schistosoma* species. It may be suggested that the duration (length) of both phases is determined by light intensity and temperature, as seen in other intermittent swimming cercariae (reviewed in Haas 1994). At least in *S. mansoni* and *S. spindale* cercariae increasing light intensity leads to shortened passive sinking and prolonged active swimming and this contributes to the distribution in upper water levels at high light intensities. Surprisingly, this effect depends only on the intensity of the light, whereas the direction of the light radiation has no effect, whether in the vertical, nor in the horizontal direction (Haas *et al.* 1990; Brachs and Haas 2008; Haas *et al.* 2008a).

In contrast to other digenean species, little is known on the interplay of the different swimming patterns in *Schistosoma* species (reviewed by Haas 1992, 1994). In *S. mansoni* cercariae the varying durations of successive active and passive phases do not correlate and this suggests that they may be controlled by separate mechanisms, as in other intermittently swimming species (Brachs and Haas 2008).

Schistosoma cercariae show, in addition to these movement patterns, different resting postures which clearly enable them to save energy, live longer and eventually reduce the risk of predation. For example, *S. mansoni* cercariae sometimes adhere motionless with the tips of their furcae at the water surface film and *S. japonicum* cercariae show long waiting adhering at the surface film with their ventral sucker.

8.4.3 Swimming Behavior: Responses to Environmental and Host Cues

Intermittently swimming cercariae which infect fish and birds (e.g., *Trichobilharzia* spp.) respond sensitively to environmental and host cues such as light stimuli, shadow, water currents and touch (reviewed by Haas 1994). However the cercariae of different *Schistosoma* species show such behavior not at all or only in a much reduced form (Haas 1976; Saladin 1982; Haas *et al.* 1987; Brachs and Haas 2008). These poor responses of *Schistosoma* species seem hardly to contribute to the host-finding success and it has been discussed whether they can be considered as relicts of formerly effective behavior patterns.

Schistosoma japonicum cercariae do not respond at all to touch, water currents, shadowing and chemical signals (Haas *et al.* 1987). *Schistosoma mansoni* cercariae show only a very weak tendency to start swimming when exposed to shadow stimuli or water currents and light stimuli tend to inhibit the start of swimming (Haas 1976; Saladin 1982; Brachs and Haas 2008). This can hardly contribute to transmission success. However, they respond to defined skin emitted chemicals with shifts between forward and backward swimming and this may support an approach towards the skin surface. Swimming cercariae of the bovid parasite *S. spindale* respond sensitively only during their active phase; they stop swimming immediately in response to shadow and touch stimuli (Haas *et al.* 1990). This behavior is absent in other human *Schistosoma* species. It might prevent the parasites from damaging themselves by colliding with obstacles. *Schistosoma haematobium* cercariae show only one sensitive response: they react to shadow stimuli with very long upward directed swimming movements and this has an influence on their distribution within the water column and might support an encounter with a potential host (reviewed by Haas 1994).

8.4.4 Distribution of Cercariae within the Aquatic Environment

The intermittent swimming behavior seems to be an adaptation of cercariae to save energy, to reduce the risk of predation and to adjust the position vertically in the water column. A disadvantage of this strategy is that the swimming movements are mainly restricted to the vertical direction and horizontal swimming occurs only sporadically.

Horizontal distribution. Human schistosome cercariae only seldom swim in a horizontal direction. Even the direction of light radiation has no effect on the swimming direction, at least in *S. mansoni* and *S. spindale* cercariae (Haas *et al.* 1990; Brachs and Haas 2008). An active distribution in the horizontal direction is only supported when the swimming path deviates occasionally by collisions with the water surface film or other obstacles. Nevertheless, *S. mansoni* cercariae have been found to spread horizontally up to 1 m within 15 min in the laboratory (Negron-Aponte and Jobin 1977) and in the field they infected caged mice at distances of up to 3.1 m in standing waters (Upatham 1974). However this is not a big distance when compared with the spreading capacity of the much smaller miracidia, which in contrast to the cercariae swim actively in all directions; *S. mansoni* miracidia infected snail hosts up to 18 m (Upatham 1973). Nevertheless, the cercariae can be transported over longer distances by wind-moved surface currents and this effect is supported by their attachment to the water surface film. Furthermore, in running water habitats they infected caged mice at distances up to 97.5 m (Upatham 1974).

Vertical distribution. The environmental factors determining the distribution of *Schistosoma* cercariae in a vertical direction are poorly studied so far and little is known on the vertical distribution under natural conditions.

S. mansoni cercariae prefer upper water levels. In cuvettes they accumulate near the water surface and in Plexiglas cylinders vertically placed in a pond, around 90% of the cercariae were found at water depths between 0 and 35 cm (Haas *et al.* 2008a). Under sunny conditions the cercariae left the uppermost water layers and this may reflect an adaptation to avoid UV radiation under tropical conditions. The concentration of the cercariae in upper water levels enabled numerous epidemiological studies, as it was sufficient to collect water samples near the water surface to detect the presence of *S. mansoni* cercariae employing various filtration techniques (Théron 1986).

In contrast to *S. japonicum* cercariae showing only sporadic vertical swimming, the *S. haematobium* cercariae perform a period of high swimming activity immediately after emergence from the snail; swimming of more than 30 mm in distance is combined with short passive phases of only a few milliseconds and in cuvettes they reach the water surface. This seems to reflect an adaptation to leave the snail's habitat and to distribute. Then they tend to sink and a periodic change in their swimming activity leads to a distribution throughout the water column in cuvettes. They respond sensitively to shadow stimuli with an intense swimming activity guiding them back to the water surface (Haas *et al.* 1994). In an endemic area the cercariae showed vertical distributions similar to those observed in cuvettes. In the morning 79% of the cercariae were collected near the water surface, at midday they tended to accumulate at about 25 cm depth and thereafter at the 50 cm deep bottom (Kimura *et al.* 1994).

8.4.5 One Step Before Host Infection

The life span of *Schistosoma* cercariae after leaving the snail host varies greatly depending on individual energy stores and physiology, parasite species/strain properties and environmental conditions.

Longevity and infectivity of cercariae. In experiments with *Schistosoma* cercariae it was recorded that temperature has a great effect on survival. In *S. mansoni* 50% of the cercariae survived for 30 h at 15°C, for 16 h at 25°C and for 8 h at 35°C and the life span was found to be related to the rate at which the glycogen reserves of the cercariae were utilized (Lawson and Wilson 1980). In addition to the depletion of glycogen reserves the loss of ions seems also to contribute to cercarial deaths. This is supported by findings that cercarial survival time is reduced in distilled water and that it increases with increasing salinity up to salt concentrations of 5% (Asch 1975; Becker 1971).

The life span of *Schistosoma* cercariae after shedding corresponds to the time in which the parasites can infect their hosts under laboratory conditions. Data on the infection success of cercariae differ greatly depending on the *Schistosoma* species or strain used, as well as the type of animal host and the method of infection (reviews: Wilson 1987; Coulson 1997). In mice infected with freshly shed *S. mansoni* cercariae, between 20 and 50% of the cercariae develop to adult worms. This maximal infection success was also achieved by 3–6 h old cercariae, but it declined with increasing age of the cercariae to half or even 10% of the initial values in 16 h old cercariae (Olivier 1966; Lawson and Wilson 1980). Relation to e.g., glycogen consumption is presented above, see section 8.3).

Also the environmental conditions have a significant effect on cercarial infectivity; the infection success was optimal at temperatures between 15 and 35°C, distilled water inhibited and addition of electrolytes augmented infections (Stirewalt and Fregeau 1965; Asch 1975; Christensen *et al.* 1979).

It is hard to test longevity and infectivity of cercariae under natural conditions. McCreesh and Booth (2014) developed a model for estimation of infection risk by *S. mansoni* cercariae. For humans the highest risk is between 16–18°C and 15–17:00 h in calm water and 20–25°C and 12–16:00 h in flowing water. Infection risk rapidly increases when temperatures increase above the minimum necessary for sustained transmission (McCreesh and Booth 2014).

Nevertheless, the infection success of cercariae under natural conditions will be probably lower than after experimental exposure to the host's skin under optimal conditions. Cercariae have to invest in optimal distribution within the host's habitats, in host-finding processes, in avoiding predation etc. and this may shorten their period of infectivity for the host. In fact, studies on their shedding patterns show that they are adapted to use a short period of maximal infectivity after shedding for optimal transmission.

Finding and approaching host. Cercariae of most *Schistosoma* species actively search their mammalian hosts just by entering and residing in their host's active space and by being present in their host's time (Combes *et al.* 1994; Haas 1994). The contact with the skin occurs, in contrast to cercariae of most other actively host invading species, merely by chance. Upon being touched by the host they recognize it and perform attachment responses of very different specificity and start their sequences of invasion behaviors.

Schistosoma mansoni is the only *Schistosoma* species known to approach actively towards its host, a strategy hitherto found only in cercariae of snail-invading echinostome species (Haberl *et al.* 2000). A direct chemotactic orientation of the swimming *S. mansoni* cercariae towards the skin could be excluded. However, they respond to chemical cues of human skin surface by increasing their shift between backward and forward swimming. They increase their rate of change of direction and this response may guide the cercariae towards the host's skin surface when they are already in close proximity to it, for example between hairs of furred mammals (Brachs and Haas 2008). The stimulating chemical cues of the human skin have been identified as free fatty acids with chemical characteristics similar to those stimulating penetration behavior, the amino acid L-arginine and small peptides with terminally located L-arginine (Haeberlein and Haas 2008). These chemical cues are also used by the cercariae in subsequent behavioral phases. However the cercariae respond only to these small molecular water soluble signals and not to cues to which they respond in other behavioral phases of host invasion which are unsuitable as chemo-attractants due to their chemical properties or distribution within the skin layers.

Among the *Schistosoma* species studied so far in more detail, the cercariae of *S. japonicum* show the lowest degree of host-specificity in their host-finding behavior (Haas *et al.* 1987). After leaving the snail the cercariae swim freely for a short time and then fix themselves to the water surface by their hydrophobic ventral sucker. In this quiescent resting posture they do not respond to any mechanical, chemical or temperature stimuli. They merely adhere to solid and hydrophobic surfaces and then start to crawl. The free swimming cercariae perform the typical attachment movements of schistosomes when encountering a solid surface. Only 30% of the cercariae contacting a substrate attach to it, fully independently of the mechanical, chemical and thermal properties of the substrate. That 70% of the cercariae do not attach spontaneously might ensure wider dispersal of the parasites after emergence from the snail.

The host-finding behavior of *S. haematobium* cercariae differs in many details from the well-known behavior of *S. mansoni* cercariae (Haas 1994). Whereas *S. mansoni* responds mainly to chemical host signals, *S. haematobium* is more sensitive to thermal host cues.

There are also sporadic data on other *Schistosoma* species; the cercariae of the bovine parasite *S. spindale* show little host-specificity after contact with the host (Haas *et al.* 1990). They attach with similar movement patterns as *S. mansoni* to the host in response to host-specific higher temperatures of the skin (Fig. 8.14), but not to chemical host signals.

To summarize host-finding behavior, it should be underlined that each of the *Schistosoma* species studied so far has evolved its individual strategy; the benefits of such diversification are not yet understood. They are subjects of the following hypotheses, which still have to be substantiated by experimental data:

S. mansoni cercariae find and recognize their human host with relatively high specificity and precision as they respond mainly to chemical host cues. This might be an adaptation to infect the human host in clear water habitats or near the water surface, where the detection of chemical cues is not disturbed by mud components.

S. haematobium cercariae respond only poorly to chemical host cues, but seem to be specialized on thermal host signals. This could reflect an adaptation to water habitats containing disguising suspended sediments and to invade the hosts in cooler habitats, where thermal host signals are especially contrasting.

S. japonicum and *S. spindale* cercariae do not respond to chemical host cues except during penetration and *S. japonicum* even performs the host-finding steps like attachment and remaining spontaneously by chance. This nonspecific host-finding strategy may be interpreted as an adaptation to the muddy habitats in which these parasites actually invade their hosts and in which numerous components of mud could interfere with a chemical host identification.

Behavior after contact with the host. Also the behavior of cercariae after contact with the host shows species-specific patterns (Table 8.1). The biological background of this high diversity of host-recognition and invasion patterns is poorly understood.

TABLE 8.1 Host recognition and invasion phases of *Schistosoma* cercariae and the stimulating host signals. Signals in [ ] show low stimulating activity. Modified from Haas W. 2003. Zoology 106(4): 349–364. Table 1.

Species	Attachment	Enduring contact	Directed creeping in gradients of	Penetration
S. mansoni[1]	L-arginine Warmth [Water currents] [Touch]	Ceramides Warmth	L-arginine Warmth	Fatty acids (Not warmth)
S. haematobium[2]	[L-arginine] Warmth	No stimuli?	L-arginine Warmth	Fatty acids (Not warmth)
S. spindale[3]	Warmth (Not chemical stimuli)	Warmth (Not chemical stimuli)	Warmth (Not chemicals)	Fatty acids (Not warmth)
S. japonicum[4]	No stimuli	No stimuli (Favoured by solid hydrophobic surfaces)	Warmth (Not chemicals)	Fatty acids Warmth

References: [1]Austin *et al.* 1972, 1974; Shiff *et al.* 1972; Haas 1976; Haas and Schmitt 1982a, b; Granzer and Haas 1986; Haas *et al.* 1994, 1997, 2002; [2]Haas *et al.* 1994; [3]Haas *et al.* 1990; [4]Haas *et al.* 1987.

Spontaneously swimming cercariae of *S. mansoni* usually approach substrates tail-first. After contact with the human skin they perform a typical attachment response (Fig. 8.14). This response of *S. mansoni* cercariae is relatively nonspecific. The cercariae prefer mammalian skin tissues, but they tend also to attach to the epidermis of fish, frog and even the snail *Biomphalaria glabrata* (Haas 1976). The chemical cue of the human skin surface stimulating attachment has been identified as L-arginine and this amino acid is recognized by the cercariae with extremely high sensitivity and specificity (Granzer and Haas 1986; Haas *et al.* 2002). However, L-arginine is an unsuitable signal for a specific identification of mammalian skin. In fact, data from a series of experiments suggest that the specialization on L-arginine may have been evolved for a function during migration of the schistosomula within the human skin. Then, later in evolution, the L-arginine receptor may have obtained its functions for identification of skin surface (Haas *et al.* 2002).

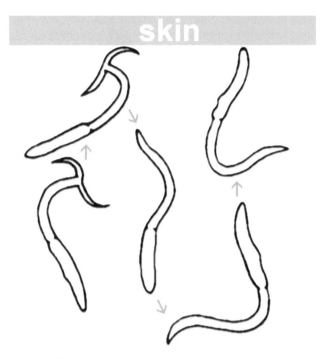

FIG. 8.14 Attachment response of *S. mansoni* cercariae. When the cercariae encounter human skin surface they shift from their backward (tail first) swimming mode to forward swimming, swim in a curve back to the skin surface and attach to it using the penetration organ, then they attach also with the ventral sucker. All responses with a shift from backward to forward swimming are typical after contact with the human skin and are defined as "attachment attempts". Other *Schistosoma* species studied so far perform similar attachment movement patterns. Drawing based on slow motion pictures of Haas 1976.

Warmth of the skin is also a cue for cercarial attachment. The attachment responses upon contact with living human finger skin decreased from 52.5% to 5.5%, when the temperature of the finger was cooled to 25°C (Haas 1976). This effect is also seen upon contact with artificial substrates (Fig. 8.15A). Furthermore, water currents and touching the cercariae during their active or passive swimming phases increases their tendency to attach to encountered substrates (Haas 1976).

After attachment the *S. mansoni* cercariae remain on the skin surface and tend to creep to entry sites. This enduring contact with the skin is stimulated by ceramides, specific lipids of the human uppermost skin layers (Haas *et al.* 2008b). Also warmth supports the tendency to remain on the skin (Fig. 8.15B).

On the surface of the skin, the cercariae creep in a leech-like manner by attaching alternately with the anterior (head) organ and ventral sucker. They select various entry sites when they are exposed to different non-host mammals and isolated human skin samples (Hackey and Stirewalt 1956; Stirewalt 1959, 1966). On the living human skin they explored the surface within a mean duration of 23 s (15 s–6.58 min) before penetrating the skin and 74% penetrated into wrinkles, 22% into the smooth skin surface and 4% into hair follicles (Haas and Haeberlein 2009). It may be suggested that the cercariae find these entry sites in response to mechanical cues. However, they are also able to follow chemical concentration gradients of human skin surface extracts and the signaling cue therein has been identified as L-arginine (Haas *et al.* 2002). Under experimental conditions *S. mansoni* cercariae tend to enter mammalian skin samples in groups and it has been suggested that the acetabular gland contents act as the attracting pheromone (Stirewalt 1971). In fact, it was reported that penetrating *S. mansoni* cercariae secrete arginine with the contents of their postacetabular glands (Stirewalt and Evans 1960; Stirewalt and Walters 1973). However, a more recent study found only very low amounts of bound or free arginine in *S. mansoni* acetabular gland secretions and these did not attract creeping cercariae (Haas *et al.* 2002). In fact, evolution of a sophisticated cooperation among the creeping parasites seems unlikely as, under natural conditions, only few cercariae may creep on the relatively large expanse of human skin surface and the chance that they meet each other seems to be minimal. Therefore, the common penetration under crowded conditions might be brought about by skin compounds which are liberated by the penetrating cercariae. The selective benefit of the specialization of the creeping cercariae on L-arginine as skin-attractant is not yet understood. It may have been evolved for a function during migration in the skin and might only subsequently be used as skin-surface signal.

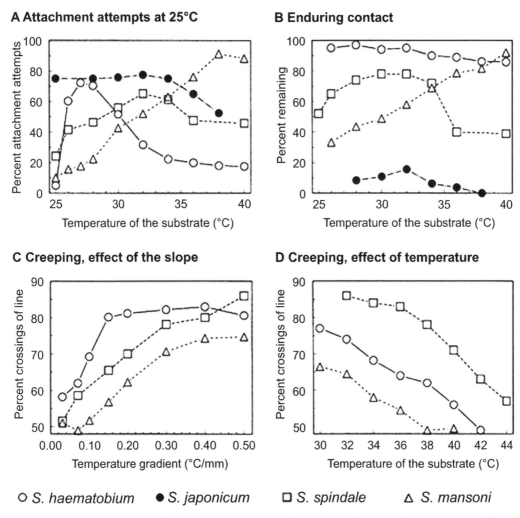

FIG. 8.15 Diversity of responses of *Schistosoma* cercariae to the host warmth. **A**. Attachment attempts; effect of the substrate temperature in a cercarial medium of 25°C. **B**. Remaining on a substrate after attachment; effect of the substrate temperature in a cercarial medium of 25°C (*S. japonicum* 28°C). **C, D**. Creeping in temperature gradients; percentage of creeping cercariae crossing a line in the direction of the higher temperature. **C**. Effect of the slope of the gradient at 32°C. **D**. Effect of the absolute temperature of the substrate on the orientation in a gradient of 0.8°C/mm in *S. spindale* and 0.5°C/mm in *S. haematobium* and *S. mansoni*. Haas W. 1994. Parasitology 109 Suppl: S15–29. Fig. 1.

Another stimulus for creeping *S. mansoni* cercariae is warmth. They follow temperature gradients as low as 0.15°C/mm (Fig 8.15C) and stop this thermo-orientation at unnatural high temperatures (Fig. 8.15D). This response may guide the cercariae from hairs towards the skin.

S. japonicum cercariae attach very nonspecifically to any substrate when free swimming (Section 8.4.5, finding and approaching host). However the normal transfer to the host will occur, when the cercariae in their resting posture just adhere to the lipophilic skin surface. They remain on substrates depending exclusively on the degree of hydrophobicity of the surface; chemical signals or warmth of the substrate have no effect. Also when creeping on a surface, the *S. japonicum* cercariae do not respond to any mechanical or chemical cues. However, in thermal gradients they migrate in the direction to the higher temperature and at temperatures of 32–40°C they try to penetrate into any substrate, independently of its mechanical or chemical properties. The warm substrate stimulates a subsequent penetration.

S. haematobium cercariae perform attachment movements similar to those of *S. mansoni* cercariae. They attach very sensitively to warm surfaces, but avoid attachments to surfaces with unphysiologically high temperatures (Fig. 8.15A). The exclusive chemical attachment stimulus in skin surface extracts is also L-arginine; however, the response is drastically lower than that of *S. mansoni* cercariae.

The cercariae of *S. haematobium* tend to remain on many substrates. Hydrophobicity of the surface seems to be the main cue and chemical and thermal factors have no effect (Fig. 8.15B). When creeping on surfaces, the cercariae orientate towards warmth with high sensitivity to gradients as low as 0.03°C/mm (threshold for *S. mansoni* 0.14°C/mm) (Fig. 8.15C). However, whereas *S. mansoni* cercariae stop their thermo-orientation at unphysiologically high temperatures, *S. haematobium* migrates towards warmth at temperatures as high as 40°C (Fig. 8.15D). Creeping *S. haematobium* cercariae respond also to chemical host cues. They migrate chemotactically along increasing concentration gradients of skin surface extracts and the exclusive cue therein is, as in *S. mansoni*, L-arginine.

The cercariae of *S. spindale* attach to and remain on the skin, responding to host-specific higher temperatures of the skin surface (Fig. 8.15A, B) and not to chemical host cues. Then the cercariae creep towards warmth in thermal gradients with a sensitivity to gradients as low as 0.07°C/mm (Fig. 8.15C) up to lethal high temperatures (Fig. 8.15D). This may guide the parasites from hairs towards the skin surface. However, they do not follow gradients of chemical host cues. Penetration of the skin is then stimulated by fatty acids with characteristics similar to those stimulating penetrations in the other schistosomes.

8.4.6   Penetration into the Skin

The infection of the definitive host, human or animal, occurs *via* schistosome cercariae that penetrate into the skin.

Schistosoma mansoni cercariae penetrate into the skin of very different mammals, applying complex movement patterns (Hackey and Stirewalt 1956). When exposed to the living human skin they need on average 6.8 min (range of 94 s to 13.1 min) for full entry (Haas and Haeberlein 2009). The chemical cues of the skin surface stimulating cercarial penetrations are free fatty acids (Austin *et al.* 1972; Shiff *et al.* 1972; Austin *et al.* 1974; Haas *et al.* 1997). Fatty acids (FA) may function by direct membrane permeation and biochemical processes. However studies on cercarial penetration responses to 230 different chemicals suggest that specific FA receptors also contribute to the response via neural mechanisms (Haas and Schmitt 1982a). The analogues stimulate cercarial penetration independently of their polarity and water solubility and experimental data suggest that receptor sites respond to hydrocarbon chains which have defined characteristics such as a carboxylic end group, a lipophilic end group, defined chain lengths and *cis*- double bonds (Haas and Schmitt 1982b). As a rule, at neutral pH, the stimulatory effect of saturated FA is restricted to chain lengths between 10 and 14 carbon atoms. Unsaturated FA stimulate penetrations in addition at longer chain lengths and their stimulatory activity increases with increasing number of *cis* double bonds. That FA receptors might be involved in cercarial penetration is also supported by the findings that the penetration response of *S. mansoni* cercariae to linolenic acid is reversibly inhibited when the supposed cercarial chemoreceptors are treated with silver ions (King and Highashi 1992) and furthermore by the fact that exposure of *S. mansoni* cercariae to linoleic acid activates protein kinase C and extracellular signal-regulated kinase in sensory papillae and in the nervous system (Ressurreicao *et al.* 2015).

The response of *S. mansoni* cercariae to this defined spectrum of FA suggests an adaptation to invade specifically humans, as the human skin surface contains a unique high concentration of free FA and 65 to 80% of the FA isolated from human skin have the chemical characteristics known to stimulate *S. mansoni* cercarial penetration (Haas and Schmitt 1982b). However, other digenean cercariae, which penetrate other mammals, birds and fish, use FA with similar characteristics as penetration signals (reviewed by Haas and Haberl 1997; Haas 1994, 2003) and this suggests that this FA specificity is a more general property within digenean cercariae. In fact, *S. mansoni* cercariae synthesize various eicosanoids (signaling molecules made by oxidation of 20-carbon FA) when they are incubated in essential FA and it was suggested that these might function in immune evasion and tegument transformation processes (Fusco *et al.* 1985, 1986; Salafsky and Fusco 1987).

Schistosoma mansoni cercariae secrete the contents of their acetabular glands when creeping on the skin surface and when penetrating the skin. It was observed that they can secrete the contents of circum- and postacetabular glands independently of one another with complex movement patterns (Hackey and Stirewalt 1956; Stirewalt and Kruidenier 1961; Stirewalt 1966), suggesting that the secretion processes are controlled by their nervous system. The chemical cues stimulating acetabular gland secretion are contained in human and pig skin surface extracts. Postacetabular gland secretion is stimulated by unknown hydrophilic compounds of the skin surface, but is often combined with secretion of the circumacetabular glands (Hackey and Stirewalt 1956; Stirewalt and Kruidenier 1961; Stirewalt 1966). Circumacetabular gland secretion is stimulated in combination with penetration behavior and probably tegument transformation by the FA fraction of the skin lipids. As free FA are restricted to the uppermost skin surface layers, these responses may be stimulated there. In addition to this behavior, cercariae secrete circumacetabular gland contents responding also to the skin lipid fractions glucosylceramides and phospholipids. These cues are typical for deeper epidermis layers, where no free FA are present and may stimulate enzyme secretion there (Haas *et al.* 1997) (Fig. 8.16).

The penetration behavior of *S. japonicum* cercariae differs significantly from that of *S. mansoni*. After attachment to a substrate or passive transfer to it they respond to warmth with vigorous penetration movements, shed their tails and die within a few minutes unless they have a chance to enter skin. The damage is probably due to osmotic sensitivity after tegument transformation. Penetration is, in addition, stimulated by the FA fraction of human skin surface lipids and the stimulating properties of the FA are the same as in other schistosomes.

The unique penetration response of *S. japonicum* cercariae to thermal cues may bear the risk that cercariae are killed by osmotic damage when they try to penetrate into warm non-animal materials. This danger may be reduced in the field, as persistent creeping, the prerequisite for penetration responses, is restricted to solid and hydrophobic surfaces. The penetration response to heat alone enables *S. japonicum* cercariae to infect a broad spectrum of mammals, which, in contrast to humans, usually have low levels of free FA on the skin surface.

The penetration behavior of *S. haematobium* cercariae resembles that of *S. mansoni*. Heat alone gives no penetration stimulus and the exclusive stimulating cue within skin surface extracts are FA with similar chemical characteristics as in the penetration response of *S. mansoni* (Shiff *et al.* 1972; Haas *et al.* 1994).

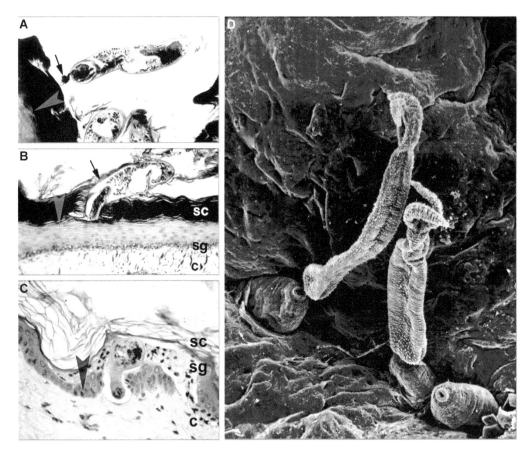

FIG. 8.16 Penetration of schistosome cercariae into the skin. **A**, **B**. Histology sections of bird schistosome *Trichobilharzia szidati* cercariae penetrating into the foot skin of a duck. **A**. Cercaria adhering to the skin surface via secretions of the circum- and postacetabular penetration glands (arrow). **B**. Cercaria penetrating the stratum corneum. The tissue is lysed and secretions of the postacetabular glands are distributed around the cercaria (arrows). Arrowhead shows a move direction. From Haas W, van de Roemer A. 1998. Parasitology Research 84: 787–795. Figs. 8, 9. **C**. *Schistosoma* cercariae penetrating the stratum corneum, stratum germinativum and corium of human skin. Cercaria in the corium after penetration of the stratum germinativum. c, corium; sc, stratum corneum; sg, stratum germinativum. Arrowhead shows a move direction. From McKerrow JH, Caffrey C, Kelly B, Loke P, Sajid M. 2006. Annual Review of Pathology 1: 497–536. Fig. 3. **D**. Scanning electron micrographs of *Schistosoma* cercariae penetrating into the skin. From cover page Parasitology Today 1990, vol. 6, no. 12.

The animated life cycle of *Schistosoma* cercariae (including also shedding of cercariae and their penetration) is downloadable at http://www.sciencedirect.com/science/article/pii/S1471492202023097 (McKerrow and Salter 2002).

Histolytic enzymes involved in the penetration process. Invasive larvae (cercariae) of schistosomes infect their hosts by active penetration of the skin. The mechanism of skin penetration, especially at the molecular level, is still not understood in some schistosome species (e.g. *Schistosoma mansoni* and *S. japonicum*); nevertheless it is agreed that the process involves release of specific proteolytic enzymes (peptidases) (Dalton and Heffernan 1989; Brindley *et al.* 1997; Whitfield *et al.* 2003; Chlichlia and Bahgat 2004; He *et al.* 2005; McKerrow *et al.* 2006; Kašný *et al.* 2009). These peptidases are present in two groups of large penetration glands (postacetabular and circumacetabular glands, see above) (see Fig. 8.11). After attachment of the cercaria to the host skin, the contents of these glands are released and the enzymes participate in disruption of the protein components of the skin and underlying tissues (McKerrow *et al.* 2006; Kašný *et al.* 2009).

The essential role of lytic enzymes in the life cycle of trematodes was revealed in the 1950s, when various researchers demonstrated proteolytic activity and developed hypotheses regarding the importance of proteolytic activity in parasite biology, especially in the mostly studied trematode species *Schistosoma mansoni* (e.g., Lee and Lewert 1956; Timms and Bueding 1959; Stirewalt and Kruidenier 1961). Dresden and Asch (1972) reported that cercarial extract is capable of hydrolyzing gelatin, casein, elastin and azo-dye-impregnated collagen (Azocoll), but not non-denatured soluble or fibrous collagen. Some authors investigated directly the gland products – a gelatinolytic activity was demonstrated in secretions from the circumacetabular glands (Stirewalt and Austin 1973). Subsequently, the content of glands was partially characterized by Campbell *et al.* (1976), having the highest proteolytic activity against azocoll and gelatin, pH optimum 8.5–8.8 and molecular weight of the analyzed proteins 25–28 kDa. At the same time peptidases responsible for host protein degradation have been demonstrated to belong to serine peptidase class (Landsperger *et al.* 1982) and the enzyme was named cercarial elastase, because of its ability to cleave a unique substrate, an insoluble component of the dermis – elastin.

Due to the continuous peptidase research a number of enzymes possibly involved in skin penetration have been reported, e.g., serine peptidases of 25 kDa (Landsperger *et al.* 1982), 28 kDa (Marikovsky *et al.* 1990), 30 kDa (McKerrow *et al.* 1985), 47 kDa (Chavez-Olortegui *et al.* 1992) and 60 kDa (Marikovsky *et al.* 1988). Salter *et al.* (2000) identified and characterized in cercarial secretions trypsin-like and chymotrypsin-like serine peptidase activities. They also revealed that only the chymotrypsin-like peptidase, elastase, is the major factor responsible for degradation of the skin by *S. mansoni*.

Schistosoma mansoni cercarial elastases have been identified by Salter *et al.* (2000) and these enzymes can be divided into two classes by amino acid and promoter sequence homologies. The two most highly expressed *S. mansoni* isoforms SmCE-1a and SmCE-1b comprised more than 90% of the released activity and have very similar biochemical properties. The gene family for these isoforms was highly conserved among several species of schistosomes, including *S. mansoni*, *S. haematobium* and *Schistosomatium douthitti*. Surprisingly, in the Asian relative, *S. japonicum*, any evidence of a similar cercarial elastase (gene product or activity) is missing. The transcripts coding for elastase-like serine peptidase were neither detected in the expression sequence tag databases (ESTs) of all life cycle stages of *S. japonicum* (Fan *et al.* 1998; Fung *et al.* 2002; Hu *et al.* 2003; Peng *et al.* 2003), nor in the *S. japonicum* cercarial mRNA (Dvořák *et al.* 2008). In addition, no cercarial elastase was identified in the acetabular glands of *S. japonicum* using mouse antisera specific to the major SmCE isoform (Chlichlia *et al.* 2005). At that time, *S. japonicum* seemed to be atypical with respect to the enzymes involved in skin penetration (Ruppel *et al.* 2004). Nevertheless, in 2009, after complete annotation of the *S. japonicum* genome, an elastase (SjCE – isoform 2b) in cercariae (and not in schistosomula) of *S. japonicum* was discovered (Zhou *et al.* 2009); it is suggested that the enzyme is responsible for histolytical activity in cercarial penetration process (Liu *et al.* 2015).

In contrast, using a selective and sensitive active-site affinity probe DCG-04, the representatives of a second important group of peptidolytic penetration enzymes, cysteine peptidases were detected in all four species of schistosomes (*S. mansoni*, *S. haematobium*, *Schistosomatium douthitti* and *S. japonicum*) (Dvořák *et al.* 2008). In *S. japonicum*, cercarial secretions had a 40-fold higher activity of cysteine peptidases than secretions of *S. mansoni*. Among the most important cercarial cysteine peptidases of schistosomes are papain-like peptidases, e.g., cathepsin B.

Important data on peptidases has also been obtained from the research on bird schistosomes – *Trichobilharzia regenti* and *T. szidati* (Dvořák *et al.* 2005; Kašný *et al.* 2007; Dolečková *et al.* 2009; Kašný *et al.* 2010). In *T. regenti*, the major activities are of cysteine peptidase origin – cathepsin B and particularly cathepsin L (Kašný *et al.* 2007). A full-length cDNA sequence coding for cathepsin B1/B2 was retrieved from a mixed cDNA library based on intramolluscan stages (sporocysts and developing cercariae) of *T. regenti* (Dolečková *et al.* 2007). Elastinolytic activity of TrCB2 indicates that this enzyme could functionally substitute for the role of chymotrypsin-like serine peptidases, namely *S. mansoni* "cercarial elastase".

Besides the peptidolytic enzymes, the presence of other active molecules such as hyaluronidase have been recognized in the secretions of schistosome cercariae (Levine and Garzoli 1948; Evans 1953). Hyaluronidase, as a glycosidase, is able to enzymatically cleave hyaluronic acid (hyaluronan) – a high molecular weight glycosaminoglycan belonging to major components of the extracellular matrix in vertebrates. The action of hyaluronidase possibly plays a role in penetration of cercariae into the skin (Levine and Garzoli 1948; Yamada *et al.* 2011).

8.5 "-OMICS"

The accelerating "-omics" research is daily generating comprehensive datasets saturating the databases by protein/peptide mass spectrum profiles (proteomics) and RNA/DNA sequences coding for particular proteins (transcriptomics/genomics) (e.g., *Schistosoma mansoni* GeneDB' website (http://www.genedb.org/genedb/smansoni/, coordinated by The Wellcome Trust Sanger Institute). In regard to cercariae this information broadens our knowledge of proteins produced by cercariae e.g. penetration gland secretions, other excretory-secretory products (Wang and Hu 2014), developmentally regulated expression of particular genes (e.g., cercariae vs. schistosomula) and total DNA composition of schistosomes.

8.5.1 Cercarial Proteome

Proteomic methods serving for characterization of complex protein mixtures provide an ideal tool to characterize larval secretions. For example, excretory-secretory proteins isolated from schistosome cercariae have been analyzed by proteomic methods many times (mass spectrometry, MALDI TOF, LC MS/MS etc.) (e.g., Knudsen *et al.* 2005; Curwen *et al.* 2006; Liu *et al.* 2015). Knudsen *et al.* (2005) and Curwen *et al.* (2006) applied proteomic techniques to *S. mansoni* cercarial proteins and identified predominantly the SmCE isoforms and other proteins associated with penetration glands. As for *S. japonicum*, protein composition of cercariae and schistosomula has been characterized in regard to stage-specific differences, including those in peptidase composition (Liu *et al.* 2015). In total, 1972 proteins were identified in association with ten main biological processes – 46 peptidases were detected in cercariae and among them, 25 peptidases disappeared after penetration and transformation to schistosomula. In the Asian relative, *S. japonicum*, the evidence of a similar cercarial elastase (gene product or activity) is missing. The transcripts coding for elastase-like serine peptidase were neither detected in the expression sequence tag databases (ESTs) of all life cycle stages of *S. japonicum* (Fan *et al.* 1998; Fung *et al.* 2002; Hu *et al.* 2003; Peng *et al.* 2003), nor in the *S. japonicum* cercarial mRNA (Dvořák *et al.* 2008). The sequence coding cercarial genes SjCE2b have been confirmed only in *S. japonicum* genome (Huang *et al.* 2007; Zhou *et al.* 2009). As in *S. mansoni*, the *S. japonicum* cysteine peptidases (cathepsin SjCB2) were found abundant and differentially expressed.

8.5.2 Cercarial Transcriptome

To understand the relationship between the information coded in genome and the functional molecules representing the proteome, the various high-throughput sequencing techniques have been developed enabling identify and characterize transcripts of messenger RNA of particular cells, tissues or whole organisms.

During penetration, the nonfeeding cercaria is converting to the schistosomulum, which is accompanied by transcription of genes involved in body remodeling, including production of a new outer surface and gut activation (even before ingestion of red blood cells starts) (Manuel-Parker *et al.* 2011). So far, approaches used to describe gene expression during transition from cercariae to schistosomula have involved quantitative analysis of Expressed Sequence Tags (ESTs) (Farias *et al.* 2011) and studies using microarrays (Dillon *et al.* 2006; Fitzpatrick *et al.* 2009; Chai *et al.* 2006; Gobert *et al.* 2009; Parker-Manuel *et al.* 2011). Gobert *et al.* (2010) compared the cercariae vs. 3-hour old schistosomula and identified 2,791 differentially expressed genes (1,608 up-regulated and 1,183 down-regulated) between these two life cycle stages. The up-regulated dataset included genes related to the structure of the tegument (e.g., tetraspanins), stress response proteins (e.g., HSP70), development (e.g., frizzle related protein) and enzymes involved in the blood meal digestion (e.g., a range of cathepsins). Another study covering 15 different life cycle time points including cercariae, 3-hour old and 24-hour old schistosomula showed that 159 genes were differentially expressed in 3 hour schistosomula vs. cercariae (114 up-regulated and 45 down-regulated) and 321 genes in 24 hour schistosomula vs. cercariae (202 up-regulated and 119 down-regulated) (Fitzpatrick *et al.* 2009).

Leontovyč *et al.* (2016) peformed robust high-throughput analysis (Illumina HiSeq 2500 platform) of transcripts of the bird schistosome *Trichobilharzia regenti* – cercariae vs. 7 days old schistosomula. The authors identified key biological and metabolic pathways specific to each of these two developmental stages and also undertook comparative analyses using data available for taxonomically related blood flukes of the genus *Schistosoma*. The comparative analyses revealed the significant differences in transcription of stage-specific genes predominantly related to carbohydrate metabolism, translation and amino acid metabolism, microaerobic metabolism (citrate cycle and oxidative phosphorylation), calcium function during the invasion of cercariae, further growth and development, function of adhesion molecules, peptidases (cathepsins) and other histolytic and lysosomal proteins playing role during further schistosomula migration.

8.5.3 Cecarial Genome

To date, the genomic studies of schistosomatids have been mostly focused on *S. mansoni, S. japonicum* and *S. haematobium* (Berriman *et al.* 2009; Zhou *et al.* 2009; Young *et al.* 2012) (see also Littlewood and Webster, Chapter 1).

All genomic data for schistosomes, such as *Schistosoma mansoni* (Berriman *et al.* 2009) *S. japonicum* (Zhou *et al.* 2009) and *S. haematobium* (Young *et al.* 2012) have been consolidated in a searchable format under the GeneDB (www.genedb.org) and SchistoDB (www.schistodb. net, Zerlotini *et al.* 2013) databases. The other species of schistosomes (*Schistosoma curassoni, S. margrebowiei, S. mattheei, S. rodhaini, T. regenti, T. szidati*) have been sequenced within the framework of The 50 Helminth Genomes Initiative and can be reached at http://parasite.wormbase. org/ (Howe *et al.* 2016).

The information generated by genomic data (for *S. mansoni*, at least 11,809 genes are in the 363 megabase nuclear genome) provides an invaluable resource for an advanced research; especially in combination with proteomic and transcriptomic data an updated platform can be defined, leading to new modes of treatment, control and eradication of schistosomosis and other neglected parasitic diseases (Berriman *et al.* 2009; Protasio *et al.* 2012).

8.6 CHAPTER SUMMARY

Cercaria certainly represents the most well known larval stage of human schistosomes, as it is responsible for parasite/disease transmission from the snail intermediate host to the vertebrate definitive host. It is a mobile stage of the parasite that actively leaves the snail host, swims in the aquatic environment and subsequently, actively penetrates the human skin. All these activities are reflected at the morphological (e.g., tail with muscles for swimming, muscular ventral sucker and anterior/head organ for attachment and penetration), physiological (e.g., aerobic metabolism consuming glycogen/glucose as energy store, histolytic secretions of penetration glands), behavioral (e.g., geo-negative and photo-positive reactions, changes of movements based on recognition of host physical and chemical cues, penetration behavior stimulated by host fatty acids) and other levels. Nowadays, adaptations of cercariae to recognize and attack their vertebrate hosts can be thoroughly studied at the molecular level and genomic, transcriptomic and proteomic methods represent powerful tools to understand these processes.

8.7 ACKNOWLEDGMENTS

The research of M.K. was financially supported by the Masaryk University, Brno (MUNI/A/1325/2015) and research of P.H. and M.K. by the Charles University in Prague (PRVOUK P41, UNCE 204017 and SVV 260202/2015).

8.8 LITERATURE CITED

Agrawal MC. 2012. *Schistosomes and Schistosomiasis in South Asia.* Springer. pp. 1–6.

Anon. 2009. The *Schistosoma japonicum* genome reveals features of host-parasite interplay. Nature 460(7253): 345–351.

Asch HL. 1972. Rhythmic emergence of *Schistosoma mansoni* cercariae from *Biomphalaria glabrata*: control by illumination. Experimental Parasitology 31(3): 350–355.

Asch HL. 1975. Effect of selected chemical agents on longevity and infectivity of *Schistosoma mansoni* cercariae. Experimental Parasitology 38(2): 208–216.

Austin FG, Frappaolo P, Gilbert B, Landis W, Da Rosa MN, Stirewalt MA. 1974. Further studies of *Schistosoma mansoni* cercarial stimulation by crude egg lecithin and other lipids. Parasitology 69: 455–463.

Austin FG, Stirewalt MA, Danzinger RE. 1972. *Schistosoma mansoni*: Stimulatory effect of rat skin lipid fractions on cercarial penetration behavior. Experimental Parasitology 31: 217–224.

Basch PF. 1991. *Schistosomes: Development, Reproduction and Host Relations.* New York: Oxford University Press. p. 248.

Becker W. 1971. Untersuchungen zur Osmo- und Ionenregulation bei Cercarien von *Schistosoma mansoni.* Zeitschrift für Parasitenkunde 35: 282–297.

Berriman M1, Haas BJ, LoVerde PT, Wilson RA, Dillon GP, Cerqueira GC, Mashiyama ST, Al-Lazikani B, Andrade LF, Ashton PD, Aslett MA, *et al.* 2009. The genome of the blood fluke *Schistosoma mansoni.* Nature 460(7253): 352–358.

Brachs S, Haas W. 2008. Swimming behaviour of *Schistosoma mansoni* cercariae: responses to irradiance changes and skin attractants. Parasitology Research 102: 685–690.

Brindley PJ, Kalinna BH, Dalton JP, Day SR, Wong JY, Smythe ML, McManus DP. 1997. Proteolytic degradation of host hemoglobin by Schistosomes. Molecular and Biochemical Parasitology 89(1): 1–9.

Bruce JI, Ruff MD, Hasegawa H. 1971. *Schistosoma mansoni*: endogenous and exogenous glucose and respiration of cercariae. Experimental Parasitology 29(1): 86–93.

Bruckner DA, Vage M. 1974. The nervous system of larval *Schistosoma mansoni* as revealed by acetylcholinesterase staining. The Journal of Parasitology 60(3): 437–446.

Bruckner DA. 1974. Differentiation of pre-and postacetabular glands of schistosome cercariae using apomorphine as a stain. Journal of Parasitology 60(5): 752–756.

Capron A, Deblock S, Biguet J, Clay A, Adenis L, Vernes A. 1965. Contribution to the experimental study of bilharziasis caused by *Schistosoma haematobium* English summary. Bulletin of the World Health Organization 32(6): 755–778.

Cavalcanti MGS, Araujo HRC, Paiva MHS, Silva GM, Barbosa C, Silva LF, Brayner FA, Alves LC. 2009. Ultrastructural and cytochemical aspects of *Schistosoma mansoni* cercaria. Micron 40(3): 394–400.

Chai M, McManus DP, McInnes R, Moertel L, Tran M, Loukas A, Jones MK, Gobert GN. 2006. Transcriptome profiling of lung schistosomula, *in vitro* cultured schistosomula and adult *Schistosoma japonicum.* Cellular and Molecular Life Sciences 63(7-8): 919–929.

Chavez-Olortegui C, Resende M, Tavares CA. 1992. Purification and characterization of a 47 kDa protease from *Schistosoma mansoni* cercarial secretion. Parasitology 105(Pt 2): 211–218.

Cheng TC, Bier JW. 1972. Studies on molluscan schistosomiasis: an analysis of the cercaria of *Schistosoma mansoni.* Parasitology 64: 129–141.

Chlichlia K, Schauwienold B, Kirsten C, Doenhoff MJ, Fishelson Z, Ruppel A. 2005. *Schistosoma japonicum* reveals distinct reactivity with antisera directed to proteases mediating host infection and invasion by cercariae of *S. mansoni* or *S. haematobium.* Parasite immunology 27(3): 97–102.

Christensen NØ, Frandsen F, Nansen P. 1979. The effect of some environmental conditions and final-host- and parasite-related factors on the penetration of *Schistosoma mansoni* cercariae into mice. Zeitschrift für Parasitenkunde 59(3): 267–275.

Coles GC. 1972. Carbohydrate metabolism of larval *Schistosoma mansoni*. International Journal for Parasitology 2(3): 341–352.

Coles GC. 1973. Further studies on the carbohydrate metabolism of immature *Schistosoma mansoni*. International journal for parasitology 3(6): 783–787.

Coles GC, Hill GC. 1972. Cytochrome C of *Schistosoma mansoni*. Journal of Parasitology 58(6): 1046.

Collins JJ, King RS, Cogswell A, Williams DL, Newmark PA. 2011. An atlas for *Schistosoma mansoni* organs and life cycle stages using cell type-specific markers and confocal microscopy. PLoS Neglected Tropical Diseases 5(3): e1009.

Combes C, Fournier A, Moné H, Théron A. 1994. Behaviours in trematode cercariae that enhance parasite transmission: patterns and processes. Parasitology 109: S3–S13.

Conde-del Pino E, Annexy-Martinez AM, Perez-Vilar M, Cintron-Rivera AA. 1968. Studies in *Schistosoma mansoni*. II. Isoenzyme patterns for alkaline phosphatase, isocitric dehydrogenase, glutamic oxalacetic transaminase and glucose 6-phosphate dehydrogenase of adult worms and cercariae. Experimental Parasitology 22(3): 288–294.

Coulson PS. 1997. The radiation-attenuated vaccine against schistosomes in animal models: paradigm for a human vaccine? Advances in Parasitology 39: 271–336.

Cousin CE, Stirewalt MA, Dorsey CH. 1981. *Schistosoma mansoni*: ultrastructure of early transformation of skin- and shear-pressure-derived schistosomules. Experimental Parasitology 51(3): 341–365.

Cummings RD, Nyame AK. 1996. Glycobiology of Schistosomiasis. FASEB Journal: Official Publication of the Federation of American Societies for Experimental Biology 10(8): 838–848.

Curwen RS, Ashton PD, Sundaralingam S, Wilson RA. 2006. Identification of novel proteases and immunomodulators in the secretions of schistosome cercariae that facilitate host entry. Molecular & Cellular Proteomics 5(5): 835–844.

Curwen RS, Wilson RA. 2003. Invasion of skin by schistosome cercariae: some neglected facts. Trends in Parasitology 19(2): 63–68.

Dalton JP, Heffernan M. 1989. Thiol proteases released *in vitro* by *Fasciola hepatica*. Molecular and Biochemical Parasitology 35(2): 161–166.

Dillon Gary P, *et al.* 2006. Microarray analysis identifies genes preferentially expressed in the lung schistosomulum of *Schistosoma mansoni*. International Journal for Parasitology 36(1): 1–8.

Dolecková K, Kasný M, Mikes L, Mutapi F, Stack C, Mountford AP, Horák P. 2007. Peptidases of *Trichobilharzia regenti* (Schistosomatidae) and its molluscan host *Radix peregra* s. lat. (Lymnaeidae): construction and screening of cDNA library from intramolluscan stages of the parasite. Folia Parasitologica 54(2): 94–98.

Dolečková K, Kasný M, Mikes L, Cartwright J, Jedelský P, Schneider EL, Dvorák J, Mountford AP, Craik CS, Horák P. 2009. The functional expression and characterisation of a cysteine peptidase from the invasive stage of the neuropathogenic schistosome *Trichobilharzia regenti*. International Journal for Parasitology 39(2): 201–2011.

Dorsey CH 1976. *Schistosoma mansoni*: description of the head gland of cercariae and schistosomules at the ultrastructural level. Experimental Parasitology 39(3): 444–459.

Dorsey CH, Stirewalt MA. 1977. *Schistosoma mansoni*: localization of calcium-detecting reagents in electron-lucent areas of specific preacetabular gland granules. Zeitschrift für Parasitenkunde (Berlin, Germany) 54(2): 165–173.

Dorsey CH, Cousin CE, Lewis FA, Stirewalt MA. 2002. Ultrastructure of the *Schistosoma mansoni* cercaria. Micron 33(3): 279–323.

Dresden MH, Asch HL. 1972. Proteolytic enzymes in extracts of *Schistosoma mansoni* cercariae. Biochimica et Biophysica Acta 289(2): 378–384.

Dresden MH, Lewis JC. 1977. Proteolytic action of *Schistosoma mansoni* cercarial proteases on keratin and basement membrane proteins. Journal of Parasitology 63: 941–943.

Dresden MH, Edlin EM. 1974. *Schistosoma mansoni*: effect of some cations on the proteolytic enzymes of cercariae. Experimental Parasitology 35(2): 299–303.

Dvořák J, Delcroix M, Rossi A, Vopálenský V, Pospíšek M, Sedinová M, Mikeš L, Sajid M, Sali A, McKerrow JH, Horák P, Caffrey CR. 2005. Multiple Cathepsin B isoforms in schistosomula of *Trichobilharzia regenti*: identification, characterisation and putative role in migration and nutrition. International Journal for Parasitology 35(8): 895–910.

Dvořák J, Mashiyama ST, Braschi S, Sajid M, Knudsen GM, Hansell E, Lim KC, Hsieh I, Bahgat M, Mackenzie B and others. 2008. Differential use of protease families for invasion by schistosome cercariae. Biochimie 90(2): 345–358.

Evans AS. 1953. Quantitative demonstration of hyaluronidase activity in cercariae of *Schistosoma mansoni* by the streptococcal decapsulation test. Experimental Parasitology 2(4): 417–427.

Faghiri Z, Skelly PJ. 2009. The role of tegumental aquaporin from the human parasitic worm, *Schistosoma mansoni*, in osmoregulation and drug uptake. FASEB J 23: 2780–2789.

Fan J, Minchella DJ, Day SR, McManus DP, Tiu WU, Brindley PJ. 1998. Generation, identification and evaluation of expressed sequence tags from different developmental stages of the Asian blood fluke *Schistosoma japonicum*. Biochemical and Biophysical Research Communications 252(2): 348–356.

Farias LP, Tararam CA, Miyasato PA, Nishiyama MY Jr, Oliveira KC, Kawano T, Verjovski-Almeida S, Leite LC. 2011. Screening the *Schistosoma mansoni* transcriptome for genes differentially expressed in the schistosomulum stage in search for vaccine candidates. Parasitology Research 108: 123–135.

Faust EC. 1920. Notes on trematodes from the Philippines. Philippines Journal of Science 17: 617–633.

Fitzpatrick JM, Peak E, Perally S, Chalmers IW, Barrett J, Yoshino TP, Ivens AC, Hoffmann KF. 2009. Anti-schistosomal intervention targets identified by life cycle transcriptomic analyses. PLoS Neglected Tropical Diseases 3(11): e543. doi: 10.1371/journal.pntd.0000543.

Frandsen F, Christensen NO. 1984. An introductory guide to the identification of cercariae from African freshwater snails with special reference to cercariae of trematode species of medical and veterinary importance. Acta Tropica 41(2): 181–202.

Fung M-C, Lau M-T, Chen X-G. 2002. Expressed Sequence Tag (EST) analysis of a *Schistosoma japonicum* cercariae cDNA library. Acta Tropica 82(2): 215–224.

Fusco AC, Salafsky B, Delbrook K. 1986. *Schistosoma mansoni*: production of cercarial eicosanoids as correlates of penetration and transformation. Journal of Parasitology 72: 397–404.

Fusco AC, Salafsky B, Kevin MB. 1985. *Schistosoma mansoni*: eicosanoid production by cercariae. Experimental Parasitology 59: 44–50.

Galaktionov KV, Dobrovolskij AA. 2003. *The Biology and Evolution of Trematodes*. 1st Ed. Boston, Dordrecht, London: Kluwer Academic Publishing. p. 592.

Glaudel RJ, Etges FJ. 1973. The effect of photoperiod inversion upon *Schistosoma mansoni* cercarial emergence from *Biomphalaria glabrata*. International Journal for Parasitology 3(5): 619–622.

Gobert GN, Moertel L, Brindley PJ, McManus DP. 2009. Developmental gene expression profiles of the human pathogen *Schistosoma japonicum*. BMC Genomics 10: 128.

Gobert GN, Tran MH, Moertel L, Mulvenna J, Jones MK, McManus DP, Loukas A. 2010. Transcriptional changes in *Schistosoma mansoni* during early schistosomula development and in the presence of erythrocytes. PLoS One Neglected Tropical Diseases 4(2): e600. doi: 10.1371/journal.pntd.0000600.

Graefe G, Hohorst W, Dräger H. 1967. Forked tail of the cercaria of *Schistosoma mansoni* – a rowing device. Nature 215: 207–208.

Granzer M, Haas W. 1986. The chemical stimuli of human skin surface for the attachment response of *Schistosoma mansoni* cercariae. International Journal for Parasitology 16: 575–579.

Haas W, Beran B, Loy C. 2008a. Selection of the host's habitat by cercariae: from laboratory experiments to the field. Journal for Parasitology 94: 1233–1238.

Haas W, Diekhoff D, Koch K, Schmalfuss G, Loy C. 1997. *Schistosoma mansoni* cercariae: stimulation of acetabular gland secretion is adapted to the chemical composition of mammalian skin. Journal for Parasitology 83: 1079–1085.

Haas W, Grabe K, Geis C, Päch T, Stoll K, Fuchs M, Haberl B, Loy C. 2002. Recognition and invasion of human skin by *Schistosoma mansoni* cercariae: the key-role of L-arginine. Parasitology 124: 153–167.

Haas W, Granzer M, Brockelman C. 1990. Finding and recognition of the bovine host by the cercariae of *Schistosoma spindale*. Parasitology Research 76: 343–350.

Haas W, Granzer M, Garcia EG. 1987. Host identification by *Schistosoma japonicum* cercariae. Journal for Parasitology 73: 568–577.

Haas W, Haberl B, Schmalfuss G, Khayyal MT. 1994. *Schistosoma haematobium* cercarial host-finding and host-recognition differs from that of *S. mansoni*. Journal for Parasitology 80: 345–353.

Haas W, Haberl B, Syafruddin, Idris I, Kallert D, Kersten S, Stiegeler P. 2005. Behavioural strategies used by the hookworms *Necator americanus* and *Ancylostoma duodenale* to find, recognize and invade the human host. Parasitology Research 95: 30–39.

Haas W, Haberl B. 1997. Host recognition by trematode miracidia and cercariae. pp. 197–227. *In*: Fried B, Graczyk TK, editors. *Advances in Trematode Biology*. Boca Raton: CRC Press.

Haas W, Haeberlein S, Behring S, Zoppelli E. 2008b. *Schistosoma mansoni*: human skin ceramides are a chemical cue for host recognition of cercariae. Experimental Parasitology 120: 94–97.

Haas W, Haeberlein S. 2009. Penetration of cercariae into the living human skin: *Schistosoma mansoni* vs. *Trichobilharzia szidati*. Parasitology Research 105: 1061–1066.

Haas W, van de Roemer A. 1998. Invasion of the vertebrate skin by cercariae of *Trichobilharzia ocellata*: penetration processes and stimulating host signals. Parasitology Research 84: 787–795.

Haas W, Schmitt R. 1982a. Characterization of chemical stimuli for the penetration of *Schistosoma mansoni* cercariae. I. Effective substances, host specificity. Zeitschrift für Parasitenkunde 66: 293–307.

Haas W, Schmitt R. 1982b. Characterization of chemical stimuli for the penetration of *Schistosoma mansoni* cercariae. II. Conditions and mode of action. Zeitschrift für Parasitenkunde 66: 309–319.

Haas W, Wulff C, Grabe K, Meyer V, Haeberlein S. 2007. Navigation within host tissues: cues for orientation of *Diplostomum spathaceum* (Trematoda) in fish towards veins, head and eye. Parasitology 134: 1013–1023.

Haas W. 1976. Die anheftung (Fixation) der cercarie von *Schistosoma mansoni*. Einfluss natürlicher substrate und der temperatur. Zeitschrift für Parasitenkunde 49: 63–72.

Haas W. 1992. Physiological analysis of cercarial behavior. Journal of Parasitology 78: 243–255.

Haas W. 1994. Physiological analyses of host-finding behaviour in trematode cercariae: adaptations for transmission success. Parasitology 109: S15–S29.

Haas W. 2003. Parasitic worms: strategies of host finding, recognition and invasion. Zoology 106: 349–364.

Haberl B, Körner M, Spengler Y, Hertel J, Kalbe M, Haas W. 2002. Host-finding in *Echinostoma caproni*: miracidia and cercariae use different signals to identify the same snail species. Parasitology 120: 479–468.

Hackey JR, Stirewalt MA. 1956. Penetration of host skin by cercariae of *Schistosoma mansoni*. I. Observed entry into skin of mouse, hamster, rat, monkey and man. Journal of Parasitology 42(6): 565–580.

Haeberlein S, Haas W. 2008. Chemical attractants of human skin for swimming *Schistosoma mansoni* cercariae. Parasitology Research 102: 657–662.

He Y-X, Salafsky B, Ramaswamy K. 2005. Comparison of skin invasion among three major species of *Schistosoma*. Trends in Parasitology 21(5): 201–203.

Hockley DJ. 1973. Ultrastructure of the tegument of *Schistosoma*. Advances in Parasitology 11(0): 233–305.

Howe KL, Bolt BJ, Cain S, Chan J, Chen WJ, Davis P, Done J, Down T, Gao S, Grove C, *et al.* 2016. Worm Base 2016: expanding to enable helminth genomic research. Nucleic Acids Research 44(D1): D774–780.

Hu W, Yan Q, Shen DK, Liu F, Zhu ZD, Song HD, Xu XR, Wang ZJ, Rong YP, Zeng LC, Wu J, *et al.* 2003. Evolutionary and biomedical implications of a *Schistosoma japonicum* complementary DNA resource. Nature Genetics 35(2): 139–147.

Huang CY, Lu Y, Wang W, Ju C, Feng Z, Yang Z, Wang SY, Hu W. 2007. Cloning, expression of *Schistosoma japonicum* elastase gene and its stage-specific transcription. Chinese Journal of Parasitology and Parasitic Diseases 25(5): 359–363. (In Chinese).

Hyman L. 1951. *The Invertebrates: Platyhelminthes and Rhynchocoela the Acoelomate Bilateria*. New York: McGraw-Hill Book Company, Inc. p. 550.

Jolly ER, Chin CS, Miller S, Bahgat MM, Lim KC, DeRisi J, McKerrow JH. 2007. Gene expression patterns during adaptation of a helminth parasite to different environmental niches. Genome Biology 8(4): R65.

Kašný M, Mikeš L, Dalton JP, Mountford AP, Horák P. 2007. Comparison of cysteine peptidase activities in *Trichobilharzia regenti* and *Schistosoma mansoni* cercariae. Parasitology 134(Pt 11): 1599–1609.

Kašný M, Mikes L, Hampl V, Dvořák J, Caffrey CR, Dalton JP, Horák P. 2009. Peptidases of trematodes. Advances in Parasitology 69: 205–297.

Kašný M, Mikeš L, Dolečková K, Hampl V, Dvořák J, Novotný M, Horák P. 2011. Cathepsins B1 and B2 of *Trichobilharzia* spp., bird schistosomes causing cercarial dermatitis. Advances in Experimental Medicine and Biology 712: 136–154.

Kemp WM. 1970. Ultrastructure of the cercarienhullen reaktion of *Schistosoma mansoni*. Journal of Parasitology 56(4): 713–723.

Khoo KH, Huang HH, Lee KM. 2001. Characteristic structural features of schistosome cercarial n-glycans: expression of Lewis x and core xylosylation. Glycobiology 11(2): 149–163.

Kimura E, Uga S, Migwi DK, Mutua WR, Kiliku FM, Muhoho ND. 1994. Hourly change in cercarial densities of *Schistosoma haematobium* and *S. bovis* at different depths in the water and distances from the shore of a dam in Kwale District, Kenya. Tropical Medicine and Parasitology 45: 112–114.

King CL, Higashi GI. 1992. *Schistosoma mansoni*: silver ion (Ag$^+$) stimulates and reversibly inhibits lipid-induced cercarial penetration. Experimental Parasitology 75: 31–39.

Knudsen GM, Medzihradszky KF, Lim K-C, Hansell E, McKerrow JH. 2005. Proteomic analysis of *Schistosoma mansoni* cercarial secretions. Molecular & Cellular Proteomics 4(12): 1862–1875.

Landsperger WJ, Stirewalt MA, Dresden MH. 1982. Purification and properties of a proteolytic enzyme from the cercariae of the human trematode parasite *Schistosoma mansoni*. The Biochemical Journal 201(1): 137–144.

Lawson JR, Wilson RA. 1980. The survival of the cercariae of *Schistosoma mansoni* in relation to water temperature and glycogen utilization. Parasitology 81(02): 337–348.

Lee CL, Lewert RM. 1956. Quantitative studies of the collagenase-like enzymes of cercariae of *Schistosoma mansoni* and the larvae of *Strongyloides ratti*. Journal of Infectious Diseases 99(1): 1–14.

Leontovyč R, Young ND, Korhonen PK, Hall RS, Tan P, Mikeš L, Kašný M, Horák P, Gasser RB. 2016. Comparative transcriptomic exploration reveals unique molecular adaptations of neuropathogenic Trichobilharzia to invade and parasitize its avian definitive host. PLoS Neglected Tropical Diseases 10(2): e0004406.

Levin MD, Garzoli RF. 1948. On the demonstration of hyaluronidase in cercariae of *Schistosoma mansoni*. Journal of Parasitology 34(2): 158–161.

Lewert RM, Para BJ. 1966. The Physiological incorporation of carbon 14 in *Schistosoma mansoni* cercariae. Journal of Infectious Diseases 116(2): 171–182.

Ligasová A, Bulantová J, Sebesta O, Kašný M, Koberna K, Mikeš L. 2011. Secretory glands in cercaria of the neuropathogenic schistosome *Trichobilharzia regenti* – ultrastructural characterization, 3-D modelling, volume and pH estimations. Parasites & Vectors 4: 162.

Linder E. 1985. *Schistosoma mansoni*: visualization with fluorescent lectins of secretions and surface carbohydrates of living cercariae. Experimental Parasitology 59(3): 307–312.

Liu M, Ju C, Du XF, Shen HM, Wang JP, Li J, Zhang XM, Feng Z, Hu W. 2015. Proteomic analysis on cercariae and schistosomula in reference to potential proteases involved in host invasion of *Schistosoma japonicum* larvae. Journal of Proteome Research 14(11): 4623–4634.

Mair GR, Maule AG, Day TA, Halton DW. 2003. Organization of the musculature of schistosome cercariae. Journal of Parasitology 89: 623–625.

Marikovsky M, Arnon R, Fishelson Z. 1990. *Schistosoma mansoni*: localization of the 28 kDa secreted peptidase in cercaria. Parasite Immunology 12(4): 389–401.

Marikovsky M, Fishelson Z, Arnon R. 1988. Purification and characterization of proteases secreted by transforming schistosomula of *Schistosoma mansoni*. Molecular and Biochemical Parasitology 30(1): 45–54.

McCreesh N, Booth M. 2014. The effect of increasing water temperatures on *Schistosoma mansoni* transmission and *Biomphalaria pfeifferi*. Population dynamics: an agent-based modelling study. PLoS One. 2014 Jul 2; 9(7): e101462. doi: 10.1371/journal.pone.0101462. eCollection 2014. Erratum in: PLoS One. 2014; 9(8):e105917. PLoS One 9(7): e101462.

McKerrow JH, Jones P, Sage H, Pino-Heiss S. 1985. Proteinases from invasive larvae of the trematode parasite *Schistosoma mansoni* degrade connective-tissue and basement-membrane macromolecules. Biochemical Journal 231(1): 47–51.

McKerrow JH, Caffrey C, Kelly B, Loke P, Sajid M. 2006. Proteases in parasitic diseases. Annual Review of Pathology 1: 497–536.

McKerrow JH, Salter J. 2002. Invasion of skin by *Schistosoma* cercariae. Trends in Parasitology 18(5): 193–195.

Meuleman EA, Holzmann PJ. 1975. The development of the primitive epithelium and true tegument in the cercaria of *Schistosoma mansoni*. Zeitschrift für Parasitenkunde (Berlin, Germany) 45(4): 307–318.

Mikes L, Zidkova L, Kasny M, Dvorak J, Horak P. 2005. *In vitro* stimulation of penetration gland emptying by *Trichobilharzia szidati* and *T. regenti* (Schistosomatidae) cercariae quantitative collection and partial characterization of the products. Parasitology Research 96(4): 230–241.

Modha JC, Redman A, Thornhill JA, Kusel JR. 1998. Schistosomes: unanswered questions on the basic biology of the host-parasite relationship. Parasitology Today 14(10): 396–401.

Morris GP. 1971. The fine structure of the tegument and associated structures of the cercaria of *Schistosoma mansoni*. Zeitschrift für Parasitenkunde (Berlin, Germany) 36(1): 15–31.

Mouahid A, Théron A. 1986. *Schistosoma bovis*: patterns of cercarial emergence from snails of the genera *Bulinus* and *Planorbarius*. Experimental Parasitology, 62(3): 389–393.

Mouahid G, Idris MA, Verneau O, Théron A, Shaban MMA, Moné H. 2012. A new chromotype of *Schistosoma mansoni*: adaptive significance. Tropical Medicine and International Health 17(6): 727–732.

Nanduri J, Dennis JE, Rosenberry TL, Mahmoud AA, Tartakoff AM. 1991. Glycocalyx of bodies versus tails of *Schistosoma mansoni* cercariae. Lectin-binding, size, charge and electron microscopic characterization. Journal of Biological Chemistry 266(2): 1341–1347.

Negron-Aponte H, Jobin WR. 1977. Guidelines for spacing and timing of samples to detect populations of *Schistosoma mansoni* cercariae in the field. International Journal for Parasitology 7(2): 123–126.

Nojima H, Sato A. 1978. The emergence of schistosome cercariae from snails. I. Hourly response of cercarial emergence of *Schistosoma mansoni* and *S. haematobium* and effect of light-cut on their emergence. Japanese Journal of Parasitology 27(3): 197–213.

Nojima H, Sato A. 1982. *Schistosoma mansoni* and *Schistosoma haematobium*: emergence of schistosome cercariae from snails with darkness and illumination. Experimental Parasitology 53(2): 189–198.

Nyame K, Cummings RD, Damian RT. 1988. Characterization of the N- and O-linked oligosaccharides in glycoproteins synthesized by *Schistosoma mansoni* schistosomula. Journal of Parasitology 74(4): 562–572.

Olivier LJ. 1966. Infectivity of *Schistosoma mansoni* cercariae. The American Journal of Tropical Medicine and Hygiene 15(6): 882–885.

Pages JR, Théron A. 1990. Analysis and comparison of cercarial emergence rhythms of *Schistosoma haematobium*, *S. intercalatum*, *S. bovis* and their hybrid progeny. International Journal for Parasitology 20(2): 193–197.

Para J, Lewert RM, Ozcel MA. 1970. *Schistosoma mansoni*: distribution of 14C in isotopically labeled cercariae and its loss during early infection. Experimental Parasitology 27(2): 273–280.

Parker-Manuel SJ, Ivens AC, Dillon GP, Wilson RA. 2011. Gene expression patterns in larval *Schistosoma mansoni* associated with infection of the mammalian host. PLoS Neglected Tropical Diseases 5(8): e1274.

Peng H-J, Chen X-G, Wang XZ, Lun Z-R. 2003. Analysis of the gene expression profile of *Schistosoma japonicum* cercariae by a strategy based on expressed sequence tags. Parasitology Research 90(4): 287–293.

Protasio AV, Tsai IJ, Babbage A, Nichol S, Hunt M, Aslett MA, De Silva N, Velarde GS, Anderson TJ, Clark RC, Davidson C, Dillon GP, Holroyd NE, LoVerde PT, Lloyd C, McQuillan J, Oliveira G, Otto TD, Parker-Manuel SJ, Quail MA, Wilson RA, Zerlotini A, Dunne DW, Berriman M. 2012. A systematically improved high quality genome and transcriptome of the human blood fluke *Schistosoma mansoni*. PLoS Neglected Tropical Diseases 6(1): e1455. doi: 10.1371/journal.pntd.0001455. Epub 2012 Jan 10.

Raymond K, Probert AJ. 1991. The daily cercarial emission rhythm of *Schistosoma margrebowiei* with particular reference to dark period stimuli. Journal of Helminthology 65(03): 159–168.

Ressurreicao M, Kirk RS, Rollinson D, Emery AM, Page NM, Walker AJ. 2015. Sensory protein kinase signaling in *Schistosoma mansoni* cercariae: host location and invasion. Journal of Infectious Diseases 212(11): 1787–1797.

Richard J. 1971. La chétotaxie des cercaires: valeur systématique et phylétique. Mémoires du Muséum National d'Histoire Naturelle. Série A(67): 1–179.

Robijn ML, Wuhrer M, Kornelis D, Deelder AM, Geyer R, Hokke CH. 2005. Mapping fucosylated epitopes on glycoproteins and glycolipids of *Schistosoma mansoni* cercariae, adult worms and eggs. Parasitology 130(Pt 1): 67–77.

Ruppel A, Chlichlia K, Bahgat M. 2004. Invasion by schistosome cercariae: neglected aspects in *Schistosoma japonicum*. Trends in Parasitology 20(9): 397–400.

Saladin KS. 1982. *Schistosoma mansoni*: cercarial responses to irradiance changes. Journal of Parasitology 68: 120–124.

Salafsky B, Fusco A. 1987. Eicosanoids as immunomodulators of penetration by schistosome cercariae. Parasitology Today 3: 279–281.

Salter JP, Lim KC, Hansell E, Hsieh I, McKerrow JH. 2000. Schistosome invasion of human skin and degradation of dermal elastin are mediated by a single serine protease. Journal of Biological Chemistry 275(49): 38667–38673.

Samuelson JC, Caulfield JP. 1985. The cercarial glycocalyx of *Schistosoma mansoni*. Journal of Cell Biology 100(5): 1423–1434.

Shiff CJ, Cmelik SH, Ley HE, Kriel RL. 1972. The influence of human skin lipids on the cercarial penetration responses of *Schistosoma haematobium* and *Schistosoma mansoni*. Journal of Parasitology 58: 476–480.

Short RB, Cartrett M. 1973. Argentophilic 'papillae' of *Schistosoma mansoni* cercariae. Journal of Parasitology 59: 1041–1059.

Skelly PJ, Shoemaker CB. 2000. Induction cues for tegument formation during the transformation of *Schistosoma mansoni* cercariae. International Journal for Parasitology 30(5): 625–631.

Smyth JD 1994. *Introduction to Animal Parasitology*. 3rd Ed. Cambridge University Press, Cambridge. p. 549.

Smyth JD, Halton DW. 1983. *The Physiology of Trematodes*. 2nd. Edition. Cambridge University Press, Cambridge. 445 p.

Stirewalt MA. 1959. Isolation and characterization of deposits of secretion from the acetabular gland complex of cercariae of *Schistosoma mansoni*. Experimental Parasitology 8(3): 199–214.

Stirewalt MA. 1966. Skin penetration mechanisms of helminths. pp. 41–60. *In*: Soulsby EJL, editor. *The Biology of Parasites*. New York: Academic Press.

Stirewalt MA. 1971. Penetration stimuli for schistosome cercariae. pp. 1–23. *In*: Cheng TC, editor. *Aspects of the Biology of Symbiosis*. Baltimore: University Park Press.

Stirewalt MA. 1974. *Schistosoma mansoni*: cercaria to schistosomule. Advances in Parasitology 12: 115–182.

Stirewalt MA, Austin BE. 1973. Collection of secreted protease from the preacetabular glands of cercariae of *Schistosoma mansoni*. Journal of Parasitology 59(4): 741–743.

Stirewalt MA, Fregeau WA. 1965. Effect of selected experimental conditions on penetration and maturation of *Schistosoma mansoni* in mice. I. Environmental. Experimental Parasitology 17(2): 168–179.

Stirewalt MA, Evans AS. 1960. Chromatographic analysis of secretions from the acetabular glands of cercariae of *Schistosoma mansoni*. Experimental Parasitology 10: 75–80.

Stirewalt MA, Kruidenier FJ. 1961. Activity of the acetabular secretory apparatus of cercariae of *Schistosoma mansoni* under experimental conditions. Experimental Parasitology 11: 191–211.

Stirewalt MA, Walters M. 1973. *Schistosoma mansoni*: histochemical analysis of the postacetabular gland secretion of cercariae. Experimental Parasitology 33(1): 56–72.

Théron A, Pointier JP. 1995. Ecology, dynamics, genetics and divergence of trematode populations in heterogeneous environments: the model of *Schistosoma mansoni* in the insular focus of Guadeloupe. Research and Reviews in Parasitology (55): 49–54.

Théron A. 1986. Cercariometry and the epidemiology of schistosomiasis. Parasitology Today 2(3): 61–63.

Théron A. 1989. Hybrids between *Schistosoma mansoni* and *S. rodhaini*: characterization by cercarial emergence rhythms. Parasitology 99(02): 225–228.

Théron A, Combes C. 1988. Genetic analysis of cercarial emergence rhythms of *Schistosoma mansoni*. Behavior Genetics 18(2): 201–209.

Théron A. 2015. Chronobiology of trematode cercarial emergence: from data recovery to epidemiological, ecological and evolutionary implications. Advances in Parasitology 88: 123–164.

Timms AR, Bueding E. 1959. Studies of a proteolytic enzyme from *Schistosoma mansoni*. British Journal of Pharmacology and Chemotherapy 14(1): 68–73.

Tyler S, Hooge M. 2004. Comparative morphology of the body wall in flatworms (Platyhelminthes). Canadian Journal of Zoology 82: 194–210.

Upatham ES. 1973. Location of *Biomphalaria glabrata* (Say) by miracidia of *Schistosoma mansoni* Sambon in natural standing and running waters on the West Indian Island of St. Lucia. International Journal for Parasitology 3(3): 289–297.

Upatham ES. 1974. Dispersion of St. Lucian *Schistosoma mansoni* cercariae in natural standing and running waters determined by cercaria counts and mouse exposure. Annals of Tropical Medicine and Parasitology 68(3): 343–352.

Valle CM, Pellegrino J, Alvarenga N. 1973. Rhythmic emergence of *Schistosoma mansoni* cercariae from *Biomphalaria glabrata*: influence of the temperature. Revista del Instituto de Medicine Tropical Sao Paulo 15(4): 195–201.

Wagner A. 1961. Papillae on three species of schistosome cercariae. Journal of Parasitology 47: 614–18.

Wang S, Wei H. 2014. Development of '-omics' research in *Schistosoma* spp. and -omics-based new diagnostic tools for schistosomiasis. Frontiers in Microbiology 5: 313.

Whitfield PJ, Bartlett A, Brown MB, Marriott C. 2003. Invasion by schistosome cercariae: studies with human skin explants. Trends in Parasitology 19(8): 339–340.

Williams CL, Gilbertson DE. 1983. Effects of alterations in the heartbeat rate and locomotor activity of *Schistosoma mansoni*-infected *Biomphalaria glabrata* on cercarial emergence. Journal of Parasitology 69(4): 677–681.

Williams CL, Wessels WS, Gilbertson DE. 1984. Comparison of the rhythmic emergence of *Schistosoma mansoni* cercariae from *Biomphalaria glabrata* in different lighting regimens. Journal of Parasitology 70: 450–452.

Wilson RA. 1987. Cercariae to liver worms: development and migration in the mammalian host. pp. 115–146. *In*: Rollinson D, Simpson AJG, editors. *The Biology of Schistosomes. From Genes to Latrines*. London, San Diego: Academic Press.

Wilson RA, Webster LA. 1974. Protonephridia. Biolological Reviews of the Cambidge Philosophical Society 49: 127–160.

Wuhrer M, Dennis RD, Doenhoff MJ, Lochnit G, Geyer R. 2000. *Schistosoma mansoni* cercarial glycolipids are dominated by Lewis X and Pseudo-Lewis Y structures. Glycobiology 10(1): 89–101.

Xu X, Stack RJ, Rao N, Caulfield JP. 1994. *Schistosoma mansoni*: ractionation and characterization of the glycocalyx and glycogen-like material from cercariae. Experimental Parasitology 79(3): 399–409.

Yamada S, Sugahara K, Ozbek S. 2011. Evolution of glycosaminoglycans: comparative biochemical study. Communicative and Integrative Biology 4(2): 150–158.

Yamaguti S. 1971. *Synopsis of Digenetic Trematodes of Vertebrates*. Tokyo: Keigaku Publishing. p. 1074.

Young ND, Jex AR, Li B, Liu S, Yang L, Xiong Z, Li Y, Cantacessi C, Hall RS, Xu X, Chen F, Wu X, *et al.* 2012. Whole-genome sequence of *Schistosoma haematobium*. Nature genetics 44(2): 221–225.

Zerlotini A, Aguiar ER, Yu F, Xu H, Li Y, Young ND, Gasser RB, Protasio AV, Berriman M, Roos DS, Kissinger JC, Oliveira G. 2013. SchistoDB: an updated genome resource for the three key schistosomes of humans. Nucleic Acids Research 41(Database issue): D728–731.

Zhou Y, Zheng H, Chen X, Zhang L, Wang K, Guo J, Huang Z, Zhang, B, Huang W, Jin K. 2009. The *Schistosoma japonicum* genome reveals features of host-parasite interplay. Nature 460(7253): 345–351.

CHAPTER

Schistosomula

Geoffrey N Gobert[1, 2,]* and Sujeevi SK Nawaratna[1]

9.1 INTRODUCTION

Penetration of the skin of the definitive host by a schistosome parasite results in a cercaria losing its tail and becoming transformed into a schistosomulum, the whole process being known as the cercarial/schistosomulum transformation (Stirewalt 1974; He *et al.* 2005). This morphological transformation is accompanied by dramatic changes in the physiology and biochemistry of the larva (Stirewalt 1974). Contraction and extension of the schistosomulum and the release of enzymes from its pre- and post-acetabular glands cause extensive tissue damage to all host structures traversed, until the larva breaches a blood vessel wall and enters a blood vessel. The timing of these early skin-schistosomulum stage events varies in schistosome species. Ninety percent of *Schistosoma mansoni* and *S. haematobium* parasites still remain in the host epidermis after 24 hours while within the same period a similar number of *S. japonicum* schistosomula will have reached the dermis or dermal vessels of the host (He *et al.* 2002).

The schistosomula then travel via the pulmonary artery to the lungs of the host where they are located intra-vascularly and are now referred to as lung schistosomula (Wilson and Coulson 1989; Dean and Mangold 1992). Again, timing of the initial arrival and period within the lung varies between species, with peak appearance being day 5-6 post-infection for *S. mansoni* with some parasites detected up to day 20. *S. japonicum* schistosomula peak in the lungs at day 3 after penetration (Gui *et al.* 1995). After exiting the lungs, schistosomula re-enter the venous circulation (Gui *et al.* 1995). The pre-adult schistosome can then reside in the liver for up to 15 days post-penetration (Mitchell *et al.* 1991). In the liver, the parasites grow and develop for 8 to 10 days before forming pairs and finally lodge in the portal and mesenteric vessels of the gut from day 11 (for *S. japonicum*) (He 1993). From this point the parasite is now referred to as an adult.

---

[1] Department of Infectious Diseases, QIMR Berghofer Medical Research Institute, Brisbane, Queensland 4006, Australia.

[2] Current Address: School of Biological Sciences, Queen's University Belfast, Belfast, Northern Ireland BT7 1NN, United Kingdom.

* Corresponding author

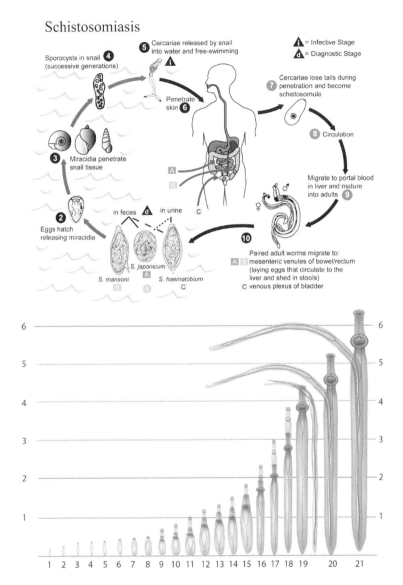

FIG. 9.1 Upper panel. Schistosome life cycle, both the figure and the following legend are from CDC website (http://www.cdc.gov/parasites/schistosomiasis/biology.html).

Eggs are eliminated with feces or urine ❶. Under optimal conditions the eggs hatch and release miracidia ❷, which swim and penetrate specific snail intermediate hosts ❸. The stages in the snail include 2 generations of sporocysts ❹ and the production of cercariae ❺. Upon release from the snail, the infective cercariae swim, penetrate the skin of the human host ❻ and shed their forked tail, becoming schistosomula ❼. The schistosomula migrate through several tissues and stages to their residence in the veins (❽, ❾). Adult worms in humans reside in the mesenteric venules in various locations, which at times seem to be specific for each species ❿. For instance, *S. japonicum* is more frequently found in the superior mesenteric veins draining the small intestine Ⓐ and *S. mansoni* occurs more often in the superior mesenteric veins draining the large intestine Ⓑ. However, both species can occupy either location and they are capable of moving between sites, so it is not possible to state unequivocally that one species only occurs in one location. *S. haematobium* most often occurs in the venous plexus of bladder Ⓒ, but it can also be found in the rectal venules. The females (size 7 to 20 mm; males slightly smaller) deposit eggs in the small venules of the portal and perivesical systems. The eggs are moved progressively toward the lumen of the intestine (*S. mansoni* and *S. japonicum*) and of the bladder and ureters (*S. haematobium*) and are eliminated with feces or urine, respectively Ⓐ.

Lower panel. Key developmental stages of schistosomula to adult schistosome (Madeleine Flynn, QIMR Berghofer Medical Research Institute adapted from Basch 1981; Clegg 1965). Length is shown in millimeters (y-axis) and the growth stages (x-axis).

9.2   SCHISTOSOMULA AND THE SCHISTOSOME LIFE CYCLE

Schistosome species vary considerably in their spectrum of definitive hosts. *Schistosoma japonicum* matures in over 30 species of mammals with domestic cattle and buffalo forming an important animal reservoir, whereas in the field, humans are the only definitive host of *S. haematobium* and *S. mansoni* (Ross *et al.* 2001; Agatsuma 2003). Cercariae use proteolytic enzymes to penetrate human skin (Rheinberg *et al.* 1997). The fore bodies of cercariae transform into schistosomula, which migrate through the dermis and connective tissues to enter blood vessels, travelling via the heart to the lungs from day 2 post infection and appearing in the liver from day 3 for *S. japonicum* or day 5 for *S. mansoni* (Sobhon *et al.* 1988; He 1993). The parasite undergoes profound morphological and biochemical changes during this migration, beginning with the loss of the bifurcated cercarial tail upon penetration, rapid modulation of its membrane, a switch from internal to external nutrient supply and the acquisition of resistance to host immune attack (Rumjanek *et al.* 1984; Lawson *et al.* 1993; Skelly and Shoemaker 2000; Dorsey *et al.* 2002; Ridi *et al.* 2003). Within the lungs, schistosomula are situated intravascularly (Dean and Mangold 1992; Wilson and Coulson 1989). In the liver, the parasites grow and develop for 8 to 10 days then form pairs which finally lodge in the portal and mesenteric vessels from day 11 (for *S. japonicum*) (He 1993). Here the 10-20 mm long adult male worm embraces the longer and more cylindrical female worm within the gynaecophoric canal and the pair begins to sexually reproduce with the formation of eggs (Jordan *et al.* 1993). See Figure 9.1 for further details and Ross and Yuesheng, Chapter 3).

9.3   GROSS MORPHOLOGY

A diagrammatic representation of the development of *Schistosoma mansoni* is shown in Figure 9.2 adapted from Basch (1981). The growth stages (Clegg 1965; Basch 1981) of the growing schistosomes are shown where stage 7 is lung schistosomula (1–7 days), stage 11 is when the ceca joined (15 days) stage 15 is 3 weeks, stage 19 is 6 weeks. However the size of the parasite in culture is 20–40% smaller than its *in vivo* counterpart (Basch 1981). Light microscopic observation of schistosomula in *in vitro* cultures show evidence of red cell digestion by day 4–5 with the appearance of brown colored pigment in the intestine (See Sections 9.5 *In vitro* schistosomes in research and 9.6.1 Gene responses of the host lung to migrating schistosomula). They gradually increase in size and the intestinal ceca start fusing around day 12. The ventral sucker becomes clearly visible by day 10. Interestingly the mechanically separated cercarial tails was noted to stays motile until about day 4 in *in vitro* culture.

9.4   KEY ULTRASTRUCTURAL FEATURES

The following section presents an ultrastructural survey of a *Schistosoma japonicum* schistosomula isolated from the lungs of a murine host three days post cercarial challenge. An excellent companion reference in terms of *S. mansoni* schistosomula and other life cycle stages was published by Basch and Basch (1982) and presents a comprehensive scanning electron microscopy (SEM) study. Furthermore the ultrastructural effects of eosinophils on *S. mansoni* schistosomula has been reported including the use of transmission electron microscopy (TEM) (Caulfield *et al.* 1985), with a focus on the parasite tegument.

When examined by SEM, lung schistosomula measure ~115 µm in length and 18 µm in width. The anterior end of the parasite often appears bulbous and demarcated from the body, while the posterior end tends to be tapered (Fig. 9.2A). Backward-facing spines are concentrated in these regions with sporadically scattered spines in the mid-region (Fig. 9.2A). A spine-free dome is sometimes observed at the anterior end, reminiscent of the cercarial head gland duct openings (Fig. 9.2B) (Sobhon *et al.* 1988; Dorsey *et al.* 2002). The mouth is located sub-terminally on the anterior end, posterior to the concentration of spines and on the ventral aspect (Fig. 9.2B). The ventral sucker is located in the same plane as the mouth but in the posterior quarter of the parasite (Figs. 9.2A, C). The excretory pore is terminal, at the posterior end (Fig. 9.2D). Prominent in parasites that appeared contracted, sagittal folds were evident in the tegument and these could be identified via SEM and TEM (Fig. 9.2E, 9.3A, B). Due to the relative sparseness of gross morphological features, the orientation of schistosomula viewed by TEM is difficult to accurately ascertain in ultrathin sections, making identification of regional differentiation difficult. No evidence of the cercarial features of pre- and post-acetabular glands, head gland or bladder was observed.

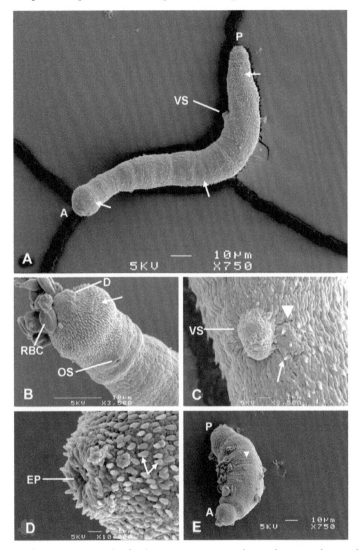

FIG. 9.2 A. Scanning electron micrograph of *Schistosoma japonicum* lung schistosomulum (3 days post cercarial infection). Note that the spines (arrows) are concentrated on the anterior (A) and posterior (P) surfaces of the parasite and are scattered at the mid region. The ventral sucker (VS) is located in the posterior quarter of the parasite. Scale bar 10 μm. **B.** Scanning electron micrograph of the anterior region of a lung schistosomulum. A spine free dome (D) is visible projecting from a concentrated region of backward-facing spines (arrow). The oral sucker (OS) is sub-terminal in a spine-free region. Some mouse red blood cells (RBC) can be seen adhering to the schistosomulum. Scale bar 10 μm. **C.** Scanning electron micrograph of the ventral sucker (VS) of a lung schistosomulum. Spines (arrow) are relatively sparse in this region. Cytoplasmic folds of the tegument are evident as deep fissures in the surface (arrowhead). Scale bar 1 μm. **D.** Scanning electron micrograph of the excretory pore (EP) which is located terminally at the posterior end of the lung schistosomulum. Spines (arrows) are abundant in this region. Scale bar 1 μm. **E.** Scanning electron micrograph of a contracted lung schistosomulum, indicating the anterior (A) and posterior (P) ends and prominent sagittal folds of the tegument (arrowhead). Scale bar 10 μm.

9.4.1 Oral and Ventral Suckers

The oral sucker or mouth is visible only as a sagittal compressed pore (1.2 μm in length) in a spine-free region, surrounded by a slightly elevated region approximately 2.6 μm in diameter (Fig. 9.2B). Although appearing small (about a quarter the diameter of a mouse red blood cell) and non-protuberant, it is possible that features visible *in vivo* may have been internalized as a result of fixation. The ventral sucker is a vestigial structure as compared to that of both adult and cercariae, visible only as a protuberance measuring about 4 μm in diameter at the base and protruding about 2 μm (Fig. 9.1A, C). The ventral sucker is spineless, covered in pits but less deeply ridged than the surrounding tegument. No central pore is visible. The sucker is situated on an underlying bulge of tegument, which is often thicker on the anterior side of the sucker so that the sucker appears to be posteriorly oriented rather than perpendicularly protruding (Fig. 9.2A).

9.4.2 Tegument

The schistosome tegument acts as a multifunctional epithelium providing physical and immunological protection; it is involved in nutrient uptake (Skelly *et al.* 1994; Dalton *et al.* 2004; Skelly and Alan Wilson 2006), sensory input via touch sensors (Gobert *et al.* 2003; Jones *et al.* 2004) and expresses receptors to monitor host hormones including insulin (Dissous *et al.* 2006; You *et al.* 2012). The tegument matrix is generally 500–700 nm thick but can be as thin as 30–40 nm in areas of deep invaginations. The tegument is extensively pitted, often organized into sagittal ridges, giving it a scalloped appearance in longitudinal section (Figs. 9.3A, B). A common approach to identify exposed proteins of the tegument is to biotin-label live parasites and isolate the biotinylated proteins for MS/MS (Braschi and Wilson 2006; Mulvenna *et al.* 2010; Zhang *et al.* 2013; Sotillo *et al.* 2015). This interface has been analyzed proteomically in adult *S. mansoni* (Braschi *et al.* 2006; Braschi and Wilson 2006) and *S. japonicum* (Mulvenna *et al.* 2010; Zhang *et al.* 2013) and more recently in *S. mansoni* schistosomula (Sotillo *et al.* 2015). As reviewed by Wilson (2012), proteomic analysis of the schistosome tegument has been an important research focus since 2004.

Further details of the schistosomula tegument, including the contrasting features and biology between life cycle stages are presented in Gobert *et al.*, Chapter 10.

9.4.3 The Sub-Tegument and Parenchyma

Two muscle layers are present beneath the tegument; transverse muscle bundles overlying longitudinal muscle (Fig. 9.3B). Muscle bundles are mostly between 2 and 5 μm in diameter. Beneath muscle layers, cells comprising the excretory and digestive systems can be identified, however the highly dynamic state of cell differentiation, reorganization and development of other nascent organ systems make identification of cells within the parenchyma difficult. Large tracts of intact lipid droplets are nestled between internal cells and individual cells appear contracted away from each other, although this may have been an artefact of processing (Figs. 9.3A, 9.4). The lack of distinguishing gross morphological features of the lung schistosomula also make orientation of internal features difficult. Several germinal cell types are identified within the parenchymal matrix, e.g., large solitary nucleated cells (8 μm) with small regions of pigmentation and small electron opaque cells (1–2 μm) in nested groups; however, cell types of the lung schistosomulum are largely undescribed and their development and organization are likely to be highly dynamic, making postulation of their differentiated fate (if any) in the adult parasite difficult (Figs. 9.3A, 9.4).

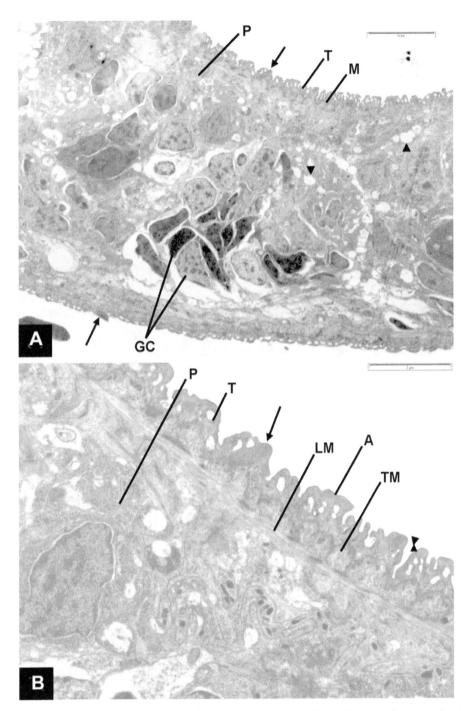

FIG. 9.3 A. Transmission electron micrograph of a *Schistosoma japonicum* lung schistosomulum cut in longitudinal section. Spines (arrows) are visible projecting through the tegument (T) and muscle (M) can be seen underlying the tegument. Nested groups of germinal cells (GC) can be seen, however their potential role in nascent organ formation is difficult to ascertain. Lipids (arrowheads) can be seen embedded in the parenchymal matrix (P). Scale bar 10 μm. **B.** A higher magnification of Figure 9.3A showing the highly folded and pitted tegument (T), a spine (arrow) projecting through the tegument matrix, layers of transverse (TM) and longitudinal muscle (LM) and the underlying parenchyma (P). Folds in the tegument create fissures (arrowhead). The apical membrane (A) of the tegument is indicated. Scale bar 2 μm.

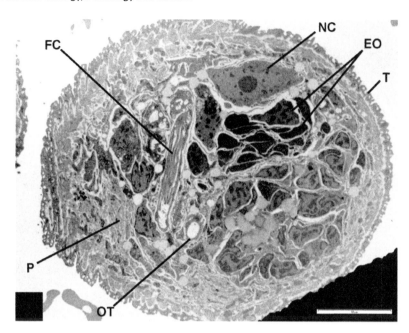

FIG. 9.4 Transmission electron micrograph of a *Schistosoma japonicum* lung schistosomulum in glancing longitudinal section. The pitted tegument (T) creates a sponge-like appearance and nests of germinal cells of variable morphology and appearance such as large, solitary nucleated cells (NC) and small electron opaque cells (EO) are visible embedded within the parenchymal matrix (P). A flame-cell (FC) and osmoregulatory tubule (OT) are labelled. Individual cells within the parenchyma appear contracted away from each other. Scale bar 10 μm.

9.4.4 Digestive Tract

In the adult parasite, the digestive tract consists of the oral cavity, esophagus and the bilateral arms of the blind-ended cecum (Senft 1976). The oral cavity was not observed by TEM in the lung parasite due to its small size and difficulty in achieving the correct orientation during sectioning, however major features of the esophagus and cecum of the adult parasite were observed in the lung-stage parasite.

The surface of the proximal esophagus (Fig. 9.5) appears convoluted by parallel folds of the apical surface which is trilaminate with some pentalaminate regions (Figs. 9.7A, C). This creates an undulating surface, forming crypts in which luminal contents are partially isolated (Fig. 9.5C). From this undulating surface, multi-laminar invaginations (approximately 50 nm in diameter) of the apical membrane extend back into the ground substance of the esophageal lining. Circular multi-laminated vesicles are also observable with a consistent diameter of approximately 50 nm. These "vesicles" may represent the apical invaginations viewed in cross-section, suggesting a tubular structure (Fig. 9.7B). The undulating apical membrane caused the esophageal matrix to vary in thickness, ranging from 50 nm to 600 nm. The proximal esophageal lining resembles the outer tegument of the lung parasite, featuring numerous laminated vesicles and elongated bodies in the matrix (Fig. 9.7A). The basement membrane often invaginates into the ground substance appearing as finger-like projections (approximately 20 nm in width and up to 350 nm long), however the absence of any circular structures of similar diameter suggests these are lateral folds of the basement membrane (Figs. 9.7A, C). These basement invaginations extend far into the esophageal matrix, associating closely with the apical surface at times, although no clear connections have been observed. The esophagus is surrounded by a layer of muscle approximately 300 nm in thickness which is closely associated with numerous mitochondria. Membrane bound organelles with granular contents are also present, perhaps glandular in nature (Fig. 9.7D). The distal esophagus (Fig. 9.6A) narrows, forming a demarcation at the junction with the cecum as in the adult parasite of *S. mansoni* (Senft 1976). Here the esophagus consisted of dense folds that dramatically reduced the luminal diameter.

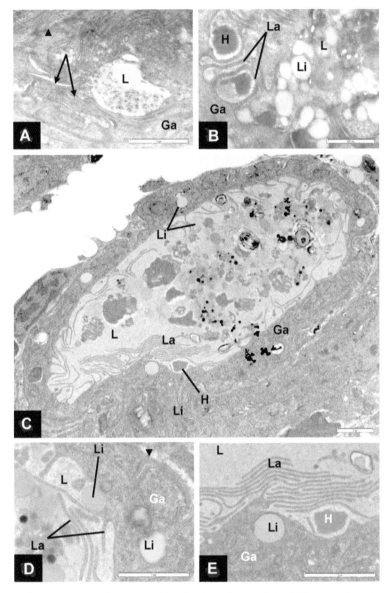

FIG. 9.5 A. Transmission electron micrograph of the distal esophagus of the *Schistosoma japonicum* lung schisto-somulum (shown in cross section), close to the junction with the cecum. The lumen (L) of the esophagus is greatly reduced by dense folds (arrows) of the gastrodermis (Ga). The basal membrane of the gastrodermis is indicated by the arrowhead. Scale bar 1 μm. **B**. Transmission electron micrograph showing the sequestration of hem-like material (H) and lipids (Li) within isolated crypts formed by lamellae (La) which project into the lumen (L) and sometimes rejoin the gastrodermis (Ga). Scale bar 500 nm. **C**. Transmission electron micrograph of the cecum of the lung stage schistosomulum. Lamellae (La) project from random thickenings of the gastrodermis (Ga) sequestering luminal contents such as hem-like material (H). Lipids (Li) can be observed within the lumen (L) and also within the gastro-dermis. Scale bar 2 μm. **D**. A higher magnification of Fig. 9.5C showing lipids (Li) in various stages of sequestration by lamellae (La) and incorporation into the gastrodermis (Ga), where lipid which is no longer intact can be seen, potentially in a state of breakdown. The lumen of the cecum is labelled (L) and an invagination of the basal mem-brane is indicated with an arrowhead. Scale bar 1 μm. **E**. A higher magnification of Figure 9.5C showing an intact lipid (Li) within the gastrodermis (Ga) of the cecum. The lumen (L), lamellae (La) and hem-like material (H) are also labeled. Scale bar 2 μm.

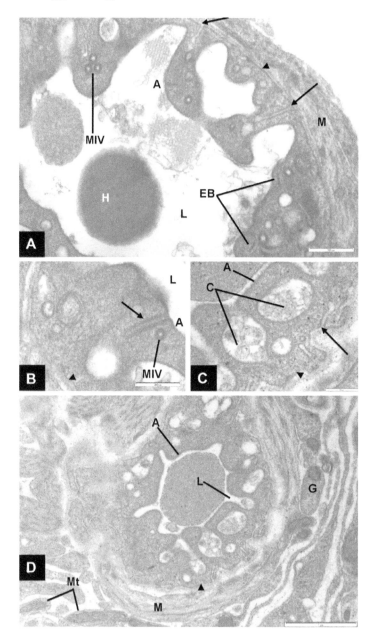

FIG. 9.6 A. Transmission electron micrograph showing the esophagus of a *S. japonicum* lung schistosomulum in cross section. The arrowhead indicates the basal membrane of the esophagus which frequently invaginates (arrows) toward the lumen (L). The ground substance of the esophagus contains elongated bodies (EB) and multilaminated vesicles (MlV). Various luminal contents can be observed. Electron-dense material (H) may potentially be haem or other products of blood digestion. The apical membrane (A) undulates. A layer of muscle (M) surrounds the esophagus. Scale bar 500 nm. **B**. Transmission electron micrograph of the esophagus indicating the basement membrane (arrowhead) and lumen (L). Circular multilaminate vesicles (MlV) may represent tubular invaginations (arrow) of the apical membrane (A) observed in cross section. Scale bar 200 nm. **C**. A higher magnification electron micrograph of Fig. 9. 6D, showing isolated crypts (C) of luminal contents formed by undulations of the apical membrane (A) of the esophagus. The apical membrane appears trilaminate. Invaginations (arrow) of the basement membrane (arrowhead) are common. Scale bar 200 nm. **D**. Transmission electron micrograph of the esophagus showing a thick layer of surrounding muscle (M) and closely associated mitochondria (Mt). Granular membrane-bound organelles (G) may be glandular in nature. The apical membrane (A), lumen (L) and basement membrane (M) are indicated. Various unidentified luminal contents are present. Scale bar 1 μm.

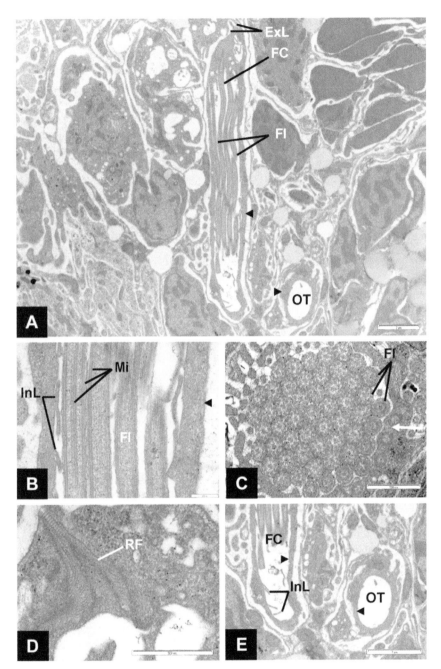

FIG. 9.7 A. *Schistosoma japonicum.* A higher magnification of Figure 9.6, showing a flame-cell (FC) and osmoregulatory tubule observed in longitudinal section (containing the flame-cell) and cross section (OT). Several flagella of the cell can be seen (Fl) and external leptotriches (ExL) are labelled. The wall of the osmoregulatory tube (made up of tubule cells) is indicated by arrowheads. Scale bar 2 μm. **B.** A higher magnification of panel a. showing the flagella (Fl) of a flame-cell. Microtubules (Mi) are visible in longitudinal section and internal leptotriches (InL) are labelled. The epithelial lining of the osmoregulatory tubule is indicated by the arrowhead. Scale bar 200 nm. **C.** Transmission electron micrograph of the flagella (Fl) of a flame-cell in cross section. The 9+2 arrangement of microtubules can be made out. The osmoregulatory tubule wall is indicated by the arrow. Scale bar 500 nm. **D.** Transmission electron micrograph of rootlet fibers (RF) near the junction of flagella to the cap cell of a flame-cell. Scale bar 500 nm. **E.** A higher magnification of panel a. showing a flame-cell (FC) and osmoregulatory tubule observed in longitudinal section (containing the flame cell) and cross section (OT). Internal (InL) leptotriches are labelled. The wall of the osmoregulatory tube (made up of tubule cells) is indicated by arrowheads. Scale bar 2 μm.

The structure of the lung-parasite cecum is also very similar to that of the adult parasite of *S. mansoni* (Senft 1976; Halton 1997; Smyth and Halton 1983b). The cecum is elliptical in shape, approximately 14 µm in length and 5–6 µm in width. The gastrodermis appears to be a syncytium about 600 nm in thickness, with random thicker projections, from which long parallel folds or lamellae extend into the lumen (Fig. 9.5C). Lamellae extend as far as 3 µm into the lumen, trapping gut contents and occasionally rejoining the gastrodermis, forming isolated crypts of contents including electron-dense granulated material and lipid droplets (Fig. 9.5B). Lipid droplets are observed in various stages of engulfment by the epithelium and are present in the matrix of the syncytium as intact and disintegrating droplets (Figs. 9.5D, 9.7E). This interaction of the gut syncytium with lipid droplets and other gut contents is very similar to that described in the active cecum of the adult parasite (*S. mansoni*) (Senft 1976). Invaginations of the basement membrane also extend into the matrix of the syncytium (Fig. 9.6D). Bifurcation of the cecum has not been observed.

9.4.5 Excretory System

Although the osmoregulatory ducts which contain flame-cells could occasionally be identified (with and without flame-cells), the exact orientation of these ducts and the number and orientation of flame-cells within the parenchyma of the lung schistosomula could not clearly be identified due to the difficulty in achieving and orientating serial sections during ultramicrotomy (Fig. 9.6). Several flame-cells are present per schistosomula and each is observed as a cap cell (not shown) from which a bunch of $\cong 60$ flagella extend into the osmoregulatory tubule (Fig. 9.7A). The microtubules of these flagella have been observed in longitudinal section (Fig. 9.7B) and in cross section, where the $9+2$ arrangement of microtubules is evident (Fig. 9.7C). Internal and external leptotriches are present (Fig. 9.7A, B, E) and rootlet fibers at the junction of the flagella and cap cell are seen (Fig. 9.7D). Tubule cells comprise the relatively thick walls of the osmoregulatory ducts, which are approximately 1.8 µm in diameter (Fig. 9.7E). Although these ducts (if there are more than one) presumably converge before joining the excretory pore, no structure resembling a bladder (as in the cercariae and adult parasite) has been observed (Smyth and Halton 1983b; Dorsey *et al.* 2002), nor was the excretory pore observed by TEM.

9.5 *IN VITRO* SCHISTOSOMES IN RESEARCH

The necessity to produce large yields of schistosomula for research has motivated the development of many techniques to transform cercariae into schistosomula *in vitro,* avoiding the time consuming and labor intensive *in vivo* technique, which produces comparatively low yields (Gold and Fletcher 2000). *In vitro* schistosomes have been successfully maintained in culture throughout all stages of the mammalian phase, even to the point of sexual pairing, although the eggs of *in vitro* worm pairs are reportedly grossly abnormal (Basch 1981; Basch and Basch 1982). *In vitro* schistosomes are now commonly used to represent various developmental stages of the schistosome, in research including vaccine-related research, ultrastructural studies, genetics and gene manipulation, development, behavior, metabolic activity, biochemistry, immunology and molecular biology (Cousin *et al.* 1981; Salafsky *et al.* 1988; Marikovsky *et al.* 1990; Harrop and Wilson 1993; Harrop *et al.* 1998; Gold and Fletcher 2000; Loukas *et al.* 2001). There seems to be a lack of universal standardization associated with the use of the various forms of *in vitro* schistosomes and culturing durations/conditions used to achieve developmental stages are varied and sometimes arbitrary (Harrop and Wilson 1993; Lawson *et al.* 1993; Curwen *et al.* 2004). There is some evidence that *in vitro* schistosomula forms are structurally comparable to their *in vivo* equivalents, nevertheless the comparative molecular biology and biochemistry of *in vitro* and *in vivo* schistosomes is not well characterized (Brink *et al.* 1977; Samuelson *et al.* 1980; Cousin *et al.* 1981; Gold and Fletcher 2000). Key morphological features of mechanically transformed schistosomula maintained in culture are presented in Figure 9.8.

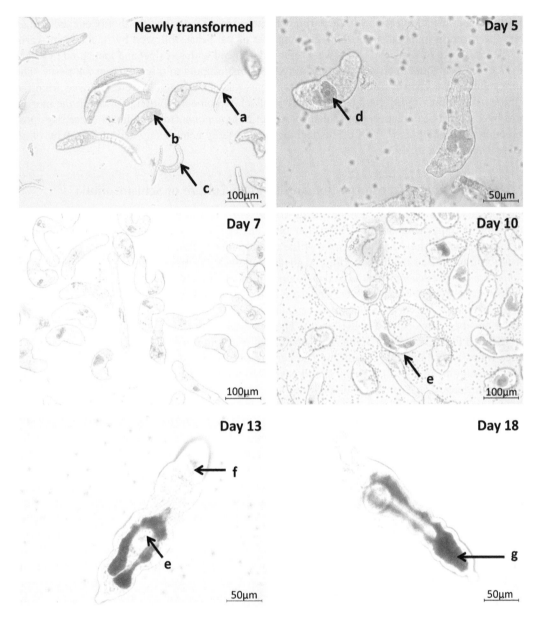

FIG. 9.8 Light microscopy of the *in vitro* development of mechanically transformed schistosomula. a, Cercariae; b, schistosomula; c, separated cercarial tail; d, brown pigment in ceca; e, ventral sucker; f, oral sucker; g, fused intestinal ceca.

9.5.1   Recovery of Lung Stage Schistosomula

Lung schistosomula were isolated using the modified technique of Gui *et al.* (1995). The *Schistosoma japonicum* schistosomula used in these experiments were supplied by QIMR. Miracidia from the Anhui (Chinese) strain of *S. japonicum* were used to infect *Oncomelania hupensis hupensis* snails. After a period of three months the snails were induced to shed cercariae by exposure to bright light. Collected cercariae were applied in water to the shaved abdomens of anaesthetized stock mice (female, Swiss, outbred) and allowed to penetrate for 60 minutes. Approximately 1000 cercariae pooled from several snails were used per mouse. Three days after the initial infection, mice were given intraperitoneal injection with heparin, euthanased by CO_2 gas and the thorax opened. Lungs were perfused by puncturing of the right ventricle of the heart, followed by injection of warm RPMI into the right ventricle. Lungs were removed, minced with a scalpel and incubated in RPMI at 37°C for 3–4 hours upon a rocker, enabling the schistosomula to migrate from the tissue. The lung solution was sieved to remove larger fragments and schistosomula were removed from the filtrate using a fine-tipped glass pipette. Studies of schistosome migration patterns in the murine host have shown peak recovery of schistosomula of *S. japonicum* from the lungs, three days post infection (Rheinberg *et al.* 1997). This technique reportedly achieves a yield of 30% (Gui *et al.* 1995; Rheinberg *et al.* 1997).

9.5.2   Mechanical Transformation and *In Vitro* Culture of Schistosomula

Methods used for obtaining schistosomula for *in vitro* research are summarized in Fig. 9.9. Mechanical transformation and cultivation of schistosomula can be carried out according to Basch (1981) with modifications from Milligan and Jolly (2011). The following is an account of a commonly used methods performed in our laboratory. Three months after infection with *Schistosoma japonicum* miracidia of the Anhui (Chinese) strain, *Oncomelania hupensis hupensis* snails were shed and cercariae were collected into a 50 ml conical tube containing RPMI 1640 medium with 10% FCS (heat inactivated) (media with FCS is not required for *Schistosoma mansoni* cercariae since they disperse in the liquid). The cercariae were placed on ice for 45 minutes then centrifuged for 10 minutes at 500 g. The supernatant was removed and the cercariae were washed three times with cold RPMI medium. The cercariae were resuspended in warm schistosomula culture medium 169 (37°C) (Table 9.1) and passed 30 times between two 10 ml syringes through a 120 mm long, 22-gauge double ended luer lock microemulsifying needle. The cercariae were then transferred to a 50 ml conical tube where the solution underwent a series of sedimentations at 37°C, each time removing the upper-portion of cercarial tail-rich supernatant. The mechanically transformed schistosomula were then transferred into a 10 ml conical tube using a Pasteur pipette and allowed to sediment at 37°C. The supernatant containing tails was removed and resuspended in 10 ml of fresh schistosomula media. Alternatively the cercariae were transferred to a petri dish, added extra media and swirled to collect the schistosomules in the center. The broken tails floating on top were then removed using a Pasteur pipette. This procedure was repeated 2–3 times after adding new media. The schistosomula collected in the center were removed using a Pasteur pipette. The newly transformed schistosomula were cultured at 37°C with 5% CO_2 gas in 24 well plates. Five hundred schistosomules in 1 ml of media were added per well which was supplemented with 10 µl of packed human RBC (diluted 1:100 in media) for long term culturing. The culture medium was changed twice a week.

Separation of cercarial tail has also been achieved by mechanical agitation in a vortex mixer with or without the use of transformation needle (Brink *et al.* 1977; Protasio *et al.* 2013). Less tail contamination is seen in schistosomules separated using a density gradient (Brink *et al.* 1977; Protasio *et al.* 2013). Transformation of cercariae through skin has also been used in research (Protasio *et al.* 2013). However the gene expression of schistosomules obtained by mechanically transformation and skin transformation differ in 38 genes only (Protasio *et al.* 2013).

A non-mechanical transformation of cercariae by leaving them undisturbed in RPMI to shed the tails has been reported with no significant difference in the number or viability of transformed schistosomules, as compared to the mechanical transformation. This method is considered to involve less risk of sample contamination and accidental infections (Coultas and Zhang 2012). Interestingly the growth of the separated cercarial tail with development of unknown structures has been described with the potential possibility of development of a *Schistosoma* cell line from this structure (Coultas and Zhang 2012).

Cercarial Transformation Methods

Cercariae shed *by exposure to light* from infected snails
Biomphalaria glabrata (Schistosoma mansoni)
or Oncomelania hupensis hupensis (Schistosoma japonicum)

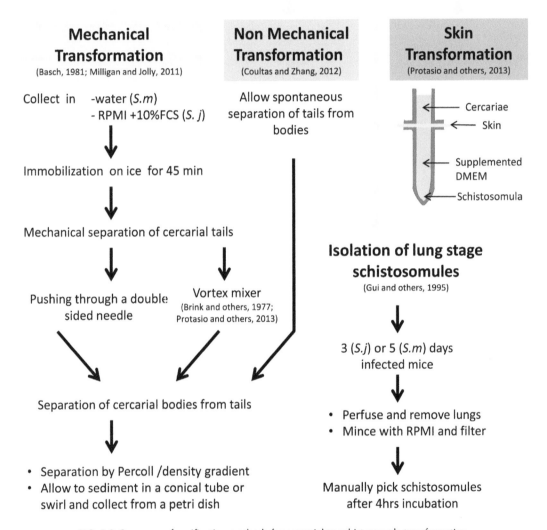

FIG. 9.9 Summary of purification methods for cercarial to schistosomula transformation.

It should be noted that the relatively few prior studies of schistosomula (skin and lung stages) relating to ultrastructure, metabolism, behavior, development and immunogenicity have mostly been carried out on *in vitro* schistosomula which have been artificially transformed and cultured, never entering a definitive host organism (Brink *et al.* 1977; Cousin *et al.* 1981; Basch and Basch 1982; Bogitsh 1989; Harrop and Wilson 1993; Harrop *et al.* 1998). The ultrastructure and biology of the cercaria (see Haas and Jamieson, Chapter 8) and adult stage parasite are much more widely researched and understood, as compared to the lung schistosomulum, which is not surprising given the relative ease of the two former life cycle stages in terms of isolation and their obvious importance to the human host regarding pathology and infectivity.

TABLE 9.1 Schistosomula culture medium 169 (Basch 1981).

Component	Concentration	100 ml Medium
Basal Medium Eagle	–	Make up to 100 ml
Lactalbumin hydrolysate	1 g/L	2 ml
Glucose	1 g/L of 100x	1 ml
Hypoxanthine	0.5 μM	100 μl
Serotonin	1 μM	100 μl
Insulin (crystalline)	8 μg/ml	100 μl
Hydrocortisone	1 μM	100 μl
Triiodothyronine	0.2 μM	100 μl
MEM vitamins	0.5x	500 μl
Schneider's medium	5%	5 ml
Human serum/fetal calf serum	10%	10 ml
Hepes	1 ml	1 ml
Penicillin/Streptomycin	300 U/ml	3 ml
L-glutamine	100x	1 ml

9.6 IMMUNOBIOLOGY OF THE LUNG STAGE SCHISTOSOMULUM

After the cercaria has infected the human host, the developing schistosomula undergoes morphological, transcriptional and biochemical alterations to adapt to the new microenvironment. This dynamism potentially presents a changing array of antigens that may stimulate immune response of the host (Mountford and Harrop 1998). Initially the skin phase schistosomula parasites are susceptible to humoral and cellular immunity, but resistance to host immune mechanisms rapidly develops (Dean 1977; Ridi *et al.* 2003). This is reflected in the sensitivity of skin schistosomula to complement mediated killing as well as antibody dependent cellular cytotoxicity; while in the lungs, migrating schistosomula and subsequently adult parasites are refractory to both modes of immunological challenge (McLaren 1980; Lawson *et al.* 1993; Parizade *et al.* 1994). Acquisition of immunological resistance may be linked to the shedding of the cercarial glycocalyx, which is previous required for osmotic protection during aquatic phase of the development of the parasite. This is replaced at the site of the apical parasite surface with the formation of a heptalaminate surface membrane several hours after penetration (Hockley and McLaren 1973; Gobert *et al.* 2003). The observed immune evasion was originally considered to be due to the masking of parasite surface antigen; however, resistance at least at the lung stage schistosomula to immune effector mechanisms was reported to be independent of host antigen adsorption (Dean 1977). These observations indicated instead that the biochemical or structural composition of the membrane surface takes place to prevent host antibody binding or in reducing parasite antigen availability (Ridi *et al.* 2003). In support of these assertions it has been suggested that lung stage schistosomulum second surface molecules in discrete, lipid-rich sites of the outer surface membrane and that in addition to this physical barrier, biochemical and immunological barriers must exist to protect the parasite from host immunity (Parizade *et al.* 1994; Ridi *et al.* 2003).

Despite these well-developed immunological stealth techniques of schistosomula, this life cycle stage is still considered the central focus of host immune elimination in response to radiation-attenuated (RA) cercariae vaccination (Wilson and Coulson 1989). It has been shown that the lungs are the principal site of worm elimination in both naive and RA cercariae immunized mice and that the increase in future protection of the parasite is proportional to the time the migrating parasites RA spends in the lungs (Dean and Mangold 1992). The basis of this protection may be reflected in the observation that parasite antigens obtained from lung stage schistosomula are particularly efficient lymphocyte proliferation stimulators and this stimulates secretion of Th1 cytokine secretion, when compared to antigen from cercarial or skin-stage sources (Mountford *et al.* 1995). Overall, this and other research has indicated that the lung stage schistosomulum, should be considered an important target for an effective protective vaccine against schistosomiasis (Mountford *et al.* 1995; Gobert *et al.* 1997; Harrop *et al.* 1998; Mountford and Harrop 1998).

9.6.1   Gene Responses of the Host Lung to Migrating Schistosomula

While the lungs have been found to be the primary site of worm elimination (Dean and Mangold 1992) little is known about the pulmonary immune responses induced in response to the migrating lung schistosomula of any schistosome species. The presence of lung schistosomula in the lung of the host results in responses both at the gene expression and cellular levels (Fig. 9.10).

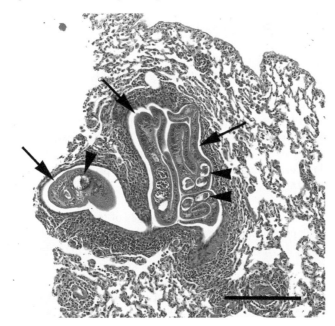

FIG. 9.10 Migrating *Schistosoma japonicum* schistosomula in the lung of a rat host three days post cercarial challenge. The presence of schistosomula is clearly seen (arrows) as is the active feeding on host blood (arrowheads). Scale bar 200 μm. (Courtesy Dr Patrick Driguez QIMR Berghofer Medical Research Institute).

The innate immune response induced in the host lung during *Schistosoma japonicum* (or any schistosome species) infection remains largely uncharacterized with most knowledge coming from infections of *Nippostrongylus brasiliensis* and the artificial schistosome pulmonary granuloma model (Fallon and Mangan 2007). Reece *et al.* detected in the host lung the transcriptional elevation of IL-4 and IL-13 and increased populations of alternatively activated macrophages (AAMs) two to four days after an initial *N. brasiliensis* infections (Reece *et al.* 2006). Unlike classically activated macrophages whose primary function is pathogen destruction via the generation of reactive nitrogen and oxygen intermediates (Loke *et al.* 2000), AAMs and are thought to function in tissue remodeling, debris scavenging, wound healing, the induction of Th2 immune responses and the activation of arginase-1 (Varin and Gordon 2009).

Several studies have reported negative correlations between the development of allergic pulmonary responses and several helminth infections (Mangan *et al.* 2004; Wilson *et al.* 2005; Reece *et al.* 2008). Specifically, humans with schistosome infections have decreased allergic responses compared to non-infected individuals (Medeiros *et al.* 2003). Both asthmatic diseases and helminth infections are characterized by eosinophilia, Th2 cytokines and IgE (Pacifico *et al.* 2009). Recruitment of eosinophils and leukocytes into the lungs in response to experimental asthma is significantly reduced in mice previously exposed to *S. japonicum* infection (Mo *et al.* 2008). This immunomodulatory effect has, in part, been correlated with elevated levels of IL-10 derived from Tregs and regulatory B-cells as well as regulatory functions of Tregs that are independent of IL-10 that act to suppress Th1 and Th2 immune responses (Hesse *et al.* 2004; Mangan *et al.* 2004; Pacifico *et al.* 2009). It is likely that the immunomodulatory effects induced by helminth infections serve to limit the extent of pulmonary inflammation and to repair damage produced by new infections (Fallon and Mangan 2007). This finding suggests that exposure to helminth parasites can result in significant alterations to the host lung environment, leading to down-modulation of pulmonary inflammatory responses induced by both allergens and new infections.

A gene expression profile of the lung of the host, obtained by Burke and colleagues (Burke *et al.* 2011) shed some light on the molecular events corresponding to this point of the parasite's development within the mammalian host. They focused on genes associated with innate immune responses induced by the presence of the migrating lung schistosomula of *S. japonicum* within a mouse host. A similar approach and some comparable responses have been demonstrated in the lungs of mice infected with *N. brasiliensis* (Reece *et al.* 2006). Genes identified as transcriptionally elevated during the active *S. japonicum* (Burke *et al.* 2011) infection within the host lung included SAA3 (Serum amyloid A3), a potent chemotactic factor for various inflammatory cells. Other pro-inflammatory genes seen as elevated included cytokine IL-1β, a member of the IL-1 super family and with SAA3 both constituents of the acute-phase response (Ahmed *et al.* 1996; Uhlar and Whitehead 1999). Furthermore the stimulation of the inflammatory mediator cascade by the acute-phase reaction, forms the core of an effective innate immune response (Uhlar and Whitehead 1999). IL-1 is a major pro-inflammatory mediator produced primarily by neutrophils, macrophages and monocytes within the inflammatory site (Arend *et al.* 1998; Uhlar and Whitehead 1999). IL-1 also induces the production of cell adhesion molecules (CAMs) on endothelial cell and platelet surfaces, which enables leukocyte extravasation to sites of infection (Ferroni *et al.* 1999; Oh and Diamond 2008). In support of this proposed mechanism Burke and colleagues (2011) also noted the transcriptional increase in cell adhesion molecule P-Selectin, in the lungs of infected mice. The presence of P-Selectin on the surface of platelets is a key requirement for pulmonary eosinophils and lymphocytes recruitment during allergic inflammation (Pitchford *et al.* 2005). This occurs since ligands for P-Selectin are present on the surface of neutrophils and eosinophils during inflammatory cells recruitment to sites of infection (Pitchford *et al.* 2005; Oh *et al.* 2009). Accordingly, several genes encoding chemokines for eosinophils and neutrophils were also up-regulated in the lungs of infected mice with migratory *S. japonicum* schistosomula (Burke *et al.* 2011).

Other key features of the host lung during the migration of schistosomula included the up-regulation of eosinophil chemokines (Burke *et al.* 2011). This was reflected in the eosinophil chemokines CCL11 and CCL24, (Eotaxin-1 and -2 (Ying *et al.* 1999)). Eosinophils form an integral part of the innate immune response in general, but are particularly prominent in the immune response to helminths and act by releasing secretory granular contents of cytotoxic cationic proteins (Blanchard and Rothenberg 2009). Eosinophilic infiltration of tissues and elevated levels within the blood are central features of the immune response mounted against a helminth infection; however, these features are also seen in lung associated diseases including emphysema and asthma (Menzies-Gow *et al.* 2002). It is thought that eosinophils contribute to the regulation of immune responses to helminths by promoting the production of chemokines that recruit Th2 lymphocytes (Fabre *et al.* 2009). In the *S. japonicum* study of Burke and colleagues (Burke *et al.* 2011) the direct detection of eosinophil populations was not, however, demonstrated by flow cytometry.

The chemokine CXCL1 is primarily associated with neutrophils and NKCs (Natural Killer Cells), was also transcriptionally up-regulated (Burke *et al.* 2011). As a major cellular component of an innate immune response neutrophils are one of the first leukocytes to infiltrate towards inflammatory sites with tissue (Menzies-Gow *et al.* 2002) and have proposed roles in fibrosis resolution and pro-inflammatory responses. Since fibrosis develops in a Th2 immune response, it is likely that up-regulation of CXCL1 in this study was associated with an inflammatory response. Flow cytometry identified the significant increases in the population of neutrophils in the lungs of infected mice, supporting the micro array data CXCL1.

In additional to investigating the role of innate immunity and the migration of lung schistosomula, the development of immunological memory through the presentation of antigen was determined to be not as prominent in the lung of infected mice. The proportion of antigen presenting cells (APCs), including alveolar macrophages (AM) and infiltrating macrophages, were not significantly up-regulated in mice during the migration of schistosomula in the host lung (Burke *et al.* 2011). Alveolar macrophages are the most abundant lung APCs and are considered less effective stimulators of immune responses when compared to other APCs (Thepen *et al.* 1989; Guth *et al.* 2009). This did not support the transcriptional data which indicated the elevation of chemokines chemotactic for macrophages, DCs (dendritic cells) and monocytes in the lungs of infected mice (Burke *et al.* 2011). This was apparent when the presence of DC were not detectible in the lungs of infected mice. This finding was in contrast to a study of Gonzalez-Juarrero *et al.* who reported increases in DC numbers in the lungs of mice infected with *Mycobacterium tuberculosis* (Gonzalez-Juarrero *et al.* 2003). These discrepancies may be related to known immunogenicity of bacterial infections that are often strongly inflammatory and therefore may recruit more DCs. Another possibility could involve a delayed activation of DCs in response to helminth infections.

While the immunological responses to migratory schistosomula are now more clear, the molecular pathways associated with tissue remodeling was also discussed by Burke and colleagues (2011). Transcriptional response to tissue damage and the promotion of healing, is a clear component of innate immunity (Reece *et al.* 2006). During the migration of lung schistosomula, counter intuitively, genes associated with tissue repair/remodeling were generally unchanged in the lungs during schistosomula migration when compared to other studies associated with early pulmonary responses to infection, namely in *N. brasiliensis* (Reece *et al.* 2006; Marsland *et al.* 2008). However two genes associated with repair, TIMP-1 and CXCL1 were detected as elevated. TIMP-1 is considered to act by reducing the activity of matrix metallopeptidase-12 (MMP-12) and as such increase the relative abundance of extracellular matrix components (Homer and Elias 2005), while CXCL1 as chemokine for neutrophils, has also been in implicated in promoting wound healing with delayed wound healing is seen in *CXCR2* (the receptor for CXCL1) knockout mice (Devalaraja *et al.* 2000; Gillitzer and Goebeler 2001). These transcriptional observations corresponded to the relative lack of histological damage seen in the lungs of mice during the schistosomula migration (Burke *et al.* 2011).

9.7 COMPARING *IN VITRO* AND *IN VIVO* SCHISTOSOMULA AS MODELS

The molecular and biochemical differences between *in vitro* and *in vivo* forms of schistosomula have not been thoroughly investigated. Commonly used *in vitro* cercarial transformation techniques include mechanical detachment of the tail (centrifuge/vortex/shear pressure), penetration of cercariae through isolated skin and incubation of cercariae in sera, chemicals and/or media varying in glucose concentration, osmolarity, lipid content etc. (Ramalho-Pinto *et al.* 1974, 1975; Brink *et al.* 1977; Cousin *et al.* 1981; Salafsky *et al.* 1988; Horemans *et al.* 1992). However, comprehensive comparison of these various forms with each other and with *in vivo* forms is lacking. After transformation, schistosomes are determined to be of a particular developmental stage (lung/liver schistosomula etc.) by gross morphology and/or the duration of culturing (Basch 1981). The mechanically transformed schistosomula for experiments in section 4.2 were harvested after three days of culturing. This time period represents the three days during which *in vivo Schistosoma japonicum* schistosomula migrate though the host, before they appear in the lungs (Rheinberg *et al.* 1997). Previous studies which have used mechanically transformed *in vitro* schistosomula in research of protein/antigen release or antigenicity of lung stage schistosomula, have cultured the schistosomula for 3 hours (Lawson *et al.* 1993), 7 days (Harrop and Wilson 1993), 8 days (Curwen *et al.* 2004) and up to 11 days (Mountford *et al.* 1995) prior to use in experiments as "lung" schistosomula. This variation in protocol exemplifies the arbitrary nature of culturing durations for *in vitro* schistosomes, although Harrop and Wilson (1993) state that the absence of erythrocytes in Medium 169 prevents transformation of parasites to the liver stage of development. Cousin *et al.* investigated the ultrastructure of mechanically transformed schistosomula and *in vivo* skin phase schistosomula and concluded that their developmental patterns achieved the same end-point but occurred at different rates (Cousin *et al.* 1981). Samuelson *et al.* (1980) also found that gross morphology of *in vitro* "lung" schistosomula and *in vivo* lung schistosomula observed by SEM was the same. One comparative study of various artificially transformed forms of schistosomula concluded their equivalence to *in vivo* forms, but used questionable criteria to define a "true schistosomulum" (Brink *et al.* 1977). Newly transformed *in vitro* schistosomula have been shown to be viable and infective by their ability to survive when introduced into the blood stream of a host and there is some evidence that *in vitro* schistosomula forms are structurally comparable to their *in vivo* equivalents, Nevertheless, the comparative molecular biology and biochemistry of *in vitro* and *in vivo* schistosomes is not well characterized (Cousin *et al.* 1981; Gold and Fletcher 2000).

9.7.1 Immune Evasion and Stress Responses in the Lung Schistosomulum

Immuno-modulating excretory/secretory products. During pulmonary migration, schistosomula interact intimately with pulmonary microvasculature endothelial cells (ECs), which play a pivotal role in initiating an inflammatory response (Carlos and Harlan 1994; Trottein *et al.* 1999a). A study of *in vitro* lung transformed schistosomula co-cultured with mouse microvascular lung endothelial cells (MLEs) and bovine brain capillary endothelial cells (BBCECs) has shown that the schistosomula actively and specifically bind to ECs resulting in decreased transendothelial permeability (Trottein *et al.* 1999a). This creates an anti-inflammatory phenotype caused by the closure of endothelial junctions, decreasing trafficking of molecules and cells such as leukocytes. Additionally, *in vitro* lung transformed schistosomula have been shown to reduce the expression of adhesion molecules E-selectin and vascular cell adhesion molecule 1 (VCAM-1) mRNA in human lung microvascular ECs (HLMECs) under inflammatory stimulation (Trottein *et al.* 1999b). This may reduce the binding and transmigration of leukocytes among these cells. It was hypothesized that the EC anti-inflammatory phenotype was induced by an excretory or secretory (ES) product of the schistosomulum (Trottein *et al.* 1999a, b).

Anti-inflammatory, immunomodulatory protein Sm16. Subsequent to the discovery that schistosomula cause anti-inflammatory responses, a potent immunomodulatory, anti-inflammatory secretory protein of infective stages of *Schistosoma mansoni* was cloned and partially characterized (Rao and Ramaswamy 2000). The 16.8 kDa protein (Sm16) has been shown to down-regulate the expression of interleukin-1α (IL-1α) in human keratinocytes, decreasing keratinocyte motility. Sm16 was also shown to prevent lymphoproliferation and suppress ICAM-1 (intracellular adhesion molecule-1) expression on ECs (Ramaswamy *et al.* 1995; Rao and Ramaswamy 2000). Rao and Ramaswamy (2000) reported expression of Sm16 in sporocysts, cercariae, *in vitro* transformed schistosomula and adult schistosomula of *S. mansoni*, with highest expression evident in schistosomula as determined by immuno-blot analysis of ES protein products. The Sm16 homolog in *S. japonicum* up regulated in lung isolated schistosomulum when compared to the mechanically transformed schistosomulum (Chai *et al.* 2006). These findings suggest an enhanced role of this immunomodulatory protein for the parasite during its stay in the lung environment where the schistosomulum must interact with immune reactive lung microvasculature endothelial cells. This may indicate that as well as the developmental pattern of stage-specific Sm16 expression observed by Rao and Ramaswamy (2000), Sm16 expression may additionally be regulated by environmental factors. It seems likely that exposure of lung schistosomula to lung microvascular ECs and various immune factors (cytokines, immmunoglobulin E) and immune cells such as mast cells, macrophages, eosinophils and basophils may affect the expression of genes involved in immune evasion (Maizels *et al.* 2004). If so, it is likely that the expression of many genes relating to immune evasion may be down-regulated in the mechanically transformed schistosomulum due to the lack of host-specific triggers or immune challenge, posed by culturing media.

Prostaglandins and 28 kDa glutathione-S-transferase. Prostaglandins (PG) are another excretory/secretory product exploited by schistosomes to modulate the host immune system. The production of PGD_2 by *S. mansoni* mediates various regulatory functions of inflammation and inhibits the migration of epidermal Langerhans cells (which play a key role in immune defense mechanisms) to the draining lymph nodes (Angeli *et al.* 2001; Herve *et al.* 2003). The parasite enzyme responsible for PGD_2 synthesis is a 28 kDa glutathione-S-transferase (Sm28GST) and this enzyme is excreted by skin-phase schistosomula (Herve *et al.* 2003; Mountford and Trottein 2004). A transcriptional comparison of *S. japonicum* mechanically transformed adult and lung *in vivo* schistosomula (Chai *et al.* 2006) found Sj28GST was up-regulated in the lung schistosomulum, again suggesting a potential role for triggers specific to the host environment in regulating the expression of this gene. The significant expression levels of GST-28 in lung schistosomula suggest a possible role for this GST and PGD_2 in the control of inflammation during the lung phase of the parasite.

Evasion of complement and antibody-mediated attack. Both cercariae and schistosomula activate complement and bind antibodies at their surface, however unlike cercariae, lung schistosomula are neither damaged nor killed by complement and are not susceptible to antibody-dependent cellular cytotoxicity (McLaren 1980; Lawson *et al.* 1993). Additionally, it has been demonstrated that antibody-mediated macrophage adherence to lung schistosomula is transient compared to mechanically transformed schistosomula and schistosomula recovered *in vivo* from host skin (Lawson *et al.* 1993). This suggests that there may exist some mechanism of immune evasion specific to schistosomula which have migrated to the lung, perhaps in response to stimuli found in the host blood or the lung itself.

Paramyosin is up-regulated in the lung schistosomulum. In addition to ES products, tegument bound molecules may also assist the lung schistosomulum in immune evasion. Paramyosin, a myofibrillary protein found in the muscle of adult schistosomes, cercariae and lung schistosomula, has also been identified on the tegument surface of the lung parasite (Gobert *et al.* 1997). A surface exposed non-filamentous form of paramyosin has been shown to inhibit the classical complement pathway by binding and inhibiting complement proteins (Deng *et al.* 2003). Additionally, tegument expressed paramyosin also acts as an Fc receptor, possibly absorbing antibodies onto the parasite surface (Laclette *et al.* 1992, 2001). The various isoforms of paramyosin have not yet been fully identified, sequenced and characterized.

Stress proteins and heat shock protein. Stress proteins are encoded by a group of genes whose expression can be induced by exposure of the organism to stress factors such as heat, nutrient deprivation and metabolic disruption (Scott and McManus 1999). The synthesis of heat shock proteins (HSPs) occurs in response to abrupt increases in environmental temperatures and correlates with the acquisition of tolerance to various stresses that would otherwise be lethal (Maresca and Kobayashi 1994). Such stresses occur during the schistosome life cycle when cercariae emerge from the fresh-water snail host, enter an aquatic environment, then penetrate the skin of the mammalian host (Hedstrom *et al.* 1987; Scott and McManus 1999). HSPs are highly conserved across many species and have been associated with the maintenance of mRNA maturation machinery, membrane structures, protein folding, repair and transport (Maresca and Kobayashi 1994). The schistosome 70 kDa HSP (HSP70) group is inducible by heat-stress and proteins are transiently expressed in high levels during cercarial transformation and constitutively expressed in the adult worm (Neumann *et al.* 1993). Various genes belonging to this group have been sequenced including *S. mansoni* HSP70 (Acc. No. L02415) and *S. japonicum* P18 HSP70 mRNA (Acc. No. AF044413), although their differing functions have not yet been fully characterized (Scott and McManus 1999). Neumann *et al.* (1993) reported higher levels of HSP70 mRNA in 24 hr mechanically transformed schistosomula when compared to adult parasites of *S. mansoni*, as indicated by northern blotting for HSP70 sequences. The up-regulation of this HSP sequence in the lung schistosomulum may indicate that during this developmental stage, environmental stresses exist that require molecular responses in the lung schistosomulum, which may have lesser emphasis in the adult parasite. Up-regulation of HSP70-related sequences in the *S. japonicum* lung schistosomulum (Chai *et al.* 2006) may be associated with acquisition of heat tolerance (the schistosomula have so far only had three days to adapt to the 37°C host temperature) or perhaps tolerance to other perhaps multifactorial stresses such as those imposed by host immune attack.

The differential gene expression profiles of lung schistosomula and mechanically transformed schistosomula, suggest that the use of *in vitro* schistosomes may potentially produce misleading or inaccurate results that could allow under- or over-estimations of actual parasite processes. *In vitro* schistosomes are utilized on the premise that schistosome development can be simulated independently of the host environment and that observed morphological similarities confer overall equivalence of *in vitro* with *in vivo* forms. However it seems likely that factors in the varying host environments encountered by *in vivo* schistosomes, manifest in their ultrastructure and molecular biology. Although perhaps useful in studies of gross morphology or pre-determined patterns of schistosome development, the application of *in vitro* schistosomes in molecular or biochemical studies should be carefully considered. A comprehensive ultrastructural comparison of mechanically transformed and lung schistosomula would be a useful complement to their expression profiles.

9.8   LUNG SCHISTOSOMULA AND NUTRITIONAL UPTAKE

Whereas cercariae are non-feeding and derive energy from endogenous glycogen reserves, adult schistosomes are dependent on glucose absorbed from the host's blood (Bruce *et al.* 1969; Halton 1997). Glucose and other small molecular weight solutes are transported across the tegument, with adult parasites reportedly consuming their dry weight in glucose every 5 hours using schistosome glucose transporter proteins (SGTPs) (Skelly *et al.* 1998; Brouwers *et al.* 1999). Cercarial transformation into schistosomula involves a complex set of metabolic and physiological changes, including a switch from aerobic to anaerobic energy metabolism, loss of the glycocalyx and complete reconstruction of the tegumental membranes (Skelly and Shoemaker 2000). Glucose transporter proteins SGTP1 and SGTP4 have been localized to the tegument of adult parasites and newly *in vitro* transformed schistosomula, but SGTP4 is absent in cercariae and other extra-mammalian parasite stages, all of which lack the heptalaminate membrane of the intra-mammalian stages (Skelly and Shoemaker 1996). The presence of SGTPs has not been reported in *in vivo* lung schistosomula and it is unknown how schistosomula in the lung supplant their depleted internal stores of glycogen (Skelly and Shoemaker 1995; Ridi *et al.* 2003).

In the lung parasite, the trilaminate cercarial membrane has been replaced by a heptalaminate membrane characteristic of the adult parasite, but the overall tegument thickness has not increased dramatically during cercarial transformation, remaining a fraction of that of the adult parasite at 500–700 nm in thickness, compared to 4–5 μm in thickness for the male parasite and 1–2 μm for the female (Gobert *et al.* 2003). The consistency in tegument thickness between specimens of lung schistosomula may indicate no sexual differentiation in this respect at the lung stage and indicates that the tegument is not yet functioning to the capacity of the adult tegument. Despite the comparative thinness of the tegument, it seems likely that nutrient absorbing machinery is already present across the lung parasite tegument, given the localization of SGTP1 and SGTP4 to the tegument of newly *in vitro* transformed schistosomula (Skelly and Shoemaker 2000). Other indications that the tegument of the lung schistosomula is used as an absorptive surface include the great increase in surface area of the lung parasite by the formation of sagittal ridges and invaginations, creating a sponge-like pitted surface (see Gobert and Jones, Chapter 10) as compared to the mildly undulating tegument of the cercariae (Sobhon *et al.* 1988). It is thought that the similarly pitted tegument in the adult parasite may trap plasma in the surface channels, facilitating absorption and/or the diffusion of glucose inward by reducing the distance between the host bloodstream and the worm's internal tissues (Hockley 1973; Dalton *et al.* 2004).

One study of schistosomula *in situ* within the lung vasculature, observed that when constricted within the pulmonary vasculature, the tegumental surface of the parasites appeared flattened and unridged with few pits (Crabtree and Wilson 1985). This may suggest extensibility and flexibility as the function of tegument folds rather than the creation of a large absorptive surface. In support of this theory, a subtegumentary interstitial fibrous layer present in the cercariae, young schistosomula and adult parasites is absent in the lung schistosomulum, perhaps permitting the flexibility necessary for migration out of the lungs (Crabtree and Wilson 1985). Conversely, a similar *in situ* study found the tegument of the lung parasites not only invaginated but seemingly entrapping and eventually absorbing the adjacent lung tissue in narrow-necked flask-shaped infoldings of the tegument (Bruce *et al.* 1974).

9.8.1   Lung Schistosomula may have a Functional Gut

There are conflicting views regarding the ability of the lung-stage schistosomula to feed from the host blood stream. There are some accounts of the lung schistosomula lacking a functional gut, while others claim the gut may be partially functional even prior to the lung-stage (Bennett and Caulfield 1991; Ridi *et al.* 2003). Male and female adult parasites ingest an estimated 39,000 to 330,000 erythrocytes per hour respectively (for *Schistosoma mansoni)* and these are digested and absorbed in the esophagus and gastrodermis (Halton 1997). Four-day-old artificially transformed cultured schistosomula of *S. mansoni* have been shown to ingest red blood cells and can apparently ingest dextrans and serum albumin prior to this (*S. mansoni* schistosomula appear in the lungs on day three, peaking at day 6), however it remains unknown whether *in vivo* lung schistosomula of *S. japonicum* can actively ingest and digest red blood cells or proteins (Bennett and Caulfield 1991; Bogitsh 1993).

In both adult schistosomes and cercariae, the lining of the oral cavity and proximal esophagus is characteristically similar to the surface tegument, featuring several similar inclusion bodies such as elongated and membranous bodies as well as numerous outfoldings and this seems in keeping with observations of the esophagus in the lung stage schistosomulum (Fig. 9.6A) (Senft 1976; Dorsey *et al.* 2002). In the adult parasite, the esophagus is glandular and associated modified tegumental cell-bodies secrete digestive enzymes for the lysis and digestion of blood cells (Dalton *et al.* 2004). However, the presence or absence of glandular activity in the lung schistosomulum esophagus or cecum is difficult to prove without histochemical localization of digestive enzymes.

In adult schistosomes, cercariae and lung schistosomula, a demarcation separating the gastrodermis and esophagus is present as the distal region of the esophagus becomes increasingly convoluted into long dense folds which evidently entrap material in the lumen (Fig. 9.5A) (Senft 1976; Dorsey *et al.* 2002). Despite being a non-feeding stage, the cecal lumen of the cercariae often contains several bodies of varying shapes and electron densities, however an increase in the electron density of esophageal granules in (vortex-transformed cultured) schistosomula as compared to cercariae has been attributed to the activity of protease in these schistosomula (Bennett and Caulfield 1991; Dorsey *et al.* 2002). Indeed, the morphology and contents of the lung schistosomulum cecum seem to exhibit the major features of the functioning cecum of the adult parasite.

The gastrodermis of the adult schistosome is described as a syncytial absorptive tissue featuring irregular syncytial projections and numerous lamellae which project into the lumen and this description also matches observations of the lung schistosomulum cecum (Smyth and Halton 1983a; Dalton *et al.* 2004). In the adult, the flaps created by lamellae sequester hemoglobin substrate into discrete masses which are either fused to the lamellae or embraced in the crypts of the folds where lamellae branch and rejoin the gut surface to form loops (Morris 1968; Senft 1976). Although the electron-dense material observed in the cecum of the lung schistosomulum cannot be proven to be hemoglobin substrate without further studies, its processing in the schistosomulum cecum is very similar to that described for the adult. Additionally, lipid-like droplets thought to be an intermediate of hemoglobin degradation similarly are seen embraced by lamellae in the adult parasite and are thought to be taken up into the epithelium, as intact droplets can be observed in various stages of incorporation into the cecal syncytium of the adult (Morris 1968). Lipids (of unknown origin) were similarly observed in various stages of incorporation into the syncytium of the lung schistosomulum cecal lining (Fig. 9.5B, C, D, E). Further brown color pigment which could be a product of breaking down hemoglobin is also noted in lung schistosomula by light microscopy (Fig. 9.8). The morphological evidence combined with the treatment of haem substrate-like material and lipids in the cecum seems to indicate that ingestion and digestion of blood cells or other blood proteins may occur in the lung stage schistosomulum.

9.9 CONCLUSIONS

The schistosomulum is a critical stage in the intra-mammalian phase of the schistosome life cycle, as defense mechanisms against host immunological attack are not yet fully established in the parasite (Wilson 2009). In a recent study by Sotillo *et al.* (2015) *Schistosoma mansoni* schistosomula were cultured after mechanical transformation for 3 hours and 2 and 5 days, their surfaces were biotinylated, the teguments were removed by freeze thawing and processed by iTRAQ labeling, OFFGEL electrophoresis and mass spectrometry. The study findings demonstrated the dynamic nature of the schistosomula surface, with the appearance of proteins associated with important functions including nutrient uptake, anti-oxidant activity and heat shock responses, as well as more general enzymatic and proteolytic activities. The latter functions could have been attributable to the functions of the parasite gastrodermis as previous studies using laser microscopy microdissection and transcriptomics of adult schistosome parasite gut identified transcripts of similar genes in this region (Gobert *et al.* 2009; Nawaratna *et al.* 2011). The Sotillo *et al.* study (Sotillo *et al.* 2015), however, involved extensive analysis by immunofluorescence of the biotinylated surface products of the parasite and eliminated the gastrodermis as a potential source of these proteins. While these results demonstrated how schistosomes adapt during the migration from the aquatic cercarial phase to the intra-mammalian schistosomulum, there are potential limitations that need to be considered. These concerns relate to the use of mechanical transformation of schistosomula followed by *in vitro* maintenance of the parasites, which are not subjected to any immunological pressure that would normally occur *in vivo*. Indeed, a complete genome transcriptional comparison of mechanically transformed *S. japonicum* schistosomula versus parasites isolated *in vivo* showed that many genes associated with avoiding the host immune response were up regulated only in worms obtained *in vivo* (Chai *et al.* 2006). However, other transcriptional studies, such as that by Protasio and colleagues (Protasio *et al.* 2013), which compared skin and mechanically transformed *S. mansoni* schistosomula, showed that the differences are, in fact, minimal. It is not clear whether any differences at the transcriptional level, when taking into account different isolation or maintenance methods for schistosomula, would be reflected at the protein level, especially when considering the host-parasite interaction.

Further insight into the biology of the early intra-mammalian schistosomulum stage of schistosomes was provided in a comparative proteomics study of *Fasciola hepatica* (liver fluke) juveniles and *S. bovis* schistosomula, both obtained *in vivo* (De la Torre Escudero *et al.* 2011). Again the focus was the host-parasite interaction with the tegument being used for analysis. The advantage of undertaking such a comparative study is clearly the distinctive biology of the two parasites used and the two very different microenvironments they inhabit in the mammalian host. While both parasites utilize ruminants as definitive hosts, *F. hepatica* when mature resides within the bile duct, whereas *S. bovis* is found within the portal and mesenteric veins. Proteins observed only on the surface of juvenile *F. hepatica* and *S. bovis* included functional groups associated with gene ontologies, proteolysis, cytoskeleton, signaling and energy/metabolism, while unique proteins were associated with each species and most were common to both. The abundance of proteins between the two species varied, when compared with all of the proteins identified, with *S. bovis* enriched in the areas of transcription and translation and signaling, while in *F. hepatica* the function of proteolysis was more prominent. What these results demonstrate is that flatworm proteins expressed at the host-parasite interface reflect the way in which these parasites adapt to the host microenvironment.

9.10 ACKNOWLEDGEMENTS

The support of the National Health and Medical Research Council of Australia (NHMRC) is acknowledged.

9.11 LITERATURE CITED

Agatsuma T. 2003. Origin and evolution of *Schistosoma japonicum*. Parasitology International 52(2003): 335–340.

Ahmed N, Thorley R, Xia D, Samols D, Webster RO. 1996. Transgenic mice expressing rabbit C-reactive protein exhibit diminished chemotactic factor-induced alveolitis. American Journal of Respiratory and Critical Care Medicine 153(3): 1141–1147.

Angeli V, Faveeuw C, Roye O, Fontaine J, Teissier E, Capron A, Wolowczuk I, Capron M, Trottein F. 2001. Role of the parasite-derived prostaglandin D-2 in the inhibition of epidermal langerhans cell migration during schistosomsiasis infection. Journal of Experimental Medicine 193: 1135–1147.

Arend WP, Malyak M, Guthridge CJ, Gabay C. 1998. Interleukin-1 receptor antagonist: role in biology. Annual Review of Immunology 16: 27–55.

Basch PF. 1981. Cultivation of *Schistosoma mansoni in vitro*. 1. Establishment of cultures from cercariae and development until pairing. Journal of Parasitology 67(2): 179–185.

Basch PF, Basch N. 1982. *Schistosoma mansoni*: scanning electron microscopy of schistsomula, adults and eggs grown *in vitro*. Parasitology 85: 333–338.

Bennett MW, Caulfield JP. 1991. *Schistosoma mansoni*: ingestion of dextrans, serum albumin and IgG by schistosomula. Experimental Parasitology 73: 52–61.

Blanchard C, Rothenberg ME. 2009. Biology of the eosinophil. Advances in Immunology 101: 81–121.

Bogitsh BJ. 1989. Observations on digestion in schistosomes or "Blood and Guts". Transactions of the American Microscopical Society 108: 1–5.

Bogitsh BJ. 1993. The feeding of A type red blood cells *in vitro* and the ability of *Schistosoma mansoni* schistosomules to acquire a epitopes on their surfaces. American Society of Parasitologists 79: 946–948.

Braschi S, Curwen RS, Ashton PD, Verjovski-Almeida S, Wilson A. 2006. The tegument surface membranes of the human blood parasite *Schistosoma mansoni*: a proteomic analysis after differential extraction. Proteomics 6(5): 1471–1482.

Braschi S, Wilson RA. 2006. Proteins exposed at the adult schistosome surface revealed by biotinylation. Molecular and Cellular Proteomics 5(2): 347–356.

Brink LH, McLAren DJ, Smithers SR. 1977. *Schistosoma mansoni*: a comparative study of artificially transformed schistosomula and schistosomula recovered after cercarial penetration of isolated skin. Parasitology 74: 73–86.

Brouwers JFHM, Skelly PJ, Golde LMGV, Tielens GM. 1999. Studies on phospholipid turnover argue against sloughing of tegumental membranes in adult *Schistosoma mansoni*. Parasitology 119: 287–294.

Bruce JI, Pezzlo F, Yajima Y, McCarty JE. 1974. *Schistosoma mansoni*: Pulmonary phase of schistosomule migration studied by electron microscopy. Experimental Parasitology 35: 150–160.

Bruce JI, Weis EM, Stirewalt MA, Lincicome DR. 1969. *Schistosoma mansoni*: glycogen content and utilization of glucose, pyruvate, glutamate and citric acid cycle intermediates by cercariae and schistosomules. Experimental Parasitology 26: 29–40.

Burke ML, McGarvey L, McSorley HJ, Bielefeldt-Ohmann H, McManus DP, Gobert GN. 2011. Migrating *Schistosoma japonicum* schistosomula induce an innate immune response and wound healing in the murine lung. Molecular Immunology 49(1-2): 191–200.

Carlos TM, Harlan JM. 1994. Leukocyte-endothelial adhesion molecules. Blood 84: 2068–2101.

Caulfield JP, Lenzi HL, Elsas P, Dessein AJ. 1985. Ultrastructure of the attack of eosinophils stimulated by blood mononuclear cell products on schistosomula of *Schistosoma mansoni*. American Journal of Pathology 120(3): 380–390.

Chai M, McManus DP, McInnes R, Moertel L, Tran M, Loukas A, Jonesa MK, Gobert GN. 2006. Transcriptome profiling of lung schistosomula, *in vitro* cultured schistosomula and adult *Schistosoma japonicum*. Cellular and Molecular Life Sciences 63(7-8): 919–929.

Clegg JA. 1965. *In vitro* cultivation of *Schistosoma mansoni*. Experimental Parasitology 16: 133–147.

Coultas KA, Zhang SM. 2012. *In vitro* cercariae transformation: comparison of mechanical and nonmechanical methods and observation of morphological changes of detached cercariae tails. Journal of Parasitology 98(6): 1257–1261.

Cousin CE, Stirewalt MA, Dorsey CH. 1981. *Schistosoma mansoni*: ultrastructure of early transformation of skin- and shear-pressure-derived schistosomules. Experimental Parasitology 51: 341–365.

Crabtree JE, Wilson RA. 1985. *Schistosoma mansoni*: an ultrastructural examination of pulmonary migration. Parasitology 92: 343–354.

Curwen RS, Ashton PD, Johnston DA, Wilson RA. 2004. The *Schistosoma mansoni* soluble proteome: a comparison across four life cycle stages. Molecular and Biochemical Parasitology 138(1): 57–66.

Dalton JP, Skelly P, Halton DW. 2004. Role of the tegument and gut in nutrient uptake by parasitic platyhelminths. Canadian Journal of Zoology 82: 211–232.

De la Torre Escudero E, Manzano-Roman R, Valero L, Oleaga A, Perez-Sanchez R, Hernandez-Gonzalez A, Siles-Lucas M. 2011. Comparative proteomic analysis of *Fasciola hepatica* juveniles and *Schistosoma bovis* schistosomula. Journal of Proteomics 74(9): 1534–1544.

Dean DA. 1977. Decreased binding of cytotoxic antibody by developing *Schistosoma mansoni*. Evidence for a surface change independent of host antigen adsorption and membrane turnover. The Journal of Parasitology 63(3): 418–426.

Dean DA, Mangold BL. 1992. Evidence that both normal and immune elimination of *Schistosoma mansoni* take place at the lung stage of migration prior to parasite death. American Journal of Tropical Medicine and Hygiene 47(2): 238–248.

Deng J, Gold D, LoVerde PT, Fishelson Z. 2003. Inhibition of complement membrane attack complex by *Schistosoma mansoni* paramyosin. Infection and Immunity 71: 6402–6410.

Devalaraja RM, Nanney LB, Du J, Qian Q, Yu Y, Devalaraja MN, Richmond A. 2000. Delayed wound healing in CXCR2 knockout mice. Journal of Investigative Dermatology 115(2): 234–244.

Dissous C, Khayath N, Vicogne J, Capron M. 2006. Growth factor receptors in helminth parasites: signalling and host-parasite relationships. FEBS Letters 580(12): 2968–2975.

Dorsey CH, Cousin CE, Lewis FA, Stirewalt MA. 2002. Ultrastructure of the *Schistosoma mansoni* cercaria. Micron 33(3): 279–323.

Fabre V, Beiting DP, Bliss SK, Gebreselassie NG, Gagliardo LF, Lee NA, Lee JJ, Appleton JA. 2009. Eosinophil deficiency compromises parasite survival in chronic nematode infection. Journal of Immunology 182(3): 1577–1583.

Fallon PG, Mangan NE. 2007. Suppression of TH2-type allergic reactions by helminth infection. Nature Reviews Immunology 7(3): 220–230.

Ferroni P, Basili S, Vieri M, Martini F, Labbadia G, Bellomo A, Gazzaniga PP, Cordova C, Alessandri C. 1999. Soluble p-selectin and proinflammatory cytokines in patients with polygenic type IIa hypercholesterolemia. Haemostasis 29(5): 277–285.

Gillitzer R, Goebeler M. 2001. Chemokines in cutaneous wound healing. Journal of Leukocyte Biology 69(4): 513–521.

Gobert GN, McManus DP, Nawaratna S, Moertel L, Mulvenna J, Jones MK. 2009. Tissue specific profiling of females of *Schistosoma japonicum* by integrated laser microdissection microscopy and microarray analysis. PLoS Neglected Tropical Diseases 3(6): e469.

Gobert GN, Stenzel DJ, Jones MK, Allen DE, McManus DP. 1997. *Schistosoma japonicum*: immunolocalization of paramyosin during development. Parasitology 114(Pt 1): 45–52.

Gobert GN, Stenzel DJ, McManus DP, Jones MK. 2003. The ultrastructural architecture of the adult *Schistosoma japonicum* tegument. International Journal for Parasitology 33(14): 1561–1575.

Gold D, Fletcher E. 2000. Influence of mechanical tail-detachment techniques of schistosome cercariae on the production, viability and infectivity of resultant schistosomula: a comparative study. Parasitology Research 86: 570–572.

Gonzalez-Juarrero M, Shim TS, Kipnis A, Junqueira-Kipnis AP, Orme IM. 2003. Dynamics of macrophage cell populations during murine pulmonary tuberculosis. Journal of Immunology 171(6): 3128–3135.

Gui M, Kusel JR, Shi RE, Ruppel A. 1995. *Schistosoma japonicum* and *S. mansoni*: comparison of larval migration patterns in mice. Journal of Helminthology 69: 19–25.

Guth AM, Janssen WJ, Bosio CM, Crouch EC, Henson PM, Dow SW. 2009. Lung environment determines unique phenotype of alveolar macrophages. American Journal of Physiology Lung Cellular and Molecular Physiology 296(6): L936–946.

Halton DW. 1997. Nutritional adaptations to parasitism within the platyhelminthes. International Journal for Parasitology 27(6): 693–704.

Harrop R, Coulson PS, Wilson RA. 1998. Characterization, cloning and immunogenicity of antigens released by lung-stage larvae of *Schistosoma mansoni*. Parasitology 118: 583–594.

Harrop R, Wilson RA. 1993. Protein synthesis and release by cultured schistosomula of *Schistosoma mansoni*. Parasitology 107: 265–274.

He YX. 1993. Biology of *Schistosoma japonicum*. From cercaria penetrating into host skin to producing egg. Chinese Medical Journal 106(8): 576–583.

He YX, Chen L, Ramaswamy K. 2002. *Schistosoma mansoni*, *S. haematobium* and *S. japonicum*: early events associated with penetration and migration of schistosomula through human skin. Experimental Parasitology 102(2): 99–108.

He YX, Salafsky B, Ramaswamy K. 2005. Comparison of skin invasion among three major species of *Schistosoma*. Trends in Parasitology 21(5): 201–203.

Hedstrom R, Culpepper J, Harrison RA, Agabian N, Newport G. 1987. A major immunogen in *Schistosoma mansoni* infections is homologous to the heat-shock protein Hsp70. Journal of Experimental Medicine 165: 1430–1435.

Herve M, Angeli V, Pinzar E, Wintjens R, Faveeuw C, Narumiya S, Capron A. 2003. Pivotal roles of the parasite PGD2 synthase and of the host D prostanoid receptor 1 in schistosome immune evasion. European Journal of Immunology 33: 2764–2772.

Hesse M, Piccirillo CA, Belkaid Y, Prufer J, Mentink-Kane M, Leusink M, Cheever AW, Shevach EM, Wynn TA. 2004. The pathogenesis of schistosomiasis is controlled by cooperating IL-10-producing innate effector and regulatory T cells. Journal of Immunology 172(5): 3157–3166.

Hockley D, McLaren DJ. 1973. *Schistosoma mansoni*: changes in the outer membrane of the tegument during development from cercaria to adult worm. International Journal of Parasitology 3: 13–25.

Hockley DJ. 1973. Ultrastructure of the tegument of *Schistosoma*. Advances in Parasitology 11: 233–305.

Homer RJ, Elias JA. 2005. Airway remodeling in asthma: therapeutic implications of mechanisms. Physiology (Bethesda) 20: 28–35.

Horemans AMC, Tielens A, van den Bergh SG. 1992. The reversible effect of glucose on the energy metabolism of *Schistosoma mansoni* cercariae and schistosomula. Molecular and Biochemical Parasitology 51: 73–80.

Jones MK, Gobert GN, Zhang L, Sunderland P, McManus DP. 2004. The cytoskeleton and motor proteins of human schistosomes and their roles in surface maintenance and host-parasite interactions. BioEssays 26(7): 752–765.

Jordan P, Webbe G, Sturrock RF. 1993. *Human Schistosomiasis*. Wallingford, UK: Cab International. 465 p.

Laclette JP, Shoemaker CB, Richter D, Arcos L, Pante N, Cohen C, Bing D, Nicholson-Weller A. 1992. Paramyosin inhibits complement C1. Journal of Immunology 148: 124–128.

Lawson BWL, Bickle QD, Taylor MG. 1993. Mechanisms involved in the loss of antibody-mediated adherence of macrophages to lung-stage schistosomula of *Schistosoma mansoni in vitro*. Parasitology 106: 463–469.

Loke P, MacDonald AS, Robb A, Maizels RM, Allen JE. 2000. Alternatively activated macrophages induced by nematode infection inhibit proliferation via cell-to-cell contact. European Journal of Immunology 30(9): 2669–2678.

Loukas A, Jones MK, King LT, Brindley PJ, McManus DP. 2001. Receptor for Fc on the surface of schistosomes. Infection and Immunity 69: 3646–3651.

Maizels RM, Balic A, Gomez-Escobar N, Nair M, Taylor MD, Allen JE. 2004. Helminth parasites – masters of regulation. Immunological Reviews 201: 89–116.

Mangan NE, Fallon RE, Smith P, van Rooijen N, McKenzie AN, Fallon PG. 2004. Helminth infection protects mice from anaphylaxis via IL-10-producing B cells. Journal of Immunology 173(10): 6346–6356.

Maresca B, Kobayashi GS. 1994. Hsp70 in parasites: as an inducible protective protein and as an antigen. Experientia 50: 1067–1074.

Marikovsky M, Parizade M, Arnon R. 1990. Complement regulation on the surface of cultured schistosomula and adult worms of *Schistosoma mansoni*. European Journal of Immunology 20(1): 221–227.

Marsland BJ, Kurrer M, Reissmann R, Harris NL, Kopf M. 2008. *Nippostrongylus brasiliensis* infection leads to the development of emphysema associated with the induction of alternatively activated macrophages. European Journal of Immunology 38(2): 479–488.

McLaren DJ. 1980. *Schistosoma mansoni*: The parasite surface in relation to host immunity. *In*: Brown KN, editor. *Tropical Medicine Research Studies*. Chichester, John Wiley. 229 p.

Medeiros M, Jr, Figueiredo JP, Almeida MC, Matos MA, Araujo MI, Cruz AA, Atta AM, Rego MA, de Jesus AR, Taketomi EA and others. 2003. *Schistosoma mansoni* infection is associated with a reduced course of asthma. Journal of Allergy and Clinical Immunology 111(5): 947–951.

Menzies-Gow A, Ying S, Sabroe I, Stubbs VL, Soler D, Williams TJ, Kay AB. 2002. Eotaxin (CCL11) and eotaxin-2 (CCL24) induce recruitment of eosinophils, basophils, neutrophils and macrophages as well as features of early- and late-phase allergic reactions following cutaneous injection in human atopic and nonatopic volunteers. Journal of Immunology 169(5): 2712–2718.

Milligan JN, Jolly ER. 2011. Cercarial transformation and *in vitro* cultivation of *Schistosoma mansoni* schistosomules. Journal of Visualized Experiments (54): e3191, doi: 10.3791/3191.

Mitchell GF, Tiu WU, Garcia EG. 1991. Infection characteristics of *Schistosoma japonicum* in mice and relevance to the assessment of schistosome vaccines. Advances in Parasitology 30: 167–200.

Mo HM, Lei JH, Jiang ZW, Wang CZ, Cheng YL, Li YL, Liu WQ. 2008. *Schistosoma japonicum* infection modulates the development of allergen-induced airway inflammation in mice. Parasitology Research 103(5): 1183–1189.

Morris GP. 1968. Fine structure of gut epithelium of *Schistosoma mansoni*. Experientia 24(5): 480–482.

Mountford AP, Harrop R, Wilson A. 1995. Antigens derived from lung-stage larvae of *Schistosoma mansoni* are efficient stimulators of proliferation and gamma interferon secretion by lymphocytes from mice vaccinated with attenuated larvae. Infection and Immunity 63(5): 1980–1986.

Mountford AP, Harrop R. 1998. Vaccination against Schistosomiasis: the case for lung-stage antigens. Parasitology Today 14(3): 109–114.

Mountford AP, Trottein F. 2004. Schistosomes in the skin: a blance between immune priming and regulation. Trends in Parasitology 20: 221–226.

Mulvenna J, Moertel L, Jones MK, Nawaratna S, Lovas EM, Gobert GN, Colgrave M, Jones A, Loukas A, McManus DP. 2010. Exposed proteins of the *Schistosoma japonicum* tegument. International Journal for Parasitology 40(5): 543–554.

Nawaratna SS, McManus DP, Moertel L, Gobert GN, Jones MK. 2011. Gene Atlasing of digestive and reproductive tissues in *Schistosoma mansoni*. PLoS Neglected Tropical Diseases 5(4): e1043.

Neumann S, Ziv E, Lantner F, Schechter I. 1993. Regulation of HSP70 gene expression during the life cycle of the parasite helminth *Schistosoma mansoni*. European Journal of Biochemistry 212: 589–596.

Oh H, Diamond SL. 2008. Ethanol enhances neutrophil membrane tether growth and slows rolling on P-selectin but reduces capture from flow and firm arrest on IL-1-treated endothelium. Journal of Immunology 181(4): 2472–2482.

Oh H, Mohler ER, 3rd, Tian A, Baumgart T, Diamond SL. 2009. Membrane cholesterol is a biomechanical regulator of neutrophil adhesion. Arteriosclerosis Thrombosis and Vascular Biology 29(9): 1290–1297.

Pacifico LG, Marinho FA, Fonseca CT, Barsante MM, Pinho V, Sales-Junior PA, Cardoso LS, Araujo MI, Carvalho EM, Cassali GD and others. 2009. *Schistosoma mansoni* antigens modulate experimental allergic asthma in a murine model: a major role for CD4+ CD25+ Foxp3+ T cells independent of interleukin-10. Infection and Immunity 77(1): 98–107.

Parizade M, Arnon R, Lachmann PJ, Fishelson Z. 1994. Functional and antigenic similarities between a 94-kD protein of *Schistosoma mansoni* (SCIP-1) and human CD59. Journal of Experimental Medicine 179: 1625–1636.

Pitchford SC, Momi S, Giannini S, Casali L, Spina D, Page CP, Gresele P. 2005. Platelet P-selectin is required for pulmonary eosinophil and lymphocyte recruitment in a murine model of allergic inflammation. Blood 105(5): 2074–2081.

Protasio AV, Dunne DW, Berriman M. 2013. Comparative study of transcriptome profiles of mechanical- and skin-transformed *Schistosoma mansoni* schistosomula. PLoS Neglected Tropical Diseases 7(3): e2091.

Ramalho-Pinto FJ, Gazzinelli G, Howells RE, Mota-Santos TA, Figueiredo EA, Pellegrino J. 1974. *Schistosoma mansoni*: a defined system for the step-wise transformation of the cercaria to schistsomule. Experimental Parasitology 36: 360–372.

Ramalho-Pinto FJ, Gazzinelli G, Howells RE, Pellegrino J. 1975. Factors affecting surface changes in intact cercariae and cercarial bodies of *Schistosoma mansoni*. Parasitology 71: 19–25.

Ramaswamy K, Salafsky B, Lykken M. 1995. Modulation of IL-1a, IL-1b and IL-1RA production in human keratinocytes by schistosomula of *Schistosoma mansoni*. Immunology and Infectious Diseases 5: 100–107.

Rao KV, Ramaswamy K. 2000. Cloning and expression of a gene encoding Sm16, an anti-inflammatory protein from *Schistosoma mansoni*. Molecular and Biochemical Parasitology 108(1): 101–108.

Reece JJ, Siracusa MC, Scott AL. 2006. Innate immune responses to lung-stage helminth infection induce alternatively activated alveolar macrophages. Infection and Immunity 74(9): 4970–4981.

Reece JJ, Siracusa MC, Southard TL, Brayton CF, Urban JF, Jr, Scott AL. 2008. Hookworm-induced persistent changes to the immunological environment of the lung. Infection and Immunity 76(8): 3511–3524.

Rheinberg CE, Mone H, Caffrey CR, Imbert-Establet D, Jourdane J, Ruppel A. 1997. *Schistosoma haematobium*, *S. intercalatum*, *S. japonicum*, *S. mansoni*, and *S. rodhaini* in mice: relationship between patterns of lung migration by schistosomula and perfusion recovery of adult worms. Parasitology Research 84: 338–342.

Ridi RE, Mohamed SH, Tallima H. 2003. Incubation of *Schistosoma mansoni* lung-stage schistosomula in corn oil exposes their surface membrane antigenic specificities. Journal of Parasitology 89(5): 1064–1067.

Ross AGP, Sleigh AC, Li YS, Davis GM, Williams GM, Jiang Z, Feng Z, McManus DP. 2001. Schistosomiasis in the People's Republic of China: Prospects and challenges for the 21st century. Clinical Microbiology Reviews 14(2): 270–295.

Rumjanek FD, Aparecida M, Pereira C, Silveira AMV. 1984. The interaction of human serum with the surface membrane of schistosomula of *Schistosoma mansoni*. Molecular and Biochemical Parasitology 14: 63–73.

Salafsky B, Fusco AC, Whitley K, Nowicki D, Ellenberger B. 1988. *Schistosoma mansoni*: analysis of cercarial transformation methods. Experimental Parasitology 67: 116–127.

Samuelson JC, Caulfield JP, David JR. 1980. *Schistosoma mansoni*: post-transformational surface changes in schistosomula grown *in vitro* and in mice. Experimental Parasitology 50: 369–383.

Scott JC, McManus DP. 1999. Identification of novel 70-kDa heat shock protein-encoding cDNAs from *Schistosoma japonicum*. International Journal for Parasitology 29: 437–444.

Senft AW. 1976. Observations in the physiology of the gut of *Schistosoma mansoni*. pp. 335–342. *In*: Van den Bosch H, editor. *Biochemistry of Parasites and Host-Parasite Relationships*. Oxford: Elsevier/North-Holland Biomedical Press.

Skelly PJ, Alan Wilson R. 2006. Making sense of the schistosome surface. Advances in Parasitology 63: 185–284.

Skelly PJ, Kim JW, Cunningham J, Shoemaker CB. 1994. Cloning, characterization and functional expression of cDNAs encoding glucose transporter proteins from the human parasite *Schistosoma mansoni*. Journal of Biological Chemistry 269(6): 4247–4253.

Skelly PJ, Shoemaker CB. 1995. A molecular genetic study of the variations in metabolic function during schistosome development. Memorias do Instituto Oswaldo Cruz 90(2): 1678–8060.

Skelly PJ, Shoemaker CB. 1996. Rapid appearance and asymmetric distribution of glucose transporter SGTP4 at the apical surface of intramammalian-stage *Schistosoma mansoni*. Proceedings of the National Academy of Sciences of the United States of America 93: 3642–3646.

Skelly PJ, Shoemaker CB. 2000. Induction cues for tegument formation during the transformation of *Schistosoma mansoni* cercariae. International Journal for Parasitology 30: 625–631.

Skelly PJ, Tielens AGM, Shoemaker CB. 1998. Glucose transport and metabolism mammalian-stage schistosomes. Parasitology Today 14(10): 402–406.

Smyth JD, Halton DW. 1983. Fine structure of the gut epithelium of *Schistosoma mansoni. The Physiology of Trematodes*. Cambridge: Cambridge University Press. 446 p.

Sobhon P, Anupunpisit V, Yuan HC, Upatham ES, Saitongdee P. 1988. *Schistosoma japonicum* (Chinese): changes of the tegument surface in cercariae, schistosomula and juvenile parasites during development. International Journal for Parasitology 18(8): 1093–1104.

Sotillo J, Pearson M, Becker L, Mulvenna J, Loukas A. 2015. A quantitative proteomic analysis of the tegumental proteins from *Schistosoma mansoni* schistosomula reveals novel potential therapeutic targets. International Journal for Parasitology 45(8): 505–516.

Stirewalt MA. 1974. *Schistosoma mansoni*: cercaria to schistosomule. Advances in Parasitology 12: 115–182.

Thepen T, Van Rooijen N, Kraal G. 1989. Alveolar macrophage elimination *in vivo* is associated with an increase in pulmonary immune response in mice. Journal of Experimental Medicine 170(2): 499–509.

Trottein F, Descamps L, Nutten S, Dehouck M-P, Angeli V, Capron A, Cecchelli R, Capron M. 1999a. *Schistosoma mansoni* activates host microvascular endothelial cells to acquire an anti-inflammatory phenotype. Infection and Immunity 67(7): 3403–3409.

Trottein F, Nutten S, Angeli V, Delerive P, Teissier E, Capron A, Staels B, Capron M. 1999b. *Schistosoma mansoni* schistosomula reduce E-selectin and VCAM-1 expression in TNF-a-stimulated lung microvascular endothelial cells by interfering with the NF-c-B pathway. European Journal of Immunology 29: 3691–3701.

Uhlar CM, Whitehead AS. 1999. Serum amyloid A, the major vertebrate acute-phase reactant. European Journal of Biochemistry 265(2): 501–523.

Varin A, Gordon S. 2009. Alternative activation of macrophages: Immune function and cellular biology. Immunobiology 214: 630–641.

Wilson MS, Taylor MD, Balic A, Finney CA, Lamb JR, Maizels RM. 2005. Suppression of allergic airway inflammation by helminth-induced regulatory T cells. Journal of Experimental Medicine 202(9): 1199–1212.

Wilson RA. 2009. The saga of schistosome migration and attrition. Parasitology 136(12): 1581–1592.

Wilson RA. 2012. Proteomics at the schistosome-mammalian host interface: any prospects for diagnostics or vaccines? Parasitology 139(9): 1178–1194.

Wilson RA, Coulson PS. 1989. Lung-phase immunity to schistosomes: a new perspective on an old problem. Parasitology Today 5(9): 274–278.

Ying S, Meng Q, Zeibecoglou K, Robinson DS, Macfarlane A, Humbert M, Kay AB. 1999. Eosinophil chemotactic chemokines (eotaxin, eotaxin-2, RANTES, monocyte chemoattractant protein-3 (MCP-3), and MCP-4), and C-C chemokine receptor 3 expression in bronchial biopsies from atopic and nonatopic (Intrinsic) asthmatics. Journal of Immunology 163(11): 6321–6329.

You H, Gobert GN, Duke MG, Zhang W, Li Y, Jones MK, McManus DP. 2012. The insulin receptor is a transmission blocking veterinary vaccine target for zoonotic *Schistosoma japonicum*. International Journal for Parasitology 42(9): 801–807.

Zhang M, Hong Y, Han Y, Han H, Peng J, Qiu C, Yang J, Lu K, Fu Z, Lin J. 2013. Proteomic analysis of tegument-exposed proteins of female and male *Schistosoma japonicum* worms. Journal of Proteome Research 12(11): 5260–5270.

Tegument and External Features of *Schistosoma* (with Particular Reference to Ultrastructure)

Geoffrey N Gobert[1, 2,]*, *Leigh Schulte*[1, 3] and *Malcolm K Jones*[3]

10.1   INTRODUCTION

The integument (hereafter tegument) of adult and larval schistosomes has attracted substantial interest from the schistosomiasis research community. This thin layer, approximately 1–5 μm thick, is the primary physical delineating interface of the parasite, with other borders being the lining of the internal digestive canal (Wilson and Xiaohong Li, Chapter 11), the excretory system (Kusel, Chapter 16) and the reproductive system (Jones *et al.*, Chapter 13). This syncytial layer provokes interest because it is considered a primary target for drug and vaccine studies (Loukas *et al.* 2007). Further, the interest arises from the intense remodeling of the layer that occurs as the parasite invades a new host (Hockley and McLaren 1973), because of multifaceted functions (Skelly and Wilson 2006) and exquisite structure. In this chapter, we will examine each of these relevant facets and will build on the strong structural knowledge that commenced during the 1960's and 1970's corresponding with the expansion of electron microscopy techniques for biological applications. The tegument has been the subject of numerous reviews (Hockley 1973; Hockley and McLaren 1973; Jones *et al.* 2004; Kusel *et al.* 2007; Loukas *et al.* 2007; Skelly and Wilson 2006; Wilson 2012a, 2012b) among many others) and the reader is referred to these others for further details and structural and functional aspects of the tissue.

10.2   NEODERMATA

The schistosomes belong to a clade of platyhelminths, the Neodermata (Ehlers 1985), a group that encompasses all tapeworms, trematodes and monogeneans. The name refers to the post-larval epidermis that develops in all species of these groups and to varying degrees in some turbellarians (Tyler and Tyler 1997). The epidermis (the tegument) in neodermatans has two primary features: it is syncytial and the nucleated regions of this cellular entity lie below the basement membrane that underlies the lining proper of the layer. Thus the external protective layer of the adult neodermatan

[1] QIMR Berghofer Medical Research Institute, Herston, Queensland 4006, Australia.
[2] Current Address: School of Biological Sciences, Queen's University Belfast, Belfast, Northern Ireland BT7 1NN, United Kingdom.
[3] University of Queensland, School of Veterinary Science, Gatton, Queensland 4343, Australia.
* Corresponding author

is a single, thin cytoplasmic entity, without lateral cellular partitions, but ornamented apically with a range of subcellular surface decorations and ornamentations, such as spines, hooks, pits, microvilli, microtriches and other elaborations and punctured in places by sensory receptors that emerge through the layer. The tegument of adult schistosomes is typically neodermatan, but bears special adaptations, no doubt correlates of the intravascular location of all post-invasive forms of the parasite in its mammalian host (Skelly and Wilson 2006).

10.3   SCHISTOSOME TEGUMENT

10.3.1   Continuity with other Epithelia

The tegument is a thin layer of anucleate cytoplasm that provides a variety of needs for the parasite including protection, nutrition, excretion, osmoregulation, as well as sensory and signaling functions (Jones *et al.* 2004). The tegument is a single syncytium that surrounds the entire worm, and is in continuity with other epithelia that line the esophagus as well as reproductive and excretory ducts (Jones *et al.* 2004).

10.3.2   Planar Polarization

One hallmark feature of the schistosome tegument is the distinctive organization of the lining cytoplasmic region (hereafter the apical cytoplasm) that forms the actual external lining. As with other epithelia, the tegument is highly polarized in its apical-basal plane, with distinctive structural and functional partitioning of cells, seen most vividly in the disparate morphology of the apical and basal membranes of the apical cytoplasm. The polarization is emphasized further by the zonal partitioning of the nucleated portions of the tegument; these pleomorphic nucleated cell bodies lie embedded in the parenchyma and connect to the apical cytoplasm through thin cytoplasmic bridges.

There exists another form of polarization that occurs in the lateral plane of the syncytium. The term planar polarization was coined for this phenomenon in other organisms (reviewed by Goodrich and Strutt 2011). Planar polarization is a common feature of tissues of all metazoans and has major functional significance in guiding development and patterning in animals. Planar polarization can occur within cells and it is certainly a feature of the tegument and lining epithelia of Platyhelminths. The phenomenon is most clearly seen in the cirrus of the cestode, *Cylindrotaenia hickmani,* in which the structure of microtriches vary markedly from filamentous to spinose with a few microns (Jones 1998).

Among the schistosomes, planar polarization is seen most clearly in males. In *Schistosoma mansoni* the dorsal surface is marked with elevated regions, the tubercles, which are ornamented with spines. The spines are true cytoplasmic organelles, bound by the apical plasma membrane and composed of a polymerized actin (Cohen *et al.* 1982). The tubercles are absent along the ventral surface, although the spines are abundant. *Schistosoma japonicum* males lack tubercles and their dorsal surface is not spinose, while the ventral surface, the region termed the gynaecophoric canal is richly ornamented with spines (Gobert *et al.* 2003). Gobert and colleagues also noted substantial variations in the organization of the apical membrane in different regions of the tegument.

The variations seen in different zones of the parasites are indicative of different functional specializations in the tegument. Sometime the advantages conferred by possession of these specialization are apparent. Abundant spines in the gynaecophoric canal of males undoubtedly serve to anchor the female. At other times, the reasons for the variations remain obscure. These variations in the tegument architecture over the surface of the parasites emphasizes the remarkable adaptations in cell biology that give rise to the biologically robust syncytial tegument of schistosomes.

10.4   EXTERNAL MORPHOLOGY

The external surface of adult schistosomes and cercariae have been studied extensively by scanning electron microscopy. This chapter will focus largely on the morphology of adult parasites, but there are many studies of the surface of different life stages and species (Hockley 1968; Silk *et al.* 1970; Race *et al.* 1971; LoVerde 1975, 1976a, b; Sakamoto and Ishii 1976; Hicks and Newman 1977; Kuntz *et al.* 1977; Tulloch *et al.* 1977). As tegument integrity is crucial for survival of the parasite within the host, there have been many studies investigating tegument integrity after perturbation with drug treatments and vaccinations. A difficulty in interpreting the structural integrity of the tegument in these largely *in vitro* studies has been the inherently labile nature of the layer once removed from the vascular micro environment. The tegument is evidently robust in its natural environment. Once removed and placed in artificial media, it can disintegrate rapidly (Jones *et al.* 2004; Skelly and Wilson 2006). It can be immensely difficult to distinguish what are true effects of treatment and what are effects of artificial culture in studying drug effects on schistosomes.

Male worms appear in life as pale, semi-transparent organisms, approximately 10–20 mm in length depending on species and host and with slightly distended lateral margins. These margins are curved ventrally and can overlap, thereby forming the gynaecophoric canal in which the female resides. The females are cylindrical and elongate, with the posterior half slightly thicker and darker than the anterior half. The denser appearance of the posterior regions is a result of the presence of a single distended cecum filled with digesta of the blood meal and the presence of vitelline follicles in that region. Worms of either gender bear an oral and ventral sucker in the anterior. Suckers of the male worm are much larger than those of the females. The genital openings of males and females occur immediately posterior to the ventral sucker, while the excretory system opens posteriorly.

Males of *S. mansoni* have numerous raised papillae or tubercles on the dorsal surface of their bodies (Silk *et al.* 1970; Skelly and Wilson 2006). Each tubercle is ornamented with numerous spines. The spines are cytoplasmic structures and hence, intracellular, but their semicrystalline nature and regular shape imparts a density to them that renders them prominent by scanning electron microscopy. Between the tubercles and elsewhere on the dorsal surface, the apical membrane is ornamented with spines or regularly arranged pits. The ventral surface of *S. mansoni* males displays externally irregular pits and abundant spines. Females display the ridges and pits seen in males, but are drastically less spinose. The females can have a small number of tubercles and these are primarily located in the posterior region (Hockley 1973). Uniciliate sense receptors emerge through the tegument in many places, but are abundant around the mouth, the margins of the suckers and openings of the gonadoducts (gonoducts).

The surface anatomy of *S. haematobium* is similar to that of *S. mansoni*. Males bear tubercles dorsally (Leitch *et al.* 1984). The tubercles are spinose except at their apex. The regions among the tubercles are heavily pitted. Uniciliate sensory receptors are found between the pits. The gynaecophoric canal is lined by forward projecting spines. The surface of the female shows regular corrugations and is heavily pitted. Spines are present (Leitch *et al.* 1984), but less apparent in scanning electron microscopy. The suckers of both genders are ringed by spines. The ventral sucker has a honeycomb appearance due to extensive pitting of its surface.

Not all species have spinose tubercles. A study of *S. bovis*, *S. currassoni* and *S. haematobium*, showed that *S. bovis* had tubercles devoid of spines, while the other two species had tubercles that were completely spinose or lacked spines at the apex (Southgate *et al.* 1986). Interestingly, samples of male *S. bovis* obtained from Spain has spinose tubercles (Fournier *et al.* 1989), indicating substantial intraspecies variation in the surface morphology. The degree of development of spines is also age dependent. Younger males of three schistosomes species consistently has aspinose tubercles, whereas these in older males were heavily spinose (Fournier *et al.* 1989).

Male *Schistosoma japonicum* lacks tubercles or spines on the dorsal surface (Sakamoto and Ishii 1977). Spines are present in the gynaecophoric canal (Gobert *et al.* 2003). The oral and ventral suckers are spinous, with the spines adorning the oral suckers. The spines point inwards and outwards and vary in sharpness (Sakamoto and Ishii 1977). The lining of the tegument displays characteristic pits and folds. At times the folding can be quite pronounced (Gobert *et al.* 2003). Females are less spinous than males. Gobert and colleagues (2003) noted spines emerging from the anterior region of the parasite. The surface of the females is much less folded and pitted than the males.

10.5 ULTRASTRUCTURE OF THE TEGUMENT OF ADULT SCHISTOSOMES

10.5.1 General Structure

The features of the tegument are shown graphically in the schematic illustration below (Fig. 10.1) and in accompanying transmission electron micrographs of *S. japonicum* and *S. mansoni* (Figs 10.2, 10.12). These features will be described systematically from the apical region of the apical cytoplasm to the basal region, the cytoplasmic bridges and the nucleated cell bodies.

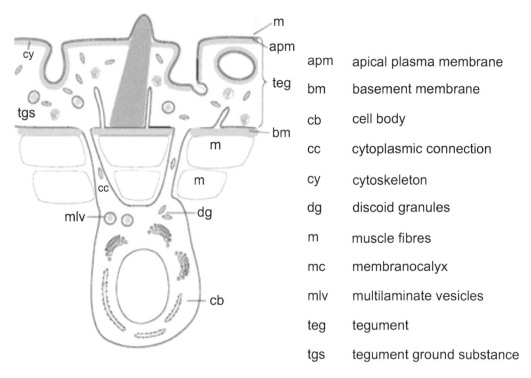

apm	apical plasma membrane
bm	basement membrane
cb	cell body
cc	cytoplasmic connection
cy	cytoskeleton
dg	discoid granules
m	muscle fibres
mc	membranocalyx
mlv	multilaminate vesicles
teg	tegument
tgs	tegument ground substance

FIG. 10.1 Diagram of the adults schistosome tegument and associated cell bodies. From Skelly and Wilson 2006. Advances in Parasitology 63: 185–284. Fig. 1.

Apical Membrane Complex. The tegument is lined apically by a complex of membranes (Fig. 10.2). This membrane system has been described as being of multi laminate or heptalaminate appearance. The heptalaminate nature is most apparent after schistosome tissues are fixed in uranyl acetate, either after the traditional sequential fixation steps of glutaraldehyde and osmium tetroxide (Wilson and Barnes 1974b) or after uranyl acetate alone in a cryo-substitution procedure (Schulte *et al.* 2013). Uranyl acetate prevents dissolution of some labile components of cells during dehydration, notably some lipids and glycoproteins. Wilson and Barnes proposed that the heptalaminate complex is a pair of unit membranes closely apposed to each other, an inner membrane proper and an outer membrane, reminiscent of a glycocalyx of other animal cells (Wilson and Barnes 1974a). They proposed the term membranocalyx for this outer lipid bilayer.

The apical membrane complex is much folded forming deep invaginations that run almost to the basal regions of the tegument. The invaginations are lined by the heptalaminate membranes, although evidence of loss or peeling of the membranocalyx can be observed (Gobert *et al.* 2003). Noticeably in *S. japonicum* and *S. mekongi*, small cave-like invaginations emerge from the base of the invaginations (Gobert *et al.* 2003) (Fig. 10.2). These give rise to small vesicles within the tegument cytoplasm. There is little direct evidence for an endocytotic pathway in the schistosomes tegument (Kusel *et al.* 2007).

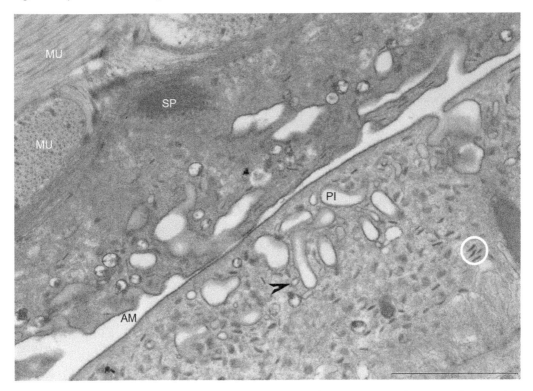

FIG. 10.2 *Schistosoma mansoni.* Juxtaposed teguments of male and female pair. The male tegument lies toward the top of the image. Arrowhead-cave-like invagination from surface invagination. The white circle shows a group of discoid bodies. AM-apical membrane complexes of the teguments. MU-myofibrils; PI- surface invagination; Scale bar 2 μm.

10.5.2   Adornments of the Apical Membrane Complex

Spines. Common features of the tegument, as observed by either light or electron microscopy, are the spines. The spines (Figs. 10.3, 10.4) are composed of a polymerized actin bound by meromyosin components (Cohen *et al.* 1982). The margins of the spines interact with cytoskeletal components associated with basal and apical membranes, thereby anchoring the structures. Spines within the same species may vary in shape from blunt to sharply pointed structures.

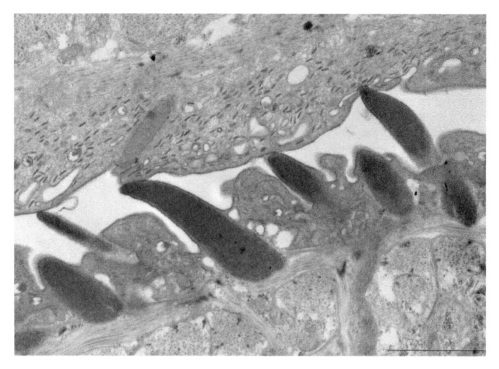

FIG. 10.3 *Schistosoma mansoni.* Electron dense spines of the tegument of a male juxtaposed with the tegument of a female (top of the micrograph). Scale bar 2 μm.

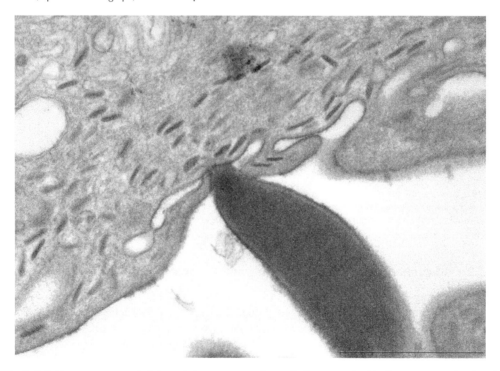

FIG. 10.4 *Schistosoma mansoni.* Spine of the tegument of a male juxtaposed with the tegument of a female. Enlargement of Fig. 10.3. Scale bar 1 μm.

Sensory Receptors. In platyhelminths, the primary sensory structure is a small dendritic extension that may either pierce through the tegument lining or abut against the lining (Fig. 10.11). The terminus is bulbous and may be uniciliate or multiciliate or it may lack cilia. In schistosomes, the predominant form of receptor is a uniciliate receptor. These structures abound in regions associated with movement, feeding and reproduction, but are also found in other locations of the tegument. Hemidesmosomes connect the receptors with the apical cytoplasm. Internally, the distinctive microtubular array of the cilium is evident. Mitochondria are at times observed within the cytoplasm of the receptor.

Tubercles. These are structures seen in males of African species of schistosomes, but not Asian species. The tubercle is a raised region of the apical cytoplasm tegument supported by intrusion of the parenchymal tissues. The tubercles may be spinous and can carry ciliated sensory receptors (Kruger *et al.* 1986a, b). The apical cytoplasm overlying the tubercles has been studied for susceptibility to drugs and vaccination. Some compounds administered to schistosomes, such ascard amonin, a chalcone extracted from the plant *Piper aduncum,* alters the numbers of tubercles (de Castro *et al.* 2015), but it is difficult to assess the underlying mechanism of this phenotypical change. Similarly, administration of the principal anti-schistosome compound praziquantel, induces disruption of the tegument, resulting in host antibody binding to the tegument particularly in the region of the tubercles (Brindley and Sher 1987).

Apical Cytoplasm. The apical cytoplasm of the tegument contains small mitochondria, as well assecretory bodies termed discoid bodies (also known as elongate bodies) (Fig. 10.2) and multilaminate vesicles (also known as membranous vesicles) (Hockley 1973; Wilson and Barnes 1974b; Skelly and Wilson 2006). In *Schistosoma mansoni*, discoid bodies are approximately 200 nm by 40 nm and are around 20-fold more abundant than multilaminate vesicles (Skelly and Wilson 2006). The tri-dimensional shape of the discoid bodies has been shown elegantly from electron tomography reconstructions (Schulte *et al.* 2013). Discoid bodies have a single lipid bilayer that contains a dense granular intracellular content. Multilaminate vesicles are 150 nm to 200 nm in diameter and contain tightly packed membranes arranged concentrically. Discoid bodies are found in the tegument of all parasitic flatworms, but multilaminate vesicles are known only in the tegument of blood flukes (Skelly and Wilson 2006).

It was originally thought that discoid bodies contributed to the maintenance of the apical membrane while the multilaminate vesicles contribute to the maintenance of the membranocalyx. However, multilaminate vesicles have been shown to contribute to both tegument membranes (Skelly and Wilson 2006). Discoid bodies have been suggested to transport material to the apical surface and release their contents into the tegument cytoplasm (Wilson and Barnes 1977). The material released from these vesicles extends over the apical plasma membranes and is observed as blobs of muco-polysaccharide (Skelly and Wilson 2006).

Cytoskeleton of the Apical Cytoplasm. Microtubules are poorly represented within the apical cytoplasm, except as the point of entry of the cytoplasmic bridges (MacGregor and Shore 1990; Duvaux-Miret *et al.* 1991) (Fig. 10.5). The microtubules in the cytoplasmic bridges are stabilized, a feature common for microtubules in the Platyhelminthes, notably those present in the nervous system (Younossi-Hartenstein *et al.* 2001; Hartenstein and Jones 2003). This microtubular array undoubtedly is used for anterograde transport through the cytoplasmic bridges, as chemical disruption of the microtubular cytoskeleton of schistosomes also disrupts surface expression of major components (Wiest *et al.* 1988).

TABLE 10.1 Comparison of structural features of the tegument of *Schistosoma japonicum*, *Schistosoma mansoni* and *Schistosoma haematobium* Gobert *et al.* (2003).

Species	Ultrastructural feature General description	*S. japonicum* (China/Philippines)	*S. mansoni* (Middle East, South America)	*S. haematobium* (South Africa)
Tubercles and/or surface spines	Raised mound on tegument, in many cases spinous Spines are arrays of actin embedded in the apical cytoplasm	Tubercles not present Spines on gynaecophoric canal and ♀ surface	Prominent tubercles on ♂, spines common on ♂ and ♀ surface	Tubercles present Spines small than in *S. mansoni*
Tegument thickness	Distance from the basal membrane to the membranocalyx	♂ 3–5 μm ♀ 1–2 μm gynaecophoric canal.	1–3 μm	1–3 μm
Apical membrane complex	The complex of membranes that lines the tegumentary cytoplasm. This complex forms the external surface of the parasite	Variety of appearance–heptalaminate, but sometimes trilaminate	heptalaminate	heptalaminate
Discoid bodies	Dominant organelles of the tegument. These are saucer-shaped, electron opaque organelles, bound by a trilaminate membrane	150 nm in length.	200 × 40 nm	130 nm in length
Multilaminate vesicles	Spherical vesicles, much less abundant than discoid bodies. The vesicles are bound by a trilaminate membrane and contain whorls or stacks of membranes	150–200 nm. Contents appear homogenous	150–200 nm. Contents appear as membrane whorls	300 nm in diameter
Mitochondria		Common basally in apical cytoplasm	Uncommon, small organelles with 1–2 cristae	Similar to those of *S. mansoni*
Key references		Gobert *et al.* 2003	Hockley 1973, Wilson and Barnes 1974a, Skelly and Wilson 2006	Leitch 1994

Actin is an abundant component of the musculature and tegument (Abbas and Cain 1987; Matsumoto *et al.* 1988; Schulte *et al.* 2013). Actin is released after the tegument is damaged by praziquantel (Matsumoto *et al.* 1988). The actin of the spines is formed into arrays with filaments having identical polarity (Cohen *et al.* 1982; Zhou and Podesta 1989; Zhou and Podesta 1992). Evidence of a cortical array of actin beneath the apical membrane complex membrane is not forthcoming, but actin is likely to be present in these regions. The strong association of dynein light chains, which link with dynein or myosin motors, with the apical membrane complex in *Schistosoma japonicum* (Zhang *et al.* 2005) suggests that a cortical actin array is present, as is common in animal cells. Further, molecules that bind actin, such as tropomyosin and α-actinin are present in the apical cytoplasm of schistosomes, (MacGregor and Shore 1990), which would suggest that actin is present in the cytoplasmic layer.

Myosin and paramyosin are muscle-associated proteins. Paramyosin occurs in lophotrochozoans (a clade to which schistosomes belong) and the ecdysozoans (nematodes and arthropod lineages) and some deuterostomes (Hooper *et al.* 2008). Immunolocalization studies have suggested that myosin may be present at the surface of *S. mansoni* schistosomula (Soisson *et al.* 1993) but not those of *S. japonicum* (Zhang *et al.* 1998). Similarly, paramyosin has been suggested to be present at the surface of cercariae and schistosomules (Matsumoto *et al.* 1988; Gobert 1998) and observation not supported by proteomic analyses of the tegument (Wilson 2012b).

Basal membrane. The basement membrane, which separates the tegument cytoplasm from the underlying muscles layer, also invaginates into the tegument cytoplasm. The degree of infolding varies with the region of the body of the parasite (Gobert *et al.* 2003), which suggest a highly absorptive surface but may also contribute to tegument flexibility (Hockley 1973).

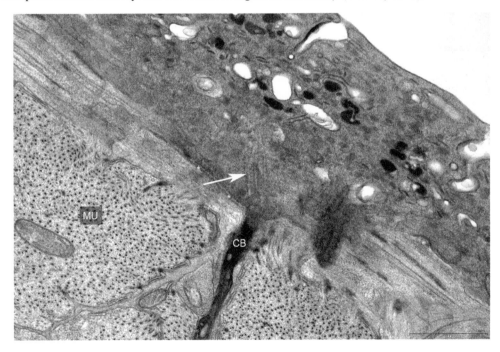

FIG. 10.5 *Schistosoma mansoni.* Tegument of a female, showing a cytoplasmic bridge that connects the apical cytoplasm with the in sunken cell bodies of the tegument. Microtubules emerging from the cytoplasmic bridge are arrowed. The basal lamina of the basal membrane is shown (asterisk). CB, cytoplasmic bridge; MU, myofibrils of the tegumentary musculature. Scale bar 1 μm.

Cytoplasmic bridges and nucleated cell bodies. The nucleated cell bodies (or cytons) (Fig. 10.6) of the tegument are withdrawn into the body of the worm, being located in the parenchymal region, below the basement membrane and myofibrils. The cytons are connected to the tegument by cytoplasmic connections, here termed cytoplasmic bridges (Fig. 10.5). These possess a cortical ring of longitudinal microtubules and allow the transport of vesicles between the tegument and cytons. The cytons, which may be multinucleate, contain endoplasmic reticulum, Golgi bodies, mitochondria and are the source of the discoid bodies and multilaminate vesicles (Wilson and Barnes 1974b). Protein synthesis occurs within the cytons, the products of which are packaged in secretory bodies for transport to the apical cytoplasm.

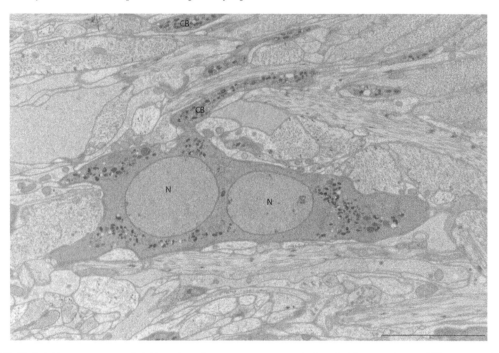

FIG. 10.6 *Schistosoma mansoni.* A bi-nucleated tegumentary cyton of a female. The prominent cell lies surrounded by parenchymal tissues. Thin cytoplasmic bridges are seen to traverse through the parenchymal tissues. These bridges will connect with the apical cytoplasm. CB, cytoplasmic bridge; N, nucleus of tegumentary cell body. Scale bar 5 μm.

10.6   CERCARIAL TRANSFORMATION AND TEGUMENT RENEWAL

A significant event in schistosome infection is the transformation of the infective larva, the cercaria, to the host-adapted larva, the schistosomulum. The ultrastructure of the *S. mansoni* cercaria has been described comprehensively by Dorsey and others (Dorsey *et al.* 2002; see also Kašný *et al.*, Chapter 8). Cercariae have a tegument with a typical bilipid membrane. This is adorned with an extensive glycocalyx (Fig. 10.7). This glycocalyx is observed only with the use of cytochemical stains ruthenium red or potassium-ferricyanide reduced osmiumtetroxide. These stains reveal a glycocalyx that is substantially thicker than the apical cytoplasm itself (Dorsey *et al.* 2002; Jones *et al.* 2004).

When cercariae penetrate host skin, the membrane quickly transforms from a simple membrane to a bi-laminar membrane complex. The cercariae shed the glyocalyx and the tegument membrane transforms to a mostly multilaminate membrane by 60 minutes post-infection (Hockley and McLaren 1973; Skelly and Shoemaker 2001; Skelly and Wilson 2006). This transformation is accompanied by extensive molecular synthesis and transport to the surface. This process is exemplified most strongly by the rapid production and transport of a surface linked glucose transporter within hours of cercaria transformation (Skelly and Shoemaker 2001) (see also Gobert and Nawaratna, Chapter 9).

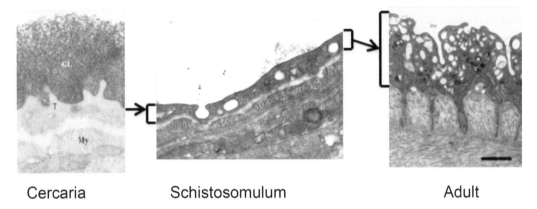

| Cercaria | Schistosomulum | Adult |

FIG. 10.7 *Schistosoma mansoni.* Cercarial transformation. The simple cercariae tegument (T) (left) is overlain by a glycocalyx (GL) and underlain by muscles (My). In the schistosomulum the tegument begins to thicken and develop (middle), loses the glycocalyx and finally becomes the thicker and more elaborate adult tegument (right). Different panels not to scale. From Jones *et al.,* 2004, BioEssays, 26(7): 752–765 and Malcolm Jones, unpublished.

The ultra structure of the schistosomulum tegument (see also Gobert and Nawaratna, Chapter 9), some 3 days after this transforming event has the following features, based on Chai *et al.* (2006). The outer membrane is heptalaminate with regions of trilaminate or pentalaminate membrane (Fig. 10.8a). No evidence of sloughing or shedding of membranes is observed at this early stage. The heptalaminate membrane measures 17 nm in thickness. Spines can be seen projecting through the syncytium from the basement membrane (Fig. 10.8b). Spines measure over 1 μm in length and 300 nm in width and the paracrystalline array of actin filaments is clearly observed (Fig. 10.8b). The basal lamina consists of the basal membrane, lamina lucida and lamina densa. Sheet-like invaginations of the basement membrane are common, extending up to 250 nm into the tegument matrix (Fig. 10.9a). Cytoplasmic bridges join subtegumental cell bodies in the underlying parenchyma to the tegument matrix (Fig. 10.8c). Discoid bodies and multilaminate vesicles are present in the cytoplasm (Fig. 10.9b, d). Discoid bodies measured approximately 80–150 nm in length and 15–20 nm in width and are shaped like biconcave discs. They are limited by a trilaminate membrane and often observed in close proximity with the apical membrane of the tegument (Fig. 10.9d). Large membranous bodies are circular or elliptical in shape and measured 150 to 200 nm in diameter. They consist of a trilaminate outer membrane, containing concentric membranous whorls which sometimes appear partially extracted. Membranous bodies are commonly seen in very close association with the apical membrane of the tegument, apparently merging with it, perhaps contributing their contents for incorporation into the apical membrane (Fig. 10.9c). Large mitochondria with many cristae are common in the tegument matrix (Fig. 10.10b). Flask-like structures, formed by narrow-necked infoldings of the apical membrane of the tegument, were observed rarely (Fig. 10.9f). From the bulb-like base of the infolding, the membrane invaginates again forming smaller cave- or flask-like invaginations, which appear to be blind-ended. Electron opaque material is observed within the cavity of the structure.

Cytoplasmic bridges are observed traversing muscle layers to form connections between cytons and the tegument (Fig. 10.8c). Mitochondria and various bodies (elongated and membranous) are observed in close association with the cytoplasmic regions of cytoplasmic bridges, seemingly banked in numbers at the narrow juncture of cyton to bridge. Although in longitudinal section, no bodies are observed within the cytoplasmic bridges themselves, cytoplasmic connections viewed in cross section revealed mitochondria contained within the peripheral ring of microtubules ($\cong 25$

nm in diameter) lining the isthmus (Fig. 10.10d). In longitudinal section, microtubules ($\cong 25$ nm in diameter) could be seen to extend in parallel structure for some distance into the tegument cytoplasm (Fig. 10.9e). Other populations of microtubules observed within the tegument were confined to the cilia of sensory organelles.

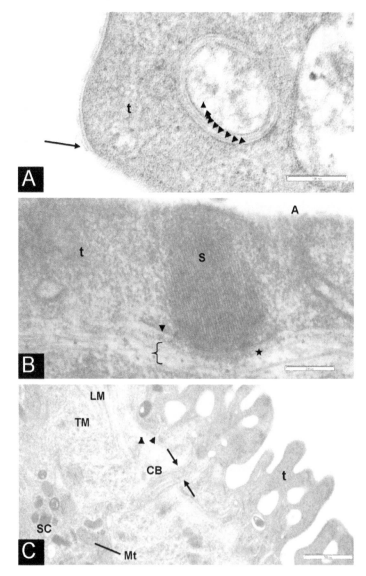

FIG. 10.8 *Schistosoma japonicum.* **A.** Transmission electron micrograph of the lung schistosomula tegument, showing regions of pentalaminate membrane (arrow) and heptalaminate membrane (individual membrane layers indicated with arrowheads) forming the apical membrane and associated invaginations, overlying the tegument matrix (t). The region of heptalaminate membrane is an apical membrane infolding observed in cross section. Scale bar 100 nm. **B.** Transmission electron micrograph of lung schistosomula tegument showing a spine (S) projecting through the tegument matrix (t) toward the apical membrane (A). The base of the spine is flush with the basement membrane of the tegument which consists of the lamina lucida (arrowhead), lamina densa (star) and lamina reticularis (bracket). The spine is formed of a paracrystalline array of actin. Scale bar 200 nm. **C.** Transmission electron micrograph of a lung schistosomulum showing a cytoplasmic bridge (CB). The boundaries of the cytoplasmic bridge are indicated by arrows. The bridge passes through transverse (TM) and longitudinal muscle (LM) to connect a subtegumental cell (SC) to the tegument matrix (t). Mitochondria (Mt) and various organelles can be seen at the juncture of the subtegumental cell to the cytoplasmic bridge. An invagination of the basement membrane of the tegument is indicated with an arrowhead. Scale bar 500 nm.

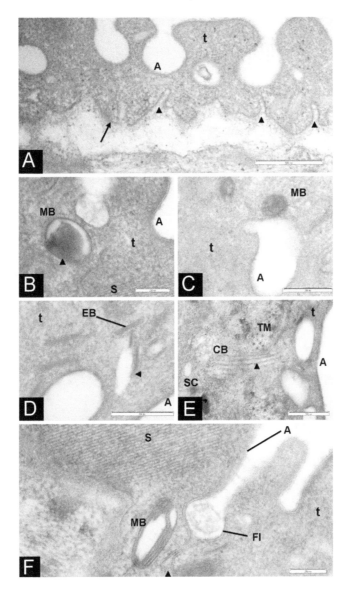

FIG. 10.9 *Schistosoma japonicum*. **A**. Transmission electron micrograph of lung schistosomula showing the basement membrane of the tegument (arrow) forming frequent sheet-like invaginations (arrowheads) which project into the tegument matrix (t). The apical membrane (A) is indicated. Scale bar 500 nm. **B**. Transmission electron micrograph of the tegument matrix (t), limited by a multilaminate apical membrane (A). A membranous body (MB) containing partially extracted concentric laminations (arrowhead) is indicated. A spine (S) with paracrystalline actin array can also be seen. Scale bar 200 nm. **C**. Transmission electron micrograph of the tegument matrix (t) showing a membranous body (MB) apparently merging with an invagination (viewed in cross section) of the apical membrane (A) of the tegument. Scale bar 200 nm. **D**. Transmission electron micrograph of the tegument matrix (t) showing elongated bodies (EB), several of which are apparently merging (arrowhead) with an invagination (viewed in cross section) of the apical membrane (A). Scale bar 200 nm. **E**. Transmission electron micrograph of the tegument matrix (t) and adjoining subtegumental cell (SC). A cytoplasmic bridge (CB) is cut in longitudinal section, showing several parallel microtubules (arrowhead) which constitute part of the peripheral ring of microtubules within the cytoplasmic bridge. The microtubules span the length of the cytoplasmic bridge and extend some distance into the tegument matrix. Transverse muscle (TM) and the apical membrane (A) are labelled. Scale bar 500 nm. **F**. Transmission electron micrograph showing a flask-like invagination (Fl) formed by an infolding of the apical membrane (A) of the tegument. A second apparently blind-ended flask-like structure (arrowhead) invaginates from the bulb-like base of the infolding. A spine (S) and a membranous body (MB) can also be observed within the tegument matrix (t). Scale bar 200 nm.

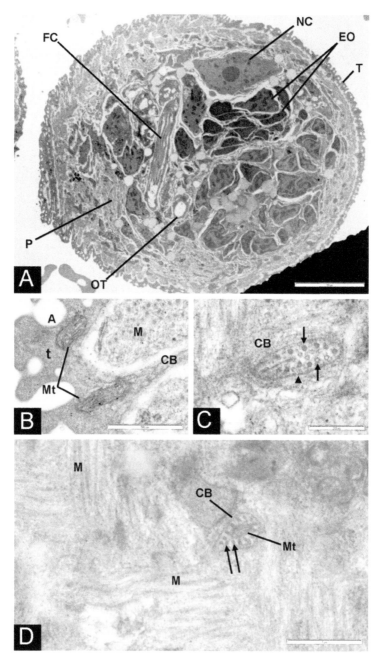

FIG. 10.10 *Schistosoma japonicum.* **A.** Transmission electron micrograph of a lung schistosomulum in glancing longitudinal section. The pitted tegument (T) creates a sponge-like appearance and nests of germinal cells of variable morphology and appearance such as large, solitary nucleated cells (NC) and small electron opaque cells (EO) are visible embedded within the parenchymal matrix (P). A flame cell (FC) and osmoregulatory tubule (OT) are labelled. Individual cells within the parenchyma appear contracted away from each other. Scale bar 10 μm. **B.** Transmission electron micrograph of the tegument and underlying transverse muscle bundles (M) showing a cytoplasmic bridge (CB) and two mitochondria (Mt) within the tegument matrix (t). The apical membrane of the tegument is indicated (A). Scale bar 500 nm. **C.** Transmission electron micrograph showing a cytoplasmic bridge (CB) in cross section. A peripheral ring of microtubules is indicated by the arrows. The limiting membrane of the bridge is indicated by the arrowhead. Scale bar 200 nm. **D.** Transmission electron micrograph of a cytoplasmic bridge (CB) observed in cross section, embedded within muscle (M) beneath the tegument of a lung stage parasite. A mitochondrion (Mt) is visible within the bridge. The peripheral ring of microtubules in indicated with arrows. Scale bar 1 μm.

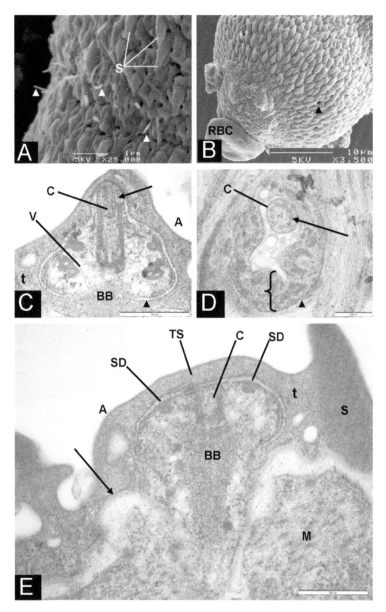

FIG. 10.11 *Schistosoma japonicum*. **A.** Scanning electron micrograph of a *S. japonicum* lung schistosomulum showing uniciliated sensory papilla (arrowheads) projecting from highly pitted tegument in the region of the ventral sucker (not pictured). Some spines (S) are also present. Scale bar 1 μm. **B.** Scanning electron micrograph of the anterior region of a lung schistosomulum showing the external opening of a multiciliated pit organelle (arrowhead). A mouse red blood cell (RBC) is labelled. Scale bar 10 μm. **C.** Transmission electron micrograph of a uniciliated papilla observed in longitudinal section. A microtubule of the cilium is indicated by the arrow. The organelle is limited by a trilaminate membrane (arrowhead). Abbreviations: A, apical membrane; BB, basal body; C, cilium; t, tegument matrix; V, vesicles. Scale bar 500 nm. **D.** Transmission electron micrograph of a uniciliated papilla observed in cross section. Microtubule (arrow) doublets can be observed within the cilium (C) however the microtubule formation is difficult to ascertain. The thickness of the wall of the sensory organelle (bracket) which is made up of the nerve-ending, is evident in cross section. The limiting membrane of the structure is indicated by an arrowhead. Scale bar 200 nm. **E.** Transmission electron micrograph of a uniciliated papilla observed in longitudinal section, projecting into the tegument matrix (t). The cilium (C) can be seen projecting from the basal body (BB) of the nerve. The cilium cannot be observed projecting through the tegument in this plane of section. The bulb of the sensory organelle is attached to the tegument sheath (TS) via a circular septate desmosome (SD). The basement membrane of the tegument is indicated with an arrow. A spine (S) and muscle (M) are also visible. Scale bar 500 nm.

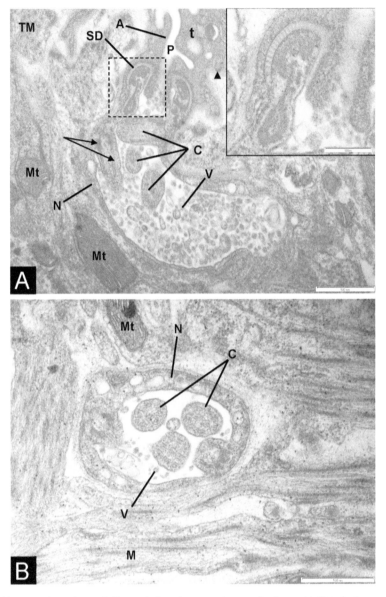

FIG. 10.12 *Schistosoma japonicum.* **A.** Transmission electron micrograph of a multiciliated pit organelle observed in longitudinal section of a lung schistosomula. The sub-surface pit is anchored to the basement membrane of the tegument (arrowhead) by a circular septate desmosome (SD). Cilia (C) project into the lumen of the pit which is filled with vesicles (V). Basal bodies (arrows) of the cilia are visible. Several mitochondria (Mt) are in close association with the nerve ending (N) making up the pit wall. Note the pit-like entrance to the organelle (P), tegument matrix (t), transverse muscle (TM) and the apical membrane (A) of the tegument. Scale bar 500 nm. Inset image shows a higher magnification of the septate desmosome. Scale bar 170 nm. **B.** Transmission electron micrograph of a multiciliated pit organelle observed in cross section. Four cilia (C) with indeterminate microtubule arrangement are present within the lumen, which also contains some vesicles (V). Note the nerve ending forming the wall of the pit (N) and closely associated mitochondria (Mt) and muscle layers (M). Scale bar 500 nm.

Two kinds of sensory organelle were identified on the surface of lung-stage schistosomula for *S. japonicum.* Observed sensory organelles were the uniciliated papilla and the multiciliated pit. The uniciliated papilla resembled that described in the cercariae and adult of *Schistosoma mansoni,* consisting of a cilium projecting from a bulbous nerve terminus, which protruded from the overlying tegument (Fig. 10.11c, d) (Dorsey *et al.* 2002; Hockley 1973). When viewed in longitudinal section, the bulb of the uniciliated papilla appeared to be entirely contained within the tegument matrix and limited by a trilaminate membrane. The bulb measured 0.9 µm in width and the bulb and tegument sheath protruded about 1 µm from the basement membrane of the tegument. A circular septate desmosome attaching the bulb to the tegument was observed and the basal body of the nerve cell could be seen extending into the sensory organelle (Fig. 10.11e). Vesicular and granular contents were observed surrounding the base of the cilium, which contained microtubules of approximately 25 nm in diameter. A uniciliated papilla observed in cross-section revealed nine pairs of microtubules circularly arranged within the cilium, however the presence of central microtubules (creating the 9 + 2 microtubule formation) was unclear (Fig. 10.11d). The thickness of the organelle wall (made up of a nerve ending) was evident in cross section. Sensory papillae with a single cilium of approximately 0.5 µm in length were observed by SEM in close association with the ventral sucker (Fig. 10.11a) however no protrusions of the tegument sheath could be seen elevating the papillae from the surface. The multiciliated pit organelle was observed by SEM, appearing as only a tiny pore in the tegument of the anterior end of the parasite, (Fig. 10.11b) however when viewed with TEM, further features were obtained. Very similar in description to that of the adult and cercariae of *S. mansoni*, the pit measured approximately 3 µm in depth and 700 nm in diameter and was anchored around its opening to the exterior by a circular septate desmosome attached to the basement membrane of the tegument (Fig. 10.12a) (Dorsey *et al.* 2002; Hockley 1973). Four cilia (450 nm in length and 260 nm in diameter) projected into the lumen of the pit, which was filled with electron lucent vesicles of various sizes (Fig. 10.12b). Microtubule arrangement could not be clearly identified and nerve attachment could not be seen at the base of the pit. Mitochondria were observed in close association with the pit.

10.7 TEGUMENT RENEWAL

Turnover. The adult tegument membranes are replaced regularly *in vivo*, with an estimated turnover rate of 5 days (Githui *et al.* 2009). Maintenance of the tegument occurs by activity of the multilaminate vesicles that migrate from the cytons along the cytoplasmic bridges to fuse with the apical membrane where they release their contents that are then integrated into membranes (Skelly and Wilson 2006; Mulvenna *et al.* 2010). The cell biology associated with tegument renewal is poorly known and studies are hampered by poor survival of schistosomes and disruption of their tegument in culture outside the host (Kusel *et al.* 2007).

In 1975, Kusel and coworkers (Kusel *et al.* 1975) postulated that sloughing of components of the complex occurred through the release of small vesicles. The work was supported by pulse chase experiments of Wilson and co-workers (1977) who noted the release of membranocalyx components into culture media, but this observation was not supported by subsequent work (Brouwers *et al.* 1999) who noted that radioactively-labelled fatty acids were incorporated into membranes. They suggested, however, that loss of radioactivity in the membranes was due to hydrolysis of the labelled acyl chains and not by sloughing of membranes. Skelly and Wilson (2006) argued that different pathways of membrane renewal exist for either component of the dual membrane system. Whereas the inner plasma membrane is recycled, they reasoned, the outer membranocalyx is sloughed. Recently, Sotillo and coworkers proposed an intriguing mode of membrane release from adult worms by a process of exosome secretion (Sotillo *et al.* 2016). Exosomes from adult worms were enriched for a range of tegument proteins including tetraspanins, heat shock protein 70, SmLy6 molecules and a saposin. The role of exosomes is thought to be in cell to cell signaling. There clearly remains much to be learned about the fate of membranes in schistosomes and the interactions between the parasite and the host micro environment.

10.8 PROTEOMICS OF THE TEGUMENT

Introduction. Proteomics has been used extensively to survey the tegument surface of schistosome parasites. The identification of unique and biologically important proteins at the host-parasite interface which is the tegument, has been a goal of many researchers for their potential use as diagnostic or biomarker targets, as well as therapeutic use for new drug applications or vaccines.

One of the earliest applications for proteomics for studying schistosomes was a high-through put study of *S. japonicum* (Liu *et al.* 2007). The complete proteomes of the parasite were examined, with developmental stages considered included miracidia, cercariae, hepatic schistosomula, adults and eggs). In addition to whole organism proteomes the tegument of the schistosomula and adult parasites also examined. The proteomes and tegument sub-proteomes were elucidated using high resolution two-dimensional-nano-liquid chromatography. This analytical approach required two sequential fractionation steps for the soluble supernatant and then for the insoluble pellet, which consisted of strong cation exchange liquid chromatography or 1 dimensional gel electrophoresis/ digestion. These fractionation steps were followed by reverse-phase liquid chromatography and tandem mass spectrometry (Liu *et al.* 2007). From this study 3260 proteins were identified, with a sub-set of 373 attributed to tegument. In contrast to this approach, less comprehensive studies of *S. mansoni* and *S. haematobium* using proteomics have mainly focused on characterizing more specific proteome fractions of these parasites including the soluble fraction of specific life cycle stages (Curwen *et al.* 2004; Higon *et al.* 2011; Neves *et al.* 2015) or surface-exposed proteins (Mulvenna *et al.* 2010; Sotillo *et al.* 2015).

The tegument of schistosome parasites, particularly that of the adult life cycle stage, is an epithelial layer providing protection from both physical and immunological sources. The tegument also is the active for nutritional uptake from the host blood stream (Skelly *et al.* 1994; Dalton *et al.* 2004; Skelly and Wilson 2006); and as the host/parasite interface it provides sensory input via touch sensors (Gobert *et al.* 2003; Jones *et al.* 2004) and molecular receptors to monitor host hormone levels, such as host insulin (Dissous *et al.* 2006; You *et al.* 2012). Biotin labeling of surface proteins of live parasites has been an effective approach in characterizing surface exposed tegument proteins (Braschi and Wilson 2006; Mulvenna *et al.* 2010; Zhang *et al.* 2013; Sotillo *et al.* 2015). The proteomic composition of this apical tegument surface has been surveyed in the adult stages of both *S. mansoni* (Braschi and Wilson 2006) and *S. japonicum* (Mulvenna *et al.* 2010; Zhang *et al.* 2013) and in *S. mansoni* schistosomula (Sotillo *et al.* 2015). As highlighted by Wilson (2012b), the utilization of proteomics to characterize schistosome tegument has been an important focus for research since the early 2000's. Particularly in the last three years there have been more comprehensive approaches to proteomics and the schistosome parasite, of which the outcomes have provided novel in sights into the schistosome tegument and its role as the primary interface between the parasite and the host micro environment.

Proteomics of the tegument of the schistosomula. The juvenile life cycle phase, schistosomulumis a critical point in the parasite development within the mammalian host and is present in the parasite life cycle as it transforms from a cercariae after leaving the aquatic environment (see Gobert and Nawaranta, Chapter 9). At this point many of the immunological defense mechanism that are employed by the adult parasite are not yet fully established (Wilson and Coulson 2009). Recently, a comprehensive analysis of the tegumentary proteome of *S. mansoni* schistosomula was performed. After mechanical transformation, schistosomula were collected at time points up to 5 days for characterization of surface proteins (Sotillo *et al.* 2015). Biotinylated tegument samples were subjected to free-thaw rupture of the tegument, followed by proteomic analysis included iTRAQ labeling, OFFGEL electrophoresis and mass spectrometry (Sotillo *et al.* 2015). Their study highlighted how the tegument of schistosomules enables the organism to be a more efficient parasite. This was made apparent due to the presence of proteins associated with dynamic functions including the uptake of nutritional sources from the host bloodstream, adaptive functions related to the microenvironment such as anti-oxidant activity and heat shock responses, as well as more general proteolytic and enzymatic activities. The last two activities could be related to the important functions performed by the gastrodermis of the parasite, a premise which is supported by previous transcriptomic studies utilizing microscopy mediated laser microdissection of adult schistosomes, where gene transcripts related to similar functions were attributed to the gut region (Gobert *et al.* 2009; Nawaratna *et al.* 2011). Immunofluorescence localization of biointylated proteins by Sotillo and colleagues indicated that the proteins they observed were derived from the tegument.

10.9   SURFACE ASSOCIATED PROTEINS OF ADULT SCHISTOSOMES

A detailed list of cytosolic proteins and membrane-linked proteins in the tegument can be found in the sources cited above. A research area of intense interest in schistosome biology lies in elucidating the organization of proteins associated with the apical membrane complex of the apical cytoplasm. Wilson (2012) has summarized the consensus information of proteomic analysis of proteins associated with the membrane complex (Fig. 10.13). Wilson's model allows for the presence of a number of proteins to be embedded within the membranocalyx, others are found in inter-membrane zones, where they are proposed to promote adhesion of the two unit membranes (the "Velcro" zone) and the apical plasma membrane. Host proteins abound at the surface of the membrane complex (Braschi and Wilson 2006; Mulvenna *et al.* 2010).

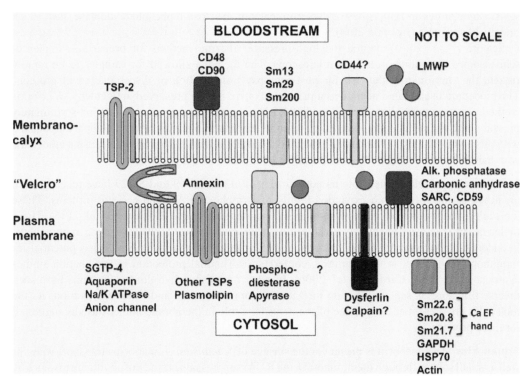

FIG. 10.13 Diagram of the putative location of proteins in the membranocalyx and apical plasma membrane of the tegument surface. Note that the proteins and lipid bilayers are not depicted to scale. Host CD44 may be anchored in the plasma membrane, but with its N-terminus protruding above the membranocalyx surface. Data suggest that calpain may be outside the plasma membrane. The location of immunoglobulins and complement factors C3 and C4 are not shown. CD-refers to 'cluster of differentiation' nomenclature for cell surface markers. LWMP-Low molecular weight proteins. Sm-refers to *Schistosoma mansoni*. TSP-tetraspanins, SGTP-schistosomes glucose transporter protein. Reproduced from Wilson RA. 2012b. Parasitology 139: 1178–1194. Fig. 2.

A number of proteins depicted by Wilson (2012) merit further comment. Two proteins have shown remarkable promise as vaccine targets; they are a tetraspanin and Sm29, a molecule with unclear affinities outside of the schistosomes. In the following paragraphs, we will give insight into the functional biology of a number of novel schistosome tegumental proteins with diverse roles in surface maintenance, homeostasis and membrane recycling.

Tetraspanins. Tetraspanins (TSPs) are a family of proteins that act as molecular facilitators through the interaction with numerous partnering proteins, resulting in the formation of "tetraspanin webs" (Boucheix *et al.* 2001; Boucheix and Rubinstein 2001). Tetraspanins interact at cellular membranes and have many varied roles proposed. Thus, tetraspanins are thought to contribute to maintenance of cell morphology, motility, invasion, fusion and signaling (Hemler 2008). Tetraspanins have been consistently found enriched on the apical tegument membranes of *S. mansoni* (van Balkom *et al.* 2005; Braschi and Wilson 2006; Castro-Borges *et al.* 2011b). Of these, SmTSP2 has been proven as a leading vaccine candidate against human schistosomiasis Tran *et al.* 2006; (Loukas *et al.* 2007). Mice vaccinated with recombinant SmTSP2 large extracellular loop showed a high efficacy with significant reductions in adult worm burden, liver egg burden and fecal egg counts (Tran *et al.* 2006). Knockdown of SmTSP2 in schistosomula and adult worms resulted in morphological changes in the tegument, most notably inducing a highly vacuolated, but thinner tegument, suggesting a role of the tetraspanins in surface membrane processing (Tran *et al.* 2010). A study of the protein interactome of *S. mansoni* TSP2 (Jia *et al.* 2014) suggested a number of common molecular partners, all of which were surface related (Fig. 10.13). These proteins included schistosome annex in B30, Sm29, a dysferlin, calpain, fructose-biphosphate aldolase, heat shock protein 70 and actin (Jia *et al.* 2014).

Schulte *et al.* (2013) adapted the high-pressure freezing method for preparing samples of schistosomes for immuno-electron microscopy. The methods allowed for samples to be rapidly frozen, thus minimizing the risk for post-necropsy translocation or disassociation of proteins. The procedure used uranyl acetate as a primary fixative, which preserved antigenicity. When this method was used to localize *S. mansoni* TSP2, it was found that the protein is not indiscriminately present along the surface membranes, but is more abundant in the membrane complex associated with pits and infoldings. These data implicated tetraspanin with biological mechanisms associated with membrane recycling on the surface of the adult schistosome parasites.

Ly6 Domain proteins. The proteins noted in Fig. 10.13 as CD59 and Sm29 have recently been identified as members of an expanded family of Lymphocyte Antigen 6 (Ly6) proteins. These molecules have a range of membrane-related functions in animals (Chalmers *et al.* 2015). These molecules in *S. mansoni* are clearly surface linked (Castro-Borges *et al.* 2011a) and are immunogenic in endemic human populations (Chalmers *et al.* 2015). SmLy6 (formerly Sm29) has potential for immunoprophylaxis of schistosomiasis, conferring substantial protection in vaccination studies (Cardoso *et al.* 2008; Cardoso *et al.* 2006). In other organisms, Ly6 domain proteins have such diverse functions as septate junction formation, odorant response and limb regeneration. As the molecules are present at the surface of the parasite, a role in membrane homeostasis is suggested for these proteins.

Aquaporins. An aquaporin is present at the surface of *S. mansoni*. Aquaporins transport water as well as small solutes through membranes. Using RNAi suppression, a study of adult schistosomes by Faghiri and Skelly (2009) was able to show that suppressed worms resisted swelling in both hyper- and hypo-osmotic solutions. The suppressed worms were further shown to be resistant to killing by the drug potassium antimonyl tartrate, which they propose passes through aquaporins channels to enter parasite cells. Further suppression studies demonstrated that aquaporin-suppressed worms were less efficient at excreting lactate into culture media (Faghiri *et al.* 2010). Lactic acid, a product of glucose catabolism, can lead to lowering of the pH of cells and therefore elaborate mechanisms are used to remove or metabolise the molecule. The role of the aquaporins in removal of lactate suggests a role for the tegument in excretion (Faghiri *et al.* 2010).

Glucose transporters. *Schistosoma mansoni* produced a number of glucose transporter proteins, two of which are expressed by the tegument cytons. One (SGTP4) is transported to the apical membrane (Skelly and Shoemaker 1996). The function of these molecules is to facilitate the uptake of glucose across the apical membrane of the parasite. Using confocal immunofluorescence, demonstrated most elegantly the synthesis and transport of SGTP4 to the surface of the developing schistosomulum during cercarial transformation. These authors demonstrated an early accumulation of the protein in tegumentary cytons, followed by the eruption of the molecules to the surface (Skelly and Shoemaker 2001). This is an example of how the nutritional needs of the invading parasite is met through uptake of low molecular weight solutes across the tegument.

10.10   THE TEGUMENT AS A TARGET IN SCHISTOSOME CONTROL

Treatment and control of schistosomiasis relies heavily on a single drug, praziquantel. Despite wide spread use of praziquantel, schistosomiasis is spreading to new regions (McManus and Loukas 2008). Treatment with the drug does not protect humans against reinfection. The potential for resistance to praziquantel clearly signifies the need for an alternative treatment for schistosomiasis. The case for an anti-schistosome vaccine is well documented (see also Bergquist and McManus, Chapter 24), however many trials have failed to produce protection against reinfection (Skelly and Wilson 2006; McManus and Loukas 2008). In addition to understanding the complex immune responses in schistosome infection, understanding the biology of the tegument is crucial in future vaccine or drug development.

Previous vaccine targets against schistosomiasis have been selected on the basis of sero-dominance of specific antigens in affected individuals and experimental hosts and more recently, after identification of putative surface antigens by proteomic analyses (Wilson 2012). While vaccine candidates that confer partial protection, resulting in significant reductions in adult worm and egg burdens have been identified, none are as yet definitive and other promising leads need to be found (McManus and Loukas 2008). The failure of sero-dominant antigens as effective vaccines may have occurred because these highly antigenic molecules were never exposed to the host immune system by living parasites, but instead only become exposed after parasites had died (Kusel *et al.* 1975; Mulvenna *et al.* 2010). The search for truly surface molecules by proteomic analyses of membrane proteins has more promise, but full development of a suitable vaccine is hampered by a very poor understanding of the roles of these molecules and how and when during the parasites development, these molecules are exposed to the host immune response.

An important aspect of understanding the biology of the tegument is how the tegument interacts with the host immune system and evades attack. Schistosomes have developed mechanisms to overcome immunological attack from the host (Gryseels *et al.* 2006). The exact molecular basis of these mechanisms of immune evasion are widely unknown, however some ideas have been proposed (Abath and Werkhauser 1996; Skelly and Wilson 2006). Schistosomes acquire host antigens and present them on their tegument, this antigenic masking tricks the host immune system and renders the worm resistant to immune attack (Abath and Werkhauser 1996). Schistosomes are able to produce host-like molecules that are presented to the host immune system and many molecules of host-origin have been found on the tegument of schistosomes. These molecules include major host histocompatibilty antigens and immunoglobulins (Skelly and Wilson 2006). Schistosomes are also thought to express receptors for these host proteins which enable the parasite to effectively camouflage itself again the host immune response (Loukas *et al.* 2001).

The surface of larval and adult schistosomes presents proteins that can bind host molecules. It is not known how these molecules are actually presented at the surface, which, as stated above is lined by the protein-depauperate membranocalyx. Fc receptors, a feature of a protective immune response, have been found on the worms surface indicating they can bind immunoglobulins and make them inaccessible for subsequent molecular interactions (Loukas *et al.* 2001). Protease activity is also thought to occur on the surface of schistosomula and adult worms. A 28 kDa serine protease has been shown to cleave complement proteins, contributing to the avoidance of complement-mediated immune attacks (Abath and Werkhauser 1996).

The expression of immunogenic molecules at the surface of schistosomes can differ over time due to developmental stage of the parasites, and secretory and repair activity in surface membranes. Over time antigens may be switched off and no longer expressed on the tegument or they may be replaced with alternative molecules (Abath and Werkhauser 1996). This switching of gene expression may also contribute to immune evasion by schistosomes. The immune evasion of the schistosome is a complex process with many different mechanisms contributing to the long term survival of schistosomes in their human hosts.

10.11   LITERATURE CITED

Abath FG, Werkhauser RC. 1996. The tegument of *Schistosoma mansoni*: functional and immunological features. Parasite Immunology 18(1): 15–20.

Abbas MK, Cain GD. 1987. Actin and intermediate-sized filaments of the spines and cytoskeleton of *Schistosoma mansoni*. Parasitology Research 73(1): 66–74.

Boucheix C, Duc GH, Jasmin C, Rubinstein E. 2001. Tetraspanins and malignancy. Expert Reviews in Molecular Medicine 2001: 1–17.

Boucheix C, Rubinstein E. 2001. Tetraspanins. Cellular and Molecular Life Sciences 58(9): 1189–1205.

Braschi S, Wilson RA. 2006. Proteins exposed at the adult schistosome surface revealed by biotinylation. Molecular and Cellular Proteomics 5(2): 347–356.

Brindley PJ, Sher A. 1987. The chemotherapeutic effect of praziquantel against *Schistosoma mansoni* is dependent on host antibody response. Journal of Immunology 139(1): 215–220.

Brouwers JF, Skelly PJ, Van Golde LM, Tielens AG. 1999. Studies on phospholipid turnover argue against sloughing of tegumental membranes in adult *Schistosoma mansoni*. Parasitology 119: 287–294.

Cardoso FC, Macedo GC, Gava E, Kitten GT, Mati VL, de Melo AL, Caliari MV, Almeida GT, Venancio TM, Verjovski-Almeida S, SC O. 2008. *Schistosoma mansoni* tegument protein Sm29 is able to induce a Th1-type of immune response and protection against parasite infection. PLoS Neglected Tropical Diseases 2: e308.

Cardoso FC, Pacifico RN, Mortara RA, Oliveira SC. 2006. Human antibody responses of patients living in endemic areas for schistosomiasis to the tegumental protein Sm29 identified through genomic studies. Clinical and Experimental Immunology 144(3): 382–391.

Castro-Borges W, Dowle A, Curwen RS, Thomas-Oates J, Wilson RA. 2011a. Enzymatic shaving of the tegument surface of live schistosomes for proteomic analysis: a rational approach to select vaccine candidates. PLoS Neglected Tropical Diseases 5(3): e993.

Castro-Borges W, Dowle A, Curwen RS, Thomas-Oates J, Wilson RA. 2011b. Enzymatic shaving of the tegument surface of live schistosomes for proteomic analysis: arational approach to select vaccine candidates. PLoS Neglected Tropical Diseases 5: e993.

Chai M, McManus DP, McInnes R, Moertel L, Tran M, Loukas A, Jonesa MK, Gobert GN. 2006. Transcriptome profiling of lung schistosomula, *in vitro* cultured schistosomula and adult *Schistosoma japonicum*. Cellular and Molecular Life Sciences 63(7-8): 919–929.

Chalmers IW, Fitzsimmons CM, Brown M, Pierrot C, Jones FM, Wawrzyniak JM, Fernandez-Fuentes N, Tukahebwa EM, Dunne DW, Khalife J, Hoffmann KF. 2015. Human IgG1 responses to surface localised *Schistosoma mansoni* Ly6 family members drop following praziquantel treatment. PLoS Neglected Tropical Diseases 9(7): e0003920.

Cohen C, Reinhardt B, Castellani L, Norton P, Stirewalt M. 1982. Schistosome surface spines are "crystals" of actin. Journal of Cell Biology 95(3): 987–988.

Curwen RS, Ashton PD, Johnston DA, Wilson RA. 2004. The *Schistosoma mansoni* soluble proteome: a comparison across four life cycle stages. Molecular and Biochemical Parasitology 138(1): 57–66.

Dalton JP, Skelly PJ, Halton DW. 2004. Role of the tegument and gut in nutrient uptake by parasitic Platyhelminthes. Canadian Journal of Zoology 82: 211–232.

de Castro CC, Costa PS, Laktin GT, de Carvalho PH, Geraldo RB, de Moraes J, Pinto PL, Couri MR, Pinto Pde F, Da Silva Filho AA. 2015. Cardamonin, a schistosomicidal chalcone from *Piper aduncum* L. (Piperaceae) that inhibits *Schistosoma mansoni* ATP diphosphohydrolase. Phytomedicine 22(10): 921–928.

Dissous C, Khayath N, Vicogne J, Capron M. 2006. Growth factor receptors in helminth parasites: signalling and host-parasite relationships. FEBS Letters 580(12): 2968–2975.

Dorsey CH, Cousin CE, Lewis FA, Stirewalt MA. 2002. Ultrastructure of the *Schistosoma mansoni* cercaria. Micron 33(3): 279–323.

Duvaux-Miret O, Baratte B, Dissous C, Capron A. 1991. Molecular cloning and sequencing of the alpha-tubulin gene from *Schistosoma mansoni*. Molecular and Biochemical Parasitology 49(2): 337–340.

Ehlers U. 1985. *Das Phylogenetische System der Platyhelminthes*. Stuttgart: Gustav Fischer. 317 p.

Faghiri Z, Camargo SM, Huggel K, Forster IC, Ndegwa D, Verrey F, Skelly PJ. 2010. The tegument of the human parasitic worm *Schistosoma mansoni* as an excretory organ: the surface aquaporin SmAQP is a lactate transporter. PLoS One 5(5): e10451.

Faghiri Z, Skelly PJ. 2009. The role of tegumental aquaporin from the human parasitic worm, *Schistosoma mansoni*, in osmoregulation and drug uptake. FASEB Journal 23: 2780–2789.

Fournier A, Pages JR, Touassem R, Mouahid A. 1989. Can tegumental morphology be used as a taxonomic criterion between *Schistosoma haematobium*, *S. intercalatum* and *S. bovis*? Parasitology Research 75(5): 375–380.

Githui EK, Damian RT, Aman RA, Ali MA, Kamau JM. 2009. *Schistosoma* spp.: isolation of microtubule associated proteins in the tegument and the definition of dynein light chains components. Experimental Parasitology 121(1): 96–104.

Gobert GN. 1998. Immunolocalization of schistosome proteins. Microscopy Research and Technique 42(3): 176–185.

Gobert GN, McManus DP, Nawaratna S, Moertel L, Mulvenna J, Jones MK. 2009. Tissue specific profiling of females of *Schistosoma japonicum* by integrated laser microdissection microscopy and microarray analysis. PLoS Neglected Tropical Diseases 3: e469.

Gobert GN, Stenzel DJ, McManus DP, Jones MK. 2003. The ultrastructural architecture of the adult *Schistosoma japonicum* tegument. International Journal for Parasitology 33(14): 1561–1575.

Goodrich LV, Strutt D. 2011. Principles of planar polarity in animal development. Development 138(10): 1877–1892.

Gryseels B, Polman K, Clerinx J, Kestens L. 2006. Human schistosomiasis. Lancet 368: 1106–1118.

Hartenstein V, Jones M. 2003. The embryonic development of the body wall and nervous system of the cestode flatworm *Hymenolepis diminuta*. Cell and Tissue Research 311(3): 427–435.

Hemler ME. 2008. Targeting of tetraspanin proteins—potential benefits and strategies. Nature Reviews in Drug Discovery 7: 747–758.

Hicks RM, Newman J. 1977. The surface structure of the tegument of *Schistosoma haematobium*. Cell Biology International Reports 1(2): 157–167.

Higon M, Cowan G, Nausch N, Cavanagh D, Oleaga A, Toledo R, Stothard JR, Antunez O, Marcilla A, Burchmore R, Mutapi F. 2011. Screening trematodes for novel intervention targets: a proteomic and immunological comparison of *Schistosoma haematobium*, *Schistosoma bovis* and *Echinostoma caproni*. Parasitology 138(12): 1607–1619.

Hockley DJ. 1968. Scanning electron microscopy of *Schistosoma mansoni* cercariae. Journal of Parasitology 54(6): 1241–1243.

Hockley DJ. 1973. Ultrastructure of the tegument of *Schistosoma*. Advances in Parasitology 11: 233–305.

Hockley DJ, McLaren DJ. 1973. *Schistosoma mansoni*: changes in the outer membrane of the tegument during development from cercaria to adult worm. International Journal for Parasitology 3: 13–25.

Hooper SL, Hobbs KH, Thuma JB. 2008. Invertebrate muscles: thin and thick filament structure; molecular basis of contraction and its regulation, catch and asynchronous muscle. Progress in Neurobiology 86(2): 72–127.

Jia X, Schulte L, Loukas A, Pickering D, Pearson M, Mobli M, Jones A, Rosengren KJ, Daly NL, Gobert GN, Jones MK, Craik DJ, Mulvenna J. 2014. Solution structure, membrane interactions and protein binding partners of the tetraspanin Sm-TSP-2, a vaccine antigen from the human blood fluke *Schistosoma mansoni*. Journal of Biological Chemistry 289(10): 7151–7163.

Jones MK. 1998. Structure and diversity of cestode epithelia. International Journal for Parasitology 28: 913–923.

Jones MK, Gobert GN, Zhang L, Sunderland P, McManus DP. 2004. The cytoskeleton and motor proteins of human schistosomes and their roles in surface maintenance and host-parasite interactions. BioEssays 26(7): 752–765.

Kruger FJ, Hamilton-Attwell VL, Schutte CH. 1986a. Scanning electron microscopy of the teguments of males from five populations of *Schistosoma mattheei.* Onderstepoort Journal of Veterinary Research 53(2): 109–110.

Kruger FJ, Hamilton-Attwell VL, Tiedt L, Du Preez L. 1986b. Further observations on an intratubercular sensory receptor of *Schistosoma mattheei.* Onderstepoort Journal of Veterinary Research 53(4): 239–240.

Kuntz RE, Tulloch GS, Huang TC, Davidson DL. 1977. Scanning electron microscopy of integumental surfaces of *Schistosoma intercalatum.* Journal of Parasitology 63(3): 401–406.

Kusel JR, Al-Adhami BH, Doenhoff MJ. 2007. The schistosome in the mammalian host: understanding the mechanisms of adaptation. Parasitology 134: 1477–1526.

Kusel JR, Mackenzie PE, McLaren DJ. 1975. The release of membrane antigens into culture by adult *Schistosoma mansoni.* Parasitology 71(2): 247–259.

Leitch B, Probert AJ, Runham NW. 1984. The ultrastructure of the tegument of adult *Schistosoma haematobium.* Parasitology 89(Pt 1): 71–78.

Liu F, Hu W, Cui SJ, Chi M, Fang CY, Wang ZQ, Yang PY, Han ZG. 2007. Insight into the host-parasite interplay by proteomic study of host proteins copurified with the human parasite, *Schistosoma japonicum.* Proteomics 7(3): 450–462.

Loukas A, Jones MK, King LT, Brindley PJ, McManus DP. 2001. Receptor for Fc on the surfaces of schistosomes. Infection and Immunity 69(6): 3646–3651.

Loukas A, Tran M, Pearson MS. 2007. Schistosome membrane proteins as vaccines. International Journal for Parasitology 37: 257–263.

LoVerde PT. 1975. Scanning electron microscope observations on the miracidium of Schistosoma. International Journal for Parasitology 5(1): 95–97.

Loverde PT. 1976a. Scanning electron microscopy of the ova of *Schistosoma haematobium* and *Schistosoma mansoni.* Egypt Journal of Bilharzia 3(1): 69–72.

Loverde PT. 1976b. Stereo-scan observations on the cercaria of *Schistosoma haematobium.* Egypt Journal of Bilharzia 3(1): 65–67.

MacGregor AN, Shore SJ. 1990. Immunocytochemistry of cytoskeletal proteins in adult *Schistosoma mansoni.* International Journal for Parasitology 20: 279–284.

Matsumoto Y, Perry G, Levine RJ, Blanton R, Mahmoud AA, Aikawa M. 1988. Paramyosin and actin in schistosomal teguments. Nature 333: 76–78.

McManus DP, Loukas A. 2008. Current status of vaccines for schistosomiasis. Clinical Microbiology Reviews 21(1): 225–242.

Mulvenna J, Moertel L, Jones MK, Nawaratna S, Lovas EM, Gobert GN, Colgrave M, Jones A, Loukas A, McManus DP. 2010. Exposed proteins of the *Schistosoma japonicum* tegument. International Journal for Parasitology 40(5): 543–554.

Nawaratna SS, McManus DP, Moertel L, Gobert GN, Jones MK. 2011. Gene atlasing of digestive and reproductive tissues in *Schistosoma mansoni.* PLoS Neglected Tropical Diseases 5(4): e1043.

Neves LX, Sanson AL, Wilson RA, Castro-Borges W. 2015. What's in SWAP? Abundance of the principal constituents in a soluble extract of *Schistosoma mansoni* revealed by shotgun proteomics. Parasites and Vectors 8: 337.

Race GJ, Martin JH, Moore DV, Larsh JE, Jr. 1971. Scanning and transmission electronmicroscopy of *Schistosoma mansoni* eggs, cercariae and adults. American Journal of Tropical Medicine and Hygeine 20(6): 914–924.

Sakamoto K, Ishii Y. 1976. Fine structure of schistosome eggs as seen through the scanning electron microscope. American Journal of Tropical Medicine and Hygeine 25(6): 841–844.

Sakamoto K, Ishii Y. 1977. Scanning electron microscope observations on adult *Schistosoma japonicum.* Journal of Parasitology 63(3): 407–412.

Schulte L, Lovas E, Green K, Mulvenna J, Gobert GN, Morgan G, Jones MK. 2013. Tetraspanin-2 localisation in high pressure frozen and freeze-substituted *Schistosoma mansoni* adult males reveals its distribution in membranes of tegumentary vesicles. International Journal for Parasitology 43(10): 785–793.

Silk MH, Spence IM, Buchi B. 1970. Observations of *Schistosoma mansoni* blood flukes in the scanning electron microscope. South African Journal of Medical Science 35(1): 23–29.

Skelly PJ, Kim JW, Cunningham J, Shoemaker CB. 1994. Cloning, characterization and functional expression of cDNAs encoding glucose transporter proteins from the human parasite *Schistosoma mansoni*. Journal of Biological Chemistry 269(6): 4247–4253.

Skelly PJ, Shoemaker CB. 1996. Rapid appearance and asymmetric distribution of glucose transporter SGTP4 at the apical surface of intramammalian-stage *Schistosoma mansoni*. Proceedings of the National Academy of Sciences of the United States of America 93(8): 3642–3646.

Skelly PJ, Shoemaker CB. 2001. The *Schistosoma mansoni* host-interactive tegument forms from vesicle eruptions of a cyton network. Parasitology 122(Pt 1): 67–73.

Skelly PJ, Wilson RA. 2006. Making sense of the schistosome surface. Advances in Parasitology 63: 185–284.

Soisson LA, Reid GD, Farah IO, Nyindo M, Strand M. 1993. Protective immunity in baboons vaccinated with a recombinant antigen or radiation-attenuated cercariae of *Schistosoma mansoni* is antibody-dependent. Journal of Immunology 151(9): 4782–4789.

Sotillo J, Pearson M, Becker L, Mulvenna J, Loukas A. 2015. A quantitative proteomic analysis of the tegumental proteins from *Schistosoma mansoni* schistosomula reveals novel potential therapeutic targets. International Journal for Parasitology 45(8): 505–516.

Sotillo J, Pearson M, Potriquet J, Becker L, Pickering D, Mulvenna J, Loukas A. 2016. Extracellular vesicles secreted by *Schistosoma mansoni* contain protein vaccine candidates. International Journal for Parasitology 46(1): 1–5.

Southgate VR, Rollinson D, Vercruysse J. 1986. Scanning electron microscopy of the tegument of adult *Schistosoma curassoni* and comparison with male *S. bovis* and *S. haematobium* from Senegal. Parasitology 93(Pt 3): 433–442.

Tran MH, Freitas TC, Cooper L, Gaze S, Gatton M, Jones MK, Lovas E, Pearce EJ, Loukas A. 2010. Suppression of mRNAs encoding tegument tetraspanins from *Schistosoma mansoni* results in impaired tegument turnover. PLoS Pathogens 6: e1000840.

Tran MH, Pearson MS, Bethony JM, Smyth DJ, Jones MK, Duke M, Don TA, McManus DP, Correa-Oliveira R, Loukas A. 2006. Tetraspanins on the surface of *Schistosoma mansoni* are protective antigens against schistosomiasis. Nature Medicine 12(7): 835–840.

Tulloch GS, Kuntz RE, Davidson DL, Huang TC. 1977. Scanning electron microscopy of the integument of *Schistosoma mattheei* Veglia & Le Roux, 1929. Transactions of the American Microscopical Society 96(1): 41–47.

Tyler S, Tyler MS. 1997. Origin of the epidermis in parasitic platyhelminths. International Journal for Parasitology 27(6): 715–738.

van Balkom BW, van Gestel RA, Brouwers JF, Krijgsveld J, Tielens AG, Heck AJ, van Hellemond JJ. 2005. Mass spectrometric analysis of the *Schistosoma mansoni* tegumental sub-proteome. Journal of Proteome Research 4(3): 958–966.

Wiest PM, Tartakoff AM, Aikawa M, Mahmoud AA. 1988. Inhibition of surface membrane maturation in schistosomula of *Schistosoma mansoni*. Proceedings of the National Academy of Sciences of the United States of America 85(11): 3825–3829.

Wilson RA. 2012a. The cell biology of schistosomes: a window on the evolution of the early metazoa. Protoplasma 249: 503–518.

Wilson RA. 2012b. Proteomics at the schistosome-mammalian host interface: any prospects for diagnostics or vaccines? Parasitology 139: 1178–1194.

Wilson RA, Barnes PE. 1974a. An *in vitro* investigation of dynamic processes occurring in the schistosome tegument, using compounds known to disrupt secretory processes. Parasitology 68(2): 259–270.

Wilson RA, Barnes PE. 1974b. The tegument of *Schistosoma mansoni*: observations on the formation, structure and composition of cytoplasmic inclusions in relation to tegument function. Parasitology 68(2): 239–258.

Wilson RA, Barnes PE. 1977. The formation and turnover of the membranocalyx on the tegument of *Schistosoma mansoni*. Parasitology 74(1): 61–71.

Wilson RA, Coulson PS. 2009. Immune effector mechanisms against schistosomiasis: looking for a chink in the parasite's armour. Trends in Parasitology 25: 423–431.

You H, Gobert GN, Duke MG, Zhang W, Li Y, Jones MK, McManus DP. 2012. The insulin receptor is a transmission blocking veterinary vaccine target for zoonotic *Schistosoma japonicum*. International Journal for Parasitology 42(9): 801–807.

Younossi-Hartenstein A, Jones M, Hartenstein V. 2001. Embryonic development of the nervous system of the temnocephalid flatworm craspedella pedum. Journal of Comparative Neurology 434(1): 56–68.

Zhang LH, McManus DP, Sunderland P, Lu X, Ye JJ, Loukas AC, Jones MK. 2005. The cellular distribution and stage-specific expression of two dynein light chains from the human blood fluke *Schistosoma japonicum*. International Journal of Biochemistry and Cell Biology 37: 1511–1524.

Zhang M, Hong Y, Han Y, Han H, Peng J, Qiu C, Yang J, Lu K, Fu Z, Lin J. 2013. Proteomic analysis of tegument-exposed proteins of female and male *Schistosoma japonicum* worms. Journal of Proteome Research 12(11): 5260–5270.

Zhang Y, Taylor MG, Bickle QD. 1998. *Schistosoma japonicum* myosin: cloning, expression and vaccination studies with the homologue of the *S. mansoni* myosin fragment IrV-5. Parasite Immunology 20(12): 583–594.

Zhou Y, Podesta RB. 1989. Surface spines of human blood flukes (*Schistosoma mansoni*) contain bundles of actin filaments having identical polarity. European Journal of Cell Biology 48: 150–153.

Zhou Y, Podesta RB. 1992. Ring-shaped organization of cytoskeletal F-actin associated with surface sensory receptors of *Schistosoma mansoni* – a confocal and electron-microscopic study. Tissue & Cell 24(1): 37–49.

CHAPTER

Alimentary Tract of *Schistosoma*

Xiao Hong Li[1] and *Alan Wilson*[2,]*

11.1 INTRODUCTION

Adult schistosomes are unusual in employing both the body surface and the alimentary tract for acquisition of nutrients, the two routes differing in importance in males and females. In this chapter we first consider the overall nutrient requirements of the two sexes before focusing on the contribution of the alimentary tract in fulfilling those needs. We begin with a brief description of the oral cavity where feeding is initiated. This is followed by a more substantial account detailing recent work on the esophagus, which has proved to be much more than just a tube for conducting blood from the mouth to the gut. The structure of the syncytium lining the gut, termed the gastrodermis, and its protease products that digest blood proteins are reviewed. The wealth of knowledge on these topics contrasts with the dearth of information about the molecular constitution and cellular functions of this key tissue. Finally there is brief coverage of alimentary tract differentiation and the thorny question of whether any alimentary tract products might serve as vaccine candidates.

11.2 FEEDING CAPABILITIES AND NUTRITION

There is a remarkable dichotomy between male and female worms in the respective importance of the gut and tegument in nutrient acquisition (Summarised in Table 11.1; Reviewed in (Skelly *et al.* 2014)). The greater importance of the alimentary route in females than males may reflect their respective cylindrical and flattened body forms. Much of our knowledge is based on a single pioneering study of blood feeding by *S. mansoni* adult worms in mice injected with ^{51}Cr labelled erythrocytes (Lawrence 1973). The body mass of a male worm is 2–3 times that of a female but the latter proved to be the more voracious consumer of blood, by a factor of 8:1. A comparison of plasma volume ingested per day with the volume of worm tissue fluids revealed that the female swallowed more than four times her fluid content versus one fifth swallowed by the male. Since vomiting by worms is infrequent and of low volume we must infer that there is a major flux of salts and water from the worm gut to the bloodstream across the tegument surface. The disparity

---

[1] Key Laboratory of Parasite and Vector Biology, Ministry of Health, National Institute of Parasitic Diseases, Chinese Center for Disease Control and Prevention, 207 Rui Jin Er Road, Shanghai 200025, China.

[2] Centre for Immunology and Infection, Department of Biology, University of York, Heslington, York, YO10 5DD, UK.

* Corresponding author

between male and female is equally marked for protein uptake. The female ingests more than four times her dry weight in protein per day, the male only one fifth. The situation with lipid intake is complicated by the fact that the female has proportionally much less parenchymal tissue than the male as two thirds of her body is occupied by vitellaria. Conversely, parenchyma is probably the most abundant tissue in the male and this is reflected in the larger lipid stores. In consequence, the female ingests about half her lipid content per day via blood feeding, the male only one fortieth.

TABLE 11.1 The feeding capacity and nutrient requirements of adult *Schistosoma mansoni.* Modified from Skelly *et al.* 2014. PLoS Pathogens 10(8): e1004246. Supplementary Table 1.

Parameter	Male	Female
Blood		
Erythrocytes ingested per hour	39,000	330,000
Volume of blood ingested per day (nl)	105	880
Volume of plasma fluid ingested per day (nl)	58	484
Body fluid volume (nl)	290	111
Protein		
Hemoglobin ingested per day (µg)	14.7	124
Total protein ingested per day (µg)	18	186
Heme ingested per day (µg)	0.56	4.8
Worm dry weight (µg)	95	40
Lipid		
Total lipid content (µg)	26.1	11
Total lipid ingested per day (µg)	0.63	5.28
Phospholipid (µg)	0.32	2.67
Triglycerides (µg)	0.11	0.94
Cholesterol (µg)	0.17	1.41
Carbohydrate		
Glycogen as % dry weight	10.5	2.83
Glucose ingested per day (µg)	0.11	0.88
Glucose consumed per day (µg)	426	115
Lactate secreted per day (µg)	490	139
Ratio glucose consumed: dry weight	4.5	2.9
Ratio glucose consumed: glucose ingested	3872	131

Carbohydrates are the one class of nutrients where the gut plays only a minimal role – the glucose ingested per day in blood is less than 1% of glucose consumed by both sexes. Around 10% of the male's dry weight comprises glycogen, localized in the parenchyma, while glycogen is <3% of the female's dry weight. Numerous studies have documented the enormous consumption of glucose and its conversion to lactate via the glycolytic pathway, as the principal energy source of adult schistosomes. However, here the tables are turned. The male consumes four times his dry weight in glucose, the female about 2.9 times. This is equivalent to ~3800 times more glucose consumed than ingested via the mouth by the male and underscores the dominant role of the tegument in glucose uptake. We have suggested that the male's reliance on its body surface for uptake of nutrients is related to its flattened body shape that would require solutes taken up via the gastrodermis to diffuse >400 µm to reach the furthest tissues (Skelly *et al.* 2014). The comparable maximum diffusion distance in the cylindrical female is only 80 µm. One final inference from these "balance sheet" estimates of nutrition is that the much more highly productive female, in terms of eggs versus sperm, has a lower total glucose uptake. This implies that the female may compensate by using amino acid backbones from ingested proteins as an energy source.

11.3 ORAL CAVITY

The oral cavity is delimited by the highly muscular and mobile oral sucker and is lined with a modified and much folded tegument (Dike 1971). The folds are longer with frequent bifurcations and inclusions packed in the tips. Like the body surface tegument these comprise multilaminate vesicles and discoid bodies (dense granules), which are manufactured in sub-tegument cell bodies and travel to the syncytium via cytoplasmic connections. Worms viewed feeding *in vivo* were facing against the direction of blood flow, either attaching themselves to the vessel wall by their ventral sucker or simply wedged in the vessel (Bloch 1980). The oral cavity was observed opening and closing at a rate of approximately once every 1 or 2 seconds, often accompanied by flexion and extension movements of the oral sucker. Video analysis of worms feeding *in vitro* on a diluted suspension of blood revealed that the grabbing motion of oral sucker could be as rapid as four times per second (Li *et al.* 2013). *In vivo*, feeding was observed to be continuous for up to an hour or intermittent with ten minute bursts, interspersed with rests (Bloch 1980). Worms remained at one site during feeding but between feeding periods some moved a few hundred micrometers along a vessel, usually centrally to a bifurcation and then up into an adjacent branch for a short distance (Bloch 1980).

11.4 THE ESOPHAGUS

The mouth, at the back of the oral sucker, opens into the schistosome esophagus, a short extensible tube connecting the oral cavity to the gut. The esophagus is surrounded by two balls of cells, separated by the paired central ganglia of the nervous system, which straddle the esophagus. The esophageal junction with the lining gastrodermis of the gut is clearly demarcated by a septate desmosome (Morris and Threadgold 1968; Dike 1971; Li *et al.* 2013). Morphologically, the whole esophageal region is more complex and highly organized than the uniform gut.

11.4.1 Morphology by Confocal Microscopy

The overall layout of the esophagus is consistent between *S. mansoni* and *S. japonicum*, comprising a syncytial lining, an investing lattice of muscle fibers and the two balls of cell bodies. It is clearly divided into anterior and posterior compartments, the anterior being approximately one third the volume of the posterior (Fig. 11.1A). Due to size disparities between the sexes, the esophageal region of the male is much larger than that of the female (Fig. 11.1A). The cell bodies are numerous and closely packed in the anterior esophagus (Dike 1971), giving the appearance of a glandular tissue (Li *et al.* 2014). Cells are even more densely packed in the posterior and have long been termed the esophageal gland. We have recently modified this terminology, designating the two cell masses as the anterior and posterior esophageal glands, respectively, since they have distinctive morphologies and functions (Li *et al.* 2014). Nucleoli are prominent in both masses of cells (Fig. 11.1B, C) indicating active ribosomal biogenesis. The approximate cell number in the anterior esophageal gland was estimated as 750 in both *S. japonicum* and *S. mansoni* males, from counts on consecutive slices in a Z stack. Cell bodies may number as many as 1000 in the posterior gland of *S. mansoni* and 1400 in *S. japonicum* males (Fig. 11.1C). At the limit of resolution of the confocal microscope, the anterior lining has a net-like appearance (Fig. 11.1B, D) while the posterior lining displays shadowy striations (Fig. 11.1C). The anterior esophageal corrugations are up to 20 µm long from base to tip in males (Fig. 11.1B, D) and proportionally shorter (only 5 µm) in females. In contrast, the striations in the posterior esophagus lining in *S. japonicum* are up to 25 µm from base to tip.

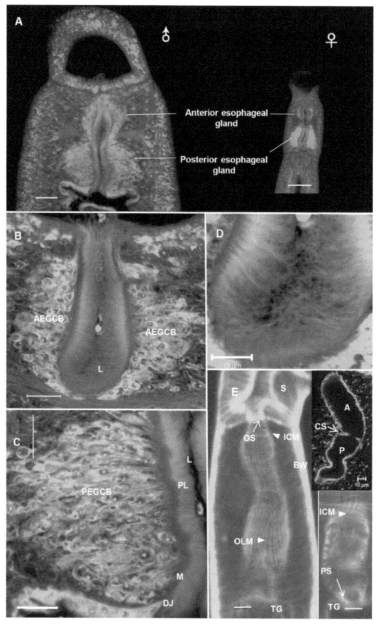

FIG. 11.1 Confocal images of esophageal region: **A.** To-scale confocal images of *S. japonicum* adult male (left) and female (right) stained with Langeron's carmine, to illustrate the large discrepancy in size of their esophageal glands. **B.** Cell bodies of the anterior esophageal gland (AEGCB) surrounding the anterior esophageal compartment. **C.** Cell bodies of the posterior esophageal gland (PEGCB) around the posterior esophageal lining. PL, Plates; L, Lumen; M, Musculature; DJ, Desmosome junction between esophageal lining and gastrodermal epithelium. **D.** Reveals the basal striated appearance of the lining and a central branched network of threads. **E.** A longitudinal side view of a female *S. mansoni*, stained with phalloidin to show only the distribution of F-actin in muscles. The minute inner circular (ICM) and outer longitudinal muscle fibers (OLM) that invest the syncytial esophageal lining appear as a fine meshwork. In comparison the larger circular and longitudinal fibers of the body wall (BW) and oral sucker (S) are intensely stained. An oral sphincter (OS, in side view) comprising a stronger circular fiber is visible at the junction between oral cavity and esophagus; a central sphincter (CS) demarcates the esophagus into anterior (A) and posterior (P) compartments (upper inset); a posterior sphincter (PS) is present at the junction between the esophagus and the transverse gut (TG) (lower inset, en face view). Scale bar: A, 50 μm; B, 20 μm; C, 50 μm; D 10 μm; E and insets, 10 μm. From Li *et al.* 2013, PLoS Neglected Tropical Diseases 7(7): e2337, Figs. 1–2. Fig. 2; Li *et al.* 2014. Parasites & Vectors 7: 565. Fig. 2.

When worms are stained with Phalloidin, a muscular network investing the syncytial lining is evident (Li *et al.* 2013). This muscular network comprises inner circular and outer longitudinal fibers (Fig. 11.1E), the spacing between which creates a lattice that works in unison to generate the peristalsis propelling food from mouth to gut. Two stronger circular fibers act as sphincters to control entry to and exit from the esophageal lumen (Fig. 11.1E and upper inset), while a central sphincter separates the anterior and posterior compartments, controlling the consecutive steps of blood processing in these two compartments (Fig. 11.1E, lower inset).

11.4.2 Surface Morphology by Scanning Electron Microscopy (SEM)

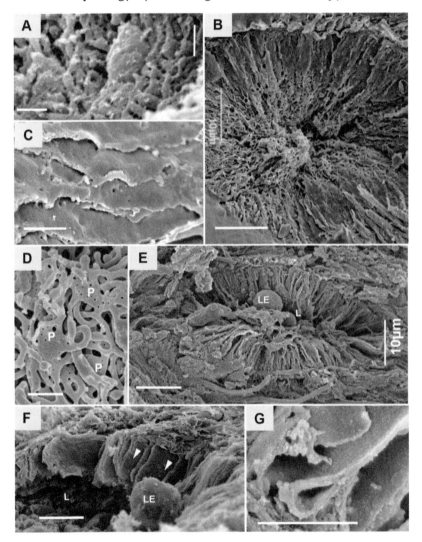

FIG. 11.2 SEM of the esophageal lining: **A.** Anterior esophageal lining of *S. mansoni* showing its highly corrugated surface. **B.** Anterior esophageal lining of *S. japonicum* showing its highly complex ultrastructure, including long, thin corrugations of cytoplasm extending from the base and central spaghetti-like cytoplasmic threads; **C.** Corrugations at higher magnification are revealed as smooth thin layers of cytoplasm with a few tiny pits in the surface. **D.** Tips of corrugation showing cytoplasmic threads connecting flattened, pitted plates of cytoplasm (P). **E., F.** Posterior esophageal lining of *S. japonicum* and *S. mansoni*, respectively, showing organized plates (arrowed), leucocytes (LE) are present in the center of the lumen (L). **G.** Posterior esophageal plates at higher magnification showing their granular matted surface. Scale bar: A, 1 μm; B, 10 μm; C, 1 μm; D, 1 μm; E, 10 μm; F, 5 μm; G, 1 μm. From Li *et al.* 2013, PLoS Neglected Tropical Diseases 7(7): e2337. Fig. 2; Li *et al.* 2014. Parasites & Vectors 7: 565. Fig. 2.

Studies on *S. mansoni* showed the anterior esophageal lining and its associated cell bodies are an extension of the body surface tegument, although the surface of the esophageal lining has more folds (Fig. 11.2A). An 'internal' examination of *S. japonicum* by SEM revealed a remarkable picture of the anterior esophageal architecture (Fig 11.2B) (Li *et al.* 2014). The basal corrugations of the anterior esophageal lining are very thin with smooth surfaces (Fig. 11.2B, C). The most striking feature in male worms is that their tips are enormously extended by threads of cytoplasm, giving the appearance of a heap of spaghetti in the esophagus lumen (Fig. 11.2B). Closer inspection shows that the cytoplasmic threads branch frequently and are anchored at both ends to the corrugations (or plates; see below), with only a few blind fingers (Fig. 11.2D). Interspersed at different levels within the tangled mass of threads are flattened plates of cytoplasm, which at high resolution are seen to be dotted with minute pits (30–40 nm diameter) spaced approximately 182 nm apart (Fig. 11.2D).

In both *S. japonicum* (Fig. 11.2E) and *S. mansoni* (Fig. 11.2F), the posterior esophageal lining is thrown into long, regular plates (Fig. 11.2E, F). SEM of worm slices reveals that the plates are orientated longitudinally, running in parallel from anterior to posterior of the gland (Fig. 11.2E, F). The plates are also very thin, reminiscent of the narrow corrugations in anterior esophageal lining, but the surface is not smooth (Fig. 11.2G).

11.4.3   Ultrastructure by Transmission Electron Microscopy (TEM)

Consistent with the SEM images of *S. mansoni*, TEM revealed that the syncytial cytoplasm lining the anterior esophagus has irregular (Dike 1971; Li *et al.* 2013), longer folds, at the centre of which are long invaginations originating from the basal lamina (Morris and Threadgold 1968). Cytoplasmic channels from the basal region of the esophageal lining connect to large nucleated cell bodies which contain the normal discoid bodies and multilaminate vesicles, exactly the same as those present in the tegument (Morris and Threadgold 1968; Spence and Silk 1970; Dike 1971; Ernst 1975); however, they are relatively much larger and more numerous. The luminal tips of the anterior folds were described as greatly expanded, containing numerous small mitochondria and vesicles, but further detailed information of such vesicles was not provided (Dike 1971). Based on the early TEM studies, the prevailing view of the anterior compartment of *S. mansoni* was that its lining and associated cell bodies were simply an extension of the body surface tegument. However, in unpublished work we have found that there is a transition in the second half of the anterior compartment, with novel vesicles manufactured and secreted. This brings it more into line with the morphology of the anterior esophagus in *S. japonicum*.

By contrast, the entire lining of the anterior esophagus in *S. japonicum* differs from the body surface tegument (Li *et al.* 2014). In the surface tegument cytoplasm, two main types of inclusion, typical discoid bodies and dark bodies are evident, in addition to mitochondria. These two inclusions are also abundant in the tegument cell bodies, where rough endoplasmic reticulum and Golgi apparatus are both frequently seen (Li *et al.* 2014). The cell bodies in the *S. japonicum* anterior esophageal region are not identical to those beneath the tegument and do not manufacture the same proportions of cytoplasmic inclusions for export to the esophageal lining. These cell bodies are very elongate (up to 45 µm) tapering down to the point where their cytoplasmic connections run through the muscle lattice and are 4.5 µm at their widest, to accommodate the nucleus (Fig. 11.3A). Similar to those in *S. mansoni*, rough endoplasmic reticulum is abundant and Golgi bodies are frequent, indicative of protein synthesis and packaging for export (Fig. 11.3B). The cytoplasm contains large numbers of vesicles (mean size 0.35 × 0.24 µm in males and 0.42 × 0.39 µm in females) quite distinct in appearance from the dark bodies of the tegument, with sparse contents, hereafter termed 'light vesicles' (Fig. 11.3C). Close inspection reveals that the light vesicles are not all identical. Some are almost devoid of contents; others contain varying amounts of granular material (Fig. 11.3D). A proportion contains membranous lamellae, which may be either adherent to the bounding membranes (Fig. 11.3E, F) or present as loose whorls in the center accompanied by a prominent single dense granule (63 nm diameter; Fig. 11.3D). These light vesicles are also different from the crystalline vesicles in the posterior esophagus (described below). In this regard, the anterior esophageal region is indeed a distinct secretory organ.

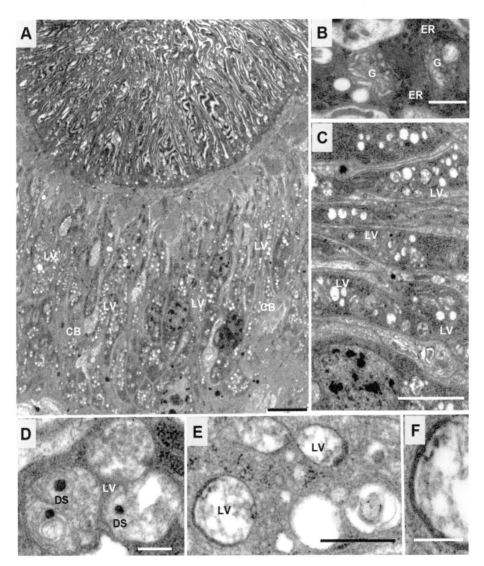

FIG. 11.3 TEM of cell bodies in the anterior esophageal gland of *S. japonicum*: **A.** Transverse section of anterior esophageal region, showing greatly extended lining cytoplasm with massive numbers of light vesicles (LV) present in the elongated cell bodies (CB). **B.** Higher magnification of the cell bodies with abundant rough endoplasmic reticulum (ER) and Golgi apparatus (G), indicative of light vesicle manufacture. **C.** Anterior esophageal gland cell bodies showing light vesicles of variable size and appearance. **D.** Light vesicles with central dense spheroidal (DS) material. **E.** Light vesicles with material applied to the outer bounding membrane. **F.** High magnification of a light vesicle reveals that this material is composed of membranous lamellae. Scale bar: A, 5 μm; B, 500 nm; C, 2 μm; D, 200 nm; E, 500 nm; F, 200 nm. From Li *et al.* 2013. PLoS Neglected Tropical Diseases 7(7): e2337. Li *et al.* 2014. Parasites & Vectors 7: 565. Fig. 3.

The lining of the anterior esophagus of *S. japonicum* comprises a thin basal layer of cytoplasm from which numerous corrugations extend; light vesicles and mitochondria are also present. This syncytial layer is not comparable in appearance to the tegument that lines the oral sucker and covers the worm body. In places, the lining corrugations are extremely narrow (50 to 70 nm), barely wide enough to accommodate two unit membranes and a little intervening cytoplasm (Fig. 11.4A). In wider regions (180 to 500 nm, mean 275 nm) central parallel membranes 28 nm apart, denote the presence of basal invaginations (Fig. 11.4B; these invaginations run for 40 to 60% of the distance up each corrugation from its base.

When viewed by TEM, the leading edge of many of the corrugations in male worms is sufficiently expanded to hold a cluster of vesicles of varying size (mean 0.4×0.29 μm) and appearance, spaced 0.4 μm apart (Fig. 11.4C, D), plus occasional mitochondria.

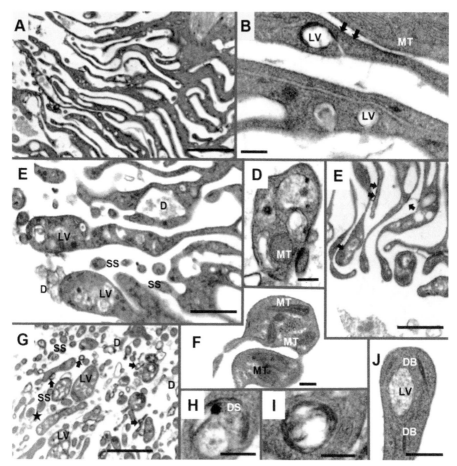

FIG. 11.4 TEM of anterior esophageal lining. A to D and G to I are from males; E, F and J are from females; C and E are to scale. **A.** Anterior esophageal lining showing clusters of light vesicles distributed all the way down to the tips of the corrugations. **B.** Light vesicles near the base of corrugations apparently emptying their cargo to the exterior; mitochondria are also evident whilst the surface is coated by membranous material (arrowed). **C.** Clusters of light vesicles (LV) are evident inside expanded corrugation tips; debris (D) is frequently seen either attached to the surface of corrugations or present between them. Small spheroids (SS), representing sections of cytoplasmic threads, are devoid of vesicles. **D.** High magnification of the tip of a male corrugation showing four vesicles in different profiles, together with a single mitochondrion (MT). **E.** Smaller expanded tips of female corrugations, most of which contain only one light vesicle; membranous material (arrowed), evident between corrugations, is attached to the outer surface. **F.** High magnification of female corrugation tips, showing presence of one or two light vesicles but more prominent mitochondria (MT). **G.** A general view of male corrugation tips. Small spheroids represent transverse section of cytoplasmic threads lacking vesicles but containing the occasional mitochondrion (starred). Other tips are expanded to accommodate a cluster of vesicles; debris (D) is common in the lumen, much of it membranous (arrowed) and adherent to the surface or to other debris to form irregular shapes. **H.** Light vesicle with whorls of membrane and a dense spheroid (DS). **I.** Light vesicle with layers of membranous material attached to the inner surface of the bounding membrane. **J.** Tip of a corrugation with discoid bodies (DB) and a single light vesicle (LV). Scale bar: A, 2 μm; B, 200 nm; C, 1 μm; D, 200 nm; E, 1 μm; F, 200 nm; G, 2 μm; H, 200 nm; I, 200 nm; J, 200 nm. Li *et al.* 2014. Parasites & Vectors 7: 565, Figs. 4 and 5.

However, in females this expansion is less obvious with only one or two light vesicles present (mean size 0.45×0.33 μm), but a predominance of mitochondria (Fig. 11.4E, F). The approximate surface area of the tip expansions is >1 μm² in males and 0.3 μm² in females (Fig. 11.4D, F). In contrast, the numerous narrow cytoplasmic threads appear circular in profile (diameter 170 nm; Fig.

11.4G) but still contain occasional mitochondria. Light vesicles containing whorls of membrane and those with a dense central granule are both evident (Fig. 11.4H, I); discoid bodies also reach as far as the expanded tips (Fig. 11.4J).

S. mansoni and *S. japonicum* share similar characteristics in the morphological aspects of the posterior esophagus. Viewed in the TEM, the plates in the posterior esophageal lining are more regularly arranged, compared with the irregular folds in anterior esophagus and revealed as thin cytoplasmic extensions tapering towards their tips. While running in parallel from anterior to posterior of the gland, they occasionally appear to bifurcate/merge, but we could not establish that individual plates ran the whole length. The plates appear loosely separated, not tightly packed together as in previous descriptions (Dike 1971). A central invagination originating from the basal lamina is prominent in each plate (Shannon and Bogitsh 1969; Dike 1971; Li *et al.* 2013), extending nearly to the outermost tips.

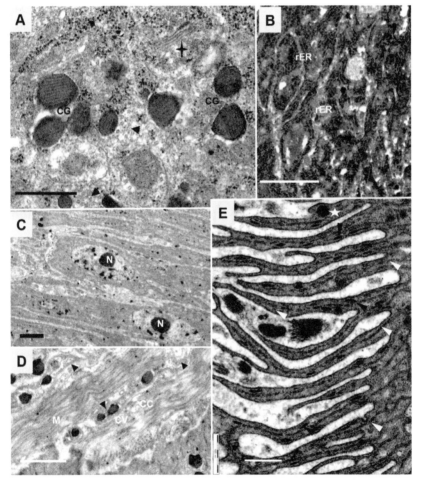

FIG. 11.5 TEM of posterior esophageal region. **A.** An assemblage of crystalloid vesicles (CV) in proximity to a Golgi body (starred) in an esophageal gland cell of *S. japonicum*. Microtubles (black arrows) are present in the vicinity of granules, suggesting that their transport occurs using kinesin motors. **B.** Esophageal gland of *S. mansoni* showing several cell bodies in section with large amounts of rough endoplasmic reticulum (rER) indicating intense protein synthesis for export. **C.** The same region in *S. japonicum* at lower magnification showing the very elongated cell bodies that convey crystalloid vesicles towards the esophageal lining; the nuclei contain very prominent nucleoli (N). **D.** Crystalloid vesicle (CV) traffic from the cell bodies, through the muscle layers (M), to the syncytium, via narrow cytoplasmic connections (CC) containing microtubules (arrowed). **E.** TEM of the thin plates; a central double line (white arrows) is evident in each plate and discoid bodies (black arrows) are numerous. A single crystalloid vesicle (starred) is located close to a potential docking site. Scale bars: A, 500 nm; B, 2 μm; C, 2 μm; D, 1 μm; E, 500 nm. Li *et al.* 2013. PLoS Neglected Tropical Diseases 7(7): e2337, Fig. 2 and Fig. S1.

The posterior esophageal gland of *S. mansoni* was described more than 40 years ago (Dike 1971). Cell bodies are closely packed, which means some are positioned up to 80 μm from the lining syncytium but each is connected by microtubule-lined extensions. The internal structure of these cells is distinct; while discoid bodies are still present, the multilaminate vesicles are no longer seen. Instead, they are replaced by a solid dense granule which contains a crystalline core arranged in parallel layers and sometimes circular profiles (Dike 1971; Bogitsh and Carter 1977; Li *et al.* 2013); hereafter we refer to them as 'crystalloid vesicles'. These vesicles contain parallel arrays of electron-dense material in evenly spaced repeating units 19 nm apart, separated by an intervening electron-lucent layer (Fig. 11.5A). The cell bodies are very obviously engaged in synthesis of protein for export, indicated by extensive rough endoplasmic reticulum (Fig. 11.5B), prominent nucleoli (Fig. 11.5C), Golgi bodies and assemblages of crystalloid vesicles in the cytoplasm (Fig. 11.5A). Where the extensions from the cell bodies pass between the esophageal wall muscles, they are only just wide enough to accommodate a single crystalloid vesicle (Fig. 11.5D). After the vesicles reach the thin lining syncytium, for the most part, they are confined to the base and a short distance up into the plates, while the discoid bodies appear capable of traveling much further towards the tips (Fig. 11.5E). The complex cellular architecture of the posterior esophageal gland can be summarized in diagrammatic form (Fig. 11.6).

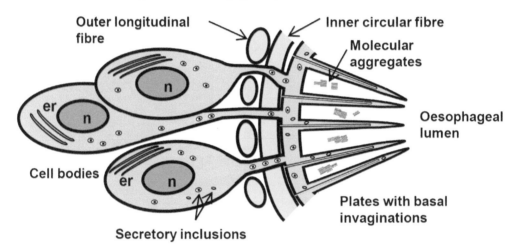

FIG. 11.6 Diagrammatic representation of the posterior esophageal gland in transverse section: A single cytoplasmic extension is shown running from each nucleated cell body, through the muscle layer to the lining syncytium, as we have no evidence that branching occurs, cf. tegument cell bodies. The secretory inclusions are crystalloid vesicles and discoid bodies. From Li *et al.* 2013. PLoS Neglected Tropical Diseases 7(7): e2337. Fig. 3.

11.4.4 Biological Functions

Videos of *S. mansoni* feeding *in vitro* have revealed that blood ingestion is a multistep process (Li *et al.* 2013). Blood, swept into the mouth by the grabbing motion of the oral sucker accumulates in the anterior compartment which expands to form a bulge. The central sphincter then relaxes allowing the ingested material to pass as a bolus to a posterior compartment. Finally the posterior sphincter relaxes and the material enters the transverse gut. The remarkable cellular architecture of the anterior esophageal lining, revealed by SEM and TEM, clearly expands enormously the surface area available for interactions while the fringe of threads extends still further the potential interaction surface with ingested blood. Indeed it is possible, especially in the males, that the spaghetti-like threads intruding into the lumen are intended to entangle cells and retard their passage, allowing more time for interaction with esophageal secretions.

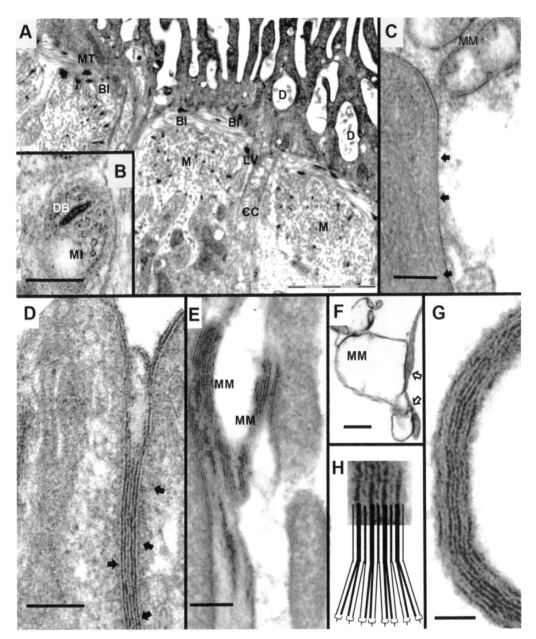

FIG. 11.7 Light vesicles supply the membranous coating on the corrugation surface in *S. japonicum*: **A.** Cytoplasmic connection (CC) between cell bodies of the anterior esophageal gland and basal corrugations showing light vesicles (LV) in transit; debris (D) is evident between the corrugations presumably derived from released contents of light vesicles (BI, basal invagination; M, muscle layer; MT, mitochondria). **B.** Transverse section of cytoplasmic connection reveals microtubules (MI) potentially involved in vesicle transportation (DB, discoid body). **C.** Membranous material (MM and arrowed) covers the whole surface of the corrugations and is also evident free in the lumen. **D.** Membranous material trapped between two adjacent corrugations, the alternating black and white stacks giving a tramline appearance. **E.** Layers of membranous material peeling off the surface of two adjacent corrugation surfaces to form the free aggregates in the lumen. **F.** Whorls of membranous material shedding from a corrugation tip (arrowed) into the esophagus lumen. **G.** High magnification of a free membranous stack showing the self-adherent properties of the constituents that produce myelin-like figures in the anterior lumen. **H.** A diagrammatic interpretation of the way in which the shed membranous layers combine to form an alternating pattern of black and white units. Scale bar: A, 5 μm; B, 500 nm; C to G, 200 nm. Li *et al.* 2014. Parasites & Vectors 7: 565. Figs. 4 & 6.

The mass of cell bodies surrounding the anterior esophagus are clearly a factory, primarily for the manufacture of light vesicles and to a lesser extent discoid bodies. After manufacture in the cell bodies, the light vesicles are transported to the corrugation via narrow cytoplasmic connections and reach all the way to the leading edge (Fig. 11.7A). Transverse sections through these connections reveal microtubules, which are probably part of the molecular-motor machinery for vesicle transport (Fig. 11.7B). This traffic must occur along designated channels wide enough to accommodate the vesicles, given the thinness of the corrugations in many places. The expanded corrugation tips of males containing clusters of vesicles, versus usually only a single vesicle in females, may indicate that they are a temporary storage site for vesicles. This accords with the recent suggestion that males feed intermittently but females continuously (Skelly *et al.* 2014). Mitochondria are more prominent in the tips of females than males, suggesting a greater need for energy at this location.

Some light vesicles can release their cargoes in basal corrugations (Fig. 11.4B) where they can be seen fusing with the bounding membranes. Debris and membranous material present between the corrugations (Fig. 11.7A) plausibly originates from the light vesicles. More commonly, the light vesicle contents are discharged from the leading edge although direct evidence of an act of vesicle fusion is difficult to capture. The heterogeneous appearance of the light vesicles themselves in the cell bodies, containing both membranous and granular material, implies a multiplicity of functions although precise function has yet to be established.

The whole surface of the corrugations from base to tip is covered with membrane-like material, displaying the multilamellar characteristics of the tegument, although the dark vesicles that supply the body surface tegument with its membrane-like coating are replaced by the light vesicles in the anterior esophagus. From this we infer that in spite of originating from a morphologically different vesicle, this secreted membranocalyx functions as an inert protective barrier in the anterior esophagus, in the same way as it does over the tegument surface.

We also infer that the membranous contents of the light vesicles are released in quantity to coat the surface of the corrugations, and are then shed into the anterior esophageal lumen. The membranous material on the corrugation surface can be seen both peeling off from the leading edge and free in the esophagus lumen. This material has proved difficult to stabilize in all schistosome species using normal fixation and tissue processing procedures that rely on a progressive series of organic solvent extractions prior to resin embedding (Fig. 11.7C). Fortuitously, it is more readily visualized when trapped between two adjacent corrugations (Fig. 11.7D). Here the alternating black and white stacks are evident and in some places layers are captured in the process of peeling off (Fig. 11.7E). It is also apparent that the released material is self-adherent since it accumulates in even larger membranous stacks, several layers thick, in the esophagus lumen (Fig. 11.7F). (This adherent property is apparent in the material when it is still within the parent vesicle; Fig. 11.4H). Careful inspection of the released membranous stacks at high magnification shows they are not of uniform electron density. Rather, there is a pattern of alternating broad and narrow dark lines (Fig. 11.7G). This suggests an asymmetry in the individual units that comprise the membrane-like secretion. One leaflet of the secreted bilayer, probably the inner-facing one, is denser than the other that faces the external environment. When the membranous secretions peel off from the surface and associate, they apparently do so thick layer to thick and thin layer to thin, to create the observed alternating pattern of broad and narrow dark lines (Fig. 11.7H).

Besides supplying the material for refreshing the membranocalyx on the corrugation surface to protect the worms from being attack, the secretions from anterior esophagus can also interact with the blood components. Such interaction might have several functions, e.g., to neutralize the 'toxic' blood components like antibody, as well as to prepare for further actions taking place in posterior esophagus.

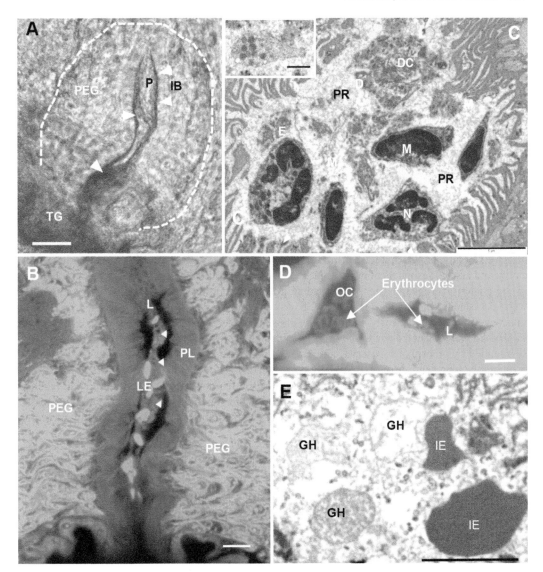

FIG. 11.8 Blood processing in posterior esophagus: **A.** Ingested blood (IB, arrowed) entering the lumen of posterior esophageal gland (PEG, outlined, based on 40 consecutive video images filmed at x 40) flows as dark line around the plug (P) of material as it passes to the transverse gut (TG). **B.** Confocal image of adult male *S. mansoni* stained with Langeron's carmine, showing a string of host leucocytes (LE) tethered in the lumen (L), surrounded by extracellular material (arrowed). The esophageal plates (PL), at the limit of resolution, appear as fine striations. **C.** TEM of posterior esophagus in *S. mansoni* showing damaged leucocytes (neutrophil, N; eosinophil, E; mononuclear cell, M; disintegrated cell, DC, a platelet with characteristic granules (inset) and debris (D) present in the centre of the lumen, surrounded by a dense precipitate (PR) of protein. **D.** Confocal image of a female worm fixed while feeding *in vitro* then stained with Langeron's carmine. Intact erythrocytes are visible within the oral cavity (OC) and the lumen (L) of the anterior esophageal compartment. The erythrocytes can only be seen when the laser intensity is greatly amplified as they do not react well with the stain. **E.** TEM of the posterior esophageal lumen from a male worm fixed *in vivo*. Three ghosts (GH) are present, showing different stages of hemoglobin leakage, plus two intact erythrocytes (IE). Scale bars: A, 25 μm; B, 10 μm; C, 5 μm; D, 10 μm; E, 5 μm. Li *et al.* 2013. PLoS Neglected Tropical Diseases 7(7): e2337. Figs. 4–6.

When the blood enters the posterior esophagus as a bolus from the anterior compartment, material is already apparent in the lumen, reminiscent of a plug, which at high magnification appears to contain cells (Fig. 11.8A). As the bolus arrives, this bulge expands with material streaming as a dark current around the plug (Li *et al.* 2013). When the stationary plug in the lumen of the posterior esophagus (Fig. 11.8A) was examined by confocal microscopy it proved to be a string of host leucocytes trapped within extracellular material (Fig. 11.8B). This interpretation was confirmed by TEM when the plug could be seen to comprise polymorphonuclear and mononuclear leucocytes in varied states of disintegration (Fig. 11.8C); indeed, some were barely recognizable as leucocytes. In addition, occasional platelets were visible, identifiable by their size and complement of characteristic granules (Fig 11.8C, inset). Using Z-stacks of consecutive confocal images through the lumen of the posterior esophagus we estimated the number of tethered leucocytes present in male *S. mansoni* fixed *in vivo* to be 28.6 (+S.E. 4.3; n = 15) and in females, 10.36 (+S.E. 3.12; n = 11). Such numbers greatly exceed the theoretical predictions based on lumen volume and density of cells in blood, indicating that a tethering mechanism functions in the posterior esophagus. More than that, the battered state of leucocytes suggests that cell killing must also occur in the posterior compartment.

Erythrocyte lysis appears to be initiated in the anterior compartment and completed in the posterior so intact erythrocytes are seldom seen there. When worms were fixed whilst being observed feeding *in vitro*, intact erythrocytes could be imaged with confocal microscopy in the oral cavity and anterior esophageal bulge (Fig. 11.8D) of both males and females but were not observed in the posterior esophagus or the transverse gut. When viewed by TEM, obvious erythrocytes were also scarce. In a single male worm, fixed while feeding *in vitro*, we observed a mixture of intact erythrocytes and clearly recognizable ghosts in different stages of hemoglobin leakage (Fig. 11.8E). All were located close to the tips of esophageal plates. These erythrocytes and ghosts were surrounded by amorphous material, similar in appearance to the residual contents of the ghosts, presumably representing released hemoglobin. Further evidence that confirms the uncoating is completed in the posterior esophagus was provided by feeding experiments with PKH2-labelled erythrocytes. As they passed down the esophagus, the lipophilic dye was transferred to the esophageal lining; this provides direct evidence for the interaction of gland products with incoming blood (Hall *et al.* 2011).

Similar to the anterior esophagus, the posterior esophageal syncytium extends its surface area greatly with thin plate-like extensions of the luminal membrane and the very elongate basal invaginations that penetrate almost the whole length of each plate. Ostensibly, this increased surface area would enhance interactions with incoming cells but the spaces between plates are too narrow to admit erythrocytes or leucocytes and quite possibly platelets, leaving only plasma proteins able to enter. The basal invaginations are typical of epithelial cells such as those of the proximal and distal tubules of the mammalian kidney (Berridge and Oschman 1972), where ion transporters are located. The Na-K-Cl cotransporter is one such protein in the basolateral membranes of secretory epithelia in many species (Haas and Forbush 2000). This raises the possibility that the enormous surface area is an adaptation to generate massive ion fluxes into or out of the posterior esophagus lumen and this in turn might be linked to erythrocyte lysis.

The location of crystalloid vesicles in the base of the syncytium and a short distance up the plates indicates that secretion of contents occurs only in this region. The contained parallel array of material is released intact to the lumen, at specific docking sites, by fusion of the vesicle membrane with the surface plasma membrane (Bogitsh and Carter 1977). Thereafter the arrays do not dissipate, as would be expected for normal secretory vesicle contents, but persist intact, as attested in several studies (Shannon and Bogitsh 1969; Spence and Silk 1970; Dike 1971; Bogitsh and Carter 1977). Indeed our observations show that they cluster into larger aggregates, thus confirming the self-affinity of their molecular constituents. However, the 40% greater repeating unit in the aggregates, compared to the vesicles, indicates that expansion of the electron lucent layers occurs after release. These observations provide another potential function of the plates, namely to sequester the crystalloid aggregates, allowing them both to interact more efficiently with the incoming blood and to prevent their wash-out (and waste) when the worm vomits. This hypothesis envisages the longitudinal plates of the posterior esophagus serving as a slow release reservoir for gland products that must interact instantaneously when blood is ingested.

The abundance of the membranous aggregates on the surface of the anterior lining and in the lumen may indicate a fast turnover of this secreted layer, which could reflect the effect of antibody binding and sloughing. In this respect, the strong reactivity of the anterior esophagus with host IgG is surprising (Li *et al.* 2013; Li *et al.* 2014; Li *et al.* 2015). In contrast, the body surface tegument of *S. japonicum* reacts only weakly for IgG implying that it is protected from antibody binding by its membranocalyx. However, if the worm has no option but to secrete proteins from specific sites on the tips of the esophageal corrugations, then such strong reactivity with antibody might be anticipated. Indeed, high resolution images of worm head cryosections revealed that the pattern of intrinsic antibody binding on the anterior lining was punctuate, not continuously uniform (Li *et al.* 2014). This indicates a marked heterogeneity of composition with only some points in the lining strongly recognized. Observations made on worms from murine and hamster permissive hosts indicate this bound antibody appears to cause no harm (Li *et al.* 2013), perhaps precisely because there is rapid sloughing of targets. By contrast, in self-cured hosts like rhesus macaques, antibody binding involved interruption of vesicle release, which resulted in the formation of swollen expanded tips with giant vesicles inside. We have speculated that impaired and ultimately lost functions of the esophagus, eventually led to feeding failure and caused the worms' demise (Li *et al.* 2015).

11.4.5 Molecular Products of the Esophageal Glands

The first application of whole mount *in situ* hybridization (WISH) to the study of gene expression in larval and adult schistosomes identified the first constituent in the esophageal glands, namely SmMEG-4.1 (Fig. 11.9A, B, previously known as Ag 10.3; (Dillon *et al.* 2007)) encoded by a micro-exon gene (MEG). Later immunocytochemistry on whole adult worms fixed and permeabilized according to the protocol of Mair *et al.* (Mair *et al.* 2000) using a rat antibody against a synthetic peptide derived from its protein sequence, confirmed the expression of the MEG-4.1 protein exclusively in the posterior gland (DeMarco *et al.* 2010). With these two powerful techniques, more constituents in the posterior esophageal gland were identified including two micro-exon genes, SmMEG-4.2 and SmMEG-14 (Fig. 11.9C) (Li *et al.* 2013) and SmVAL-7 (Rofatto *et al.* 2012). Seven proteins (six MEGs and VAL-7) (Fig. 11.9E, F, G) were also localised to the posterior esophageal gland of *S. japonicum* by immunocytochemistry (Li *et al.* 2013, 2015).

Clearly, to obtain an insight into those genes expressed in the distinctive tissues of the schistosome esophagus that encode the proteins involved in the initial processing of ingested blood, a more comprehensive method had to be adopted. The difficulties in characterizing patterns of gene expression that occur in the discrete organ systems of an acoelomate metazoan with a solid body plan should not be underestimated. Laser capture microdissection (Nawaratna *et al.* 2014) has been applied but the amount of tissue obtained and the precision needed to excise the organ of choice without contamination are just two of the limitations.

The advent of cheaper technologies has made comparative transcriptome analysis by direct sequencing feasible, providing a wide dynamic range of detection with low background and no saturation of signal, no gaps in the repertoire and the ability to identify new genes. We have used the massive parallel capacity of ion semiconductor sequencing on an Ion Torrent instrument to investigate differential gene expression in the esophageal region of adult male *S. mansoni* (Wilson *et al.* 2015). Schistosomes possess epithelia (tegument, gastrodermis) and rudimentary organ systems (muscles, nerves and sense organs, alimentary tract, protonephridial system, parenchyma) present throughout the whole body but the solid acoelomate body plan means they are not readily isolated for analysis. However, the cell masses surrounding the anterior and posterior esophageal compartments, plus the paired cerebral ganglia of the nervous system are unique to the esophageal region. We therefore reasoned that a subtractive comparison of the patterns of gene expression in heads and tails would delineate this unique 'head' subset. The core of this strategy was the isolation by microdissection of the entire esophageal region and matching tails from adult male bodies stabilized with RNA Later.

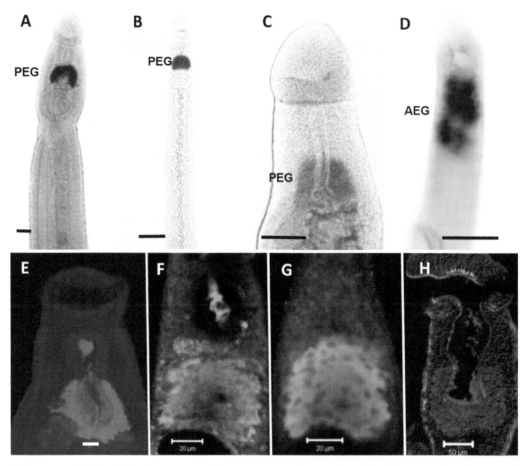

FIG. 11.9 Localization of MEG transcripts and MEG proteins to the esophageal glands: A to D show detection of mRNA transcripts in *S. mansoni* using WISH, E to H show detection of proteins by immunocytochemistry on *S. japonicum*, using monospecific antibodies. **A.** and **B.** Male and female worms, respectively, at low magnification to show the specific expression of MEG-4.1 solely in the posterior esophageal gland (PEG). **C.** Higher magnification of a male showing MEG-14 expression in the posterior esophageal gland. **D.** MEG-12 Expression of in the anterior esophageal gland of a female. **E., F.** and **G.** Localisation of MEGs 4.1, 4.2 and 14, respectively, in the posterior esophageal gland of permeabilised male worms. **H.** Localisation of MEG-8.2 in the posterior esophageal gland on a cryosection of a male. Scale bars: A-C, 100 μm; D, 25 μm; E, 50 μm; F and G, 20 μm; H, 50 μm. Li *et al.* 2013. PLoS Neglected Tropical Diseases 7(7): e2337. Fig. 7; Li *et al.* 2015. PLoS Neglected Tropical Diseases, 9(7): e0003925. Fig. 7; Wilson *et al.* 2015. PLoS Neglected Tropical Diseases, 9(12): e0004272. Fig. 8.

The millions of transcripts were mapped by Tophat and Cufflinks programmes to approximately 5000 genes, half more highly expressed in the heads and half in the tails, with a dynamic range between four and five orders of magnitude. Expression of the vast majority of ~250 signature genes for the major schistosome tissues lay within a very narrow range either side of the equivalence line, allowing us to define differential expression of outliers by setting a generous four fold margin either side of the equivalence line. On that basis 92 annotated genes were differentially expressed in the head samples. The remarkable feature of the top 20 genes sorted by abundance (Table 11.2), was that MEGs accounted for more than half the total, including the three already known to be expressed in the esophageal gland (MEGs 4.1, 4.2 and 14), plus a further eight (MEGs 8.1, 8.2, (Fig. 11.9D) 9, 11, 12, 15, 16 & 17), whose site of expression was not previously recorded. Nine lysosomal hydrolases comprised a second prominent group of differentially expressed genes (Table 11.3). These included a group of six proteases, annotated as subfamily A1A unassigned peptidases (A01 family), encoding closely related Cathepsin D aspartyl proteases. Two Phospholipase A2 enzymes were almost exclusively expressed in the heads, while the final hydrolase, palmitoyl protein thioesterase 1 enzyme, removes thioester-linked fatty acyl groups from modified cysteine residues in proteins or peptides.

TABLE 11.2 Microexon (MEG) genes differentially expressed in the male *S. mansoni* esophageal glands, which encode secreted or membrane anchored proteins. From Wilson *et al.* 2015. PLoS Neglected Tropical Diseases 9(12): e0004272. Table S5.

MEGs mapped to the genome by top Hat/Cufflinks				Novel MEG genes from trinity assembly			
Rank out of 92	Annotation in Gene DB	Description	Abundance (RPKM score)	Rank out of 51	Trinity gene id	Description	Abundance (RPKM score)
1	Smp_085840	MEG-4.2	66833.85	1	c9796_g1	MEG-8.3	3954.4
2	Smp_172180	MEG-8.2	49111.40	2	c8054_g1	MEG-22	2559.7
3	Smp_010550	MEG-15	44058.30	3	c7574_g1	MEG-31	1947.5
4	Smp_124000	McG-14	40535.45	4	c7993_g1	MEG-27	1764.4
5	Smp_176020	MEG-11	25935.05	5	c9222_g1	MEG-29	1351.7
6	Smp_163630	MEG-4.1	22409.80	6	c8532_g1	MEG-20	1261.8
7	Smp_123200	MEG-32.2	21064.70	8	c12069_g2	MEG-3.4	1071.1
8	Smp_171190	MEG-8.1	16785.50	11	c6870_g1	MEG-4.3	762.4
11	Smp_152630	MEG-12	11620.65	12	c10941_g3	MEG-8.4	536.9
12	Smp_125320	MEG-9	10265.61	14	c10894_g2	MEG-10.2	426.4
13	Smp_123100	MEG-32.1	9963.52	18	c10348_g3	MEG-19	257.1
14	Smp_158890	MEG-16	6500.60	20	c586_g1	MEG-30	229.2
16	Smp_180620	MEG-17	5468.51	22	c10014_g1	MEG-28	203.7
				23	c10048_g1	MEG-26	195.2

TABLE 11.3 Lysosomal hydrolase and VAL genes expressed in the male *S. mansoni* esophageal glands. Modified from Wilson *et al.* 2015. PLoS Neglected Tropical Diseases 9(12): e0004272. Table S3D.

Rank out of 92	Annotation in Gene DB	Description	Abundance (RPKM score)
9	Smp_199890	VAL 7 protein	16190.15
10	Smp_070240	VAL 7 protein	14618.00
17	Smp_136830	Aspartyl protease, Cathepsin D	1934.89
18	Smp_142970	palmitoyl protein thioesterase 1	1379.78
21	Smp_132480	Aspartyl protease, Cathepsin D	1058.14
25	Smp_031190	phospholipase A	623.88
35	Smp_031180	phospholipase A	245.07
47	Smp_018800	Aspartyl protease, Cathepsin D	99.33
55	Smp_132470	Aspartyl protease, Cathepsin D	81.07
59	Smp_205390	Aspartyl protease, Cathepsin D	70.57
80	Smp_136720	Aspartyl protease, Cathepsin D	27.46

The reads not mapped by Tophat and Cufflinks were assembled de novo using Trinity and a further 51 unannotated differentially expressed contigs was filtered out for manual analysis. Sixteen of the novel genes were identified as encoding unannotated MEGs (Table 11.2) including six previously described (MEGs 8.3, 8.4, 19, 20, 22, 24 (Almeida *et al.* 2012), a further eight that were entirely new, namely MEGs 26–31, plus two new members of existing families, MEGs 10.2 and 4.3 and two improved gene models for annotated genes now designated as MEGs 32.1 and 32.2. Subsequent work has shown that MEG-26 is a representative of a family with six closely related members (unpublished data). In summary, a total of 32 transcripts from 22 out of 32 MEG families (two-thirds) was detected in the schistosome head region, making it the most intense site for the expression of this enigmatic group of genes so far discovered. Based on WISH and immunocytochemistry we can now be confident that three MEGs (12, 16 and 17) plus a phospholipase A2 are expressed in the anterior gland and nine MEGs (4.1, 4.2, 8.1, 8.2, 9, 11, 14, 15 and 22) plus aspartyl protease and palmitoyl thioesterase in the posterior gland (Wilson *et al.* 2015).

A major focus of current research is to relate the functions of these esophageal secretions to the processing of blood that we have shown occurs in the anterior and posterior esophageal compartments. For the numerous MEG encoded proteins this is made more difficult by their lack of homology to any known protein. Nine MEGs (4.1, 8.1, 8.2, 14, 15, 19, 20, 29 and 32.1) show putative disordered regions that are predicted to be heavily O-glycosylated, while a further five MEGs (8.3, 10.2, 22, 32.1, 32.2) are predicted to be O-glycosylated proteins but without extensive regions of disorder. Both the MEG-4 and MEG-8 family proteins possess a conserved C-terminus between schistosome species (Li *et al.* 2013; Wilson *et al.* 2015). MEG-15 also displays a relatively hydrophobic C-terminus. Potentially this C terminal region might target host leucocytes, e.g., by binding to a pan-leucocyte marker such as CD45 (Li *et al.* 2013). Equally, plasma proteins or leucocyte secretions could be the intended ligands. An alternative possibility is that the O-glycosylated proteins may use the C-terminal motifs to organize themselves in a similar way to secreted mucins, creating a net that confers a gel-like consistency with viscoelastic proprieties (Cone 2005). The immunolocalization of both SjMEG-4.1 (Li *et al.* 2013) and SjMEG-8.2 (Li *et al.* 2015) in a cocoon-like association with tethered leukocytes in the lumen of the *S. japonicum* esophagus provides visual evidence that a mucus-like complex traps incoming leukocytes.

The most likely role for the MEGs anchored by a transmembrane helix (e.g., MEGs 14, 29, 32.1, 32.2) and also predicted to be O-glycosylated, is to provide a protective lining coat of O-glycans for the entire esophagus. This suggestion is corroborated by the detection of a thin layer of neutral muco-substance lining the esophagus of the related blood fluke *Schistosomatium douthitti* (Shannon and Bogitsh 1969). Another group of MEGs (9, 12, 26 family, 27 and 28) encode a small peptide, predicted to contain an amphipathic helix with a hydrophobic interaction face. Such peptides are widespread in animals and have both anti-microbial and hemolytic properties (e.g., Kozlov *et al.* 2006; de la Salud Bea *et al.* 2015). In the context of the schistosome esophagus they may interact with incoming erythrocytes and leucocytes to destabilize their membranes. Our recent observations on the localization of *S. japonicum* MEG-9 confirm its association with the surface of leucocytes *in situ* in the esophagus lumen (Li *et al.* 2015). That still leaves approximately one third of the proteins encoded by esophageal MEGs that have no distinguishing features to provide a clue to putative function, other than a signal peptide.

In contrast to the MEG proteins, the lysosomal hydrolases all have orthologues in mammals with well characterised properties, which provide pointers to their potential function in blood processing. The secretion of two phospholipases, one at least from the anterior gland, strongly suggests a role in the lysis of erythrocytes as these cells pass through the two compartments. The palmitoyl thioesterase may also participate in the process via its ability to cleave the lipid anchor from proteins on the cytoplasmic face of plasma membranes. One such is p55/MPP1 (Ruff *et al.* 1991), an important component of the ternary complex that attaches the spectrin-based skeleton to the erythrocyte membrane (Lach *et al.* 2012). Thus we can envisage a cascade where short amphipathic MEG peptides such as MEG-9 or MEG-12 bind to and destabilize the erythrocyte membrane, enhancing the interaction of the two phospholipases with their plasma membrane substrates. Increased permeability then permits the palmitoyl thioesterase to enter and disrupt the cytoskeleton; the erythrocyte loses shape, leaks hemoglobin and is destroyed. Judging from videos of worm feeding (Li *et al.* 2013) the whole process takes only seconds.

The secretion of six aspartyl proteases, at least one from the posterior compartment, indicates a powerful attack is also made on proteins in the plasma or on the external surface of host blood cells. Note that these enzymes are distinct from the one already described for the *S. mansoni* gut (see below) (Wong *et al.* 1997). The number of homologs suggests either redundancy of function, potentially as a means of immune evasion, or the existence of subtly different substrate specificities in the target proteins. The most obvious candidates for proteolysis are the components of the clotting cascade since clot formation does not occur in the worm esophagus.

11.5   THE GUT

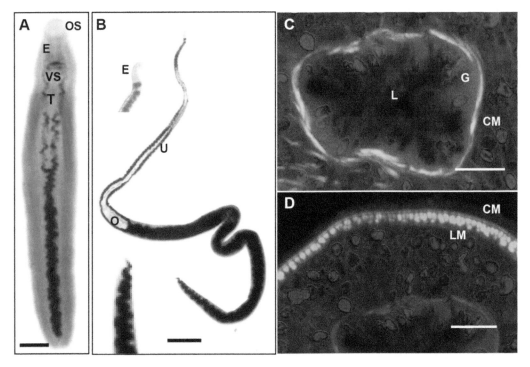

FIG. 11.10 Layout of the alimentary tract: **A.** Day 28 *S. mansoni* male worm from a hamster host. The gut lumen is clearly demarcated by the black hemozoin breakdown product of hemoglobin digestion. The oral sucker (OS) and short esophagus (E) lead to the transverse gut from which two branches descend either side of the ventral sucker (VS) and testes (T) before uniting as a single tube that ends blindly almost at the extreme posterior. **B.** Twenty one week female *S. japonicum* worm from a rabbit host. The much greater amount of hemozoin reflects both the larger volume of the female gut and the amount of blood ingested. The magnified upper inset (x2.5) emphasizes the small size of the female esophagus compared to the male and relative to the volume of blood ingested per day. The branches of the female gut run either side of the uterus (U) and ovary (O) before uniting to run as a single tube that to the extreme posterior (lower inset, x2.5). The gut area comprises 10% of a female cross section but only 2% of the male body. **C.** Cryosection of a female *S. japonicum* worm showing muscle (orange) and nuclei (blue). The gastrodermal syncytium (G) surrounding the lumen (L) contains numerous prominent nuclei and is invested by a ring of very thin circular muscle fibers (CM). **D.** The body wall of the same female, showing the outer ring of thin circular fibers (CM) overlying the much larger and stronger longitudinal muscle fibers (LM). Scale bars: A, 0.5 mm; B, 1 mm, C & D, 20 μm. All images are originals.

The organisation of the gut cecum in both sexes is easily discerned from the distribution of dark hematin pigment comprising polymers of the detoxified heme moieties released when hemoglobin is digested. The short esophagus empties into an initial transverse region of gut that immediately bifurcates, the two arms passing either side of the seminal vesicle and testes in the male (Fig. 11.10A) or ovary, ootype and uterus in the female (Fig. 11.10B). Approximately halfway down the body the two arms reunite and continue to the extreme posterior as a single tube which ends blindly. In the female this layout can be regarded as a way of transporting nutrients close to the tissues that need them but this is not the case for the male (see above/below). Compared to the esophagus, the cecum has a remarkably uniform structure along it whole length, comprising the syncytial gastrodermal layer of cytoplasm containing numerous nuclei. This is overlain by a thin layer of circular smooth muscle fibers (Fig. 11.10C), but unlike the esophagus and body wall (Fig. 11.10D), longitudinal fibers are absent. Coordinated contraction and relaxation of the circular fibers is responsible for the waves of peristalsis that sweep to and fro along short sections of the cecum to move and mix the blood meal. However, there is no information about the innervations from the peripheral nerve net that control these movements. *In vitro* feeding experiments with fluorescently labelled dextran (Hall *et al.* 2011) and bovine serum albumin (Delcroix *et al.* 2006) have revealed that by one hour (the earliest sampling time in both studies) the label had reached the furthest extremity of both males and females and was uniformly mixed. The dextran performs as a space marker since it is not hydrolysed by normal intestinal carbohydrases. Twenty four hours after ingestion by adult worms much of the label had entered the gastrodermal cells where it accumulated in numerous small discrete and intense spots, the largest approximately 2 µm in diameter (Hall *et al.* 2011). This was the first demonstration of endocytosis of material into the gastrodermal syncytium.

11.5.1 Ultrastructure of Gastrodermis

The first report by Morris that described the principal structural features of the gastrodermis, was a model of brevity and accuracy that has not been surpassed (Morris 1968). The syncytium varies in thickness from approximately 1.5 to 4 um, with some irregular protrusions of cytoplasm into the gut lumen; it is possible that the variations in thickness simply reflect the state of extension or contraction of the worm body. The luminal surface is not extended by regular villi, as in many animals. Rather, numerous sheet-like lamellae project a short distance from the surface into the lumen (Fig. 11.11A, B, C). A finely filamentous coat present external to the surface plasma membrane, is rich in glycans, representing a glycocalyx (Fig. 11.11E) (Wilson and Barnes 1974). Small surface depressions or caveolae, on the luminal surface have been interpreted as indicative of micropinocytosis. The entire gastrodermis is separated from the investing circular muscle fibers by a basal lamina and fibrous interstitial material but is firmly attached to them by numerous junctional complexes. The basal plasma membrane of the gastrodermis is thrown into numerous long invaginations which penetrate almost the entire thickness of the syncytium, potentially indicating fluxes of water and solutes that are characteristic of transporting epithelia (Fig. 11.11C). Lipid droplets, numerous mitochondria and an extensive endoplasmic reticulum are present in the syncytium (Fig. 11.11C). The cisternae of the endoplasmic reticulum are often distended and contain an amorphous material of appreciable density, confirming the synthesis of protein for export to the lumen (Fig. 11.11D). Use of the enzyme marker thiamine pyrophosphatease has confirmed the presence of numerous Golgi bodies throughout the syncytium (Bogitsh 1975) but surprisingly for such an active protein-exporting tissue, secretory vesicles are not obvious (Fig. 11.11D).

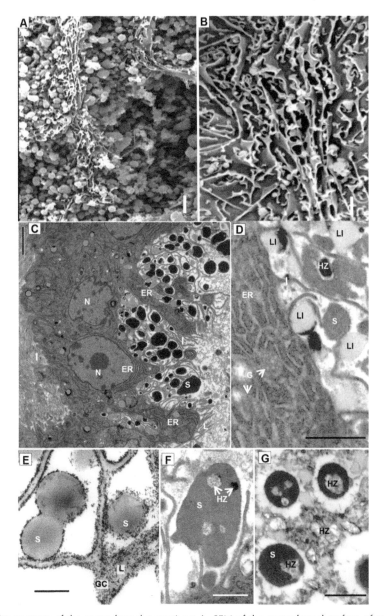

FIG. 11.11 Ultrastructure of the gastrodermal syncytium: **A.** SEM of the gastrodermal surface of an *ex vivo* adult worm showing the lamellate surface liberally coated with spherical globules of hemoglobin. **B.** SEM of an *ex vivo* 28 day worm cultured *in vitro* for 72 hours. The hemoglobin spheres have all been digested, revealing the extent of the gastrodermal surface lamellae. **C.** TEM of the gastrodermal syncytium showing nuclei (N), invaginations of the basal plasma membrane (I), spheres of hemoglobin (S), thin lamellae projecting into the gut lumen (L) and extensive rough endoplasmic reticulum (ER). **D.** Gastrodermal syncytium showing the lumen of the endoplamic reticulum (ER) cisternae distended by amorphous material. The lumen contains spheres of hemoglobin (S), hemozoin granules (HZ) and four lipid droplets (LI). Two have a peripheral accumulation of hemozoin and two do not; G, Golgi bodies (arrowed). **E.** The glycocalyx (GC) of the gastrodermis visualised by the PATCO technique. The surface lamellae (L) have a glycan-rich coat that potentially equates to CAA and CCA and the spheres (s) of hemoglobin are also covered in glycan reaction product. **F.** Hemoglobin sphere (S) anchored to the gastrodermis via the tip of a lamella. Dense aggregates of hemozoin (HZ) are arrowed. **G.** Three spheres (S) of hemoglobin with granules of hemozoin (HZ) in an advanced state of formation. A cluster of free hemozoin granules lies between the spheres. Scale bars: A, 2 μm; B, 1 μm; C, 2 μm; D, 1 μm; E & F, 0.5 μm; G, 1 μm. Images A, B and F modified from Hall *et al.* 2011. Molecular and Biochemical Parasitology 179(1): 18–29. Figs. 2 & 3.C from Delcroix *et al.* 2007. Molecular and Biochemical Parasitology 154: 95–97. Fig. 1. Image E from Wilson and Barnes 1974. Parasitology 68: 239–258. Fig. 7. Images D and G are originals.

We can also make some inferences about gut processes and function from the ultrastructural studies. Hemoglobin is present in large masses in the center of the gut lumen, but in close proximity to the gastrodermal surface forms into compact almost spherical globules by an unknown mechanism (Fig. 11.11A, C–F). The edges of the gut lamellae are often observed inserted into depressions in the matrix of the luminal hemoglobin globules (Fig. 11.11F). This could indicate contact digestion by the gut lamellae or simply an anchoring function. Lipid droplets are also present in quantity (Fig. 11.11D) and their similar size has led to them being mistaken for the hemoglobin globules. They are easily distinguished by their relative electron densities, that of hemoglobin spheres being greater. The distribution of lipid droplets in the gut lumen, the gastrodermal syncytium and the surrounding parenchyma, is suggestive of transcytosis (the process of uptake into vesicles on one side of an epithelium followed by their release on the other side, unaltered) but has not been formally demonstrated by feeding experiments. A third inclusion type is found only in the gut lumen (Fig. 11.11G) and termed dense wrinkled particles by Morris (Morris 1968). They are the granules of hemozoin, soluble in alcoholic picric acid, which derive from the excess heme released from hemoglobin by proteolysis (Tsiftsoglou *et al.* 2006). The hemozoin is envisioned as a lattice of hydrogen-bonded dimers linked in a head-to-tail manner (Toh *et al.* 2010), thus avoiding cell damage due to the potentially pro-oxidant effects of heme. It has been proposed that hemozoin formation occurs at the surface of lipid droplets in the gut lumen (Oliveira *et al.* 2005). However, hemozoin particles are frequent within luminal hemoglobin aggregates undergoing digestion (Fig. 11.11F, G). An explanation for the presence of hemozoin in lipid droplets (Fig. 11.11D) is that it naturally partitions there due to its lipophilic properties (Skelly *et al.* 2014). In this situation, it has been suggested that the luminal lipid droplets (some ringed with Hz) continue to be degraded by gut enzymes until eventually all that is left is an insoluble core, adding to the dark, inert hemozoin accumulations in the schistosome gut (Skelly *et al.* 2014).

The cellular organization of the gastrodermis is illustrated in Fig. 11.12.

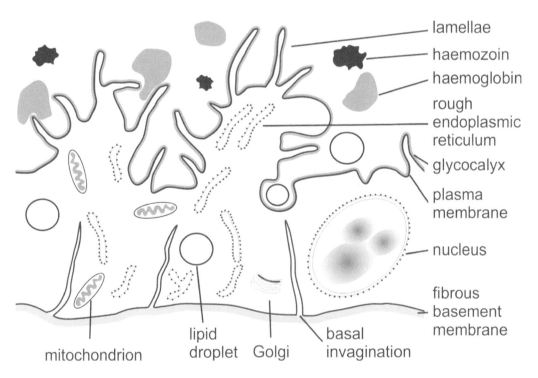

lamellae
haemozoin
haemoglobin
rough endoplasmic reticulum
glycocalyx
plasma membrane
nucleus
fibrous basement membrane

mitochondrion   lipid droplet   Golgi   basal invagination

FIG. 11.12 Diagrammatic representation of the gastrodermal syncytium. From Wilson and Coulson (2009). Trends in Parasitology 25: 423. Fig. 1C.

The fine filamentous coat covering the luminal surface (Fig. 11.11E) was shown to be the source of a polysaccharide antigen (circulating anodic antigen; CAA) that could be detected in the host circulation (Berggren and Weller 1967; Nash 1974). A second circulating antigen was subsequently identified (cathodic antigen; CCA) (Nash and Deelder 1985); the mucin-like coat they form over the gastrodermis potentially protects it against the low luminal pH and the hydrolytic environment. The major O-linked chains of CAA consist of long, negatively charged repeats of 1–6-linked N-acetylgalactosamine residues substituted with 1–3-linked glucuronic acid (Bergwerff *et al.* 1994). The glycan moiety of CCA consists of linear repeats of the Lewis x (Lex) trisaccharide that comprise galactose β1–4 linked to N-acetlyglucosamine, the latter substituted with α1–3 linked fucose (Van Dam *et al.* 1994). The polypeptide backbone of CCA is reputed to be a 39 kDa protein (Abdeen *et al.* 1999) but that of CAA is not known. However, the CCA protein lacks a signal peptide and is currently annotated on Gene DB as a 'cleavage and polyadenylation factor' so the original attribution must be suspect. The detection of CAA in serum and CCA in urine, using dipstick tests, has become the principal tool for immunodiagnosis of schistosomiasis (van Lieshout *et al.* 2000; Colley *et al.* 2013; Coulibaly *et al.* 2013). Furthermore, the sensitivity of the CAA test has now been increased almost to the point where single worm infections can be detected (Corstjens *et al.* 2014).

11.5.2   The Gastrodermis and its Products

The way in which schistosomes digest blood and take up nutrients has attracted much attention, but primarily this has involved the piecemeal characterisation of single proteins. Obtaining an overview of hydrolytic processes occurring in the gut lumen, transport into the gastrodermis and further hydrolysis within intracellular compartments has proved more difficult. While development of methods for the isolation of the tegument syncytium have led to quite a detailed appreciation of it molecular composition and of potential secreted products, no similar methods are available for the gastrodermis so researchers have had to use alternative approaches. One has been to collect and analyse vomitus on the assumption that it will contain the secreted products of the syncytial layer (Hall *et al.* 2011). Another has been the use of laser capture microdissection to excise the epithelium for analysis of gene expression (Nawaratna *et al.* 2011). Both approaches have technical limitations so their results must be treated with caution. A major limitation of vomitus collection is to persuade the worms to release their gut contents *in vitro*. Incubation in distilled water (Chappell and Dresden 1986; Delcroix *et al.* 2007) is not recommended for subsequent proteomic analysis due to massive leakage of cytosolic proteins. Applying a cold/warm shock to worms in culture medium results in a steady stream of vomitus in the warm-up phase (Planchart *et al.* 2007). However, while proteomic analysis of such medium revealed enrichment of known gut enzymes there was still contamination with cytosolic and cytoskeletal constituents (Hall *et al.* 2011). The most successful method involved the *in vitro* culture of pre-adult 28-day worms in protein-free medium for four days (Hall *et al.* 2011). These worms are very active, robust and less easily damaged than 49-day paired adults.

Tandem mass spectrometry was used to identify the major proteins highly enriched in the vomitus preparations over those present in the soluble adult worm extract known as SWAP (Curwen *et al.* 2004; Neves *et al.* 2015). Several host proteins were identified in one or more vomitus preparations, including the alpha and beta chains of mouse hemoglobin. The most abundant plasma protein, serum albumin, was detected but there was no evidence for any immunoglobulin isotypes, suggesting that they are rapidly degraded in the acidic hydrolytic conditions of the gut lumen. A single host membrane protein, CD44 and two murine protease inhibitors, a contrapsin and a serpin, were identified. The latter two were presumably present as their inhibitory properties prevented their hydrolysis by the gut cathepsins. A total of eight parasite proteases was identified in the various preparations, five previously implicated in digestion of blood proteins (Caffrey *et al.* 2004) and three that were novel (Table 11.4). The first group included two isoforms of cathepsin.

TABLE 11.4 Secretory proteins, mostly of lysosomal origin identified by proteomic analysis of worm vomitus. From Wilson (2012b). Parasitology, 139(9): 1178–94. Table 4.

Acid hydrolases and inhibitors	Annotation	Lysosome and transport protein	Annotation
Asparaginyl endopeptidase (Sm32)	Smp_179170	ferritin-2 heavy chain, isoform 1	Smp_047660
Cathepsin B1 isotype 1 (Sm31)	Smp_103610	ferritin-2 heavy chain, isoform 2	Smp_047650
Cathepsin B1 isotype 2	Smp_067060	Apoferritin	Smp_063530
Dipeptylpeptidase 1 (Cathepsin C)	Smp_019030	Saposin B domain-containing 1	Smp_194910
Cathepsin K/S	Smp_139240	Saposin B domain-containing 2	Smp_014570
Lysosomal Pro-X carboxylpeptidase	Smp_002600	Saposin B domain-containing 3	Smp_105450
Lysosomal Pro-X carboxylpeptidase	Smp_071610	Saposin B domain-containing 4	Smp_130100
DJ-1/PARK7-like protease	Smp_082030	NPC-like cholesterol binding protein	Smp_194840
Ester hydrolase	Smp_010620	Calumenin, EF-hand Ca binding protein	Smp_147680
Glucan 1, 4 beta-glucosidase	Smp_043390	Lysosome membrane-associated glycoprotein	Smp_167770
Long chain acyl-coenzyme thioesterase 1	Smp_150820	Vesicle associated membrane protein	Smp_136240
Serpin	Smp_090080		
Alpha-2-macroglobulin	Smp_089670		

B, cathepsin C, a cathepsin L (annotated as cathepsin S in the *S. mansoni* genome) and asparaginyl endopeptidase, the pro-enzyme convertase for cathepsins. The second group comprised two serine proteases, Pro-X carboxyl peptidase and dipeptidyl peptidase II (also a potential proline carboxypeptidase), plus a DJ-1/PARK7 orthologue. Glucan 1, 4 beta-glucosidase/beta-xylosidase, an enzyme capable of hydrolyzing carbohydrates linkages and an ester hydrolase were also detected. All the hydrolases apart from the ester hydrolase, which appears to lack a signal peptide, have a potential association with lysosomes.

A second class of vomitus constituents comprised "carrier" proteins potentially involved in the sequestration of molecules that may be crucial for parasite metabolism (Table 11.4). Three variants of ferritin, a protein that catalyzes the uptake and transport of $Fe2+$ ions, were evident. Four proteins possessing the characteristic saposin domain, which can bind sphingolipids to facilitate their degradation by ceramidases, were identified predominantly in the culture vomit. The Niemann–Pick C2-like protein (NPC-2), capable of sequestering cholesterol and calumenin, a calcium-binding protein, were also found. Like the enzymes, some of the carrier proteins are also typical constituents of lysosomes. The demonstration of dextran uptake into the gastrodermis (Hall *et al.* 2011) highlights a potential route and mechanism for the uptake of the carrier proteins like ferritin and the saposins, hypothesized to load their ligands in the gut lumen. As no dextran reached the interstitial spaces beyond the gastrodermis over a 68 h period, it seems that the ligands are unloaded in the syncytium. Whether uptake involves the formation of coated pits and the internalization of specific receptors or macropinocytosis, the non-selective endocytosis of solute macromolecules in uncoated vesicles from 0.2 μm to 5 μm in diameter, remains to be established. As the endocytosed dextran aggregates range up to 2 μm in diameter, their size argues for the macropinocytosis at the luminal surface of the gut epithelium (Swanson and Watts 1995).

The identification of numerous acid hydrolases in the vomitus, together with lysosomal carrier proteins and the acid pH of the gut lumen, certainly <pH 6.2 and inferred as pH 4.3–4.9, the pH optimum of several gut proteases (Chappell and Dresden 1986) strongly suggests the secretion of lysosomal contents into the gut lumen. In this context, the identification of two membrane proteins, lysosome-associated membrane protein (LAMP) and vesicle associated membrane protein (VAMP), adds further weight to the idea that digestion of the hemoglobin and plasma proteins that comprise the schistosome diet is effected by the secreted contents of lysosomes. The long chain acyl-coenzyme thioesterase 1 identified, is normally a constituent of peroxisomes and may indicate that these organelles also make a contribution to luminal digestion. Lastly, the detection of schistosome serpin and alpha-2 macroglobulin in the vomitus is intriguing. The former group of inhibitors target serine proteases, the latter has more generalized protease inhibitory properties. It has been speculated that they might act of prevent clotting of ingested blood in the gut lumen (Hall *et al.* 2011).

A major limitation of the laser capture method is the amount of material that can be excised from sections of worm tissue for extraction of mRNA to determine patterns of gene expression in the tissue of interest. Nevertheless, it has been attempted for the gastrodermis (Nawaratna *et al.* 2011) and a few more genes with elevated expression have been added to the list. Saposin Smp_014570, NPC-2 and LAMP were common to both approaches, but genes encoding three Cathepsin proteases (B, Smp_085010; D2, Smp_136730; L, Smp_139160) did not match the four Cathepsins identified in vomitus. This reinforces the idea of differential hydrolysis occurring between lumen and intracellular gastrodermal compartments (Brindley *et al.* 2001). Additional genes encoding LAMP-2 (CD36 scavenger receptor; Smp_128430), Multidrug resistance protein (Smp_151190), Zinc metalloprotease (Smp_173160), plasma membrane calcium-transporting ATPase (Smp_137170), cationic amino acid transporter (Smp_004190) and a trafficking kinesin-binding protein 1 (Smp_127380) were unique to the laser capture study. Another gene, Smp_166170 is a chimeric model, one part of which encodes an egg shell protein presumably expressed in vitellaria adjacent to the gastrodermis; this highlights the need for absolute precision in tissue excision by the laser capture method.

An inevitable conclusion is that, compared with the knowledge we have about differential gene expression in the schistosome esophagus, we are only scratching the surface with the gastrodermis. This tissue participates not just in the synthesis and secretion of proteins from the syncytium into the gut lumen but also in the uptake of many classes of nutrients into the syncytium, via endocytosis and membrane transporters, as well as further processing within the cytoplasm. We might anticipate some hundreds of differentially expressed genes to be involved. All the necessary components for receptor-mediated endocytosis have been identified in the *S. mansoni* transcriptome or genome, but none has yet been localized to any tissue (Wilson 2012a). Similarly, the multiple constituents of the vacuolar ATPase responsible for acidifying compartments are all encoded in the *S. mansoni* genome but none has been localized. However, an orthologue of the epsilon vATPase subunit in the human liver fluke *Clonorchis* has been localized to the gastrodermis of that species (Lv *et al.* 2014). Thus we might anticipate finding the complete vATPase proton pump machinery in the gastrodermis. Certainly transcripts of all the genes encoding the complex were identified in a recent subtractive RNASeq investigation of male *S. mansoni* heads and tails but with similar levels of expression (Wilson *et al.* 2015). It is thus conceivable that luminal acidification occurs in both the esophagus (heads) and gut (tails).

11.5.3   Proteolysis in the Gut

The way in which schistosomes degrade the constituents of their protein-rich diet, particularly hemoglobin and serum albumin, has attracted a lot of attention. The early work inevitably focused on the functional properties of homogenates, rather than individual proteins purified and characterised biochemically. For example, a thiol proteinase of MW 27-32 kDa and pH optimum of 4.0 was described in homogenates of males and females with the properties of a Cathepsin B (Dresden and Deelder 1979) and loosely described as a hemoglobinase. The real advances came ten years later with the cloning of two transcripts encoding mature proteins of 31 and 32 kDa (Davis *et al.* 1987; Klinkert *et al.* 1989). The hemoglobinase term was then applied to the second enzyme of MW32 kDa, which later proved to be an asparaginyl endopeptidase. (Asparginyl endopeptidases, Smp_075800 and Smp_179170 are still annotated as hemoglobinase in the *S. mansoni* database on Gene DB).

   Expression of Cathepsin B in functional form and its use to digest hemoglobin and albumin, showed that the process was not complete, even after 24 hours, suggesting that other enzymes must be involved (Ghoneim and Klinkert 1995). Subsequently a Cathepsin D aspartyl protease was cloned and characterized, first in *S. japonicum* (Becker *et al.* 1995) and then in *S. mansoni* (Wong *et al.* 1997). A Cathepsin L1 was added to the growing list (Brady *et al.* 2000; Brady *et al.* 1999) and both cathepsins were localised to the gastrodermis (Brady *et al.* 1999; Brindley *et al.* 2001). Cathepsin C activity was also detected in adult worm homogenates (Hola-Jamriska *et al.* 1999) and the *S. japonicum* enzyme then expressed and characterised (Hola-Jamriska *et al.* 2000). This may not be the whole story as a leucine amino peptidase (McCarthy *et al.* 2004) and second Cathepsin L (SmCL3) have been characterized and localised to the gastrodermis (Dvorak *et al.* 2009). Publication of the genomes of *S. mansoni* and *S. japonicum* has also provided the means to create an inventory of the "degradome" of schistosomes (Berriman *et al.* 2009; Consortium 2009). To summarize the above findings, digestion in the schistosome gut involves a battery of acid proteases from different C1 and A1 families usually associated with lysosomes and acting at an acid pH optimum.

   Asparaginyl endopeptidase is unlike the other gut proteinases in belonging to a group of enzymes first identified in leguminous plants and referred to as legumains. Cathepsins B, C, D and L are synthesized as larger proproteins and must activated by cleavage. It was proposed that the schistosome asparaginyl endopeptidase was not the hemoglobinase involved in hemoglobin digestion, but acted as the required convertase (Dalton *et al.* 2009) to activate the Cathepsins. However, this function was recently challenged when expression of the asparginyl endopeptidase gene in adult worms was inhibited by RNAi and Cathepsin B was still converted to the functionally active form (Krautz-Peterson and Skelly 2008). Most recently a novel scheme for Cathepsin B activation involving sulphated polysaccharides such as heparin or dextran sulphate has been described (Dalton and Dvorak 2014; Jilkova *et al.* 2014). It was proposed that the sulphated polysaccharides acted as a molecular switch between the autocatalytic and asparaginyl endopeptidase activation routes (Jilkova *et al.* 2014). It was speculated that the source of the sulphated polysaccharides may be intrinsic, on the lining of the gastrodermis, or from ingested host blood (Dalton and Dvorak 2014).

   Not all approaches have relied on molecular biology to advance our understanding of protein degradation. Judicious use of protease inhibitors on a preparation designated as "gastro-intestinal contents" (GIC) has helped to define the functions of Cathepsins B1, L1, D and Asparaginyl endopeptidase in a cooperative network (Delcroix *et al.* 2006). For the digestion of albumin, Cathepsins B and L initiated substrate cleavage with some redundancy of function, while inhibition of Cathepsin D had only a minor effect. However, Cathepsin D played a greater role in the primary cleavage of hemoglobin (Brindley *et al.* 2001) releasing two peptides of molecular mass between 6 and 16 kDa, which were, in turn, degraded by Cathepsins B and L. Exopeptidases such as Cathepsin C (dipeptidyl peptidase I) (Hola-Jamriska *et al.* 2000) and speculatively, aminopeptidase, further degraded peptides released by the action of the endopeptidases B1, L1 and D.

There is an unresolved question about where albumin and hemoglobin degradation take place, particularly as hemoglobin is only degraded by the GIC preparation below pH 6. There are two sources for the confusion. The first is the actual measured pH of the schistosome gut contents, variously inferred or determined as anywhere between 4.0 and 6.0. Given the availability of a range of non-toxic fluorescent pH indicators, it is surprising that there have been no recent attempts to use them in worm feeding experiments to settle the question. The second is the localization of the various enzymes in the cascade by immunocytochemistry with monospecific antibodies to the gastrodermal tissues. As an example, Cathepsin D has been localized to the gastrodermis in both *S. mansoni* and *S. japonicum* (Bogitsh *et al.* 1992; Brindley *et al.* 2001). Such a result is to be expected since that is where synthesis of proteases for export takes place, but it does not mean that they interact with their substrates only in the gastrodermis. Immunocytochemical techniques are not well suited to localization of targets in the gut lumen since the soluble proteases can easily leach out of the sections during processing. This is where the proteomic analyses of vomitus are invaluable since they provide a profile of the proteins actually secreted from the gastrodermis (Hall *et al.* 2011). Cathepsins B, C, L and asparaginyl endopeptidase were common to the molecular studies and proteomic analyses. The further proteomic identification of two prolylcarboxypeptidases is a useful addition to the proteolysis repertoire since the dipeptidyl peptidase activity of Cathepsin C is blocked by proline residues in the polypeptide chain. The prolylcarboxypeptidases are serine proteases that cleave a peptide bond adjacent to a proline residue; there are 24 such bonds in human serum albumin and 7 each in the α and β chains of hemoglobin. Cathepsin S, another C1 cysteine peptidase needs to be factored into the luminal digestion mix but the detection of a DJ-1 type protease must remain enigmatic as it does not fit any obvious scheme of protein degradation. The one glaring absence among components of the luminal digestion scheme in the proteomic analyses is the *S. mansoni* Cathepsin D (Wong *et al.* 1997). There are a number of possible explanations. It seems unlikely that this enzyme is sequestered only within internal gastrodermal compartments as later enzymes in the projected cascade are found in the lumen. The formation of hemozoin in the gut lumen also argues for the breakdown of hemoglobin there and not within a gastrodermal compartment. A theoretical tryptic digest of Cathepsin D indicates that peptides in the range suitable for detection by tandem mass spectrometry exist although that number may be reduced by the predicted existence on N and O-linked glycosylation sites that would alter the peptide molecular mass and so prevent detection. It is of course possible that the Cathepsin D has a very short half-life in the presence of the other proteases in the gut lumen and is rapidly hydrolyzed.

11.6 DEVELOPMENT OF THE ALIMENTARY TRACT

It has long been known that feeding on blood cells only begins after the migrating schistosomulum arrives in the hepatic portal system and transforms from an elongate to a short stumpy body shape (Clegg 1965; Crabtree and Wilson 1980). Indeed, premature transformation would abolish the ability of the migrating parasite to traverse capillary beds. Ingestion of the first erythrocytes is rapidly followed by the appearance of hemozoin in the gut lumen and triggers a rapid growth phase into the adult body form over a period of about 4 weeks (Lawson and Wilson 1980). What is most surprising is that expression of the genes encoding both esophageal and gastrodermal proteins is dissociated from the morphological development and onset of erythrocyte feeding. A microarray analysis of the gene expression patterns in larval *S. mansoni* associated with invasion of the mammalian host revealed that by day 3 transcripts for many of the proteins highlighted above were detected, relative to the preceding intramolluscan germ ball and infective cercarial life cycle stage (Parker-Manuel *et al.* 2011). (At day-3 the vast majority of schistosomula are still in the skin.) These genes included numerous MEGs subsequently localised to the esophagus, plus proteases and saposins of gastrodermal origin. The differentiation of the anterior esophageal gland at day 3 was confirmed by upregulation of MEGs 12, 16 and 17 and of the posterior gland by upregulation of MEGs 4.1, 4.2, 8.1, 8.2, 9, 14 and 15 (Parker-Manuel *et al.* 2011; Wilson *et al.* 2015). In addition, expression of five of the six esophageal aspartyl proteases was also elevated. Evidence for differentiation of the gastrodermis was provided by the striking upregulation of 10 of the 12 saposins encoded in the genome (Parker-Manuel *et al.* 2011), compared to the four identified by proteomic analysis of vomitus (Hall *et al.* 2011). Five gut cathepsins and an asparaginyl endopeptidase were similarly upregulated up to 50-fold, compared to the germ balls. It is notable that most of these genes were also enriched in the cercaria, revealing that alimentary tract differentiation is a very early event in parasite development. Intriguingly the previously described gut cathepsin D aspartyl protease (Wong *et al.* 1997) was not found to be upregulated at Day 3. Both cathepsin B and asparaginyl endopeptidase proteins have been localised to the cercarial gut by immunocytochemistry (Skelly and Shoemaker 2001), while cathepsin L mRNA has been localised to the gastrodermis of lung stage schistosomula by WISH (Dillon *et al.* 2007). The demonstration of Lucifer Yellow ingestion by newly transformed schistosomula (Thornhill *et al.* 2009) suggests that the gut may be active early after infection in processing plasma fluids, although not yet capable of dealing with blood cells.

11.7 ALIMENTARY TRACT PROTEINS AS VACCINE TARGETS

On the face of it, the gastrodermal secretions do not appear to present a promising source of vaccine targets because of their acidic and proteolytic environment. Indeed no immunoglobulins were detected by proteomic analysis of vomitus, which contrasts both with the detection of albumin and hemoglobin alpha and beta chains in vomitus (Hall *et al.* 2011) and with the detection of IgG and IgM on the tegument surface by biotinylation (Braschi and Wilson 2006). It suggests that ingested immunoglobulins have a very short half-life in the gut lumen. The rapid destruction of mixed IgG isotypes by Cathepsin D aspartyl protease from *S. japonicum* at pH 3.6–4.5 confirms the supposition (Verity *et al.* 2001). Nevertheless, gut secretions have been proposed as vaccine targets (Figueiredo *et al.* 2015), although the evidence for protection is sparse. One reported success has been with administration of enzymatically active recombinant Cathepsin B to mice, where it had both adjuvant and protective properties (El Ridi *et al.* 2014). However, this study has been criticised both for its experimental design and for potentially operating via non-specific effects on the intravascular migration of schistsomula. The gut-secreted saposins provide a different angle on the vaccine potential of gut proteins. The ratio of synonymous to non-synonymous substitution rates in different schistosome species has been used as a measure of exposure to selection pressure during the evolution of the Genus *Schistosoma* (Philippsen *et al.* 2015). Three genes coding for saposins and one coding for Niemann Pick C2 protein (NPC2) displayed dN/dS values higher than 0.5 in comparisons between *S. mansoni* and *S. haematobium* genes. In contrast the various gut proteases displayed much lower ratios. These data suggest that some gut proteins may be susceptible to neutralisation by antibodies. However, protection experiments with one of the three saposins, SmSLP-1 (equivalent to Smp_105450), failed to elicit any reduction in worm burden (Don *et al.* 2008).

The identification of the anterior and posterior esophageal glands as a hotspot for the secretion of numerous MEG-encoded proteins and lysosomal hydrolases distinct from those of the gastrodermis, provides an entirely novel group of potential vaccine targets (Li *et al.* 2015; Wilson *et al.* 2015). They have one possible advantage over gastrodermal targets as the acidic hydrolytic environment is likely to be less severe in the esophagus lumen. Indeed, the binding of host antibodies in the esophagus has been demonstrated in several host species (Li *et al.* 2013; Li *et al.* 2014; Li *et al.* 2015). Targeting of esophageal secretions by host IgG has also been linked to self-cure in the rhesus macaque host infected with *Schistosoma japonicum* (Li *et al.* 2015). One *S. mansoni* MEG has so far been tested for protection with modest results (Martins *et al.* 2014). However, a note of caution is called for since almost all MEG genes show very high dN/dS ratios indicating that they had been subject to strong selective pressure during schistosome evolution (Philippsen *et al.* 2015). It will take some time to evaluate whether any of the newly defined esophageal secretions has protective potential.

11.8   CHAPTER SUMMARY

The alimentary tract is one of the two major interfaces between the adult schistosome and the host bloodstream. Its development and differentiation begins at the cercarial stage, with rapid amplification of gene expression after infection but the full capacity for processing whole blood is only achieved after the parasite arrival in the hepatic portal system. The cascade of proteases that hydrolyse blood proteins has been well researched although proteomic analysis of vomitus has added further potential participants, including exopeptidases. By contrast, the molecular composition and functions of the gastrodermis are poorly understood and await a robust method for isolation of the tissue to facilitate transcript and protein analysis. The role of the esophagus in blood processing is now much better appreciated as a result of recent studies. It is structurally and functionally more complex than the gastrodermis, with anterior and posterior compartments demarcated by sphincters. Videos of blood feeding have revealed a two-step ingestion process, the anterior compartment first filling before the material passes as a bolus to the posterior compartment. This allows it to mix with the distinct secretions of the anterior and posterior glands that surround their respective compartments. A combination of transcript analysis, immunocytochemistry and WISH has revealed that these two glands secrete approximately 40 proteins to interact with the ingested blood; MEG-encoded proteins and lysosomal hydrolases are particular prominent. The roles these secretions play in erythrocyte lysis, tethering and killing of leucocytes and prevention of clot formation are presently speculative. Some esophageal secretions appear to be targets of host antibody, especially in rhesus monkeys, but it is not clear whether they would make good vaccine targets.

11.9   LITERATURE CITED

Abdeen HH, Attallah AF, Mansour MM, Harrison RA. 1999. Molecular cloning and characterization of the polypeptide backbone of *Schistosoma mansoni* circulating cathodic antigen. Molecular and Biochemical Parasitology 101(1–2): 149–159.

Almeida GT, Amaral MS, Beckedorff FC, Kitajima JP, DeMarco R, Verjovski-Almeida S. 2012. Exploring the *Schistosoma mansoni* adult male transcriptome using RNA-seq. Experimental Parasitology 132(1): 22–31.

Becker MM, Harrop SA, Dalton JP, Kalinna BH, McManus DP, Brindley PJ. 1995. Cloning and characterization of the *Schistosoma japonicum* aspartic proteinase involved in hemoglobin degradation. Journal of Biological Chemistry 270(41): 24496–24501.

Berggren WL, Weller TH. 1967. Immunoelectrophoretic demonstration of specific circulating antigen in animals infected with *Schistosoma mansoni*. American Journal of Tropical Medicine and Hygiene 16(5): 606–612.

Bergwerff AA, van Dam GJ, Rotmans JP, Deelder AM, Kamerling JP, Vliegenthart JF. 1994. The immunologically reactive part of immunopurified circulating anodic antigen from *Schistosoma mansoni* is a threonine-linked polysaccharide consisting of --> 6)-(beta-D-GlcpA-(1 --> 3))-beta-D-GalpNAc-(1 --> repeating units. Journal of Biological Chemistry 269(50): 31510–31517.

Berridge MJ, Oschman JL. 1972. *Transporting Epithelia*. New York and London: Academic Press. 462 p.

Berriman M, Haas BJ, LoVerde PT, Wilson RA, Dillon GP, Cerqueira GC, Mashiyama ST, Al-Lazikani B, Andrade LF, Ashton PD and others. 2009. The genome of the blood fluke *Schistosoma mansoni*. Nature 460(7253): 352–358.

Bloch EH. 1980. *In vivo* microscopy of schistosomiasis. II. Migration of *Schistosoma mansoni* in the lungs, liver and intestine. American Journal of Tropical Medicine and Hygiene 29(1): 62–70.

Bogitsh BJ. 1975. Cytochemistry of gastrodermal autophagy following starvation in *Schistosoma mansoni*. Journal of Parasitology 61(2): 237–248.

Bogitsh BJ, Carter OS. 1977. *Schistosoma mansoni*: ultrastructural studies on the esophageal secretory granules. Journal of Parasitology 63(4): 681–686.

Bogitsh BJ, Kirschner KF, Rotmans JP. 1992. *Schistosoma japonicum*: immunoinhibitory studies on hemoglobin digestion using heterologous antiserum to bovine cathepsin D. Journal of Parasitology 78(3): 454–459.

Brady CP, Brinkworth RI, Dalton JP, Dowd AJ, Verity CK, Brindley PJ. 2000. Molecular modeling and substrate specificity of discrete cruzipain-like and cathepsin L-like cysteine proteinases of the human blood fluke *Schistosoma mansoni.* Archives of Biochemistry and Biophysics 380(1): 46–55.

Brady CP, Dowd AJ, Brindley PJ, Ryan T, Day SR, Dalton JP. 1999. Recombinant expression and localization of *Schistosoma mansoni* cathepsin L1 support its role in the degradation of host hemoglobin. Infection and Immunity 67(1): 368–374.

Braschi S, Wilson RA. 2006. Proteins exposed at the adult schistosome surface revealed by biotinylation. Molecular and Cellular Proteomics: MCP 5(2): 347–356.

Brindley PJ, Kalinna BH, Wong JY, Bogitsh BJ, King LT, Smyth DJ, Verity CK, Abbenante G, Brinkworth RI, Fairlie DP and others. 2001. Proteolysis of human hemoglobin by schistosome cathepsin D. Molecular and Biochemical Parasitology 112(1): 103–112.

Caffrey CR, McKerrow JH, Salter JP, Sajid M. 2004. Blood 'n' guts: an update on schistosome digestive peptidases. Trends in Parasitology 20(5): 241–248.

Chappell CL, Dresden MH. 1986. *Schistosoma mansoni*: proteinase activity of "hemoglobinase" from the digestive tract of adult worms. Experimental Parasitology 61(2): 160–167.

Clegg JA. 1965. *In vitro* cultivation of *Schistosoma mansoni.* Experimental Parasitology 16: 133–147.

Colley DG, Binder S, Campbell C, King CH, Tchuem Tchuente LA, N'Goran EK, Erko B, Karanja DM, Kabatereine NB, van Lieshout L and others. 2013. A five-country evaluation of a point-of-care circulating cathodic antigen urine assay for the prevalence of *Schistosoma mansoni.* American Journal of Tropical Medicine and Hygiene 88(3): 426–432.

Cone RA. 2005. Mucus. pp. 49–72. *In*: Mestecky J, Lamm JR, McGhee JR, Bienenstock J, Mayer W, Strober W, editors. *Mucosal Immunology.* 3rd ed. New York & London: Academic Press.

Consortium SjG. 2009. The *Schistosoma japonicum* genome reveals features of host-parasite interplay. Nature 460(7253): 345–351.

Corstjens PL, DeDood CJ, Kornelis D, Fat EM, Wilson RA, Kariuki TM, Nyakundi RK, Loverde PT, Abrams WR, Tanke HJ and others. 2014. Tools for diagnosis, monitoring and screening of *Schistosoma* infections utilizing lateral-flow based assays and upconverting phosphor labels. Parasitology 141(14): 1841–1855.

Coulibaly JT, N'Gbesso YK, Knopp S, N'Guessan NA, Silue KD, van Dam GJ, N'Goran EK, Utzinger J. 2013. Accuracy of urine circulating cathodic antigen test for the diagnosis of *Schistosoma mansoni* in preschool-aged children before and after treatment. PLoS Neglected Tropical Diseases 7(3): e2109.

Crabtree JE, Wilson RA. 1980. *Schistosoma mansoni:* a scanning electron microscope study of the developing schistosomulum. Parasitology 81(Pt 3): 553–564.

Curwen RS, Ashton PD, Johnston DA, Wilson RA. 2004. The *Schistosoma mansoni* soluble proteome: a comparison across four life cycle stages. Molecular and Biochemical Parasitology 138(1): 57–66.

Dalton JP, Brindley PJ, Donnelly S, Robinson MW. 2009. The enigmatic asparaginyl endopeptidase of helminth parasites. Trends in Parasitology 25(2): 59–61.

Dalton JP, Dvorak J. 2014. Activating the cathepsin B1 of a parasite: a major route with alternative pathways? Structure 22(12): 1696–1698.

Davis AH, Nanduri J, Watson DC. 1987. Cloning and gene expression of *Schistosoma mansoni* protease. Journal of Biological Chemistry 262(26): 12851–12855.

de la Salud Bea R, Ascuitto MR, de Johnson LE. 2015. Synthesis of analogs of peptides from *Buthus martensii* scorpion venom with potential antibiotic activity. Peptides 68: 228–232.

Delcroix M, Medzihradsky K, Caffrey CR, Fetter RD, McKerrow JH. 2007. Proteomic analysis of adult *S. mansoni* gut contents. Molecular and Biochemical Parasitology 154(1): 95–97.

Delcroix M, Sajid M, Caffrey CR, Lim KC, Dvorak J, Hsieh I, Bahgat M, Dissous C, McKerrow JH. 2006. A multienzyme network functions in intestinal protein digestion by a platy helminth parasite. Journal of Biological Chemistry 281(51): 39316–39329.

DeMarco R, Mathieson W, Manuel SJ, Dillon GP, Curwen RS, Ashton PD, Ivens AC, Berriman M, Verjovski-Almeida S, Wilson RA. 2010. Protein variation in blood-dwelling schistosome worms generated by differential splicing of micro-exon gene transcripts. Genome Research 20(8): 1112–1121.

Dike SC. 1971. Ultrastructure of the esophageal region in *Schistosoma mansoni.* American Journal of Tropical Medicine and Hygiene 20(4): 552–568.

Dillon GP, Illes JC, Isaacs HV, Wilson RA. 2007. Patterns of gene expression in schistosomes: localization by whole mount *in situ* hybridization. Parasitology 134(Pt 11): 1589–1597.

Don TA, Bethony JM, Loukas A. 2008. Saposin-like proteins are expressed in the gastrodermis of *Schistosoma mansoni* and are immunogenic in natural infections. International journal of infectious diseases: IJID: official publication of the International Society for Infectious Diseases 12(6): e39–47.

Dresden MH, Deelder AM. 1979. *Schistosoma mansoni*: thiol proteinase properties of adult worm "hemoglobinase". Experimental Parasitology 48(2): 190–197.

Dvorak J, Mashiyama ST, Sajid M, Braschi S, Delcroix M, Schneider EL, McKerrow WH, Bahgat M, Hansell E, Babbitt PC and others. 2009. SmCL3, a gastrodermal cysteine protease of the human blood fluke *Schistosoma mansoni*. PLoS Neglected Tropical Diseases 3(6): e449.

El Ridi R, Tallima H, Dalton JP, Donnelly S. 2014. Induction of protective immune responses against schistosomiasis using functionally active cysteine peptidases. Frontiers in Genetics 5: 119.

Ernst SC. 1975. Biochemical and cytochemical studies of digestive-absorptive functions of esophagus, cecum and tegument in *Schistosoma mansoni*: acid phosphatase and tracer studies. Journal of Parasitology 61(4): 633–647.

Figueiredo BC, Ricci ND, de Assis NR, de Morais SB, Fonseca CT, Oliveira SC. 2015. Kicking in the guts: *Schistosoma mansoni* digestive tract proteins are potential candidates for vaccine development. Frontiers in Immunology 6: 22.

Ghoneim H, Klinkert MQ. 1995. Biochemical properties of purified cathepsin B from *Schistosoma mansoni*. International Journal for Parasitology 25(12): 1515–1519.

Haas M, Forbush B, 3rd. 2000. The Na-K-Cl cotransporter of secretory epithelia. Annual Review of Parasitology 62: 515–534.

Hall SL, Braschi S, Truscott M, Mathieson W, Cesari IM, Wilson RA. 2011. Insights into blood feeding by schistosomes from a proteomic analysis of worm vomitus. Molecular and Biochemical Parasitology 179(1): 18–29.

Hola-Jamriska L, Dalton JP, Aaskov J, Brindley PJ. 1999. Dipeptidyl peptidase I and III activities of adult schistosomes. Parasitology 118(Pt 3): 275–282.

Hola-Jamriska L, King LT, Dalton JP, Mann VH, Aaskov JG, Brindley PJ. 2000. Functional expression of dipeptidyl peptidase I (Cathepsin C) of the oriental blood fluke *Schistosoma japonicum* in *Trichoplusia ni* insect cells. Protein Expression and Purification 19(3): 384–392.

Jilkova A, Horn M, Rezacova P, Maresova L, Fajtova P, Brynda J, Vondrasek J, McKerrow JH, Caffrey CR, Mares M. 2014. Activation route of the *Schistosoma mansoni* cathepsin B1 drug target: structural map with a glycosaminoglycan switch. Structure 22(12): 1786–1798.

Klinkert MQ, Felleisen R, Link G, Ruppel A, Beck E. 1989. Primary structures of Sm31/32 diagnostic proteins of *Schistosoma mansoni* and their identification as proteases. Molecular and Biochemical Parasitology 33(2): 113–122.

Kozlov SA, Vassilevski AA, Feofanov AV, Surovoy AY, Karpunin DV, Grishin EV. 2006. Latarcins, antimicrobial and cytolytic peptides from the venom of the spider *Lachesana tarabaevi* (Zodariidae) that exemplify biomolecular diversity. Journal of Biological Chemistry 281(30): 20983–20992.

Krautz-Peterson G, Skelly PJ. 2008. Schistosome asparaginyl endopeptidase (legumain) is not essential for cathepsin B1 activation *in vivo*. Molecular and Biochemical Parasitology 159(1): 54–58.

Lach A, Grzybek M, Heger E, Korycka J, Wolny M, Kubiak J, Kolondra A, Boguslawska DM, Augoff K, Majkowski M and others. 2012. Palmitoylation of MPP1 (membrane-palmitoylated protein 1)/p55 is crucial for lateral membrane organization in erythroid cells. Journal of Biological Chemistry 287(23): 18974–18984.

Lawrence JD. 1973. The ingestion of red blood cells by *Schistosoma mansoni*. Journal of Parasitology 59(1): 60–63.

Lawson JR, Wilson RA. 1980. Metabolic changes associated with the migration of the schistosomulum of *Schistosoma mansoni* in the mammal host. Parasitology 81(2): 325–336.

Li XH, de Castro-Borges W, Parker-Manuel S, Vance GM, Demarco R, Neves LX, Evans GJ, Wilson RA. 2013. The schistosome oesophageal gland: initiator of blood processing. PLoS Neglected Tropical Diseases 7(7): e2337.

Li XH, Stark M, Vance GM, Cao JP, Wilson RA. 2014. The anterior esophageal region of *Schistosoma japonicum* is a secretory organ. Parasites & Vectors 7: 565.

Li XH, Xu YX, Vance G, Wang Y, Lv LB, van Dam GJ, Cao JP, Wilson RA. 2015. Evidence that rhesus macaques self-cure from a *Schistosoma japonicum* infection by disrupting worm esophageal function: A new route to an effective vaccine? PLoS Neglected Tropical Diseases 9(7): e0003925.

Lv X, Huang L, Chen W, Wang X, Huang Y, Deng C, Sun J, Tian Y, Mao Q, Lei H and others. 2014. Molecular characterization and serological reactivity of a vacuolar ATP synthase subunit epsilon-like protein from *Clonorchis sinensis*. Parasitology Research 113(4): 1545–1554.

Mair GR, Maule AG, Day TA, Halton DW. 2000. A confocal microscopical study of the musculature of adult *Schistosoma mansoni*. Parasitology 121(Pt 2): 163–170.

Martins VP, Morais SB, Pinheiro CS, Assis NR, Figueiredo BC, Ricci ND, Alves-Silva J, Caliari MV, Oliveira SC. 2014. Sm 10.3, a member of the micro-exon gene 4 (MEG-4) family, induces erythrocyte agglutination *in vitro* and partially protects vaccinated mice against *Schistosoma mansoni* infection. PLoS Neglected Tropical Diseases 8(3): e2750.

McCarthy E, Stack C, Donnelly SM, Doyle S, Mann VH, Brindley PJ, Stewart M, Day TA, Maule AG, Dalton JP. 2004. Leucine aminopeptidase of the human blood flukes, *Schistosoma mansoni* and *Schistosoma japonicum*. International Journal for Parasitology 34(6): 703–714.

Morris GP. 1968. Fine structure of the gut epithelium of *Schistosoma mansoni*. Experientia 24(5): 480–482.

Morris GP, Threadgold LT. 1968. Ultrastructure of the tegument of adult *Schistosoma mansoni*. Journal of Parasitology 54(1): 15–27.

Nash TE. 1974. Localization of the circulating antigen within the gut of *Schistosoma mansoni*. American Journal of Tropical Medicine and Hygiene 23(6): 1085–1087.

Nash TE, Deelder AM. 1985. Comparison of four schistosome excretory-secretory antigens: phenol sulfuric test active peak, cathodic circulating antigen, gut-associated proteoglycan and circulating anodic antigen. American Journal of Tropical Medicine and Hygiene 34(2): 236–241.

Nawaratna SS, Gobert GN, Willis C, Chuah C, McManus DP, Jones MK. 2014. Transcriptional profiling of the oesophageal gland region of male worms of *Schistosoma mansoni*. Molecular and Biochemical Parasitology 196(2): 82–89.

Nawaratna SS, McManus DP, Moertel L, Gobert GN, Jones MK. 2011. Gene atlasing of digestive and reproductive tissues in *Schistosoma mansoni*. PLoS Neglected Tropical Diseases 5(4): e1043.

Neves LX, Sanson AL, Wilson RA, Castro-Borges W. 2015. What's in SWAP? Abundance of the principal constituents in a soluble extract of *Schistosoma mansoni* revealed by shotgun proteomics. Parasites and Vectors 8: 337.

Oliveira MF, Kycia SW, Gomez A, Kosar AJ, Bohle DS, Hempelmann E, Menezes D, Vannier-Santos MA, Oliveira PL, Ferreira ST. 2005. Structural and morphological characterization of hemozoin produced by *Schistosoma mansoni* and *Rhodnius prolixus*. FEBS Letters 579(27): 6010–6016.

Parker-Manuel SJ, Ivens AC, Dillon GP, Wilson RA. 2011. Gene expression patterns in larval *Schistosoma mansoni* associated with infection of the mammalian host. PLoS Neglected Tropical Diseases 5(8): e1274.

Philippsen GS, Wilson RA, DeMarco R. 2015. Accelerated evolution of schistosome genes coding for proteins located at the host-parasite interface. Genome Biology and Evolution 7(2): 431–443.

Planchart S, Incani RN, Cesari IM. 2007. Preliminary characterization of an adult worm "vomit" preparation of *Schistosoma mansoni* and its potential use as antigen for diagnosis. Parasitology Research 101(2): 301–309.

Rofatto HK, Parker-Manuel SJ, Barbosa TC, Tararam CA, Alan Wilson R, Leite LC, Farias LP. 2012. Tissue expression patterns of *Schistosoma mansoni* venom allergen-like proteins 6 and 7. International Journal for Parasitology 42(7): 613–620.

Ruff P, Speicher DW, Husain-Chishti A. 1991. Molecular identification of a major palmitoylated erythrocyte membrane protein containing the src homology 3 motif. Proceedings of the National Academy of Sciences of the United States of America 88(15): 6595–6599.

Shannon WA Jr, Bogitsh BJ. 1969. Cytochemical and biochemical observations on the digestive tracts of digenetic trematodes. V. Ultrastructure of *Schistosomatium douthitti* gut. Experimental Parasitology 26(3): 344–353.

Skelly PJ, Da'dara AA, Li XH, Castro-Borges W, Wilson RA. 2014. Schistosome feeding and regurgitation. PLoS Pathogens 10(8): e1004246.

Skelly PJ, Shoemaker CB. 2001. *Schistosoma mansoni* proteases Sm 31 (cathepsin B) and Sm 32 (legumain) are expressed in the cecum and protonephridia of cercariae. Journal of Parasitology 87(5): 1218–1221.

Spence IM, Silk MH. 1970. Ultrastructural studies of the blood fluke—*Schistosoma mansoni*. IV. The digestive system. South African Journal of Medical Science 35(3): 93–112.

Swanson JA, Watts C. 1995. Macropinocytosis. Trends in Cell Biology 5(11): 424–428.

Thornhill J, Coelho PM, McVeigh P, Maule A, Jurberg AD, Kusel JR. 2009. *Schistosoma mansoni* cercariae experience influx of macromolecules during skin penetration. Parasitology 136(11): 1257–1267.

Toh SQ, Glanfield A, Gobert GN, Jones MK. 2010. Heme and blood-feeding parasites: friends or foes? Parasites and Vectors 3: 108.

Tsiftsoglou AS, Tsamadou AI, Papadopoulou LC. 2006. Heme as key regulator of major mammalian cellular functions: molecular, cellular and pharmacological aspects. Pharmacology and Therapeutics 111(2): 327–345.

Van Dam GJ, Bergwerff AA, Thomas-Oates JE, Rotmans JP, Kamerling JP, Vliegenthart JF, Deelder AM. 1994. The immunologically reactive O-linked polysaccharide chains derived from circulating cathodic antigen isolated from the human blood fluke *Schistosoma mansoni* have Lewis x as repeating unit. European Journal of Biochemistry/FEBS 225(1): 467–482.

van Lieshout L, Polderman AM, Deelder AM. 2000. Immunodiagnosis of schistosomiasis by determination of the circulating antigens CAA and CCA, in particular in individuals with recent or light infections. Acta Tropica 77(1): 69–80.

Verity CK, Loukas A, McManus DP, Brindley PJ. 2001. *Schistosoma japonicum* cathepsin D aspartic protease cleaves human IgG and other serum components. Parasitology 122(Pt 4): 415–421.

Wilson RA. 2012a. The cell biology of schistosomes: a window on the evolution of the early metazoa. Protoplasma 249(3): 503–518.

Wilson RA. 2012b. Proteomics at the schistosome-mammalian host interface: any prospects for diagnostics or vaccines? Parasitology 139(9): 1178–1194.

Wilson RA, Barnes PE. 1974. The tegument of *Schistosoma mansoni*: observations on the formation, structure and composition of cytoplasmic inclusions in relation to tegument function. Parasitology 68(2): 239–258.

Wilson RA, Coulson PS. 2009. Immune effector mechanisms against schistosomiasis: looking for a chink in the parasite's armour. Trends in Parasitology 25(9): 423–431.

Wilson RA, Li XH, MacDonald S, Neves LX, Vitoriano-Souza J, Leite LC, Farias LP, James S, Ashton PD, DeMarco R and others. 2015. The schistosome esophagus is a 'Hotspot' for microexon and lysosomal hydrolase gene expression: implications for blood processing. PLoS Neglected Tropical Diseases 9(12): e0004272.

Wong JY, Harrop SA, Day SR, Brindley PJ. 1997. Schistosomes express two forms of cathepsin D. Biochimica et Biophysica Acta 1338(2): 156–160.

CHAPTER

Nervous and Sensory System
of *Schistosoma*

Barrie GM Jamieson

12.1 INTRODUCTION

12.1.1 General Neural Morphology

This chapter deals with the adult nervous system. For the miracidium see Jamieson and Haas, chapter 6 and for the cercaria Haas and Jamieson, chapter 8. The schistosome central nervous system (CNS), consists of two cerebral ganglia (the 'brain') located above the esophagus, immediately anterior to the ventral sucker. These are conjoined by a short commissure (Silk and Spence 1969; Skuse *et al.* 1990, Fig. 12.1; Halton 2004, Fig. 12.2, upper panel; Collins *et al.* 2011, Fig. 12.2, lower panel), giving the brain a 'butterfly-shape' (Schmidt-Rhaesa *et al.* 2015). The ganglia can be revealed by various staining procedures, including lectin and immunofluorescence labeling. From the ganglia arise small immunoreactive nerve fibers which innervate the suckers. Paired longitudinal nerve cords arise from the ganglia and are cross-linked at regular intervals by transverse commissures. The ventral nerve cords are thicker than the dorsal pair (Gustafsson 1992). A similar neural morphology has been described for *Opisthorchis viverrini* (Leksomboon *et al.* 2014).

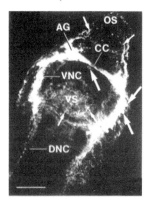

FIG. 12.1 *Schistosoma mansoni*. Innervation of the oral (OS) and ventral (VS) suckers of adult. Ag, anterior ganglion; CC, central commissures; VNC, ventral nerve cord; DNC, dorsal nerve cord. Nerve cell bodies are associated with the ganglia and the central commissure (large arrows). Immunoreactive nerve fibers (small arrows) innervate the suckers. Confocal scanning laser micrograph of whole mount preparation. Scale bar 200 µm. From Gustafsson MKS. 1992. Advances in Neuroimmunology 2(3): 267–286. Fig. 9A. After Skuce PJ, Johnston CF, Fairweather I, Halton DW, Shaw C, Buchanan KD. 1990. Cell and Tissue Research 261(3): 573–581. Fig. 1.

Department of Zoology and Entomology, School of Biological Sciences, University of Queensland, Brisbane 4072, Australia.

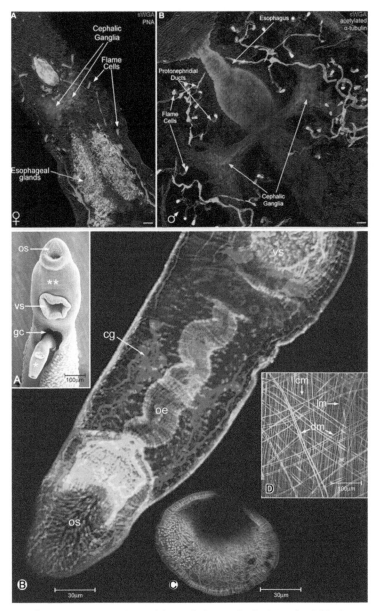

FIG. 12.2 Upper panel. *S. mansoni.* Two views of the cephalic ganglia (brain) and neighboring structures. **A.** Lectin PNA staining of the esophageal gland cells. Lectins WGA labeling of cephalic ganglia and protonephridial ducts in a female. Note the commissure joining the two ganglia. **B.** Lectins WGA labeling of cephalic ganglia and non-ciliated protonephridial ducts of a male. Also shown is anti-acetylated α-tubulin staining the ciliated regions of the protonephridial system. From Collins JJ, King RS, Cogswell A, Williams DL, Newmark PA. 2011. PLoS Neglected Tropical Diseases 5(3). Fig. 6A, B.

Lower panel. A. SEM image of adult *Schistosoma mansoni* pair in permanent copula from human mesenteric veins, where the male worm (*) holds the female (**) in the gynaecophoric canal (gc). Note the well-developed oral (os) and ventral suckers (vs). **B.** Confocal image of the forebody of a whole-mount of a female schistosome showing immunofluorescence (red) for FaRP neuropeptide in the cerebral ganglion/brain (cg) and associated neurons, including those innervating the oral (os) and ventral (vs) suckers. The muscle fibers of the suckers, esophagus (oe) and body wall (bw) are stained green using FITC-labelled phalloidin. **C.** Confocal image of the neural plexus that innervates the musculature of the oral sucker of a male worm, immunostained for FaRP neuropeptide. **D.** Confocal image of trematode body wall musculature revealing a lattice-like arrangement of outer circular (cm) and inner longitudinal fibers (lm), below which are bundles of diagonal muscle fibers (dm) that cross each other at an angle of approximately 120 degrees. From Halton DW. 2004. Micron 35(5): 361–390. Fig. 15.

A peripheral nervous system is linked to the CNS and consists of smaller nerve cords and nerve plexuses that supply all the major body structures, in particular the tegument, the somatic musculature, oral and ventral suckers, the alimentary tract and reproductive organs (see Halton and Gustafsson 1996; Halton 2004; Hatton and Maule 2004; Ribeiro and Geary 2010; Collins *et al.* 2011) (Figs. 12.1, 12.2). The combination of central and peripheral neuronal elements provides a complex network of communication that effectively links every region of the body (Ribeiro and Geary 2009). As the latter authors note, schistosomes, as acoelomates and lacking a circulatory system, do not have the capacity for endocrine signaling through a circulating body fluid. Thus much of the signal transduction that occurs in these animals is presumed to involve the nervous system, probably through a combination of synaptic and paracrine mechanisms.

12.2 CYTOLOGY

We owe to the comprehensively illustrated ultrastructural work of Silk and Spence (1969) most of our knowledge of the cytology or histology of the schistosome nervous system. This is briefly reviewed here.

The circumesophageal ganglia contain cells with large nuclei and prominent osmiophilic nucleoli which gave rise to groups of closely packed non-myelinated axons. Small osmiophilic clusters and particles approximately 45–60 nm in diameter are scattered throughout the chromatin and the eccentric nucleolus consists of a dense aggregation of 4–13 nm osmiophilic granules. The perikaryon contains numerous clusters of ribosomes, round to elongate mitochondria with sparse well-defined cristae, osmiophilic granules approximately 62–68 nm in diameter, stellate clusters of granules resembling α-glycogen, membrane-bound multivesicular bodies, a Golgi zone and poorly defined endoplasmic reticulum.

Adjacent axolemmas of the closely packed axons are separated by a uniform layer of weakly osmiophilic material approximately 12.5 nm wide. The organelles within the predominantly electron-lucent axoplasm (details in Silk and Spence 1969) comprise:

1. Collections of clear synaptic vesicles approximately 20–50 nm in diameter bounded by about 10–15 osmiophilic 5 nm granules fused together along the periphery.
2. Stellate osmiophilic clusters resembling α-glycogen and individual 20 nm osmiophilic putative β-glycogen granules.
3. Dense 32–90 nm circular too ovoid, sometimes membrane-bound osmiophilic granules. This type of granule is sometimes associated with the clear synaptic type 1 vesicles.
4. A larger type of 100–160 nm membrane-bound dense granule, apparently neurosecretory in nature, in single rows or small aggregations within the nerve processes investing muscle and other cells. The granules are accompanied by large type 5 axoplasmic vesicles frequently devoid of content.
5. Clear axoplasmic 50–150 nm long vesicles associated with the larger variety of dense granule (type 4) in a ground substance comprising finely amorphous material and putative β-glycogen particles. The vesicles frequently contain small particles of osmiophilic material and are bounded by a membrane composed of linked 5 nm granules.
6. Circular to elongate mitochondria with well-developed but randomly orientated cristae sometimes accompanied by small osmiophilic granules.
7. Parallel arrays and fragments of microtubules approximately 25 nm in diameter.

Axons in the nerve trunks extending from the ganglia contain fewer inclusions of types 1–7 than those forming the circumesophageal commissure. The nerve trunks ramify throughout the fluke and are demarcated from the surrounding musculature and other tissues by an 11 nm layer of amorphous material identifiable with the cement substance separating apposed axolemmas in the nerve process.

Synapses between axons are present within the circumesophageal commissure and in other regions. These are characterized by an accumulation of clear type 1 vesicles in an axon closely applied to another in which the post-synaptic space has few inclusions. The apposing axolemmas are spaced approximately 9–17 nm apart and clear vesicles are attached to the pre-synaptic membrane which appear less thickened and osmiophilic than the post-synaptic. Type 3 osmiophilic granules are sometimes associated with the type 1 clear vesicles in the pre-synaptic axoplasm.

Neuromuscular junctions similar to the axo-axonal synapses are seen. In these areas the axon is packed with clear type 1 synaptic vesicles which are occasionally accompanied by type 3 osmiophilic granules and glycogen. Clear type 1 vesicles are sometimes seen attached to the slightly thickened axolemma which is apposed to the broadened osmiophilic sarcolemma of the adjoining muscle cell at a distance of about 10 nm.

Osmiophilic type 4 neurosecretory granules and their associated clear type 5 axoplasmic vesicles are present in the sarcoplasm of muscle cells and in close association with the sarcolemma and in rows or small clusters associated with the plasma membranes of flame-cells and esophageal integumentary cells. Close association between the esophageal integument and a nerve process containing type 4 neurosecretory granules has been observed but direct innervation of the integument has not been seen. Nerve trunks are seen in close proximity to the deep-seated nucleated regions of the integument but nerve processes are not directly associated with the outer syncytial layer of the integument except in the esophageal region or where they penetrate the outer layer to form sensory bulbs (Silk and Spence 1969).

12.3 SENSORY RECEPTORS

Silk and Spence (1969) briefly described putative sensory receptors in the adult *S. mansoni* and they are discussed by Jones *et al*., Chapter 10, who recognize uniciliated, multiciliate or aciliate receptors, the predominant type being the uniciliate receptor. As noted by Silk and Spence (1989) a number of nodular protuberances with an apical finger-like projection are present on both the male and female integument. These distensions cover uniciliate sensory receptors consisting of a bulbous of nerve tissue from which a single cilium extends into the apical integumentary projection overlying the bulb (Fig. 12.3). The nerve process and its bulbous portion are demarcated from the integument by an 8–12.5 nm layer of cement substance continuous with the basement membrane separating the integument from the underlying musculature.

The base of the cilium is surrounded by two osmiophilic rib-like structures which are described in detail by Silk and Spence (1969). The sensory bulb contains mitochondria with well-defined cristae, stellate clusters resembling α-glycogen and clear ovoid vesicles with a major axis 40–135 nm in length. Microtubules are also observed within the axoplasm of the bulb and in the nerve fibers leading to the sensory receptors.

The cilium appears to originate at a level slightly beneath that of the rib-like structures. The basal body has a thickened osmiophilic wall comprising nine groups of conjoined tubular fibers. Spoke-like arms extend radially from the wall of the basal body at two levels and these converge to form small osmiophilic satellite structures at a distance of about 50 nm from the cilium. A number of these satellite bodies surround the thickened basal fibers of the cilium (Silk and Spence 1989).

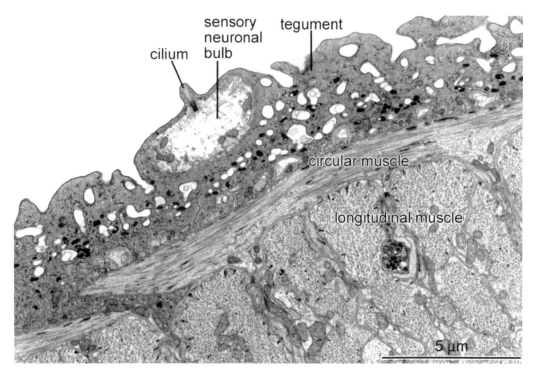

FIG. 12.3 *Schistosoma mansoni.* Transmission electron microscope section of a unicliate sensory receptor in the adult tegument, showing the sensory neuronal bulb and cilium. From Jones MK (unpublished).

12.4 NEUROACTIVE SUBSTANCES

12.4.1 Introduction

Neurotransmitter receptors and other neuronal proteins are attractive potential targets for chemotherapeutic intervention, including elimination of the parasite by drug application. A variety of activities that are essential to schistosome survival are controlled by neuronal signaling. These include host attachment and penetration, the control of movement and migration, feeding, excretion and egg-laying (Ribeiro and Geary 2010). Schistosomes have a rich diversity of neuroactive substances and receptors, some of which have no vertebrate or invertebrate orthologues and thus could be schistosome-specific (Hu *et al.* 2003; Verjovski-Almeida *et al.* 2003). This opens avenues for development of selective anti-schistosomal drugs (Geary *et al.* 1992; Ribeiro and Geary 2010) and vaccines (Bergquist and McManus, Chapter 24).

The neuroactive substances include a well-developed peptidergic system, consisting of several families of neuropeptides, in particular NPF/Y-like and FMRFamide-like peptides. In addition, the nervous system is rich in small "classical" transmitters, the most abundant of which are acetylcholine (ACh), glutamate and biogenic amines. The reader is referred to Ribeiro *et al.* (2005), Ribeiro and Geary (2010), Ribeiro (2015), for a comprehensive coverage of biogenic amines and other agents from which the following brief account is largely drawn.

12.4.2 Biogenic Amines

Biogenic amines make up the largest subset of classical transmitters. Most familiar is serotonin (5-hydroxytryptamine or 5HT). It is an indoleamine derived from the amino acid tryptophan. Other examples are histamine (HA) which is a histidine derivative, dopamine (DA) and noradrenaline (NA), both derived from tyrosine. Recent findings suggest that BAs may play an important role in the interaction with the snail intermediate host. This new evidence adds an important piece of information to our understanding of this complex system (Ribeiro 2015). Biogenic amines interact with cell surface receptors which are beyond the scope of this chapter.

Serotonin. Serotonin (5 HT) is widespread throughout the schistosome nervous system. It is a strong muscle stimulant but appears to require potentiation from other myoactive substances, particularly neuropeptides. It has also been shown to stimulate carbohydrate metabolism. Genomic searches have demonstrated at least two HT receptors, suggesting that 5 HT acts primarily through cAMP and associated signaling pathways. Schistosomes may regulate endogenous production of 5 HT, decreasing production when supply by the host is sufficient (references in Ribeiro and Geary 2010).

Catecholamines. In addition to HT, the catecholamines DA (dopamine) and NA (noradrenaline) have been implicated as modulators of neuromuscular activity in schistosomes. Experimental results suggest that DA acts directly on the musculature or some of the associated nerve endings rather than through centrally located neurons. DA causes muscle relaxation possibly by activating SmD_2 and elevating cAMP within the myofiber. Further proteins of relevance to catecholamine signaling are discussed by Ribeiro and Geary (2010).

In addition, El-Shhabi *et al.* (2012) have described a *Schistosoma mansoni* G protein-coupled receptor (named SmGPR-3) that was cloned, expressed heterologously and shown to be activated by dopamine, a well-established neurotransmitter of the schistosome nervous system. SmGPR-3 belongs to a new clade of "orphan" amine-like receptors that, importantly, exist in schistosomes but not the mammalian host. Further analysis of the recombinant protein showed that SmGPR-3 can also be activated by other catecholamines, including the dopamine metabolite, epinine and it has an unusual antagonist profile when compared to mammalian receptors. SmGPR-3 is abundantly expressed in the nervous system of schistosomes, particularly in the main nerve cords and the peripheral innervation of the body wall muscles. Dopamine, epinine and other dopaminergic agents have strong effects on the motility of larval schistosomes in culture. SmGPR-3 appears to be an important neuronal receptor and is probably involved in the control of motor activity in schistosomes. There are potentially important differences between SmGPR-3 and host dopamine receptors that could be exploited to develop new, parasite-selective anti-schistosomal drugs (El-Shehabi *et al.* 2012).

Histamine is present in all examined flatworms. In *S. mansoni*, exogenous application of HA causes a dose-dependent increase in motor activity and treatment with antihistamines causes paralysis. Some of its receptors appear to have no mammalian orthologue.

Novel amine-like receptors. The reader is referred to Ribeiro and Geary for a discussion of additional 15 potential biogenic amines receptors, all of which have a characteristic heptahelical topology and conserved signature peptide receptors.

12.4.3   Neuropeptides

Schistosoma mansoni and *S. japonicum* may have as many as 16–17 neuropeptides. These are major transmitters of the flatworm nervous system. The cerebral ganglia, longitudinal nerve cords, transverse connectives, peripheral plexuses, innervations of the somatic musculature, oral and ventral suckers and the reproductive tract are all rich in peptidergic neurons (references in Ribeiro and Geary 2010). Halton (2004) demonstrated FaRP neuropeptide by red immunofluorescence in the cerebral ganglion/brain and associated neurons, including those innervating the oral and ventral suckers. The neural plexus that innervates the musculature of the oral sucker of a male worm was also immunostained for this neuropeptide (Fig. 12.2).

12.4.4 Acetylcholine (ACh)

In schistosomes, cholinergic neurons have been detected in the cerebral ganglia, longitudinal nerve cords, transverse commissures and the innervation of all major bodies of muscle, including attachment organs, somatic and subtegumental muscle layers and muscles of reproductive organs, often co-localized with neuropeptide immunoreactivity (Halton and Gustafsson 1969; Ribeiro and Geary 2010). In contrast with vertebrate neuromuscular junctions, ACh is a major inhibitory neuromuscular transmitter in schistosomes. It appears to act contrary to peptidergic transmitters in controlling the body wall musculature and movement. It is also implicated in glucose transport from the host across the tegument (references in Ribeiro and Geary 2010).

12.4.5 Glutamate

Glutamate, the major excitatory neurotransmitter in vertebrates is little studied in flatworms. However, a glutamate binding site has been found in membrane preparations and evidence produced for glutamate receptors and for an important role of glutamate in the control of motor activity in *Schistosoma*. The truncated mGluR-like receptor of *S. mansoni* can bind glutamate. Antibody testing has shown that the receptor is enriched in the adult tegument, particularly of the male (references in Ribeiro and Geary 2010).

More recently, Taman and Ribeiro (2011) have described a novel glutamate receptor in *S. mansoni* (SmGluR). The receptor was shown to be activated by glutamate, whereas aspartate and the glutamate derivative gamma-aminobutyric acid (GABA) had no significant effect. The pharmacological profile of SmGluR was shown to be substantially different from that of receptors in the host. Confocal immunolocalization studies revealed that SmGluR is strongly expressed in the nervous system of adult worms and larvae. In the adults, the receptor was detected in the longitudinal nerve cords and cerebral commissures, as well as the peripheral nerve fibers and plexuses innervating the acetabulum and the somatic musculature. SmGluR was also detected along the length of the female reproductive system, including the oviduct, ootype and the uterus. It is expressed at about the same level in cercaria and adult stages. The results identified SmGluR as an important neuronal receptor and provided the first molecular evidence for a glutamate signaling system in schistosomes (Taman and Ribeiro 2013).

12.5 CHAPTER SUMMARY

General neuronal morphology of adult *Schistosoma* is described, chiefly from literature depicting SEM and confocal images with various staining procedures including immunofluorescence and lectin staining. Neuroactive substances are described, including biogenic amines (serotonin, catecholamines and histamine) and neuropeptides. Acetylcholine and glutamate, with their receptors, are also discussed.

12.6 LITERATURE CITED

Collins JJ, King RS, Cogswell A, Williams DL, Newmark PA. 2011. An atlas for *Schistosoma mansoni* organs and life cycle stages using cell type-specific markers and confocal microscopy. PLoS Neglected Tropical Diseases 5(3): e1009.

El-Shehabi F, Taman A, Moali LS, El-Sakkary N, Ribeiro P. 2012. A novel G protein-coupled receptor of *Schistosoma mansoni* (SmGPR-3) is activated by dopamine and is widely expressed in the nervous system. PLoS Neglected Tropical Diseases 6(2): e1253.

Geary TG, Klein RD, Vanover L, Bowman JW, Thompson DP. 1992. The nervous systems of helminths as targets for drugs. Journal of Parasitology 78(2): 215–230.

Gustafsson MKS. 1992. The neuroanatomy of parasitic flatworms. Advances in Neuroimmunology 2(3): 267–286.

Halton DW. 2004. Microscopy and the helminth parasite. Micron 35(5): 361–390.

Halton DW, Gustafsson MKS. 1996. Functional morphology of the platyhelminth nervous system. Parasitology 113: S47–S72.

Halton DW, Maule AG. 2004. Flatworm nerve-muscle: structural and functional analysis. Canadian Journal of Zoology 82(2): 316–333.

Hu W, Yan Q, Shen DK, Liu F, Zhu ZD, Song HD, Xu XR, Wang SJ, Rong YP, Zeng LC, *et al.* 2003. Evolutionary and biomedical implications of a *Schistosoma japonicum* complementary DNA resource. Nature Genetics 35(2): 139–147.

Leksomboon R, Chaijaroonkhanarak W, Arunyanart C, Umka J, Jones MK, Sripa B. 2012. Organization of the nervous system in *Opisthorchis viverrini* investigated by histochemical and immunohistochemical study. Parasitology International 61(1): 107–111.

Ribeiro P. 2015. Exploring the role of biogenic amines in schistosome host-parasite interactions. Trends in Parasitology 31(9): 404–405.

Ribeiro P, Geary TG. 2010. Neuronal signaling in schistosomes: current status and prospects for postgenomics. Canadian Journal of Zoology 88(1): 1–22.

Ribeiro P, El-Shehabi F, Patocka N. 2005. Classical transmitters and their receptors in flatworms. Parasitology 131: S19–S40.

Schmidt-Rhaesa A, Harzsch S, Purschke G. 2015. *Structure and Evolution of Invertebrate Nervous Systems.* Oxford UK: Oxford University Press. 768 p.

Silk MH, Spence IM. 1969. Ultrastructural studies of the blood fluke – *Schistosoma mansoni* III. The nerve tissue and sensory structures. South African Journal of Medical Sciences 34: 93–104.

Skuce PJ, Johnston CF, Fairweather I, Halton DW, Shaw C, Buchanan KD. 1990. Immunoreactivity to the pancreatic-polypeptide family in the nervous-system of the adult human blood fluke, *Schistosoma mansoni.* Cell and Tissue Research 261(3): 573–581.

Taman A, Ribeiro P. 2011. Glutamate-mediated signaling in *Schistosoma mansoni*: A novel glutamate receptor is expressed in neurons and the female reproductive tract. Molecular and Biochemical Parasitology 176(1): 42–50.

Verjovski-Almeida S, deMarco R, Martins EAL, Guimaraes PEM, Ojopi EBP, Paquola ACM, Piazza JP, Nishiyama MY, Kitajima JP, Adamson RE, *et al.* 2003. Transcriptome analysis of the acoelomate human parasite *Schistosoma mansoni.* Nature Genetics 35(2): 148–157.

CHAPTER

Reproductive System of *Schistosoma*

Malcolm K Jones[1],*, Barrie GM Jamieson[2] and Jean-Lou Justine[3]

13.1 INTRODUCTION

Schistosomes are dioecious parasites, a rare phenomenon within the Playhelminthes. The female *Schistosoma* lies within the gynaecophoric canal, a groove formed by folding of the lateral extremities of the male worm. Males hold the female worms in this groove throughout adult life of the parasite pairs (Figs. 13.1, 13.2). The females are held rather loosely by the males and it would appear that they can be mobile (or can be manipulated by males) within the canals, so that the two members of the worm pair can lie with their genital pores opposing each other or in tandem.

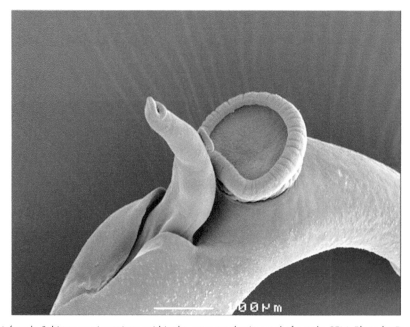

FIG. 13.1 A female *Schistosoma japonicum* within the gynaecophoric canal of a male, SEM. Photo by Breanna Dunn and Malcolm Jones.

[1] University of Queensland, School of Veterinary Science, Gatton, Qld 4343, Australia.
[2] Department of Zoology and Entomology, School of Biological Sciences, University of Queensland, Brisbane, Qld 4072, Australia.
[3] ISYEB, Institut de Systématique, Évolution, Biodiversité (UMR7205 CNRS, EPHE, MNHN, UPMC), Muséum National d'Histoire Naturelle, Sorbonne Universités, CP 51, 55 rue Buffon, 75231 Paris cedex 05, France.
* Corresponding author

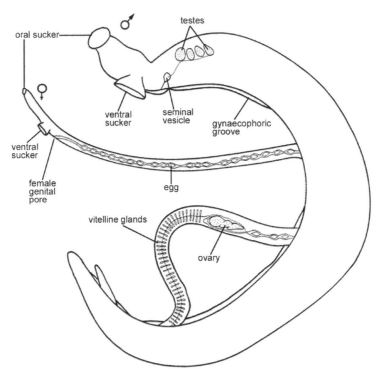

FIG. 13.2 *Schistosoma bovis*. Male and female in coitus (semidiagrammatic). Adapted from Justine J-L. (1980). Thèse de Troisième Cycle. France: Université des Sciences et Techniques du Languedoc (Montpellier II). Fig. A.

Observations of *Schistosoma mansoni* and the related species *Austrobilharzia variglandis* of chickens, indicate that female worms are quite motile within the male, often pushing their way into the small venules near the intestine where they deposit eggs against the endothelium (Bloch 1980; Wood and Bacha 1983). Single females could be observed in blood vessels, which at least in the case of *A. variglandis*, are thought to migrate to the narrowest blood vessels to lay their eggs.

Pairing of schistosomes adults has important implications for sexual maturation and development of the parasites, particularly of the female worm. Females in single sex infections cannot develop to full sexual maturity. This has been attributed to a lack of endocrine signal from the male as well as undernourishment (Pearce and Huang 2015) but several elements including Src kinases and TGFβ pathway-associated and neural processes may also contribute to the male-induced sexual development of the female (Grevelding *et al.*, Chapter 15). Intersex forms may occur (Erasmus 1987).

A functional consequence of the pairing behavior of schistosomes is that interspecies pairs can form, especially where two species are sympatric. Substantial research has been deployed to investigate the genetic consequences of these cross-species matches. The situation is complex and dependent on the species involved in hybridization, the host selection of the pairs and gender of hybridizing individuals. Thus, a recent study (Webster *et al.* 2013) in Senegal demonstrated the occurrence of *S. haematobium* introgressions with either *S. bovis* or *S. curassoni* in parasite progeny obtained from infected humans. No *S. haematobium* or *S. haematobium* hybrids were found in parasite progeny discharged from natural ruminant hosts. In some experimental infections, the result of pairing can be parthenogenesis (Tchuenté *et al.* 1994), although genotyping may be required for confirmation of this. Molecular studies have shown that *S. mansoni* and *S. bovis* have hybridized in Corsica where *Schistosoma* is a new incursion into Europe (Boissier *et al.* 2015; see also Madsen, Chapter 3). For reviews of sex determination and female reproductive development in *Schistosoma*, see Loverde and Chen (1991), Ribeiro-Paes and Rodrigues (1997) and Grevelding *et al.* (Chapter 15). For an extensive review of sexual biology in schistosomes see Moné and Boissier (2004).

13.2 MALE GENITAL SYSTEM

13.2.1 General Genital Anatomy

Aspects of the anatomy of the reproductive system of *Schistosoma* have been described for *S. mansoni* (Gönnert 1955; Spence and Silk 1971: Erasmus 1973, 1975; Nollen *et al.* 1976; Otubanjo 1980; Shaw and Erasmus 1981; Loverde and Chen 1991; Machado-Silva *et al.* 1998; Southgate *et al.* 1998; Neves *et al.* 2005; Collins *et al.* 2011; and deWalick *et al.* 2012). Details of the female system and eggs are reviewed for seven species of *Schistosoma* by Southgate and Bray (2003) (Table 13.1). Testicular structure was described by Lindner (1914) for *S. haematobium.* Detailed descriptions of the anatomy and gametogenesis are given for *Schistosoma bovis* by Justine (1980).

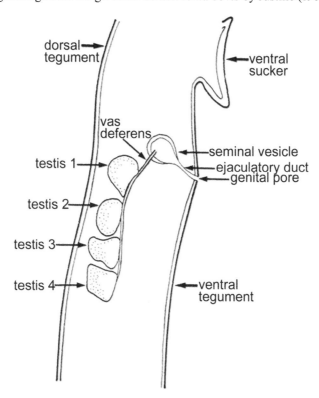

FIG. 13.3 *Schistosoma bovis.* Longitudinal section of the male genital system (semidiagrammatic). Adapted from Justine J-L. 1980. Thèse de Troisième Cycle. France: Université des Sciences et Techniques du Languedoc (Montpellier II). Fig. B.

In *Schisosoma bovis* the male genital organs (Figs. 13.2, 13.3, 13.5) are situated anteriorly in the body, immediately behind the ventral sucker. The testis lobes (alternatively regarded as individual testes), numbering 3 to 6 but predominantly 4, are arranged in an irregular row. The number of testis lobes varies among species (Table 13.1), but generally number from 4 to 6. The testis lobes are joined to a seminal vesicle by a fine duct (vas deferens) which enters the vesicle according to Justine (1980) but in *S. mansoni* the vesicle appears to join the duct as a pear-shaped diverticulum as shown by confocal scanning microscopy (Silva-Leitao *et al.* 2009). The latter authors found that a testis lobe supernumerary to the normal four was amorphous, without connection to the normal set. The male system discharges at the male pore as an ejaculatory duct. The ejaculatory duct is eversible as a narrow cirrus (Collins *et al.* 2011). Awad and Probert (1990) deny the presence of a cirrus in *S. margrebowiei.*

Testis wall. The testes of *S. bovis* are surrounded by a parenchyma which is found throughout the body with no particular differentiation. From the exterior to the interior, the wall consists of the following components (Fig 13.4A, B): some muscle cells seen in transverse and longitudinal sections and therefore probably "crisscrossed"; a clear space of uniform thickness, about 0.1 µm, with some fibrous elements less than 10 nm thick; a basal lamina of uniform thickness, about 30 nm and a space of about 15 nm containing the cytoplasm of more internal cells; a cytoplasmic bed of variable thickness containing in places mitochondria and some membranous elements. It is joined by processes of the sustentacular cells. This bed is present over almost the entire surface of the testis but seems to be absent in some regions (Fig. 13.4C).

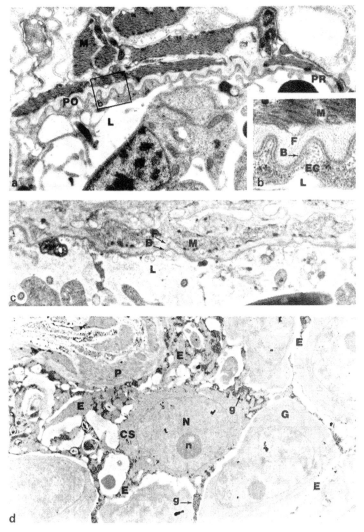

FIG. 13.4 *Schistosoma bovis.* **A.** Testis with lumen (L) surrounded by germinal cells. The testis wall is rectilinear (PR) or undulating (PO). x 11000. **B.** Enlargement of the undulating wall from A. The wall is formed from muscle (M), a clear zone flanked by fibers (F), by a basal lamina (B) and by cytoplasmic extensions of the sustentacular cells (EC) surrounding the lumen (L). x 33000. **C.** Different region of the testicular wall, lacking cytoplasmic extensions of the sustentacular cells. The lumen (L) is directly bordered by the basal lamina (B) to which is attached the fibrous zone (F) and the musculature (M). x 11000. **D.** Section showing the relationship between the wall (P) and a sustentacular cell (CS). This cell has a nucleus (N) containing a large nucleolus (n). Glycogen (G) is present in the peripheral region of the cell body and in the extensions (E). These extensions join the wall or surround the germinal cells (G). x 7000. From Justine J-L. 1980. Thèse de Troisième Cycle. France: Université des Sciences et Techniques du Languedoc (Montpellier II). Fig. 2.

In certain places the basal lamina presents undulations with an irregular amplitude of ca 0.3 μm (Fig. 13.4A, B). These undulations are not seen in the basal membrane nor in the clear fibrous layer (Justine 1980).

Lindner (1914) described the testis of *S. haematobium* as merely a cavity in the parenchyma, without a proper wall but as the wall described above was less than 0.5 μm thick, this may explain why it was not observed by light microscopy by that author. He observed only the basal lamina and the fibrous layer. The muscle fibers were seen only as a discontinuous trellis and the cytoplasmic layer produced by the sustentacular cells was not observed.

Three distinct regions of the male ducts are recognized ultrastructurally by Otubanjo (1980) in *S. mansoni*: the vas deferens, the seminal vesicle and the cirrus; they merge into each other without prominent demarcation but long septate desmosomes are present at the junctions. The short vas efferens originates from the mid-ventral region of each testis and leads into the single undulating vas deferens. This enters the wider seminal vesicle, 35 μm wide, at a dorsolateral angle. The vesicle lies directly posterior to the ventral sucker. It opens into the cirrus tube which lies posteriorly and ventrally, measures 8–9 μm in diameter and turns at approximately 90° to open at the anterior end of the gynaecophoric canal. No glandular structures were observed around the cirrus tube.

Vas efferens/vas deferens. The epithelium of the vas efferens/deferens consists of a single layer of flat cells which lie on a fibrous basal lamina, of variable thickness, which is continuous with that of the multiple testes. The basal membranes form deep infoldings or tubular invaginations, which extend deeply into the epithelial cells. Circular and longitudinal muscle bundles are in intimate contact with the basal lamina. The cells forming the epithelium possess an irregularly shaped nucleus (5–8 × 3–4 μm) with a heterogeneous nucleolus and a dense layer of chromatin adjacent to the nuclear envelope. In close association with the nucleus are numerous strands of granular endoplasmic reticulum. Clusters of glycogen particles, oval mitochondria with few cristae and dense matrix, as well as free microtubules and ribosomes, are distributed throughout the cytoplasm. Golgi complexes and large inclusions are present. Sometimes these granular inclusions appear to lack limiting membranes. The luminal surface of the epithelium is covered with a granular material. The apical cytoplasm of the cells is extended to form highly convoluted lamellae, the ends of which appear to be fused. Those portions of the epithelium without nuclei appear relatively flat, although they still possess numerous lamellae. Numerous septate desmosomes arising from the apical surface are located between the epithelial cells. The luminal surface of the vas deferens possesses longer and more abundant lamellae and dense secretory bodies as compared to the vas efferens. The anterior tips of spermatids and spermatozoa present in the lumen may be closely associated with the lamellae.

Seminal vesicle. The seminal vesicle is differentiated into four distinct morphological layers: (1) the anucleate syncytial epithelium, which lies on (2) the basal lamina, (3) the smooth muscle fibers and (4) the nucleated cytoplasmic region within the parenchyma. The wall of the seminal vesicle resembles the tegument, but it lacks spines, sensory papillae, multilaminate bodies and abundant channels as observed in the surface tegument of schistosomes. The luminal surface of the seminal vesicle is covered by an extraneous coat of dense, granular material. The lumen increases in diameter nearer the cirrus tube, at which point the lumen becomes interrupted by the infolded surface to form channels. Germinal cells, mainly spermatozoa and spermatids at varying stages of development, occur in its lumen. The spermatozoa are orientated in the direction of the cirrus tube and associate with the luminal surface of the duct. The commencement of the cirrus tube is marked by a decrease in luminal volume.

Cirrus tube and genital pore. The cirrus tube is also tegumentary in structure and is continuous with the surface gynaeophoric canal. It differs from the seminal vesicle by its restricted lumen, its highly folded luminal surface, the absence of spermatids and spermatozoa in the channels, a rather thick coat of circular muscle fibers and the presence of multilaminate vesicles in its luminal wall near the region of the genital pore. The external opening of the cirrus tube forms the male genital pore. The region around the pore lacks spines and accessory structures. The genital pore lies at the beginning of the gynaeophoric canal and is approximately disc-shaped whereas the cirrus tube in its evaginated condition appears sponge-like, possibly allowing for passage of spermatozoa but not of spermatids. Abundant rows of sensory papillae surround the male opening and are surmised to aid in locating the female genital pore. Anterior to the genital pore, short blunt spines are present (Otubanjo 1980).

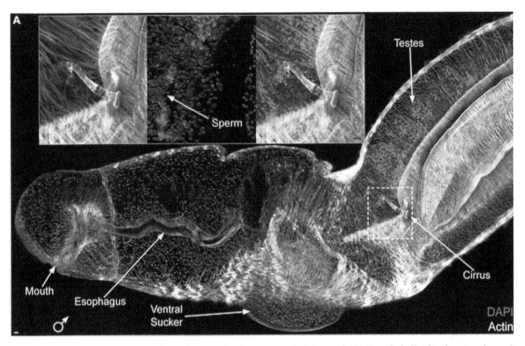

FIG. 13.5 *Schistosoma mansoni.* The male reproductive system. Staining with DAPI and phalloidin showing the male head and various parts of the male reproductive system. Inset, magnified view of male cirrus. Anterior to left, dorsal towards top. Collins JJ, King RS, Cogswell A, Williams DL, Newmark PA. 2011. PLoS Neglected Tropical Diseases 5(3): e1009. Fig. 7.

13.2.2 Sustentacular Cells

Gametogenesis is discussed in Chapter 15 but an important component of the testis, the sustentacular cells, as described by Justine (1980), for *S. bovis* will be described here. They are seen in all cut sections of the testis but in small numbers. The cell body, with a diameter of 5–6 μm, is provided with very long radiating prolongations which insinuate between the germinal cells and progressively ramify (Fig. 13.4D). All stages of spermatogenesis are involved, including the cytoplasmic rejects of the spermatids, but rarely the spermatozoa. They are sometimes situated in proximity to the testis wall, with which their prolongations associate closely (Fig. 13.4A, B, D). The ultrastructure of schistosome spermatozoa is described in Justine and Jamieson (Chapter 14).

The nucleus is globular, 4 μm wide. The nucleoplasm is clear and finely granular. The dense chromatin is present only as small masses apposed to the nuclear envelope. The single nucleolus is very large and often of a diameter exceeding 1 μm. The nuclear membrane is only slightly undulating. The cytoplasm forms a very thin layer (0.2–0.5 μm) around the nucleus; it is abundantly provided with ribosomes grouped as polysomes. Endoplasmic reticulum is represented by short elements covered by large numbers of ribosomes and very compact. The mitochondria are generally rounded, with a diameter of ca 0.2 μm and a dense matrix crossed by clear cristae. The cytoplasmic extensions are sometimes dilated by the mitochondria, even at a considerable distance from the cell-body, especially where they line the testicular wall. Neither Golgi bodies nor centrioles have been observed in these cells. Their cytoplasm has abundant β-glycogen predominantly localized against the plasma membrane of the cell-body and its branches. Pronase digestion indicates the presence of proteins in the cytoplasm.

The extensions or the cell-bodies of the sustentacular cells are pressed against the germinal cells and follow their contours at a distance of ca 0.5 μm. In some cases the extensions are zigzagged and seem to contact germinal cells alternately. In other cases the extensions remain central: the space that separates them from the germinal cells is penetrated by fine rays of cytoplasm. Similar rays are sometimes present between germinal cells in the absence of cytoplasm of the sustentacular cells at the center of the intercellular space. It is proposed that the fine cytoplasmic rays are formed by contraction of the cells on fixation.

It is deduced that the sustentacular cells have a nutritive function for the spermatogenetic stages and the wall of the testis, as evidenced by the presence of β-glycogen and the intimate association of the prolongations with the germinal cells and this extends to fusion of the cell membranes.

13.3 FEMALE GENITAL SYSTEM

13.3.1 General Anatomy

The female genital system has been described for *Schistosoma bovis* by Justine (1980). It extends for almost the entire length of the body and consists principally, from anterior to posterior, of the common genital pore, uterus, ootype which receives Mehlis'gland, the oviduct, the seminal receptacle and the ovary and the vitelline glands which join the anterior end of the oviduct, via the vitelline duct, forming the common ovovitelline duct (vitello-oviduct) (Fig. 13.6). The female system of *S. mansoni*, as illustrated by Gönnert (1955), Smyth (1966), Spence and Silk (1971) (Fig. 13.7A); Loverde and Chen (1991); Collins *et al.* (2011) (Fig. 13.15); deWalick *et al.* (2012) (Fig. 13.7B) and by confocal laser scanning microscopy (tomography) by Neves *et al.* (2005) (Fig. 13.8), resembles that of *S. bovis*, though with some differences noted in Table 13.1. The table may be consulted for details of number of testes, position of the ovary, length of the uterus and number of eggs it contains and mode of discharge of the eggs, whether urinary or fecal.

TABLE 13.1 Features of the female reproductive system and eggs of species of *Schistosoma.* Based on Southgate and Bray (2003).

Character	Haematobium	Mansoni	Japonicum	Intercalatum	Mekongi	Bovis	Malayensis
Number of testes	4 or 5	2–14	6–8	4–6	6–7	3–6	
Ovary	In posterior third	In anterior half	Central	In posterior half	In anterior 5/8	midbody	In anterior half
Uterus	Anterior, long. Holds 10–100 eggs at any one time. Produces 20290 daily	Anterior, short, Holds few eggs at one time. Produces 100–300 daily	Anterior, long. Holds 50 or more eggs at one time. Produces 1500–3500 daily	Anterior, long. Holds 5–50 eggs at one time.	Anterior, long	Anterior, long, holds many eggs at same time	Contains many eggs
Egg	83–187 × 60 μm; Terminal Spine. Discharged in urine	112–175 × 45–70 μm; Lateral spine. Discharged in feces	70–100 × 50-65 μm Rudimentary lateral spine. Discharged in feces	140–240 × 50–85 μm; long terminal spine. Discharged in feces	30–55 × 5065 μm; Small lateral knob. Discharged in feces	?	52–90 × 33–62 μm; small knob usually located laterally

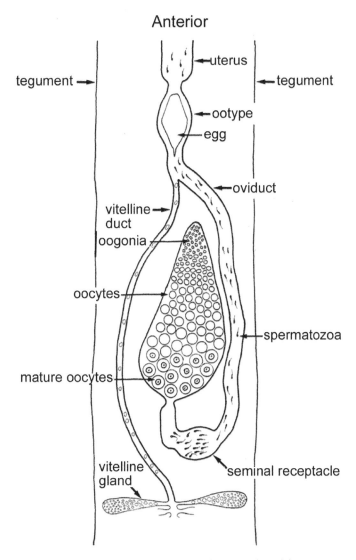

FIG. 13.6 *Schistosoma bovis*. Female genital apparatus. Dorsal view. Adapted from Justine J-L. 1980. Thèse de Troisième Cycle. France: Université des Sciences et Techniques du Languedoc (Montpellier II). Fig. H.

13.3.2   Vitellaria

Vitelline glands. The vitelline glands are racemose and when viewed by confocal microscopy, consist of columnar cells with a central nucleus of irregular size (Neves *et al.* 2005) (Fig. 13.8A). The vitelline follicles (Fig. 13.11A) form a large mass which occupies the whole of the width of the body behind the ovary and surrounds the cecum. Each follicle opens independently into the long vitelline duct. It debouches into the oviduct (thereafter the ovovitelline duct or vitello-oviduct) (Spence and Silk 1971, Fig. 13.7A; Justine 1980, Fig. 13.6; Neves *et al.* 2005; deWalick *et al.* 2012, Fig. 13.7B). Only the anterior region of the vitelline duct is ciliated (Spence and Silk 1971; Erasmus 1973). An oocyte receives spermatozoa from the seminal receptacle and when fertilized becomes surrounded by 30 to 40 vitelline cells shed by the vitelline glands (Smyth and Clegg 1959; deWalick *et al.* 2012).

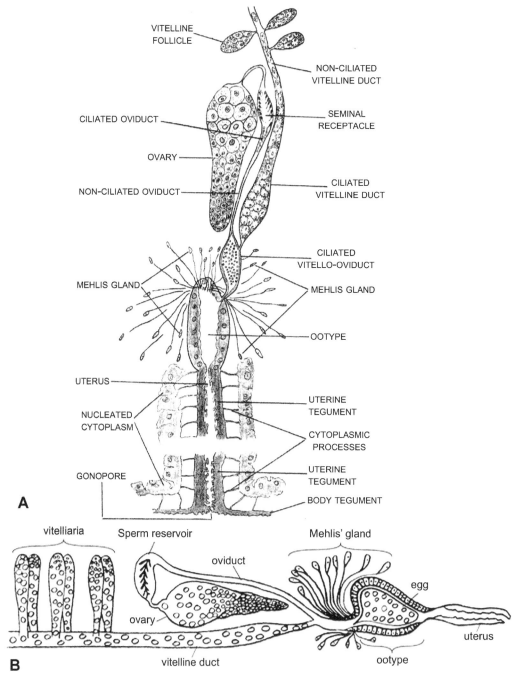

FIG. 13.7 *Schistosoma mansoni.* Two semidiagrammatic representations of the female system. **A.** In ventral view. From Spence IM, Silk MH. 1971. The South African Journal of Medical Sciences 36(3): 41–50. Fig. A. **B.** In lateral view. From deWalick S, Tielens AGM, van Hellemond JJ. 2012. Experimental Parasitology 132: 7–13. Fig. 2. After Gönnert, R., 1955. Schistosomiasis-Studien. II. Uber die Eibildung bei *Schistosoma mansoni* und das Schicksal der Eier im Wirtsorganismus. Zeitschrift für Tropenmedizin und Parasitologie 6: 33–52 and Smyth (1966). *The Physiology of Trematodes.* Oliver and Boyd. Edinburgh and London.

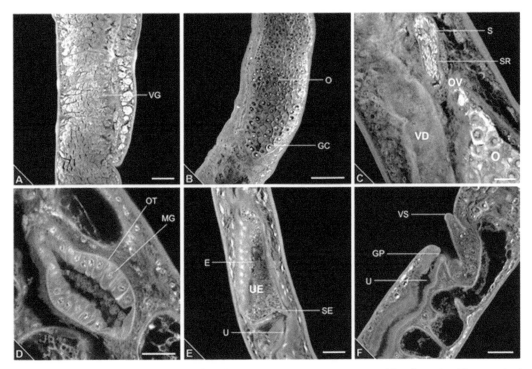

FIG. 13.8 *Schistosoma mansoni.* A–F Confocal laser scanning microscope images of females isolated from a mixed infection. **A.** Vitelline glands (VG), bar 50 μm. **B.** Ovary (O) and germ cells (GC), bar 100 μm. **C.** Ovary (O), oviduct (OV), seminal receptacle (SR), spermatozoa (S) and vitelline duct (VD), bar 50 μm. **D.** Ootype (OT). Supposed Mehlis' gland (MG) is the endothelium of the ootype, bar 100 μm. **E.** Uterus (U), uterine egg (UE), embryo (E) and spine egg (SE), bar 50 μm. **F.** Uterus (U), genital pore (GP) and ventral sucker (VS), bar 50 μm. From Neves RH, de Lamare Biolchini C, Machado-Silva JR, Carvalho JJ, Branquinho TB, Lenzi HL, Hulstijn M, Gomes DC. 2005. Parasitology Research 95(1): 43–49. Fig. 1.

Vitelline cells (vitellocytes) in different stages of maturation are found within the vitelline follicles (Fig. 13.10A, 13.11). The most mature cells (Fig. 13.12) are usually found proximal to the vitelline duct. Nuclear and cytoplasmic volumes increase during maturation while the ribosomal and mitochondrial complements become more abundant and synthesis of proteinaceous vitelline droplets, concentrated mainly at the periphery of the cell, together with shell precursor material (SPM) is initiated. Progression to maturity is associated with the production of rosettes of SPM, the appearance of lipid globules scattered through the cytoplasm and conglomerates of proto-SPM, the formation of concretions within distended sacs of rough endoplasmic reticulum (RER) and plaques of degenerating RER. The lipid droplets have peripheral clusters of α-glycogen. The nucleus has scattered patches of heterochromatin and a large nucleolus; nuclear pores are obvious and the outer nuclear membrane bears ribosomes. The cytoplasm is electron-dense owing to closely packed ribosomes and RER. The fully mature cell is discharged into the vitelline duct for transport to the vitello-oviduct [ovo-vitelline duct]. The granular endoplasmic reticulum is present up to the time that the vitelline cell is enclosed in the egg capsule (Spence and Silk 1971; Erasmus 1973, 1975).

The vitelline glands are enriched for a number of secretory products, most notable of which are the precursor proteins for the egg shell (SPM), as noted above. Additionally, the cells contain a yolk material, identified as ferritin-containing yolk platelets (Schübler *et al.* 1995) as well as abundant stores of the divalent metal, iron (Jones *et al.* 2007). More information on the nature of the formation of egg shells is provided in Chapter 5 and in detail in deWalick *et al.* (2012), but a brief resumé of the cellular activity and fate of vitelline cells will be provided here. Under influence of the Mehlis gland, the numerous vitelline cells within the ootype are induced to release their secretory content. The electron opaque egg-shell material coalesces to form the thin shell, presumably under direct influence of the Mehlis gland secretions. Once entrapped in the egg, the spent vitelline cells apoptose and gradually become incorporated into the developing embryo (Khampoosa *et al.* 2012). By virtue of this, the vitelline cells can be seen as true yolk cells.

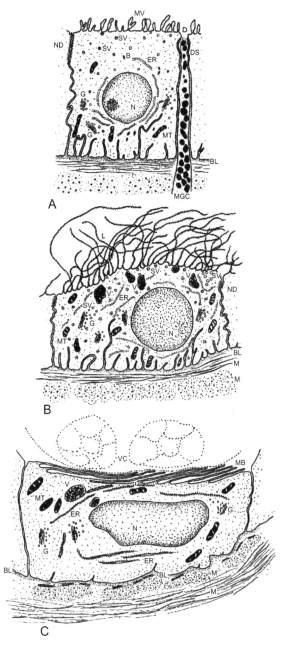

FIG. 13.9 *Schistosoma mansoni.* Semidiagrammatic representations of sections through A. The ootype, B. The oviduct and C. the vitelline duct. **A.** Transverse section though the epithelium at the posterior end of the ootype. **B.** Transverse section through the non-ciliated region of the oviduct. **C.** Longitudinal section through a vitelline duct cell. B, small dense bodies; BL, basement layer; D, duct from the cell of Mehlis' gland; DS, septate desmosome; ER, granular endoplasmic reticulum; G, Golgi complexes; M, muscle; MB, stack of membranes; MGC, cell of Mehlis' gland, MT, mitochondria; MV, spatulate microvilli; N, nucleus; ND, non-septate desmosomes; L, lamellae; SV, small vesicles; V, large or irregular vesicles containing electron-dense granular material. VC, vitelline cells. From Erasmus DA. 1973. Parasitology 67: 165–183. Figs. 1–3. Explanation in text.

Vitelline duct. The oviduct and the vitelline ducts separately join the ovovitelline duct, the latter being an expansion of the oviduct (Fig. 13.6). In the direction of the female genital pore, the vitelline duct is at first narrow and non-ciliated but it widens greatly before joining the ovovitelline duct to contain the post-mature vitelline cells. The basement membrane of the vitelline duct is separated by fibrillar interstitial tissue from a single surrounding layer of circular myofiber bundles. Invaginations of the basal plasma membrane penetrate into the cytoplasm which contains mitochondria, Golgi zones, ribosomes and regions of active secretory granule production. Large accumulations of glycogen are present in cells in the more anterior regions of the duct, which is syncytial in nature. According to Spence and Silk (1971) the apical cytoplasm is subdivided into narrow looped extensions characteristic of all tegumental tissues and the luminal border is re-inforced at intervals by septate desmosomes associated with zonulae occludentes. However, Erasmus (1973) reports that the luminal surface is elevated into stacks of membranes (Fig. 13.9C).

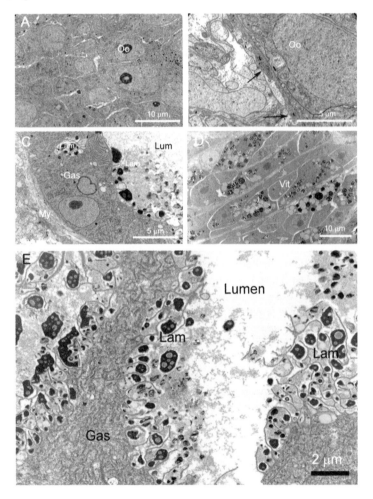

FIG. 13.10 *S. japonicum*. Ultrastructural morphology of tissues of a female. **A**. Ovary. Numerous oocytes with a high nucleo-cytoplasmic ratio are present as the sole cell type. **B**. A thin cellular layer, incorporating myofibrils, is present as the limiting margin of the ovary. Arrow indicates margin of ovary. **C**. Gastrodermis. The gastrodermis is a unilaminate syncytial layer forming the absorptive lining of the gut. **D**. Vitelline cells. These accessory cells of the female reproductive system secrete egg-shell precursors and possibly yolk. Electron opaque lipid droplets are prominent features of this region. **E**. Luminal surface of gastrodermis showing multiple stacked lamellae. Gas, Gastrodermis epithelium, Lam, Lamellae, Lum, Lumen of gut; My, Smooth muscle fibers; Oo, Oocyte; Vit, Vitelline cell. From Gobert *et al.* 2009. PLoS Neglected Tropical Diseases, 3(6): e469. doi: 10.1371/journal. pntd. 0000469. Fig. 3.

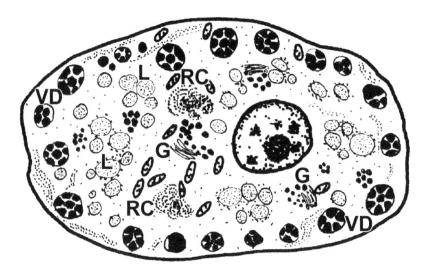

FIG. 13.11 *Schistosoma mansoni.* A mature vitelline cell. G, Golgi complexes; L, lipid; RC, ribosomal complexes; VD, viteline globules. From Erasmus DA. 1975. Experimental Parasitology 38(2): 240–256. Fig. 1D.

The ovovitelline duct is a widely distended and profusely ciliated chamber connecting posterolaterally to the ootype via a narrow non-ciliated duct. The structure of the main portion is similar to that of the oviduct but in the regions proximal to the ootype, the cytoplasm becomes more electron-dense, secretory granule formation is less marked and the structure is intermediate between that of the oviduct and the ootype (Spence and Silk 1971).

13.3.3   Ovary and Oviduct

Ovary. The ovary in *S. bovis* is situated anteriad of the region of confluence of the paired ceca, approximately in the middle of the body. The ovary is elongated and spiral in form, with a narrow anterior extremity (Fig. 13.6). The posterior region is larger and gives rise to the oviduct. According to Nollen *et al.* (1976), ovaries in the three predominant species of schistosomes differ from each other in shape but are similar in organization: *S. mansoni* and *S. haematobium* are reported to have serpentine ovaries while in *S. japonicum* the ovary is oval. A large, elongate ovary with posterior egress of the oviduct is well illustrated for *S. mansoni* by DAPI labeling in Collins *et al.* (2011) (Fig. 13.14), though a serpentine form is not evident. They note that spermatozoa passed to a female migrate up the female reproductive tract and are stored in the seminal receptacle located posterior to the ovary; tightly packed sperm were observed in this receptacle by staining with DAPI and anti-acetylated α-tubulin. From the seminal receptacle sperm are able to begin the fertilization process as oocytes emerge from the ovary. Oocytes then pass anteriorly through the oviduct (Fig. 13.14), to reach Mehlis' gland (Collins *et al.* 2011).

The ovary of *S. mansoni* is enclosed by a thin layer of circular muscle attached to the fibrous basement layer. Internal to this is the basal epithelium of the ovary. The epithelial cells have flattened nucleus, mitochondria, granular endoplasmic reticulum, several Golgi complexes and irregular dense bodies and vacuoles with variable contents. Numerous ribosomes but no cortical granules are present. The narrower, anterior region of the ovary contains closely packed, small oogonia and oocytes at all stages of development and a central lumen is absent. Posteriorly there are large ova (Erasmus 1973).

In a longitudinal section of the ovary of *S. bovis* it is possible to define three zones in anterior to posterior succession, based on the ultrastructure of the component cells. The anterior zone has a striking resemblance to certain zones of a testis, having small polyhedric cells. The intercellular spaces are occupied by cytoplasmic extensions resembling those of male sustentacular cells. The midregion contains cells of a different appearance. They are large with a large clear nucleus and increase in size posteriad. The cytoplasm is weakly electron-dense. Interceullular spaces are reduced or absent and sustentacular cells are not seen. The nucleus of germinal cells displays characteristic synaptonemal complexes. In the posterior zone of the ovary the intercellular spaces reappear progressively but are free of sustentacular cells. The germinal cells are large and polyhedric. The cytoplasm is electron-dense and the nucleus has a very large nucleolus. The sustentacular cells of the ovary differ from those of the testes in being flattened against the ovarian wall and emitting lateral, centripetal prolongations whereas in the testis the cell bodies are more internal and extend processes in all directions with some flattened against the testis wall. β-glycogen is more abundant in the testis cells than in those of the ovary (Justine 1980). The ultrastructural morphology of tissues of the adult *S. japonicum* is illustrated in Fig. 13.10A–E; including that of developing oocytes (Figs. 13.10A, B).

An ovarian wall similar to that of *S. bovis* was described by Erasmus (1973) for *S. mansoni*. In particular, a parietal position of the cell bodies was reported.

Oogonia. Oogenesis is covered by Grevelding *et al.*, in Chapter 15, chiefly from a biochemical standpoint, but we may note some features of the contents of the ovary. The smallest cells in the ovary are the oogonia.

Primary and secondary oocytes are the next largest cells in the ovary and are found in the center of the anterior region of the ovary, diminishing in numbers posteriorly. The ooctyes are hexagonal or irregular cells arranged in cords and having a central nucleus, at different stages of maturation. The smaller oocytes are usually irregular in shape, with a relatively small nuclear-cytoplasmic ratio. The nucleus contains a nucleolus and dense patches of chromatin; also present are mitochondria, long strings of granular endoplasmic reticulum and centrioles but few Golgi complexes and no cortical granules (Erasmus 1973; Neves *et al.* 2005) (Fig. 13.7B). The maturation of oogonia to primary and secondary oocytes involves a sequence of changes in both the nuclear and cytoplasmic components (see Chapter 15).

Immature ova are larger than the primary and secondary oocytes but vary in size according to their stage of maturation. They are absent from the anterior region of the ovary but are found in the central and posterior areas of the mid-section. A distinguishing feature is production of cortical granules.

Mature ova. In contrast to the smaller oocytes, the nucleus does not contain dense patches of chromatin but usually has a well-defined nucleolus. The ova have the cortical granules arranged peripherally. Annulate lamellae, an electron-dense reticulum, centrioles and a lipid-like body are present in the cytoplasm which also has a very abundant complement of ribosomes mostly associated with the porous nuclear membrane, mitochondria and Golgi zones. The oolemma of a mature ovum is in close contact with the confining cytoplasmic extensions of the oviduct. Numerous tight junctions are formed between the plasma membrane of the oviduct and the contiguous sarcolemma (Spence and Silk 1971; Erasmus 1973).

Oviduct and seminal receptacle. The oviduct has three distinct regions (Erasmus 1973). Arising from the posterior face of the ootype is a common duct which is ciliated and receives the anterior end of the oviduct as well as the anterior end of the vitelline duct. The anterior third of the oviduct is lined by cells which, in addition to lamellae, bear cilia. The second region is lined with cells which bear lamellae but no cilia (Fig. 13.9B); the lamellae are highly plicate. The epithelial cells contain mitochondria, polyribosome, granular endoplasmic reticulum, glycogen masses and numerous Golgi complexes with vesicles. The vesicles become associated with the external plasma membrane and are possibly secreted (Erasmus 1971). The third region, nearest to the ovary, is similar in structure but is distended for storage of spermatozoa; most authors, including Spence and Silk (1971) and Justine (1980), refer to this region as the seminal receptacle. It is described as an outpocketing by Neves *et al.* (2005). The oviduct then turns anteriorly and after extending well beyond the anterior tip of the ovary discharges into the ootype. The oviduct is considered to be syncytial as each nuclear component is in cytoplasmic continuity with its neighbors (Spencer and Silk 1971).

13.3.4   Stages of Egg Formation

Two stages in egg formation can be recognized in *S. mansoni* (Jurberg *et al.* 2009). In the brief prezygotic stage, single oocytes leave the ovary by projecting their cytoplasm as pseudopods (Fig. 13.12a). Each oocyte is fertilized in the seminal receptacle and develops to the zygotic stage (also named as stage 0, see Chapter 5). In this stage, the zygote passes through the oviduct and the ovovitelline duct, where it is surrounded by approximately 40–45 vitelline cells (vitellocytes). Vitelline cells are smaller than the zygote and present a central, eosinophilic nucleus and azurophilic cytoplasm full of blue-stained yolk granules by Giemsa. In PASH and Masson's trichrome-stained sections, the vitelline cell cytoplasm is slightly eosinophilic with weakly red and green-blue tinge, respectively. The zygote plus the surrounding vitelline cells are pushed by the female muscular contractions to enter the ootype, where the eggshell is formed (Fig. 13.12b). The fully formed egg with its characteristic conspicuous lateral spine is then released into the uterus (Fig. 13.12c). Finally, the egg is released through the gonopore to the external environment, without presenting any cleavage as observed in newly laid eggs *in vitro* (Fig. 13.12d) (Jurberg *et al.* 2009). The subsequent states of cell division and embryogenesis are detailed in Chapters 5 and 15.

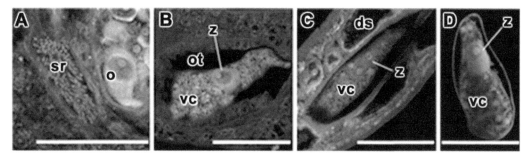

FIG. 13.12 Intrafemale egg development. **A–C.** Confocal laser scanning microscope images of entire females stained with hydrochloric carmine. **A.** Detail of a mature oocyte leaving the ovary by projecting a pseudopod. **B.** Eggshell formation in the ootype (ot). This zygote surrounded by 40 vitelline cells. **C.** Egg in the uterus. This egg has 45 vitelline cells counted in virtual three-dimensional tomographies. **D.** *In vitro* laid egg stained with hydrochloric carmine, laid in culture. Notice no cleavage during intrafemale egg development. Plasma membranes seem to disappear progressively. z, zygote; ds, digestive system; sr, seminal receptacle full of sperm. Scale bars 50 μm. From Jurberg A. *et al.* 2009. Development, Genes and Development 219(5): 219–234.

13.3.5   Mehlis' Gland and the Ootype

The ootype, a dilatation of the oviduct, is composed of cylindrical cells which were identified as Mehlis's gland by Neves *et al.* (2005) (Fig. 13.8D). However, Collins *et al.* (2011) correctly dispute this identification and state that the oviduct (Fig. 13.14), merges with the vitelline duct (now the ovovitelline duct), before reaching a mass of secretory cells, which collectively they refer to as Mehlis' gland. The ootype contains the developing embryo around which an egg shell is secreted. Mehlis' gland is depicted as radiating glands in line drawings by Spence and Silk (1971), Loverde and Chen (1991) and deWalick *et al.* (2012) (Fig. 13.7A, B) and the presence of the radiating canals was clearly demonstrated by Moczon *et al.* (1992) (Fig. 13.13). Mehlis' gland and the ootype are regarded as a syncytium by Spence and Silk (1971) and Moczon *et al.* (1992) but Erasmus (1973) (Fig. 13.9) clearly demonstrates separate epithelial cells of the ootype joined by non-septate desmosomes and unicellular Mehlis' gland cells. The passage of a duct, filled with secretory granules, from a cell of the ootype through the epithelium to the lumen is illustrated in this figure, as in Figure 13.13.

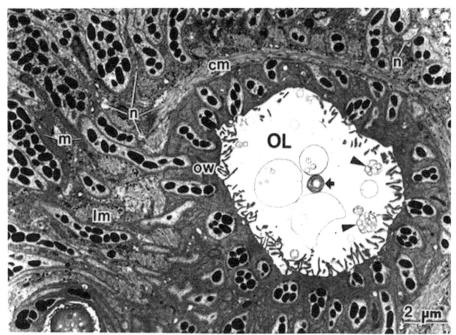

FIG. 13.13 *Schistosoma mansoni*. Section through the proximal region of the ootype. This shows clearly the ducts of Mehlis' gland opening through the endothelial wall into the lumen of the ootype. Elliptical black bodies are secretory granules. cm, circular muscle; lm, longitudinal muscle; m, mitochondrion; ne, nerve fiber; ow, ootype wall. From Moczon T, Swiderski Z, Huggel H. 1992. International Journal for Parasitology 22(1): 65–73. Fig. 6.

The luminal plasma membrane in the anterior and mid-region of the ootype projects as blunt, finger-like folds; these become finer, more leaf-like and microvillous towards the posterior end of the ootype (Fig. 13.9A), particularly where the ducts of Mehlis' gland enter. They are about 1 μm long and some are bifid. The lateral cell membranes are straight except basally where they are infolded. The basal plasma membrane is attached to a fibrous basement layer external to which is a layer of circular and outer longitudinal muscle. The cytoplasm contains a finely granular nucleus with an eccentric nucleolus, mitochondria, a few Golgi complexes associated with small granular vesicles and sparse, basal granular endoplasmic reticulum. Anteriorly, the ootype is continuous with the uterine lining, the junction being marked by a muscular sphincter. The secretory processes of the Mehlis' gland cells are holocrine and are filled with secretory bodies; in cells at this stage most of the cytoplasmic organelles have disappeared. The long slender necks form ducts, each supported by about 40 microtubules, running between the cells of the ootype. Septate desmosomes join the ducts to the surrounding ootype cells. The cells appear to be radially arranged around the ootype (Fig. 13.9A) (Erasmus 1973).

The mature secretory granules of Mehlis' gland exhibit a very compact, pseudocrystalline structure and stain strongly for polysaccharides and glycoproteins. The largest granules found in the canals of the gland measure approximately 1000 × 400 nm (Fig. 13.13). They are embedded in the apical cytoplasm in which mitochondria, β-glycogen particles and periodate-unreactive vesicular structures. The canals of the gland pass between the bundles of the longitudinal and circular muscles, which surround the ootype and they penetrate the wall of the proximal ootype to open into the ootype lumen (Fig. 13.13). Groups of closely packed, unmyelinated axons are present between the canals in the vicinity of the ootype. The axoplasm of the axons contains large neurosecretory granules (100–150 nm in diameter), numerous α-glycogen particles and parallel arrays of microtubules. The lumen bears straight and bifurcated microvilli the surfaces of which are lined by a periodate-reactive substance, presumably a glycoconjugate. The mature secretory granules of the Mehlis' gland are secreted into the lumen of the proximal ootype together with the apical cytoplasm. When they reach the ootype lumen, they assume the appearance of Golgi multivesicular bodies loaded with lipoprotein particles. They now measure ca 600 nm in diameter and their contents are periodate-unreactive. In the ootype lumen the multivesicular bodies expand and decompose (Moczon *et al.* 1992) (Fig. 13.13).

A variety of functions has been proposed for Mehlis' gland, including providing lubrication for the reproductive tract, activating sperm and providing materials for egg shell biosynthesis (references in Collins *et al.* 2011). These authors observed anti-acetylated α-tubulin to label microtubules of the canals before they entered the ootype. Within the ootype the eggshell coalesces to surround a single fertilized egg and 30 to 40 vitelline cells. This egg, containing the zygote, is passed anteriorly through the uterus (see Chapter 5).

Egg biosynthesis is examined by Moczon *et al.* 1992, deWalick *et al.* (2012) and authors there referenced (see also Chapter 5).

13.3.6   Uterus

The ootype opens into the long uterus, with a cylindrical epithelium, into which the fully formed eggs are released (Fig. 13.8E). In *S. bovis* it contains numerous fusiform eggs and opens at the posterior of the ventral sucker through the female genital pore. Spermatozoa are observed in the uterus, ootype, oviduct and seminal receptacle (Justine 1980). In *S. mansoni*, unlike *S. bovis*, only one to a few eggs is observed in the uterus at a time. The uterus again opens at the genital pore located shortly below the ventral sucker (Fig. 13.8F) (e.g., Neves *et al.* 2005). Despite the paucity of eggs in the uterus, *S. mansoni* sheds approximately 100–300 eggs per day (Moore and Sandground 1956; Southgate and Bray 2003). The output of other species is summarized in Table 13.1.

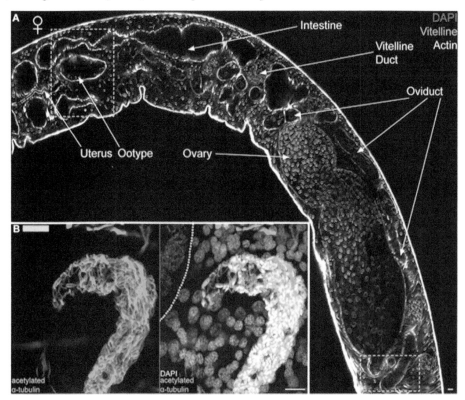

FIG. 13.14 *Schistosoma mansoni.* The female reproductive system. **A.** Top, female labeled with DAPI. Autofluorescence from the vitelline cells is shown in green. **B.** Left, anti-acetylated a-tubulin labeling microtubules of sperm present in the seminal receptacle. Right, large nuclei of oocytes in the ovary can be seen stained with DAPI to the left. Dashed line represents the position of the muscle layer surrounding the ovary. Anterior faces left in panel A and up in panels B. From Collins JJ, King RS, Cogswell A, Williams DL, Newmark PA. 2011 PLoS Neglected Tropical Diseases 5(3): e1009. Fig. 8.

Knowledge of early development of schistosomes has been aided recently by detailed transcriptomic analyses of the female and male reproductive tissues (Gobert *et al.* 2009; Nawaratna *et al.* 2011). Gonadal tissues of *S. japonicum* and *S. mansoni* were isolated from parasites using a laser microdissection method, coupled with microarray analyses. More recently, a combination of detergent and enzymatic digestion of tissues has been used to purify viable ovaries from female *S. mansoni* (Hahnel *et al.* 2013). This intriguing method will allow more detailed analyses of oocyte function and may open new protocols for studies in schistosome transgenesis (see Chapter 5).

The transcriptomics surveys of Gobert, Nawaratna and colleages (Gobert *et al.* 2009; Nawaratna and others 2011) revealed distinct functions for the ovary and vitelline cells. Oocytes are transcriptionally active and transcripts encoding molecules with signaling function, regulatory and cell cycle function and receptors for sperm docking were abundant. Abundant transcripts in vitelline cells included those encoding egg-shell precursors and signaling molecules (Nawaratna *et al.* 2011). Some gene transcripts enriched in the reproductive tissues had no clear homology with molecules in other organisms, drawing attention to the need for continuing studies on oogenesis and reproductive biology of these parasites. One such molecule is a recently characterized hem transporter, found to be enriched in the ovary of schistosomes (Toh *et al.* 2015). Suppression of the transporter, either by transcriptional suppression or drug treatment, led to reduced egg production indicating an important role of this molecule and the hem it transports, in maintaining the high fecundity of schistosomes.

13.4 RECEPTORS AND SIGNALING MOLECULES REGULATING SCHISTOSOMES REPRODUCTION

Recent research has given insight into regulatory control of reproduction and cell migration of gametes within the reproductive tract. Taman and Ribeiro (2011) have shown that the glutamate receptor SmGluR is present along the length of the female reproductive system, including the oviduct, ootype and the uterus. Protein kinases C (PKCs) and extracellular signal-regulated kinases/mitogen-activated protein kinases (ERKs/MAPKs), enzymes involved in a range of cellular processes are expressed in the vitellaria, ootype and Mehlis' gland (Ressurreição *et al.* 2014).

13.5 CHAPTER SUMMARY

Schistosomes and some didymozoids are unique in the digenetic Trematoda in having separate sexes. In *Schistosoma* the male holds the female in the gynaeocophoric groove. Pairing of the sexes is essential for normal development and reproduction of the female and factors affecting this are described. The male genital system consists of a row of testes (or testicular lobes) attached to the vas efferens, which continues as the vas deferens. This joins a seminal vesicle which discharges through an ejaculatory duct and an eversible cirrus at the male genital pore. The female system consists of a large ovary which discharges oocytes via the oviduct to an expansion, the ootype, which also receives the long duct from the vitelline glands; the short common duct, entering the ootype, is the ovovitelline duct. The ootype receives the ducts of unicellular glands making up Mehlis'gland a major function of which appears to be provision of materials for eggshell biosynthesis. The oocytes are fertilized by spermatozoa in the seminal receptacle, a proximal expansion of the oviduct. Vitelline cells surround the fertilized ovum and also provide materials for eggshell biosynthesis, the ootype acting as an egg mold. The ootype discharges into the wide uterus which opens at the female genital pore shortly below the ventral sucker. The vitelline cells will provide yolk to the developing miracidium after egg laying. The histology, including ultrastructure of the various organs is described.

13.6 LITERATURE CITED

Awad AH, Probert AJ. 1990. Scanning and transmission electron microscopy of the female reproductive system of *Schistosoma margrebowiei* Le Roux, 1933. Journal of Helminthology 64(3): 181–192.

Bloch EH. 1980. *In vivo* microscopy of schistosomiasis. II. Migration of *Schistosoma mansoni* in the lungs, liver and intestine. American Journal of Tropical Medicine and Hygiene 29(1): 62–70.

Boissier J, Mone H, Mitta G, Bargues MD, Molyneux D, Mas-Coma S. 2015. Schistosomiasis reaches Europe. The Lancet Infectious Diseases 15(7): 757–758. doi: 10.1016/S1473-3099(15)00084-5.

Collins JJ, King RS, Cogswell A, Williams DL, Newmark PA. 2011. An atlas for *Schistosoma mansoni* organs and life cycle stages using cell type-specific markers and confocal microscopy. PLoS Neglected Tropical Diseases 5(3): e1009.

deWalick S, Tielens AGM, van Hellemond JJ. 2012. *Schistosoma mansoni*: the egg, biosynthesis of the shell and interaction with the host. Experimental Parasitology 132: 7–13.

Erasmus DA. 1973. A comparative study of the reproductive system of mature, immature and "unisexual" female *Schistosoma mansoni*. Parasitology 67: 165–183.

Erasmus DA. 1975. *Schistosoma mansoni* development of the vitelline cell its role in drug sequestration and changes induced by astiban. Experimental Parasitology 38(2): 240–256.

Erasmus DA. 1987. The adult schistosome: structure and reproductive biology, from genes to latrines. pp. 51–82. *In*: Rollinson D, Simpson AJG, editors. *The Biology of Schistosomes*. London: Academic Press.

Gobert GN, McManus DP, Nawaratna S, Moertel L, Mulvenna J, Jones MK. 2009. Tissue specific profiling of females of *Schistosoma japonicum* by integrated laser microdissection microscopy and microarray analysis. PLoS Neglected Tropical Diseases 3: e469.

Gönnert R. 1955. Schistosomiasis-studien. II. Über die eibildung bei *Schistosoma mansoni* und das schicksal der eier im wirtsorganismus. Zeitschrift für Tropenmedizin und Parasitologie 6: 33–52.

Hahnel S, Lu Z, Wilson RA, Grevelding CG, Quack T. 2013 Whole organ isolation approach as a basis for tissue-specific analyses in *Schistosoma mansoni*. PLoS Neglected Tropical Diseases 7(7): e2336.

Jones MK, McManus DP, Sivadorai P, Glanfield A, Moertel L, Belli SI, Gobert GN. 2007. Tracking the fate of iron in early development of human blood flukes. International Journal of Biochemistry and Cell Biology 39(9): 1646–1658.

Jurberg A, Gonçalves T, Costa T, de Mattos A, Pascarelli B, de Manso P, ibeiro-Alves M, Pelajo-Machado M, Peralta J, Coelho P and others. 2009. The embryonic development of *Schistosoma mansoni* eggs: proposal for a new staging system. Development, Genes and Development 219(5): 219–234.

Justine J-L. 1980. Étude ultrastructurale de la gametogénèse de *Schistosoma bovis* Sonsino, 1876 (Trematoda: Schistosomatidae) [Thèse de Troisième Cycle]. France: Université des Sciences et Techniques du Languedoc (Montpellier II). 87 + 19 plates.

Khampoosa P, Jones MK, Lovas EM, Srisawangwong T, Laha T, Piratae S, Thammasiri C, Suwannatrai A, Sripanidkulchai B, Eursitthichai V and others. 2012. Light and electron microscopy observations of embryogenesis and egg development in the human liver fluke, *Opisthorchis viverrini* (Platyhelminthes, Digenea). Parasitology Research 61: 799–808.

Lindner E. 1914. Über die spermatogenese von *Schistosoma haematobium* Bilh. (*Bilharzia haematobia* Cobb.) mit besonderer Berüksichtungen der Geschlechtschromosomen. Archiv für Zellfoschung 12: 516–538.

Loverde PT, Chen L. 1991. Schistosome female reproductive development. Parasitology Today 7(11): 303–308.

Moczon T, Swiderski Z, Huggel H. 1992. *Schistosoma mansoni*: the chemical nature of the secretions produced by the Mehlis' gland and ootype as revealed by cytochemical studies. International Journal for Parasitology 22(1): 65–73.

Moné H, Boissier J. 2004. Sexual biology of schistosomes. Advances in Parasitology 57: 89–189.

Moore DV, Sandground DH. 1956. The relative egg producing capacity of *Schistosoma mansoni* and *Schistosoma japonicum*. American Journal of Tropical Medicine and Hygiene 5(5): 831–840.

Nawaratna S, McManus DP, Moertel L, Gobert GN, Jones MK. 2011. Gene atlasing of digestive and reproductive tissues in *Schistosoma mansoni*. PLoS Neglected Tropical Diseases 5(4): e1043.

Neves RH, de Lamare Biolchini C, Machado-Silva JR, Carvalho JJ, Branquinho TB, Lenzi HL, Hulstijn M, Gomes DC. 2005. A new description of the reproductive system of *Schistosoma mansoni* (Trematoda: Schistosomatidae) analyzed by confocal laser scanning microscopy. Parasitology Research 95(1): 43–49.

Nollen PM, Floyd RD, Kolzow RG, Deter DL. 1976. The timing of reproductive cell development and movement in *Schistosoma mansoni*, *S. japonicum* and *S. haematobium*, using techniques of autoradiography and transplantation. Journal of Parasitology 62(2): 227–231.

Otubanjo OA. 1980. *Schistosoma mansoni*: the ultrastructure of the ducts of the male reproductive system. Parasitology 81(Pt 3): 565–571.

Pearce EJ, Huang SC. 2015. The metabolic control of schistosome egg production. Cellular Microbiology 17(6): 796–801.

Ressurreicao M, De Saram P, Kirk RS, Rollinson D, Emery AM, Page NM, Davies AJ, Walker AJ. 2014. Protein kinase C and extracellular signal-regulated kinase regulate movement, attachment, pairing and egg release in *Schistosoma mansoni*. PLoS Neglected Tropical Diseases 8(6): e2924.

Ribeiro-Paes JT, Rodrigues V. 1997. Sex determination and female reproductive development in the genus *Schistosoma*: a review. Revista do Instituto de Medicina Tropical de Sao Paulo 39(6): 337–344.

Schübler P, Potters E, Winnen R, Bottke W, Kunz W. 1995. An isoform of ferritin as a component of protein yolk platelets in *Schistosoma mansoni*. Molecular Reproduction and Development 41: 325–330.

Shaw JR, Erasmus DA. 1981. *Schistosoma mansoni*: an examination of the reproductive status of females from single sex infections. Parasitology 82(1): 121–124.

Silva-Leitao FW, Biolchini CL, Neves RH, Machado-Silva JR. 2009. Development of *Schistosoma mansoni* in the laboratory rat analyzed by light and confocal laser scanning microscopy. Experimental Parasitology 123(4): 292–295.

Smyth JD. 1966. *The Physiology of Trematodes*. Oliver and Boyd. Edinburgh and London. p. 256.

Smyth JD, Clegg JA. 1959. Egg-shell formation in trematodes and cestodes. Experimental Parasitology 8: 286–323.

Southgate VR, Jourdane J, Tchuente LA. 1998. Recent studies on the reproductive biology of the schistosomes and their relevance to speciation in the Digenea. International Journal for Parasitology 28(8): 1159–1172.

Southgate V, Bray R. 2003. Medical helminthology. pp. 1649–1716. *In*: Cook G, Zumla A, editors. *Manson's Tropical Diseases*. London: Saunders Elsevier.

Spence IM, Silk MH. 1971. Ultrastructural studies of the blood fluke-*Schistosoma mansoni*. V. The female reproductive system – a preliminary report. The South African Journal of Medical Sciences 36(3): 41–50.

Taman A, Ribeiro P. 2011. Glutamate-mediated signaling in *Schistosoma mansoni*: a novel glutamate receptor is expressed in neurons and the female reproductive tract. Molecular and Biochemical Parasitology 176(1): 42–50.

Tchuenté LA, Imbert-Establet D, Southgate VR, Jourdane J. 1994. Interspecific stimulation of parthenogenesis in *Schistosoma intercalatum* and *S. mansoni*. Journal of Helminthology 68(20): 167–73.

Toh SQ, Gobert GN, Martínez DM, Jones MK. 2015. Haem uptake is essential for egg production in the haematophagous blood fluke of humans, *Schistosoma mansoni*. FEBS Journal 282: 3632–3646.

Webster BL, Diaw OT, Seye MM, Webster JP, Rollinson D. 2013. Introgressive hybridization of *Schistosoma haematobium* group species in Senegal: species barrier break down between ruminant and human schistosomes. PLoS Neglected Tropical Diseases 7(4): e2110.

Wood LM, Bacha WJ, Jr. 1983. Distribution of eggs and the host response in chickens infected with *Austrobilharzia variglandis* (Trematoda). Journal of Parasitology 69(4): 682–688.

CHAPTER

Spermatozoa, Spermatogenesis and Fertilization in *Schistosoma*

Barrie GM Jamieson[1],* and Jean-Lou Justine[2]

14.1 INTRODUCTION

The spermatozoa of schistosomes are exceptional in the Digenea in having a single axoneme (two in other Digenea) which extends as a free flagellum (incorporated in others); in possessing a single diffuse central structure in the 9 + "1" axoneme (instead of the highly structured central element otherwise constant for the 9 + "1" axonemes normal in Digenea and other Trepaxonemata); in the short inverted conical (rather than filiform) condition of the nucleus and sperm body; location of more or less discrete mitochondria anterior to the nucleus (rather than filiform, fused and parallel to it); retention of a striated rootlet (rather than loss during spermiogenesis); and retention of a single centriole consisting of nine triplets (rather than two centrioles with nine doublets or singlets) (Justine 1980, 1982, 1985, 1991; Justine and Mattei 1981; Justine *et al.* 1993).

These characteristics of schistosome sperm have been described, in whole or in part, for *Schistosoma mansoni* (Kitajima *et al.* 1976; Irie *et al.* 1983; Otubanjo 1980, 1981; Spence and Silk 1981); *S. bovis* (Justine 1980, 1982, 1985; Justine and Mattei 1981); *S. haematobium* (Erasmus 1987); *S. margrebowiei* (Awad and Probert 1989) and *S. japonicum* (Zhou *et al.* 1993; Yang *et al.* 1998, 2003). In a comparative account, Justine *et al.* (1993), described sperm ultrastructure for six schistosome species. The species investigated were *S. curassoni* (Dakar, Senegal), *S. rodhaini* (Bujumbura, Burundi), *S. intercalatum* (Franceville, Gabon) which had not previously been studied, *S. bovis* (Tambacounda, Senegal), *S. margrebowiei* (Lochinvar National Park, Zambia) and *S. mansoni* (Medinha, Saudi Arabia). Triplet centrioles were confirmed for *S. bovis* and *S. margrebowiei*; in all species the features of schistosome sperm listed above were found to be constant (Fig. 14.1A–Q).

A detailed account of sperm ultrastructure is given in Section 14.3, below. We will first deal with the process of spermatogenesis.

---

[1] Department of Zoology and Entomology, School of Biological Sciences, University of Queensland, Brisbane, Qld 4072, Australia.

[2] ISYEB, Institut de Systématique, Évolution, Biodiversité (UMR7205 CNRS, EPHE, MNHN, UPMC), Muséum National d'Histoire Naturelle, Sorbonne Universités, CP 51, 55 rue Buffon, 75231 Paris cedex 05, France.

* Corresponding author

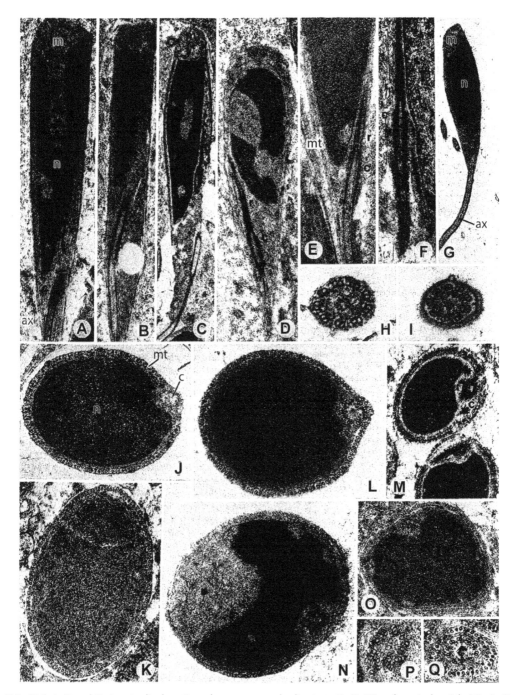

FIG. 14.1 A–D and **G.** Longitudinal section of mature sperm bodies in testis (**A–D**) and seminal vesicle (**G**); **A., B.** *S. rodhaini*, x 14100; **C.** *S. margrebowiei*, x 9400; **D.** *S. curassoni*, x 19500; **G.** *S. bovis*, x 13000. **E–F** Basal part of axoneme in mature sperm (**E**) and spermatid (**F**) with associated rootlet; **E.** *S. curassoni*, x 18800; **F.** *S. margrebowiei*, x 18800. **H–I.** Transverse sections of 9 + "1" (but non-trepaxonematan) flagella in mature sperm; dynein arms are absent; **H.** *S. rodhaini*, x 72300; **I.** *S. bovis*, x 3800. **J–O.** Transverse sections of mature sperm bodies. **J.** *S. mansoni*, x 29000; **K.** *S. intercalatum*, x 29000; **L.** *S. bovis*, x 37600; **M.** Section with centriole, *S. margrebowiei*, x 28900; **N.** Section with centriole, *S. bovis*, x 28200; **O.** *S. curassoni*, x 28900. **P–Q** Transverse sections of basal part of sperm body. **P.** Centriole, *S. bovis*, x 54200; **Q.** base of axoneme and encircling microtubules, *S. curassoni*, x 37600. c, centriole; m, mitochondria; r, striated root. After Justine JL, Jamieson BGM, Southgate VR. 1993. Annales de Parasitologie Humaine et Comparée 68(4): 185–187. Fig. 1.

14.2 SPERMATOGENESIS

14.2.1 Spermatogonia

With regard to chromosome complements and karyotypes, the pioneering studies of earlier workers (e.g., Lindner 1914; Severinghaus 1928; Niyamasena 1940) have largely been superseded by more recent studies. Thus, the chromosomes of schistosomes, including *Schistosoma, Schistosomatium* and *Heterobilharzia* have been reviewed and illustrated by Short (1983). More recently Hirai (2014) has further dealt with chromosome differentiation in schistosomes. Other works are beyond the scope of this chapter. The basic [plesiomorphic] diploid number of chromosomes is deduced to be 16, as in *Schistosoma,* which has seven pairs of autosomes and one pair of heteromorphic sex chromosomes. Primary spermatocytes, spermatids and spermatozoa therefore have 7 autosomes and either the Z or the W sex chromosome. The sex chromosomes are arbitrarily named Z and W, as in birds and paralleling the latter, the female is heterogametic, with ZW and the male homogametic, with ZZ. For further references see Basch (1991), Mahmoud (2001), Secor and Colley (2005).

Schistosoma bovis is not a human parasite but the spermatozoon is similar to that of *S. mansoni* and there is no reason to believe that spermatogenesis in the two species differs in any important respects. The comprehensive account of Justine (1980) is therefore reviewed here. For a detailed comparison with other platyhelminths and other animal groups the reader is referred to that work.

In *Schistosoma mansoni,* spermatogonia are found in the peripheral zone of the testis and scattered in its interior. Resting nuclei are usually rounded. The chromatin is in the form of small bundles distributed in a network. There is one, rarely two nucleoli, 5–6 µm in diameter (Niyamasena 1940).

The spermatogonia of *S. bovis* (Fig. 14.2) are small cells (4.5 µm) with a central nucleus (3.3 µm). They are few in number and are dispersed through the testis in most cases but sometimes they form an important part of the testis and are grouped in rows. Their form is regular and no intercellular bridges are evident. The chromatin is clumped against the nuclear membrane and the remainder of the nucleoplasm is hyaline. The nucleolus is large and very dense. The nuclear membrane is dilated in places; its outer leaf is deformed towards the cytoplasm. Concentric membranes of a myeloid appearance are present in the clear space so formed. Numerous ribosomes, often as polysomes, are present.

A small amount of rough endoplasmic reticulum is present. Clear vesicles 0.3 to 1 µm in diameter contain concentric membranes. Two centrioles are situated close to the nucleus. The mitochondria of this stage are characteristic: few in number, small, elongate (0.08 µm × 0.4 µm), with a dark matrix with numerous, often longitudinal cristae. It is not possible to distinguish primary, secondary or tertiary spermatogonia (Justine 1980, 1982).

14.2.2 Primary Spermatocytes

In many animal groups, including mammals (Bruslé 1970) and oligochaetes (Jamieson 1981), the spermatocytes are grouped around and connected to a central mass, the cytophore. This condition is seen in platyhelminths in Turbellaria, Monogenea, Cestoda and Digenea including the closely related genus *Schistosomatium* (references in Justine 1980; Basch 1991) but has never been seen in *Schistosoma.*

In *S. mansoni,* the primary spermatocytes, lying in the interior of the testis, are 7–9 um in diameter. The nucleus is larger and paler than in the spermatogonia. Pachytene stages are particularly frequent, indicating that this is a prolonged stage (Niyamasena 1940).

In *S. bovis,* primary spermatocytes exceed 4.5 µm to 7 µm and more in length during meiotic prophase, when the size of the nucleus also increases. Initially the cells are rounded with central nucleus but become progressively more irregular with an eccentric nucleus which comes to almost lie against the nuclear membrane in a cytoplasmic zone rich in organelles. The nucleocytoplasmic ratio diminishes.

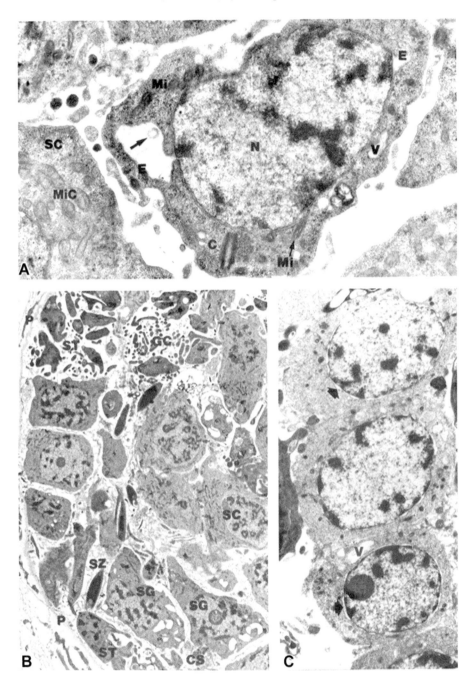

FIG. 14.2 *Schistosoma bovis*. Spermatogonia. **A.** The spermatogonium is a cell with a voluminous nucleus (N) and the nuclear membrane has expansions (E) which delimit clear spaces with concentric membranes (arrow). The small amount of cytoplasm contains vesicles (V), some mitochondria (MI) and a centriole (C). Compare the mitochondria of spermatogonia (Mi) with dark matrix and few longitudinal cristae with the abundant mitochondria (MIC) of the neighboring spermatocyte (SC), with clear matrix and transverse cristae. x 13000. **B.** General view of the wall of the testis (P), showing the successive stages of spermatogenesis. Spermatogonia (SG), primary spermatocytes (SC), spermatids (ST), spermatozoa (SZ) and their cytoplasmic droplets (GC). There are no rosette arrangements and the stages are mixed. Sustentacular cells (CS) envelope the extensions of all germinal cells. x 3600. **C.** Group of four spermatogonia. The perinuclear space is dilated (broad arrow) and vacuoles (V) are present in the cytoplasm. x 8700. After Justine J-L. 1980. Thèse de Troisième Cycle. France: Université des Sciences et Techniques du Languedoc (Montpellier II). Fig. 4.

Mitochondria, with a clear matrix, are numerous, elongated, with many transverse cristae. With a size of 0.4 µm × 1 µm, they are larger than those of the spermatogonia. Endoplasmic reticulum is at first limited in amount but progressively increases and the Golgi apparatus appears and contains well developed dictyosomes. Ribosomes are numerous. Centrioles numbering two or four are grouped in diplosomes. The internal microtubule (tubule a) of each peripheral triplet is provided with an arm. Between this microtubule and microtubule c of the neighboring triplet there is a fine connection. At the center of the circle of triplets there is a tubular structure with a diameter of ca 30 nm.

The nucleus is characteristic, with synaptonemal complexes of meiotic prophase. When fully developed these have two rather dense bands (lateral arms), 40 nm wide, surrounding a clear band 110 nm wide. This has a central dense central band 15 nm wide. The length of the synaptonemal complex can reach 1 µm.

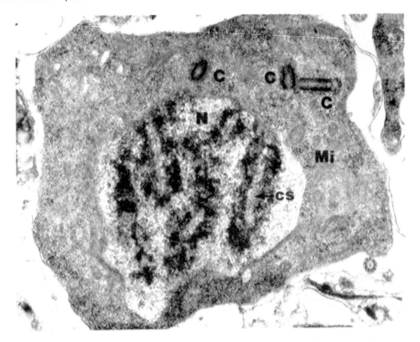

FIG. 14.3 *Schistosoma bovis.* Small type of spermatocyte. The nucleus (N) shows very dense chromosomes, containing synaptonemal complexes (CS). The section passes through three centrioles (C). There are few mitochondria (MI). After Justine J-L. 1980. Thèse de Troisième Cycle. France: Université des Sciences et Techniques du Languedoc (Montpellier II). Fig. 5a. x 9300.

There seem to be two types of cells which contain synaptonemal complexes. The first are small, nearing a spermatogonium, at 4.5 µm, with a nucleus which has clearly visible, very dense chromosomes (Fig. 14.3). The mitochondria appear intermediate in appearance between those of spermatogonia and those described above. The second type of cell is large (7 µm), with an eccentric nucleus. Cytoplasm is abundant and mitochondria are clearly developed and numerous. Golgi is well developed. The nucleus is pale, the nucleoplasm finely granular and contains long, not dense, chromosomes at which stage the nucleus is more regular than previously when the nuclear envelope undulated.

Some spermatocytes are in contact at certain regions of their membranes. Other are provided with thin cytoplasmic extensions which join them to the neighboring spermatocyte. Some extensions possibly originate from sustentacular cells. Nuclear pores are more numerous than in spermatogonia and are grouped. Secondary spermatocytes have not been distinguished.

14.3   SPERMIOGENESIS

The young spermatids are often found in groups dispersed throughout the testis. The form is rounded, as is the nucleus (Fig. 14.4A). The cytoplasm contains mitochondria, numerous free ribosomes a little rough endoplasmic reticulum and a centriole one end of which is in contact with the plasma membrane. Stages of spermiogenesis are shown diagrammatically in Figure 14.5.

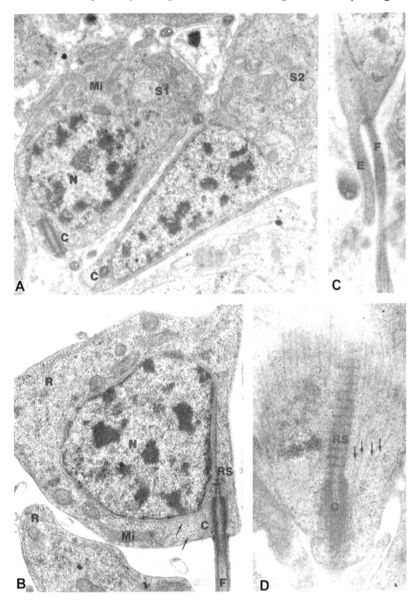

FIG. 14.4 *Schistosoma bovis.* Young spermatids. **A.** Spermatid (S1) with a round nucleus. The cytoplasm with mitochondria (Mi) surrounds the nucleus (N). The centriole (C) is pressed against the nuclear membrane. An older spermatid (S2) has a conical nucleus. x 12900. **B.** Later spermatid. The cytoplasm still abundant around the nucleus (N) which is becoming triangular in section (conical). The reticulum (R) is parallel to the plasma membrane. The centriole (C) has a striated rootlet and gives rise to a flagellum (F). Some microtubules (arrows) arise from the centriole and surround the nucleus. x 1700. **C.** In some sections a small transient cytoplasmic expansion (E) parallels the flagellum. x 17400. **D.** The centriole of the spermatid (C) is at the origin of the striated rootlet (RS) and numerous microtubules (arrows) which come to surround the nucleus longitudinally. x 32100. After Justine (1980). Thèse de Troisième Cycle. France: Université des Sciences et Techniques du Languedoc (Montpellier II). Fig. 7.

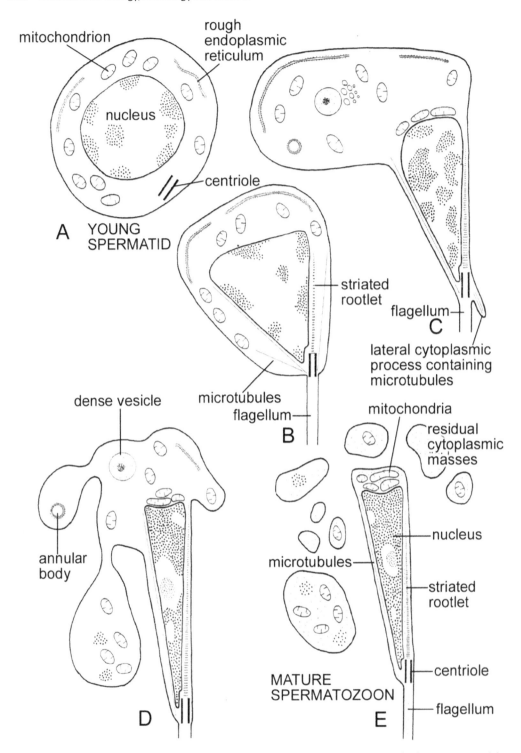

FIG. 14.5 *Schistosoma bovis.* Spermiogenesis. **A.** Young round spermatid. **B.** Spermatid adopting a conical form; the flagellum, striated rootlet and cortical microtubules are in place. **C.** Elongation of the nucleus and appearance of a transitory cytoplasmic expansion. **D.** Dislocation of the cytoplasm and shedding of cytoplasmic droplet; elongation of the nucleus and condensation of the chromatin. **E.** Mature spermatozoon surrounded by cytoplasmic fragments. Adapted from Justine J-L. 1980. Thèse de Troisième Cycle. France: Université des Sciences et Techniques du Languedoc (Montpellier II). 87 + 19 plates. Figs. A-G.

The spermatids of *S. bovis* rapidly take on a conical form, about 4 μm wide (Fig. 14.4B, S2, 14.5B). The nucleus, in the form of an inverted cone the wide base of which is at the future anterior end of the sperm, measures 2.5 to 3 μm in width. The nucleoplasm is clear with some dark chromatin masses. The nuclear cone is indented at the level of the centriole. Some microtubules emanate from this, radiating around the nucleus. The flagellum originates from the centriole. At its base a striated rootlet arises and intrudes between the nucleus and the plasma membrane. The cytoplasm at the base of the nuclear cone contains ribosomes and rough endoplasmic reticulum which form a long ribbon which runs parallel to the plasma membrane. Mitochondria are abundant, elongate (0.4×0.2 μm), with a clear matrix and numerous cristae. Some are adpressed longitudinally to the nuclear envelope at the wide end of the nuclear cone.

The nucleus becomes more closely invested by the plasma membrane (Fig. 14.5C) and the cytoplasmic organelles are rejected at the base of the nuclear cone, only the centriole and the microtubules remaining in place at the point of the cone and parallel with this. The nucleus elongates, while retaining its conical form. Its wide end is directed towards the cytoplasm and constitutes the anterior pole of the spermatid. Its length reaches 5 to 6 μm, with a basal width of 1 μm. During this elongation its chromatin condenses as large very dense strands, interrupted in places by clear patches. The flagellum elongates and at its base a small probably ephemeral cytoplasmic expansion, surrounded by longitudinal microtubules, parallels it. The striated rootlet grows and maintains the same length as the nucleus. Its width at the base is ca 100 nm, though subsequently becoming narrower. It appears to be composed of piled dense discs 20 nm thick, separated by clear spaces of 40 nm. The period is about 60 nm. At maximum development it adds a fine transverse striation (13 nm) and a longitudinal striation. The perinuclear microtubules also elongate but do not surpass the length of the nucleus. At the greatest width of the nucleus there are 100 microtubules. Each is connected to the plasma membrane by a fine perpendicular line. The organelles of the anterior cytoplasm disappear. Some clear vesicles appear and tend to unite as a single vesicle about 0.6 wide with a dense center surrounded by a zone less dense to electrons. Electron-dense granules arise and unite in a mass; they are Thiéry-positive, indicating β-glycogen.

The chromatin progressively condenses (Fig. 14.5D) and becomes electron-dense with some clear patches. EDTA testing confirms that the dense regions of the nucleus consist of DNA. The clear zones show two components: a circular electron-dense body ca 0.3 μm in diameter and an irregular, less dense zone. The cytoplasmic mass deforms and comes to lie along the length of the nucleus.

A narrow annular body, sometimes forming an open U is eliminated with the cytoplasmic droplet (Fig. 14.5D). After separation (Fig. 14.5E) the droplet takes on a complex form with digitations which intermix with the cytoplasmic droplets of other sperm. They contain Thiéry-positive α-glycogen granules ca 60 nm in diameter which group in a mass. Sparse quantities of 3 nm β-glycogen granules are also present.

Throughout the process of spermiogenesis the spermatids are enveloped in processes of the sustentacular cells (Justine 1980).

Transformation of spermatids to spermatozoa is described by Lindner (1914) for *S. haematobium* and Faust and Meleney (1924) and Severinghaus (1928) for *S. japonicum*.

Brief reference may be made to points discussed in greater detail by Justine (1980) for *S. bovis*. The sperm of most platyhelminths have two flagella, unlike schistosomes. Furthermore, restriction to a single centriole is rare in the Animalia. The short, ephemeral cytoplasmic expansion parallel to the flagellum in *Schistosoma* may be the equivalent of a median expansion between the two flagella of other platyhelminths. An intercentriolar body between the two centrioles in other platyhelminths is not seen in *Schistosoma*. Migration of the mitochondria to the anterior end of the nucleus also contrasts diametrically with the usual situation where they become grouped around the centriole as the midpiece. *Schistosoma* accords with other platyhelminths in the presence of cortical microtubules, a striated rootlet, failure of an acrosome to develop and possibly the small expansion near the flagellum comparable with the median expansion common in the group.

Worms cultured *in vivo* also produce spermatozoa in *S. haematobium* (Smith *et al.* 1976), *S. japonicum* (Yasuraoka *et al.* 1978) and *S. mansoni* (Clegg 1959; Irie *et al.* 1983).

14.4 ULTRASTRUCTURE OF THE SPERMATOZOON

14.4.1 Overview

Sperm are present in the testis and seminal vesicle. Numbers are usually small (Kitajima *et al.* 1976; Justine 1980, 1985; Justine and Mattei 1981; Justine *et al.* 1993) but Irie *et al.* (1983) reported many mature sperm associated with sustentacular cells in *Schistosoma mansoni* cultured from cercariae. In all cases, the number of spermatozoa in schistosomes is much lower than in other digeneans.

Under the phase contrast light microscope, *S. mansoni* sperm demonstrate a bulbous head 8 μm by 2 μm, with a rounded anterior end and a tapering posterior part, from which emerges a motile, relatively short tail, ca 20 μm long. Some clear patches are noticeable in the head (Fig. 14.6) (Kitajima *et al.* 1976). In *S. japonicum*, the head measures 6.2 × 1.4 μm in average size (Yang *et al.* 2003) while in *S. bovis* it measure 5.5 μm long and the width passes from 1.5 μm to 0.4 μm, at the point near the centriole, with nuclear dimensions of 4 μm by 1 μm (Justine 1980).

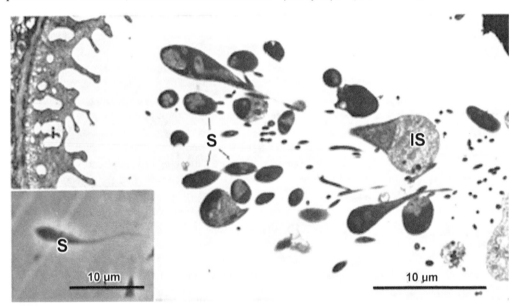

FIG. 14.6 *Schistosoma mansoni.* A group of mature (S) and immature (IS) sperm in the seminal vesicle of a male, which is lined by the integument (i). Insert shows a mature sperm as seen under phase contrast light microscope. Relabeled after Kitajima EW, Paraense WL, Correa LR. 1976. Journal of Parasitology 62(2): 215–221. Fig. 1.

The general ultrastructure of the sperm of *Schistosoma curassoni, S. rodhaini, S. intercalatum, S. bovis* (Justine 1980) and *S. margrebowiei* is similar to that of *S. mansoni* (Justine *et al.* 1993) and *S. japonicum* (Kitajima *et al.* 1976; Yang *et al.* 2003). No acrosome or Golgi-derived vesicle is present in the sperm head. The report (Zhou *et al.* 1993) of an acrosome-like structure in the spermatid has not been confirmed. Poorly distinguishable mitochondria are present at the anterior part of the sperm head, within a large indentation of the nucleus. They are however, clearly visible by acid staining or permanganate fixation (Kitajima *et al.* 1976; Justine 1980) and have numerous cristae (Justine 1980). The nucleus is electron-dense, except for some electron-transparent lacunae of various sizes which are not membrane-bound (Kitajima *et al.* 1976; see also Justine 1980; Justine *et al.* 1993; Yang *et al.* 2003). The more electron-dense areas of the nucleus of *S. bovis* sperm stain for DNA (Justine 1980). Mature sperm found in the oviduct of the female *S. mansoni*, near the ovary, are similar in morphology to those found in the male organs (Erasmus 1973; Kitajima *et al.* 1976).

For *S. bovis*, Thiéry–positive granules have been demonstrated around the mitochondria, the length of the nucleus and at the level of the centriole and also in the flagellum. Their size of ca 3 nm corresponds with β-glycogen (Justine 1980).

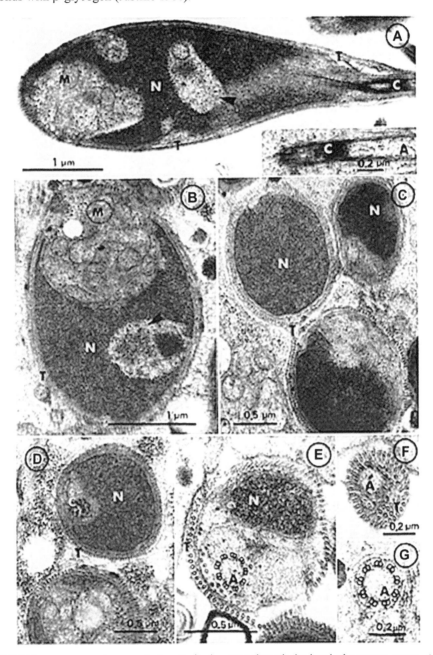

FIG. 14.7 *Schistosoma mansoni* sperm. **A.** Longitudinal section through the head of a mature sperm. Acrosome lacking and undifferentiated mitochondria (M) accumulate at anterior part of sperm head. Single row of microtubules (T) appears just beneath the plasma lemma. Dense nucleus (N) presents some electron-transparent patches (arrow). Insert shows detail of the centriole (C) and initial part of axial complex of the flagellum (A) in which absence of the central pair of tubules is noticeable. **B-G.** Transverse sections of the sperm at different levels, from anterior part of head (Fig. **B**) to sperm tail (Fig. **G**). The peripheral and longitudinal array of the microtubules (T) and an apparent 9 + 0 pattern of the flagellar axoneme (A) is evident. Fig. **F** shows the terminal part of the head where the peripheral microtubules surround the axoneme. M, mitochondrion, N, nucleus. Relabeled after Kitajima EW, Paraense WL, Correa LR. 1976. The Journal of Parasitology 62(2): 215–221. Figs. 2–8.

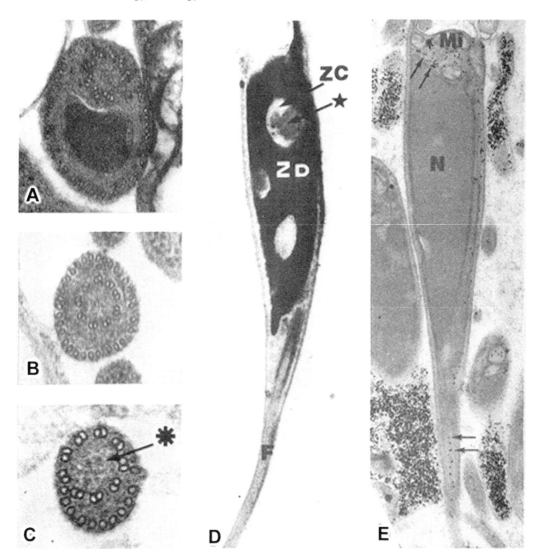

FIG. 14.8 *Schistosoma bovis.* **A.** Section through the centriole and nucleus. The nucleus has a groove in which the centriole, with 9 triplets, lodges. Some longitudinal microtubules are present under the plasma membrane. x 60000. **B.** Section just behind the centriole. The axoneme has nine doublets. Submembranar microtubules are present around the whole periphery. x 66000. **C.** Section at the departure of the flagellum. The eccentric axoneme is flanked laterally by cytoplasm garnished with submembranar microtubules. The center of the axoneme has a dense mass. x 111000. **D.** Longitudinal section of a spermatozoon. The sperm body contains the nucleus and a single flagellum (F). The nucleus has a large dense zone (ZD) interrupted by clear spaces (ZC). Glutaraldehyde fixation. x 18000. **E.** Longitudinal section of a spermatozoon. The body of the sperm contains the nucleus (N) and in an anterior position, the mitochondria (Mi). Some glycogen (arrows) is present around the mitochondria, the length of the nucleus and in the flagellum. Thiéry staining. X 17000. Relabeled after Justine J-L. 1980. Thèse de Troisième Cycle. France: Université des Sciences et Techniques du Languedoc (Montpellier II). Figs. 11 d-h.

The nucleus occupies most of the head, except for the anterior pocket containing the mitochondrial mass and the posterior region where the flagellar apparatus is posterolaterally inserted (Justine 1980; Justine and Mattei 1981; Justine *et al.* 1993; Kitajima *et al.* 1976). A single layer of microtubules, upto 100, is situated beneath the plasma lemma (Figs. 14.7 A–G), extending longitudinally along the major axis of the sperm head, from the very anterior end to the initial portion of the flagellum. Where the flagellum emerges from the sperm head; it is accompanied by an anteriorly directed striated rootlet emanating from the centriole and extending to the anterior end of the cell; each microtubule has a connection to the plasma membrane (Justine 1980; Justine and Mattei 1981; Justine *et al.* 1993) but the rootlet is said to be absent in *S. japonicum* (Yang *et al.* 2003). The axial complex (axoneme) was described by Kitajima *et al.* (1976) as being of the 9 + 0 type, lacking the central pair of the tubules (Fig. 14.7F, G). However, it has since been shown (Justine 1980; Justine and Mattei 1981, for *S. bovis*), particularly by the technique of rotation, that the axoneme in transverse section has a central non-tubular element, resulting in what they have termed a special 9 + "1" configuration distinct from the 9 + 1 general trepaxonematan condition (see details below). The 9 + "1" configuration was confirmed for *S. japonicum* for most of the length of the axoneme but the posterior part was reported to lack a central element (Yang *et al.* 2003).

In *S. mansoni*, dense, ribosome-like granules are occasionally seen near the mitochondria and the centriole (Fig. 14.7A, F). The microtubular sheath surrounds the sperm head and the anterior "mitochondrial cap". The microtubules overlap the axonemal complex at the zone where the flagellum emerges from the sperm head (Kitajima *et al.* 1976).

14.4.2   The Schistosome Sperm Flagellum

At the risk of some repetition, the detailed studies of the flagellum by Justine (1980) and Justine and Mattei (1981) requires special mention.

Within the Platyhelminthes various authors described a peculiar type of axonemal structure in the spermatozoon, termed 9 + "1". The general, almost ubiquitous structure of axonemes is designated as 9 + 2, in which 9 refers to the number of peripheral doublets and 2 to the number of central microtubules; the 9 + "1" terminology uses quotes to indicate that the "1" is not a microtubule but instead a special structure (Justine and Mattei 1981). The central core of the trepaxonematan 9 + "1" axoneme is not labelled by antitubulin antibodies (Iomini *et al.* 1995). This differentiates it from the 9 + 1 structure found in very rare cases, in which there is a single central microtubule (Justine and Mattei 1988). Ehlers (2005) named this 9 + "1" axonemal structure the "trepaxonematan" pattern (trepaxo- refers to the helical structure of the central core) and considered that the trepaxonematan axoneme in spermatozoa was a key characteristic of the thus named Trepaxonemata, a group which represents the majority of the phylum Platyhelminthes, including the free-living triclads and polyclads and the main parasitic groups, i.e. the digeneans, monogeneans and cestodes.

Among the digeneans, the schistosomes are exceptional in not having in their sperm the trepaxonematan 9 + "1" structure (with the characteristic trepaxonematan core) but, instead, a *special* 9 + "1" structure in which the central element is different. Hundreds of species of digeneans have been examined for sperm ultrastructure (e.g., Justine 1995; Bakhoum *et al.* 2015); all have 9 + "1" axonemes except the didymozoid *Didymozoon* which has two 9 + 0 axonemes (i.e. axonemes with 9 peripheral doublets and no central structure; Justine and Mattei 1983, 1984). Kitajima *et al.* (1976) designated the structure they found in *Schistosoma mansoni* as 9 + 0 but the structure in *Schistosoma* species (*S. bovis* and *S. mansoni*, Justine and Mattei 1981; *S. rodhaini*, Justine *et al.* 1993) is the same in all species, i.e. an aberrant 9 + "1" type.

The sperm of *S. bovis* presents the same structure that has already been described for *S. mansoni* (Kitajima *et al.* 1976). The cytoplasm appears uniformly electron-dense and there is a single centriole with nine triplets (Fig. 14.9A) which is associated with a striated rootlet. A cross section of the flagellum at its exit from the body of the sperm, just at the base of centriole, reveals the axoneme with nine doublets and a set of microtubules situated against the flagellar membrane (Fig. 14.9B, C). These microtubules represent the ends of longitudinal microtubules situated under the plasma membrane throughout the length of the sperm. Slightly posterior to this area microtubules are absent from the flagellum and only doublets are present along its entire length (Fig. 14.9D, E). These peripheral doublets are totally devoid of dynein arms. Tenuous connections exist between neighboring doublets and between doublets and the flagellar membrane (Fig. 14.9D). Superposition by rotation confirms these characteristics (Fig. 14.9E).

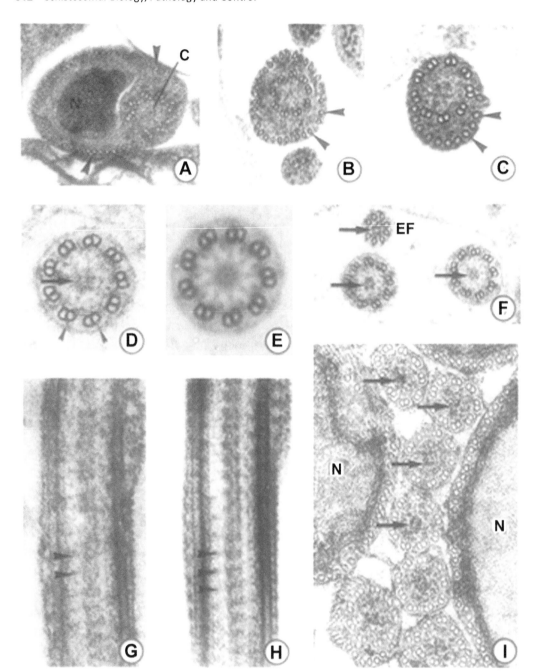

FIG. 14.9 *Schistosoma bovis.* **A.** Transverse section (TS) of the basal region of a spermatozoon. C, Centriole. N, Nucleus; the arrows indicate the submembranar microtubules. x 63600. **B.** TS of the flagellum at its exit from the body of the spermatozoon. Around the axoneme are the extremities of the submembranar microtubules (arrows). x 60100. **C.** TS flagellum posterior to the previous. Only some submembranar microtubules are still visible (arrows). x 86500. **D.** TS of the spermatozoal flagellum. The doublets lack arms. An electron opaque element is present at the center of the axoneme (arrows). The points of the arrows indicate a connection between the doublet and the flagellar membrane. x 109200. **E.** Rotation clearly demonstrates the particulars mentioned in D but the central element is an artefact of rotation. x 127400. **F.** TS of flagella, showing the axial element (arrows). This element is also present at the tip of the flagellum (EF). x 56400. **G.** TS flagellum, showing the periodicity at the level of the axial element and rays (arrows). x 100000. **H.** The technique of translation shows clearly the intraflagellar periodicity (arrows). x 100000. **I.** TS spermatozoa in the seminal receptacle of the female. N, nucleus. Arrows indicate the central element. x 77400. Relabeled after Justine J-L, Mattei x. 1981. Journal of Ultrastructure Research 76: 89–95. Figs. 1–9.

No microtubule is visible in the central region of the axoneme but, contrary to what is usually seen in the classical 9 + 0 flagellum, there is electron-dense material at its center. This structure is ill-defined but non-tubular with a maximum diameter in the order of 50 nm. It is present in sperm flagella in the testes and seminal vesicle in the male and in the seminal receptacle of the female (Fig. 14.9I). In longitudinal section it shows a periodic structure consisting of a succession of elements 50 nm wide and 30 nm thick (Figs. 14.9G, 14.10A), with a periodicity of 42 nm. This was confirmed by the technique of image building by translation (Fig. 14.9H). It is connected to the peripheral doublets by rays that run the length of the flagellum with the same periodicity. At the distal extremity of the flagellum, the doublets become nine singlets but the central structure persists (Fig. 14.9F). This 9 + "1" configuration of the axoneme and the absence of dynein arms correlate with feeble motility of the flagellum, contrasting with the active motility of the 9 + 2 flagella of the protonephridia. The sperm nevertheless are immotile in physiological saline (Justine 1980). The sperm of *S. mansoni* also show limited mobility (Kitajima *et al.* 1978).

PTAAg staining reveals particles of β-glycogen within the axoneme (Fig. 14.10B) at levels corresponding to the axial elements. In contrast, flagella of the excretory system of *S. bovis* have the 9 + 2 configuration classical in Platyhelminthes, irrespective of the structure of the sperm flagella. Cross sections of sperm flagella of *S. mansoni* reveal a structure similar to that in *S. bovis*.

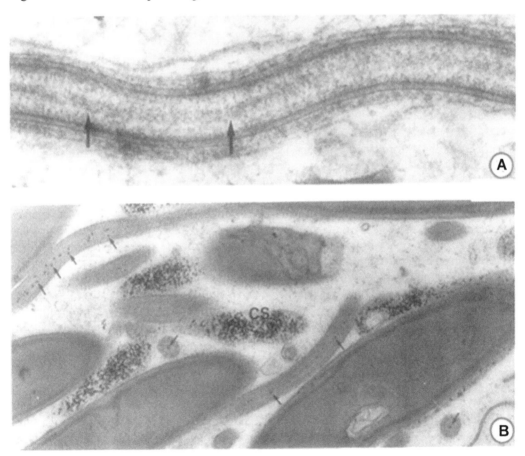

FIG. 14.10 *Schistosoma bovis*. **A.** Longitudinal section of a sperm flagellum. Arrows indicate the axial element. x 82300. **B.** Sections of spermatozoa in the testis, with Thiéry staining. CS, Cytoplasmic extension of the Sertoli cell, very rich in glycogen. Arrows indicate glycogen granules in the axoneme. x 21500. Relabeled after Justine J-L, Mattei X. 1981. Journal of Ultrastructure Research 76: 89–95. Figs. 10–11.

14.4.3 Comparison with other Digenea

Justine (1991) has compared spermatogenesis in *Schistosoma* with that of other digeneans (Table 14.1).

TABLE 14.1 Comparison between usual digenean spermatogenesis and that of *Schistosoma* (adapted from Justine 1991).

Other Digeneans	Schistosomes
Elongation of spermatid	No elongation
Migration of centriole to proximal part of dense zone of nucleus	No migration of single centriole
Lengthening of a median cytoplasmic process	Cytoplasmic process aborts
Great lengthening of two free flagella (up to 400 μm)	Moderate lengthening of single flagellum (20 μm)
Lengthening of nucleus	Nucleus remains pyriform (invert cone)
Lengthening of mitochondria	Mitochondria remain ovoid
Migration of mitochondria posterior to centrioles	Mitochondria remain anterior to nucleus and centrioles
Striated rootlets disappear	Striated rootlet retained
Centrioles simplify from 9 triplets to 9 singlets	Centriole keeps 9 triplets
Two 9 + "1" axonemes in mature sperm (sometimes 9 + 0 in spermatid)	Single axoneme with 9 + 0 configuration in immature sperm. Special 9 + "1" in mature sperm

Most digeneans (and most neodermatans, i.e. the parasitic Platyhelminthes, including digeneans, monogeneans and cestodes) have a filiform sperm morphology. Justine (1982, 1991) considered that the outstanding mature cone-shaped spermatozoon of schistosomes correspond to the immature spermatid of the other digeneans. The spermatid was considered to precociously attain maturity and its spermiogenesis was therefore considered to be case of progenesis (Justine 1991) or neoteny (Justine 1982).

The question is whether derivation of schistosomes from the general pattern of filiform neodermatan spermatozoa can be interpreted as a correlation with phylogeny or with biology. Correlation with phylogeny is questionable, since the closest relatives to the schistosomes, i.e. the other blood flukes such as Aporocotylidae (sometimes designated Spirorchidae) have a normal filiform pattern with two axonemes (Justine 1995). Correlation with the biology of fertilization is a more convincing hypothesis. Adult schistosomes are united as pairs (see Chapter 13), thus limiting sperm competition between spermatozoa from different males; absence of sperm competition allows abandonment of the characteristics of competing sperm in other digeneans, such as extreme length and active movement (see, however, section 14.3.4, below). Hundreds of species of neodermatans have been studied for sperm ultrastructure; it is striking that important derivations from the general structure have been found only in digeneans of the family Didymozoidae and in the monogenean *Diplozoon*. In the Didymozoidae, the two members of a couple are permanently together within a cyst in the flesh of their fish host; derivations from the general pattern include the very short spermatozoa of *Gonapodasmius* (Justine and Mattei 1982) and the spermatozoa with 9 + 0 axonemes in *Didymozoon* (Justine and Mattei 1983, 1984). In the monogenean *Diplozoon*, the two hermaphroditic members of the couple are permanently paired and the genital ducts are in continuity; sperm structure in *Diplozoon* is highly derived and very different from that of other polyopisthocotylean monogeneans, with the loss of the axonemes (Justine *et al.* 1985).

Although sperm structures in didymozoids, *Diplozoon* and schistosomes are highly different, these cases illustrate the drastic effect of reduction of sperm competition on sperm morphology in Platyhelminthes.

Although fertilization biology will always dictate sperm morphology, phylogenetic patterns are evident in sperm morphology of many animal groups. Of a few examples (Jamieson 1987, 1991a,b, 1999, 2007; Jamieson *et al.* 1999), from many which might be given, perhaps the most striking is the demonstration from sperm ultrastructure that the Pentastomida should be included in the Crustacea (Storch and Jamieson 1992). Furthermore, the distinctive sperm type of *Schistosoma* indicates phylogenetic relationship between its species.

14.4.4   Male-Male Competition

Steinauer (2009) appears to have convincingly demonstrated (Fig. 14.11) that a male may acquire a female in his gynaecophoric groove by pulling the female from another couple in coitus. This is here considered to have the implication that if there are sperm of the primary male in the seminal receptacle of the abducted female, there may be subsequent competition between the spermatozoa of the primary male and that of the new partner.

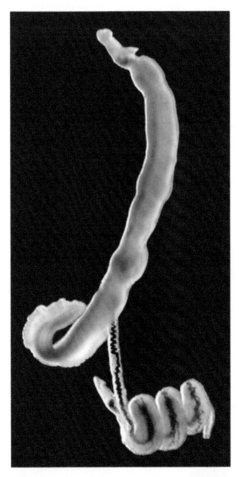

FIG. 14.11 *Schistosoma mansoni.* Photograph of putative male–male competition *in vitro.* A single male (bottom) is actively pulling a female out of the gynaecophoric groove of her primary mate (top) by grasping her with his gynaecophoric groove and coiling and twisting. From Steinauer ML. 2009. International Journal for Parasitology 39(10): 1157–1163. Fig. 1.

14.5 FERTILIZATION

Fertilization in *Schistosoma japonicum* has briefly been described by Yang *et al.* (2003) whose account is paraphrased here. Fertilization occurs in the posterior part of the oviduct near the ovary, in which area there are no cilia or lamellae on the wall of the oviduct, although lamellae are observed in the seminal receptacle. Secondary oocytes move to the oviduct and are surrounded by many mature sperm but it appears that only a single spermatozoon penetrates into the oocyte to produce the fertilized ovum. In general, one or several fertilized ova are located in the center of the oviduct lumen, surrounded by dozens or hundreds of sperm as seen in a cross section. The rounded ovum is 10.0×11.9 mm on average. The cytoplasm contains endoplasmic reticulum, glycogen granules and some vacuoles but only a few mitochondria. Some vacuoles fuse and link together, to form a net-like plasma membrane. In the process of fertilization, the main morphological changes can be attributed three stages. Spermatozoa have also been observed lining the wall of the seminal receptacle of the female in a cross section in *S. bovis* (Justine 1980) (Fig. 14.12) and in longitudinal section of the oviduct in *S. mansoni* (Erasmus 1987).

Cortical granules release and degeneration. In early stages of fertilization, there are some intact cortical granules (CG) in the ovum. The CGs are divided into three types: concentric type, partial dense type and entirely dense type. The ultrastructure of these three types is the same as those in secondary oocytes prior to fertilization. However, the total amount of CGs in the fertilized ovum is clearly less than pre-fertilization. Some of the CGs degenerated during fertilization and most of them have disappeared by the late stage of fertilization. Some fertilized ova were observed releasing CGs which ruptured on the surface of the ova with membrane fusion. Other CGs broke down or degenerated in the cytoplasm.

Pronucleus appears in cytoplasm. By the second maturation division, the secondary oocyte finally divides into a second polar body and one female pronucleus. During this period, the male pronucleus also forms. The nuclear membrane and nucleolus disappear. The shapes of both pronuclei are irregular.

Female pronucleus and male pronucleus fuse. The female pronucleus and male pronucleus approach each other, come into contact in their central region, a narrow band-like bridge appears and they finally fuse to form a zygote.

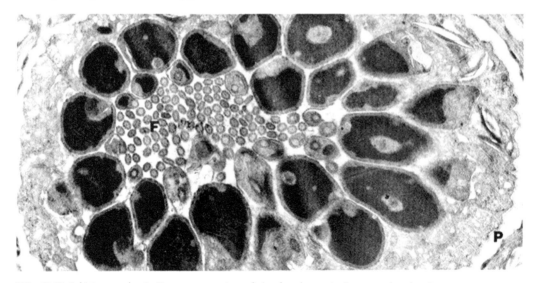

FIG. 14.12 *Schistosoma bovis.* Transverse section of the female seminal receptacle, showing numerous sperm sectioned through the sperm body (head) or the flagella (F) lining the wall (P) of the receptacle. x 1200. After Justine J-L. 1980. Thèse de Troisième Cycle. France: Université des Sciences et Techniques du Languedoc (Montpellier II). Fig. 12a.

14.6 CHAPTER SUMMARY

The spermatozoa of schistosomes are exceptional in the Digenea in having a single axoneme (two in other Digenea) which extends as a free flagellum (incorporated in others); in possessing a single diffuse central structure in the special 9 + "1" axoneme (instead of the highly structured central element otherwise constant for the 9 + "1" axonemes normal in Digenea and other Trepaxonemata); in the short inverted conical (rather than filiform) condition of the nucleus and sperm body; location of more or less discrete mitochondria anterior to the nucleus (rather than filiform, fused and parallel to it); retention of a striated rootlet (rather than loss during spermiogenesis); and retention of a single centriole consisting of nine triplets (rather than two centrioles with nine doublets or singlets). Six species: *S. curassoni*, *S. rodhaini*, *S. margrebowiei*, *S. mansoni* and *S. bovis*, of which only *S. mansoni* and *S. intercalatum* are human parasites, have homogeneous spermatozoa, the characteristics of which are described. The sperm flagella have limited motility. The limited information on fertilization is presented. It is hypothesized that the derived morphology of spermatozoon in schistosomes (compared to other digeneans) is correlated with the reduction of sperm competition due to the intimate pairing of adult schistosomes.

14.7 LITERATURE CITED

Awad AHH, Probert AJ. 1989. Transmission and scanning electron microscopy of the male reproductive system of *Schistosomama rgrebowiei* Le Roux 1933. Journal of Helminthology 63(3): 197–205.

Bakhoum AJS, Quilichini Y, Justine J-L, Bray RA, Bâ CT, Marchand B. 2015. Ultrastructural study of sperm cells in Acanthocolpidae: the case of *Stephanostomum murielae* and *Stephanostomoides tenuis* (Digenea). Peer J 3: e744.

Basch PF. 1991. *Schistosomes: Development, Reproduction and Host Relations*. New York, Oxford: Oxford University Press. 248 p.

Bruslé, J. 1970. Les apports récents de la microscopie électronique à la connaissance de l'ultrastructure des cellules germinales mâles précoces. Symbioses 2: 185–200.

Clegg JA. 1959. Development of sperm by *Schistosoma mansoni* cultured *in vitro*. Bulletin of the Research Council of Israel Section E Experimental Medicine 8E: 1–6.

Erasmus DA. 1973. A comparative study of the reproductive system of mature, immature and "unisexual" female *Schistosoma mansoni*. Parasitology 67: 165–183.

Erasmus DA. 1987. The adult schistosome: structure and reproductive biology. pp. 51–82. *In*: Rollinson D, Simpson AJG, editors. *The Biology of Schistosomes from Genes to Latrines*. London etc.: Academic Press.

Faust EC, Meleney HE. 1924. Studies on schisosomiasis japonica. American Journal of Hygiene Monographic Series 3: 1–99.

Hirai H. 2014. Chromosomal differentiation of schistosomes: what is the message? Frontiers in Genetics 5: 301.

Irie Y, Basch PF, Beach [Basch] N. 1983. Reproductive ultrastructure of adult *Schistosoma mansoni* grown *in vitro*. Journal of Parasitology 69(3): 559–566.

Jamieson BGM. 1981. *The Ultrastructure of the Oligochaeta*. London: Academic Press. 462 p.

Jamieson BGM. 1987. *The Ultrastructure and Phylogeny of Insect Spermatozoa*. Cambridge: Cambridge University Press. 320 p.

Jamieson BGM. 1991a. *Fish Evolution and Systematics: Evidence from Spermatozoa*. Cambridge: Cambridge University Press. 319 p.

Jamieson BGM. 1991b. Ultrastructure and phylogeny of crustacean spermatozoa. Memoirs of the Queensland Museum 31: 109–142.

Jamieson BGM. 1999. Spermatozoal phylogeny of the Vertebrata. pp. 303–331. *In*: Gagnon C, editor. *The Male Gamete: From Basic Science to Clinical Applications*. Vienna (USA): Cache River Press.

Jamieson BGM. 2007. Avian spermatozoa: structure and phylogeny. pp. 349–511. *In*: Jamieson BGM, editor. *Reproductive Biology and Phylogeny of Birds*. Enfield, New Hampshire, NH, Plymouth, UK Science Publishers. Volume 6A.

Jamieson BGM, Dallai R, Afzelius BA. 1999. *Insects: Their Spermatozoa and Phylogeny.* Enfield (NH) USA; Plymouth UK: Science Publishers. 555 p.

Justine J-L. 1980. Étude ultrastructurale de la gametogenèse de *Schistosoma bovis* Sonsino, 1876 (Trematoda: Schistosomatidae) [Thèse de Troisième Cycle]. France: Université des Sciences et Techniques du Languedoc (Montpellier II). 87 + 19 plates. doi: 10.6084/m9.figshare.154986.

Justine J-L. 1982. Étude ultrastructurale de la gametogenèse de *Schistosomabovis* Sonsino, 1876 (Trematoda: Schitosomatidae). Afrique Médicale 21: 287–293.

Justine J-L. 1985. Étude ultrastructurale comparée de la spermiogenèse des Digènes et des Monogènes (Plathelminthes). Relations entre la morphologie du spermatozoide, la biologie de la fecondation et la phylogenie. Thèsed'Etat, University of Montpellier, France. doi: 10.6084/m9.figshare.154987.

Justine J-L. 1991. The spermatozoa of the schistosomes and the concept of progenetic spermiogenesis. pp. 977–979. *In*: Baccetti B, editor. *Comparative Spermatology 20 years after.* New York: Raven Press.

Justine J-L. 1995. Spermatozoal ultrastructure and phylogeny of the parasitic Platyhelminthes. pp. 55–86. *In*: Jamieson BGM, Ausio J, Justine J-L, editors. *Advances in Spermatozoal Phylogeny and Taxonomy.* Paris: Mémoires du Muséum National d'Histoire Naturelle.

Justine J-L, Le Brun N, Mattei X. 1985. The aflagellate spermatozoon of Diplozoon (Platyhelminthes: Monogenea: Polyopisthocotylea). A demonstrative case of relationship between sperm ultrastructure and biology of reproduction. Journal of Ultrastructure Research 92: 47–54.

Justine J-L, Mattei X. 1981. Etude ultrastructurale du flagelle spermatique des schistosomes (Trematoda: Digenea). Journal of Ultrastructure Research 76: 89–95.

Justine J-L, Mattei X. 1983. A spermatozoon with two 9+0 axonemes in a parasitic flatworm, *Didymozoon* (Digenea: Didymozoidae). Journal of Submicroscopic Cytology 15: 1101–1105.

Justine J-L, Mattei X. 1984. Atypical spermiogenesis in a parasitic flatworm, *Didymozoon* (Trematoda: Digenea: Didymozoidae). Journal of Ultrastructure Research 87: 106–111.

Justine J-L, Mattei X. 1988. Ultrastructure of the spermatozoon of the mosquito *Toxorhynchites* (Diptera, Culicidae). Zoologica Scripta 17: 289–291.

Justine J-L, Jamieson BGM, Southgate VR. 1993. Homogeneity of sperm structure in six species of schistosomes (Digenea, Platyhelminthes). Annales de Parasitologie Humaine et Comparée 68(4): 185–187.

Kitajima EW, Paraense WL, Correa LR. 1976. The fine structure of *Schistosoma mansoni* sperm (Trematoda: Digenea). The Journal of Parasitology 62(2): 215–221.

Lindner E. 1914. Über die spermatogenese von *Schistosoma haematobium* Bilh. (*Bilharzia haematobia* Cobb.) mit besonderer Berüksichtungen der Geschlechtschromosomen. Archiv für Zellfoschung 12: 516–538.

Mahmoud AAF. 2001. *Tropical Medicine Science and Practice, Volume 3. Schistosomiasis.* Singapore. London: World Scientific. 526 p.

Niyamasena SG. 1940. Chromosomen und geslchechtbei *Bilharzia mansoni.* Zeitschrift für Parasitenkunde 11: 690–710.

Otubanjo OA. 1980. *Schistosoma mansoni* the ultrastructure of the ducts of the male reproductive system. Parasitology 81(3): 565–572.

Otubanjo OA. 1981. *Schistosoma mansoni*: astiban-induced damage to tegument and the male reproductive system. Experimental Parasitology 52: 161–170.

Secor WE, Colley DG [eds]. 2005. *Schistosomiasis.* Boston, USA: Springer. 235 p.

Severinghaus AE. 1928. Sex studies on *Schistosoma japonicum.* Quarterly Journal of Microscopical Science 71(4): 653–702.

Short RB. 1983. Sex and the single schistosome. Journal of Parasitology 69: 4–22.

Smith M, Clegg JA, Webb G. 1976. Culture of *Schistosoma haematobium in vivo* and *in vitro.* Annals of Tropical Medicine and Parasitology 70: 101–107.

Spence IM, Silk MH. 1981. Ultrastructural studies of the blood fluke-*Schistosoma mansoni* – a preliminary report. South African Journal of Medical Science 36: 41–50.

Steinauer ML. 2009. The sex lives of parasites: investigating the mating system and mechanisms of sexual selection of the human pathogen *Schistosoma mansoni.* International journal for parasitology 39(10): 1157–1163.

Storch V, Jamieson BGM. 1992. Further spermatological evidence for including the Pentastomida (Tongue worms) in the Crustacea. International Journal for Parasitology 22: 95–108.

Yang M-Y, Dong H-F, Jiang M-S. 2003. Ultrastructural observation of spermatozoa and fertilization in *Schistosoma japonicum*. Acta tropica 85(1): 63–70.

Yang M, Jiang M, Li Y, Dong H, Zhou S. 1998. Ultrastructure of *Schistosoma japonicum* sperm. Chinese Journal of Parasitology and Parasitic Diseases 16(4): 264–268. (In Chinese).

Yasuraoka K, Irie Y, Hata H. 1978. Conversion of schistosome cercariae to schistosomula in serum-supplemented media and subsequent culture *in vitro*. Japanese Journal of Experimental Medicine 48: 53–60.

Zhou SL, Yang MX, Li Y, Kun CH, Liang HL, Fang P, Rei SL. 1993. Ultrastructural study on spermatogenesis and sustentacular cell of the testes of *Schistosoma japonicum*. Journal of Parasitology and Parasitic Diseases 11(1): 50–52. (In Chinese).

CHAPTER

Ova and Oogenesis in *Schistosoma*

Christoph G Grevelding*, Steffen Hahnel and *Zhigang Lu*

15.1 INTRODUCTION

15.1.1 The Male-Female Interaction Leads to Egg Production

First reports about the remarkable reproductive biology of schistosomes date back about 90 years showing these blood-inhabiting endoparasites to be the only trematodes that have evolved separate sexes (Cort 1921; Severinghaus 1928; Basch 1991; Platt and Brooks 1997). Cort (1921) observed the extreme sexual dimorphism that exists in the adult stage, whereas earlier stages in the life cycle exhibit no sexual dimorphism (Jamieson and Haas, Chapter 6; Yoshino *et al.*, Chapter 7; Haas and Jamieson, Chapter 8). Later Severinghaus (1928) discovered by a series of final host infections with different sets of cercariae, the life stage of schistosomes infectious for final hosts (Ross and Yuesheng, Chapter 3), that unpaired females not only fail to develop normal body size and form but also lack fully developed reproductive organs.

Following copulation in the liver, adult male and female schistosomes migrate as couples to their final destination in the host body. Depending on the species the couples settle within the mesenteric veins of the gut (e.g., *Schistosoma mansoni*, *S. japonicum*) or the venous plexus surrounding the bladder (*S. haematobium*). Here the couples produce hundreds (e.g., *S. mansoni*) to thousands of eggs (e.g., *S. japonicum*) each day (Cheever *et al.* 1994), of which about 50% reach the gut lumen (most species) or the bladder (*S. haematobium*) (Moore and Sandground 1956). The remaining eggs tramp through the vasculature to liver and spleen, where they become trapped and induce the pathologic consequences of schistosomiasis (Ross *et al.* 2002; Yuesheng and Ross, Chapter 17; Olveda and Ross, Chapter 18). The structure of the egg and its embryological development are treated in Jones *et al.*, Chapter 5.

Although lipophilic substances (Eveland and Haseeb 1986) and male body size (Steinauer 2009) have been considered to be important for mate attraction and pairing, the secrets of partner finding are not yet disclosed as are the mechanisms leading to female sexual maturation.

15.1.2 The Male Governs Female Sexual Maturation

The most spectacular biological feature of schistosomes is the male-dependent sexual maturation of the schistosome female. As a consequence of a constant pairing contact, mitoses and differentiation processes are induced leading to the development of ovary and vitellarium, which are the main reproductive organs of the female (Popiel and Basch 1984; Den Hollander and Erasmus 1985;

Institute for Parasitology, Biomedical Research Center, Justus-Liebig-University, 35392 Gießen, Germany.
* Corresponding author

Erasmus 1987; Basch 1991; Kunz 2001; Knobloch *et al.* 2002a; Jones *et al.*, Chapter 13). Cells provided by these organs, oocytes and vitelline cells, are needed for egg formation. Upon pairing, spermatozoa from the male are received by the female at its genital pore and migrate via the uterus down to the seminal receptacle (Gönnert 1955; Spence and Silk 1971a; Basch 1991; Nollen 1997; Jones *et al.*, Chapter 13; Jamieson and Justine, Chapter 14). Fertilization occurs at the seminal receptacle, an enlarged posterior part of the oviduct. The oviduct connects the ovary with the ootype to transport oocytes, which complete meiosis upon fertilization. The ootype represents the egg-forming organ and receives oocytes via the oviduct but also vitelline cells delivered via the vitelloduct. Egg formation finally includes the amalgamation of the zygote with the cellular content of 30–40 vitelline cells per egg. The latter have a "nursing" character providing energy resources for the embryo and egg-shell precursor proteins needed for egg-shell formation (Spence and Silk 1971a; Basch 1991; Nollen 1997; deWalick *et al.* 2012). This process of egg-shell synthesis is supported by secretion products of the Mehlis' gland, a group of secretory cells which are also connected to the ootype (Spence and Silk 1971b; Nollen 1997; Jones *et al.*, Chapter 13).

Without pairing schistosome females remain in a virgin-like, sexually immature status (Shaw 1987; Basch 1991). Furthermore, pairing is not only necessary for the initiation of the sexual development of the female, but also for the maintenance of its mature state. Pairing-induced developmental changes are reversible since egg production stops and the female regresses to an immature state following separation from its partner. These processes can be re-initiated upon re-pairing, which includes the control of transcriptional activities of female-specifically expressed genes with functions in the gonads (Clough 1981; Popiel *et al.* 1984; Grevelding *et al.* 1997; Knobloch *et al.* 2002a; Hoffmann 2004). Many experimental approaches have been performed to uncover the underlying principles. However, their nature still remains unknown although in contrast to earlier work (Armstrong 1965; Popiel *et al.* 1984) recent studies favor the idea that not one key factor of the male but several elements including Src kinases and TGFβ pathway-associated and neural processes may contribute to the male-induced sexual development of the female (Hoffmann 2004; Knobloch *et al.* 2007; Beckmann *et al.* 2010; Buro *et al.* 2013; Leutner *et al.* 2013; Lu *et al.* submitted).

15.2 THE OVARY, ONE OBJECTIVE OF PAIRING-DEPENDENT PROCESSES

15.2.1 Differentiation Processes in the Female Gonads

Besides the vitellarium, the ovary is central to pairing-induced developmental processes in females. Ontogenetically, oocytes and vitelline cells have arisen from common precursor cells (Kunz 2001). Selection forces towards the need for mass production of robust eggs on the one side and energy supply for the developing embryo and the evolving miracidium on the other hand may have influenced the development of more specialized subtypes of gonad cells in the female (Adiyodi and Adiyodi 1988). Thus the vitelline cells produce lipids, yolk and egg-shell precursor proteins while the oocytes harbor the genetic material for sexual reproduction and maintain the property for conducting meiosis. Studies in the past indicated that unpaired females showed some degree of variation with respect to vitelline cells in worms 30–200 days post-infection; however, in general a low level of differentiation and turnover of vitelline cells was observed. In contrast, the ovary was always found in an immature stage in unpaired females (Severinghaus 1928; Erasmus 1973; Shaw and Erasmus 1981). Therefore, it was concluded that the role of the male may not necessarily be to initiate sexual maturation of the female upon pairing but to increase the rate of vitelline cell differentiation and to co-ordinate the final development of the entire reproductive system (Shaw and Erasmus 1981). The latter includes the ovary, which is located in the anterior part of the female body. In unpaired, immature females, the ovary is small and bean-shaped. In the mature, paired female the size of the ovary has considerably increased and it appears as a bulb-shaped, twisted organ in *Schistosoma mansoni* as in *S. haematobium* and *S. Bovis* whereas in *S. japonicum* it is oval. With respect to structure and cell content, the ovary of an unpaired female looks more homogeneous, whereas remarkable differences are observed in the ovary following pairing. In this situation, two parts can be distinguished: the smaller anterior part containing stem cell-like oogonia while the larger posterior part contains mature, primary oocytes. There is a kind of line of demarcation that clearly separates both parts. Structurally, this is accompanied by a twist within the bulb-shaped ovary which in a constriction-like manner separates the small part from the large part. Novel methods such as carmine-red staining combined with confocal laser scanning microscopy (CLSM) or gonad isolation (Hahnel *et al.* 2013) followed by microscopy have allowed the easy visualization of the morphological differences among and within oocytes (Neves *et al.* 2005; Beckmann *et al.* 2010; Fig. 15.1).

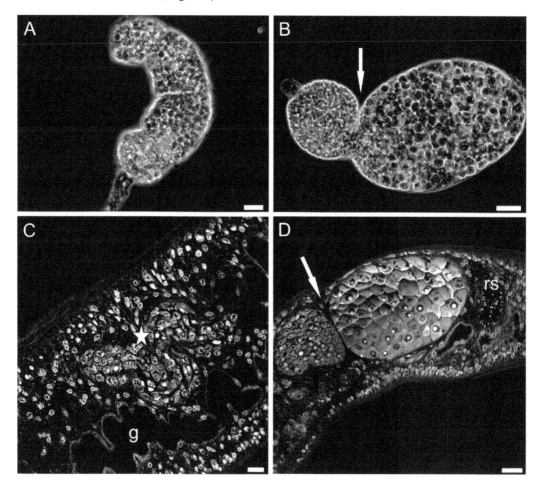

FIG. 15.1 Microscopic analyses (**A**, **B**: phase contrast; **C**, **D**: CLSM) of ovaries that were obtained by the organ-isolation procedure (**A**, **B;** Hahnel *et al.* 2013) or focused on in an unpaired and a paired female (**C**, **D**). **A.** Ovary isolated from an unpaired female. This ovary has a length of approximately 100 µm. **B.** Ovary isolated from a paired female with a length of approximately 200 µm. The smaller part (left; about 60 µm) of the ovary contains stem cell-like oogonia, while the larger part (right, about 140 µm) contains mature, primary oocytes. Both parts are separated by a constriction within the anterior third of the ovary (see arrow), which results from a structural twist in this area. **C.** Ovary (convoluted structure, marked by a star) within an unpaired female; g, gut. **D.** Ovary within a paired female; rs, receptaculum seminis (filled with some sperm). Scale bars: A, 10 µm; B, 20 µm; C, 10 µm; D, 20 µm.

15.2.2   Common and Uncommon Signaling Molecules Control Oocyte Differentiation

In eukaryotes proliferation and differentiation events are controlled by complex signaling networks. Based on studies on the development of model organisms such as *Drosophila* or *Caenorhabditis elegans*, signal transduction processes were investigated also in schistosomes already in the pre-genomic era. This was possible due to the high degree of functional but also sequence conservation of especially such signaling molecules that govern important biological processes such asmitosis and differentiation (Krauss 2014). Phylogenetic analyzes of these molecules, which in many cases represent enzymes, exhibited a remarkable degree of sequence conservation, particularly within their catalytic domains. Among these were protein tyrosine kinases (PTKs) and protein serine/threonine kinases (PS/TKs), key regulatory molecules for cell proliferation and differentiation processes. PTKs and PS/TKs possess highly conserved kinase domains, which catalyze the phosphorylation of tyrosine or serine/threonine residues in target molecules. The superfamily of PTKs and PS/TKs comprises membrane-spanning receptor tyrosine kinases (RTKs) or receptor serine/threonine kinases (RS/TKs) and cytoplasmic tyrosine or serine/threonine kinases (CTKs, CS/TKs) that often directly interact in signaling cascades (Hubbard and Till 2000; Tatosyan and Mizenina 2000; Bromann *et al.* 2004).

Based on their sequence similarity representatives of both families could be successfully cloned and subsequently characterized by reverse genetic approaches in schistosomes. Besides gene structural analyses, gene expression studies were performed including localization approaches, the identification of interaction partners by yeast-two-hybrid (Y2H) and/or co-immunoprecipitation studies, knock-down approaches by RNA interference (RNAi) and inhibitor studies *in vitro*. Furthermore, the functional conservation of schistosome genes coding for signaling proteins was also shown by replacing endogenous signaling proteins with their schistosome orthologs in yeast or mammalian cell-culture systems (Knobloch *et al.* 2007; Freitas *et al.* 2007; LoVerde *et al.* 2009; Beckmann *et al.* 2010; Swierczewski and Davies 2010; You *et al.* 2010; Andrade *et al.* 2014; Dissous *et al.* 2014; Walker *et al.* 2014). Along with ongoing genome/transcriptome sequencing projects, further evidence has been obtained for the existence of conserved signaling molecules in schistosomes (Verjovski-Almeida *et al.* 2003; Hu *et al.* 2004; Berriman *et al.* 2009; Gobert *et al.* 2009; *Schistosoma japonicum* Genome Sequencing and Functional Analysis Consortium 2009; Andrade *et al.* 2011; Nawaratna *et al.* 2011; Protasio *et al.* 2012; Young *et al.* 2012; Stroehlein *et al.* 2015).

Although the majority of studies dealing with the characterization of kinases in schistosomes have been performed with *S. mansoni*, conclusions from the results obtained may also hold true for the closely related species. With respect to the ovary, signaling molecules were discovered that fulfill additional roles in other tissues such as integrins (Beckmann *et al.* 2012), RTKs of the epidermal growth factor (EGFR) family (Shoemaker *et al.* 1992; Buro, personal communication), transforming growth factor β (TGFβ) receptors (LoVerde *et al.* 2007), the fibroblast growth factors receptors SmFGFR-A and SmFGFR-B (Collins *et al.* 2013; Hahnel *et al.* 2014; Lu *et al.* submitted), G protein-coupled receptors (GPCRs; Wang *et al.* 2011; Hahnel *et al.* 2014), the calcium channel ORAI (Strange *et al.* 2007; Buro *et al.* 2013; Hogan *et al.* 2015), polo kinases (PLKs; Dissous *et al.* 2011), Abl kinases (Beckmann *et al.* 2010c) or the Src kinase SmTK3 (Kapp *et al.* 2004) and the Src/Fyn-like kinase SmTK5 (Kapp *et al.* 2001) (Fig. 15.2). However, ovary-specifically expressed signaling molecules were also detected in females such as the Syk kinase SmTK4 (Knobloch *et al.* 2002b; Beckmann *et al.* 2010b), the Src/Abl hybrid kinase SmTK6 (Beckmann *et al.* 2011) and the RTK-like venus kinase receptors (VKRs), SmVKR1 and SmVKR2 (Gouignard *et al.* 2012). SmVKR1 was originally identified by a classical degenerate primer RT-PCR approach in the context of identifying insulin receptors (IRs) in *S. mansoni* (Vicogne *et al.* 2003). Indeed, sequencing of cloned amplification products revealed a molecule with a high degree of similarity to RTKs of the IR class, especially with respect to its intracellular TK domain. Unexpected, however, was the finding of a Venus FlyTrap (VFT) module in its extracellular part. VFTs had been identified before in GPCRs of the C class (Morill *et al.* 2015). Thus VKRs represent hybrid RTKs that are closely related to IRs as well as GABA-receptors. Later studies also detected VKRs in other species and phylogenetic analyses revealed their singularity. This justified their classification as a new RTK family and it was shown that it occurs exclusively in invertebrates from at least five phyla of the Bilateria branch (Platyhelminthes, Arthropoda, Annelida, Mollusca, Echinodermata) as well as in Cnidaria (Ahier *et al.* 2009; Dissous 2015).

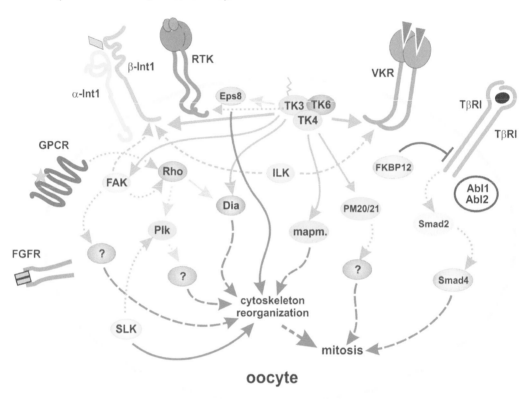

FIG. 15.2 Extended schematic model of signaling molecules and their interactions on the basis of a previous review for *S. mansoni* (Beckmann *et al.* 2010a) about signaling molecules and cascades in oocytes regulating differentiation-associated processes such as cytoskeleton reorganization and mitotic activity [continuous lines: direct interactions confirmed by YTH analyses and/or co-immunoprecipitation; dotted lines: direct interactions known from other organisms, not proven for schistosomes yet; grey dashed lines: indirect interactions via adapter molecules known from other organisms; green dashed lines: activating function; red lines: inhibitory function; brown circle, localized Abl kinases from *S. mansoni*, interactions yet unknown]. For further details and explanation of acronyms, see text.

15.2.3 Venus Kinase Receptors are Key Factors for Oogenesis

Functional analyses in *S. mansoni* showed SmVKR1 cooperation with a CTK complex consisting of the Src-kinase SmTK3, the Syk-kinase SmTK4 and the Src/Abl hybrid kinase SmTK6 (Beckmann *et al.* 2010, 2011) (Fig. 15.2). Analogous to the SmVKRs, SmTK6 exhibits unique characteristics because of its hybrid character, which was shown by structural and biochemical studies and its evolutionary role as potential precursor of Abl kinases of higher eukaryotes as was suggested by phylogenetic analyses (Beckmann *et al.* 2011). Targeting SmVKR1 and SmVKR2 for RNAi-mediated knock-down approaches led to a strong disorganization of the ovary and defects in egg production as well as disturbed spermatogenesis in males. The ovary phenotype was strongest when a double knock-down of both schistosome VKRs was performed (Vanderstraete *et al.* 2014). Since SmTK4 knock-down by RNAi resulted in similar gonad phenotypes, it appears likely that SmVKRs in cooperation with SmTK4 and the other members of the kinase complex govern important differentiation processes in the gonads. This view was reinforced by inhibitor experiments blocking these molecules at the protein level. Both, the Syk kinase inhibitor Piceatannol targeting SmTK4 (Beckmann *et al.* 2010b) as well as the IR inhibitors tyrphostin AG1024 or HNMP-A3 targeting SmVKRs negatively affected gonad integrity, egg production, but also viability of adult schistosomes (Vanderstraete *et al.* 2013).

How are VKRs activated? Biochemical assays using *Xenopus* oocytes as expression system and germinal vesicle breakdown (GVBD) assays for monitoring showed that calcium ions in the case of SmVKR2 and L-Arginine in the case of SmVKR1 were able to induce autophosphorylation of the respective receptor and thus to activate it (Vanderstraete *et al.* 2014). Although this represents a first breakthrough to de-orphanizing SmVKRs with respect to their activation, ions and calcium may not represent the only potential ligands. A recent study showed that the VKR ortholog of *Aedes aegypti*, AaeVKR, was preferentially expressed in the ovaries of blood-fed adult females (Vogel *et al.* 2015). RNAi against AaeVKR affected egg formation but only when the ovary ecdysteroidogenic hormone (OEH) was added. OEH and insulin-like peptides (ILPs) are neurohormones that are released from neurosecretory cells in the mosquito brain when blood feeding occurs (Brown *et al.* 1998, 2008). Ecdysteroid production was not affected upon RNAi indicating a very specific role of AaeVKR for egg formation in the mosquito (Vogel *et al.* 2015). Blood consumption is well known for schistosomes due to its importance for egg production (Schüssler *et al.* 1995; Glanfield *et al.* 2007; Toh *et al.* 2015). Furthermore, the first evidence for schistosome ecdysteroids such as ecdysone and 20-hydroxy ecdysone (Nirde *et al.* 1983) was found as well as an ortholog of the *Drosophila* ecdysone receptor SmE78, which was shown to be highly expressed in *S. mansoni* eggs (Wu *et al.* 2008). However, our knowledge about hormonal mechanisms regulating oogenesis and further reproductive processes in schistosomesis in its infancy but will be an interesting topic for future research activities.

15.2.4   A Closer Look into Schistosome Reproduction – Isolation and Characterization of Gonads

Many efforts have been made to unravel the male-female interaction at the molecular level. However, studies such as transcriptome analyses and single gene characterization have suffered from the limitation that whole worms had to be used impeding the analyses of tissue-specific effects. At the morphological level this is explained by the compact body plan of Platyhelminthes lacking an inner visceral cavity. Instead, inner organs and tissues are embedded in the parenchyma, a supporting tissue filling the whole worm body (Reissig 1970; Mehlhorn 1988). Furthermore, a syncytial layer, the tegument, forms a robust outer surface protecting the parasite from external impact (Skelly and Wilson 2006; Gobert and Jones, Chapter 10).

To overcome these limitations, different approaches were performed to isolate reproductive tissues from adult schistosomes. Among others, laser microdissection microscopy (LMM) was successfully established by Gobert *et al.* (2009) and Nawaratna *et al.* (2011) to obtain specific tissue from sections of adult schistosomes. In combination with subsequent microarray analyses this approach allowed for the first time the study of tissue-specific transcription patterns for a large set of genes.

Taking the time- and cost-intensive effort of LMM into account a more straight forward method was established by Hahnel *et al.* (2013) which allowed the isolation of complete, intact gonads from adult worms by a combinatory detergent/protease treatment of adult worms. In a first step the use of a detergent mixture allowed the removal of the tegument followed by incubation with elastase which breaks up sub-tegumental muscle layers. By incubation under mechanical forces this treatment led to the disintegration of the parenchyma and the release of testes, ovaries (Hahnel *et al.* 2013), vitellarium tissue fragments and vitelline cells (Lu *et al.* 2015), which were then enriched by pipetting. Organ and cell samples obtained by this method are a source of tissue-specific RNA and protein. Their quality and quantity was sufficient for RT-PCR and Western blot analyses which were used to verify former localization results obtained by *in situ* hybridization and to investigate transcript profiles of other genes of interest (Hahnel *et al.* 2013; Lu *et al.* 2015). Furthermore, gonad isolation was successfully applied to study pairing-dependent transcriptional processes at the level of specific tissues. Using RNA from testes and ovaries as template for qRT-PCRs analyses allowed the identification of the pairing-dependent transcription of genes coding for signaling proteins such as SmFz1 (frizzled 1) and SmFGFRs (Hahnel *et al.* 2014). Besides demonstrating that these genes are involved in gonad-associated processes, the gonad RNA-based approach exhibited its great advantage over conventional gene transcription analyses. The latter are routinely performed taking RNA of complete worms for analysis. With respect to SmFz1 and further genes analyses (Hahnel *et al.* 2014), their transcription seemed to be unaffected by pairing when transcript levels were compared using RNA of whole worms. Thus "zooming-in" into specific tissues can unmask features of pairing-regulated gene expression that otherwise could remain undiscovered.

For obvious reasons, cell isolation of schistosomes has been a matter of research for decades (Weller and Wheeldon 1982; Hobbs *et al.* 1993; Bayne *et al.* 1994, 1997; Dong *et al.* 2002; Quack *et al.* 2010; Ye *et al.* 2013). Former studies used trypsin-based enzymatic digestion and similar biochemical approaches to dissect complete larval or adult stages for cell recovery and *in vitro*-cultivation approaches. Although these efforts demonstrated that cells can be obtained, their maintenance in culture and their dividing capacities *in vitro* was limited. Today, we have to state that neither primary cell lines exist nor was the immortalization of cells yet achieved. Among many reasons these limitations may have originated from the diversity of cell types within such schistosome cell extracts and their varying demands for long-term cultivation. Former studies with insect cells showed that their requirements varied between cell lines from different species, but also between cell lines from the same species (Hink *et al.* 1973; Hink 1979). Furthermore, the numbers of specific cell types within a mixed-cell population of schistosomes may have been too low to initiate cell-culture establishment. Against this background, the gonad isolation protocol provides a novel basis for access and enrichment of specific, uniform cells of the germ line. Even FACS sorting of such cells is possible as was shown for vitelline cells (Lu *et al.* 2015). This represents a promising starting point for new efforts to establish schistosome cell cultures, which are needed in the post-genomic era of schistosome research (Quack *et al.* 2010).

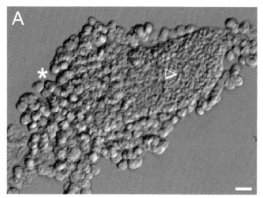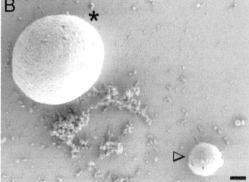

FIG. 15.3 Microscopic analyses of isolated oocytes from *S. mansoni*. **A.** Light microcopy (Hoffman modulation contrast; Olympus IX 81) of oocytes obtained by trypsin treatment of ovaries that were isolated by the organ isolation method (Hahnel *et al.* 2013). **B.** Scanning electron microscopy (Zeiss DSM 982) showing oocytes of different stages of development. Mature, primary oocytes are marked by a star, germline stem-cell-like oogonia are indicated by a triangle. Scale bars: A, 20 µm; B, 2 µm.

Besides cell culture options, the access to gonads also allows the pursuit of other research activities. By mechanical or enzymatic disruption of the epithelium surrounding the ovary, cells can be released, which are accessible for e.g., morphological analyses like electron microscopy (Fig. 15.3). This opens the possibility of following oogenesis-associated processes in the ovary at the level of distinct cell populations from germline stem-cell-like oogonia, which may originate from primordial germ cells (Dunlop *et al.* 2014; Greenspan *et al.* 2015), to primary oocytes.

15.2.5   Pairing-Induced Changes of Gene Transcription in the Ovary – Molecular insights with Evolutionary Implication

In the context of studying pairing-dependent transcriptional processes we recently compared the transcriptomes of isolated ovaries from mature, paired *S. mansoni* females with those from immature females that were not paired before by RNAseq (Lu *et al.* submitted). Among the findings was that the majority of genes known to be present in *S. mansoni* were transcribed in the ovary. This may be explained by the special role of the oocyte to be prepared for zygote formation and post-zygotic processes to organize development and early embryogenesis. To manage this complex biological program, part of the synthesized mRNAs has to be stored, a process previously shown for oocytes of other organisms (Seydoux 1996; Zhou and King 2004; Barckmann and Simonelig 2013).

Many of the ovary-transcribed genes in *S. mansoni* were found to be transcribed in a pairing-influenced manner. This confirms our previous findings (Hahnel *et al.* 2014) and demonstrates the remarkable influence of the male on transcriptional processes in the female including its gonads. In contrast, transcript profiles of only a very small number of genes in male testes were influenced by pairing. This points to an, albeit biased, bidirectional molecular communication between males and females during the pairing contact. Furthermore, it substantiates the fundamental impact of males on gonad developmental processes in the female. This may have evolutionary implication with respect to the origin of the family Schistosomatidae. On the basis of phylogenetic analyses it has been hypothesized that this parasite family is closely related to Spirorchidae, which parasitize poikilotherms including turtles (Platt and Brooks 1997; Snyder 2004; Brant and Loker 2005). Following the principle of sequential hermaphroditism, Spirorchidae are protandric, the male gonad develops ahead of that of the female. The same principle is realized in cestodes, in which the testes develop first in the anterior part of the body and produce sperm, whereas ovary development, the production of eggs and fertilization occur later (Johnson *et al.* 1970). In a sense, the situation of dioecious schistosomes resembles that of sequential, protandric hermaphrodites. The male gonad including differentiated sperm is present in males independent of pairing, whereas the development of the female gonad depends on the physical presence of the male. Thus the RNAseq data support this evolutionary view showing a clear bias of more transcripts influenced by pairing in the ovary compared to the testes. Further indirect evidence has been provided by the observation of pseudo-ovaries and vitellaria, which occasionally occur in schistosome males in addition to testicular lobes (Shaw and Erasmus 1982; Hulstijn *et al.* 2006; Beckmann *et al.* 2010a). In contrast, to our knowledge, pseudo testes have not been found yet in schistosome females. Moreover, microarray and SuperSAGE data correspondingly showed the occurrence of transcripts of female-specific genes associated with egg synthesis in males (Leutner *et al.* 2013). Therefore, Schistosomatidae represent an exceptional family of platyhelminths in evolutionary terms as the principle of sequential hermaphroditism is still maintained, though split into two sexes (Bach 1990; Lu *et al.* submitted).

15.3   LITERATURE CITED

Adiyodi KH, Adiyodi RG. 1988. *Reproductive Biology of Invertebrates. Vol. III, Acessory Sex Glands.* New York: John Wiley and Sons. 542 p.

Ahier A, Rondard P, Gouignard N, Khayath N, Huang S, Trolet J, Donoghue DJ, Gauthier M, Pin JP, Dissous C. 2009. A new family of receptor tyrosine kinases with a venus flytrap binding domain in insects and other invertebrates activated by amino acids. PLoS One 4(5): e5651.

Andrade LF, Nahum LA, Avelar LG, Silva LL, Zerlotini A, Ruiz JC, Oliveira G. 2011. Eukaryotic protein kinases (ePKs) of the helminth parasite *Schistosoma mansoni*. BMC Genomics 12: 215.

Andrade LF, MourãoMde M, Geraldo JA, Coelho FS, Silva LL, Neves RH, Volpini A, Machado-Silva JR, Araujo N, Nacif-Pimenta R, Caffrey CR, Oliveira G. 2014. Regulation of *Schistosoma mansoni* development and reproduction by the mitogen-activated protein kinase signaling pathway. PLoS Neglected Tropical Diseases 8(6): e2949.

Armstrong JC. 1965. Mating behavior and development of schistosomes in the mouse. Journal of Parasitology 51: 605–616.

Barckmann B, Simonelig M. 2013. Control of maternal mRNA stability in germcells and early embryos. Biochimica Biophysica Acta 1829(6–7): 714–724.

Basch PF. 1990. Why do schistosomes have separate sexes? Parasitology Today 6(5): 160–163.

Basch PF. 1991. *Schistosomes: Development, Reproduction and Host Relations*. New York: Oxford University Press. 264 p.

Bayne CJ, Menino JS, Hobbs DJ, Barnes DW. 1994. *In vitro* cultivation of cells from larval *Schistosoma mansoni*. Journal of Parasitology 80(1): 29–35.

Bayne CJ, Barnes DW. 1997. Culture of cells from two life stages of *Schistosoma mansoni*. Cytotechnology 23(1–3): 205–210.

Beckmann S, Quack T, Burmeister C, Buro C, Long T, Dissous C, Grevelding CG. 2010a. *Schistosoma mansoni*: signal transduction processes during the development of the reproductive organs. Parasitology 137: 497–520.

Beckmann S, Buro C, Dissous C, Hirzmann J, Grevelding CG. 2010b. The Syk kinase SmTK4 of *Schistosoma mansoni* is involved in the regulation of spermatogenesis and oogenesis. PLoS Pathogens 6(2): e1000769.

Beckmann S, Grevelding CG. 2010c. Imatinib has a fatal impact on morphology, pairing stability and survival of adult *Schistosoma mansoni in vitro*. International Journal for Parasitology 40(5): 521–526.

Beckmann S, Hahnel S, Cailliau K, Vanderstraete M, Browaeys E, Dissous C, Grevelding CG. 2011. Characterization of the Src/Abl hybrid kinase SmTK6 of *Schistosoma mansoni*. Journal of Biological Chemistry 286(49): 42325–42336.

Beckmann S, Quack T, Dissous C, Cailliau K, Lang G, Grevelding CG. 2012. Discovery of platyhelminth-specific α/β-integrin families and evidence for their role in reproduction in *Schistosoma mansoni*. PLoS One 7(12): e52519.

Berriman M, Haas BJ, LoVerde PT, Wilson RA, Dillon GP, Cerqueira GC, Mashiyama ST, Al-Lazikani B, Andrade LF, Ashton PD, *et al.* 2009. The genome of the blood fluke *Schistosoma mansoni*. Nature 460(7253): 352–358.

Brant SV, Loker ES. 2005. Can specialized pathogens colonize distantly related hosts? Schistosome evolution as a case study. PLoS Pathogens 1(3): e38.

Bromann PA, Korkaya H, Courtneidge SA. 2004. The interplay between Src family kinases and receptor tyrosine kinases. Oncogene 23: 7957–7968.

Brown MR, Graf R, Swiderek KM, Fendley D, Stracker TH, Champagne DE, Lea AO. 1998. Identification of a steroidogenic neurohormone in female mosquitoes. Journal of Biological Chemistry 273: 3967–3971.

Brown MR, Clark KD, Gulia M, Zhao Z, Garczynski SF, Crim JW, Suderman RJ, Strand MR. 2008. An insulin-like peptide regulates egg maturation and metabolism in the mosquito *Aedesaegypti*. Proceedings National Academy of Sciences USA 105: 5716–5721.

Buro C, Oliveira KC, Lu Z, Leutner S, Beckmann S, Dissous C, Cailliau K, Verjovski-Almeida S, Grevelding CG. 2013. Transcriptome analyses of inhibitor-treated schistosome females provide evidence for cooperating Src-kinase and TGFβ receptor pathways controlling mitosis and eggshell formation. PLoS Pathogens 9(6): e1003448.

Cheever AW, Macedonia JG, Mosimann JE, Cheever EA. 1994. Kinetics of egg production and egg excretion by *Schistosoma mansoni* and *S. japonicum* in mice infected with a single pair of worms. American Journal Tropical Medicine and Hygiene 50(3): 281–295.

Clough ER. 1981. Morphology and reproductive organs and oogenesis in bisexual and unisexual transplants of mature *Schistosoma mansoni* females. Journal of Parasitology 67(4): 535–539.

Collins JJ 3rd, Wang B, Lambrus BG, Tharp ME, Iyer H, Newmark PA. 2013. Adult somatic stem cells in the human parasite *Schistosoma mansoni*. Nature 494(7438): 476–479.

Cort WW. 1921. Sex in the trematode family Schistosomidae. Science 53(1367): 226–228.

Den Hollander JE, Erasmus DA. 1985. *Schistosoma mansoni*: male stimulation and DNA synthesis by the female. Parasitology 91(Pt 3): 449–457.

deWalick S, Tielens AG, van Hellemond JJ. 2012. *Schistosoma mansoni*: the egg, biosynthesis of the shell and interaction with the host. Experimental Parasitology 132(1): 7–13.

Dissous C, Grevelding CG, Long T. 2011. *Schistosoma mansoni* polo-like kinases and their function in control of mitosis and parasite reproduction. Anais Da Academia Brasileira de Ciências 83(2): 627–635.

Dissous C, Morel M, Vanderstraete M. 2014. Venus kinase receptors: prospects in signaling and biological functions of these invertebrate kinases. Frontiers in Endocrinology 5: 72.

Dissous C. 2015. Venus kinase receptors at the crossroads of insulin signaling: their role in reproduction for helminths and insects. Frontiers Endocrinology 6: 118.

Dong HF, Chen XB, Ming ZP, Zhong QP, Jiang MS. 2002. Ultrastructure of cultured cells from *Schistosoma japonicum*. Acta Tropica 82(2): 225–234.

Dunlop CE, Telfer EE, Anderson RA. 2014. Ovarian germline stem cells. Stem Cell Research and Therapy 5(4): 98.

Erasmus DA. 1973. A comparative study of the reproductive system of mature, immature and "unisexual" female *Schistosoma mansoni*. Parasitology 67(2): 165–183.

Erasmus DA. 1987. The adult schistosome: structure and reproductive biology, from genes to latrines. pp. 51–82. *In*: Rollinson D, Simpson AJG, editors. *The Biology of Schistosomes*. London: Academic Press.

Eveland LK, Haseeb MA. 1986. Schistosome behavior *in vitro*. Journal of Chemical Ecology 12(8): 1687–1698.

Freitas TC, Jung E, Pearce EJ. 2007. TGF-beta signaling controls embryo development in the parasitic flatworm *Schistosoma mansoni*. PLoS Pathogens 3(4): e52.

Glanfield A, McManus DP, Anderson GJ, Jones MK. 2007. Pumping iron: a potential target for novel therapeutics against schistosomes. Trends Parasitology 23(12): 583–588.

Gobert GN, McManus DP, Nawaratna S, Moertel L, Mulvenna J, Jones MK. 2009. Tissue specific profiling of females of *Schistosoma japonicum* by integrated laser microdissection microscopy and microarray analysis. PLoS Neglected Tropical Diseases 3(6): e469.

Gönnert R. 1955. Schistosomiasis-studien. I, II. Zeitschrift für Tropenmedzin und Parasitologie 6: 18–52.

Gouignard N, Vanderstraete M, Cailliau K, Lescuyer A, Browaeys E, Dissous C. 2012. *Schistosoma mansoni*: structural and biochemical characterization of two distinct venus kinase receptors. Experimental Parasitology 132(1): 32–39.

Greenspan LJ, de Cuevas M, Matunis E. 2015. Genetics of gonadal stem cell renewal. Annual Review of Cell and Developmental Biology 31: 291–315.

Grevelding CG, Sommer G, Kunz W. 1997. Female-specific gene expression in *Schistosoma mansoni* is regulated by pairing. Parasitology 115(Pt 6): 635–640.

Hahnel S, Lu Z, Wilson RA, Grevelding CG, Quack T. 2013. Whole-organ isolation approach as a basis for tissue-specific analyses in *Schistosoma mansoni*. PLoS Neglected Tropical Diseases 7(7): e2336.

Hahnel S, Quack T, Parker-Manuel SJ, Lu Z, Vanderstraete M, Morel M, Dissous C, Cailliau K, Grevelding CG. 2014. Gonad RNA-specific qRT-PCR analyses identify genes with potential functions in schistosome reproduction such as SmFz1 and SmFGFRs. Frontiers Genetics 5: 170.

Hink WF, Richardson BL, Schenk DK, Ellis BJ. 1973. Utilization of amino acids and sugars by two codling moth (*Carpocapsa pomonella*) cell lines (CP-1268 and CP-169) cultured *in vitro*. pp. 195–208. *In*: Rehacek J, Blaskovic D, Hink WF, editors. *Proceedings Third International Colloquium on Invertebrate Tissue Culture*. Bratislava: Publishing House of the Slovak Academy of Sciences.

Hink WF. 1979. Cell lines from invertebrates. Methods Enzymology 58: 450–466.

Hobbs DJ, Fryer SE, Duimstra JR, Hedstrom OR, Brodie AE, Collodi PA, Menino JS, Bayne CJ, Barnes DW. 1993. Culture of cells from juvenile worms of *Schistosoma mansoni*. Journal of Parasitology 79(6): 913–921.

Hoffmann KF. 2004. An historical and genomic view of schistosome conjugal biology with emphasis on sex-specific gene expression. Parasitology 128 Suppl 1: S11–22.

Hogan PG, Rao A. 2015. Store-operated calcium entry: mechanisms and modulation. Biochemical and Biophysical Research Communication 460(1): 40–49.

Hu W, Brindley PJ, McManus DP, Feng Z, Han ZG. 2004. Schistosome transcriptomes: new insights into the parasite and schistosomiasis. Trends in Molecular Medicine 10: 217–225.

Hubbard SR, Till JH. 2000. Protein tyrosine kinase structure and function. Annual Reviews of Biochemistry 69: 373–398.

Hulstijn M, Barros LA, Neves RH, Moura EG, Gomes DC, Machado-Silva JR. 2006. Hermaphrodites and supernumerary testicular lobes in *Schistosoma mansoni* (Trematoda: Schistosomatidae) analyzed by bright field and confocal microscopy. Journal of Parasitology 92(3): 496–500.

Johnson AD, Gomes WR, Vandermark NL. 1970. *The Testis; Biochemistry*, Vol. II. Heidelberg: Academic Press. p. 457.

Kapp K, Schussler P, Kunz W, Grevelding CG. 2001. Identification, isolation and characterization of a Fyn-like tyrosine kinase from *Schistosoma mansoni*. Parasitology 122: 317–327.

Kapp K, Knobloch J, Schussler P, Sroka S, Lammers R, Kunz W, Grevelding CG. 2004. The *Schistosoma mansoni* Src kinase TK3 is expressed in the gonads and likely involved in cytoskeletal organization. Molecular and Biochemical Parasitology 138: 171–182.

Knobloch J, Kunz W, Grevelding CG. 2002a. Quantification of DNA synthesis in multicellular organisms by a combined DAPI and BrdU technique. Development Growth and Differentiation 44: 559–563.

Knobloch J, Winnen R, Quack M, Kunz W, Grevelding CG. 2002b. A novel Syk-family tyrosine kinase from *Schistosoma mansoni*, which is preferentially transcribed in reproductive organs. Gene 294: 87–97.

Knobloch J, Beckmann S, Burmeister C, Quack T, Grevelding CG. 2007. Tyrosine kinase and cooperative TGF beta signaling in the reproductive organs of *Schistosoma mansoni*. Experimental Parasitology 117: 318–336.

Krauss G. 2014. *Biochemistry of Signal Transduction and Regulation*. 5th edition. Weinheim: Wiley-VCH. 844 p.

Kunz W. 2001. Schistosome male-female interaction: induction of germ-cell differentiation. Trends in Parasitology 17(5): 227–231.

Leutner S, Oliveira KC, Rotter B, Beckmann S, Buro C, Hahnel S, Kitajima JP, Verjovski-Almeida S, Winter P, Grevelding CG. 2013. Combinatory microarray and Super SAGE analyses identify pairing-dependently transcribed genes in *Schistosoma mansoni* males, including follistatin. PLoS Neglected Tropical Diseases 7(11): e2532.

LoVerde PT, Osman A, Hinck A. 2007. *Schistosoma mansoni*: TGF-β signaling pathways. Experimental Parasitology 117: 304–317.

LoVerde PT, Andrade LF, Oliveira G. 2009. Signal transduction regulates schistosome reproductive biology. Current Opinion in Microbiology 12(4): 422–428.

Lu Z, Quack T, Hahnel S, Gelmedin V, Pouokam E, Diener M, Hardt M, Michel G, Baal N, Hackstein H, Grevelding CG. 2015. Isolation, enrichment and primary characterisation of vitelline cells from *Schistosoma mansoni* obtained by the organ isolation method. International Journal for Parasitology 45(9-10): 663–672.

Mehlhorn H. 1988. *Parasitology in Focus*. Berlin: Springer Publishing. 924 p.

Moore DV, Sandground JH. 1956. The relative egg production capacity of *Schistosoma mansoni* and *Schistosoma japonicum*. American Journal Tropical Medicine and Hygiene 5: 831–840.

Morrill GA, Kostellow AB, Gupta RK. 2015. Computational analysis of the extracellular domain of the Ca^{2+}-sensing receptor: an alternate model for the Ca^{2+} sensing region. Biochemical Biophysical Research Communication 459(1): 36–41.

Nawaratna SS, McManus DP, Moertel L, Gobert GN, Jones MK. 2011. Gene atlasing of digestive and reproductive tissues in *Schistosoma mansoni*. PLoS Neglected Tropical Diseases 5(4): e1043.

Neves RH, de Lamare Biolchini C, Machado-Silva JR, Carvalho JJ, Branquinho TB, Lenzi HL, Hulstijn M, Gomes DC. 2005. A new description of the reproductive system of *Schistosoma mansoni* (Trematoda: Schistosomatidae) analyzed by confocal laser scanning microscopy. Parasitology Research 95(1): 43–49.

Nirde P, Torpier G, De Reggi ML, Capron A. 1983. Ecdysone and 20 hydroxyecdysone: new hormones for the human parasite *Schistosoma mansoni*. FEBS Letters 151: 223–227.

Nollen PM. 1997. Reproductive physiology and behavior of digenetic trematodes. pp. 117–148. *In*: Fried B, Graczyk T, editors. *Advances in Trematode Biology*. USA: CRC Press.

Platt TR, Brooks DR. 1997. Evolution of the schistosomes (Digenea: Schistosomatidea): the origin of dioecy and colonization of the venous system. Journal of Parasitology 83: 1035–1044.

Popiel I, Basch PF. 1984. Reproductive development of female *Schistosoma mansoni* (Digenea: Schistosomatidae) following bisexual pairing of worms and worm segments. Journal of Experimental Zoology 232: 141–150.

Protasio AV, Tsai IJ, Babbage A, Nichol S, Hunt M, Aslett MA, De Silva N, Velarde GS, Anderson TJ, Clark RC, *et al.* 2012. A systematically improved high quality genome and transcriptome of the human blood fluke *Schistosoma mansoni*. PLoS Neglected Tropical Diseases 6(1): e1455.

Quack T, Wippersteg V, Grevelding CG. 2002. Cell cultures for schistosomes – chances of success or wishful thinking? International Journal for Parasitology 40(9): 991–1002.

Reissig M. 1970. Characterization of cell types in the parenchyma of *Schistosoma mansoni*. Parasitology 60(2): 273–279.

Ross AG, Bartley PB, Sleigh AC, Olds GR, Li Y, Williams GM, McManus DP. 2002. Schistosomiasis. New England Journal of Medicine 346(16): 1212–1220.

Schistosoma japonicum Genome Sequencing and Functional Analysis Consortium. 2009. *The Schistosoma japonicum* genome reveals features of host-parasite interplay. Nature 460(7253): 345–351.

Schüssler P, Pötters E, Winnen R, Bottke W, Kunz W. 1995. An isoform of ferritin as a component of protein yolk platelets in *Schistosoma mansoni*. Molecular Reproduction and Development 41(3): 325–330.

Severinghaus AE. 1928. Memoirs: sex studies on *Schistosoma japonicum*. Quarterly Journal of Microscopical Science 71: 653–702.

Seydoux G. 1996. Mechanisms of translational control in early development. Current Opinion in Genetics and Development 6(5): 555–561.

Shaw JR, Erasmus DA. 1981. *Schistosoma mansoni*: an examination of the reproductive status of females from single sex infections. Parasitology 82(1): 121–124.

Shaw MK, Erasmus DA. 1982. *Schistosoma mansoni*: the presence and ultrastructure of vitelline cells in adult males. Journal of Helminthology 56(1): 51–53.

Shoemaker CB, Ramachandran H, Landa A, dos Reis MG, Stein LD. 1992. Alternative splicing of the *Schistosoma mansoni* gene encoding a homologue of epidermal growth factor receptor. Molecular and Biochemical Parasitology 53(1–2): 17–32.

Skelly PJ, Wilson AR. 2006. Making sense of the schistosome surface. Advances in Parasitology 63: 185–284.

Snyder SD. 2004. Phylogeny and paraphyly among tetrapod blood flukes (Digenea: Schistosomatidae and Spirorchiidae). International Journal for Parasitology 34: 1385–1392.

Spence IM, Silk MH. 1971a. Ultrastructural studies of the blood fluke—*Schistosoma mansoni*. V. The female reproductive system—a preliminary report. The South African Journal of Medical Sciences 36(3): 41–50.

Spence IM, Silk MH. 1971b. Ultrastructural studies of the blood fluke—*Schistosoma mansoni*. VI. The mehlis gland. The South African Journal of Medical Sciences 36(3): 69–76.

Steinauer ML. 2009. The sex lives of parasites: investigating the mating system and mechanisms of sexual selection of the human pathogen *Schistosoma mansoni*. International Journal for Parasitology 39(10): 1157–1163.

Strange K, Yan X, Lorin-Nebel C, Xing J. 2007. Physiological roles of STIM1 and orai1 homologs and CRAC channels in the genetic model organism *Caenorhabditis elegans*. Cell Calcium 42(2): 193–203.

Stroehlein AJ, Young ND, Jex AR, Sternberg PW, Tan P, Boag PR, Hofmann A, Gasser RB. 2015. Defining the *Schistosoma haematobium* kinome enables the prediction of essential kinases as anti-schistosome drug targets. Science Reports 5: 17759.

Swierczewski BE, Davies SJ. 2010. Conservation of protein kinase a catalytic subunit sequences in the schistosome pathogens of humans. Experimental Parasitology 125(2): 156–160.

Tatosyan AG, Mizenina OA. 2000. Kinases of the Src family: structure and functions. Biochemistry (Mosc.) 65: 49–58.

Toh SQ, Gobert GN, Malagón Martínez D, Jones MK. 2015. Haem uptake is essential for egg production in the haematophagous blood fluke of humans, *Schistosoma mansoni*. FEBS Journal 282(18): 3632–3646.

Vanderstraete M, Gouignard N, Cailliau K, Morel M, Lancelot J, Bodart JF, Dissous C. 2013. Dual targeting of insulin and venus kinase receptors of *Schistosoma mansoni* for novel anti-schistosome therapy. PLoS Neglected Tropical Diseases 7(5): e2226.

Vanderstraete M, Gouignard N, Cailliau K, Morel M, Hahnel S, Leutner S, Beckmann S, Grevelding CG, Dissous C. 2014. Venus kinase receptors control reproduction in the platy helminth parasite *Schistosoma mansoni*. PLoS Pathogens 10(5): e1004138.

Verjovski-Almeida S, DeMarco R, Martins EA, Guimaraes PE, Ojopi EP, Paquola AC, Piazza JP, Nishiyama Jr MY, Kitajima JP, Adamson RE, *et al.* 2003. Transcriptome analysis of the acoelomate human parasite *Schistosoma mansoni.* Nature Genetics 35: 148–157.

Vicogne J, Pin JP, Lardans V, Capron M, Noel C, Dissous C. 2003. An unusual receptor tyrosine kinase of *Schistosoma mansoni* contains a venus flytrap module. Molecular and Biochemical Parasitology 126: 51–62.

Vogel KJ, Brown MR, Strand MR. 2015. Ovary ecdysteroidogenic hormone requires a receptor tyrosine kinase to activate egg formation in the mosquito *Aedesaegypti.* Proceeding National Academy of Sciences USA 112: 5057–5062.

Walker AJ, Ressurreição M, Rothermel R. 2014. Exploring the function of protein kinases in schistosomes: perspectives from the laboratory and from comparative genomics. Frontiers Genetics 5: 229.

Wang X, Li H, Qi X, Shi Y, Xia Y, Yang J, Yuan C, Feng X, Lin J. 2011. Characterization and expression of a novel Frizzled 9 gene in *Schistosoma japonicum.* Gene Expression Patterns 11(3–4): 263–270.

Weller TH, Wheeldon SK. 1982. The cultivation *in vitro* of cells derived from adult *Schistosoma mansoni.* I. Methodology: criteria for evaluation of cultures; and development of media. American Journal of Tropical Medicine and Hygiene 31: 335–348.

Wu W, Tak EY, LoVerde PT. 2008. *Schistosoma mansoni:* SmE78, a nuclear receptor orthologue of *Drosophila* ecdysone-induced protein 78. Experimental Parasitology 119: 313–318.

Ye Q, Dong HF, Grevelding CG, Hu M. 2013. *In vitro* cultivation of *Schistosoma japonicum*-parasites and cells. Biotechnology Advances 31(8): 1722–1737.

You H, Zhang W, Jones MK, Gobert GN, Mulvenna J, Rees G, Spanevello M, Blair D, Duke M, Brehm K, McManus DP. 2010. Cloning and characterisation of *Schistosoma japonicum* insulin receptors. PLoS One 5(3): e9868.

Young ND, Jex AR, Li B, Liu S, Yang L, Xiong Z, Li Y, Cantacessi C, Hall RS, Xu XP. 2009. Whole-genome sequence of *Schistosoma haematobium.* Nature Genetics 44(2): 221–225.

Zhou Y, King ML. 2004. Sending RNAs into the future: RNA localization and germ cell fate. IUBMB Life 56(1): 19–27.

CHAPTER

Excretory System of Schistosomes

John Kusel

16.1 INTRODUCTION

The flame-cell, which is the terminal and immediately and strikingly visible component of the excretory system, is one of the strangest and most intriguing microscopic structures that one encounters as a student of biology. Almost every student exhales and then takes a deep breath when first observing the flame-cell and many have undertaken a career in biology after witnessing this phenomenon!

The morphology of flame-cells and the excretory organs in invertebrates has been described by Goodrich (1945) and Hyman (1951); scholarly chapters on this topic can be found in Smyth and Halton (1983) and Erasmus (1972). The now classical review on protonephridia was published by Wilson and Webster (1974).

The fine structure of the flame-cell of *Schistosoma mansoni* in the miracidium has been referred to by Jamieson and Haasin Chapter 6 of this volume and a longitudinal section for electron microscopy has been included. Images of the distribution of flame-cells and excretory tubules in schistosomes are shown in Figures 16.1 to 16.8. Further details of the structure and function of the flame-cells are given later in this Chapter.

The flame-cell structure seems to have been preserved in the development of the schistosome from the miracidium into the later stages and is also found as a model structure in all trematodes and cestodes. Although these two groups appear to have similar structures, there are subtle differences, as shown by Howells (1969), where pores were observed in the tubule cell to form a nephridial funnel (nephrostome). This would suggest the presence of a nephrostome in *Moniezia*. However, in *Fasciola* miracidia, Wilson (1969) could find no evidence for organized pores, which might suggest there is no nephrostome in this species. Rohde *et al.* (1989) have concluded that "flame-cells appear to be a very conservative element and their more or less identical structure in all major groups of parasitic platyhelminths might suggest that parasitic platyhelminths have originated from one or a few closely related groups of free-living Turbellaria with the same type of flame-cell. "However, an extensive discussion of the structures of the various kinds of flame-cell (Wilson and Webster 1974) concludes that the very varied structures may not be homologous and may have arisen in evolution in separate lines of development. The evolution of the protostome excretory system is thus a very interesting and controversial area though it will not be considered further here. The interested reader might consult Rohde *et al.* (1992) and Bartolomaeus and Ax (1992).

Division of Infection and Immunity, Institute of Biomedical and Life Sciences, Glasgow Biomedical Research Centre, University of Glasgow, 120 University Place, Glasgow G12 8TA, UK.

Excellent recent morphological studies have been published by Collins *et al.* (2011) using fluorescent lectins and antibodies. These striking images and movies of the nephridial system of *Schistosoma mansoni* enhance the pictures inherited from the conventional histology and confirm these earlier findings. Other beautiful images of the nephridial system can be seen through the use of monoclonal antibodies and fluorescence microscopy (Figs. 16.2–16.5). A diagrammatic overview of the distribution of the elements of the excretory system of adult paired *Schistosoma mansoni* is show in Figure 16.1.

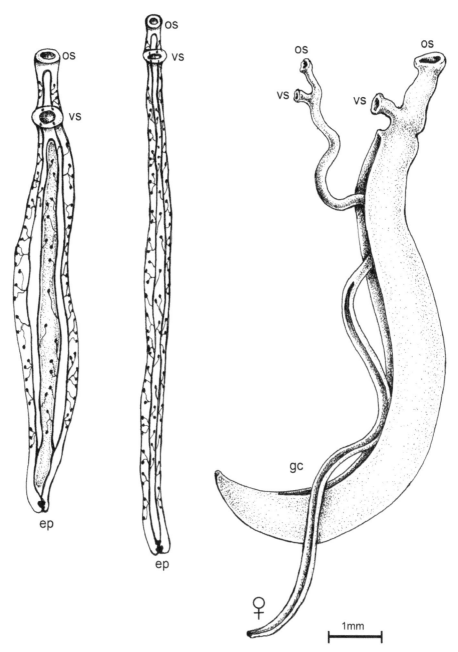

FIG. 16.1 *Schistosoma mansoni.* A diagrammatic overview of the positions of flame-cells, proximal, distal tubules and collecting ducts in adult male and female parasites. os, oral sucker; vs, ventral sucker; gc, gynaecophoric canal; ep, excretory pore. Original.

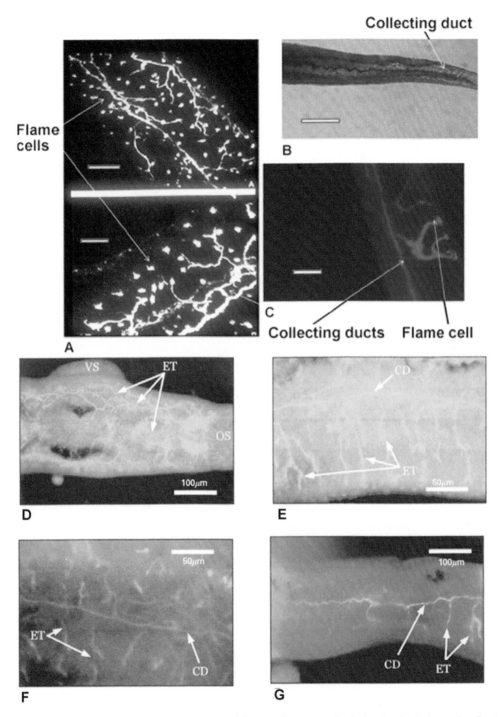

FIG. 16.2 *Schistosoma mansoni.* The distribution of flame-cells, proximal tubules distal tubules and collecting ducts in fixed or living parasites. **A.** Staining of fixed adult worm with a monoclonal antibody against carbohydrate epitopes. From Bogers JJ, Nibbeling HA, van Marck EA, Deelder AM. 1994. American Journal of Tropical Medicine and Hygiene 50(5): 612–619. Fig. 4. **B.** Bright field light microscope of posterior of adult male showing distended collecting ducts. **C.** Staining of living worm with the fluorescent dye resorufin. Scale bars: A 100 μm B 500 μm. C 50 μm. B, C Original. **D–G** *Schistosoma mansoni.* Showing ramifications of tubule systems in living male and female adult worms stained with the fluorescent calcium-specific indicator Fluo-3 AM. From Sato H, Kusel JR, Thornhill JA. 2004. Parasitology 128: 43–52. Fig. 4.

Other images of the distribution of flame-cells and tubules shown by a variety of fluorescent substrates are shown in Figures 16.2–16.6.

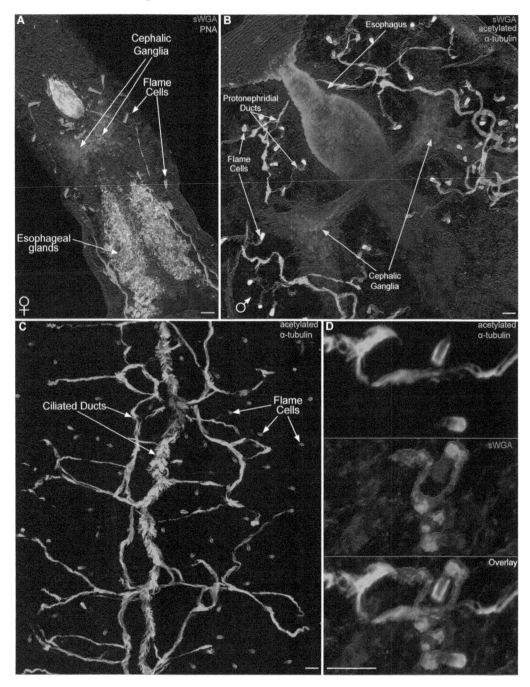

FIG 16.3 *Schistosoma mansoni.* Excretory tubules and flame-cells revealed by immunostaining of fixed worms **A.** Lectin PNA staining of oesophageal gland cells. Lectin sWGA staining of cerebral ganglia and protonephridial ducts of female. **B.** Lectin labelling of cephalic ganglia and non-ciliated protonephridial ducts of male. Anti-acetylated tubulin staining of ciliated region of protonephridial system. **C.** Anti-acetylated tubulin staining of ciliated regions of protonephridia and collecting ducts. **D.** Anti-acetylated tubulin (Ciliated flame-cells and ducts) and sWGA labeling (nonciliated tubules) of a protonephridial unit. Scale bar 10 μm. From Collins JJ, King RS, Cogswell A, Williams DL, Newmark PA. 2011. PLoS Neglected Tropical Diseases 5(3): e1009. Fig. 6.

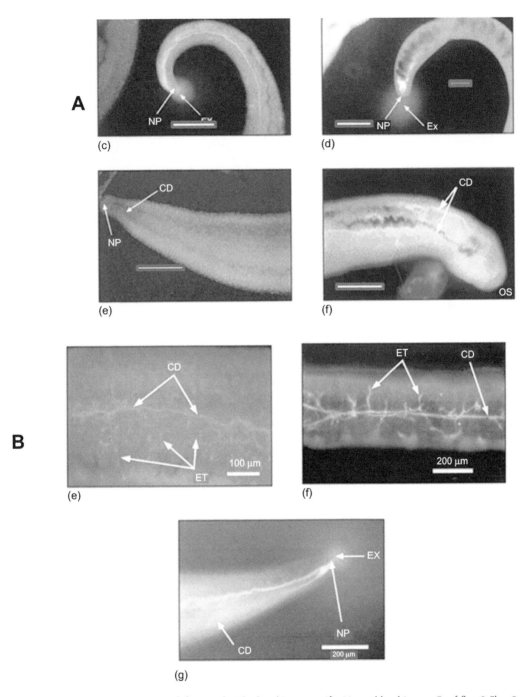

FIG. 16.4 *Schistosoma mansoni* **A.** cdef, stained with glutathione specific Monochlorobimane. **B.** ef fluo-3 Fluo-3 AM. g. carboxy fluorescein diacetate. Scale bars A 500 μm. From Sato H, Kusel JR, Thornhill JA. 2004. Parasitology 128(1): 43–52. Figs. 1, 2.

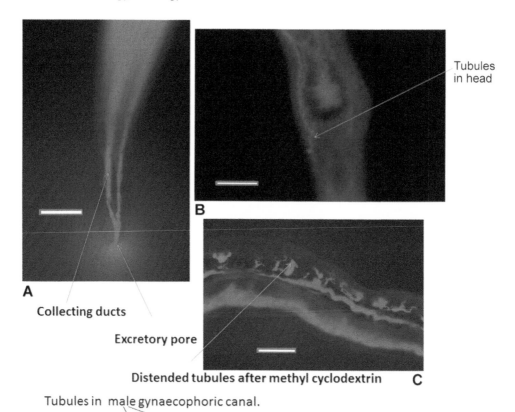

Tubules
in head

B

A

Collecting ducts

Excretory pore

Distended tubules after methyl cyclodextrin **C**

Tubules in male gynaecophoric canal.

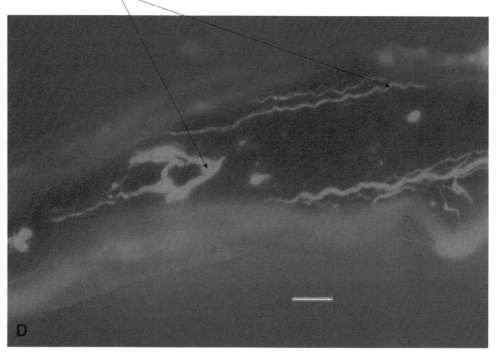

D

FIG. 16.5 *Schistosoma mansoni.* **A.** Staining of adult worms with resorufin to show excretion through excretory pore (nephridiopore). **B.** Tubules in the head region. **C.** Massive activity seen after cholesterol removal from membrane by methyl cyclodextrin. Scale bars 200 μm. **D.** Staining with resorufin to show distribution in gynaecophoric groove. Scale bars 200 μm except left middle 50 μm. Original.

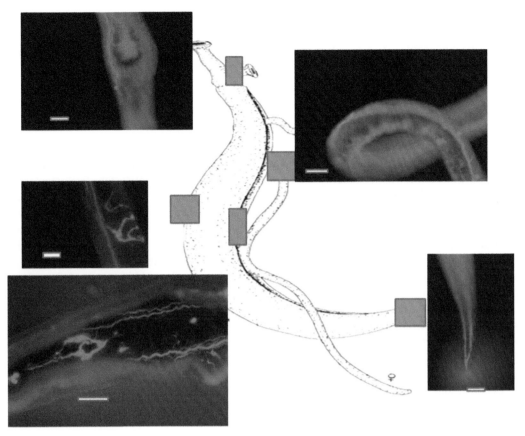

FIG. 16.6 *Schistosoma mansoni.* Staining with resorufin to show staining in various regions of the paired male and female worms Scale bars 100 µm. Original.

16.2 EMBRYOLOGY

The development of the cercarial protonephridial system has been studied in the sporocyst in careful histological work by Cheng and Bier (1972). They sectioned sporocysts during their growth in infected *Biomphalaria glabrata* and demonstrated the presence of stem cells and the appearance of the excretory pore, but the origin of the flame-cells and tubule cells is unclear. A detailed study on the germinal cells of a variety of digenetic trematodes was carried out by Cort *et al.* (1954) but no indication was given of the development of the excretory system in the early embryos studied.

Recent work extending the detailed studies of Cheng and Bier (1972) has been described (Wang *et al.* 2013) and is relevant to the development of protonephridia. These workers have used the fluorescent POPO-1 to label the germinal cells of sporocysts and cercariae. Using this method and also the incorporation of the thymidine analogue EdU during growth from miracidium to mother sporocyst to identify the germinal cells, they demonstrated, using the RNAi method, that several genes were essential for their development. These genes were also found to be important for the development of stem cells (neoblasts) in *Schmidtea*, a free living planarian (see below). They were identified as *arg2-1 and vlg3*. The similarity of these stem cells in *Schmidtea* to neoblasts in adult *S. mansoni* (Collins *et al.* 2013) led those authors to use RNAi methods to deplete fgfrA, the fibroblast growth factor receptor gene in sporocysts and it was indeed found that this gene was essential for the growth of a proportion of the sporocyst germinal cells.

The similarity between the function of germinal cells in sporocysts and that of neoblasts in adult *S. mansoni* and *Schmidtea* allows the speculation that the germinal cell/neoblast/stem cells are a very ancient phenomenon in evolution and links the evolution of parasitic flatworms to the free-living species. The authors conclude:

> *"These (germinal cells of sporocysts) cells possess a molecular signature similar to that of neoblasts in free-living flatworms, as well as stem cells from diverse organisms. This conserved molecular context opens access to understanding these cells and may lead to strategies for intervening and blocking the transmission of the disease".*

Recent studies on the free living flatworm *Schmidtea mediterranea* (Scimone *et al.* 2011, 2014) have real potential for understanding the growth of the protonephridial system from stem cells and have given valuable insights into the development of the flame-cell and the excretory tubules. By the use of RNAi techniques and immuno-localization it has been shown that several genes are responsible for the production of tubule cells, flame-cells and tubules during regeneration of wound regions of the worm (Scimone *et al.* 2011). These workers took anterior and posterior blastemas (undifferentiated tissue appearing at the wound surface) and used RNAi inhibition of six genes. The importance of these genes was concluded because the RNAi treatment led to edema formation in the worm, which indicated lack of osmoregulation by the flame-cells of the excretory system. The genes *Six1/2-2*, *POU2/3* and *hunchback* were essential for the development and the regeneration of the protonephridial system. These genes were also important for the maintenance of the intact system. Homologs of these genes are important for the development of the vertebrate kidney. Microarray experiments on RNAi-treated animals also showed loss of genes coding for ion transport, protein clearance and acid-base balance, further emphasizing the similarity to the vertebrate kidney. Since many of these genes are important in the formation of vertebrate kidney tubules, it is concluded that these genes were active during the formation of protonephridia before the evolution of the bilateria (Scimone *et al.* 2011). Regeneration in planarians requires adult dividing cells called neoblasts. These can be isolated by flow cytometry because of their greater than 2N DNA content. When examining neoblasts, Scimone *et al.* (2011) verified that those destined by their gene content to form nephridia also stained with the *cubilin* gene, a protonephridial marker for intrinsic factor/vitamin B12 complexes. This proved that specific neoblasts were destined to become protonephridia and that the regenerating protonephridia were not formed from preformed nephridial fragments.

In the work of Rink *et al.* (2011), *Schmidtea* has also been used to study the regeneration of the flame-cells and associated tubules in blastemas taken from the head region and the tail region. During regeneration in growth medium it was shown that flame-cells developed first followed by proximal and then distal tubules. The very first group of cells developing into the protonephridia were named the proto-tubule.

The flame-cells were identified by immunofluorescence using an anti-tubulin and anti-desmosome antibody and the proximal tubules using anti-tubulin antibodies. The tubulin is present in abundance in cilia and this antibody showed the presence of ciliated tubules also. Molecular markers for the diversity of cell types were used in *in situ* hybridization, co-localizing with the tubulin immunofluorescence. These genes included the gap junction gene *Smed-innexin 10 (inx10)* and the carbonic anhydrase gene *Smed-CAVII-1 (CAVII-1)*. Distal tubules were particularly rich in *CAVII-1* staining, which may indicate a role in bicarbonate transport (see below).

Thus the flame-cells were identified by anti-tubulin antibodies and expression of dynein heavy chain gene (*Smed-DNAH-beta3*). These were connected to *inx-10* expressing ciliated tubules (proximal tubules) which transition into *CAVII-1* positive, non-ciliated tubules (distal tubules). No excretory pore could be identified.

The complete regeneration of the system took about 6 days after blastema formation. When RNAi experiments were carried out with blastema and whole worms it was found that one inhibition in particular prevented the development of the system. This was the inhibition of the expression of epidermal growth factor receptor (EGFR). There are five different genes in the EGFR family and the inhibition of the expression of only one of these (*EGFR-5*) prevented protonephridial development. Whole worms treated with RNAi (*EGFR-5*) became waterlogged (edema formation), suggested to be due to the lack of flame-cell growth and function in osmoregulation.

Co-expression of *EGFR-5* with *inx10* confirmed the presence of the epidermal growth factor receptor in the proximal tubules. The EGFR seems to be important both in branching morphogenesis and flame-cell formation. It was suggested that the flame-cell unit might initiate the branching morphogenesis. The RNAi (*EGFR-5*) effect on the blastema was not quite as efficient as with the whole organisms, perhaps because of lack of access of RNAi to the cells owing to poor endocytosis. It is unclear whether the proto-tubules are formed in the blastemal by invagination of cells from the surface or from condensation within the blastemal cell mass. Nevertheless, the dramatic effect on the need for the EGFR was demonstrated.

The work of Scimone *et al.* (2011) and Rink *et al.* (2011) gives a powerful indication of how the development of protonephridia may be investigated. *Schmidtea* is a valuable model system but there will be differences from the development and structure of parasitic trematodes. For example, there was no demonstrable excretion of resorufin in *Schmidtea* (Kusel, unpublished). The precise histological work of Cheng and Beir (1972) and Cort *et al.* (1954) needs to be linked in histology with the development of the excretory system as has been shown in careful work of Wang *et al.* (2013).

16.3 FUNCTION OF THE PROTONEPHRIDIAL SYSTEM

From an experimental point of view, the following methods have been used to ascertain the function of the nephridial tubules.

1. Micropuncture methods to draw fluids from the tubules (*Hymenolepis*).
2. Freezing point measurements of environmental and tissue fluids.
3. Enzyme distribution (e.g., alkaline phosphatase) and identification by histochemical methods.
4. Uptake and excretion of fluorescent compounds.
5. Inferences from ultra-structural studies, e.g., the presence of microvilli.
6. Localization of proteins such as tubulin and fluorescent lectins by confocal laser scanning microscopy.
7. Localization of P glycoproteins and Multidrug resistance proteins by immunostaining.
8. Localization, in nematodes, of glutamate-gated chloride channels.

The classical work of Webster (1971) and the subsequent review (Wilson and Webster 1974) has been the most influential work on protonephridia in the present era. Very little definitive work on the function of the protonephridial system of any organism has been published until very recently (Vu *et al.* 2015), for *Schmidtea*. From early work it is not clear that the protonephridial system is involved in osmoregulation (Wilson and Webster 1974) and reports of increase of flame-cell activity due to water influx during egg hatching may be due to general activation prior to hatching. It can also be concluded that there is little evidence for modification of the tubular fluid from the early experiments (Wilson and Webster 1974). In ingenious experiments with *Hymenolepis*, tubules were injected with both C14 inulin, an impermeable marker for tissue space and H3 glucose. Any increase in ratio of C14 and H3 indicated a reabsorption of glucose. This was found and this reabsorption could be inhibited by the glucose transport inhibitor phloridzin (Webster 1971;

Webster 1972). It is particularly difficult to analyze the contents of the protonephridial tubules. Methods using a refined stop flow perfusion system have been recommended (Webster 1971) and recent use of fluorescent pH and other ion and solute markers will prove of great value (Vu *et al.* 2015). Recent work using *Schmidtea* as a model indicates that edema formation can occur if flame-cell activity is compromised in RNAi experiments (Rink *et al.* 2011; Scimone *et al.* 2014).

The work of Rink *et al.* (2011) and Scimone *et al.* (2014) suggests the role of the flame-cell in osmoregulation. However the RNAi technique which results in edema in the worms could have affected other processes involved in ion transport. This needs to be investigated.

Vu *et al.* (2015) have reported work with *Schmidtea* which has greatly increased our knowledge and understanding of the protonephridial system. Fluorescent dextrans of 10 kDa and 500 kDa carrying different fluorescent labels were injected into the body tissues of adult worms and after washing it was shown that the 10 kDa dextran had entered the proximal tubule by filtration, while the 500 kDa was excluded.

They analyzed those genes expressed in the proximal tubules by searching for homologues in the vertebrate kidney proximal tubule. Numerous inorganic and organic ion transporters (solute carrier series) were identified, which suggests that modification of the tubule contents does take place. These solute carriers overlap in their function with the P glycoproteins (Hediger *et al.* 2013) but at present no localization of P glycoproteins in the excretory systems of schistosomes or *Schmidtea* has been demonstrated.

These experiments with *Schmidtea* indicate to us the kinds of questions and the methods by which we may obtain answers for trematodes using clear experimental techniques.

The questions we can ask are:

1. What is the mechanism of filtration? What powers the ciliary body in the flame-cell? Is it mitochondrial ATP or some other ion gradient?
2. What are the functions of the various tubule cells? What are the distribution and function of enzymes such as carbonic anhydrase and alkaline phosphatase in the cells?
3. How is the ultra-filtrate altered during its passage to the excretory pore from the flame-cell?
4. What macromolecules are excreted?

Early work has reported that the excretory system of a variety of trematodes contains macromolecules. Detailed studies of *Cyathocotyle bushiensis* (Erasmus 1972) using histochemical techniques for the light microscope and followed by electron-microscopy showed the likely secretion of lipid globules from the excretory tubule epithelium into the tubule lacuna near the excretory bladder of metacercariae. These lipid globules were shown by light microscopy to be released through the nephridiopore into the saline incubation medium. The lipid is excreted in large amounts on excystation of the metacercaria and secretion is resumed three days after infection of the final host (Dixon 1966). Calcareous bodies, believed to contain calcium carbonate, have been detected in the excretory tubules of the metacercaria of *Fasciola hepatica* (Martin and Bils 1964). The function of these is not known but may be related to removal of carbon dioxide or to act as a proteoglycan modifying agent during protease activity in the host as is found in the pre-acetabular gland of the cercariae of *Schistosoma mansoni* (Landsperger *et al.* 1982). The finding of carbonic anhydrase in the distal tubules may strengthen the idea for the role of these tubules in bicarbonate excretion. The secretory nature of the tubule epithelium is supported by its content of Golgi apparatus and the presence at the excretory pore of particulate material, as seen with electron microscopy (Wilson 1969).

These observations are valuable since they alert us to the great flexibility in function of the excretory system and its ability to respond to the environment at particular developmental stages. We can conclude that this system performs a number of functions in the life of the parasite, depending on environmental conditions and on the developmental stage.

16.4   FILTRATION

Howells (1969) described pores adjacent to and associated with the flame-cell surface (a fenestrated wall) and suggested that a nephrostome system exists in *Moniezia*. Permanent pores were not found in the flame-cell system of the miracidium of *Fasciola hepatica* by Wilson (1969). Whether or not nephrostomes are observable may depend on the species and the environmental conditions. The definition of meta- and protonephridia of Goodrich (1945) and Hyman (1951) may not be clear cut due to a more dynamic nature of the flame-cell, which may be able to alter its properties according to conditions. If a nephrostome system-operated, passage of macromolecules could occur without filtration. The dye Howells used to stain the protonephridial system was monolite fast blue (indanthrone blue, anthraquinone blue). This was an ingenious way to stain the protonephridial system and no reference can be found of its use since. It had been directly injected into the longitudinal vessels of *Moniezia*. The structure of the flame-cell is very complex and has been elucidated over the years and described by Wilson and Webster (1974), Ruppert and Smith (1988) and more recently by Vu *et al.* (2015). It is comprised of two cells, a cap cell and a tubule cell, the interdigitated projections from which make up the barrel. This is well described in Chapter 6 on the miracidium: the proximal quarter of the barrel of the flame-cell forms a filtration grille, fenestra or weir. It contains an outer row of thick bars (filaments) and an inner row of thin filaments. In cross section, thick filaments (160 nm across) alternate with thin filaments (100 nm across) forming a zigzag line. These filaments are cytoskeletal supports which strengthen this membrane complex which forms the filter. The mechanism by which molecules enter the excretory tubules through the flame-cell is still a matter for speculation. The flame-cell is thought to create a negative pressure within the barrel and proximal tubule by the beating of the flame cilia. The extent of the pressure was carefully calculated by Wilson and Webster (1974) as 12.1 dynes per cm^2, considered adequate for filtration of fluid through the flame-cell. The tissue extracellular fluid appears to move though the "filter" or "weir" which consists of the cell membrane of the barrel and extracellular macromolecules (glycocalyx) forming a matrix. Ions can pass though this filter (Wilson and Webster 1974) and so can macromolecules such as iron labelled dextran (Ruppert and Smith 1988) and fluorescent dextrans (Vu *et al.* 2015) of a size up to 10 kDa but excluding 500 kDa. The filter must contain some kind of pore to allow such molecules through, but other processes such as endocytosis certainly occur within the tubules and perhaps the flame-cell, so this process could also be involved.

The integrity of the filter is crucial for these processes to occur without leakage or loss of selective filtration. Vu *et al.* (2015) showed by RNAi knockdown studies in *Schmidtea*, that trans-membrane proteins NEPH1 and NPHS1, homologous with those found in the human nephron filtration barrier, were essential for the maintenance of the selectively permeable filtration of the fluorescent dextrans. The structural basis for the maintenance of size-selective filtration is unknown.

There are a number of signaling molecules and calcium binding proteins which have been localized in the excretory tubules of adult *Schistosoma mansoni*. In the schistosomula secretory-excretory products have been suggested to be secreted both to immunize the host and to modify the immune response by immune-regulation. It has, however, never been shown that any of those proteins localized in the adult or those clearly crucial in the life of the schistosomula are secreted from the excretory system via the nephridiopore. Fluid would need to be collected from the nephridiopore to definitely establish this point.

Good evidence for the presence and excretion of Bone morphogenetic protein (BMP) was shown by Freitas *et al.* (2009). BMP was localized in the collecting ducts of adult male *S. mansoni* and appeared in the excretory products collected from the worm in culture. However, Leutner *et al.* (2013) have localized the BMP in the testes and shows that it has a role in reproduction and it was not found in the excretory system. Such inconsistencies may be due to differences in sample preparation.

In a very ingenious paper, Moreno *et al.* (2010) showed that in the nematode *Brugia malayi* the excretory bladder of the microfilariae contains glutamate-gated chloride channels which are blocked by ivermectin. This has the effect of preventing the secretion of a variety of macromolecules (e.g., cysteine protease inhibitor) which are known to modulate the immune response of the host. Thus in the absence of these molecules the immune response can act to remove the microfilariae. This provides an explanation of the ability of ivermectin to clear the microfilariae very rapidly from the infected host after its administration. Thus in nematodes there is good evidence for the excretion of immuno-modulatory macromolecules through the excretory pore. A recent publication (Chehayeb *et al.* 2014) on *Ascaris* fluid compartments and secretory products supported this concept. This work suggested the possible presence of exosomes in excretory/secretory fluid (ES) which isan attractive possibility when considering the dramatic effects that exosomes may have on the immune system and other tissue cells in the host environment.

In *Fasciola* and *Echinococcus*, the exosomes collected in the ES are thought to come from the tegumental membrane. In the trematodes (Marcilla *et al.* 2012), however, it is not easy to determine the origin of ES molecules and whether the protonephridial system is involved in the release of exosomes.

The nematode excretory system is very different from that of the helminth, but these elegant experiments (Moreno *et al.* (2010) encourage the hypothesis that schistosomes excrete a variety of macromolecules via the nephridiopore.

The P glycoproteins and multidrug resistance proteins that are likely to participate in such excretion are known to have an ability to transport a wide variety of compounds (Greenberg 2014). The activity of these proteins is assumed in the work of Sato *et al.* (2002, 2004), but definite demonstration of their activity has only recently been obtained. The demonstration of Kasinathan *et al.* (2010) that the P glycoprotein SMDR2 is active in transporting rhodamine 123 and fluorescent praziquantel by transfected CHO cell membrane vesicles shows that these proteins can be active in membrane transport.

The role of ABC transporters in schistosomes is unknown. They are likely to be found in a large variety of regions in the parasite and they would be expected to be localized in the excretory system. However, it has been shown that there is localization of PgP (SMDR2) in the gut. Surface membrane localization has not been demonstrated but the PgP inhibitor tariquidar was shown to cause surface damage (Kasinathan *et al.* 2014) and therefore an effect on PgP might be expected in this site.

Experiments were carried out on the inhibition of the variety of ABC transporters by tariquidar and with three combinations of inhibitors such as zosuquidar, Ko143 and MK571. This latter combination is known to inhibit PgP, respectively (ABCB1), BRCP (ABCG2) and MRP1 (ABCC1) Two other combinations were also used. It was found by enhancement of adult worm activity in the presence of sub-lethal amounts of praziquantel (PZQ), that these inhibitors potentiated the activity of PZQ. They also enhanced the retention of fluorescent PZQ by the adult worms. Thus this is evidence that the ABC transporters do have an efflux ability in normal adult worms, which can be inhibited to make the PZQ more effective. This work was supported by RNAi experiments which suppressed the activity of SMDR2, SmMRP1, ABCA4 and ABCB6. This suppression led to the increase in retention of fluorescent PZQ and a greater sensitivity to sub-lethal PZQ (Kasinathan *et al.* 2014).

16.5 STUDIES WITH THE INTACT EXCRETORY SYSTEM IN LIVING *SCHISTOSOMA MANSONI*

We first became aware of the excretory tubules in schistosomula, when a very small proportion of *in vitro* transformed schistosomula cultured with fluorescent 10 kDa dextrans showed clear labeling of the tubules and excretory pore after washing the schistosomula, which were expected to have excluded this macromolecule (Kusel *et al.* 2009).

How did the dextran enter the schistosomula? We subsequently found that damage to the surface membrane appeared to increase labeling of the excretory area of a large proportion of schistosomula with Texas-red Bovine serum albumin (Tan *et al.* 2003). It was thought that much of this label had entered the excretory tubules via the nephridiopore. Although there is evidence that this may be a route during skin transformation (Thornhill *et al.* 2010a), this hypothesis needs to be re-evaluated after evidence that the excretory epithelium can exhibit rapid phagocytosis (Kusel *et al.* 2007). When we labeled living adult worms with Amplex Red (Invitrogen), in order to test for hydrogen peroxide, the resorufin dye released was concentrated into the excretory tubules of the adult male and female worm. Resorufin is a highly fluorescent redox dye of structure 7-Hydroxy-3H-phenoxazin-3-one sodium salt. This alerted us to a possible function for the excretory system, that of drug excretion. We extended the range of fluorescent substrates to which the adult worm pairs were exposed and found labeling of excretory tubules by many of the substrates of P glycoproteins (PgP) and multi-resistance proteins (MRP). Inhibitors of these proteins known from work with mammalian cells inhibited the concentration of fluorescent substrates into the tubules (Oliviera *et al.* 2006; Sato *et al.* 2002, 2004; Kusel *et al.* 2006; Couto *et al.* 2010).

Preliminary results with *Echinococcus* (Sato unpublished) showed very similar excretory properties for these substrates as was seen in the schistosome but no evidence for excretion of resorufin could be found in *Schmidtea* (Kusel, unpublished). This must make us aware that very great differences in function may be seen in similar structures!

16.6 *IN VITRO* STUDIES WITH ADULT *S. MANSONI* PAIRS

16.6.1 Interaction of Male and Female Worms

It was found that resorufin (a redox dye of structure 7-Hydroxy-3H-phenoxazin-3-one sodium salt) labeling of the male tubules could be intense at the same time as a total lack of labeling of the female tubules during incubation in resorufin and washing (Oliveira *et al.* 2006). This lack of labeling of the female could be reversed if the male surface was damaged (e.g., by a pin prick) or if the living parasite pair was cooled to 8°C. It seemed as if the male worm could inhibit the activity of the female even under *in vitro* conditions, unless damaged or by having its metabolism disrupted.

16.6.2 Disruption of Surface Membrane Rafts

The activity of the tubules of the male worms in particular could be greatly enhanced by prior extraction of the living parasites with methyl-cyclodextrin, a reagent which removes cholesterol and may disrupt the raft structure of the surface (Kusel and Gordon 1989). The effect described emphasizes the close correspondence of the state of the surface membrane and the activity of the excretory tubules. The effect observed (Fig. 16.5C) may be due to increased ion permeability after raft disruption or a change in endocytosis.

16.6.3 Disruption of Surface Membrane by Poly-l-lysine

After treatment with poly-l-lysine, the adult worms show enhanced activity of the nephridial tubules owing to membrane damage, but also show a great phagocytic activity of Texas-red BSA into the proximal tubules, particularly in the regions adjacent to areas of damage. This enhanced endocytosis may be a protective device to prevent leakage of internal contents and thus avoid alerting the host's immune system.

16.6.4 Disruption of the Worm's Musculature due to Praziquantel (PZQ)

PZQ treatment of adult worms *in vitro* causes immediate cessation of worm activity due to muscular contraction and an inhibition of the uptake of resorufin by the excretory tubules. This inhibition can be reversed by washing in culture medium (Kusel *et al.* 2006). The influx of calcium ions is thought to disrupt the tubular enzyme activity in the excretory system and the intense muscular contraction might inhibit the movement and function of the tubules.

16.6.5 The Medium Influences Tubule Activity

Labeling of tubules by resorufin in a phosphate free or calcium free medium is inhibited. This may be due to internal ion balance or due to changes in the surface membrane (e.g., membrane potential, Kusel *et al.* 2007).

16.6.6 Activity of Excretory System During Skin Penetration of Cercariae

It was observed that there was a very rapid uptake of a variety of membrane-impermeable molecules by schistosomula derived from cercariae as they penetrated mouse skin (Thornhill *et al.* 2010a). This unexpected increase in permeability of the parasite was explained initially as an uptake through the nephridiopore. Although this is still a viable hypothesis as far as penetrating cercariae are concerned (Thornhill *et al.* 2010b) migrating schistosomula taken from skin after 24 hr show a greatly enhanced permeability (Jeremias *et al.* 2015) with no staining of the nephridiopore. Thus it is possible for the surface membrane to show very flexible properties of permeability given the particular *in vivo* conditions.

This work reveals the properties of the excretory system of the schistosomulum as being an organelle that may have endocytic activity and also secretory activity as we have seen in earlier sections. In considering function, we must consider the whole worm and other organ systems, including gut, surface membrane and nervous system during their interaction with the protone-phridial system. Little studied properties, such as the importance of surface membrane potential, must also be considered as influencing the activity of the excretory system.

16.6.7 PgP Activity as Sensitive to Environment

There are many reports of the regulation of Pglycoproteins by inflammatory mediators (Heemskerk *et al.* 2010). Cholesterol in the membrane can be influenced by PgP activity (Zhu *et al.* 2010). Levels of calcium ions (Sulova *et al.* 2009), pH change (Lu *et al.* 2008) and phosphate (Prie *et al.* 2001) have all been shown to influence PgP activity. We propose a very speculative scheme based on some of these reports to show how PgP and MRP and other environmental factors could be central to the regulation of host-parasite interactions through their effects on the protonephridial system (Fig. 16.7). The possible role of PgP in host-parasite interactions has been well reviewed by Greenberg (2014). The possible role of PgP in transporting signaling molecules is particularly intriguing and is described in this review.

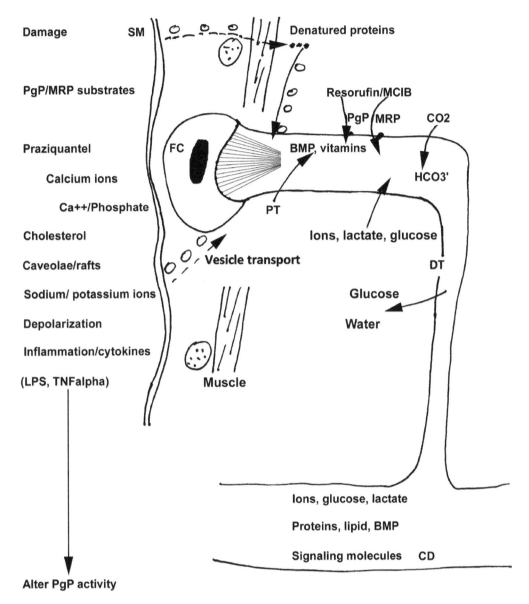

FIG. 16.7 Summary diagram showing activities of the schistosome excretory system in adult worm. SM, surface membrane; FC, flame-cell; PT, proximal tubule; DT distal tubule; CD collecting duct; PgP, P glycoprotein. The figure shows a diagrammatic longitudinal section of the adult tegument (male or female), showing a summary of activities of the flame-cell (FC) and proximal (PT) and distal (DT) excretory tubules. Damage to the surface membrane leads to endocytosis and vesicle transport of macromolecules and denatured proteins to the flame-cells and proximal tubules. PgP and MRP substrates can be excreted into tubules which can also excrete BMP (bone morphogenetic protein), vitamins, bicarbonate and a variety of lipids, proteins and carbohydrates into the collecting ducts (CD). Environmental changes can activate the P glycoproteins (PgP), such as membrane damage, removal of cholesterol, change in calcium ion concentration and inflammatory cytokines. Inhibition of PgP can also occur by membrane depolarization, praziquantel and low calcium and phosphate ions. Original.

16.7 CONCLUSIONS

We have attempted to describe the morphology and distribution and some of the properties of the excretory system of schistosomes. We have described some of the stimulating work being carried out with the free-living flatworm *Schmidtea* in order to gain vital understanding of the schistosome but the adaptation of the parasite may require changes in the activities of the excretory system out of all recognition to that seen in the non-parasitic species. The earlier studies had no evidence for a role in osmoregulation, but RNAi studies with *Schmidtea* could lead to "waterlogging" or edema formation, indicating a role for the flame-cells in osmoregulation. The flame-cells and proximal tubules have been shown by Vu *et al.* (2015) to filter and concentrate dextran molecules of 10 kDa but not 500 kDa. Protein transport and ion transport have also been suggested for these tubules on the basis of microarray studies. The distal tubule contains carbonic anhydrase and therefore may be involved in regulation of pH in a bicarbonate/CO2 equilibrium. Schistosomes excrete PgP and MRP substrates and may concentrate praziquantel for excretion. Waste products such as lactate and some fatty acids may be excreted but there is no firm evidence for this except in *Hymenolepis*. Protein, lipids and carbohydrates may be excreted, especially those involved in regulating the host's immune response. Other functions of excreted materials may be to make nutrients available from the host cells. Damage repair is very evident in *Schmidtea* and in schistosomes. Surface membrane damage can also be accompanied by endocytosis of membrane fragments and environmental proteins by the flame-cells and proximal tubules. Similar pathways may exist in both free-living and parasitic species for the generation of stem cells and the function of the tubules. The discovery in *Schmidtea* of the cubilin gene as a marker for the flame-cell may give a vital clue to the function of the excretory system. Cubilin is the receptor for the intrinsic factor/Vitamin B12 complex. Vitamin B12 is essential for the parasite, as it is to the host in all tissues, for growth and division. Thus the excretory system could concentrate and distribute Vitamin B12 as in a circulatory system. Other growth factors such as bone morphogenetic protein have also been found in the schistosome excretory system.

Much of this is speculative and is summarized in Figure 16.7 which attempts to show the dynamic interaction of the flame-cell with many other intra-cellular organelles and environmental factors.

The manifestations *in vivo* of schistosome excretory activity are unknown. It is essential to discover ways of examining this, perhaps with the appropriate fluorescent substrates. One thing is certain. Unexpected discoveries will be made in the future!

16.8   ACKNOWLEDGEMENTS

I should like to thank Barrie Jamieson for constant support and advice during the writing of this chapter. Jochen Rink gave some very valuable ideas and very recent data about tubule function in *Schmidtea*. Philip Newmark made me aware of many beautiful images of both schistosomes and *Schmidtea* and gave permission for the use of some of them. Richard Martin continued our friendship by advising on *Ascaris* and *Brugia* and families. Robert Greenberg shared the latest results of his laboratory and read the Chapter very carefully. Christoph Grevelding told us of a grandchild, football, rock bands and signaling molecules. Alan Wilson shared his expertise on protonephridia with us. I am most grateful to Caroline Morris for Figure 16.1 and Heather Collin for Figure 16.7. Many thanks to all of you.

16.9   LITERATURE CITED

Bartolomaeus T, Ax P. 1992. Protonephridia and metanephridia – their relation within the bilateria. Journal of Zoological Systematics and Evolutionary Research 30: 21–45.

Bogers JJ, Nibbeling HA, van Marck EA, Deelder AM. 1994. Immunofluorescent visualization of the excretory and gut system of *Schistosoma mansoni* by confocal laser scanning microscopy. American Journal of Tropical Medicine and Hygiene 50(5): 612–619.

Chehayeb JF, Robertson AP, Martin RJ, Geary TG. 2014. Proteomic analysis of adult *Ascaris suum* fluid compartments and secretory products. PLoS Neglected Tropical Diseases 8(6): e2939. doi: 10.1371/journal.pntd0002939.

Collins JJ, King RS, Cogswell A, Williams DL, Newmark PA. 2011. An atlas for *Schistosoma mansoni* organs and life cycle stages using cell type-specific markers and confocal microscopy. PLoS Neglected Tropical Diseases 5(3): e1009.

Cort WW, Ameel DJ, Van der Woude A. 1954. Germinal development in the sporocysts and rediae of the digenetic trematodes. Experimental Parasitology 3: 185–225.

Couto FF, Coelho PM, Araujo NJ, Kusel JR, Katz N, Mattos AC. 2010. Use of fluorescent probes as a useful tool to identify resistant *Schistosoma mansoni* isolates to praziquantel. Parasitology 137(12): 1791–1797.

Cheng TC, Beir JW. 1972. Studies on molluscan schistosomiasis: an analysis of the development of the cercaria of *Schistosoma mansoni*. Parasitology 64: 129–141.

Dixon KE. 1966. The physiology of excystment of the metacercaria of fasciola hepatica. Parasitology 56: 431–456.

Erasmus DA. 1972. *The Biology of the Trematodes*. Edward Arnold. p. 312.

Freitas TC, Jung E, Pearce EJ. 2009. A bone morphogenetic protein homologue in the parasitic flatworm, *Schistosoma mansoni*. International Journal for Parasitology 39(3): 281–287.

Goodrich ES. 1945. Nephridia and genital ducts since 1895. Quarterly Journal of Microscopical Science S2-86: 303–392.

Greenberg RM. 2014. Schistosome ABC multidrug transporters: from pharmacology to physiology. International Journal for Parasitology: Drugs and Drug Resistance 4: 301–309.

Hediger MA, Clemoncon B, Burrier RE, Brufford EA. 2013. The ABCs of membrane transporters in health and disease (SLC series): Introduction. Molecular Aspects of Medicine 34: 95–107.

Heemskerk S, Peters JG, Louisse J, Sagar S, Russel FG, Masereeuw R. 2010. Regulation of P-glycoprotein in renal proximal tubule epithelial cells by LPS and TNF-alpha. Journal of Biomedical Biotechnology 525: 180.

Howells RE. 1969. Observations on the nephridial system of the cestode *Moniezia expansa*. Parasitology 59: 449–459.

Hyman LH. 1951. *The Invertebrates. Vol. 2*. New York: McGraw-Hill.

Jeremias W de J, de Cunha Melo DA, Baba EH, Coelho PM, Kusel JR. 2015. The skin migratory stage of the schistosomulum of *Schistosoma mansoni* has a surface showing greater permeability and activity in membrane internalisation than other forms of skin or mechanical schistosomula. Parasitology 142: 1143–1151.

Kasinathan RS, Morgan WM, Greenberg RM. 2010. *Schistosoma mansoni* express higher levels of multidrug resistance-associated protein 1 (SmMRP1) in juvenile worms and in response to praziquantel. Molecular and Biochemical Parasitology 173(1): 25–31. doi: 10.1016/j.molbiopara.2010.05.003. Epub 2010 May 12.

Kasinathan RS, Greenberg RM. 2012. Pharmacology and potential physiological significance of schistosome multidrug resistance transporters. Experimental Parasitology 132(1): 2–6. doi: 10.1016/j.exppara.2011.03.004. Epub 2011 Mar 21.

Landsperger WJ, Stirewalt MA, Dresden MH. 1982. Purification and properties of a proteolytic enzyme from the cercariae of the trematode parasite *Schistosoma mansoni*. Biochemical Journal 201: 137–144.

Leutner S, Oliveira KC, Rotter B, Beckmann S, Buro C, Hahnel S, Kitajima JP, Verjovski-Almeida Winter P, Grevelding CG. 2013. Combinatory microarray and Super SAGE analyses identify pairing-dependently transcribed genes in *Schistosoma mansoni* males, including follistatin. PLoS Neglected Tropical Diseases 7(11): e2532.

Kusel JR, Gordon JF. 1989. Biophysical studies on the schistosome surface and their relevance to its properties under immune and drug attack. Parasite Immunology 11: 431–451.

Kusel JR, Al-Adhami BH, Doenhoff MJ. 2007. The schistosome in the mammalian host: understanding the mechanisms of adaptation. Parasitology 134: 1477–1526.

Kusel JR, McVeigh P, Thornhill JA. 2009. The schistosome excretory system: a key to regulation of metabolism, drug excretion and host interaction. Trends in Parasitology 25(8): 353–358.

Kusel JR, Oliveira FA, Todd M, Ronketti F, Lima SF, Mattos AC, Reis KT, Coelho PMZ, Thornhill JA, Ribeiro F. 2006. The effects of drugs, ions and poly-l-lysine on the excretory system of *Schistosoma mansoni*. Memorias Instituto do Oswaldo Cruz 101: Suppl 1: 293–298.

Lu Y, Pang T, Wang J, Xiong D, Ma L, Li Q, Wakabayashi S. 2008. Down-regulation of P-glycoprotein expression by sustained intracellular acidification in K562/Dox cells. Biochemical and Biophysical Research Communications 377(2): 441–446.

Marcilla AM, Trelis MA, Cortes AJ, Sotillo J, Cantalapiedra F, Minguez MT, Valero ML, Sanchez del Pino MM, Munoz-Antoli C, Toledo R, Bernal D. 2012. Extracellular vesicles from parasitic helminths contain specific excretory/secretory proteins and are internalized in intestinal host cells. PLoS One 7(9): e45974, 2.

Martin WE, Bils RF. 1964. Trematode excretory concretions: formation and fine structure. Journal of Parasitology 50: 337–344.

Mecozzi B, Rossi A, Lazzaretti P, Kady M, Kaiser S, Valle C, Cioli D, Klinkert MQ. 2000. Molecular cloning of *Schistosoma mansoni* calcineurin subunits and immunolocalization to the excretory system. Molecular and Biochemical Parasitology 110(2): 333–343.

Moreno Y, Nabhan JF, Solomon J, Mackenzie CD, Geary TG. 2010. Ivermectin disrupts the function of the excretory-secretory apparatus in microfilariae of *Brugia malayi*. Proceedings of the National Academy of Sciences USA 107(46): 20120–20125.

Newmark PA, Wang Y, Chong T. 2008. Germ cell specification and regeneration in planarians. Cold Spring Harbour Symposium for Quantitative Biology 73: 573–581.

Oliveira FA, Kusel JR, Ribeiro F, Coelho PMZ. 2006. Responses of the surface membrane and excretory system of *Schistosoma mansoni* to damage and to treatment with praziquantel and other biomolecules. Parasitology 132(Pt 3): 321–330.

Prie D, Couette S, Fernandes I, Silve C, Friedlander G. 2001. P-glycoprotein inhibitors stimulate renal phosphate reabsorption in rats. Kidney International 60(3): 1069–1076.

Rink J, Vu H, Sanchez TK, Alvarado AS. 2011. The maintenance and regeneration of the planarian excretory system is regulated by EGFR signalling development. Development 138: 3769–3780.

Rohde K, Watson N, Roubal F. 1989. Ultrastructure of the protonephridial system of *Dactylogyrus* sp. and an unidentified ancyrocephaline (Monogenea: Dactylogyridae). International Journal for Parasitology 19 (8): 859–864.

Rohde K, Watson NA, Roubal FR. 1992. Ultrastructure of the protonephridial system of *Anoplodiscus cirrusspiralis* (Monogenea Monopisthocotylea). International Journal for Parasitology 22(4): 443–457.

Ruppert EE, Smith PR. 1988. The functional organization of filtration nephridia. Biological Reviews 63: 231–258.

Sato H, Kusel JR, Thornhill JA. 2002. Functional visualization of the excretory system of adult *Schistosoma mansoni* by the fluorescent marker resorufin. Parasitology 125(6): 527–535.

Sato H, Kusel JR, Thornhill JA. 2004. Excretion of fluorescent substrates of mammalian multidrug resistance-associated protein (MRP) in the *Schistosoma mansoni* excretory system. Parasitology 128(1): 43–52.

Scimone ML, Srivastava M, Bell GW, Reddien PW. 2011. A regulatory program for excretory system regeneration in planarians. Development 138(20): 4387–4398. doi: 10.1242/dev.068098.

Scimone L, Kravarik KMS, Lapan SW, Reddien PW. 2014. Neoblast specialization in regeneration of the planarian *Schmidtea mediterranea*. Stem Cell Reports 3(2): 339–352.

Smyth JD, Halton DW. 1983. *The Physiology of Trematodes*. Cambridge University Press.

Sulova Z, Seres M, Barancik M, Gibalova L, Uhrik B, Polekova L, Breier L. 2009. Does any relationship exist between P-glycoprotein-mediated multidrug resistance and intracellular calcium homeostasis. General Physiology and Biophysics 28 Focus issue: F89–F95.

Tan HH, Thornhill JA, Al-Adhami BH, Akhka A, Kusel JR. 2003. A study of the effect of surface damage on the uptake of texas red-BSA by schistosomula of *Schistosoma mansoni*. Parasitology 126(3): 235–240.

Thornhill J, Kusel J, Oliviera FA, Ribeiro F, Lima SF, Coelho PMZ, McVeigh P, Mattos AC. 2010a. Uptake of macromolecules by cercariae during skin penetration and transformation to schistosomula (*Schistosoma mansoni*). Memorias Instituto do Oswaldo Cruz 105(4): 387–390.

Thornhill JA, McVeigh P, Jurberg AD, Kusel JR. 2010b. Pathways of influx of molecules into cercariae of *Schistosoma mansoni* during skin penetration. Parasitology 137: 1089–1099. doi: 10.1017/S0031182009991983.

Vu H T-K, Rink JC, McKinney SA, McClain M, Lakshmanaperumal N, Alexander R, Alvarado AS. 2015. Stem cells and fluid flow drive cyst formation in an invertebrate excretory organ. eLife: 10.7554/eLife.07405.

Wang B, Collins III JJ, Newmark PA. 2013. Functional genomic characterization of neoblast-like stem cells in larval *Schistosoma mansoni*. eLife 2: e00768.

Webster LA. 1971. *The Physiology of Protonephridia*. A thesis for the degree of Doctor of Philosophy University of York. Heslington, York, YO1 5DD.

Webster LA. 1972. The absorption of glucose, lactate and urea from the protonephridial tubules of *Hymenolepis diminuta*. Comparative Biochemistry and Physiology 41A: 861–868.

Wilson RA. 1969. The fine structure of the protonephridial system in the miracidium of *Fasciola hepatica*. Parasitology 59: 461–467.

Wilson RA, Webster LA. 1974. Protonephridia. Biological Reviews 49: 127–160.

Wippersteg V, Ribeiro F, Liedtke S, Kusel JR, Grevelding CG. 2003. The uptake of texas red-BSA in the excretory system of schistosomes and its colocalization with ER60 promoter-induced GFP in transiently transformed adult males. International Journal for Parasitology 33: 1139–1143.

Zhu MLE, Moyec L, Starzec A, Stierle V, Marbeuf-Gueye C. 2010. Caveolin-1 and doxorubicin-induced P-glycoprotein modulate plasma cholesterol membrane accessibility in erythrolymphoblastic cell line. Anticancer Research 30(9): 3451–3458.

CHAPTER

Acute Schistosomiasis

Li Yuesheng[1,2] and *Allen GP Ross*[3,]*

17.1 INTRODUCTION

Katayama syndrome is an early clinical manifestation of schistosomiasis that occurs several weeks post-infection with *Schistosoma* spp. (trematode) worms. Because of this temporal delay and its non-specific presentation, it is the form of schistosomiasis most likely to be misdiagnosed by travel medicine physicians and infectious disease specialists in non-endemic countries. Katayama syndrome appears between 14–84 days after non-immune individuals are exposed to first schistosome infection or heavy reinfection. Disease onset appears to be related to migrating schistosomula and egg deposition with individuals typically presenting with nocturnal fever, cough, myalgia, headache and abdominal tenderness. Serum antibodies and schistosome egg excretion often substantiate infection if detected. Diffuse pulmonary infiltrates are found radio logically and almost all cases have eosinophilia and a history of water contact before presentation of clinical symptoms. Patients respond well to regimens of praziquantel with and without steroids. Artemisinin treatment given early after exposure may decrease the risk of the syndrome.

17.2 CLINICAL MANIFESTATIONS

Percutaneous penetration of the cercariae can provoke a temporary urticarial rash that can manifest within hours and persist for days as maculopapular lesions (Fig. 17.1) (Appleton 1984; Visser *et al.* 1995; Lambertucci *et al.* 1997; Yasaway 2004). In temperate climate zones, a similar "swimmers itch" is also frequently seen with avian trematode cercariae (Bouree and Caumes 2004). The presentation of delayed-onset dermatitis, manifested as urticaria or angioedema, can occur within 1–12 weeks after heavy exposure to cercariae-infested water with initial symptoms subsiding within 48 hours (Ross *et al.* 2002; Corachan 2002; CDC 2005; Bottieau *et al.* 2006). The skin lesions are often puritic. They are discrete erythematous raised lesions that vary in size from 1–3 cm (Fig. 17.1) (Ross *et al.* 2002; Corachan 2002). Because of the temporal association with water exposure, the diagnosis is usually suspected clinically.

[1] Hunan Institute of Parasitic Diseases, Yueyang, Hunan Province, World Health Organisation Collaborating Centre for Research and Control on Schistosomiasis in Lake Region, Yueyang, Hunan, 414000, People's Republic of China.
[2] QIMR Berghofer Medical Research Institute, 300 Herston Rd, Brisbane QLD 4006, Australia.
[3] Tropical Medicine and Global Health, Griffith University, Logan Campus, University Drive, Meadowbrook QLD 4131, Australia.
* Corresponding author

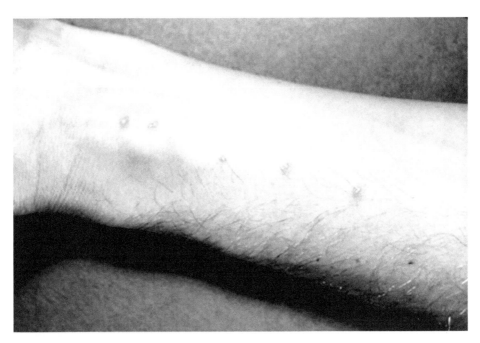

FIG. 17.1 Clinical features of acute schistosomiasis: cercarial dermatitis. Image from public domain at: https://en.wiki2.org/wiki/Schistosomiasis#/media/File:Schistosomiasis_itch.jpeg.

As noted above, Katayama syndrome is a systemic hypersensitivity reaction against the migrating schistosomula and eggs and can occur within 14–84 days after a primary infection (Ross *et al.* 2002). In many cases, acute infections are asymptomatic. Symptoms that do manifest, however, are thought to be a result of an allergic reaction during larval migration and early oviposition by adult worms and will vary in severity depending on the infecting species (Hiatt *et al.* 1979; Visser *et al.* 1995; Cooke *et al.* 1999; Ross *et al.* 2002; Bottieau *et al.* 2006). The delay in symptoms by weeks to months after exposure adds to the diagnostic difficulty. Disease onset is usually sudden, producing many non-specific symptoms, such as fever, fatigue, myalgia, malaise, urticaria, non-productive cough, eosinophilia and patchy pulmonary infiltrates on chest radiograph (Fig. 17.2) (Rocha *et al.* 1995). Examples of pronounced radiological alterations include thickening of bronchial walls and beaded micronodulation in the lower pulmonary fields (Rocha *et al.* 1995). Abdominal symptoms may develop within a few weeks, because of migration of juvenile worms and egg deposition of the mature worms (Ross *et al.* 2002). High-grade nocturnal fever and eosinophilia is generally present and should alert the clinician to the parasitic nature of the illness, once other potential causes of fever and eosinophilia have been ruled out (Ribeiro *et al.* 2002). Most patients recover spontaneously after 2–10 weeks, whereas some develop persistent and more serious, disease with weight loss, dyspnoea, diarrhea, diffuse abdominal pain, hepatomegaly and generalised rash.

Within chronically exposed populations, Katayama syndrome caused by *Schistosoma mansoni* and *S. haematobium* is rarely reported. A potential explanation for this phenomenon is that in-utero sensitization might decrease the severity of common symptoms of Katayama syndrome in chronically exposed populations; however, it may be equally likely that cases from endemic areas simply go unrecognized (King *et al.* 1998). Katayama syndrome caused by *S. japonicum* infections is not restricted to primary infection, but also occurs in people living in endemic areas with a history of previous infection. In China, for example, rebound epidemics have typically been reported in endemic communities exposed to floods (Chen 1993; Ross *et al.* 2001).

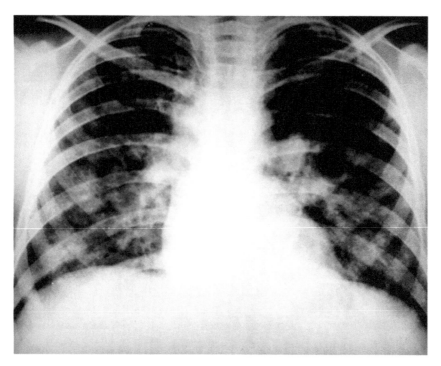

FIG. 17.2 Clinical features of acute schistosomiasis: pulmonary infiltrates. Image from public domain at: http://www.isradiology.org/tropical_deseases/tmcr/chapter2/lungs.htm.

17.3   DIAGNOSIS

Travelers to endemic areas, who were at risk of exposure to schistosomal cercariae, are advised to contact a tropical medicine or infectious disease specialist. Schistosomiasis is potentially a long-lived infection and acute schistosomiasis may be difficult to diagnose in lightly infected travelers, since they are likely to possess low worm burden, with low egg production and excretion (Tarp *et al.* 2000). Schistosomiasis is, however, associated with malignant transformation. The association between *S. haematobium* infection and squamous cell carcinoma of the bladder is better established than the risk of hepatocellular carcinoma in the two hepatic forms (*S. mansoni* and *S. japonicum*) (Ross *et al.* 2002). In either case, it is advisable to find and treat asymptomatic cases. For these reasons, it is worth enquiring about freshwater exposure dating back many years (Blanchard 2004).

In previously non-exposed patients, such as travelers, diagnosis of Katayama syndrome can often be confirmed by either detecting the eggs of the different schistosome species or by identifying the presence of antischistosomal antibodies (Doenhoff *et al.* 2004; Bierman *et al.* 2005; Bottieau *et al.* 2006). Work in the 1990s demonstrated that detection of antischistosomal antibodies with adult worm microsomal antigens was an effective means of diagnosis in previously unexposed individuals with high sensitivity and specificity, since seropositivity will likely occur before the onset of symptoms (Cetron *et al.* 1996). Although egg detection and serology can be effective for confirming an infection among travelers, there may not be detectable antibody levels or eggs being excreted at the time of the first clinical presentation (Hamilton *et al.* 1998; Yin-Chang 2005). Neither a negative serological test, nor the absence of eggs is sufficient evidence to rule out a diagnosis of Katayama syndrome. It often takes time for oviposition and seroconversion to occur (Bottieau *et al.* 2006). As a consequence, these tests are often repeated several weeks after the diagnosis is suspected clinically.

Routine screening following freshwater exposure in an endemic area should consist of a full blood count, absolute eosinophil count, serology (particularly the presence of IgE antibodies), urinanalysis and fecal microscopy (Rocha *et al.* 1995; Blanchard 2004; Bierman *et al.* 2005). In asymptomatic travelers, this might be best done at 3–6 months post travel, at a time when skin tests for tuberculosis would be advised, to decrease the number of false-negative tests. Schistosomal serology results frequently remain positive for life despite implementing adequate treatment regimens. As a result, serology is often not helpful in individuals born and raised in schistosome endemic countries (Blanchard 2004). Urinanalysis will often reveal microscopic haematuria, particularly in patients from endemic areas and fecal samples should be examined using a concentration (or centrifugal) method (Blanchard 2004).

In field studies and when the diagnosing clinician already suspects schistosome infection, the Kato-Katz thick smear technique is used to quantify egg numbers in fecal matter, as well as to determine egg morphology (Ross *et al.* 2002). Semen microscopy may also reveal the presence of schistosome ova and red cells (Blanchard 2004). The most sensitive test for the hepatic forms of schistosomiasis are superficial rectal biopsies, but this invasive procedure is rarely indicated. With *S. haematobium* the most sensitive examination is to centrifuge an entire morning urine collection and examine its sediment.

Despite its expediency and previously recorded success, issues surrounding the sensitivity, specificity and cross-reactivity still prevail and are problematic with the serological diagnosis of acute schistosomiasis (Blanchard 2004). Additionally, indirect haemagglutination (IHA) sensitivity to Katayama syndrome is estimated to be approximately 70–80% (van Gool *et al.* 2002; Bottieau *et al.* 2006). ELISAs provide no added diagnostic value to samples already analyzed by IHA. This is largely because of the fact that most commercially available ELISAs use egg antigens and not adult worm antigens (Bottieau *et al.* 2006). A four year investigation of IHA-confirmed acute schistosomiasis cases (without ELISA) among travelers reported a positive diagnosis in 15 out of 23 (65%) of patients at first presentation and by ova detection in only 5 out of 23 (22%) of patients (Bottieau *et al.* 2006). Previous research using antigen detection as an alternative to antibody detection in travelers with acute schistosomiasis has been reported (van Lieshout *et al.* 1997; van Lieshout *et al.* 2000). However, the reliability of this method in detecting early schistosome infections in human beings remains to be determined.

The occurrence of other clinical indicators (e.g., eosinophila) might not be sufficiently specific to diagnose Katayama syndrome since eosinophilia is common with other helminth infections. It may also be helpful to rule out other common travel-related infections, such as tuberculosis, strongyloidiasis or trichuriasis (Blanchard 2004). Some viral infections (such as hepatitis B, hepatitis C and HIV) can also cause an increase in polyclonal immunoglobins (specifically IgM), thus leading to false-positive serological tests (Blanchard 2004).

17.4   TREATMENT

Treatment of Katayama syndrome is often initiated on clinical grounds and infection should be confirmed retrospectively through serological analysis or stool examination or both. Safe and effective oral drugs are readily available for treating schistosomiasis (Ross *et al.* 2002).

Praziquantel is the foremost-prescribed drug and is quite effective against the adult worms of all *Schistosoma* species (Doherty *et al.* 1996; Ross *et al.* 2002; Fenwick *et al.* 2003; Magnussen 2003; Olds 2003; Chen 2005; Xiao 2005). As mentioned, patients who present with Katayama syndrome may receive additional treatment with corticosteroids to reduce the severity of the immunological reaction (Ross *et al.* 2002). Because of the lack of clinical trials in non-immune populations, the exact treatment regimen for Katayama syndrome with praziquantel is not known. Standard treatment of chronic schistosomiasis is 60 mg/kg of praziquantel in divided doses. Treatment failures with this dose are reported, particularly in endemic areas and among children (Botros *et al.*

2005; Alonso *et al.* 2006). It is unknown whether this is because of praziquantel-resistant worms or because the drug works in concert with the host immune response, which takes time to develop or because of the presence of migrating larvae that are not susceptible to praziquantel. Depending on the species of schistosome responsible for infection, treatment with 40 mg/kg for three days (for *S. mansoni* or *S. haematobium*) and 60 mg/kg for 6 days (for *S. japonicum*) has been reported to work well. The latter dose may also be effective if the standard single-dose of 40 mg/kg of praziquantel fails to clear *S. haematobium* infections.

A derivative of the antimalarial drug, artemisinin (artemether), has recently been studied for the control of schistosomal infections. The addition of this drug to the treatment of schistosomiasis holds the potential for increasing the cure rate for acute schistosomiasis. Artemether is effective against young schistosomula (e.g., *S. japonicum*) and also has a prophylactic effect against *S. mansoni* (Utzinger *et al.* 2000; Xiao *et al.* 2000; Xiao *et al.* 2002; Li *et al.* 2005; Xiao 2005). Although there is promising evidence on the effectiveness of this drug against acute schistosomiasis, artemether is currently used to treat malaria and it is feared that its widespread use in endemic countries, particularly in Africa, could create selective pressures for resistant forms of *Plasmodium* spp. (Uzinger *et al.* 2000). As a result, this drug is likely to be used sparingly in field applications, but is being investigated for the treatment of Katayama syndrome, particularly in travelers returning from endemic areas.

Katayama syndrome can recur, even after treatment. This might be because—as referred to previously—praziquantel is relatively ineffective against migrating larvae that eventually mature and begin egg deposition at a later time. Clinically, a patient appears to improve, often with resolution in fever, only to have symptoms (particularly fever) recur days or weeks later. If steroids are used in the initial treatment, recurrence can take place during the steroid taper. Recurrences are normally self-limiting, even without repeat treatment.

Developing novel regimens for treating Katayama syndrome may be an effective method of pharmacological prevention in non-immune travelers. A study of 18 travelers has shown that early treatment (i.e., 10–15 days post-exposure) with praziquantel was less effective than later treatment (i.e., 28–40 days post-exposure) and 17 (94%) of the study participants developed chronic schistosomiasis (Grandiere-Perez *et al.* 2006). It is reasonable to postulate that prophylactic efficacy will increase and relapse rates might be lowered if primary treatment with praziquantel is given with artemether, because this drug kills immature parasites. Alternatively, combination treatments of both praziquantel and artemether are often used in this situation.

Because no treatment regimen for schistosomiasis is 100% effective and because chronic schistosomiasis is both asymptomatic and has a small risk of malignant transformation, we recommend re-screening individuals with Katayama fever 6–8 weeks after their symptoms subside. Individuals who remain infected should be re-treated with a standard 40–60 mg/kg praziquantel dose.

17.5 PREVENTION

Exposure to schistosomiasis is a health risk to those who travel to endemic areas. It is a growing problem for tourists, travelers and other people who are accidentally exposed to schistosome cercariae (Jelinek *et al.* 1996; Hatz 2005). Most cases in western travel clinics are imported from Sub-Saharan Africa. Frequent sources of infection include Lake Malawi, Lake Victoria and Lake Volta, the Zambesi and Niger deltas and lake resorts in South Africa. Activities that can lead to contact with infested water range from bathing and swimming to scuba diving, water skiing and rafting.

There are many myths regarding exposure to schistosomiasis (e.g., fast-flowing water is snail free and wetsuits protect against cercarial penetration) (Blanchard 2004; CDC 2005). Studies that have described the burden of acute schistosomiasis among travelers have rarely outlined effective recommendations for prevention, besides avoidance of wading, swimming or any other contact with fresh water in endemic areas. It should be noted that the US Centers for Disease Control and Prevention (CDC) publishes an annual guide for international travel that provides recommendations to prevent infection with specific infectious diseases and would be enlightening for both physicians and travelers (CDC 2005).

Fresh water that comes directly from canals, lakes, rivers or springs in endemic countries could contain schistosome cercariae. Travelers cannot become infected from contact with salt water (CDC 2005). Other than assuming that all fresh water is infested, there is often no convenient way for travelers to distinguish between water that is infested and water that is not. To reduce the risk of infection, water that is to be used for washing can be heated to 50°C (122°F) for five minutes, killing any cercariae present or can be collected and left to stand for two days since cercariae are often not infective for more than 12–24 hours (CDC 2005). If these measures are unavailable and exposure to infested fresh water is suspected, vigorous towel drying and application of insect repellents containing diethyltoluamide have been reported to remove and help prevent cercariae from entering the skin (CDC 2005). However, the CDC warns that these are not rigorously supported preventive measures and are only recommended if accidental exposure has occurred. Returning travelers, who may have been exposed to schistosome-infested water, are advised to visit a physician for screening (CDC 2005).

17.6   CHAPTER SUMMARY

The recent increase in professed "adventure tourism" has resulted in increased exposure to schistosomiasis in non-immune populations. The most challenging presentation is Katayama syndrome because of the delay in the onset of symptoms following exposure as well as the diverse and non-specific symptomatology of the syndrome. Both these issues often lead to misdiagnosis. A better knowledge of this syndrome should lead the astute physician to the correct diagnosis. For reasons not yet fully clear, the recurrence of Katayama symptoms (even after treatment) is not unusual and can complicate the medical management of this disease. Additional studies are needed to better define optimum treatment options and strategies. However, artemether might improve our ability to treat active schistosomiasis, particularly in travelers. Further clinical and epidemiological research in the field of travel medicine will aid in resolving issues related to the pathogenesis of Katayama syndrome and the relation between exposure to schistosomes and the acquisition of infection.

17.7   LITERATURE CITED

Alonso D, Munoz J, Gascon J, Valls ME, Corachon M. 2006. Failure of standard treatment with praziquantel in two returned travelers with *Schistosoma haematobium* infection. American Journal Tropical Medicine Hygiene 74: 342–44.

Appleton CC. 1984. Schistosome dermatitis—an unrecognised problem in South Africa. South African Medical Journal 65: 467–69.

Bierman WF, Wetsteyn JC, van Gool T. 2005. Presentation and diagnosis of imported schistosomiasis: relevance of eosinophilia, microscopy for ova and serology. Journal of Travel Medicine 12: 9–13.

Blanchard TJ. 2004. Schistosomiasis. Travel Medicine and Infectious Diseases 2: 5–11.

Botros S, Sayed H, Amer N, El-Ghannam M, Bennett JL, Day TA. 2005. Current status of sensitivity to praziquantel in a focus of potential drug resistance in Egypt. International Journal Parasitology 35: 787–91.

Bottieau E, Clerinx J, de Vega MR, Van den Enden E, Colebunders R, Van Esbroeck M, Vervoort T, Van Gompel A, Van den Ende J. 2006. Imported Katayama fever: clinical and biological features at presentation and during treatment. Journal of Infection 52: 339–45.

Bouree P, Caumes E. 2004. Cercarialdermatitis. La Presse Médicale 33: 490–93 (in French).

CDC. Schistosomiasis. In: Travelers' health: yellow book, health information for international travel, 2005–2006. Atlanta, GA: Centers for Disease Control and Prevention, 2005.

Cetron MS, Chitsulo L, Sullivan JJ, Pilcher J, Wilson M, Noh J, Tsang VC, Hightower AW, Addiss DG. 1996. Schistosomiasis in Lake Malawi. Lancet 348: 1274–78.

Chen MG. 1993. *Schistosoma japonicum* and *S japonicum*-like infections: epidemiology, clinical and pathological aspects. *In*: Jordan P, Webbe G, Sturrock FS, editors. *Human Schistosomiasis.* Wallingford: CAB International 1993: 237–70.

Chen MG. 2005. Use of praziquantel for clinical treatment and morbidity control of Schistosomiasis japonica in China: a review of 30 years' experience. Acta Tropica 96: 168–76.

Cooke GS, Lalvani A, Gleeson FV, Conlon CP. 1999. Acute pulmonary schistosomiasis in travellers returning from Lake Malawi, sub-Saharan Africa. Clinical Infectious Diseases 29: 836–39.

Corachan C. 2002. Schistosomiasis and international travel. Clinical Infectious Diseases 35: 446–50.

Doenhoff MJ, Chiodini PL, Hamilton JV. 2004. Specific and sensitive diagnosis of schistosome infection: can it be done with antibodies? Trends Parasitology 20: 35–39.

Doherty JF, Moody AH, Wright SG. 1996. Katayama fever: an acute manifestation of schistosomiasis. British Medical Journal 313: 1071–72.

Fenwick A, Savioli L, Engels D, Bergquist NR, Todd MH. 2003. Drugs for the control of parasitic diseases: current status and development in schistosomiasis. Trends in Parasitology 2003; 19: 509–15.

Grandiere-Perez L, Ansart S, Paris L, Faussart A, Jaureguiberry S, Grivois JP, Klement E, Bricaire F, Danis M, Caumes E. 2006. Efficacy of praziquantel during the incubation and invasive phase of *Schistosoma haematobium* schistosomiasis in 18 travelers. American Journal of Tropical Medicine and Hygiene 74: 814–18.

Hamilton JV, Klinkert M, Doenhoff MJ. 1998. Diagnosis of schistosomiasis: antibody detection, with notes on parasitological and antigen detection methods. Parasitology 117(suppl): S41–57.

Hatz C. 2005. Schistosomiasis: an underestimated problem in industrialized countries? Journal of Travel Medicine 12: 1–2.

Hiatt RA, Sotomayor ZR, Sanchez G. 1979. Factors in pathogenesis of acute schistosomiasis. Journal of Infectious Diseases 172: 1336–42.

Jelinek T, Nothdurft HD, Loscher T. 1996. Schistosomiasis in travellers and expatriates. Journal of Travel Medicine 3: 160–64.

King CL, Malhotra I, Mungai P, Wamachi A, Kioko J, Ouma JH, Kazura JW. 1998. B cell sensitization to helminthic infection develops in utero in humans. Journal of Immunology 160: 3578–84.

Lambertucci JR, Rayes AA, Barata CH, Teixeira R, Gerspacher Lara R. 1997. Acute schistosomiasis: report on five singular cases. Memórias do Instituto Oswaldo Cruz 92: 631–35.

Li YS, Chen HG, He HB, Hou XY, Ellis M, McManus DP. 2005. A double-blind field trial on the effects of artemether on *Schistosoma japonicum* infection in a highly endemic focus in southern China. Acta Tropica 96: 184–90.

Magnussen P. 2003. Treatment and re-treatment strategies for schistosomiasis control in different epidemiological settings: a review of 10 years' experiences. Acta Tropica 86: 243–54.

Olds RG. 2003. Administration of praziquantel to pregnant and lactating women. Acta Tropica 86: 185–95.

Ribeiro de Jesus A, Silva A, Santana LB, Magalhães A, de Jesus AA, Almeida RP, Rego MAV, Burattini MN, Pearce EJ, Carvalho EM. 2002. Clinical and immunologic evaluation of 31 patients with acute *Schistosomiasis mansoni*. Journal of Infectious Diseases 185: 98–105.

Rocha MO, Rocha RL, Pedroso ER, Greco DB, Ferreira CS, Lambertucci JR, Katz N, Rocha RS, Rezende DF, Neves J. 1995. Pulmonary manifestations in the initial phase of *Schistosomiasis mansoni*. Revista do Instituto de Medicina Tropical de São Paulo 37: 311–18.

Ross AG, Bartley PB, Sleigh AC, Olds GR, Li YS, Williams GM, McManus DP. 2002. Schistosomiasis. New England Journal of Medicine 346(16): 1212–20.

Ross AG, Sleigh AC, Li YS, Davis GM, Williams GM, Jiang Z, Feng Z, Jingping G, McManus DP. 2001. Schistosomiasis in the People's Republic of China: prospects and challenges for the 21st century. Clinical Microbiology Reviews 14(2): 270–295.

Tarp B, Black FT, Petersen E. 2000. The immunofluorescence antibody test (IFAT) for the diagnosis of schistosomiasis used in a non-endemic area. Tropical Medicine International Health 5: 185–91.

Utzinger J, N'Goran EK, N'Dri A, Lengeler C, Shuhua X, Tanner M. 2000. Oral artemether for prevention of *Schistosoma mansoni* infection: randomised control trial. Lancet 355: 1320–25.

van Gool T, Vetter H, Vervoort T, Doenhoff MJ, Wetsteyn J, Overbosch D. 2002. Serodiagnosis of imported schistosomiasis by a combination of commercial indirect hemagglutination test with *Schistosoma mansoni* adult worm antigens and an enzyme-linked immunosorbent assay with *S. mansoni* egg antigens. Journal of Clinical Microbiology 40: 3432–37.

van Lieshout L, Polderman AM, Visser LG, Verwey JJ, Deelder AM. 1997. Detection of the circulating antigens CAA and CCA in a group of dutch travellers with acute schistosomiasis. Tropical Medicine International Health 2: 551–57.

van Lieshout L, Polderman AM, Deelder AM. 2000. Immunodiagnosis of schistosomiasis by determination of the circulating antigens CAA and CCA, in particular in individuals with recent or light infections. Acta Tropica 77: 69–80.

Visser LG, Polderman AM, Stuiver PC. 1995. Outbreak of schistosomiasis among travellers returning from Mali, West Africa. Clinical Infectious Diseases 20: 280–85.

Xiao SH, Booth M, Tanner M. 2000. The prophylactic effect of artemether against *Schistosoma japonicum* infection. Parasitology Today 16: 122–26.

Xiao SH, Tanner M, N'Goran EK, Utzinger J, Chollet J, Bergquist R, Chen M, Zheng J. 2002. Recent investigations of artemether, a novel agent for the prevention of schistosomiasis japonica, mansoni and haematobia. Acta Tropica 82: 175–81.

Xiao SH. 2005. Development of antischistosomal drugs in China, with particular consideration to praziquantel and the artemisinins. Acta Tropica 96: 153–67.

Yasaway MI. 2004. Katayama syndrome. Saudi Medical Journal 25: 234–36.

Yin-Chang Z. 2005. Immunodiagnosis and its role in schistosomiasis control in China: a review. Acta Tropica 96: 130–36.

CHAPTER

Chronic Schistosomiasis

David U Olveda[1] and *Allen GP Ross*[2,*]

18.1 INTRODUCTION

Schistosomiasis is a chronic enteropathogenic disease caused by blood flukes of the genus *Schistosoma*. Chronic infections with morbidity and mortality occur as a result of granuloma formation in the intestine, liver or in the case of *S. haematobium*, the bladder. Various methods are utilized to diagnose and evaluate liver fibrosis due to schistosomiasis. Liver biopsy is still considered the gold standard but it is invasive. Diagnostic imaging has proven to be an invaluable method in assessing hepatic morbidity in the hospital setting but has practical limitations in the field. The potential of non-invasive biologic markers, serum antibodies, cytokines and circulating host microRNAs to diagnose hepatic fibrosis are presently undergoing evaluation and are discussed.

18.2 UROGENITAL DISEASE

The classic sign of urogenital schistosomiasis is haematuria and is specifically noted with *S. haematobium* (Ross *et al.* 2002, 2013). Bladder, ureter fibrosis and kidney damage are sometimes seen in advanced cases. The urogenital form may present with genital lesions (e.g., vulvar nodules), vaginal bleeding, dyspareunia and fallopian tube damage (in the late stages) in females. Genital infection in males may result in damage to seminal vesicles, prostate and other related organs; this may lead to irreversible infertility (Ross *et al.* 2002). Urogenital schistosomiasis in both sexes is a significant risk factor for Human Immunodeficiency Virus (HIV) infection due to both local genital tract and systemic immunological effects (Ross *et al.* 2002). Schistosomal co-infection may hasten HIV disease progression in individuals already infected with HIV and facilitate viral transmission to sexual partners (Ross *et al.* 2013).

The link between *S. haematobium* and urinary bladder cancer has been documented. For example, squamous cell carcinoma of the urinary bladder has been associated with *S. haematobium* infection in studies in many areas of Africa (Ross *et al.* 2002, 2013). In addition, studies from Africa have shown that the estimated incidence of urinary bladder cancer is higher in areas with a high prevalence of infection with *S. haematobium*, compared to areas with a low prevalence. Urinary bladder cancer as a proportion of all cancers appears to be 10 times more common among men in Egypt than among men in Algeria (Ross *et al.* 2002, 2013). Furthermore, the estimated incidence of urinary bladder cancer is related to the proportion of cancerous urinary bladder

[1] School of Medical Sciences, Griffith University, Gold Coast Campus, Parklands Drive, Southport, Qld 4222, Australia.

[2] Tropical Medicine and Global Health, Griffith University, Logan Campus, University Drive, Meadowbrook QLD 4131, Australia.

* Corresponding author

specimens containing *S. haematobium* eggs. It is more common in men, who are more involved in agricultural work, than in women. A higher proportion of urinary bladder cancers are seen in areas where there is histological evidence of infection compared to areas without these characteristics (Ross *et al.* 2002, 2013).

18.3   INTESTINAL DISEASE

Intestinal schistosomiasis is caused by *Schistosoma mansoni, S. japonicum, S. mekongi* and *S. intercalatum* (Jenkins-Holick 2013). Some intestinal cases of *S. haematobium* and *S. guineensis* infection have also been reported (Ata, el-Raziky *et al.* 1970; Betson *et al.* 2010). The pathology associated with intestinal schistosomiasis is due to egg deposition and granuloma formation which eventually leads to acute then chronic schistosomal colitis and polyp formation (Elbaz and Esmat 2013). Although areas in both the small and large intestine may be involved, most severe lesions are found in the large intestine. It is theorized that the adult worms have a tendency to inhabit the branches of the inferior mesenteric vein and superior haemorrhoidal vein; hence, more eggs are deposited in the large intestine, especially in the rectum, sigmoid and descending colon (Chen 1991).

Ova are generally distributed in the loose submucosa of the large intestine and to a lesser extent in the subserosa. The muscularis mucosa becomes involved subsequently and the underlying mucosa may either undergo hyperplastic changes or be denuded and form small superficial ulcers. When the submucosa becomes heavily thickened with fibrous tissue containing massive amounts of calcified eggs, atrophy of the overlying mucosa ensues and it acquires a granular dirty yellowish appearance (Strickland 1994). Polyps, which are said to be the most common among the spectrum of intestinal lesions, may result from immune-mediated inflammatory process associated with continued egg deposition and ova entrapment leading to a foreign body reaction with progressive inflammation and fibrosis (Ismail *et al.* 1994; Delgado *et al.* 2004). Schistosomal eggs are deposited in the superficial layers of the submucosa where reactive cellular debris and vascular granulation tissue accumulate. Eggs will then produce a cell mediated inflammatory response with granuloma formation and necrosis. The subsequent healing of the necrotic foci will lead to the formation of fibrous connective tissue and hypertrophy of the muscularis mucosa. The fibrous connective tissue in the submucosa and the hypertrophied muscularis mucosa form a barrier to the ova transiting from the mesenteric veins to the gut lumen. The trapped ova then elicit further inflammation and fibrosis. This continuous process elevates the hypertrophied muscularis mucosa to form a nodule which is the earliest detectable polyp (Mostafa 1997).

Clinical manifestations of intestinal schistosomiasis include abdominal pain, altered bowel habits and bloody stools (Ross *et al.* 2002). Iron-deficiency anemia and eosinophilia are also present (Strickland *et al.* 1982). Polyposis from intestinal schistosomiasis does not appear to be related with colorectal carcinoma but a recent study has shown that a history of colonic schistosomiasis japonica is a probable independent risk factor for the development of colorectal neoplasias (Barsoum 1953; Nebel *et al.* 1974; Liu *et al.* 2013).

Appendiceal schistosomiasis was first documented in 1909 and the most frequent species associated with this condition are *S. haematobium* and *S. mansoni.* In one case report, schistosomiasis hematobia presented as acute appendicitis in a 26 year old Israeli male who developed symptoms two years after visiting Africa; tissue sections showed extensive inflammatory areas and fibrosed granulomas (Weber *et al.* 1998). This rare condition was also reported recently in a 30 year old male UK resident from Ghana; histologic sections of his appendix revealed luminal pus associated with numerous *S. mansoni* egg masses transmurally and within the subserosal adipose tissue. The usual granulomatous response around the eggs was evident. Eggs in the submucosa produce an obstructive type of appendicitis, while serosal lesions produce inflammation and adhesion formation (Madavo 2006).

In Saudi Arabia, an unusual case of disseminated peritoneal *S. japonicum* has also been reported in a 32-year old Filipino female who presented with signs and symptoms of acute appendicitis.

However, a right iliac fossa mass was also seen on diagnostic laparoscopy. Microscopic sections of both the appendiceal wall and the adherent omental mass showed suppurative inflammation and multiple foci of schistosomal ova highly indicative of the *S. japonicum* species. Interestingly, a granulomatous response was not seen in the sections examined (Al-Waheeb *et al.* 2009).

18.4   HEPATOSPLENIC DISEASE

Hepatic schistosomiasis represents the best known form of chronic disease with a wide range of clinical manifestations and its pathogenesis is related to the host cellular immune response (Elbaz and Esmat 2013). The mechanisms involved in granuloma formation and fibrosis have been documented extensively in experimental models and humans infected with *S. mansoni* and *S. japonicum*. Eggs trapped in the pre-sinusoidal portal venules secrete soluble egg antigens which are taken up by antigen-presenting cells such as macrophages. Subsequently, antigen presentation stimulates Th1 cells (CD4+ve T lymphocytes) to secrete interleukin (IL)-2, interferon (IFN)-γ and tumor necrosis factors (TNF) which in turn drive a cell-mediated response and attract more immune cells around the ova (Boros 1999). As the granuloma becomes more organized, the Th1 cells are gradually replaced by Th2 cells which produce IL-4, IL-5, IL-10 and IL-13 completing granuloma maturation (Boros 1999). Towards the late stage of granuloma formation, the fibroblasts are stimulated by egg products and by T lymphocyte cytokines to proliferate, replacing most of the cellular elements and mediating fibrotic collagenous material deposition around the portal vein tributaries. The pathogenesis of hepatic fibrosis leading to hepatosplenic schistosomiasis is illustrated for *S. japonicum* in Figure 18.1.

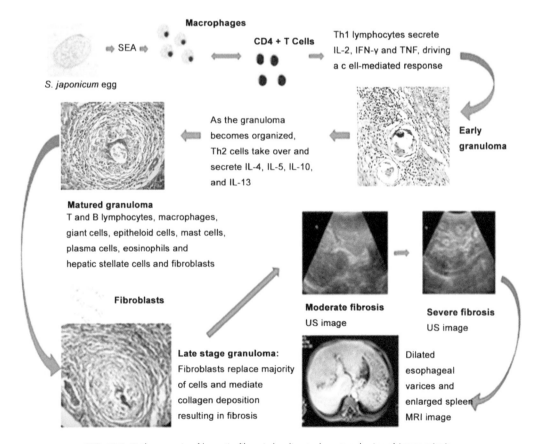

FIG. 18.1 Pathogenesis of hepatic fibrosis leading to hepatosplenic schistosomiasis.

Fibrosis, leading to portal hypertension, is the major cause of disease morbidity and mortality. Grossly, whitish plaques, known as "clay-pipestem" fibrosis, are evident on cut sections contrasting with the intact liver acinar architecture (Andrade 2009). Lesions commence as eosinophilic infiltrates surrounding trapped eggs which may subsequently lead to abscess formation. A Hoeppli-Splendore reaction (asteroid body formation) may sometimes occur (Colley *et al.* 2014). Periovular granulomas develop and over time, the eggs inside degenerate and calcify. As older granulomas involute, macrophage-predominant granulomas begin to form (Andrade 2009). Eventually, granulomas are replaced by surrounding fibrous tissue. As more new eggs arrive, resultant damage to larger diameter veins occurs, along with periportal granulomatous inflammation and inter-granulomatous fibrosis. In the absence of coexisting hepatotropic viral infection, the liver of patients with periportal fibrosis secondary to schistosomiasis retains its hepatocellular function, differentiating the disease from cirrhosis and other liver diseases (Colley *et al.* 2014). Severe schistosome-induced hepatic fibrosis causes portal vein obliteration leading to the development of portal hypertension and lethal complications of the hepatosplenic form of the disease include pulmonary hypertension, glomerulopathy, splenomegaly and thrombocytopenia (Brito *et al.* 1999; Butrous *et al.* 2008; Rodrigues *et al.* 2010; Carvalho *et al.* 2011; Luiz *et al.* 2013).

Despite the general mechanism of granuloma formation leading to hepatic fibrosis in schistosomiasis, studies have shown some peculiarities regarding this pathogenesis among schistosome species. For instance, in a study involving mice models, tissue studies were done to shed light on the characteristics of granulomas caused by *S. mekongi* compared with *S. japonicum*-induced granulomas. In the murine livers, it was shown that *S. mekongi*-induced granulomas were initially cellular, formed by foam cells and continuously appeared in the intralobular areas, whereas *S. japonicum*-induced granulomas were fibrous and did not continuously appeared in the intralobular areas. Portal fibrosis is also not observed in *S. mekongi*-infected murine livers whereas this lesion appeared later in livers infected with *S. japonicum*. It was thought that absence of portal fibrosis in *S. mekongi* infection allowed the eggs to infiltrate the interlobular areas continuously which may account for the absence of echogenic pattern on ultrasonographic evaluation (a feature noted with *S. japonicum* species) (Shimada *et al.* 2010). On the other hand, a study on the intensity of liver fibrosis among patients with *S. mekongi* infection was assessed by liver biopsy. Liver biopsies revealed complete disorganization of hepatic architecture with fibrous enlargement of portal tracts and some portal-portal bridging fibrosis, but there was no cirrhosis. There was blood vessel congestion and thrombosis with inflammation in the portal areas. Numerous eggs of *Schistosoma mekongi* were observed mostly in fibrous areas and more rarely in the parenchyma. Some eggs were surrounded by epithelioid and giant cell reaction. It was noted in this study that a high degree of fibrosis is observed among young adults and this contrasts the findings observed with other schistosome species (Monchy *et al.* 2006).

Clinical manifestations with *S. mekongi* are similar to those of *S. mansoni* and *S. japonicum* infections (Ohmae *et al.* 2004). Clinical hepatomegaly, splenomegaly, reported blood in stool, abdominal pain, diverted circulation and ascites are the frequent reported clinical signs in studies assessing hepatosplenic morbidity from *S. mekongi* (Stich *et al.* 1999; Keang *et al.* 2007).

Although several studies have clearly shown that schistosome egg-induced granulomas are pathogenic to hosts, there are some current studies suggesting that the lesions may have some protective functions. In an *S. mansoni* study involving transgenic mice, it was thought that granuloma formation around the egg offered some protection against schistosome-related hepatotoxicity (an occurrence that is currently observed only in *S. mansoni*-infected mice) (Fallon *et al.* 2000). Mice models that were not able to produce the necessary cytokines to form a functional granuloma die early due to egg-induced endotoxemia. The granuloma is not only thought to act as a physical barrier, but also functions to sequester the antigenic products secreted by the egg. Nevertheless, it is also thought that the parasite takes advantage of this functional granuloma by facilitating excretion of its eggs, without killing the host and therefore continuing its life cycle (Hams *et al.* 2013).

Recently, hepatic stellate cells (HSCs) have been acknowledged to be major players in the liver fibrotic process (Anthony *et al.* 2012). The interaction of the HSCs between *S. japonicum* and *S. mansoni* eggs have been investigated and both studies supported the contributory role of these cells in schistosome-induced hepatic fibrosis (Bartley *et al.* 2006; Chang *et al.* 2006). For example, it has been demonstrated that the activated HSCs myofibroblastic phenotype can be reversed to its quiescent phenotype (manifested by lipoprotein storage in cells) by *S. mansoni* egg antigens, an interaction not observed with *S. japonicum* eggs. This inability of the *S. japonicum* eggs to induce a quiescent myofibroblastic phenotype has been suggested to partly explain why *S. japonicum* is more pathogenic to the liver compared with *S. mansoni* (Anthony *et al.* 2013).

18.5 EVALUATION OF HEPATIC FIBROSIS

The pathology of schistosome-induced liver disease has been studied extensively since its first description in 1904 (Symmers 1904). The term 'Symmers fibrosis' was originally adopted to describe the unusual pattern of collagen and glycosaminoglycan (GAG) deposition observed in an autopsy. This term was changed in 1947 to clay-pipestem fibrosis in recognition of the fact that the hepatic parenchyma was spared in schistosomal-induced liver disease and appeared fundamentally different from "cirrhotic" deposits of collagen and GAGs (Hashem 1947).

Currently there are a number of methods that can be used to diagnose and evaluate liver fibrosis severity. Liver biopsy is considered the gold standard but causes significant discomfort and possible post-procedural risk. Morphologic tissue evaluation may be informative but does not have sufficient sensitivity to diagnose periportal fibrosis (PPF) and cannot explain the disease dynamics occurring between sampling periods (Hou *et al.* 2011). Imaging modalities like ultrasound (US), computed tomography (CT) scan and magnetic resonance imaging (MRI), on the other hand, are used not only to support the diagnosis of schistosomiasis but also to accurately assess and detect target organ damage which can develop due to chronic infection with schistosomiasis. Research to identify non-invasive markers for hepatic fibrosis is also underway and the combination of these biomarkers along with comprehensive history and physical examination, basic laboratory tests and imaging methods seem to offer the best approach for evaluating patients with this disease (Lambertucci 2014).

18.5.1 Diagnostic Imaging

The reliability of ultrasound (US) has made it the routine imaging method in the evaluation of hepatosplenic schistosomiasis for the past 30 years (Gerspacher-Lara *et al.* 1997; Martins *et al.* 1998; Marinho *et al.* 2006; Pinto-Silva *et al.* 2010). In the diagnosis of schistosomiasis mansoni, US can demonstrate periportal fibrosis appearing as echogenic tubular shadows with anechoic lumen which radiates from the porta hepatis. When the tubular structure is viewed crosswise, it appears as a ring of concentric fibrosis surrounding portal venous vasculature and is known as the "Bull's eye lesion" (Cerri *et al.* 1984; Fataar *et al.* 1984; Hussain *et al.* 1984). Other ultrasonographic findings include hypertrophy of the left hepatic lobe, atrophy of the right hepatic lobe, gallbladder wall thickening, granulomas and splenic nodules. These above lesions can also be seen in liver pathology associated with *S. japonicum* infection. However, the demonstration of a septal formation by high echogenic bands like a mosaic or network pattern in the liver by US is typical only for *S. japonicum* infection (Ohmae *et al.* 2003; Chigusa *et al.* 2006).

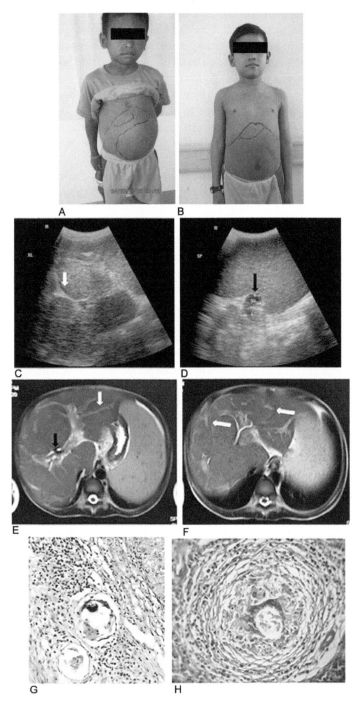

FIG. 18.2 Clinical outcome of hepatosplenic schistosomiasis. The abdomen of a 12-year-old Filipino boy with severe schistosomiasis. **A.** Before. **B.** After splenectomy. **C.** Ultrasound of the patient showing a markedly thickened branch of the main portal vein (white arrow). **D.** A markedly enlarged spleen with dilated splenic vein (black arrow). **E.** MRI depicting a cavernous transformation of the right portal vein (black arrow) and periportal fibrosis running along the second branch of the portal vein (white arrow). **F.** Curvilinear tracts scattered throughout the liver parenchyma consistent with periportal fibrosis (arrows). **G.** Histopathology sections showing early *Schistosoma japonicum* egg granuloma. **H.** Late granuloma in the liver.

CT, on the other hand, is not routinely used in the evaluation of schistosomiasis due to its cost and the utilization of ionizing radiation. CT scan shows similar imaging findings to that of US including atrophy of the right hepatic lobe, hypertrophy of the left hepatic lobe, splenomegaly and ascites. Periportal fibrosis in seen in CT as a band of low attenuation around portal vein branches throughout the liver, with enhancement following intravenous administration of contrast. The "Bull's Eye" lesion in the liver demonstrated by ultrasound is seen as concentric layers of periportal enhancement and is thought by some authors to be a useful, specific indicator of schistosomiasis (Hammerman *et al.* 1991). The network pattern seen in the liver in *S. japonica* appears in CT as "turtleback" or "tortoiseshell" lesion. This lesion is thought to be due to septal calcification of schistosomal ova (Araki *et al.* 1985).

In the diagnosis of hepatosplenic schistosomiasis using MRI, the most frequent findings are accentuation of periportal signal in T2-weighted sequences and hypointense signal in relation to the normal liver parenchyma in T1-weighted sequences with fat suppression. The periportal signal is accentuated on T1-weighted sequences following contrast administration. It has been suggested that periportal inflammation may be differentiated from fibrosis by the hyperintense signal observed in T2-weighted sequences (Lambertucci *et al.* 2004). Portal vein thrombosis (PVT), which may be due to hepatosplenic schistosomiasis, is best diagnosed by MRI (Olveda *et al.* 2014). Cavernous portal vein transformation due to PVT is described in MRI as small enhancing flow voids around the right portal vein in a T1 weighted image. Although MRI uses a gadolinium-based contrast, its higher cost precludes it from routine use.

The utility of imaging techniques in the diagnosis of schistosomiasis is demonstrated in Fig. 18.2. A 12-year-old Filipino boy from a known schistosomiasis endemic area in the central Philippines was diagnosed by portable gray scale US to have moderate periportal fibrosis with severe splenomegaly due to schistosomiasis japonica. Further examination with MRI confirmed the periportal fibrosis and showed an additional finding of portal vein thrombosis and cavernous transformation of the right branch of the main portal vein. Massive splenomegaly was also demonstrated by MRI. The patient was successfully treated by splenectomy.

18.5.2   Serum Biologic Markers

The supporting framework of the normal and fibrotic liver comprises a group of macromolecules called extracellular matrix (ECM). In advanced stages of fibrosis, the liver contains approximately six times more ECM than normal, including collagens (I, III and IV), fibronectin, undulin, elastin, laminin, hyaluronan and proteoglycan (Bataller and Brenner 2005). Qualitative and quantitative ECM changes in liver fibrosis can be measured in the blood or urine using indirect and direct biomarkers. Direct biomarkers (see also Bergquist and Van Dam, Chapter 21) are classified into three groups: (a) those that measure matrix deposition: procollagen I carboxy terminal peptide (PICP), procollagen III amino terminal peptide (PIIINP), tissue inhibitor of metalloproteinase (TIMP), transforming growth factor beta (TGF-β) tenascin; (b) those that reflect matrix removal or degradation: procollagen IV C peptide, procollagen IV N peptide (7-S collagen), collagen IV, metalloproteinase MMP, undulin, urinary desmosine and hydroxylysyl pyridinoline and; (c) those which cannot clearly determine the relationship to the matrix deposition or removal: hyaluronic acid, YKL-40 (Chondrex) and laminin (Bataller *et al.* 2005; Marinho *et al.* 2010). Many of these biomarkers have been evaluated and have been found useful in the diagnosis and grading of liver fibrosis caused by several conditions, including chronic viral hepatitis, alcoholic cirrhosis, non-alcoholic steatohepatitis and schistosomiasis (Domingues *et al.* 2011). However, their ability to identify and grade liver fibrosis in schistosomiasis at the community level in endemic areas needs further evaluation. The performance of some direct serum markers in the evaluation of schistosome-induced hepatic fibrosis is discussed below and is summarized in Table 18.1.

TABLE 18.1 The performance of serum markers in the evaluation of hepatic fibrosis due to schistosomiasis.

Serum marker	Species	Studies undertaken	Results	Reference
Hyaluronic acid (HA)	*S. mansoni*	Correlation of levels of HA in patients with intense, moderate, light and without fibrosis by ultrasound in 79 subjects.	HA was able to separate individuals with fibrosis from those without and light from intense fibrosis. No sense The HA diagnostic accuracy for fibrosis was 0.89. With a cut-off level of 115.4 ng/ml, sensitivity and specificity were 0.98 and 0.64, respectively.	Marinho *et al.* 2010.
		Serum levels of HA assessed in 122 patients with PPF and 12 schistosomiasis patients without fibrosis.	Serum levels of HA in the nonfibrotic group was 23.9 micrograms per litre (mcg/L), 51.9 mcg/L in the mild fibrotic group and 64.3 mcg/L in those with advanced fibrosis. Cut off value of 27.8 mcg/L had a sensitivity of 78.2% and a specificity of 83.7%.	Ana Lúcia Coutinho Domingues *et al.* 2011.
		A study of 61 patients with schistosomiasis mansoni and 16 healthy individuals.	A serum HA level of 20.2 mcg/L was observed that differentiated between patients with milder PPF (patterns C+D) and those with more severe PPF by ultrasound with a sensitivity of 60% and a specificity of 65%.	Silva *et al.* 2011.
		Correlation of HA level in hepato-splenic schistosomiasis and with mild to advanced fibrosis.	Higher levels of HA were noted in more advanced forms of schistosomiasis.	Ricard-Blum *et al.* 1999.
		Serum levels in advanced forms of the disease.	Serum HA levels were increased.	Pascal *et al.* 2000; Eboumbou *et al.* 2005.
		Levels of HA in 153 patients in Senegal with schistosomiasis hepatic fibrosis.	No changes in patient levels of HA. However 60% of patients showed early stages of hepatic involvement with US, while only 3% presented with HS disease.	Burchard *et al.* 1998.
	S. japonicum	HA levels in 38 cases with portal hypertension.	Levels of HA are higher in those with portal hypertension than the normal range.	Guangjin S *et al.* 2002.
		HA levels in 193 individuals exposed to endemic *Schistosoma japonicum.*	hyaluronic acid levels correlated with ultrasound findings.	Sleigh YS *et al.* 2000.
		Patients were identified with either mild (n = 30) or severe (n = 30) hepatic fibrosis due to *S. japonicum* infection.	HA levels in normal, mild and severe cases of hepatic fibrosis were 83.0 ± 35.7, 216.1 ± 77.9 and 212.6 ± 80.9 μg/ml, respectively. HA levels did not make good correlation with degree.	Min Zheng *et al.* 2005.

Table 18.1 Contd.

Serum marker	Species	Studies undertaken	Results	Reference
Laminin	*S. mansoni*	Correlation of levels with portal hypertension.	Laminin levels correlated with portal hypertension.	Ricard-Blum *et al.* 1999.
		Field studies evaluating the levels of laminin in individuals with milder forms of schistosomiasis.	Results did not reveal a correlation between laminin levels with milder form of periportal fibrosis.	Kardorff *et al.* 1997, 1999.
		Correlation of serum laminin levels with infection (egg-positive patients).	Serum laminin level was significantly higher in egg-positive infected patients than in endemic controls.	Tanabe *et al.* 1989; Ricard-Blum *et al.* (1999).
		Serum levels of laminin in Hepatosplenic and hepatointestinal patients.	Serum levels of laminin were higher in Hepatosplenic than in hepatointestinal patients and were also higher in hepatointestinal patients than the controls.	Parise & Rosa 1992.
		Laminin levels were determined in cases with initial stages of disease in hepatointestinal cases and in advanced disease.	Progressive increase of laminin levels in those with initial to advanced cases of hepatointestinal schistosomiasis were observed.	Wyszomirska *et al.* 2005.
	S. japonicum	Laminin levels in 193 individuals exposed to endemic *Schistosoma japonicum*.	Levels of laminin are correlated with re-infection.	Sleigh YS *et al.* 2000.

Table 18.1 Contd.

Serum marker	Species	Studies undertaken	Results	Reference
Collagen IV	*S. mansoni*	Correlation of Procollagen IV peptide levels intensity of schistosomiasis, liver fibrosis, splenomegaly, portal vein dilatation and the presence of portosystemic collaterals.	Levels of Procollagen IV peptide did not correlate with the presence or intensity of schistosomiasis infection, but was significantly correlated with liver fibrosis and signs of portal hypertension. Type IV collagen had a good specificity (over 90%) but poor sensitivity because more than half of those with severe liver involvement exhibited normal levels of type IV collagen).	Kardorff *et al.* 1997, 1999.
		Correlation between type IV collagen and advanced forms of schistsomiasis mansoni.	A positive correlation between type IV collagen and advanced forms of schistosomiasis mansoni. However, no correlation to the grade of PPF assessed by ultrasound examination as gold-standard method.	Wyszomirska *et al.* 2005, 2006.
	S. japonicum	Collagen IV levels in 193 individuals exposed to endemic *Schistosoma japonicum*.	Levels of collagen IV correlated with re-infection.	Sleigh YS *et al.* 2000.
Procollagen type III	*S. mansoni*	Levels of PIIIP in 82 individuals infected with *S. mansoni*, without liver enlargement and 20 with hepatic or hepatosplenic disease.	PIIINP was elevated in patients with hepatic disease, but normal in uncomplicated cases.	Zwingenberger *et al.* 1988.
		Correlation of levels of PIIIP with the presence or intensity of infection or fibrosis scores.	No significant correlation evident between PIIIP levels and presence or intensity of infection or fibrosis scores.	Kardorff *et al.* 1999; Tanabe *et al.* 1989; Kardorff *et al.* 1997; Burchard *et al.* 1998; Ricard-Blum *et al.* 1999.
	S. japonicum	PIIIP levels in 38 cases with portal hypertension.	Levels of PIIIP were higher in those with portal hypertension than the normal range.	Guangjin S *et al.* 2002.
YKL-40	*S. japonicum*	Levels of YKL-40 were determined in 60 patients with mild, moderate or severe hepatic fibrosis due to *S. japonicum* infection.	Levels of YKL-40 were 49.0 ± 10.4, 92.3 ± 18.5 and 172.1 ± 35.9 µg/ml, respectively for mild, moderate and severe heaptic fibrosis. Serum levels of YKL-40 correlated with the stage of hepatic fibrosis.	Min Zheng *et al.* 2005.

Collagen. Collagen and its metabolic products have been examined in the blood and urine of patients with different aetiologies of fibrotic liver disease (Rojkind and González 1974). Collagen types I, III and IV have been extensively used in hepatic fibrosis evaluation. Serum PIIINP has also gained wide acceptance as a blood test for collagen metabolism (Bienkowsk *et al.* 1978). Type III collagen is secreted in the extracellular space as a pro-collagen molecule. The amino terminal portion of the molecule is then cleared by enzymatic cleavage by two specific endopeptidases (McCullough *et al.* 1987). This releases the helical type III collagen molecule to combine with other extracellular matrix macromolecules to form a collagen fibril. This short peptide fragment (PIIIP) diffuses from fibrotic loci, circulates in the blood and is metabolized further to form a fragment termed Col-1 (Procollagen type I peptide [PIP]) (Bruckner *et al.* 1978). The PIIIP assay appears to have its greatest utility in the early stages of fibrosis when Type III collagen biosynthesis predominates (Friedman 1997). In late stages of most cirrhotic conditions, PIP is the dominant peptide synthesized (Zwingenberger *et al.* 1988). A study on PICP in schistosomiasis mansoni showed that the levels of this biomarker were higher in infected patients than in uninfected controls and that serum levels decreased during the first year post-treatment with the anti-schistosome drug praziquantel; however, no correlation was established between PIP and fibrosis scores by US examination (Ricard-Blum *et al.* 1999). On the other hand, PIIINP levels in schistosomiasis mansoni were elevated in patients with hepatic disease compared with normal or uncomplicated control subjects. It has also been shown that PIIINP levels returned to normal levels 18 months after patients received praziquantel (Zwingenberge *et al.* 1908). Higher levels of PIIINP were also observed in subjects with advanced PPF diagnosed by histology or with more severe hepatic disease (Shahin *et al.* 1992).

Serum type IV collagen has also been examined in schistosomiasis patients. Levels of procollagen IV peptide did not correlate with the presence or intensity of infection, but they correlated significantly with liver fibrosis, splenomegaly, portal vein dilatation and the presence of portosystemic collaterals (Kardorff *et al.* 1997, 1999). In another study, a positive correlation between type IV collagen and advanced schistosomiasis was noted, with a significant reduction being observed in serum levels following splenectomy, but there was no correlation with the grade of PPF as determined by US (Wyszomirska *et al.* 2005, 2006).

Hyaluronic acid. Hyaluronic acid (HA), a high molecular weight GAG (Glycosaminoglycan), is an essential extracellular matrix component. It is synthesized by hepatic stellate cells and is degraded by the sinusoidal endothelial cells (Lindqvist *et al.* 1992; Guechot *et al.* 1995). Its increased levels are either due to decreased hepatic removal or to increased production by stellate cells or to both processes (Grigorescu 2006). Several studies have examined HA serum levels in a number of chronic liver diseases and have suggested that it may be a candidate marker for detecting fibrosis in cirrhosis, chronic hepatitis B and C and alcoholic and non-alcoholic fatty liver disease (Pares *et al.* 1996; Wong *et al.* 1998; Pontinha *et al.* 1999; Santos *et al.* 2005). Studies in patients with different degrees of fibrosis due to schistosomiasis mansoni suggested a correlation between HA serum levels and disease severity. More severe cases of PPF had much higher levels of HA while subjects with mild fibrosis had low HA levels (Pascal *et al.* 2000; Kopke-Aguiar *et al.* 2002; Silva *et al.* 2011). Evaluation of serum levels of HA and type IV collagen with respect to US patterns of PPF showed that only the former was capable of distinguishing patients with mild fibrosis from those with intense fibrosis. Moreover, HA correlated positively with portal hypertension and PPF and collateral circulation predicted HA increase (Marinho *et al.* 2010). It is also noteworthy that, as with serum type I collagen, the high HA levels in patients with hepatic fibrosis were reduced following praziquantel treatment. In another study involving advanced schistosomiasis japonica patients and controls, serum HA and TIMP-1 levels were elevated in the former, with HA outperforming TIMP-1 (Hou *et al.* 2011).

Chitinase-3-like protein 1. Chitinase-3-like protein 1 (YKL-40) is a novel liver fibrosis marker. This recently described glycoprotein belongs to the chitinase family and is expressed in human liver and arthritic articular cartilage (Hakala *et al.* 1993; Hu *et al.* 1996; Saitou *et al.* 2005). Although its precise physiologic function is not known, YKL-40 is thought to contribute to tissue remodeling or degradation of the extracellular matrix in liver fibrosis. In a study of patients with HCV-associated liver disease comparing Type IV collagen, PIIINP, HA, YKL-40 and biochemical parameters in the assessment of hepatic fibrosis, it was concluded that HA and YKL-40 were more useful than other markers for assessing the fibrosis stage; in particular, YKL-40 was the most useful for monitoring the fibrosis of liver disease and for distinguishing extensive from mild stage of liver fibrosis assessing the fibrosis stage. Additionally, YKL-40 was more sensitive than HA in measuring the degree of hepatic fibrosis due to schistosomiasis, the serum levels increased in patients infected with *S. japonicum* and correlated with hepatic fibrosis stage (Zheng *et al.* 2005).

Matrix metalloproteinases and inhibitors. The matrix metalloproteinases (MMPs) and their inhibitors are involved in controlling matrix degradation. MMPs are enzymes that are produced intracellularly and secreted as proenzymes that require cleavage by cell surface mechanisms for functional activity. The action of MMPs is counteracted by tissue inhibitors of metalloproteinases (TIMPs) (Kelleher 2005). As such, they act to degrade ECM and to permit new matrix deposition. It has been suggested that the imbalance between MMPs and TIMPs affects the rate of fibrosis progression and that their levels correlate with the fibrosis stage but the results have been variable and dependent upon the MMP/TIMP being assessed (Murawaki *et al.* 1999a, b, Murawaki *et al.* 1999c, Murawaki *et al.* 2012).

The diagnostic accuracy of TIMP-1 and pro-MMP-2 (the free precursor molecule of MMP-2) and their relationship to histological inflammatory scores have been evaluated in patients with hepatitis C infection; both assays performed either as well as or better than HA for the diagnosis of cirrhosis but only TIMP-1 showed diagnostic value for the identification of patients with early stages of fibrosis whereas the levels of MMP-2 exhibited no relationship with the stage of fibrosis thereby limiting its value for staging liver disease (Walsh *et al.* 1999; Boeker *et al.* 2002). Serum MMP-1 and MMP-3 levels have also not been shown to be of any diagnostic value. On the other hand, one study has demonstrated that TIMP-1 can predict the risk of hepatic fibrosis in patients after one year of Praziquantel treatment. This highlights the importance of identifying biomarkers for fibrosis risk because fibrosis can occur despite effective treatment (Fabre *et al.* 2011).

Laminin. Laminin is a major noncollagenous glycoprotein of basement membranes. Synthesised by hepatic stellate cells, its serum levels correlate with portal hypertension. Field studies evaluating individuals with milder forms of schistosomiasis did not, however, reveal a correlation between laminin and PPF fibrosis or the presence or intensity of infection (Burchard *et al.* 1998). However, some studies noted that the mean value of serum laminin was significantly higher in egg-positive schistosome infected patients than in endemic controls (Tanabe *et al.* 1989). In another study, serum laminin levels were shown to be higher in hepatic schistosomiasis subjects than in hepatointestinal patients and were also higher in the latter compared with controls (Parise and Rosa 1992). A progressive increase of laminin has also been reported in the initial stages of hepatointestinal schistosomiasis and in advanced cases (Nelson *et al.* 1997).

Cytokines. Cytokines have been evaluated as biomarkers for hepatic fibrogenesis (HF) in a limited number of studies in schistosomiasis with conflicting results. Transforming growth factor β (TGF-β) is the dominant stimulus for producing extracellular matrix by hepatic stellate cells. In a study of 88 patients who had chronic hepatitis C, there was a correlation between total TGF-β 1 and the degree of hepatic fibrosis (Nelson *et al.* 1997). The role of cytokines in hepatic fibrosis development before the hepatosplenic and early hepatosplenic stages of schistosomiasis mansoni was evaluated in a group of patients with different degrees of HF determined by US. Peripheral blood mononuclear cells (PBMC) from schistosomiasis japonica patients were stimulated by *S. japonicum* antigens and the levels of IL-5, IL-10, IL-13, gamma interferon, tumor necrosis factor alpha and TGF-β determined in the PBMC supernatants. Significantly, higher levels of IL-5, IL-10 and IL-13 were found in the supernatants of soluble egg antigen-stimulated PBMC from subjects with stage III hepatic fibrosis compared with patients with stage I or II fibrosis. Significant increases in IL-5 and IL-13 levels were also observed in some of the subjects who remained untreated for 1 year following initial assessment and who developed more serious fibrosis during this period (Magalhães *et al.* 2004). In a study of hepatosplenic schistosomiasis mansoni, there was no significant difference in the mean serum concentrations of IL-10 and IL-13 between the different categories of hepatosplenic disease (Brandt *et al.* 2010).

Circulating miRNAs. MicroRNAs comprise a family of conserved small non-coding RNAs (~22nt) which can be detected in a wide range of body fluids, including blood plasma/serum. The high level of stability of miRNAs in biofluids has been attributed to two mechanisms: (1) formation of a protein-miRNA complex with argonaute proteins (mainly Ago2) or high-density lipo-proteins and (2) incorporation into exosomes, macrovesicles or apoptotic bodies (Valadi *et al.* 2007; Arroyo *et al.* 2011; Esteller *et al.* 2011; Vickers *et al.* 2011). MicroRNAs are being developed as novel biomarkers for various cancers and other diseases (Brase *et al.* 2010; Gandhi *et al.* 2013; Rani *et al.* 2013).

For example, serum levels of liver-specific miR-122 and miR-34a were suggested to be correlated with fibrosis, steatosis and inflammatory activities in a study of chronic hepatitis C viral (HCV) infection and non-alcoholic fatty liver disease (NAFLD) (Cermelli *et al.* 2011). Liver fibrosis and/ or cirrhosis were also shown to be associated with increased serum levels of miR-571 and miR-513-3p and reduced serum levels of miR-29 and miR-652 (Roderburg *et al.* 2012). A recent study in HCV-infected patients has suggested that the serum miR-20a may serve as a predictive biomarker for HCV mediated fibrosis (Shrivastava *et al.* 2013). It is noteworthy that the serum exosomal miRNA expression profile was linked to grade and stages of liver fibrosis in patients with chronic HCV infection. The expression levels of two miRNAs (miR-483-5p and miR-671-5p) significantly increased and the expression levels of 14 miRNAs (let-7a, miR-106b, miR-1274a, miR-130b, miR-140-3p, miR-151-3p, miR-181a, miR-19b, miR-21, miR-24, miR-375, miR-548l, miR-93 and miR-941) were progressively reduced as liver fibrosis increased (Murakami *et al.* 2012).

As schistosomal-egg induced immunopathology is an unusual type of chronic liver disease distinguishable from many other liver disease types, it is tempting to speculate that a unique set of circulating miRNAs may be defined to potentially serve as sensitive molecular signatures for assessment of the severity of fibrotic pathology of schistosomiasis. To date, a set of miRNAs of parasite origin have been identified in both *S. japonicum* and *S. mansoni* and miRNA expression profiles at different developmental stages of *S. japonicum* have been characterized (Hao *et al.* 2010; Wang *et al.* 2010; Cai *et al.* 2011; de Souza *et al.* 2011; Simoes *et al.* 2011; Cai *et al.* 2013a). However, the abundance of schistosome-specific miRNAs in the plasma of the definitive mammalian host has been shown to be relatively low, which will likely limit its utility for the diagnosis and staging of schistosomiasis (Cheng *et al.* 2013). Encouragingly, however, alterations of specific host miRNAs within different tissues have been shown to be associated with *S. japonicum* infection. The expression of several host miRNAs was shown to be altered rapidly in lungs, liver and spleen of BALB/c mice as early as 10 days post infection with *S. japonicum* (Han *et al.* 2013). A broad array of miRNAs in liver was dysregulated and associated with the progression of *S. japonicum* infection (Cai *et al.* 2013b). These results raise the possibility that circulating host miRNAs may be dynamically altered during the course of schistosome infection and this provides an avenue for their development as useful diagnostic and pathogenic biomarkers for schistosomiasis, because of the strong correlation between the expression profile of miRNAs and the status/progression of the disease.

18.6 CHAPTER SUMMARY

Schistosomiasis can causes significant pathology and chronic morbidity in humans infected with numerous egg granulomas and this typically results in fibrosis in target organs. The intestinal form of the disease is caused by the deposition of eggs within the bowel wall while the hepatosplenic form of the disease, is due to the eggs trapped in the liver pre-sinusoids. In the former, the eggs cause bowel lesions ranging from colitis to polyp formation whereas in the latter, granuloma formation and subsequent fibrosis involve immune responses initially mediated by Th1 and later by Th2 lymphocytes. Severe fibrosis leads to portal hypertension-related complications causing significant illness or death.

There are various methods to diagnose and evaluate schistosome-induced liver fibrosis. Liver biopsy is the still considered the gold standard but the procedure is clinically impractical in the field. Ultrasonography is invaluable in assessing pathology but is not readily available in many endemic communities and the results among users can vary widely. Although costly and limited to the hospital setting, CT and MRI show distinct imaging features associated with hepatosplenic schistosomiasis and aid in diagnosis and clinical management of patients.

Attention has recently been given to non-invasive biologic markers that can determine hepatic fibrosis severity and monitor qualitative and quantitative changes following treatment. The most promising serum markers for the evaluation of schistosome-induced hepatic fibrosis appear to be hyaluronic acid, collagen type III, YKL-40 and laminin. More studies are needed to evaluate the utility of matrix metalloproteinases and inhibitors and cytokines. Circulating host miRNAs show promise as useful biomarkers for schistosomiasis diagnosis due to the strong correlation between the expression profile of miRNAs and the disease status/progression. Biological markers are presently limited to research investigations and may prove too costly for broader clinical application but their potential in role in the diagnosis of schistosomiasis warrants further investigation.

18.7   LITERATURE CITED

Al-Waheeb S, Al-Murshed M, Dashti F, Hira PR, Al-Sarraf L. 2009. Disseminated peritoneal *Schistosoma japonicum*: a case report and review of the pathological manifestations of the helminth. Annals of Saudi Medicine 29: 149–152.

Andrade ZA. 2009. Schistosomiasis and liver fibrosis. Parasite Immunology 31: 656–663.

Anthony BJ, Ramm GA, McManus DP. 2012. Role of resident liver cells in the pathogenesis of schistosomiasis. Trends in Parasitology 28: 572–579.

Anthony BJ, James KR, Gobert GN, Ramm GA, McManus DP. 2013. *Schistosoma japonicum* eggs induce a proinflammatory, anti-fibrogenic phenotype in hepatic stellate cells. PLoS One 8(6): e68479.

Araki T, Hayakawa K, Okada J, Hayashi S, Uchiyama G, Yamada K. 1985. Hepatic schistosomiasis japonica identified by CT. Radiology 157: 757–760.

Arroyo JD, Chevillet JR, Kroh EM, Ruf IK, Pritchard CC, Gibson DF, Mitchell PS, Bennett CF, Pogosova-Agadjanyan EL, Stirewalt DL. 2011. Argonaute 2 complexes carry a population of circulating microRNAs independent of vesicles in human plasma. Proceedings of the National Academy of Sciences 22; 108(12): 5003–5008.

Ata AA, el-Raziky SH, el-Hawey AM, Rafla H. 1970. A clinicopathological study of schistosomal colonic polyposis and their pathogenesis. Journal of the Egyptian Medical Association 53: 762–772.

Barsoum H. 1953. Cancer in Egypt: its incidence and clinical forms. Acta-Unio Internationalis Contra Cancrum 9: 241–250.

Bartley PB, Ramm GA, Jones MK, Ruddell RG, Li Y, McManus DP. 2006. A contributory role for activated hepatic stellate cells in the dynamics of *Schistosoma japonicum* egg-induced fibrosis. International Journal for Parasitology 36: 993–1001.

Bataller R, Brenner DA. 2005. Liver fibrosis. The Journal of Clinical Investigation 115: 209–218.

Betson M, Sousa-Figueiredo JC, Rowell C, Kabatereine NB, Stothard JR. 2010. Intestinal schistosomiasis in mothers and young children in Uganda: investigation of field-applicable markers of bowel morbidity. American Journal Tropical Medicine Hygiene 83: 1048–1055.

Bienkowski R, Cowan M, McDonald J, Crystal R. 1978. Degradation of newly synthesized collagen. Journal Biological Chemistry 253: 4356–4363.

Boeker KHW, Haberkorn CI, Michels D, Flemming P, Manns MP, Lichtinghagen R. 2002. Diagnostic potential of circulating TIMP-1 and MMP-2 as markers of liver fibrosis in patients with chronic hepatitis C. Clinica Chimica Acta 316: 71–81.

Boros DL. 1999. T helper cell populations, cytokine dynamics and pathology of the schistosome egg granuloma. Microbes and Infection 1: 511–516.

Brandt CT, Rino M, Pitta MG, Muniz JS, Silveira Dde O, Castro CM. 2010. Evaluation of the cytokines IL-10 and IL-13 as mediators in the progression of symmers fibrosis in patients with hepatosplenic schistosomiasis mansoni. Revista do Colégio Brasileiro de Cirurgiões 37: 333–337.

Brase JC, Wuttig D, Kuner R, Sültmann H. 2010. Serum microRNAs as non-invasive biomarkers for cancer. Molecular Cancer 9: 306.

Brito TD, Nussenzveig I, Carneiro CRW, Silva AMG. 1999. *Schistosoma mansoni* associated glomerulopathy. Revista do Instituto de Medicina Tropical de São Paulo 41: 269–272.

Bruckner P, BÄChinger HP, Engel J, Timpl R. 1978. Three conformationally distinct domains in the amino-terminal segment of type III procollagen and its rapid triple helix $\rightleftarrows$ coil transition. European Journal of Biochemistry 90: 595–603.

Burchard GD, Guissé-Sow F, Diop M, Ly A, Lanuit R, Gryssels B, Gressner AM. 1998. *Schistosoma mansoni* infection in a recently exposed community in Senegal: lack of correlation between liver morphology in ultrasound and connective tissue metabolites in serum. Tropical Medicine International Health 3: 234–241.

Butrous G, Ghofrani HA, Grimminger F. 2008. Pulmonary vascular disease in the developing world. Circulation 118: 1758–1766.

Cai P, Hou N, Piao X, Liu S, Liu H, Yang F, Wang J, Jin Q, Wang H, Chen Q, *et al*. 2011. Profiles of small non-coding RNAs in *Schistosoma japonicum* during development. PLoS Neglected Tropical Diseases 5: 2.

Cai P, Piao X, Hao L, Liu S, Hou N, Wang H, Chen Q. 2013a. A deep analysis of the small non-coding RNA population in *Schistosoma japonicum* eggs. PLoS One 8(5): e64003.

Cai P, Piao X, Hao L, Liu S, Hou N, Wang H, Chen Q. 2013b. MicroRNA-gene expression network in murine liver during *Schistosoma japonicum* infection. PLoS One 8(6): e67037.

Carvalho VT, Barbosa MM, Nunes MCP, Cardoso YS, de Sá Filho IM, Oliveira FR, Antunes CM, Lambertucci JR. 2011. Early right cardiac dysfunction in patients with schistosomiasis mansoni. Echocardiography 28: 261–267.

Cermelli S, Ruggieri A, Marrero JA, Ioannou GN, Beretta L. 2011. Circulating microRNAs in patients with chronic hepatitis C and non-alcoholic fatty liver disease. PLoS One 6: 23.

Cerri GG, Alves VA, Magalhães A. 1984. Hepatosplenic schistosomiasis mansoni: ultrasound manifestations. Radiology 153: 777–780.

Chang D, Ramalho LNZ, Ramalho FS, Martinelli ALC, Zucoloto S. 2006. Hepatic stellate cells in human schistosomiasis mansoni: a comparative immunohistochemical study with liver cirrhosis. Acta Tropica 97: 318–323.

Chen MG. 1991. Relative distribution of *Schistosoma japonicum* eggs in the intestine of man: a subject of inconsistency. Acta Tropica 48: 163–171.

Cheng G, Luo R, Hu C, Cao J, Jin Y. 2013. Deep sequencing-based identification of pathogen-specific microRNAs in the plasma of rabbits infected with *Schistosoma japonicum*. Parasitology 40: 1751–1756.

Chigusa Y, Otake H, Ohmae H, Kirinoki M, Ilagan EJ, Barzaga NG, Kawabata M, Hayashi M, Matsuda H. 2006. Determination of the period for establishment of a liver network echogenic pattern in *Schistosoma japonicum* infection. Parasitology International 55: 33–37.

Colley DG, Bustinduy AL, Secor WE, King CH. 2014. Human schistosomiasis. The Lancet 383(9936): 2253–2264.

de Souza Gomes M, Muniyappa MK, Carvalho SG, Guerra-Sa R, Spillane C. 2011. Genome-wide identification of novel microRNAs and their target genes in the human parasite *Schistosoma mansoni*. Genomics 98: 96–111.

Delgado J, Delgado B, Sztarkier I, Baer A, Depsames R. 2004. Schistosomal rectal polyp—an unusual cause of rectal bleeding. Israel Medical Association Journal 6: 114–115.

Domingues AL, Medeiros TB, Lopes EP. 2011. Ultrasound versus biological markers in the evaluation of periportal fibrosis in human *Schistosoma mansoni*. Memorias do Instituto Oswaldo Cruz 106: 802–807.

Elbaz T, Esmat G. 2013. Hepatic and intestinal schistosomiasis: review. Journal of Advanced Research 4: 445–452.

Esteller M. 2011. Non-coding RNAs in human disease. Nature Reviews Genetics 12: 861–874.

Fabre V, Wu H, PondTor S, Coutinho H, Acosta L, Jiz M, Olveda R, Cheng L, White ES, Jarilla B, McGarvey ST, Friedman JF, Kurtis JD. 2011. Tissue inhibitor of matrix-metalloprotease-1 predicts risk of hepatic fibrosis in human *Schistosoma japonicum* infection. Journal Infectious Diseases 203: 707–714.

Fallon PG, Richardson EJ, McKenzie GJ, McKenzie AN. 2000. Schistosome infection of transgenic mice defines distinct and contrasting pathogenic roles for IL-4 and IL-13: IL-13 is a profibrotic agent. Journal Immunology 164: 2585–2591.

Fataar S, Bassiony H, Satyanath S, Vassileva J, Hanna RM. 1984. Characteristic sonographic features of schistosomal periportal fibrosis. American Journal of Roentgenology 143: 69–71.

Friedman S. 1997. Molecular mechanisms of hepatic fibrosis and principles of therapy. Journal of Gastroenterology 32: 424–430.

Gandhi R, Healy B, Gholipour T, Egorova S, Musallam A, Hussain MS, Nejad P, Patel B, Khoury S, Quintana F, et al. 2013. Circulating MicroRNAs as biomarkers for disease staging in multiple sclerosis. Annals of Neurology 73: 729–740.

Gerspacher-Lara R, Pinto-Silva RA, Rayes AAM, Drummond SC, Lambertucci J. 1997. Ultrasonography of periportal fibrosis in schistosomiasis mansoni in Brazil. Transactions of The Royal Society of Tropical Medicine and Hygiene 91: 307–309.

Grigorescu M. 2006. Noninvasive biochemical markers of liver fibrosis. Journal Gastrointestinal Liver Disease 15: 149–159.

Guechot J, Loria A, Serfaty L, Giral P, Giboudeau J, Poupon R. 1995. Serum hyaluronan as a marker of liver fibrosis in chronic viral hepatitis C: effect of alpha-interferon therapy. Journal Hepatology 22: 22–26.

Han H PJ, Hong Y, Zhang M, Han Y, Liu D, Fu Z, Shi Y, Xu J, Tao J, Lin J, et al. 2013. MicroRNA expression profile in different tissues of BALB/c mice in the early phase of Schistosoma japonicum infection. Molecular and Biochemical Parasitology 188: 1–9.

Hao L, Cai P, Jiang N, Wang H, Chen Q. 2010. Identification and characterization of microRNAs and endogenous siRNAs in Schistosoma japonicum. BMC Genomics 11: 1471–2164.

Hakala BE, White C, Recklies AD. 1993. Human cartilage gp-39, a major secretory product of articular chondrocytes and synovial cells, is a mammalian member of a chitinase protein family. Journal of Biological Chemistry 268: 25803–25810.

Hammerman AM, Kotner LM, Doyle TB. 1991. Periportal contrast enhancement on CT scans of the liver. American Journal of Roentgenology 156: 313–315.

Hams E, Aviello G, Fallon PG. 2013. The Schistosoma granuloma: friend or foe? Frontiers in Immunology 2013: 4: 89.

Hashem M. 1947. Etiology and pathogenesis of endemic form of hepatosplenomegaly: Egyptian splenomegaly. Journal of the Royal Egyptian Medical Association 30: 48–79.

Hou XY EM, McManus DP, Wang YY, Li SD, Williams GM, Li YS. 2011. Diagnostic value of non-invasive biomarkers for stage-specific diagnosis of hepatic fibrosis in patients with advanced schistosomiasis japonica. International Journal of Parasitology 41: 325–332.

Hu B, Trinh K, Figueira WF, Price PA. 1996. Isolation and sequence of a novel human chondrocyte protein related to mammalian members of the chitinase protein family. Journal Biology Chemistry 271: 19415–19420.

Hussain S, Hawass ND, Zaidi AJ. 1984. Ultrasonographic diagnosis of schistosomal periportal fibrosis. Journal of Ultrasound in Medicine 3: 449–452.

Hussein MR. 2008. Mucocutaneous Splendore-Hoeppli phenomenon. Journal of Cutaneous Pathology 35: 979–988.

Ismail MM, Attia MM, el-Badawy AA, Farghaly AM, Husein MH, Metwally A. 1994. Treatment of schistosomiasis with praziquantel among school children. Journal Egypt Society Parasitology 24: 487–494.

Jenkins-Holick DS, Kaul TL. 2013. Schistosomiasis. Urologic Nursing Journal 33: 163–170.

Kardorff R, Gabone RM, Mugashe C, Obiga D, Ramarokoto CE, Mahlert C, Spannbrucker N, Lang A, Günzler V, et al. 1997. Schistosoma mansoni-related morbidity on Ukerewe Island, Tanzania: clinical, ultrasonographical and biochemical parameters. Tropical Medicine International Health 2: 230–239.

Kardorff R, Mugashe C, Gabone RM, Mahlert C, Doehring E. 1999. Diagnostic value of connective tissue metabolites in Schistosoma mansoni related liver disease. Acta Tropica 73: 153–164.

Keang H, Odermatt P, Odermatt-Biays S, Cheam S, Degremont A, Hatz C. 2007. Liver morbidity due to Schistosoma mekongi in Cambodia after seven rounds of mass drug administration. Transactions Royal Society Tropical Medicine Hygiene 101: 759–765.

Kelleher TB, Afdhal N. 2005. Noninvasive assessment of liver fibrosis. Clinics in Liver Disease 9: 667–683, vii.

Kopke-Aguiar LA, Martins JR, Passerotti CC, Toledo CF, Nader HB, Borges DR. 2002. Serum hyaluronic acid as a comprehensive marker to assess severity of liver disease in schistosomiasis. Acta Tropica 84: 117–126.

Lambertucci JR, Silva LCdS, Andrade LM, Queiroz LCd, Pinto-Silva RA. 2004. Magnetic resonance imaging and ultrasound in hepatosplenic schistosomiasis mansoni. Revista da Sociedade Brasileira de Medicina Tropical 37: 333–337.

Lambertucci JR. 2014. Revisiting the concept of hepatosplenic schistosomiasis and its challenges using traditional and new tools. Revista da Sociedade Brasileira de Medicina Tropical 47: 130–136.

Lindqvist U, Laurent TC. 1992. Serum hyaluronan and aminoterminal propeptide of type III procollagen: variation with age. Scandanavian Journal Clinical Laboratory Investigations 52: 613–621.

Liu W, Zeng HZ, Wang QM, *et al.* 2013. Schistosomiasis combined with colorectal carcinoma diagnosed based on endoscopic findings and clinicopathological characteristics: a report on 32 cases. Asian Pacific Journal Cancer Prevention 14: 4839–4842.

Luiz Arthur Calheiros L, Ana Lúcia Coutinho D, Edmundo Pessoa L, Rita de Cássia dos Santos F, Adenor de Almeida Pimenta F, Caíque Silveira Martins da Fonseca B, Santana dos Santos V, Lúcia de Menezes L. 2013. Relationship between splenomegaly and hematologic findings in patients with hepatosplenic schistosomiasis. Revista Brasileira de Hematologia 35(5): 332–336.

Madavo CaH, H. 2006. Schistosomiasis of the appendix. Journal of the Royal Society of Medicine 99: 473–474.

Magalhães A, Miranda DG, Miranda RG, Araújo MI, Jesus AA, Silva A, Santana LB, Pearce E, Carvalho EM, Jesus AR. 2004. Cytokine profile associated with human chronic schistosomiasis mansoni. Memórias do Instituto Oswaldo Cruz 99: 21–26.

Marinho CC, Voieta I, Azeredo LM, Nishi MP, Batista TS, Pereira AC, Serufo JC, Queiroz LC, Ruiz-Guevara R, Antunes CM, Prata A, Lambertucci JR. 2006. Clinical versus ultrasound examination in the evaluation of hepatosplenic schistosomiasis mansoni in endemic areas. Memórias do Instituto Oswaldo Cruz 101: 317–321.

Marinho CC, Bretas T, Voieta I, Queiroz LC, Ruiz-Guevara R, Teixeira AL, Antunes CM, Prata A, Lambertucci JR. 2010. Serum hyaluronan and collagen IV as non-invasive markers of liver fibrosis in patients from an endemic area for schistosomiasis mansoni: a field-based study in Brazil. Memórias do Instituto Oswaldo Cruz 105: 471–478.

Martins MJ, Pinto-Silva RA, Serufo JC, Rayes AA, Damasceno MP, Martins ML, Santos AP, Drummond SC, Bezerra MA, Lambertucci JR. 1998. Morbidity of schistosomiasis in an endemic area of the northeast of the state of Minas Gerais in Brazil: a clinical and sonographic study. Memórias do Instituto Oswaldo Cruz, Rio de Janeiro 93: 243–244.

McCullough A, Stassen W, Wiesner R, Czaja A. 1987. Serum type III procollagen peptide concentrations in severe chronic active hepatitis: relationship to cirrhosis and disease activity. Hepatology 7: 49–54.

Monchy D, Dumurgier C, Heng TK, Hong K, Khun H, Hou SV, Sok KE, Huerre MR. 2006. Histology of liver lesions due to *Schistosoma mekongi*. About six cases with severe portal hypertension operated in Cambodia. Bulletin de la Société de pathologie exotique et de ses filiales 99: 359–364.

Mostafa I. 1997. Schistosomal colonic polyposis. Gastrointestinal Endoscopy 46(6): 584–587.

Murawaki Y, Ikuta Y, Okamoto K, Koda M, Kawasaki H. 1999a. Serum matrix metalloproteinase-3 (stromelysin-1) concentration in patients with chronic liver disease. Journal of Hepatology 31: 474–481.

Murawaki Y, Ikuta Y, Idobe Y, Kawasaki H. 1999b. Serum matrix metalloproteinase-1 in patients with chronic viral hepatitis. Journal of Gastroenterology and Hepatology 14: 138–145.

Murawaki Y, Yamada S, Ikuta Y, Kawasaki H. 1999c. Clinical usefulness of serum matrix metalloproteinase-2 concentration in patients with chronic viral liver disease. Journal of Hepatology 30: 1090–1098.

Murakami Y, Toyoda H, Tanahashi T, Tanaka J, Kumada T, Yoshioka Y, Kosaka N, Ochiya T, Taguchi YH. 2012. Comprehensive miRNA expression analysis in peripheral blood can diagnose liver disease. PLoS One 7: e48366.

Nebel OT, el-Masry NA, Castell DO, Farid Z, Fornes MF, Sparks HA. 1974. Schistosomal disease of the colon: a reversible form of polyposis. Gastroenterology 67: 939–943.

Nelson D, Gonzalez-Peralta R, Qian K. 1997. Transforming growth factor-beta 1 in chronic hepatitis C. Journal of Viral Hepatology 4: 29–35.

Ohmae H, Sy OS, Chigusa Y, Portillo GP. 2003. Imaging diagnosis of schistosomiasis japonica—the use in Japan and application for field study in the present endemic area. Parasitology International 52: 385–393.

Ohmae H, Sinuon M, Kirinoki M, Matsumoto J, Chigusa Y, Socheat D, Matsuda H. 2004. Schistosomiasis mekongi: from discovery to control. Parasitology International 53: 135–142.

Olveda DU, Olveda RM, Lam AK, Chau TN, Li Y, Gisparil AD, 2nd, Ross AG. 2014. Utility of diagnostic imaging in the diagnosis and management of schistosomiasis. Clinical microbiology 3(2): pii: 142.

Pares A, Deulofeu R, Gimenez A, Caballería L, Bruguera M, Caballería J, Ballesta AM, Rodés J. 1996. Serum hyaluronate reflects hepatic fibrogenesis in alcoholic liver disease and is useful as a marker of fibrosis. Hepatology 24: 1399–1403.

Parise ER, Rosa H. 1992. Serum laminin in hepatic schistosomiasis. Transactions Royal Society Tropical Medicine Hygiene 86: 179–181.

Pascal M, Abdallahi OM, Elwali NE, Mergani A, Qurashi MA, Magzoub M, de Reggi M, Gharib B. 2000. Hyaluronate levels and markers of oxidative stress in the serum of Sudanese subjects at risk of infection with *Schistosoma mansoni*. Transactions of the Royal Society of Tropical Medicine and Hygiene 94: 66–70.

Pinto-Silva RA, Queiroz LCd, Azeredo LM, Silva LCdS, Lambertucci JR. 2010. Ultrasound in schistosomiasis mansoni. Memórias do Instituto Oswaldo Cruz 105: 479–484.

Pontinha N, Pessegueiro H, Barros H. 1999. Serum hyaluronan as a marker of liver fibrosis in asymptomatic chronic viral hepatitis B. Scandinavian Journal of Clinical and Laboratory Investigation 59: 343–347.

Rani S, Gately K, Crown J, O'Byrne K, O'Driscoll L. 2013. Global analysis of serum microRNAs as potential biomarkers for lung adenocarcinoma. Cancer Biological Therapy 14: 1104–1112.

Ricard-Blum S, Hartmann DJ, Grenard P, Ravaoalimalala VE, Boisier P, Esterre P. 1999. Relationships between several markers of extracellular matrix turn-over and ultrasonography in human Schistosomiasis mansoni. American Journal of Tropical Medicine and Hygiene 60: 658–663.

Roderburg C, Mollnow T, Bongaerts B, Elfimiva N, Cardenas DV, Berger K, Zimmerman H, Koch A, Vucur M, Luedde M, et al. 2012. Micro-RNA profiling in human serum reveals compartment-specific roles of miR-571 and miR-652 in liver cirrhosis. PLoS One 7: e32999.

Rodrigues VL, Otoni A, Voieta I, Antunes CMdF, Lambertucci JR. 2010. Glomerulonephritis in schistosomiasis mansoni: a time to reappraise. Revista da Sociedade Brasileira de Medicina Tropical 43: 638–642.

Rojkind M, González E. 1974. An improved method for determining specific radio activities of proline-14C and hydroxyproline-14C in collagen and in noncollagenous proteins. Analytical Biochemistry 57: 1–7.

Ross AG, Bartley PB, Sleigh AC, Olds GR, Li YS, Williams GM, McManus DP. 2002. Schistosomiasis. New England Journal of Medicine 346(16): 1212–1220.

Ross AGP, Olds GR, Cripps AW, Farrar JJ, McManus DP. 2013. Enteropathogens and chronic illness in returning travelers. New England Journal of Medicine 368: 1817–1825.

Saitou Y, Shiraki K, Yamanaka Y, Yamaguchi Y, Kawakita T, Yamamoto N, Sugimoto K, Murata K, Takeshi Nakano T. 2005. Noninvasive estimation of liver fibrosis and response to interferon therapy by a serum fibrogenesis marker, YKL-40, in patients with HCV-associated liver disease. World Journal of Gastroenterology: WJG 11: 476–481.

Santos VN, Leite-Mor MM, Kondo M, Martins JR, Nader H, Lanzoni VP, Parise ER. 2005. Serum laminin, type IV collagen and hyaluronan as fibrosis markers in non-alcoholic fatty liver disease. Brazilian Journal Medical Biological Research 38: 747–753.

Shahin M, Schuppan D, Waldherr R, Risteli J, Risteli L, Savolainen ER, Oesterling C, Abdel Rahman HM, el Sahly AM, Abdel Razek SM, et al. 1992. Serum procollagen peptides and collagen type VI for the assessment of activity and degree of hepatic fibrosis in schistosomiasis and alcoholic liver disease. Hepatology 15: 637–644.

Shimada M, Kirinoki M, Shimizu K, Kato-Hayashi N, Chigusa Y, Kitikoon V, Pongsasakulchoti P, Matsuda H. 2010. Characteristics of granuloma formation and liver fibrosis in murine schistosomiasis mekongi: a morphological comparison between *Schistosoma mekongi* and *S. japonicum* infection. Parasitology 137: 1781–1789.

Shrivastava S, Petrone J, Steele R, Lauer GM, Di Bisceglie AM, Ray RB. 2013. Up-regulation of circulating miR-20a is correlated with hepatitis C virus-mediated liver disease progression. Hepatology 58: 863–871.

Silva CC, Domingues AL, Lopes EP, Morais CN, Santos RB, Luna CF, Nader HB, Martins JR. 2011. Schistosomiasis mansoni: ultrasound-evaluated hepatic fibrosis and serum concentrations of hyaluronic acid. Annals of Tropical Medicine and Parasitology 105: 233–239.

Simões MC, Lee J, Djikeng A, Cerqueira GC, Zerlotini A, da Silva-Pereira RA, Dalby AR, LoVerde P, El-Sayed NM, Oliveira G. 2011. Identification of *Schistosoma mansoni* microRNAs. BMC Genomics 12: 1471–2164.

Stich AH, Biays S, Odermatt P, Men C, Saem C, Sokha K, Ly CS, Legros P, Philips M, Lormand JD, Tanner M. 1999. Foci of schistosomiasis mekongi, Northern Cambodia: II. Distribution of infection and morbidity. Tropical Medicine International Health 4: 674–685.

Strickland GT, Merritt W, El-Sahly A, Abdel-Wahab F. 1982. Clinical characteristics and response to therapy in Egyptian children heavily infected with *Schistosoma mansoni*. Journal of Infectious Diseases 146: 20–29.

Strickland GT. 1994. Gastrointestinal manifestations of schistosomiasis. Gut 35: 1334–1337.

Symmers W. 1904. Note of a new form of liver cirrhosis due to the presence of ova of *Bilharzia haematobium*. Journal Pathology Bacteriology 9: 237–239.

Tanabe M, Sekiguchi T, Kaneko N, Kobayashi S, Takeuchi T, Coutinho A, Tateno S, Maruyama K, Okazaki I. 1989. Elevation of laminin and beta-subunit of propyl 4-hydroxylase in the sera of human subjects with *Schistosoma mansoni*. Japanese Journal of Experimental Medicine 59: 109–119.

Valadi H, Ekström K, Bossios A, Sjöstrand M, Lee JJ, Lötvall JO. 2007. Exosome-mediated transfer of mRNAs and microRNAs is a novel mechanism of genetic exchange between cells. Nature Cell Biology 9: 654–659.

Vickers K, Palmisano B, Shoucri B, Shamburek R, Remaley A. 2011. MicroRNAs are transported in plasma and delivered to recipient cells by high-density lipoproteins. Nature Cell Biology 13: 423–433.

Wang Z, Xue X, Sun J, Luo R, Xu X, Jiang Y, Zhang Q, Pan W. 2010. An "in-depth" description of the small non-coding RNA population of *Schistosoma japonicum* schistosomulum. PLoS Neglected Tropical Diseases 4: 0000596.

Walsh KM, Timms P, Campbell S, MacSween RN, Morris AJ. 1999. Plasma levels of matrix metalloproteinase-2 (MMP-2) and tissue inhibitors of metalloproteinases-1 and -2 (TIMP-1 and TIMP-2) as noninvasive markers of liver disease in chronic hepatitis C: comparison using ROC analysis. Digestive Diseases and Sciences 44: 624–630.

Weber G, Borer A, Zirkin HJ, Riesenberg K, Alkan M. 1998. Schistosomiasis presenting as acute appendicitis in a traveler. Journal of Travel Medicine 5: 147–148.

Wong VS, Hughes V, Trull A, Wight DG, Petrik J, Alexander GJ. 1998. Serum hyaluronic acid is a useful marker of liver fibrosis in chronic hepatitis C virus infection. Journal of Viral Hepatology 5: 187–192.

Wyszomirska RM, Nishimura NF, Almeida JR, Yamanaka A, Soares EC. 2005. High serum laminin and type IV collagen levels in schistosomiasis mansoni. Archives of Gastroenterology 42: 221–225.

Wyszomirska RM, Lacet Cda C, Tenorio LR, Nishimura NF, Mesquita MA, Neto JB, Oliveira F, Balwani M do C, Almeida JR, Soares EC. 2006. Decrease of type IV collagen and TIMP-1 serum levels after splenectomy in patients with schistosomiasis mansoni. Acta Tropica 97: 301–308.

Zheng M, Wei-Min C, Jun-Kang Z, Shao-Ming Z, Rong-Hua L. 2005. Determination of serum levels of YKL-40 and hyaluronic acid in patients with hepatic fibrosis due to schistosomiasis japonica and appraisal of their clinical value. Acta Tropica 96: 148–152.

Zwingenberger K, Harms G, Feldmeier H, Müller O, Steiner A, Bienzle U. 1998. Liver involvement in human schistosomiasis mansoni: regression of immunological and biochemical disease markers after specific treatment. Acta Tropica 45: 263–275.

19

CHAPTER

Neuroschistosomiasis: Pathogenesis and Clinical Manifestations

Allen GP Ross[1,]*and *Richard J Huntsman*[2]

19.1 INTRODUCTION

Neuroschistosomiasis is one of the most severe clinical manifestations of the disease caused by *Schistosoma mansoni, S. japonicum, S. mekongi* and *S. haematobium*. Central nervous system (CNS) schistosomiasis has been described in soldiers and aid workers serving in areas where schistosomiasis is endemic and in tourists who have had relatively limited exposure to such areas (Kane and Most 1948; Cetron *et al.* 1996). Among Chinese adults hospitalized with schistosomiasis, approximately 4% have CNS disease (Chen and Mott 1989). Focal or generalized seizures are a typical presentation for *S. japonicum* infection with CNS involvement. The prevalence of seizures in communities where infections have occurred has been estimated at 1–4%–eight times higher than in the general non-endemic population (Chen and Mott 1989). Transverse myelitis is the most common neurological manifestation of *S. mansoni* or *S. haematobium* infection (Kane and Most 1948; Chen and Mott 1989). However, myeloradiculopathy occurs far less frequently than the cerebral form of the disease. To date, approximately 500 clinical cases of spinal neuroschistosomiasis have been reported since 1930, but the true prevalence is unknown (Carod-Artal 2008). Indeed, the overall burden of neuroschistosomiasis is probably grossly understated due to the lack of awareness of the disease and its CNS complications.

19.2 PATHOGENESIS

Schistosomes may reach the CNS once worms have matured, paired and laid eggs. In order for symptomatic CNS involvement to occur, eggs must reach the CNS through retrograde venous flow in the Batson vertebral epidural venous plexus which connects the portal venous system and the venae cavae to the spinal cord and cerebral veins (Batson 1940; Pittella 1997; Ferrari 2004; Carod-Artal 2008). This route permits migration of the adult worms to sites close to the CNS followed by *in situ* oviposition or massive embolization of eggs from the mesenteric-pelvic system (Batson 1940; Pittella 1997; Ferrari 2004; Carod-Artal 2008). The small round eggs of *S. japonicum* may

[1] Tropical Medicine and Global Health, Griffith University, Logan Campus, University Drive, Meadowbrook QLD 4131, Australia.

[2] Paediatric Neurology, Division of Paediatric Neurology, Department of Paediatrics, University of Saskatchewan, Saskatoon, Canada.

* Corresponding author

travel all the way to the brain but the eggs of *S. mansoni* and *S. haematobium*, which are slightly larger, are retained within the lower spinal cord (Ferrari 2004). Asymptomatic CNS involvement, as seen in advanced hepatosplenic and cardiopulmonary schistosomiasis, may result from the above-mentioned venous mechanism of invasion or via arterial invasion. During arterial migration the eggs may pass into the arterial circulation through pulmonary arteriovenous shunts or through portopulmonary anastomoses via the azygos vein, which shunts form as a consequence of portal hypertension (Batson 1940; Pittella 1997; Ferrari 2004; Carod-Artal 2008). The lack of symptoms is attributed to the sparse distribution of the eggs and to the scant periovular inflammatory reaction observed in this severe chronic form of infection (Pittella 1997; Ferrari 2004).

The presence of eggs in the CNS induces a cell-mediated CD4+ T-helper cell driven periovular granulomatous reaction as in other tissue sites. The mass effect of thousands of eggs and the large granulomas concentrated within the brain or spinal cord explains the signs and symptoms of increased intracranial pressure, radiculopathy (eggs surrounded by granulomas on the surface of the nerve roots) and subsequent clinical sequelae. The intensity and duration of infection determine the amount of antigen released and the severity of chronic fibro-obstructive disease. Neuroschistosomiasis rarely occurs during the invasive phase of the disease, but when it does the neurological involvement due to Katayama syndrome differs from that in the chronic phase. Migrating schistosomula have not reached their adult stage and hence, are unable to lay eggs. The earliest that egg laying starts is at 2 months after initial infection (Davis 2002; Jaureguiberry and Caumes 2008). Therefore, eggs are not responsible for the neurological findings seen in this subset of patients. The most likely pathophysiological phenomena to explain such sequelae is an eosinophil-mediated toxicity leading to vasculitis and small-vessel thrombosis (Granier *et al.* 2003; Jaureguiberry *et al.* 2007; Jaureguiberry and Caumes 2008).

19.3 CLINICAL MANIFESTATIONS

As mentioned, only a minority (5%) of patients will go on to develop CNS symptoms due to schistosomiasis with cerebral complications being more prevalent than spinal (Pittella 1997; Ferrari 2004; Carod-Artal 2010). The onset of neurological symptoms usually takes place within weeks after infection and progresses in an acute or subacute manner with the symptoms and signs of the disease progressively getting worse. Approximately 90% of patients develop a full neurological picture within 2 months (Ferrari 2004; Carod-Artal 2008). Patients may present with acute encephalitis or myelopathy or both during the acute phase of the disease after systemic manifestation or during the early stages of chronic disease (Ferrari 2004; Carod-Artal *et al.* 2006; Carod-Artal 2008; Lambertucci *et al.* 2009; Carod-Artal 2010). Cerebral schistosomiasis involvement occurs more frequently in *S. japonicum* infections while myelopathy is more common in *S. mansoni* and *S. haematobium* infections (Carod-Artal *et al.* 2006; Lambertucci *et al.* 2009; Carod-Artal 2010).

19.3.1 Cerebral Schistosomiasis

Yamagiwa published the first description of cerebral schistosomiasis caused by *S. japonicum* in 1889 (Yamagiwa 1889; Carod-Artal 2008). Cerebral complications were, likewise, reported in 2.3% of 1,200 diagnosed cases of acute schistosomiasis among World War II soldiers stationed in The Philippines (Kane and Most 1948). The chief neurological feature is diffuse encephalopathy and seizures (Ferrari 2004; Carod-Artal 2008). Patients typically present with any of the following symptoms: fever, headache, nystagmus, speech disturbances, some degree of motor weakness and cranial nerve abnormalities due to the formation of mass granulomatous lesions and increased intracranial pressure in the cortex, subcortical white matter, basal ganglia or internal capsule (Ferrari 2004; Carod-Artal 2008). Papilledema may be seen as a result of intracranial pressure.

Seizures, focal or generalized, are a typical presentation in *S. japonicum* infection with CNS involvement. Cerebral involvement due to *S. mansoni* or *S. haematobium* infection typically involves the cerebral and cerebellar cortex and leptomeninges (Ferrari 2004; Carod-Artal 2008). Cerebellum involvement has increasingly been reported in Brazil. The presence of granulomas and inflammation in the cerebellar vermis may raise intracranial pressure and result in hydrocephalus (Braga *et al.* 2003; Ibahoin *et al.* 2004; Raso *et al.* 2006; Carod-Artal 2010).

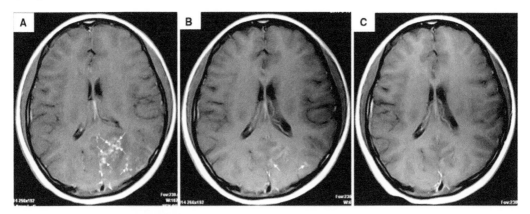

FIG. 19.1 A 14-year-old girl presenting with cerebral schistosomiasis due to *S. japonicum* in China. Enhanced MR images following intravenous administration of gadolinium-DTPA. **A.** Before PZQ treatment, showing multiple small nodular or 'silt-like' enhancements (white arrows) scattered or clustered in the cortical or subcortical area. **B.** 25 days after PZQ treatment, showing two nodular enhancements (white arrows). **C.** 3 months after PZQ treatment the enhancements have almost disappeared.

We recently described the diagnosis of a young Chinese schistosomiasis patient with cerebral involvement (Li *et al.* 2011). A 14-year-old girl presented to Xiang-Yue hospital, Yueyang City, Hunan Province, People's Republic of China, with a 4-month history of vertigo, headache, vomiting and syncope. The patient had lived in the Dongting Lake region, a highly endemic area for schistosomiasis, since birth. The patient reported no past medical history of epilepsy, tuberculosis or hepatitis B/C and no known drug allergies. Upon physical examination, the patient's vital signs were within normal limits and she had no evidence of fever, bloody stool or obvious abdominal discomfort and no other neurological signs were detected. Electrocardiography and chest radiograph were unremarkable. Hematological testing revealed evidence of eosinophilia (eosinophil count 0.8 $\times 10^9$), but white blood cells (white blood cell count 5.6×10^9; lymphocyte count 1.2×10^9; neutrophil count 3.4×10^9) and liver and renal functions were within normal limits. Both serological tests including indirect hemagglutination (IHA) and enzyme linked immunosorbent assays (ELISA) using soluble egg antigen and stool tests including the Kato-Katz thick smear stool examination, were positive for *S. japonicum* infection. Ultrasonography showed a wide echo dot pattern and increased echogenicity but no network in the liver and no abnormality in the size or texture of the spleen. An unenhanced axial section CT brain scan showed a 1.6×2.4 cm isodensity mass in the left parietal lobe with edema. Axial section MRI showed two hypointensity and hyperintensity lesions on T1- and on T2-weighted spin-echo images, respectively (Fig. 19.1). Multiple small nodular or 'silt-like' enhancements clustered in the subcortical region were evident following intravenous administration of gadolinium-DTPA. A diagnosis of 'cerebral schistosomiasis' was subsequently made and the patient was treated intravenously with 20% mannitol (125 ml daily for 5 days) to lower the intracranial pressure and orally with praziquantel (PZQ) (120 mg/kg three times a day for 6 days after meals). Follow-up MRI and CT scans 25 days after PZQ treatment showed that the edema and lesions were partially resolved and the lesions were almost completely resolved 3 months after PZQ treatment (Fig. 19.1).

19.3.2   Spinal Schistosomiasis

Myelopathy, such as transverse myelitis, is the most common neurological manifestation of *S. mansoni* or *S. haematobium* infection (Ferrari 2004; Carod-Artal 2008). The first reported case of thoracolumbar transverse myelitis was described in a German patient who visited Brazil in 1930 (Muller and Stender 1930; Carod-Artal 2008). Since then more than 500 cases have been reported in the medical literature (Ferrari 2004). Myeloradiculopathy occurs far less frequently than the cerebral form of the disease (Ferrari 2004; Carod-Artal 2008). Patients with spinal schistosomiasis usually present with lumbar pain, lower limb radicular pain, muscle weakness, sensory loss and bladder dysfunction due to egg deposition and granuloma formation in the spinal cord or cauda equina. This is typically seen at the T12 to L1 level in patients infected with *S. mansoni* and *S. haematobium* and is less frequently seen in patients infected with *S. japonicum* (Bill 2003; Carod-Artal 2008). Spinal schistosomiasis can also present as a progressive paraparesis mimicking a spinal cord neoplasm or tuberculosis (Carod-Artal 2008).

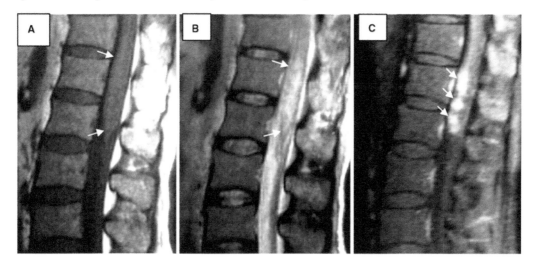

FIG. 19.2 A 40-year-old Chinese man was admitted to hospital with a 20-day history of pain and fatigue in his legs. A serological test was positive *for S. japonicum* egg antigen. Histology revealed schistosomal eggs in the spinal tissue. Intramedullary schistosomal granulomas were seen on MR images: **A.** Sagittal T1-weighted image shows moderate expansion of the distal cord isointense relative to the cord (white arrows). **B.** Sagittal T2-weighted image, showing a heterogeneous hyperintense lesion with an unclear boundary (white arrows). **C.** Sagittal T1-weighted image after contrast agent administration, showing multiple patchy lesions (white arrows) resembling a string of beads in the ventral spinal cord with significant enhancement.

Although spinal schistosomiasis is less commonly seen in China, Jiang and colleagues recently described a case from the Dongting lake region (Jiang *et al.* 2008). A 36-year-old man presented with bilateral leg pain, weakness and numbness followed by progressive bowel and bladder dysfunction for 3 months before admission. Upon physical examination there was hyperaesthesia at the level of T9 through L2 and anaesthesia at the level of L2. The cerebrospinal fluid (CSF) contained 10.5 WBCs per cubic millimetre (98% lymphocytes and 2% eosinophils), 120 mg/dl total protein and 60 mg/dl glucose. The CSF was positive for IgG against soluble egg antigen of *S. japonicum* by ELISA. MRI revealed an ill-defined lesion in the conus medullaris and lower thoracic spinal cord on sagittal T2-weighted FRFSE images (Fig. 19.2). Contrast-enhanced sagittal and T1-weighted images showed a cyst-like lesion (7 × 7 × 15 mm) and coronal T1-weighted images showed enhancement of ill-defined margins with central low signal intensity in the conus medullaris. On axial imaging the lesion showed enhancement as a mass (Fig. 19.2). The patient was treated with PZQ (60 mg/kg per day) and prednisone (60 mg per day) for 2 weeks. The patient showed no recovery from the neuronal damage. Surgery was performed to excise the intramedullary mass. PZQ (60 mg/kg per day) was administered for 1 week after surgery. No recurrence was seen 6 months after surgery and the patient had greatly improved clinically.

19.3.3 Other Manifestations

Cognitive impairment and memory deficits have also been described among children infected with *S. japonicum* or *S. mansoni* (Nokes *et al.* 1999). Recently, cognitive impairment, poor short-term memory and slower reaction times have been found among Tanzanian school children diagnosed with the disease (Jukes *et al.* 2002). Likewise, an Egyptian case-control study among school children revealed lower Wechsler Intelligence Scale Scores and poorer comprehensive, vocabulary and picture completion subtests in children with schistosomiasis versus controls (Nazel *et al.* 1999). Clearly, the ramifications of schistosomiasis cerebral involvement are much broader for communities where the disease is endemic.

19.4 DIAGNOSIS

The finding of eggs in the stool or positive serology, provide supportive but not direct evidence of schistosomal CNS involvement. However, a positive diagnostic finding coupled with neuroimaging and neurological symptoms should place neuroschistosomiasis high on the list of differentials. A definitive diagnosis can only be ascertained with histopathological study at biopsy or at necropsy showing schistosome eggs and granulomas (Carod-Artal 2008, 2010). For further discussion of diagnostics see Bergquistand van Dam (Chapter 21).

19.4.1 Direct Parasitological Methods

The detection of schistosomal eggs in the urine or feces is diagnostic of schistosomiasis. However, a number of studies examining the utility of the Kato-Katz technique in the diagnosis of schistosomes have shown variation in sensitivity with differences in the numbers of stool samples and slide preparations per stool employed. In general, as the numbers of stool samples collected and slides prepared increase, so does the sensitivity (Ebrahim *et al.* 1997; Yu *et al.* 1998). Studies in *S. japonicum* have also shown day-to-day variation in stool egg counts and also clustering of eggs within the stool (Ross *et al.* 1998). The presence of eggs in the urine or stool can be detected in less than 50% of neuroschistosomiasis patients (Balliauw *et al.* 2010; Carod-Artal 2010; Jongste *et al.* 2010). The Helmintex test for detection of schistosome eggs in feces, through their interaction with paramagnetic beads in a magnetic field, has recently been described (Fagundes *et al.* 2007). Although highly sensitive, its cost-effectiveness will need to be considered against that of the Kato-Katz technique to determine whether it has wide applicability in a variety of field settings.

19.4.2 Immunological Methods

Antibody detection of a schistosome infection is quite sensitive and useful in a few specific circumstances, but its use can be limited because antibodies persist after parasitological cure (Rabello 1997; Utzinger *et al.* 2005). Additional supportive clinical and laboratory findings for diagnosis of schistosomiasis include evidence of: peripheral blood eosinophilia; CSF (cerebrospinal fluid) eosinophilia, pleocytosis and increased protein concentration; anaemia (e.g., iron deficiency anaemia, anaemia of chronic disease or macrocytic anaemia); hypoalbuminaemia; elevated urea and creatinine levels; and hypergammaglobulinaemia (Rabello 1997). With regard to the specific immunological diagnosis of neuroschistosomiasis, ELISA (enzyme-linked immunosorbent assay) against soluble egg antigens is the most reliable immunological method and has 50% sensitivity and 95% specificity (Carod-Artal 2010). IHA (indirect hemagglutination assay) tests against worm antigens have 70–90% sensitivity (Carod-Artal 2010). The combination of ELISA and IHA has 90% sensitivity and 93% specificity (Van Gool *et al.* 2002; Carod-Artal 2010; Zhu *et al.* 2010). Recently, immune complexes containing soluble egg antigens of *S. mansoni* were reported in the CSF of four patients with spinal cord schistosomiasis (Ferrari *et al.* 2011). Further studies are needed to verify these findings, but this diagnostic method holds considerable promise for the immunological diagnosis of neuroschistosomiasis.

19.4.3　Neuroimaging

CT and MRI are of value in investigating neuroschistosomiasis. CT images in patients symptomatic for cerebral schistosomiasis usually show mass lesions with hyperdense lesions surrounded by hypodense areas of edema with variable contrast enhancement. MRI may show small nodular or 'silt-like' enhancements scattered or clustered at the cortical or subcortical areas following intravenous administration of gadolinium-DTPA (Fig. 19.1). Spinal schistosomiasis typically reveals edema of the spinal cord, conus medullaris and cauda equina (Ferrari 2004; Carod-Artal 2008). Intramedullary schistosomal granuloma may show moderate expansion of the distal cord, isointense relative to the cord, a heterogeneous hyperintense lesion with an unclear boundary or multiple patchy nodular lesions resembling a string of beads mainly in the ventral spinal cord which may be significantly enhanced (Fig. 19.2). Atrophy of the spinal cord may be found in long standing cases (Carod-Artal 2008).

19.5　TREATMENT

Early recognition of neuroschistosomiasis is imperative so that timely treatment can be initiated in order avoid severe disability. However, there is no definitive consensus or randomized controlled trials on the course of treatment once a diagnosis has been made (Carod-Artal 2008). Schistosomicidal drugs, steroids and surgery are presently used for the treatment of neuroschistosomiasis. During the 'acute phase' of infection, neuroschistosomiasis is treated with corticosteroids which are augmented with a course of PZQ once ovipositioning has occurred (Jaureguiberry and Caumes 2008).

19.5.1　Praziquantel

PZQ, a pyrazinoisoquinoline derivative, is a safe and highly effective oral drug that is active against all schistosome species (Ross *et al.* 2001; Ross *et al.* 2002). Standard treatment of chronic schistosomiasis is 60 mg/kg of PZQ in divided doses; for mass chemotherapy a single dose (40 mg/kg) is used (Ross *et al.* 2001; Ross *et al.* 2002). Treatment failures with this dose have been reported, particularly in areas where schistosomiasis has been recently introduced (Ross *et al.* 2002). Whether this is because the drug works in concert with the host immune response, which has yet to develop or due to migrating larvae not yet susceptible to PZQ is unknown. Drug susceptibility usually takes three weeks to develop. Other drugs that have been used in the treatment of schistosomiasis are oxamniquine (Vansil) for *S. mansoni* and metrifonate (trichlorfon) for *S. haematobium* (Ross *et al.* 2002).

19.5.2　Corticosteriods

PZQ is usually augmented by a course of corticosteroids (prednisone 1.5–2.0 mg/kg per day for 3 weeks) for the treatment of schistosomal encephalopathy during the oviposition stage (Ross *et al.* 2002; Ferrari 2004). Corticosteroids help to alleviate acute allergic reactions and mass effects caused by excessive granulomatous inflammation in the CNS (Fowler *et al.* 1999; Carod-Artal 2008). It is noteworthy that no double-blind randomized control trial has been conducted to determine the effects of corticosteroids on the spinal cord (Carod-Artal 2008). In contrast, PZQ is contraindicated during the acute phase of infection (Grandiere-Perez *et al.* 2006; Jaureguiberry and Caumes 2008; Caunes and Vidailhet 2010). As mentioned, PZQ cannot be used as a chemoprophylactic because of its short half-life and its inability to kill schistosomula (the migrating larvae) that are 3–21 days old. Instead corticosteroids are recommended for treatment (Grandiere-Perez *et al.* 2006; Jaureguiberry and Caumes 2008; Caunes and Vidailhet 2010). To date, no clinical trial has been undertaken to examine the role of ART alone or in combination with corticosteroids for the treatment of neurological symptoms complicating Katayama syndrome.

19.5.3   Other Therapies

In a patient suspected of having neuroschistosomiasis, priority should be given to the emergent management of seizures in particular if they are prolonged or recurrent. As most of the available anticonvulsants interfere with the metabolism of PZQ through induction or inhibition of the hepatic cytochrome P450 pathway, the PZQ dosage will need to be adjusted accordingly. Phenobarbital and phenytoin are probably the most readily available anticonvulsants in regions where schistosomiasis is endemic, but they are both known to markedly increase the metabolism and clearance of PZQ. If long-term anticonvulsant therapy is required in conjunction with PZQ, consideration should be given to the use of a non-hepatically metabolized anticonvulsant such as levetiracetam (Keppra) or agents that have less hepatic interaction with PZQ such as sodium valproate or carbamazepine. Surgery should be reserved for special cases such as inpatients with evidence of medullary compression and those who deteriorate despite clinical treatment (Ferrari 2004). Surgical intervention, ventriculoperitoneal shunt and corticosteroids are required to treat hydrocephalus and intracranial hypertension in cerebral schistosomiasis (Carod-Artal 2008).

19.6   CHAPTER SUMMARY

Neuroschistosomiasis is one of the most severe clinical outcomes associated with schistosome infection. Neurological complications early during the course of infection are thought to occur through *in situ* egg deposition following aberrant migration of adult worms to the brain or spinal cord. The presence of eggs in the CNS induces a cell-mediated Th2-driven periovular granulomatous reaction. The mass effect of thousands of eggs and the large granulomas concentrated within the brain or spinal cord explain the signs and symptoms of increased intracranial pressure, myelopathy, radiculopathy and subsequent clinical sequelae. Myelopathy (acute transverse myelitis and subacute myeloradiculopathy) of the lumbosacral region is the most common neurological manifestation of *S. mansoni* or *S. haematobium* infection, whereas, acute encephalitis of the cortex, subcortical white matter, basal ganglia or internal capsule is typical of *S. japonicum* infection. Cerebral complications include encephalopathy with headache, visual impairment, delirium, seizures, motor deficits and ataxia, while spinal symptoms include lumbar pain, lower limb radicular pain, muscle weakness, sensory loss and bladder dysfunction. The finding of eggs in the stool or a positive serology, provides supportive but not direct evidence of neuroschistosomiasis. A definitive diagnosis can only be made with histopathological study showing *Schistosoma* eggs and granulomas in the CNS. Schistosomicidal drugs (notably praziquantel), steroids and surgery are currently used for the treatment of neuroschistosomiasis.

Many questions remain unanswered with regard to neuroschistosomiasis. Its true prevalence is yet unknown but is presently estimated at between 1–5% of all diagnosed cases of schistosomiasis. No pathological studies of neuroschistosomiasis cases associated with Katayama fever are available and no double-blind randomized control trials have been undertaken to determine the effects of steroids on the spinal cord. Moreover, no clinical trial has examined the role of ART alone or in combination with corticosteroids for the treatment of neurological symptoms complicating Katayama syndrome. There is also a lack of clinical data in the literature regarding the timing of treatment, results of surgery and combination therapy. A complete or partial recovery occurs in 70% of patients with myeloradiculopathy but a less favorable outcome is seen in those treated for transverse myelitis. Early diagnosis and prompt treatment should improve the prognostic outlook.

19.7 LITERATURE CITED

Balliauw C, Martens F, Steen KVD, Bladt O, Vanhoenacker P. 2010. Spinal schistosomiasis. European Journal of Radiology Extra's 73: e49–e51.

Bartley PB, Ramm GA, Jones MK, Ruddel RG, Li Y, McManus DP. 2006. A contributory role for activated hepatic stellate cells in the dynamics of *Schistosoma japonicum* egg-induced fibrosis. International Journal Parasitology 36: 993–1001.

Batson OV. 1940. The function of the vertebral veins and their role in the spread of metastases. Annals Surgery 112: 138–149.

Bill P. 2003. *Schistosomiasis and the Nervous System.* Practical Neurology. Blackwell, Oxford.

Braga BP, Costa LB, Lambertucci JR. 2003. Magnetic resonance imaging of cerebellar schistosomiasis mansoni. Revista da Sociedade Brasileira de Medicina Tropical 36: 635–636.

Burke ML, Jones MK, Gobert GN, Li YS, Ellis MK, McManus DP. 2009. Immunopathogenesis of human schistosomiasis. Parasite Immunology 31: 163–176.

Carod-Artal FJ, Mesquita HM, Gepp Rde A, Antunes JS, Kalil RK. 2006. Brain involvement in a *Schistosoma mansoni* myelopathy patient. Journal Neurological Neurosurgery Psychiatry 77(4): 512.

Carod-Artal FJ. 2008. Neurological complications of *Schistosoma* infection. Transactions Royal Society Tropical Medicine Hygiene 102: 107–116.

Carod-Artal FJ. 2010. Neuroschistosomiasis. Expert Reviews Antimicrobial Infective Therapy 8: 1307–1318.

Caunes E, Vidailhet M. 2010. Acute neuroschistosomiasis: a cerebral vasculitis to treat with corticosteroids not praziquantel. Journal of Travel Medicine 17: 359.

Cetron MS, Chitsulo L, Sullivan JJ, Pilcher J, Wilson M, Noh J, Tsang VC, Hightower AW, Addiss DG. 1996. Schistosomiasis in Lake Malawi. Lancet 348: 1274–1278.

Chen MG, Mott K. 1989. Progress in the assessment of morbidity due to *Schistosoma japonicum* infection: a review of recent literature. Tropical Disease Bulletin 85: R1–R56.

Chen MG. 1991. Relative distribution of *Schistosoma japonicum* eggs in the intestine of man: a subject of inconsistency. Acta Tropica 48: 163–171.

Davis A. 2002. Schistosomiasis. pp. 1431–1469. *In*: Cook GC, Zumla AI, editors. *Manson's Tropical Diseases.* Saunders, London.

Ebrahim A, El-Morshedy H, Omer E, El Daly S, Barakat R. 1997. Evaluation of the Kato-Katz thick smear and formol ether sedimentation techniques for quantitative diagnosis of *Schistosoma mansoni* infection. American Journal Tropical Medicine Hygiene 57: 706–708.

Ferrari TC. 2004. Involvement of the central nervous system in the schistosomiasis. Memórias do Instituto Oswaldo Cruz 99: 59–62.

Ferrari TC, Faria LC, Vilaca TS, Correa CR, Goes AM. 2011. Identification and characterization of immune complexes in the cerebrospinal fluid of patients with spinal cord schistosomiasis. Journal Neuroimmunology 230: 188–190.

Fowler R, Lee C, Keytone JS. 1999. The role of corticosteroids in the treatment of cerebral schistosomiasis caused by *Schistosoma mansoni*: case report and discussion. American Journal Tropical Medicine Hygiene 61(1): 47–50.

Grandiere-Perez L, Ansart S, Paris L, Faussart A, Jaureguiberry S, Grivois JP, Klement E, Bricaire F, Danis M, Caumes E. 2006. Efficacy of praziquantel during the incubation and invasive phase of *Schistosoma haematobium* schistosomiasis in 18 travelers. American Journal Tropical Medicine Hygiene 74: 814–818.

Granier H, Potard M, Diraison PP, Nicolas X, Laborde JP, Talarmin F. 2003. Acute encephalitis concurrent with primary infection by *Schistosoma mansoni*. Médecine Tropicale 63: 60–63.

Gryseels B, Polman K, Clerinx J, Kestens L. 2006. Human schistosomiasis. Lancet 368: 1106–1118.

Ibahioin K, Chellaoui A, Lakhdar A, Hilmani S, Naja A, Sami A, Achouri M, Ouboukhlik A, El Kamar A, El Azhari A. 2004. Cerebellar schistosomiasis. A case report. Neurochirurgie 50: 61–65.

Jauréguiberry S, Ansart S, Perez L, Danis M, Bricaire F, Caumes E. 2007. Acute neuroschistosomiasis: two cases associated with cerebral vasculitis. American Journal of Tropical Medicine Hygiene 76: 964–966.

Jaureguiberry S, Caumes E. 2008. Neurological involvement during Katayama syndrome. Lancet Infectious Diseases 8: 9–10.

Jiang YG, Zhang MM, Xiang J. 2008. Spinal cord schistosomiasis japonica: a report of four cases. Surgery Neurological 69: 392–397.

Jongste AH, Tilanus AMR, Bax H, Willems MH, van der Feltz M, van Hellemond JJ. 2010. New insights in diagnosing *Schistosoma* myelopathy. Journal Infection 60: 244–247.

Jukes MC, Nokes CA, Alcock KJ, Lambo JK, Kihamia C, Ngorosho N, Mbise A, Lorri W, Yona E, Mwanri L, Baddeley AD, Hall A, Bundy DA. 2002. Partnership for child development. Heavy schistosomiasis associated with poor short-term memory and slower reaction times in Tanzanian schoolchildren. Tropical Medicine International Health 7: 104–117.

Kane CA, Most H. 1948. Schistosomiasis of the central nervous system: experiences in World War II and a review of the literature. Archives Neurological Psychiatry 59: 141–183.

Lambertucci JR, Souza-Pereia SR, Carvalho TA. 2009. Simultaneous occurrence of brain tumor and myeloradiculopathy in schistosomiasis mansoni: case report. Revista da Sociedade Brasileira de Medicina Tropical 42: 338–341.

Li YS, Ross AG, Hou X, Lou Z, McManus P. 2011. Oriental schistosomiasis with neurological complications: case report. Annals of Clinical Microbiology and Antimicrobials 10: 1–5.

McManus DP, Li Y, Gray DJ, Ross, AG. 2009. Conquering 'snail fever': schistosomiasis and its control in China. Expert Reviews Antimicrobial Infective Therapy 7(4): 473–485.

Muller HR, Stender A. 1930. Bilharziose des Rückenmarkes unter dem Bilde einer myelitis dorso-lumbalis transversa completa. Archiv für Schiffs- und Tropen-hygiene 34: 527–538.

Nazel MW, el-Morshedy H, Farghaly A, Shatat H, Barakat R. 1999. *Schistosoma mansoni* infection and cognitive functions of primary schoolchildren, in Kafr El Sheikh, Egypt. Journal Egyptian Public Health Association 74: 97–119.

Nokes C, McGarvey ST, Shiue L, Wu G, Wu H, Bundy DA, Olds GR. 1999. Evidence for an improvement in cognitive function following treatment of *Schistosoma japonicum* infection in Chinese primary schoolchildren. American Journal of Tropical Medicine Hygiene 60: 556–565.

Pittella JE. 1997. Neuroschistosomiasis. Brain Pathology 7: 649–662.

Rabello A. 1997. Diagnosing schistosomiasis. Memórias do Instituto Oswaldo Cruz 92: 669–676.

Raso P, Tafuri A, Lopes Nda F, Monteiro ER, Tafuri WL. 2006. The tumoral form of cerebellar schistosomiasis: case report and measure of granulomas. Revista da Sociedade Brasileira de Medicina Tropical 39: 283–286.

Ross AG, Sleigh AC, Li Y, Williams GM, Waine G, Forsyth SJ, Li Y, Hartel GF, McManus DP. 1998. Measuring exposure to *S. japonicum* in China. II. Activity diaries, pathways to infection and immunological correlates. Acta Tropica 71: 229–236.

Ross AG, Sleigh AC, Li Y, Davis GM, Williams GM, Jiang Z, Feng Z, McManus DP. 2001. Schistosomiasis in the People's Republic of China: prospects and challenges for the 21st century. Clinical Microbiology Reviews 14: 270–295.

Ross AG, Bartley PB, Sleigh AC, Olds GR, Li Y, Williams GM, McManus DP. 2002. Schistosomiasis. New England Journal Medicine 346: 1212–1220.

Schramm G, Hamilton JV, Balog CI, Wuhrer M, Gronow A, Beckmann S, Wippersteg V, Grevelding CG, Goldmann T, Weber E, *et al.* 2009. Molecular characterisation of kappa-5, a major antigenic glycoprotein from *Schistosoma mansoni* eggs. Molecular Biochemistry Parasitology 166: 4–14.

Shiff C. 2000. Epidemiology of helminth infections. pp. 917–930. *In*: Nelson KE, Masters Williams C, Graham NM, editors. *Infectious Disease Epidemiology: Theory and Practice.* Aspen, Gaithersburg.

Utzinger J, Zhou XN, Chen MG, Bergquiest R. 2005. Conquering schistosomiasis in China: the Long March. Acta Tropica 96: 69–96.

Van Gool T, Vetter H, Vervoort T, Doenhoff MJ, Wetsteyn J, Overbosch D. 2002. Serodiagnosis of imported schistosomiasis by a combination of a commercial indirect hemagglutination test with *Schistosoma mansoni* adult worm antigens and an enzyme-linked immunosorbent assay with *S. mansoni* egg antigens. Journal of Clinical Microbiology 40: 3432–3437.

Waine GJ, McManus DP. 1997. Schistosomiasis vaccine development–the current picture. Bioessays 19: 435–443.

Wilson MS, Mentink-Kane MM, Pesce JT, Ramalingam TR, Thompson R, Wynn TA. 2007. Immunopathology of schistosomiasis. Immunology Cell Biology 85: 148–514.

Wynn TA, Thompson RW, Cheever AW, Mentink-Kane MM. 2004. Immunopathogenesis of schistosomiasis. Immunology Reviews 201: 156–167.

Yamagiwa J. 1889. Beitrage zur aetiologie der jacksonschen epilepsie. Virchows Archives Pathology Anatomy 119: 449–460.

Yu JM, De Vlas SJ, Yuan HC, Gryseels B. 1998. Variations in faecal *Schistosoma japonicum* egg counts. American Journal Tropical Medicine Hygiene 59: 370–375.

Yu JM, de Vlas SJ, Jiang QW, Gryseels B. 2007. Comparison of the Kato-Katz technique, hatching test and indirect hemagglutination assay (IHA) for the diagnosis of *Schistosoma japonicum* infection in China. Parasitology International 56: 45–49.

Zhu H, Yu C, Xia X, Dong G, Tang J, Fang L, Du Y. 2010. Assessing the diagnostic accuracy of immunodiagnostic techniques in the diagnosis of schistosomiasis japonica: a meta-analysis. Parasitology Research 107(5): 1067–1073.

CHAPTER

Subtle Morbidity in Schistosomiasis

G Richard Olds[1],* and *Jennifer F Friedman*[2]

20.1 INTRODUCTION

Because of the many overt clinical manifestations of schistosomiasis, such as the swollen bellies with enlarged liver and spleen with *S. mansoni, S. mekongi* and *S. japonicum* as well as hematuria and urinary obstruction in *S. haematobium*, schistosomiasis has been known clinically for thousands of years (Ross *et al.* 2002; Olds *et al.* 1996, 2001; see also Berquist *et al.*, Chapter 2). In addition, since adult worm pairs are visible with a naked eye and even the eggs of the parasite can be seen with low level microscopy, the basic pathophysiology of the disease was determined over a century ago (Bilharz 1853; Manson 1902; Symmers 1903; Migairi *et al.* 1914). The acute manifestations of swimmers itch and Katayama syndrome (Migairi and Suzuki 1914) as well as the chronic pathologic manifestations such as clay pipe-stem fibrosis (Symmers 1903), hepatosplenomegaly, bleeding esophageal varices and even frank corpulmonale were all described more than fifty years ago. Even the link between *S. haematobium* infection and squamous cell carcinoma of the bladder was made in the middle of the last century (Elsebai 1902).

As important as these *Schistosoma*-specific pathologic events have been to the understanding of all five human species of the parasite they generally occurred in a relatively small subset of the infected population. Since, however these sequelae often led to premature deaths they quickly became the target of most schistosomiasis research during the late 20th century (Global Burden of Disease Study 2015). Active field investigations have been implemented in an effort to forestall development of morbidity (Fig. 20.1).

Most population based research on chronic human infections were based on cross sectional human population surveys where it was assumed these chronic morbid sequelae were a result of both the intensity and duration of *Schistosoma* infection. It was not until the 1980's and 90's that host factors were considered in their development, including the genetic background of the host (Abel-Salem *et al.* 1986) and co-infections with such agents as Hepatitis B and C (Lyra *et al.* 1976; Koshy 1993).

From a control standpoint, this focus on severe end organ pathology, unique to schistosomiasis, combined with the high cost of treatment soon after curative drugs were discovered, led to a public health strategy that concentrated treatment on heavily infected individuals (Grysells and Polderman 1991). The cost of screening was significantly less than the cost of treatment. As a result, during

[1] Medicine, St George's University, Grenada West Indies.
[2] Pediatrics, Alpert Medical School of Brown University and Center for International Health Research Rhode Island Hospital, Providence R.I. USA.
* Corresponding author

the early stages of schistosomiasis control in such countries as Egypt, the Philippines and Brazil, great efforts were expended on finding heavily infected cases using direct observation of eggs in the urine and stool.

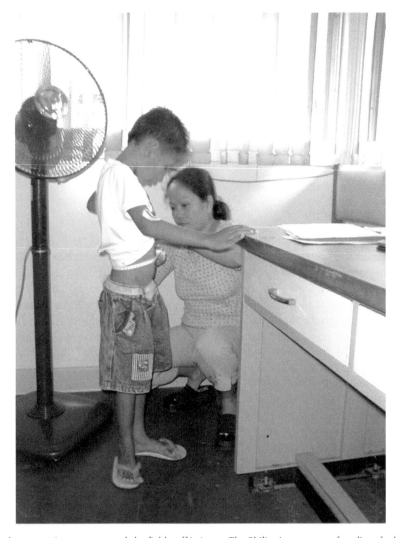

FIG. 20.1 Anthropometric measures made by field staff in Leyte, The Philippines as part of studies of schistosomiasis-related morbidity.

With the introduction of effective oral treatment for schistosomiasis in the 1980's it became possible to study the less obvious morbid sequelae of *Schistosoma* infections. Around the same time, public health scholars began to encourage the inclusion of morbidity in weighing the burden of diseases, as opposed to approaches that almost exclusively used mortality to weigh disease burden (Global Burden of Disease Study 2015). This first led to the consideration of the role of schistosomiasis in the pathogenesis of anemia, undernutrition and cognitive impairment. The specific impact that schistosomiasis had on these parameters could be determined by longitudinal studies following treatment. Such studies took advantage of the fact that Praziquantel has a very narrow spectrum of effectiveness, treating primarily human trematodes and cestodes (WHO 2006). This area of less obvious schistosomiasis-induced morbidity is now referred to as subtle morbidity, though many physicians rightly take issue with designating these debilitating morbidities, often occurring in already vulnerable populations (Fig. 20.2) as "subtle" (King 2015).

FIG. 20.2 Filipina girl collecting water for her household. Activities of daily living necessitate contact with fresh water, placing individuals at persistent risk of schistosomiasis.

Initially, studies addressed childhood growth and development and anemia (Mahmoud *et al.* 1972; Stephenson *et al.* 1987, 1989; McGarvey *et al.* 1992). As the field evolved, other parameters were studied including cognitive development (Nokes *et al.* 1999), school and work performance, functional work capacity, birth outcomes (Olds 2003) and inhibition of effective TH-1 responses to vaccines and co-infections (Al-Shamiri *et al.* 2011). All of these negative effects of a *Schistosoma* infection are potentially reversible with treatment but needed to be studied with either double blind placebo controlled trials or cohort studies that carefully controlled for potential confounders given the inter-correlations among risk for helminth infections, poverty and its attendant risks for undernutrition and anemia. Today these effects have become at least as important to our understanding of the impact that schistosomiasis has on humans as the classical studies of *Schistosoma*-specific pathologic changes (Olds *et al.* 1996; WHO 2002, 2006). This is due to the fact that the overall assessment of the DALY or disability adjusted life years attributed to schistosomiasis, was re-quantified to include these "subtle morbidities," greatly increasing its understood impact on the health of populations (King *et al.* 2005). In addition, studies demonstrated that those who were undernourished or experienced growth faltering, had the most significant improvements in those domains following treatment (Coutinho *et al.* 2006). Finally, in many endemic areas, it was recognized that schistosomiasis was responsible for significant population attributable risk for many key morbidities, most notably anemia (Leenstra *et al.* 2006b). Addressing these "subtle morbidities" has become the driving force around population based chemotherapy.

It should be noted that some population based studies have failed to show a consistent effect of treatment on specific morbidities. There are many possible reasons for this. First, these morbid sequelae of infection are likely influenced by host factors for which schistosomiasis is a contributing factor. Thus, many individuals suffer significant morbidity while others with the same or even higher intensities of infection are much less affected. It has been suggested that some individuals have a more profound pro-inflammatory response to infection such that these individuals experience schistosomiasis with greater anemia, undernutrition and perhaps fibrosis in the human host (Coutinho *et al.* 2005). In addition, some populations with better access to macro and micronutrients in their diet may be able to "compensate" for the deleterious effects of infection. Furthermore, many studies re-assess children too soon after treatment. It has been demonstrated that improvements in hemoglobin take at least 3 months to become apparent, likely due to the time it takes for inflammation to resolve such that iron absorption can improve and occult blood loss in the stool declines. With respect to cognition, susceptibility to cognitive impairment with infection may be greater in very young children and therefore might be missed in studies of older children already in school (Nokes *et al.* 1999). In addition, children may need time to recover and re-acquire or acquire cognitive processes affected by schistosomiasis such as anemia of inflammation and iron deficiency and likely undernutrition (King *et al.* 2005; Olson *et al.* 2009; Ezeamama *et al.* 2012). Finally, in the context of many other poverty related insults, treatment of a single infection may not allow the desired impact on morbidities with multiple etiologies.

Given the current very low cost of treatment and the relatively high cost of diagnosis, the focus of control efforts has switched from screening for individuals with heavy infections to periodic mass treatment of populations at-risk. In addition to these changes in the public health approach to schistosomiasis the new focus on subtle morbidity was instrumental in schistosomiasis retaining its place as the most significant helminth infection of humans. As discussed, the worldwide impact of all tropical diseases including schistosomiasis has been redefined using a new parameter; disability-adjusted life years (DALY's) (WHO 2006). If schistosomiasis were judged only by the late, *Schistosoma*-specific pathology of clay pipe-stem fibrosis, bleeding esophageal varices, renal obstruction and renal failure, then today schistosomiasis would be considered a relatively unimportant parasitic infection globally. Such insights have led directly to the current international interest in de-worming at-risk school children throughout the developing world (Olds 2013).

For the purpose of this chapter we have divided the types of subtle morbidity into specific categories. Most of the research on this topic has been done in *S. haematobium, S. mansoni* and *S. japonicum* human infections. Few studies on this topic have been on *S. mekongi* or *S. intercalatum.*

20.2 ANEMIA

Anemia in schistosomiasis has been studied the longest of the subtle morbidities (Mahmoud and Woodruff 1972). The causes of anemia, however are likely multi-factorial (Friedman *et al.* 2005). Blood loss through the stool or urine is common (Kanzaria *et al.* 2005; Bustinduy *et al.* 2013) and likely contributes to anemia through iron loss, compounding iron deficiency anemia caused by hookworm and dietary insufficiency of iron. In addition, adult schistosomes live in the human vasculature and use red blood cells as an energy source (Friedman *et al.* 2005), though this is thought to be a minor contributor. Improvements in hemoglobin with the treatment of schistosomiasis likely occurs through decreasing occult blood loss, which occurs at higher intensities of infection and amelioration of anemia of inflammation (AI) (Friedman *et al.* 2005) It is likely that anemia of inflammation is the primary contributor based on studies demonstrating higher population attributable risk for anemia of inflammation due to schistosomiasis than for iron deficiency anemia (Leenstra *et al.* 2006a, 2006b). Other work supporting this etiology includes RCTs of praziquantel which demonstrated higher hemoglobin levels (p<.001) at six month following Praziquantel treatment of children over placebo treated controls (McGarvey *et al.* 1996). The placebo treated children had further deterioration in their hemoglobin when left untreated while the Praziquantel treated children improved their hemoglobin levels over pre-treatment values without being given iron supplementation. The authors suggest this was due to release of sequestered iron stores following cure. The improvement in hemoglobin following treatment was more marked in males than females. This effect on anemia was then confirmed in a three country, double-blind placebo controlled study in which *S. haematobium, S. mansoni* and both the Philippine and Chinese strains of *S. japonicum* were studied. In all sites a significant increase in serum hemoglobin was observed in children treated with Praziquantel at six month follow-up even without iron supplementation (Olds *et al.* 1999). This improvement following Praziquantel treatment was seen even in children who were living in the endemic area but not found to have parasite eggs in their stool or urine. This finding suggested that children with very light infections (below the detection rate of two stool or urine samples) might benefit from treatment. Interestingly no effect on anemia was observed using Albendazole, despite a significant prevalence of hookworms. This observation again points out the differences in anemia between these two helminth species.

The role of AI in schistosomiasis has important treatment implications. In the context of AI iron absorption from the gut is decreased and iron is shunted from bio-available forms to storage forms. The provision of iron in this context, therefore, will have a dampened effect and treatment of schistosomiasis will have dual benefit by treating the underlying inflammation due to infection and maximizing children's ability to benefit from iron supplementation. This has led to recent calls for the inclusion of treatment for schistosomiasis in the Integrated Management of Childhood Illness (IMCI) guidelines such that young children can maximally benefit from limited contact with the health care sector and allow interventions to synergize.

This positive impact on anemia has recently been confirmed during the follow-up studies in Uganda and Burkina Faso where population based mass chemotherapy has been used for almost a decade (Kabatereine *et al.* 2007). In recent years, a major focus of the Schistosomiasis Control Initiative (SCI) and the Schistosomiasis Consortium for Operational Research and Evaluation (SCORE) has been identifying the optimal frequency of and target populations for MDA that will both decrease prevalence of infection and associated morbidities. This includes better understanding of the frequency needed to reduce population levels of anemia as well as the key age groups to include in MDA, ranging from more targeted and accessible school age children in school based campaigns to community MDA which has even started to include preschool aged children (Stothard *et al.* 2011; Verani *et al.* 2011).

20.3   CHILDHOOD GROWTH AND NUTRITIONAL STATUS

Nutritional status is considered a key "vital sign" in the assessment of the pediatric patient, integrating both long and short-term insults to health. Schistosomiasis is well known to cause deficits in linear growth leading to chronic undernutrition (stunting) as well as acute undernutrition (wasting) (Coutinho *et al.* 2006; Colley *et al.* 2014). Praziquantel has the greatest nutritional benefit for children who are undernourished at the time of treatment (Coutinho *et al.* 2006).

Historically there was evidence that schistosomiasis likely had negative effect on childhood growth. This was particularly true for *Schistosoma japonicum* although reports exist with *S. haematobium* and *S. mansoni* as well. In China, a condition known as Schistosomal Dwarfism was described (McGarvey *et al.* 1993) and in Japan, military recruits from prefectures endemic for schistosomiasis were noted to be shorter in stature as late as World War II (Olds and Dasarathy 2001).

The first epidemiologic studies were cross sectional in design in chronically infected populations. Schistosome infected school children appeared to be shorter in stature, lighter in weight and had less fat as assessed by skin fold thickness. For example, in a study from The Philippines (McGarvey *et al.* 1992), males aged 16–18 infected with *S. japonicum*, were almost 8 centimeters shorter and almost 6 kilograms lighter than uninfected individuals from the same community. They also had smaller arm circumferences and sum of skin folds. In studies from both China and The Philippines (Olds *et al.* 1996), females appeared to be more severely affected than males with reductions in stature, weight, arm muscle area and sum of skin folds. The greatest negative effect was seen during the adolescent growth spurt, which occurs earlier in females. Growth stunting was also related to intensity of infection. Multi-variable analysis suggested that these changes were independent of other parasitic infections. It is important to point out that these population based studies were primarily done prior to the wide spread use of oral treatment and thus these subjects had significantly more heavy infections and were likely infected for many years.

Because child growth is affected by a variety of factors, the most important studies came from blinded, randomized placebo controlled trials with Praziquantel and well-controlled longitudinal studies. In RCTs improvements in the sum of skin folds were observed at six month following treatment. It was found that children infected with *S. japonicum*, successfully cured and who remained free of infection for a year experienced "catch up" growth (Olds *et al.* 1996), which was faster than uninfected age matched controls. Unfortunately, children who were initially uninfected and became infected during the one year follow-up had significant deceleration of their growth. Similarly, in longitudinal treatment studies, children who were infected at baseline and remained infection free experienced the greatest longitudinal growth benefit (Coutinho *et al.* 2006). These studies therefore suggested that populations need to be treated frequently, at least once per year and more in communities with higher transmission pressure, if the impact on child growth is to be sustained.

Also of note, mechanistic studies have demonstrated that children experiencing the greatest nutritional morbidity due to schistosomiasis have elevated levels of pro-inflammatory cytokines that lead to anorexia and cachexia, which may limit interventions provided for undernutrition in this setting. This has implications for treatment of highly prevalent undernutrition, as seizing the opportunity to treat children with co-morbid undernutrition will have both direct effects on nutritional status and indirect effects by improving appetite and the ability to take advantage of nutritional interventions provided.

In summary, there is significant evidence that at least three species of schistosomiasis have an impact on child growth and nutritional status. *S. japonicum* appears to have the greatest negative impact on growth specifically of the various species. The impact seems to be greatest in girls and during the adolescent growth spurt in both sexes. Infection, treatment and then re-infection appears to slow normal growth suggesting that keeping a population of growing children free of infection should be the treatment goal. Because children can catch up to their growth before their epiphyseal plates close, particular attention should be paid to keeping prepubescent children free of infection. Given the widespread use of population based chemotherapy worldwide and the ethical considerations involved in doing placebo controlled trials today, it appears unlikely that more large scale studies will be done on this issue.

20.4 COGNITIVE DEVELOPMENT AND SCHOOL PERFORMANCE

The impact of schistosomiasis on cognitive development and school performance has been significantly less studied than growth or anemia. One of the major obstacles had been a lack of a culturally appropriate tool to assess cognition. Two studies, one in *S. haematobium* (Stephenson *et al.* 1989) and one in *S. japonicum* (Nokes *et al.* 1999), were done in conjunction with placebo controlled trials. In both cases very young children (< 5 years old) appeared to be most affected. Improved performance was documented after parasitological cure as compared to controls.

More recently, cognitive tests have been adapted that are more culturally appropriate and likely improve the ability to understand causal relationships by decreasing measurement error. Studies in The Philippines adapted the Wide Range Assessment of Memory and Learning in collaboration with a child psychologist and utilized an instrument developed and validated in The Philippines (The Philippines Non-Verbal Intelligence Test) (Ezeamama *et al.* 2012). These studies demonstrated improvements in cognitive function among children who remained infection free following treatment after adjustment for key confounders (Ezeamama *et al.* 2012). Other studies examined the contribution of AI to cognitive impairment, finding that this type of anemia did negatively impact cognition and further suggesting that treatment of schistosomiasis and consequent AI may improve cognitive function. This is important given the very limited educational opportunities afforded children in low and middle income countries and the need to maximize benefits from these.

20.5 FUNCTIONAL WORK CAPACITY

To our knowledge, no blinded placebo controlled trials have been done on functional work capacity. Most studies have compared exercise tolerance and performance of basic physical fitness activities before and after treatment. Some studies have shown improvement in these parameters but some have not (Ming-Gang 2005). In one study from China (Wu *et al.* 2002) individuals with severe *S. japonicum* infection sequelae appeared to have a 15% reduction in work capacity. A similar study from Egypt showed a 20% improvement after treatment (El Karim *et al.* 1981). Consistent with the theme that host factors are often more important to parameters such as functional work capacity it appears that infected healthy males often do not have a significant reduction in work capacity (Collins *et al.* 1976). On the other hand growing children, malnourished individuals or those with significant anemia may well have decreased work capacity related to their schistosome infection, but these populations in general have not been studied.

20.6 PREGNANT WOMEN AND BIRTH OUTCOMES

The understanding of anemia in schistosomiasis increased the concern about the potential adverse effects this infection might have on pregnant and lactating women and their offspring. When Praziquantel was introduced in the early 1980's sadly studies were not performed on pregnant women. As a result, PZQ was released as a Pregnancy Category B drug, presumed safe based on animal studies but not tested in humans. As a result pregnant or lactating women were excluded from national control programs in several countries. In fact, post pubescent girls were occasionally removed from school based treatment programs because of the difficulty in determining pregnancy in the first few months following conception (Cool 1992; WHO 1998).

This exclusion of women and girls was particularly concerning because animal studies showed that schistosome-infected and pregnant mice had a decreased number of viable litters, increased maternal deaths, increased spontaneous abortions and stunted off-spring (Amano *et al.* 1990). In cross-sectional field studies, infected women were reported to have an increased incidence of prematurity, low birth weight children, preeclampsia and fetal death (Siegrist and Siegrist-Obimpeh 1992). In addition, PZQ has been used extensively in the field throughout the world for decades without apparent negative effects on women who were inadvertently treated early in their first trimester. As a result, the WHO Expert Committee on schistosomiasis strongly recommended in their guidelines that pregnant women be treated (WHO 2002, 2006).

Despite this recommendation many women continued to be excluded from treatment, largely due to lack of data from well controlled trials addressing safety. Finally two successful phase 2, randomized, double-blind placebo controlled trails were performed in Uganda and The Philippines which supported the safety of PZQ for pregnant women and their fetuses. (Ndibazza *et al.* 2010; Olveda RM *et al.* 2015). Though the Ugandan study did not demonstrate an effect on anemia, the recently completed study in The Philippines demonstrated that women who were treated had higher iron stores as captured by ferritin at 32 weeks gestation. Neither study demonstrated a significant impact on birth weight, but study in The Philippines found that newborns of treated women had a trend toward improved iron endowment. Limitations of both studies included the low prevalence of moderate and higher intensity infections, limiting our understanding of the impact among women with more intense infections.

Results from these trials provided important data from controlled trials in support of the expansion of treatment policies to include pregnant women as already recommended by WHO. Though neither study observed an impact on birth weight, it remains possible, based on animal models, as well as human observational and mechanistic studies, that schistosomiasis adversely affects pregnancy outcomes. However, treatment during gestation is too late to be impactful. If this is the case, including pregnant women in MDA campaigns will increase the likelihood that they enter subsequent pregnancies free of schistosomiasis infection. This is important for the health of their pregnancies as well as their own experience of morbidity, particularly anemia, such they do not continue to accrue infections during their reproductive years.

20.7   FEMALE GENITAL SCHISTOSOMIASIS

Apart from the impact on anemia and pregnancy outcomes, schistosomiasis of the female genital tract has become an important area of focus. Female Genital Schistosomiasis (FGS) causes bleeding and the three distinct pathologic lesions. Studies led by Eryun Kjetland *et al.* in Tanzania and Zimbabwe found that women with FGS have a three- to four-fold increased risk of having HIV, they increased secondary infertility, contact bleeding and incontinence (Kjetland *et al.* 2012, 2014). The increased risk for HIV is thought to be the result of mucosal inflammation caused by egg granulomas along the female genital urinary tract (WHO 2010; Downs *et al.* 2011) which may increase the prevalence of HIV targeted CD4+ cells (Jourdan *et al.* 2011). These morbidities, together with the known effects of schistosomiasis on anemia and nutritional status, support greater efforts to treat women of reproductive age, particularly ensuring MDA programs worldwide include them.

20.8   IMMUNOLOGIC IMPAIRMENT AND CO-INFECTION

Most worm infections induce a state of TH-2 dominant immune responses within the infected host (Pearce *et al.* 2004). This shift is not only to parasite antigens but also to new antigens presented to the host. Emerging evidence demonstrates that both antenatal and childhood parasitic infections alter levels of protective immune response to routine vaccinations (Labeaud *et al.* 2009). Studies have demonstrated that schistosome antigens cross the placenta and prime or tolerize fetal immune responses (Malhotra *et al.* 1999). In fact, several studies have shown that schistosome-infected children have reduced immune responses to childhood vaccines, particularly those that are driven by TH-1 responses to the vaccine antigens (Sabin *et al.* 1996). This shift in TH-2 dominance modulates three to six months after successful cure. This finding has led to efforts to coordinate population based mass treatment of children (de-worming) with natural efforts to immunize children national vaccine days.

This understanding of how schistosomiasis and various other helminths alter host immune responses has led to several studies of co-infection with viruses that require an intact TH-1 response for self-cure (Gasim *et al.* 2015). The best studied of these is Hepatitis C where co-infection with schistosomiasis appears to accelerate the liver disease in those with chronic active hepatitis (Kamal *et al.* 2004). A similar negative effect of co-infection has been found with Hepatitis B, but this relationship is more controversial (Pereira *et al.* 1994). Co-infection with HIV and schistosomiasis appears to lead to increased viral load (Lawn *et al.* 2000) and co-infection with HIV and schistosomiasis effects egg excretion (Kallestrup *et al.* 2005). Genital schistosomiasis appears to increase the risk of acquiring HIV. Thus treatment of schistosomiasis will likely have a positive impact on both HIV transmission and progression (Downs *et al.* 2012).

20.9   CHAPTER SUMMARY

The field of schistosomiasis was transformed when effective oral treatment became available in the 1980's. Efforts at control and even elimination have clearly accelerated now that the cost of treatment is very low and major pharmaceutical companies have provided millions of doses of PZQ for free.

As a result the classic end organ disease induced by this parasite may someday be of only historical value. Until, however, schistosomiasis is eradicated, acute manifestations of infection and subtle morbidities will be major causes of human suffering.

The ideal time to study subtle morbidity induced by schistosomiasis was during the 1980's and 90's when populations could easily be found that were both heavily infected and without a history of prior treatment. From the studies that were done during this period it appears that at least three species of *Schistosoma* (*S. haematobium*, *S. mansoni* and *S. japonicum*) have a negative effect on childhood growth, nutritional status, anemia, cognition, pregnant women and potentially their newborns. It may also alter immune responses to key childhood vaccines and host immune responses to viral pathogens.

We are not surprised that investigators have had increasing difficulty replicating those studies today as subtle morbidities are of multi-factorial etiologies and human host genetics, including varying immunologic responses to the same infection burden, culminate in different responses to infection and treatment. Thus, double-blind, placebo controlled studies in populations at high-risk and with high intensity infections are likely to further elucidate changes in some of these parameters. Some of the success of MDA programs has fortunately led to dwindling numbers of populations with high intensity infections. In addition, ethically RCTs simply cannot now be done. Strong evidence for this perspective comes from the great difficulties encountered in doing the most recent double-blind placebo controlled trails in pregnant women.

The authors believe that schistosomiasis is detrimental to infected individuals and agree with Charlie King that there is no "asymptomatic schistosomiasis" (King 2015). Subtle morbidity is likely to be most severe in individuals who already are at-risk for these sequelae and for which schistosomiasis is just another cumulative insult. This may reflect, in part, that these individuals are more susceptible to that morbidity based on host factors, such that treatment is more beneficial. Our knowledge of subtle morbidity should, however, direct ongoing control measures. Treating only children in school is unlikely to maximize effectiveness because the children at highest risk for subtle morbidity may be too young to attend school (cognitive performance) or not attending school due to poverty, which already places them at risk for schistosomiasis and nutritional morbidities. Given the reversibility of anemia found in schistosomiasis more attention should be paid to treating women of all ages, despite the fact that the prevalence and intensity of infection is often lower than in their male counterparts. Finally, periodic treatment must be maintained, at least at a frequency to suppress subtle morbidity until public health and sanitation standards can be improved. Our greatest fear is that the current lack of schistosome-induced pathology in at-risk populations undergoing periodic treatment will prompt governments to shut down national control efforts before this infection becomes eradicated.

20.10   LITERATURE CITED

Abel-Salem E, Abdel-Khalik A, Abdel-Meguid A. 1986. Association of HLA Class 1 antigens (A1, B5, B8 and CW2) with disease manifestations and infection in human schistosomiasis mansoni in Egypt. Tissue Antigens 27: 142–146.

Allen H, Crompton D, DeSilva N, LoVerde P, Olds GR. 2002. New policies for using anti-helmintics in high-risk groups. Trends in Parasitology 18(9): 381–382.

Al-Shamiri AH, Al-Taj MA, Ahmed AS. 2011. Prevalence and co-infections of schistosomiasis/hepatitis B and C viruses among school children in an endemic area in Taiz, Yemen. Asian and Pacific Journal of Tropical Medicine May 4(5): 404–8.

Amano T, Freeman GL, Colley DG. 1990. Reduced reproductive efficiency in mice with schistosomiasis mansoni and in uninfected pregnant mice injected with antibodies against *Schistosoma mansoni* soluble egg antigen. American Journal of Tropical Medicine and Hygiene 43(2): 180–185.

Bilharz T. 1853. A study on human helminthography. Derived from information by letter from Dr. Bilharz in Cairo, along with remarks by Prof. Th. V. Srebold in Breslau. Zeitschrift für Wissenschaftliche Zoologie 4: 53–71.

Bustinduy AL, Parraga IM, Thomas CL, Mungai PL, Mutuku F, Muchiri EM, Kitron U, King CH. 2013. Impact of polyparasitic infections on anemia and undernutrition among Kenyan children living in a *Schistosoma haematobium*-endemic area. The American Journal of Tropical Medicine and Hygiene 88(3): 433–440.

Colley DG, Bustinduy AL, Secor WE, King CH. 2014. Human schistosomiasis. Lancet 383(9936): 2253–2264.

Collins KJ, Brotherhood RJ, Davies CT, Dore C, Hackett AJ, Imms FJ, Musgrove J, Weiner JS, Amin MA, El Karim M, Ismail HM, Omer AH, Sukkar MY. 1976. Physiological performance and work capacity of Sudanese came cultures with *Schistosoma mansoni* infection. The American Journal of Tropical Medicine and Hygiene 25(3): 410–21.

Cook GC. 1992. Use of anti-protozoan and anti-helmintic drugs during pregnancy: side effects and contra-indications. Journal of Infection 25: 1–9.

Coutinho HM, Acosta LP, McGarvey ST, Jarilla B, Jiz M, Pablo A, Su L, Manalo DL, Olveda RM, Kurtis JD and others. 2006. Nutritional status improves after treatment of *Schistosoma japonicum*-infected children and adolescents. Journal of Nutrition 136(1): 183–188.

Coutinho HM, McGarvey ST, Acosta LP, Manalo DL, Langdon GC, Leenstra T, Kanzaria HK, Solomon J, Wu H, Olveda RM and others. 2005. Nutritional status and serum cytokine profiles in children, adolescents and young adults with *Schistosoma japonicum*-associated hepatic fibrosis, in Leyte, Philippines. Journal of Infectious Diseases 192(3): 528–536.

Downs JA, Mguta C, Kaatano GM, Mitchell KB, Bang H, Simplice H, Kalluvya SE, Changulucha JM, Johnson WD, Fitzgerald DW. 2011. Urogenital schistosomiasis in women of reproductive age in Tanzania's Lake Victoria Region. American Journal of Tropical Medicine and Hygiene 84: 364–369.

Downs JA, van Dam GJ, Fitzgerald DW. 2012. Association of schistosomiasis and HIV infection in Tanzania. The American Journal of Tropical Medicine and Hygiene 87(5): 868–873.

El Karim MA, Collins KJ, Omer AH, Amin MA, Dore C. 1981. An assessment of anti-schistosomal treatment on physical work capacity. The Journal of Tropical Medicine and Hygiene 84(2): 67–72.

Elsebai I. 1977. Parasites in the etiology of cancer; bilharziasis and bladder cancer. CA: A Cancer Journal for Clinicians 27: 100–106.

Ezeamama AE, McGarvey ST, Hogan J, Lapane KL, Bellinger D, Acosta LP, Leenstra T, Olveda RM, Kurtis JD, Friedman JF. 2012. Treatment of *Schistosoma japonicum*, reduction of intestinal parasite load and congnitive test score improvement in school-aged children. PLoS Neglected Tropical Diseases 6(5): 1634.

Friedman JF, Kanzaria HK, McGarvey ST. 2005. Human schistosomiasis and anemia: the relationship and potential mechanisms. Trends in Parasitology 21(8): 386–392.

Friedman JF, Mital P, Kanzaria HK, Olds GR, Kurtis JD. 2007. Schistosomiasis and pregnancy: Trends in Parasitology 23(4): 159–164.

Gasim GI, Bella A, Adam I. 2015. Schistosomiasis, hepatitis B and hepatitis C co-infection. Virology Journal 12: 19–36.

Global Burden of Disease Study C. 2015. Global, regional and national incidence, prevalence and years lived with disability for 301 acute and chronic diseases and injuries in 188 countries, 1990–2013: a systematic analysis for the Global Burden of Disease Study 2013. Lancet 386(9995): 743–800.

Grysells B, Polderman AM. 1991. Morbidity due to schistosomiasis mansoni and its control in subsaharan Africa. Parasitology Today 7: 244–49.

Jourdan PM, Holmen SD, Gundersen SG, Roald B, Kjetland EF. 2011. HIV target cells in *Schistosoma haematobium*-infected female genital mucosa. The American Journal of Tropical Medicine and Hygiene 85(6): 1060–1064.

Kabatereine NB, Brooker S, Koukounari A, Kazibwe F, Tukahebwa EM, Fleming FM, Zhang Y, Webster JP, Stothard JR, Fenwick A. 2007. Impact of a national helminth control programme on infection and morbidity in Ugandan school children. Bulletin of the World Health Organization 85(2): 91–99.

Kallestrup P, Zinyama R, Gomo E, Butterworth AE, van Dam GJ, Erikstrup C, Ullum H. 2005. Schistosomiasis and HIV-1 infection in rural Zimbabwe: implications of co-infection for excretion of eggs. The Journal of Infectious Diseases 191: 1311–1320.

Kamal SM, Graham CS, HE Q, Bianchi L, Tawil AA, Rasenack JW. 2004. Kinetics of intrahepatic hepatitis C virus (HCV) specific CD4+ T Cell responses in HCV and *Schistosoma mansoni* co-infection: relationship to progression of liver fibrosis. Journal of Infectious Diseases 189: 1140–1150.

Kanzaria HK, Acosta LP, Langdon GC, Manalo DL, Olveda RM, McGarvey ST, Kurtis JD, Friedman JF. 2005. *Schistosoma japonicum* and occult blood loss in endemic villages in Leyte, The Philippines: American Journal of Tropical Medicine and Hygiene 72(2): 115–118.

King CH, Dickman K, Tisch DJ. 2005. Reassessment of the loss of chronic helmintic infection: ameta-analysis of diability-related outcomes in endemic schistosomiasis. Lancet 365: 1561–1569.

King CH. 2015. It's time to dispel the myth of "asymptomatic" schistosomiasis. PLoS Neglected Tropical Diseases 9(2): e0003504.

Kjetland EF, Leutscher PD, Ndhlovu PD. 2012. A review of female genital schistosomiasis. Trends in Parasitology 28(2): 58–65.

Kjetland EF, Norseth HM, Taylor M, Lillebo K, Kleppa E, Holmen SD, Andebirhan A, Yohannes TH, Gundersen SG, Vennervald BJ and others. 2014. Classification of the lesions observed in female genital schistosomiasis. International Journal of Gynaecology and Obstetrics 127(3): 227–228.

Koshy A, Al-Nakib B, Al-Mufti S. 1993. Anti-HCV positive cirrhosis associated with schistosomiasis. American Journal of Gastroenterology 88: 1428–1437.

Koukounari A, Gabrielli AF, Toure S, Bosque-Oliva E, Zhang Y, Sellin B, Donnelly CA, Fenwick A, Webster JP. 2007. *Schistosoma haematobium* infection and morbidity before and after large-scale administration of praziquantel in Burkina, Faso. Journal of Infectious Diseases 196: 659–669.

Labeaud AD, Malhotra I, King MJ, King CL, King CH. 2009. Do antenatal parasite infections devalue childhood vaccination? PLoS Neglected Tropical Diseases 3(5): e442.

Lawn SD, Kavanja DM, Mwinzia P, Andove J, Colley DG, Folks TM, Secor WE. 2000. The effect of treatment of schistosomiasis on blood plasma HIV-1 RNA concentration in co-infected individuals. AIDS 14: 2437–2443.

Leenstra T, Acosta LP, Langdon GC, Manalo DL, Su L, Olveda RM, McGarvey ST, Kurtis JD, Friedman JF. 2006a. Schistosomiasis japonica, anemia and iron status in children, adolescents and young adults in Leyte, Philippines 1. American Journal of Clinical Nutrition 83(2): 371–379.

Leenstra T, Coutinho HM, Acosta LP, Langdon GC, Su L, Olveda RM, McGarvey ST, Kurtis JD, Friedman JF. 2006b. *Schistosoma japonicum* reinfection after praziquantel treatment causes anemia associated with inflammation. Infection and Immunity 74(11): 6398–6407.

Lyra LG, Reboueas G, Andrade ZA. 1976. Hepatitis B surface antigen carrier state in hepatosplenic schistosomiasis. Gastroenterology 71: 641–645.

Mahmoud AF, Woodruff AW. 1972. Mechanisms involved in the anaemia of schistosomiasis. Transactions of the Royal Society of Tropical Medicine and Hygiene 66: 75–84.

Malhotra I, Mungai P, Wamachi A, Kioko J, Ouma JH, Kazura JW, King CL. 1999. Helminth-and bacillus calmette-guerin-induced immunity in children sensitized in utero to filariasis and schistosomiasis. Journal of Immunology 162(11): 6843–6848.

Manson P. 1902. Report of the case of bilharzia from the West Indies. British Medical Journal 2: 1894–1895.

McGarvey ST, Aligui G, Daniel BL, Peters P, Olveda R, Olds GR. 1992. Child growth and schistosomiasis japonicum in Northeastern Leyte, The Philippines: cross-sectional results. American Journal of Tropical Medicine and Hygiene 46(5): 571–581.

McGarvey ST, Aligui G, Graham KK, Peters P, Olds GR, Olveda R. 1996. Schistosomiasis japonica and childhood nutritional status in northeastern Leyte, The Philippeans: a randomized trial of praziquantel versus placebo: American Journal of Tropical Medicine and Hygiene 54(5): 498–502.

McGarvey ST, Wu G, Zhang S, Wang Y, Peters P, Olds GR, Wiest PM. 1993. Child growth, nutritional status and schistosomiasis japonica in Jiangxi, People's Republic of China. American Journal of Tropical Medicine and Hygiene 48(4): 547–553.

Migairi K, Suzuki M. 1914. Der Zwichenwirt der Schistosomiasis Japonicum Katsurada. M. Heilunger der Medizinishen Fakulm. Heilunger der Medizinischer Fakultat Kalrserchen Universitat Kyusohu 1: 187–97.

Ming-Gang C. 2005. Use of praziquantel for clinical treatment and morbidity control of schistosomiasis japonica in China: a review of 30 years experience. Acta Tropica 96(2): 168–176.

Ndibazza J, Muhangi L, Akishule D, Kiggundu M, Ameke C, Oweka J, Kizindo R, Duong T, Kleinschmidt I, Muwanga M, Elliot AM. 2010. Effects of deworming during pregnancy on maternal and perinatal outcomes in Entebbe, Uganda: a randomized controlled trial. Clinical Infectious Diseases 50: 531–540.

Nokes K, McGarvey ST, Shive L, Wu G, Lin C, Wu H, Bundy DAP, Olds GR. 1999. Evidence for an improvement in cognitive function following treatment of *Schistosoma japonicum* infection in Chinese primary school children. American Journal of Tropical Medicine and Hygiene 60(4): 556–565.

Olds GR. 2003. Administration of praziquantel to pregnant and lactating women Acta Tropica 86(2–3): 185–195.

Olds GR. 2013. De-worming the world. Transactions of the American Clinical and Climatological Association 124: 265–274.

Olds GR, Dasarathy S. 2001. Schistomiasis. pp. 369–405. *In*: Gillespie S, Pearson R editors. *Principles and Practice of Clinical Parasitology.* London: Wiley.

Olds GR, King C, Johnson J, Olveda R, Wu G, Ouma J, Peters P, McGarvey ST, Odhiambo O, Koech D, Lui CY, Aligui G, Gachihi G, Kombe Y, Parraga I, Ramirez B, Honton J, Horst HJ, Reeve P. 1999. A double-blind placebo-controlled study of concurrent administration of albendazole and praziquantel in school children with schistosomiasis. Journal of Infectious Diseases 179: 996–1003.

Olds GR, Olveda R, Wu G, Wiest P, McGarvey ST, Aligui G, Zhang S, Ramirez B, Daniel B, Peters P, Romulo R, Fevidal P, Tiu W, Yuan J, Domingo E, Blas B. 1996. Immunity and morbidity in schistosomiasis japonicum infection. American Journal of Tropical Medicine and Hygiene 55(5): 121–126.

Olson CL, Acosta LP, Hochberg NS, Olveda RM, Jiz M, McGarvey ST, Kurtis JD, Bellinger D, Friedman JF. 2009. Anemia of inflammation is related to cognitive impairment among children in Leyte, The Philippines. PLoS Neglected Tropical Diseases 3(10): 533.

Olveda RM, Acosta LP, Tallo V, Baltazar PI, Lesiguez JLS, Estanislao GG, Ayaso EB, Monterde DBS, Ida A, Watson N, McDonald EA, Wu HW, Kurtis JD, Friedman JF. 2015. Efficacy and safety of praziquantel for the treatment of human schistosomiasis during pregnancy: a phase 2 randomized double-blind, placebo-controlled trial. Lancet Infectious Diseases: Published online October 26, 2015.

Pearce EJ, Kane MC, Sun J, Taylor JJ, McKee AS, Cervi L. 2004. TH2 response polarization during infection with the helminth parasite *Schistosoma mansoni.* Immunologic Reviews 201: 117–126.

Pereira LMMB, Melo MCV, Lacerda C, Spinelli V, Domingues ALC, Massarolo P, Mies S, Saleh MG, McFarlane IG, Williams R. 1994. Hepatitis B virus infection in schistosomiasis mansoni. Journal of Medical Virology 42: 203–206.

Ross AGP, Bartley PB, Sleigh AC, Olds GR, Li Y, Williams GM, McManus DP. 2002. Schistosomiasis: a clinical perspective: New England Journal of Medicine 346(16): 1212–1220.

Sabin EA, Araujo MI, Carvalho EM, Pearce EJ. 1996. Impairment of tetanus toxoid – specific TH1-like immune responses in humans infected with *Schistosoma mansoni.* The Journal of Infectious Diseases 173: 269–272.

Siegrist D, Siegrist-Obimpeh P. 1992. *Schistosoma haematobium* infection in pregnancy. Acta Tropica 50(4): 317–321.

Stephenson LS. 1987. Schistosomiasis. pp. 47–88. *In*: Stephenson LS, editor. *Impact of Helminth Infections on Human Nutrition.* New York: Taylor and Francis.

Stephenson LS, Latham MC, Kurz KM, Kinoti SN. 1989. Single dose metrifonate or praziquantel treatment in Kenyan children II. Effects on growth in relation to *Schistosoma haematobium* and hookworm egg counts. American Journal of Tropical Medicine and Hygiene 41: 445–453.

Stothard JR, Sousa-Figueiredo JC, Betson M, Green HK, Seto EY, Garba A, Sacko M, Mutapi F, Vaz Nery S, Amin MA and others. 2011. Closing the praziquantel treatment gap: new steps in epidemiological monitoring and control of schistosomiasis in African infants and preschool-aged children. Parasitology 138(12): 1593–1606.

Symmers W, St C. 1903. Note on a new form of liver cirrhosis due to the presence of the ova of a bilharzia haematobia. Journal of Pathology and Bacteriology 9: 237–89.

WHO. 1998. Report of the WHO informal consultation on monitoring of drug efficiency in the control of schistosomiasis and intestinal nematodes. Geneva: World Health Organization.

WHO. 2002. Prevention and control of schistosomiasis and soil-transmitted helminthiasis. Geneva: World Health Organization.

WHO. 2002. Report of the WHO informal consultation on the use of praziquantel during pregnancy, lactation and albendazole/praziquantel in children under 24 months. Geneva: World Health Organization.

WHO. 2006. Preventative chemotherapy in human helminthiasis. coordinated use of anti-helminthic drugs in control interventions. Geneva: World Health Orgranization.

WHO. 2010. Report of an informal working group on urogenital schistosomiasis and HIV transmission. Geneva: World Health Organization.

Wu XH, Wang TP, Lu DB, Hu HT, Gao ZB, Zhu CG, Fang GR, He YC, Mei QJ, Wu WD, Ge JH, Zheng J. 2002. Studies of impact on physical fitness and working capacity of patients with advanced schistosomiasis japonica in Susong County, Anhui Province. Acta Tropica 82(2): 247–252.

CHAPTER

Diagnostic Tests for Schistosomiasis

Robert Bergquist[1],*, Govert van Dam[2] and Jing Xu[3]

21.1 INTRODUCTION

Diagnosis is not only the starting point for dealing with infectious diseases, but is also central to their control and future elimination. Schistosomiasis, one of the major diseases of poverty is no exception. The still unacceptably high global burden of disease continues to exert serious public health and socioeconomic impact (Hotez *et al.* 2014; WHO 2015) in spite of widespread, large-scale chemotherapy with praziquantel, which has remained the global control strategy for over 30 years. Although significant improvements in the morbidity associated with this disease have been made, schistosomiasis is still common in the tropics and people still die because of this infection. The mortality level is, however, notoriously difficult to pin down, which explains the large variations in the literature, e.g., the 280,000 deaths annually in sub-Saharan Africa alone reported by van der Werf *et al.* (2003) and the much smaller figure, 11,700 globally, provided by the Global Burden of Disease (GBD) study in 2010 (Lozano *et al.* 2012). Even if the general reduction of morbidity between 2003 and 2010 must have had a positive impact, it cannot explain this dramatic decrease in mortality that might rather be explained by different ways of assessment. Although schistosomiasis is often a contributing factor to premature death, it is seldom the direct cause. On the other hand, both these mortality estimates fade into insignificance when compared to the 250 million people infected (Hotez *et al.* 2014), a fact that clearly proves that schistosomiasis is primarily a chronic disease. The latest figure from the GBD 2010 study on the burden of schistosomiasis to society and individual victims is 3.3 million disability-adjusted life years (DALYs) (Murray *et al.* 2012). Although this number of healthy human days lost annually is much below those calculated for malaria and tuberculosis (TB), schistosomiasis is no. 3 on the list of DALYs caused by the 17 neglected tropical diseases (NTDs) listed by the World Health Organization (WHO) (Hotez *et al.* 2014). The DALY score for schistosomiasis has risen to this level from no. 7 in 1990 due to inclusion of mild symptoms of the disease, such as diarrhea, dysuria and anemia, which were not counted before (Hotez *et al.* 2014). However, the true impact of this disease is in actual fact higher, since even the reformed DALY metric in GBD 2010 does not give a strong weight of disability to subtle symptoms and pathology in individuals with infections too light to always be revealed by current diagnostic approaches (King 2015a). Indeed, even if the GBD 2010 study

[1] UNICEF/UNDP/World Bank/WHO Special Programme for Research and Training in Tropical Diseases (TDR), World Health Organization, Geneva, Switzerland.
[2] Department of Parasitology, Leiden University Medical Center, P.O. Box 9600, 2300 RC Leiden, The Netherlands.
[3] Department of Schistosomiasis, National Institute for Parasitic Diseases, China CDC, Shanghai, People's Republic of China.
* Corresponding author

emphasizes advanced schistosomiasis, not all effects of the infection, such as infertility and late effects of growth stunting and cognitive impairment (King 2015b) are included. Correct diagnosis is thus not only necessary for appropriately targeting those individuals or populations who require chemotherapy, but it should also support the moral duty of society to recognize the full extent of damage inflicted by this disease.

Resolution 65.21 of the World Health Assembly (WHA 65.21) is a WHO agenda item concerned with the possibility of elimination of schistosomiasis. It encourages endemic member countries to ensure the provision of essential medicines, water, sanitation and hygiene interventions as well as to intensify control interventions, to strengthen surveillance and to initiate elimination campaigns with a view to ultimately break transmission (WHO 2012a). The latest schistosomiasis fact sheet emphasizes the goal of regular treatment with praziquantel of at least 75% of children by 2020 (WHO 2015). In addition to the general measures mentioned before, accurate, sensitive diagnostic tests and surveillance are also recommended. The WHO Road Map for the global control of the NTDs (WHO 2012b) includes schistosomiasis but this disease has, however, proved hard to control, first due to lack of effective drugs and later because of rapid re-infection related to limited sanitary facilities. The diagnostic techniques, mainly stool examination (for intestinal schistosomiasis) and urine filtration (for the urogenital form), used before the introduction of praziquantel in the early 1980s (Davis and Wegner 1979), were certainly adequate for that stage as individual burdens of disease were then high in most endemic areas. These tests are still recommended even if the intensity of disease has generally come down due to good coverage of chemotherapy and is now approaching the limits of sensitivity of these microscopy-based techniques in many areas. This situation not only results in underestimation of disease prevalence, but also contributes to inaccurate assessments of praziquantel impact. The daily variations in host egg excretion, which is particularly pronounced in *Schistosoma japonicum* infections (Yu *et al.* 1998), is an added problem increasing the number of falsely negative tests in areas characterized by low-intensity infection. Thus, approaches that worked well in the past are less suitable when the target shifts from control of morbidity to interruption of transmission, the critical juncture that has been reached in some countries, notably China, where elimination of schistosomiasis within the next decade is now a distinct possibility (Lei and Zhou 2015). The solution lies in the adoption of an approach modifying the methodology according to the diagnostic need (Bergquist *et al.* 2009), e.g., by institution of more sensitive and specific assays than previously used, when the priority moves from traditional monitoring and control to surveillance and response.

The great divide in diagnostics from the methodological point of view falls between non-microbiological techniques, such as the use of questionnaires and imaging techniques on the one hand and parasitological assays on the other. Today, disease and infection are diagnosed through the expression of biological markers (http://www.biomarkersconsortium.org/). In the medical sense, however, a biomarker can be a very vague concept. For example, body temperature is associated with infection in a general way and increased levels of the C-reactive protein indicates an inflammatory response, while the schistosome egg is a specific biomarker for schistosome infection. This train of thoughts leads from a physical indicator over a general biomarker to a specific one. They are all biological markers, but in this chapter we limit ourselves with the subset that might be discovered by microscopy, immunological techniques, genomics, proteomics or technologies that visualize pathology, such as imaging approaches like ultrasonography and histology.

While the questionnaire approach in principle consists of school-based interview or forms to be completed by the pupils, the imaging techniques constitute an extension of the clinical examination carried out at the individual level. This part is relatively briefly presented at the end of the chapter as the approaches used are not as diverse as the laboratory-related group that makes up the major part of the chapter. Here, however, new or modified assay systems appear on an almost monthly basis forcing us to only indicate the general areas of approach, while focusing on the most commonly used ones.

21.2 PARASITOLOGICAL DIAGNOSIS

Detection of schistosome eggs by light microscopy in stool or urine samples is still the most common diagnostic approach in the endemic areas. The approach is highly specific though completely dependent on the competence of the observer. However, the main drawback is the relatively low sensitivity that has become a problem in areas with low-level infection intensities (Lier *et al.* 2008; Lamberton *et al.* 2014), now increasingly common thanks to better coverage of chemotherapy schedules.

The currently most widely used approach for diagnosis of schistosomiasis is based on detection of parasite eggs in feces or urine. However, oviposition does not commence until about 4–8 weeks after infection in all five schistosome species that are adapted to human infection (*S. mansoni, S. japonicum, S. mekongi, S. intercalatum* and *S. haematobium*). Only the last species mentioned takes slightly longer to produce eggs but as there are great variations for all of the species, it is safe to say that egg detection should not be attempted before three months after supposed infection. Although this is of no practical importance in the endemic areas, it would be useful for testing of tourists and people moving between non-endemic and endemic areas. Overall, the outcome of diagnostic approaches should be informative for decision-making and policy formation.

21.2.1 Kato-Katz

The tests used for eggs produced by the species causing intestinal schistosomiasis (along with those produced by soil-transmitted helminths) is the thick stool smear technique, originally described by the Japanese researchers Kato and Miura (1954) and later standardized by the introduction of a 41.7 mg template by Katz and colleagues in Brazil (Katz *et al.* 1972). Based on this test, infections resulting in ≥400 eggs per gram (EPG) of feces are categorized as high, the 101–400 EPG range as moderate and EPG ≤100 as light (WHO 1993). These standards have been followed around the world for the latest 40 years and shown a bell-shaped distribution of infection intensity at all scales. However, in the last few decades, the bell curve maximum has been displaced towards lighter infections. This is not only thanks to intensive control activities that are ongoing in most endemic areas, but also due to a general economic development that has resulted in better living standards and improved education in many places. Indeed, Chinese control managers noticed that 97% of all positive cases were found below the 100 EPG level in a recent provincial-level survey encouraging the introduction of a 1–40 EPG group that turned out to include 66% of all positive cases (Zhou *et al.* 2014) resulting in a tentative classification with 100 EPG as the lower limit for heavy infection and 40 EPG as the division point between moderate and light infection. However, this classification strains the reliability of light microscopy which can miss patients with less than 100 EPG altogether if not subjected to multiple stool examinations.

Kato-Katz is a simple, relatively inexpensive assay for the field with a high level of specificity. However, the reference position of Kato-Katz notwithstanding, egg-based techniques are not optimal as they are highly dependent on the training and capability of the staff preparing and reading the test. Sensitivity is an important issue and the risk for missing subjects with low-intensity infections argues against the Kato-Katz approach in the era of mass drug administration (MDA). Even if sensitivity improves if several smears are carried out for each person (Lin *et al.* 2008; Lamberton *et al.* 2014; Levecke *et al.* 2014), this is not affordable, neither with respect to time nor cost, in large-scale surveys and it influences also compliance (Engels *et al.* 1996). In addition, the day-to-day variability in egg excretion from the host and clumping of eggs together (Teesdale *et al.* 1985; Engels *et al.* 1996), particularly pronounced in *S. japonicum* infection (Yu *et al.* 1998), add further uncertainty to the outcome.

21.2.2   Miracidium Hatching Test

The hatching test, based on observation of the phototrophic behavior of schistosome miracidia when hatched from feces diluted in water and allowed to sediment, is widely used in China (Qiu and Xue 1990). Lofty (2009) developed and evaluated thoroughly a modified hatching technique, showing that it can take up to three days before all eggs have hatched, even under optimal conditions. The sensitivity of the hatching technique is slightly better than the Kato-Katz method, mainly due to the large volume of feces investigated, but a combination of stool examination and hatching produces even better results (Zhu *et al.* 2014). However, the result is influenced by temperature, quality of the water added and examiner experience, and the slight improvement of sensitivity achieved may not justify wider use of the hatching test due to the added time and cost required. The technique has been used also for other schistosome species (Borges *et al.* 2013) and shown considerably better sensitivity (down to 12 EPG) but the approach has not caught on outside China.

21.2.3   Urine Filtration

Urinary schistosomiasis, caused by *S. haematobium* infection, is normally diagnosed by finding parasite eggs in the urine. This can be done by microscopy after centrifugation of an arbitrary amount of urine sampled in the morning or early afternoon, but the technique has been standardized by specifying the amount, normally 10 mL. A more user-friendly approach is filtration; Mott (1982) originally used Nytrel® polyamide mesh filters, while polycarbonate filter with holes of a few micrometers in diameter (Nuclepore®) are generally used today (Talaat and Evans 1996). The procedure is simple: shaking the sample followed by drawing the specified amount in a 10 mL syringe and then discharged through a new Nuclepore® membrane to be read in a microscope using a 10x lens. Like the Kato-Katz test, urine filtration has high specificity but low sensitivity (Stete *et al.* 2012).

21.2.4   Other Tests

A number of other parasitological procedures for the detection of helminth eggs have been used for schistosomiasis diagnosis, e.g., formol-ether sedimentation (Truant *et al.* 1981; Ebrahim *et al.* 1997), Mc Master (Bekana *et al.* 2015), including some newer developments, such as the FLOTAC (Knopp *et al.* 2009) and mini-FLOTAC (Barda *et al.* 2014). The latter approach is basically a flotation technique using test solutions of various viscosities that provides higher sensitivity since up to 1 g of stool is examined. Mini-FLOTAC has the added advantage of field-application since it does not need a centrifugation step. A custom-made zinc sulphate solution is particularly suitable for the detection of *S. mansoni* eggs (Barda *et al.* 2014).

There are also two promising, novel methodologies that are clearly much more sensitive than Kato-Katz, but both need large-scale validation and field testing before a verdict can be made. One is based on the isolation of *S. mansoni* eggs under a magnetic field through the interaction with paramagnetic beads (Fagundes Teixeira *et al.* 2007; Candido *et al.* 2015) and the other utilizes differential saline gradient sedimentation of parasite eggs when subjected to a slow, continuous flux of 3% saline solution through a porous plaque (Coelho *et al.* 2009). The saline gradient method requires 500 mg, while the assay based on paramagnetic beads processes as much as 30 g of feces.

21.3   ANTIBODY DETECTION

The discovery in the 1920s by Heidelberger and Avery, chronicled by Van Epps (2006), that the serological immune response relies on antibodies that are a type of proteins with a particular chemical composition, its confirmation by Marrack (1934), followed by the lock-and-key theory by Pauling and colleagues (1943) and the understanding that specific antibodies are produced by plasma cells in response to infection (Fagraeus 1948) were all notable achievements. However, the quantization of the precipitin reaction, originally discovered by Rudolph Kraus in 1897 and developed further by Pauling *et al.* (1943), is considered an equally important accomplishment as it initiated the development of agglutination tests in a wide variety of medical disciplines. One format depending on the diffusion of antibodies and antigens towards each other in agar gel to form a precipitate in the middle, invented by Ouchterlony (1948), was much used for a time both for research and as confirmatory test. In the meantime, however, various techniques relying on other formats of antigen-antibody reactions, e.g., the hemagglutination assay (HA), were developed to assess the relative concentration of viruses and bacteria (Hirst 1942). HA depends on sialic acid receptors on the surface of animal red blood cells capable of binding glycoproteins in the coat of various viruses to produce virus-erythrocyte lattice elements that remain suspended in contrast to negative samples, in which the blood cells instead settles at the bottom of the test tube producing an easily observed red dot. Originally the assay was carried out in test tubes, while machine-readable microtiter plates are used today.

Based on the HA approach, Stavitsky and Jarchow (1954) developed the indirect hemagglutination (IHA), which also relies on erythrocytes (commonly from sheep) but they are treated with diluted tannic acid to allow antigens to attach to the surface, after which they can be agglutinated by antibodies specific for the antigen in question. Thus, in contrast to the HA test, it is not the cell wall itself but the antigens attached to it that react with antibodies in samples being investigated for this specificity. Twofold titration is used to find the HI titer of the serum investigated. This golden era of immunological methods development produced the idea of tagging a tracing antibody with a label that would visualize its reaction with the antigen sought. The first one to do so was Albert Coons, who used fluorescent labels, first for immunochemistry and later also for serology (Coons *et al.* 1941; Coons 1959). Meanwhile, Yalow and Berson (1960) carried out groundbreaking work on diabetes and insulin based on antibodies tagged with a radioactive isotope. This led eventually to the birth of the radioimmunoassay (RIA), a very sensitive *in vitro* assay for the measurement of concentrations of antigens in serum by use of specific antibodies. This radioactive technique can also be used for antibody quantification by instead tagging the corresponding antigen. Although RIA is very attractive due to its high sensitivity and low cost, the need for specialized equipment and precautions (due to the radioactive labels used) eventually eclipsed its wider use. The replacement consisted of the enzyme immunoassays (EIA), which are based on various enzyme-substrate reactions that produce a strong, easy to read color change. This innovative step was first published by Engvall and Perlmann (1971), who called it the enzyme-linked immunosorbent assay (ELISA). Thus, the idea of tagging tracing antibodies (or antigens) stayed, even if the label shifted.

A long list of tests was developed for the serologic diagnosis of schistosomiasis, essentially depending on agglutination by antibodies and thus presaging the advent of the first proper immunoassays. Assays for the detection of anti-schistosomal reactions, developed and used in the middle of last century, have been reviewed in great detail by Kagan and Pellegrino (1961) including many test systems no longer used for schistosomiasis diagnosis, such as complement-fixation, precipitin formation, cercarial agglutination, miracidial immobilization and various hemagglutination applications. However, only the most commonly assays used today are discussed at some length here.

Immunodiagnostic methods have a much higher sensitivity than direct microscopy making these assays excellent epidemiological tools for screening populations in low-endemicity areas and in areas where the strategy is to interrupt transmission. The approach has gone through yet another revolution by the advent of the hybridoma technique (Kohler and Milstein 1975), that made it possible to devise super-specific and very sensitive approaches based on monoclonal-antibodies (MAbs) reactive with specific schistosome antigens. On the other hand, it is also clear that the immunodiagnostic methods cannot replace direct methods completely because antibody levels change only slowly, preventing differentiation between current and past infections. In addition, they are not quantitative and have generally a lower specificity compared with direct techniques using microscopy and/or molecular methods. In fact, control programs the world over still rely on egg counts rather than antibody screening, except in areas with low-level transmission such as in China, where serology and direct microscopy are currently of equal importance (Zhu 2005).

21.3.1   Skin Test

The intradermal test represents the earliest immunological test for schistosomiasis. It is based on the reaction between patient specific antibodies (and immunocompetent cells) resulting from a schistosome antigen solution injected intradermally. Various antigens were tried and one consisting of the dried powdered livers of snails infected with *S. spindalis* seems to have been quite successful (Fairley and Williams 1927). These authors reported that seven out of eight patients with *S. haematobium* infection reacted with a rapidly appearing large white wheal surrounded by a zone of erythema after intradermal injection of this concoction, while three of them also showed a delayed allergic response due to cellular immunity. Antigens were later extracted from adult worms, eggs, miracidia or cercariae from different schistosome species and used in various parts of the world (Gan 1936; Pellegrino 1958; Gazzinelli *et al.* 1965; Hunter *et al.* 1982). These tests, read after 15 to 30 min, were simple, sensitive (estimated at 90%) and cost-effective and therefore widely used in prevalence studies in the early days of schistosomiasis diagnosis.

21.3.2   Immunofluorescent Assay (IFA)

The IFA is often used as an in-house test in a number of laboratories for diagnosing of imported schistosomiasis, usually in combination with one or more other assays (Kinkel *et al.* 2012). An IFA specifically targeting IgM antibodies against polysaccharide antigens in the schistosome gut of Rossman's fixed sections of adult male worm was developed by Nash (1978). Such antibodies appear to be mainly directed against a circulating cathodic antigen (CCA), which is now part of one of the most promising novel tests (see below under 21.4.2). As CCA is already secreted by immature worms, it can be demonstrated early in infection, even before egg deposition has started. Generally, IgM titers are found to be significantly higher in infected returning travelers from non-endemic areas as compared to the immigrants from endemic areas (Polderman *et al.* 1989). The tradition of using IFA for the localization of antigens in schistosome tissues can thus both used for routine testing as well as a research approach (Peng *et al.* 2008).

21.3.3   Cercarien-Hullen Reaktion (CHR)

During their studies on acquired immunity in experimental schistosomiasis, Vogel and Minning (1949) observed that schistosome cercariae developed transparent membranes with well-defined edges when placed in sera from two monkeys experimentally infected with *S. japonicum*. The acronym CHR for the German expression 'cercarien-hüllen reaktion' (cercarial envelope reaction) is generally used for this phenomenon that prompted a series of studies culminating in its introduction as a test for schistosomiasis. CHR was relatively widely used for a while (Smit 1961; Ahmed *et al.* 1993). Meanwhile, further early research proposed that CHR is partly based on antigens common to the parasite and its two hosts and that these antigens include potential candidate vaccines and as well as diagnostic targets (Kemp 1972; Kemp *et al.* 1974). However, CHR should not even have been mentioned here, had not a novel version under the name of cercarial transformation fluid (CTF) recently been published (Smith *et al.* 2012) concluding it to be superior to other commonly used antigens and also cheaper to produce. CHR, in the form of CTF, may therefore still survive as a future diagnostic for schistosomiasis.

21.3.4   Circumoval Precipitin Test (COPT)

The COPT test emanates from the observation by Oliver-Gonzalez (1954) that precipitates formed around *S. mansoni* eggs when incubated at 37°C with serum from schistosomiasis patients. The technique uses cleaned, intact, fresh or lyophilized schistosome eggs, which are allowed to react with diluted human sera at 37°C for 48 hours (Rodriguez-Molina *et al.* 1962; Tanaka *et al.* 1975). The sensitivity of COPT is well above 90% and adequate specificity. False positive reactions in healthy people from non-endemic areas is low (2–4%) (de Noya *et al.* 1992; Li 1991). The test is supposed to have originated in China, where the oldest reference found is that of Liu *et al.* (1958), while Oliver-Gonzales (1954) discusses this test as already being part of the diagnostic arsenal in Latin America. The important point, however, is that the COPT has been widely used in China in the latter part of the 20th century and to some extent also in other parts of the world, Brazil and Venezuela in particular. However, COPT is less often used now since it is comparatively complicated, time-consuming and there is a long delay in seroconversion following treatment.

21.3.5   Indirect Hemagglutination Assay (IHA)

The IHA test has found universal use since any antigen can be used for coating the erythrocyte indicator system, e.g., within the field of schistosomiasis diagnosis alone, antigens from the different life cycle stages have been used, such as the soluble egg antigen (SEA), the adult worm antigen (AWA) and various larval antigens (Van Gool *et al.* 2002; Pardo *et al.* 2007). IHA has been used extensively for serological surveys, but it should be noted that the diagnostic sensitivity and specificity depend entirely on the antigens used.

IHA was first employed for the diagnosis of *S. japonicum* in China by Huang *et al.* (1966) and is today the most commonly used antibody test for community diagnosis and population screening for targeting chemotherapy in China, second only to COPT (Zhu *et al.* 2009; Zhou *et al.* 2011). After multiple modifications by scientists, the diagnostic efficacy and stability of IHA have improved significantly. Using Kato-Katz as the reference, Zhou *et al.* (2008) reported that the specificity of the IHA varied between 60 and 77%, while sensitivity and negative predictive value were higher than 97%. Although IHA is used in the world's endemic areas, a quick look in 'PubMed' or any other search engine, it is obvious that almost all modern IHA references with respect to schistosomiasis have been produced by Chinese researchers; evidently a reflection of the generally low prevalence and intensity of disease in their country.

21.3.6   Enzyme-linked Immunosorbent Assay (ELISA)

ELISA is currently the most commonly used serologic test and exists in more modifications than any other assay. Like IHA, various fractionated antigen components can be used, which decrease the likelihood of cross-reactivity with other helminth infections, while modification of assay components can improve the diagnostic efficacy and extend the range of application.

Rather than being adsorbed onto erythrocytes, ELISA is based on the antigens coating the inner surface of tests tubes or the wells of a microtiter plate. Serum samples to be investigated are added in various dilutions followed by an anti-immunoglobulin antibody that binds to the layer of sample antibodies (if present). This tracer antibody is labeled by an enzyme, which can be visualized by adding the enzyme's specific substrate. The subsequent reaction produces a color change that can be read by eye or in an electronic reader. The procedure includes further minor steps, e.g., addition of a substance that stops the reaction after a certain time etc., which makes the assay somewhat time-consuming. To that end, the Falcon assay screening test (FAST) system was developed (Hancock and Tsang 1986). It is a kinetic test approach with polystyrene beads on sticks fixed to the lid of a microtiter plate coated with the antigen of choice with speed gained because of all steps involved treat the plate with its wells as a whole. Detection is quantitative and based on a standard curve. The FAST ELISA is sensitive and quantitative and has been modified for field use, a technique suitable for large-scale community screening operations (Richards *et al.* 1989). Abdel-Fattah *et al.* (2011) have further improved the sensitivity by adopting an improved standard curve.

Apart from the main adult, cercarial and egg antigen groups (Sarhan *et al.* 2014), there are many sub-fractions thereof, e.g., the sixth cationic antigen fraction (CEF6) of SEA (Doenhoff *et al.* 1985; Bligh *et al.* 2010), the MAMA or HAMA, i.e. the adult microsomal fractions from *S. mansoni* (Tsang *et al.* 1983) or *S. haematoubium* Abdel-Fattah *et al.* (2011), respectively, the adult worm tegumental protein (Carvalho *et al.* 2014) to mention just a few. In addition, many recombinant peptide antigens have shown substantial advantages, reporting excellent sensitivity and specificity parameters, e.g., Xu *et al.* (2014). A commercially available *S. haematobium* SEA-ELISA using sera from finger-prick blood has shown 89% sensitivity, 70% specificity, a 57% positive predictive value and a 90% negative predictive value (Stothard *et al.* 2009a), while the FA-ELISA, based on a fractionated *S. japonicum* SEA antigen provides acceptable sensitivity and specificity for case detection and has even been reported to be useful for assessment of cure (Zhu *et al.* 2012).

Antibody-based, specific methods for the diagnosis of less common schistosome species, such as *S. mekongi* and *S. intercalatum* infections, do not exist as dedicated tests, but these infections can be diagnosed using heterologous detection assays based on cross-reactions with the three main species.

21.3.7 Rapid Diagnostic Kits

There is a pressing need for simple tests applicable for population screening. Assays carried out in the field should be possible to read without microscope or other special equipment with the results preferably available at the point of care (POC), which means a time limit of a few minutes rather than hours. None of the assay systems discussed above fit these requirements. Urine filtration and reagent strips for microhematuria (see below under biomarkers) come close but are limited to the diagnosis of urinary schistosomiasis. Only diagnostic kits based on the principles of lateral flow (Fenton *et al.* 2010) with colloid dyes or metals as markers fit the bill, as they can be read with the unaided eye within 5–10 min.

Rapid field tests, particularly common in China, are exemplified by three assays: the dot immunogold filtration assay (DIGFA) (Fig. 21.1) where the probe is anti-human IgG labeled with colloidal gold (Ding *et al.* 1998; Wen *et al.* 2005; Jiang *et al.* 2009), the dipstick dye immunoassay (DDIA) where it consists of the antigen labeled with a blue colloidal dye (He *et al.* 2000; Zhu *et al.* 2002) and the latex immuno-chromatographic assay (DLIA), where the results are revealed by anti-human IgG tagged with red latex particles (Yu *et al.* 2011).

The DIGFA (Fig. 21.1, top panel) consists of a nitrocellulose membrane, onto which a spot of antigen has been added by micropipette. The control consists of a dot of normal human IgG added in the same way next to the antigen spot. The serum to be tested is added followed by the labeled anti-human IgG probe in a second step. Each move is preceded by a washing step. The probe reacts with the human IgG at the control spot, while it only reacts at the antigen spot in case the test serum contains specific antibodies that have reacted there in the first step. The slide becomes transparent when dried and the appearance of two dots, colored red by the gold, is a positive reaction, while the appearance of a single red dot is negative. Using SEA as antigen, Wen *et al.* (2005) found the level of sensitivity in a material of 1,091 subjects to be close to 100%. The cross-reaction rates for paragonimiasis and clonorchiasis patients were 14.3% and 0%, respectively, while the specificity was reported to be 100% in a control group living in an area non-endemic for any of these diseases.

Dot Immunogold Filtration Assay (DIGFA)

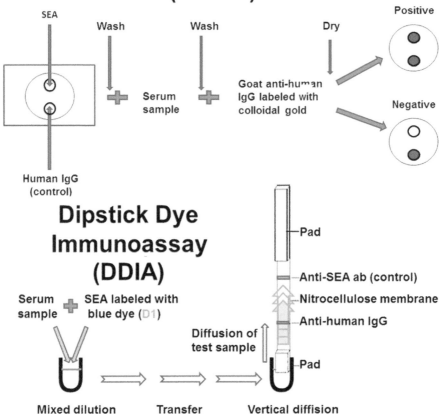

Dipstick Dye Immunoassay (DDIA)

Dipstick Latex Immuno-chromatographic Assay (DLIA)

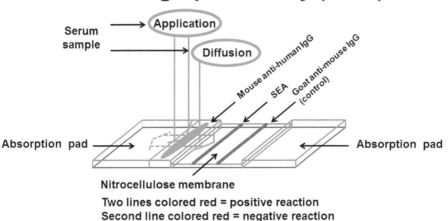

FIG. 21.1 The three most commonly used rapid diagnostic kits for schistosomiasis.

The DDIA (Fig. 21.1, middle panel) relies on the principle of immunochromatography where the antigen, normally SEA, has been conjugated with a blue colloidal dye. This probe and test serum are mixed in a well of a microtiter plate or the like followed by absorption of the contents into a nitrocellulose membrane fixed on a dipstick, where the liquid extends into the whole membrane passing two bands next to each other. The first consists of anti-human IgG and will capture human antibodies, with or without SEA, while the next consists of anti-SEA antibodies and will capture this antigen, with or without human IgG. The test is only regarded as positive if both bands turn blue. Investigating 6,285 individuals aged 6–65 years, in an area endemic for schistosomiasis of 4% prevalence according to Kato-Katz stool examination, DDIA showed a sensitivity of 91% and a mean negative predictive value of 99% (Xu *et al.* 2011). The specificity was a relatively low 53% with multivariate analysis indicating higher age, certain occupations and previous schistosome infections as significantly associated with false positive results. Cross-reactions due to *Clonorchis* and/or *Paragonimus* infection, a growing problem in China of today, were probably also an important cause.

The DLIA (Fig. 21.1, bottom panel) approach is similar to DDIA, but here, like DIGFA, the anti-human antibodies are labeled and instead of the blue dye, red latex particles are used as indicator. After being absorbed into one end of the testing device, the serum sample will pass through a layer laced with mouse anti-human IgG by lateral flow before entering the nitrocellulose membrane with the test band consisting of the antigen followed by a control band with goat anti-mouse IgG. This means that all human antibodies first combine with the anti-human IgG and that these complexes pass through the two bands by lateral flow. If the complexes include anti-SEA antibodies, they will be captured by the first band, while all remaining complexes will be captured by the control band. Thus, a positive sample will show two red bands. A trial involving 430 sera with known diagnoses, including 275 from health controls, showed both sensitivity and specificity of DLIA around 95% with no cross-reactions with *Angiostrongylus*, *Clonorchis* or intestinal nematodes. However, 42% of sera from patients with *Paragonimus* infection were found to show false positive results (Yu *et al.* 2011).

These assays are only the three most commonly used ones. The list of different labels and different testing system is long, while the variation of antigens is less impressive. Generally SEA is used, which explains the loss of specificity. Zhu *et al.* (2012), however, reached 100% using fractioned SEA in their FA-ELISA and FA-DIGFA assays. One of the fractions neither reacted with sera from healthy individuals nor showed any cross-reaction with sera from patients with *Paragonimus* or *Clonorchis* infections, which was seen in 10% and 26%, respectively, of the cases when crude SEA was used. They also discovered that this particular fraction only reacts with antibodies of short duration (7 weeks) emphasizing its potential use in population studies as surveillance tool after treatment.

The methodological capacity of the test kits is similar but, as argued by Zhu *et al.* (2012), the preparation of the antigen is important for avoiding unwanted cross-reactions. In the same vein, Yu *et al.* (2012a, b, 2014) compared SEA, 26-kDa glutathione *S*-transferase and the Sj14-3-3 tegumental protein described by Zhang *et al.* (1999) and noted absence of cross-reactions after fractionation, while Tang *et al.* (2008) showed that focusing on different antibody classes can tell more about the stage of the infection.

In summary, the results of diagnostic assays depend on methodological aspects on the one hand and the composition of the antigen on the other. Fractionation of antigen preparations usually improves specificity of the assay, while focusing on different antibody classes can also give information on the stage of an infection.

21.3.8 Other Assays

There is not only a 'never-ending' list of assays for the detection of schistosome antibodies, but there are also a wide variety of antigen preparations used. However, before leaving this area, a few novel approaches that may be of interest in the future should be mentioned. Promising results, both in the laboratory and the field, have for example been reported with *S. mansoni* cercarial transformation fluid (CTF) as a possibly superior antigen for the ELISA and lateral-flow POC formats (Chand *et al.* 2010; Coulibaly *et al.* 2013a; Nausch *et al.* 2014).

Another approach of methodological interest is magnetic affinity enzyme-linked immunoassay (MEIA), a technique for the isolation of cell organelles and molecules (Olsvik *et al.* 1994), which can also be used as a diagnostic technique by coupling antigens to magnetic micro-beads and let them capture specific antibodies from serum samples. This approach has shown as good sensitivity and specificity as ELISA, while being superior with respect to simplicity and speed. Yu *et al.* have published three papers using the MEIA technique for schistosomiasis diagnosis with three different *S. japonicum* antigen preparations: SEA (2012a), the recombinant 26-kDa glutathione *S*-transferase (2012b) and Sj14-3-3, an adult worm tegumental protein (2014). They used the alkaline phosphatase immunoassay (APIA) system together with the enzyme conjugated to the anti-human IgG as probe (Pujol *et al.* 1989). Although this series of papers does not surprise, it is both attractive and interesting to see this very clear account of how antigen fractionation improves specificity.

MEIA is not a straightforward, rapid technique but deserves to be considered as an alternative to other techniques for schistosomiasis surveillance in areas of low endemicity. However, the technique needs further validation before large-scale implementation.

21.4 ANTIGEN DETECTION

21.4.1 Deoxyribonucleic Acid Molecules (DNA)

Schistosome DNA was first identified in feces from individuals with intestinal schistosomiasis and later also in other body fluids. Its detection by the techniques described below constitutes one of the most sensitive and specific diagnostic approaches known.

21.4.1.1 Polymerase Chain Reaction (PCR)

Based on the rationale that schistosome DNA has a turnover resulting in circulating, cell-free parasite DNA (CFPD), application of the polymerase chain reaction (PCR) was successfully attempted (Sørensen *et al.* 1999; Pontes *et al.* 2002; Rabello *et al.* 2002). Indeed, *S. mansoni* CFPD could not only be identified in serum by PCR, but also in stool samples from infected individuals (Pontes *et al.* 2002) and in urine sediments (Lodh *et al.* 2013). Importantly, CFPD can be positive in cases with ectopic schistosomiasis, where conventional techniques normally present false-negative results (Härter *et al.* 2014).

Weerakoon *et al.* (2015) have reviewed current diagnostic approaches for schistosomiasis and refer to a number of various *S. japonicum* retrotransposons, which not only emphasizes the sensitivity of PCR diagnostics, but also clearly shows that the technique is useful for assessment of cure as well as early detection. Generally, infection can be found within a week of infection, while CFPD disappears 10–17 weeks after treatment (Xia *et al.* 2009; Guo *et al.* 2012).

Real-time PCR is an advanced form of the original PCR that maximizes the potential of the amplification technique by monitoring the reaction as it develops, i.e. instead of reading bands on a gel at the end of the reaction, the process is continuously recorded by a detector as originally described by Higuchi *et al.* (1993). The advantages include the possibility of (semi) quantitative measurements, as well as a substantially lower risk of contamination, resulting in virtually 100% specificity. Moreover, the procedure can be easily automated, which makes high-throughput analysis feasible. Multiple targets can be combined in a single assay format (Van Lieshout and Roestenberg 2015). In the field of schistosomiasis, real-time PCR has proven particularly valuable for monitoring therapeutic responses as the amount of CFPD declines gradually (Espírito-Santo *et al.* 2014). The new technique has also been shown to produce higher schistosomiasis prevalence figures in various parts of the world (Meurs *et al.* 2015; Gordon *et al.* 2015). Although samples can be collected in the field, transported and stored for processing in the laboratory at a later date, the requirement for dedicated expertise, expensive reagents and equipment conspire against the wider applicability of this approach outside advanced laboratories.

The earlier the diagnosis of schistosomiasis is confirmed, the earlier the transmission cycle would be counteracted. The understanding of this fact has emphasized the search for candidate assay capable of early diagnosis. Zhang *et al.* (2015) produced and analyzed transcriptional profiles of the six antigens representing different developmental stages by quantitative PCR. Of the six antigens, 50% were highly expressed in 21-day old young worms, while the other half were highly expressed in eggs. The early diagnostic validity of SjSP-216, a highly expressed gene in young worms, was further evaluated in mice and rabbits infected with *S. japonicum* (Zhang *et al.* 2015). They recorded 100% diagnostic sensitivity and specificity of SjSP-216-based ELISA three weeks after infection in both infected mice and rabbits. This study strongly suggests that SjSP-216 could serve as a potential biomarker for the early immunodiagnosis of *S. japonicum* infections in vertebrate hosts.

21.4.1.2 *Loop-mediated Isothermal Amplification (LAMP)*

LAMP builds on one-step amplification based on the addition of iron that produces a strong visual signal through a chain involving iron pyrophosphate, which forms an insoluble salt on combining with a divalent metallic iron ion (Notomi *et al.* 2000; Tomita *et al.* 2008). This enables naked-eye detection of test endpoints and hence avoids the need for electrophoresis equipment. The one-step LAMP approach does not only make the reaction more rapid than PCR, but LAMP has also a higher amplification efficiency and can be read visually. The absence of equipment for recording the visual signal makes LAMP the method of choice for field use.

A sequence from the highly repetitive retrotransposon *SjR2* gene of *S. japonicum* was successfully amplified in a LAMP assay using samples from experimentally infected rabbits and humans infected with *S. japonicum*. The results indicate that LAMP is more sensitive than PCR and capable of detecting *S. japonicum* DNA 1 week post-infection and revert to negative as early as 12 weeks post-treatment. (Xu *et al.* 2010; Wang *et al.* 2011). These results have been confirmed by a LAMP modification targeting a mitochondrial *S. mansoni* mini-satellite DNA region in a murine model detecting *S. mansoni* CFPD in stool samples (Fernández-Soto *et al.* 2014).

LAMP has thus been shown a cost-effective and feasible alternative to conventional PCR for the detection of schistosome CFPD in fecal and serum samples. However, before rolling out the techniques for large-scale implementation in the field, it would be important to validate the approach in community settings.

21.4.2 Circulating Adult Worm Antigens

Living schistosomes excrete or release a number of different antigens into the circulation of the host, among them those named the gut-associated antigens, which enter the bloodstream after regular regurgitation of the undigested contents from the parasite's gut. Ever since the early description of these antigens by Berggren and Weller (1967) and later characterized by other investigators (Gold *et al.* 1969; Nash *et al.* 1974, 1977; Deelder *et al.* 1976, 1980), research has been focused on utilizing the detection of these antigens as an immunodiagnostic approach. By far the most well-studied antigens are the circulating anodic antigen (CAA) and circulating cathodic antigen (CCA), names that are based on their electrophoretic mobility (Deelder *et al.* 1976). The presence of these antigens both in serum and urine samples and their detection by MAbs has been reported by Deelder *et al.* (1989a, b), De Jonge *et al.* (1990) and Van Lieshout *et al.* (2000). Both these antigens are highly glycosylated glycoproteins, the carbohydrate/polysaccharide parts of which constitute the antigenic elements detected in the assays (Bergwerff *et al.* 1994; Van Dam *et al.* 1994). Although CAA and CCA are mainly determined in established worm infections, they are already excreted by young schistosomula from 1 week after transformation, suggesting that early detection of schistosome infections is feasible (Van Dam *et al.* 1996).

Many versions of assays targeting many different circulating antigens have been applied, particularly in China. For example, Gao *et al.* (2015) recently reported the results of a test based on MAbs specific for enolase, the most abundant excretory-secretory antigen found so far. A sandwich ELISA using this MAb as capture antibody and the polyclonal antibody as detection antibody in the field showed a sensitivity and specificity of 85 and 96%, respectively. The cross-reaction rates for clonorchiasis and paragonimiasis were 3% and 5%, respectively (Gao *et al.* 2015). However, before the development and recent introduction of the up-converting phosphor (UCP) reporter technology in combination with the lateral flow (LF) strip platform, the performance of most of these various assays has not been reliable due to unsatisfactory sensitivity, especially in patients with light infections (Guan and Shi 1996; Zhang *et al.* 2016). The UCP-LF strip platform led to a second generation of CAA testing, by significantly improving robustness and applicability for single-case identification (Corstjens *et al.* 2008; van Dam *et al.* 2013). The novel approach has been shown to detect CAA at 10 to 1,000 times lower levels than the ELISA approach, both in serum and urine, while maintaining the high specificity (Corstjens *et al.* 2014, 2015). This would eventually allow determination of infections consisting of single or a few worms (Van Dam *et al.* 1996; Wilson *et al.* 2006).

21.4.2.1   CAA-detection Assays

The CAA-ELISA test, although useful for diagnosing all human and veterinary *Schistosoma* species with virtually 100% specificity (Deelder *et al.* 1989a, b; Flowers *et al.* 2002; Gabriel *et al.* 2002), remained for many years mainly a research tool due to its relative high complexity, need for proper laboratory facilities and cost. As such, it has been applied for many immunological, epidemiological, as well as modeling studies of all human and most veterinary schistosome species. Over 100 peer-reviewed papers and Ph.D. theses by Deelder's group and others have been published and recently been the focus of a Cochrane review group (Ochodo *et al.* 2015).

The UCP-LF CAA assay (Fig. 21.2) has recently been used to evaluate the prevalence of *S. haematobium* in a close-to-elimination setting in Zanzibar, Tanzania. It was clearly shown that prevalence measured with urine filtration and/or hemastix, was largely underestimating active schistosome infections, the levels of which were surpassed by a factor of five through measuring the presence of CAA (Knopp *et al.* 2015). Similarly, in large schistosomiasis japonica screening and control programs in China, the UCP-LF CAA assay demonstrated a prevalence of about 10 times higher than estimated by triplicate Kato-Katz thick smears (Van Dam *et al.* 2015a). The same study also presented evidence that the routine IHA, applied as first-line screening tool for antibody detection, failed to identify as many as 50% of active cases. The UCP-LF CAA test also showed its value in the diagnosis of *S. mekongi* infected cases, using banked urine samples from Cambodia (Van Dam *et al.* 2015b). The high specificity of the UCP-LF CAA assay (Fig. 21.2, upper panel) is ensured by the complete uniqueness of the antigen, by the highly specific MAbs-based detection, as well as enrichment of the CAA following sample pretreatment by trichloroacetic acid (Van Dam *et al.* 2015b). Experimental infections of mice, rabbits and baboons with *Schistosoma* unequivocally indicate the correlation of serum CAA levels with the number of living adult worms (Qian and Deelder 1983; Agnew *et al.* 1995; Wilson *et al.* 2006; Corstjens *et al.* 2014). Correlation of CAA levels with egg counts are nearly always found for all human species (Van't Wout *et al.* 1992; Polman *et al.* 1995; Leutscher *et al.* 2008; Van Dam *et al.* 2015a). The levels of CAA decrease quickly after successful treatment, in some cases even within days (De Jonge 1989; Van't Wout *et al.* 1992; Van Lieshout *et al.* 1994). In experimental infections, increase of CAA-levels over time has been shown very clearly (Van't Wout *et al.* 1995; Agnew *et al.* 1995; Van Dam *et al.* 1996; Kariuki *et al.* 2006). Finally, using sets of negative control samples and the application of a high-specificity cut-off threshold furthermore substantiated the very high specificity of the UCP-LF CAA assay (Van Dam *et al.* 2013; Corstjens *et al.* 2014).

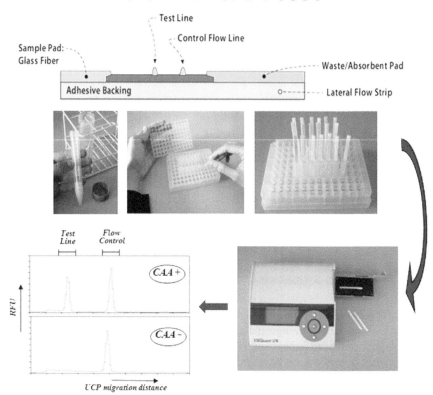

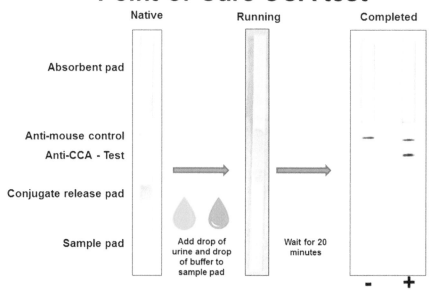

FIG. 21.2 The two highly sensitive and specific tests for *Schistosoma* CAA and CCA.

The UCP-LF CAA assay was recently adapted to a dry reagent format that allows convenient storage at ambient temperature and worldwide shipping without the need for a cold chain (Van Dam *et al.* 2013). The possibility of concentrating larger urine volumes allows the approach using pooled samples for a more cost-effective way of screening specific population groups in close-to-elimination settings (Corstjens *et al.* 2014, 2015). The use of dried blood spots for sample storage and transportation and subsequently successful elution and detection of CAA with the ELISA has already been shown before (Nilsson *et al.* 2001) and was recently confirmed using the UCP-LF CAA assay (Downs *et al.* 2015).

21.4.2.2 *CCA-detection Assays*

Similar to the CAA-ELISA, a MAb-based ELISA detecting CCA has been widely used for diagnosing all human *Schistosoma* species showing high sensitivity and specificity (Deelder *et al.* 1989a, b; De Jonge *et al.* 1990; Polman *et al.* 1995; Van Lieshout *et al.* 2000). This CCA-ELISA remained a research tool for many years and was mostly used for detection of CCA in urine samples of *S. mansoni* infected individuals. It was already at an early stage established that the concentrations in urine were much higher than those in serum, which is an obvious advantage not only from the ease-of-approach viewpoint, but also when considering patient compliance. In most studies the best diagnostic performance was achieved with the urine CCA assay, while the determination of serum CAA was seen as a better indicator of worm burden (Van Lieshout *et al.* 1995; Polman *et al.* 1995).

For *S. mansoni*, two urine-based CCA assays have been developed into commercially available test kits: the CCA carbon test (van Dam *et al.* 2004; Midzi *et al.* 2008; Obeng *et al.* 2008; Stothard *et al.* 2009a; Shane *et al.* 2011; Koukounari *et al.* 2013) and the more widely used CCA cassette test (Shane *et al.* 2011; Colley *et al.* 2013; Coulibaly *et al.* 2013b; Lamberton *et al.* 2014). The production of CCA carbon test was discontinued, while the gold-based cassette test has finally become commercially available as the POC-CCA test from Rapid Medical Diagnostics, Pretoria, South-Africa (http://www.rapid-diagnostics.com). It has been very extensively evaluated and satisfies all requirements of the 'ASSURED' criteria, i.e. Affordable, Sensitive, Specific, User-friendly, Rapid/Robust, Equipment-free and Deliverable to those who need it (Deelder *et al.* 2012; Bergquist 2013; Colley *et al.* 2013).

The POC-CCA test (Fig. 21.2, lower panel) is more rapid and sensitive than multiple Kato-Katz thick smears for the detection of *S. mansoni* and is a viable alternative for the rapid diagnosis of human infections (Coulibaly *et al.* 2011; Shane *et al.* 2011; Navaratnam *et al.* 2012; Tchuem Tchuenté *et al.* 2012; Coulibaly *et al.* 2013b; Erko *et al.* 2013) or supplement to Kato-Katz (Lodh *et al.* 2013; Colley *et al.* 2013). It has recently been recommended by the Schistosomiasis Consortium for Operational Research and Evaluation (SCORE) to WHO's NTD department as being sufficiently sensitive and specific for use as a mapping tool to determine prevalence among school-age children (Colley *et al.* 2013). At about 1–1.5 US$ per determination, the assay is cost-effective allowing its inclusion as the test of choice in currently operating, national, schistosomiasis control programs as well as in primary health care settings (Stothard *et al.* 2009a; Speich *et al.* 2010; Sousa-Figueiredo 2013; Adriko *et al.* 2014; Worrell *et al.* 2015). Last but not least, as people are more likely to provide urine samples than any other type of sample, the POC-CCA test may increase compliance as well as allowing examination of pre-school children from whom stool samples are not always easy to obtain.

The correlation between urine CCA levels and egg counts has already been extensively shown by the CCA-ELISA (De Jonge *et al.* 1989; Deelder *et al.* 1989a, b). Consequently, the color intensity of the test line on the LF strip (read semi-quantitatively) has also been shown to correlate to the number of eggs in the stool (Standley *et al.* 2010; Mwinzi *et al.* 2015), thereby being a proxy for the number of worms present in the host. Effective treatment with praziquantel would reduce these worm numbers leading to a decrease in POC-CCA positives as well as test line intensity (Coulibaly *et al.* 2013b; Mwinzi *et al.* 2015). While eggs continue to be detectible in host feces for about 2–3 weeks (Scherrer *et al.* 2009), CCA-levels drop rather quickly, sometimes turning negative within

1 week after treatment (Van Lieshout *et al.* 1994; Lamberton *et al.* 2014; Coulibaly *et al.* 2013b). This makes the POC-CCA a promising tool to monitor drug efficacy in schistosomiasis. A recent study already shows evidence that effective cure rates for praziquantel when evaluated by a more sensitive assay as the POC-CCA, are significantly much less than those inferred from fecal testing (Lamberton *et al.* 2014; Koukounari *et al.* 2013; Mwinzi *et al.* 2015).

By comparing various diagnostics, such as the CCA assays, the Kato-Katz and a soluble worm antigen preparation (SWAP)-specific IgG ELISA using statistical modeling techniques like latent class analysis (LCA), the CCA-cassette test was shown to have 96.3% sensitivity and 74.7% specificity (Shane *et al.* 2011). In individuals with low egg counts and in areas characterized by low endemicity, the sensitivity and specificity of CCA test are not as high as what DNA-based detection methods can offer (Van Lieshout *et al.* 2000; Obeng *et al.* 2008; Lodh *et al.* 2013) suggesting that the PCR is superior to CCA assay. Yet, the ultrasensitive UCP-LF format of the CAA-test (Van Dam *et al.* 1996; Wilson *et al.* 2006), which allows single-worm detection, surpass even PCR in sensitivity (these studies have yet to be published).

The recent, extensive Cochrane analysis (Ochodo *et al.* 2015) concluded that the POC-CCA test for *S. mansoni* detects a very large proportion of infections identified by microscopy, but "it misclassifies a large proportion of microscopy negatives as positives in endemic areas with a moderate to high prevalence of infection, *possibly because the test is potentially more sensitive than microscopy"*. The line above is revealing as it admits that the test under review might indeed be 'more sensitive'. A true 'gold standard' is not referred to and can only be provided by an experimental animal model. However, for the correct validation of a new diagnostic test, it is not sufficient to only compare against currently accepted techniques, particularly if their short-comings are not well-recognized. Multiple, diagnostic approaches should be compared together taking into account modeling and statistical methods in combination with knowledge of biological systems (Koukounari *et al.* 2013; Knopp *et al.* 2015).

For the diagnosis of *S. haematobium* infection, the urine POC-CCA test is usually less sensitive than microscopic urinary egg detection (Obeng *et al.* 2008; Ayele *et al.* 2008; Stothard *et al.* 2009a) thereby limiting its use as a general schistosomiasis test. However, it appears that in some regions the POC-CCA test shows satisfactory results also for *S. haematobium* infections (Midzi *et al.* 2008), which calls for further study. Recently, it has been shown to have a reasonable accurate result for detection of *S. japonicum* infections (Van Dam *et al.* 2015a), but also here further research is needed.

21.4.2.3   *Other Circulating Antigens*

A number of other circulating antigens, associated with different life cycle stages of the parasite have also been studied as targets for diagnostic assays. So far, only a few of these antigens have been more extensively characterized and the number of studies in humans is limited. Specifically, antigens providing information complementary to data on worm antigens would potentially be very interesting in view of the egg/worm dynamics. For example, circulating cercarial antigens may have an application for the diagnosis of acute schistosomiasis. However, after some initial work on infected mice and on pooled serum samples of chronically infected schistosomiasis patients (Hayunga *et al.* 1986, 1987), no further human studies have been published. Detection of circulating egg antigens (described for *S. mansoni* as well as for *S. haematobium*) seems to be more promising and may be a useful indicator for tissue egg burdens. It may also provide information on egg-induced pathology or may be used to measure worm-fecundity (Nour el Din *et al.* 1994; Bosompem *et al.* 1997; Kahama *et al.* 1998; Nibbeling *et al.* 1998). The dynamics of egg antigen concentration after successful treatment of school children has been described by Kihara *et al.* (2009), showing that after a steep decrease within 2 weeks, especially in children <5 years old, levels already start to rise after about 4 weeks. ELISA-format assays targeting these egg antigen have been described, but they have neither been applied nor further investigated in spite of the detection of specific glycan fragments by advanced, highly specific techniques utilizing mass spectrometry (Robijn *et al.* 2007, 2008; Balog *et al.* 2010).

A proteomic study of excretory/secretory proteins of adult *S. japonicum* worms identified more than 100 proteins, among which a fatty acid binding protein (FABP) (Gobert *et al.* 1997) was the most abundant. FABP has the potential to induce immunogenic reactions in the host circulation, raising the possibility of a role for an anti-FABP antibody in serodiagnosis of *S. japonicum* (Liu *et al.* 2009). In another, similar study, (Zhong *et al.* 2010) used proteomics in combination with Western blotting and identified four putative diagnostic protein candidates, namely, leucine aminopeptidase (LAP), fructose bisphosphate aldolase (FBPA), glutathione *S*-transferase and a 22.6-kDa tegumental antigen.

21.5 NON-SPECIFIC BIOMARKERS

The determination of biomarkers includes various indicators that are quite general, but it may still be possible to fine-tune them for better specificity in the future. Already, however, information on some of these biomarkers complement other tests and can thus contribute to a better description of the immunological or physiological state of an individual based on measureable indicators of current or past presence of an infectious agent. Although these biomarkers are of considerable interest as basis for further research, they are of limited diagnostic value in isolation.

21.5.1 Metabonomic Markers

Host metabolic responses to parasitic infections is a promising approach, which has yet to impinge on the field of diagnostics. Wang *et al.* (2004) carried out a pilot metabolic profiling of *S. mansoni* infection in a mouse model using proton nuclear magnetic resonance (NMR), mass spectrometric methods and multivariate pattern recognition techniques. They found a metabolic signature consisting of reduced levels of tricarboxylic acid cycle intermediates, increased pyruvate levels and a range of microbial-related metabolites, findings that together highlight metabonomics as a promising approach for the development of novel diagnostic approaches.

Interestingly, a proteonomics approach has just been developed as an accurate diagnostic test for human African trypanosomiasis (Papadopoulos *et al.* 2004) and a similar approach might be possible in the field of other parasitic diseases as well. It might even be possible to trace various clinical complications of a chronic *S. mansoni* infection in humans, such as periportal hepatic fibrosis followed by venous hypertension and splenomegaly by following disturbances in the metabolism of amino acids. Indeed, native peptides in urine samples from individuals infected with *S. mansoni* have been identified as novel biomarkers using 'peptidomic' profiling, an approach based on mass spectrometry in combination with pattern recognition algorithms (Balog *et al.* 2010, 2011). Progress in the field has, however, been slow. Wang and Hu (2014) have recently reviewed the latest developments with special reference to diagnostics.

21.5.2 Cytokines

The different stages of schistosomiasis infection are characterized by typical cytokine spectra reflecting the host parasite interplay resulting in various T_H responses (Wilson *et al.* 2012; Olveda *et al.* 2014a; Colley and Secor 2014) with the early infection stage dominated by T_H1 proinflammatory cytokines, such as TNF_α, IFNγ, IL-1, IL-2 followed by a switch to T_H2 response (IL-4, IL-5, IL-10, IL-13) coinciding with parasite egg deposition. However, despite these typical immune responses at the different stages of the infection, the T_H1 and T_H2 cell activities are not strictly separated but overlap to some extent, which make diagnostic interpretation of cytokine profiles dubious in real life (Colley and Secor 2014).

21.5.3 Circulating Ribonucleic Acids (miRNAs)

Aberrant expression of some of the highly conserved, micro-RNA molecules (miRNAs) that regulate gene expression after transcription (Bartel 2004) are associated with various types of human diseases including liver disease (Bala *et al.* 2009) and schistosomiasis (He *et al.* 2013). The miRNAs either leak into the circulation from broken cells or appear there through active secretion (Chen *et al.* 2012) and the presence of miRNAs in plasma and other body fluids opens up many possibilities including the application of circulating miRNAs as diagnostic markers (Makarova *et al.* 2015).

In the field of schistosomiasis Cai *et al.* (2015) established miRNAs for cell-free diagnosis providing evidence for the potential of utilizing circulating host miRNAs to indicate different immune responses and association with the severity of liver damage. There are thus reasons to believe that miRNAs could serve as novel biomarkers, e.g., in the early stages of schistosomiasis when eggs are not yet excreted as well as in evaluation of disease progression. However, the approach first needs to be tested and validated in clinical settings. Although miRNAs profiling gives great hope for future diagnostics, the cost of the reagents and the technical resources required limit their wide-scale application for the time being.

21.6 GENERAL APPROACHES

Here the focus is on a class of non-microbiological biomarkers that can indicate the presence of schistosomiasis, either as a sign of a certain schistosome species or of a specific stage of the disease. However, no test should be used in isolation. For example, indicators, such as macro- and microhematuria, can conveniently be applied in community studies by using rapid-detection reagent strips in combination with demographic information obtained using questionnaires (Bogoch *et al.* 2012; Stete *et al.* 2012).

21.6.1 Questionnaires

Adolescent hematuria has been noted in areas endemic for urinary schistosomiasis since ancient times but rather than signaling a serious disease, it was seen as a sign of adulthood. Although the disease association has been widely understood for a long time, its use as indirect indicator for *S. haematobium* infection is relatively recent. The targeting of chemotherapy for schistosomiasis requires rapid assessment procedures for identifying high-risk foci for the disease and as shown by Mott *et al.* (1985), the questionnaire approach was promising in showing that a few simple questions can provide good sensitivity and specificity for the presence of morbidity due *S. haematobium*. This is commonly achieved through parasitological techniques, but the use of questionnaires would lower the cost considerably (Lengeler *et al.* 1991, 2002). An association between intestinal schistosomiasis and blood in the stools has also been found though it is generally weaker than that between urinary schistosomiasis and hematuria (Lengeler *et al.* 2002; Yang *et al.* 2015). In a meta-analysis of the diagnostic efficiency of the questionnaires screening Yang *et al.* (2015) found that the best diagnostic odds ratio (OR) was obtained for *S. haematobium* followed by *S. japonicum* and *S. mansoni,* in that order. They further reported that pretested and standardized questionnaires had a better diagnostic performance and concluded that questionnaires in general can be used to diagnose schistosomiasis with moderate sensitivity and specificity (Yang *et al.* 2015).

The use of simple school questionnaires over a period of more than a decade has been validated in 10 countries with 133,880 children interviewed and 54,996 children examined for *S. haematobium* showing that this approach is well accepted, inexpensive and highly reliable (Lengeler *et al.* 2002). Questionnaires are now available for promptly defining the magnitude of schistosomiasis in a large area, which will allow limited resources for morbidity control to be allocated in an optimal way. Some years later Clements *et al.* (2008) were able to produce national risk maps of urinary schistosomiasis in mainland Tanzania based on spatial analysis and self-reported schistosomiasis data from over 2.5 million school students from all over the country, which facilitated targeting the resources of the national schistosomiasis control program.

21.6.2   Urine Reagent Strips

Hematuria is strongly associated with urinary schistosomiasis and urine reagent strips are therefore highly useful for inexpensive, rapid mapping of *S. haematobium* infection (Robinson *et al.* 2009; Emukah *et al.* 2012; King and Bertsch 2013). Indeed, hematuria has even been shown to be a better indicator of infection than excretion of *S. haematobium* eggs (Stete *et al.* 2012). A recent systematic review found that reagent strip testing for *S. haematobium* diagnosis has an overall sensitivity and specificity of 75% and 87%, respectively (Ochodo *et al.* 2015). Still, urine hemastix® proved so useful for the assessment of cure after treatment of urinary schistosomiasis that it is recommended for monitoring *S. haematobium* MDA programs (Emukah *et al.* 2012).

Albustix® or combination sticks senstive both for albumin and heme are not only practical, but in fact necessary as the presence of albumin is common in urinary schistosomiasis. Another type of urine dipsticks are the microalbustix®, also rapid and inexpensive and with the additional capability of indicating whether or not the disease has started to produce pathology in the upper urinary tract by measuring the ratio of albumin to creatinine. These strips represent an excellent example of advanced field diagnosis, which makes them a useful complement to hemastix® for screening of children, as they indicate who should be followed up by ultrasonography (Stothard *et al.* 2009b).

21.6.3   Other Potential Markers

Increased activity of eosinophil granulocytes is common in inflammation and have been noted releasing the eosinophilic cationic protein (ECP) in response to schistosome infection (Reimert *et al.* 2000, 2008). Indeed, increased levels of ECP in the body could be a biomarker of urogenital schistosomiasis in combination with related morbidity (Poggensee *et al.* 1996; Midzi *et al.* 2003). Although circulating ECP levels can vary considerably among patients, its determination could be useful for monitoring individual patients. It has also been shown that ECP levels increase to pre-treatment levels, sometimes significantly higher, following reinfection (Reimert *et al.* 2008). This response pattern may imply a rebound effect during reinfection with resolution of the immunosuppressive mechanisms in function during the chronic phase. Despite these very interesting characteristics, it is unlikely that regular ECP testing will be adopted outside the hospital environment because of cost as well as need for trained staff and adequate laboratory conditions.

Calprotectin, a protein found in neutrophils and increasingly used as a marker of bowel inflammation, is measured in feces (Tibble *et al.* 2000). Like fecal occult blood, increased calprotectin concentrations are strongly associated with high-intensity intestinal schistosomiasis and show significant decrease after treatment (Bustinduy *et al.* 2013). Both tests are inexpensive with excellent operational performance and reliability and therefore appropriate for use in the field as an approach corresponding to reagent strips in urinary schistosomiasis.

21.7   IMAGING TECHNIQUES

The introduction of computer tomography (CT) and magnetic resonance imaging (MRI) has facilitated the clinical evaluation of the pathologic changes associated with schistosomiasis. However, these techniques are costly and require the hospital environment, which limits their wider use for this purpose. Ultrasonography (US), on the other hand, is not only considerably less expensive, but the availability of mobile US equipment has also made it possible to bring the technology to the bedside in the field. The technique is an extension of the clinical examination providing guidance whether or not patients need to be transferred to a hospital for surgical treatment. Developed as far back as in the 1940s, US was not widely used for the study of schistosome-induced organ morbidity until the 1980s, when the appearance of mobile units rapidly propelled US to center stage in the examination of chronic/advanced schistosomiasis in the hospital as well as in the field. However, lack of methodological standardization delayed the integration of US into routine. Efforts made to deal with this problem resulted in the Cairo approach, that was used in the 1980s and early 1990s, only to be replaced with a tentative protocol recommended by WHO that was helpful in many ways, in particular due to the agreement on a standard for grading the severity of lesions as well as a system of weighting the scores given (Richter *et al.* 1996; WHO 2000). Even if the WHO protocol was a step forward, it became quickly obvious that work remained (King *et al.* 2003). Further discussions resulted in improved standards discussed in up-to-date reviews on the experience of US and chronic infection by *S. mansoni* (El Scheich *et al.* 2014, 2015) and *S. haematobium* (Akpata *et al.* 2015), respectively. However, even if the introduction of mobile US equipment broadened image documentation of organ lesions in endemic community settings, it is still not commonly included in large-scale control interventions as called for by many, e.g., Antwi *et al.* (2014) and Koukounari *et al.* (2006).

21.7.1 Liver Pathology

Long standing intestinal schistosomiasis, mainly caused by *S. mansoni* infection, invariably results in chronic fibrotic hepatic damage starting with focal areas of chronic inflammation around parasite eggs trapped in this organ. These granulomas are distributed at the periphery of the portal vein system giving a picture of initiation and regulation of the pathology where host immunity plays as big a role as the root cause. If not cured at an early stage, the developing fibrosis results in portal hypertension with the classical features of portal dilatation, periportal fibrosis, thickening of the portal vascular tree. Typical US findings include strongly echogenic bands along the portal vessels (Symmer's fibrosis), reduction in the size of the right lobe with hypertrophy of the left lobe, splenomegaly and finally ascites (Fig. 21.3). Difficulties observed include overestimation of the risk of portal vein dilatation and left liver lobe enlargement (Koukounari *et al.* 2006). The *S. japonicum* patterns are very similar to those caused by *S. mansoni*, but there are differences, mainly the net-like appearance of periportal fibrosis typical of advanced schistosomiasis due to *S. japonicum* (Chou *et al.* 2003).

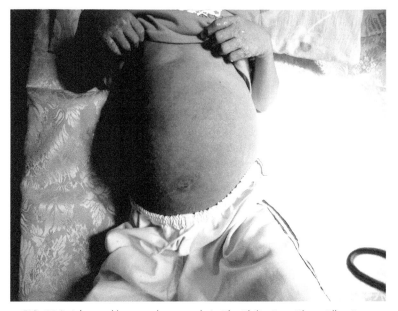

FIG. 21.3 Advanced hepatosplenomegaly in The Philippines. Photo Allen Ross.

US is not only useful in the diagnosis and differential diagnosis of advanced pathology, but also for guidance of treatment and evaluation of therapeutic effects as well as prediction of the risk for gastrointestinal hemorrhage. Advanced equipment, e.g., devices capable of color Doppler US, can emulate portal vein perfusion, which is critical for the prognosis and treatment options for complicated portal hypertension (Olveda *et al.* 2014b).

21.7.2   Urogenital Lesions

Schistosome-related pathology in the urogenital system is caused by *S. haematobium* infection that produces granulomatous inflammation provoked by egg deposition in the lower urogenital tract, mainly in the bladder wall, which results in urogenital inflammation and scarring leading to strictures as well as polyposis, ulceration and bladder carcinoma. Although cystoscopy and biopsies from the bladder wall are commonly used in the hospital environment, US has become a convenient approach to assess the morbidity in the endemic areas giving the degree of severity of bladder damage according to thickness, irregularities and shape of the bladder wall as well as presence of polyps, tumors and hydronephrosis.

Assessing morbidity indicators in the context of large-scale control intervention, evaluated according to the WHO guidelines for ultrasound in schistosomiasis, Koukounari *et al.* (2006) found the approach to be particularly useful for detection of *S. haematobium* pathology. Although lesions can appear relatively slowly depending on the level of transmission in the endemic environment, they can also grow very rapidly as reported from a case study in Ghana (Antwi *et al.* 2014). However, interpretation of echographic images are sometimes not easy, e.g., some forms of urogenital schistosomiasis in women and prostatic lesions in men (Richter *et al.* 1995; Ramarakoto *et al.* 2008).

21.7.3   Neuroschistosomiasis

Clinical signs, such as epileptic seizures, intracranial pressure and myelopathy, are relatively unusual, debilitating outcomes associated with schistosomiasis, thought to occur through ectopic egg deposition following aberrant schistosome migration to the brain or spinal cord. Demonstration of schistosome eggs in the stool or urine supports the diagnosis but biopsy and positive histopathology are needed for a definitive diagnosis (Ross *et al.* 2012; see also Ross and Huntsman, Chapter 19). However, imaging approaches, such as CT and MRI can be helpful in detecting evidence of lesions, such as expansive masses or tissue edema (Wu *et al.* 2012). Efforts have also been made to explain the neurological manifestations of schistosomal myelitis by increased blood flow within the portal venous system using Doppler US (Vidal *et al.* 2010).

21.7.4   Biopsy Studies

Two areas dominate with regard to the use of biopsy in schistosomiasis: rectum and liver. The former was used to confirm intestinal schistosomiasis in very light infections, while the aim of the latter was to better understand the extent and type of liver pathology. There are few published records of these approaches, but a report of 360 patients (Shak *et al.* 1959) is included to complete this diagnostic overview. In this study, 61 out of 360 needle biopsies done, schistosome eggs and/or granulomata were seen, but parasite eggs were recovered in the stools or urine in only 42% of the cases. The changes observed were periportal inflammation or fibrosis, schistosomal pigment, focal necroses and granulomata with and without eggs. Diagnosis of the species of schistosome involved was possible by the examination of liver tissue, but in only 18% of the cases. Successful spleen biopsy, showing diffuse fibrosis or fibrocongestive splenomegaly and/or schistosome pigments, was achieved in 31 of the 61 patients. Another paper (Dimmette 1955) points out that wedge biopsy of the liver reveals more than twice as many cases of schistosomiasis as does needle biopsy because the lesions are scattered and small and that splenomegaly due to schistosomiasis cannot be distinguished from splenomegaly due to other reasons unless schistosome pigments are found.

Today, records of needle or of surgical biopsies of any type in patients suspected to be infected with schistosomiasis cannot be found, as these techniques are invasive and have therefore been completely supplanted by other approaches.

21.8   INTERMEDIATE SNAIL HOST DIAGNOSTICS

The last human infection by schistosomiasis in Japan occurred in 1977 (Tanaka and Tsuji 1997). Although eradication is not close for the majority of the endemic areas in the world, the countries north of Sahara in Africa and China in Asia are now in the position to contemplate elimination of schistosomiasis defined as reducing prevalence to a point when it is no longer a public health threat. However, the permanent reduction of the disease in a geographical area is not only difficult to achieve in practice, it is also difficult to certify. Although tests for humans and animals having the required sensitivity are now available, absence of infection in the snail intermediate host must also be shown.

The contribution of local infection sources including humans and livestock to the distribution of infected snails might be relatively small and snail control may limit infected snails to increasingly small areas ecologically most suitable for transmission. A spatio-temporal method to identify areas and risk factors for persistent infected snail presence has been suggested by Yang *et al.* (2013). This approach would be useful for developing a surveillance system capable to detect the spread of infected snails at an early stage, including sensitive and reliable methods to identify risk factors for the establishment of infected snails.

Microscopy for sporocysts and cercariae in crushed snails or observation of cercarial shedding induced by exposure to light (Qiu and Xue 1990) have been widely used but are replaced by molecular methodology in situations characterized by low prevalence and intensity of infection. Over the years, different, sensitive indicator systems, from labeled MAbs to molecular techniques, have been applied for the detection of schistosome infection in snails. Caldeira *et al.* (2004) developed a methodology to detect the presence of *S. mansoni* in snails by DNA extraction using low-stringency PCR. Infection profiles were shown to be present in experimental shell specimens as far as the eighth week after being removed from the aquarium. This approach has also been tried for host snails both with respect to *S. mansoni* and to *S. haematobium* (Amarir *et al.* 2014). Most reported results concern the application of PCR in advanced laboratories, while recent results with LAMP-based assays show that this approach works equally well in low-technology parasitology laboratories in the field (Abbasi *et al.* 2012; Hamburger *et al.* 2013; Tong *et al.* 2015).

Detection of cercariae in natural water bodies is just a step from detection of the infection in snails but just as important in certifying areas as non-endemic from the intermediate host point of view. Various cercarial concentrations experimentally added to natural water followed by a filtration process produced positive detection down to the level of a single cercaria by PCR (Hamburger *et al.* 1998; Hung and Remais 2008). A different approach for detection of cercariae in water based on geographical information systems (GIS) and sentinel mice has been successfully used in China (Yang *et al.* 2012) but comparisons cannot be made as the dilution factor is unknown in this latter case. In addition, the resulting worm load of sentinel rodents may not have a linear relationship with cercarial density in the water (Aoki *et al.* 2003). Overall, however, cercariometry is a rapid, reliable assay for cercarial location in endemic areas and provides also information on diurnal fluctuation, seasonal and spatial distribution of cercariae in natural water bodies.

21.9   CONCLUDING REMARKS

In contrast to diseases caused by viruses or bacteria, the diagnostic objective with respect to schistosomiasis is as much aimed at quantity as at quality. The latter is due to the need to know the geographical distribution of the infection (risk mapping), while the former is directly correlated with infection levels, including the degree of pathology at the individual level (ultrasonography). The degree of infection is of considerable importance as this governs the number of parasite eggs released, which not only contributes to the severity of lesions, but also ensures that transmission continues.

The use of specific diagnostic assays, in contrast to clinical examination and ultrasonography, are geared at providing data on prevalence and intensity of the infection (number of worms present), which are critically linked to the effect of chemotherapy. Sensitivity and specificity therefore need to be close to 100%, preferably above 95%.

Together with convenience, speed and cost, definition of sensitivity, specificity, predictive values, etc. are central to any discussion of diagnostics. These latter issues are not specific for schistosomiasis and have been extensively dealt with elsewhere (https://en.wikipedia.org/wiki/Positive_and_negative_predictive_values). Suffice it to say that the importance of all these different features varies with the progress of control activities, i.e. the prevailing epidemiological situation. For example, in highly endemic areas, sensitivity is less important than specificity and cost, while sensitivity and specificity play a more prominent role than cost at the other end of the spectrum. For this reason, tests such as Kato-Katz are useful at the start of a control program, while highly sensitive and specific approaches, e.g., PCR and LAMP are needed at the elimination step. Antibody-detection, although not always the most specific approach, is good enough for long-term surveillance due to its high sensitivity and low cost. Novel techniques for urine testing, such as POC-CCA and UCP-LF CAA, fit almost the whole spectrum thanks to high sensitivity and specificity in combination with moderate cost and expected good compliance.

A summary schedule of tests for schistosomiasis, most of which are currently available, is presented below together with some critical parameters (Table 21.1). Sensitivity and specificity are perceived as essential but for large-scale operations, time and cost can be as important.

Individual diagnosis is as important as the mapping of endemic areas, but the approaches differ. While a positive result is adamant for the individual case before treatment can be prescribed, a few such outcomes of sentinel testing in the field would suffice to begin MDA of a whole population. This latter approach is only possible because the drug treatment is not only highly effective but also very safe. However, from the perspective of control and elimination, also very light infections must be found, indeed even if only consisting of one worm pair, when the objective changes from morbidity control to interruption of transmission.

The individual approach varies over a wide range of diagnostic activities from confirming the diagnosis to finding out the extent of any pathological process present and if so, deciding on the possibility of surgery, while the critical factors in mapping exercises mainly deal with prevalence, intensity of disease, monitoring of cure after chemotherapy. In the individual case, cost is normally not a limiting factor, but expenses become a very important criterion when large-scale, population-based interventions are under consideration. If the goal is morbidity control, highly sensitive, yet inexpensive, methodologies must be used. Control activities focus on surveillance and preparation for response. They should include monitoring of the efficacy of chemotherapy and the diagnostic assays should be user-friendly, field-applicable, inexpensive, have a long shelf life and above all, have a sensitivity/specificity ratio adapted to the endemic situation at hand. The requirements can be summarized as follows:

- Methods based on egg counts are sufficiently sensitive and specific for targeting MDA in highly endemic areas, but they do not allow evidence-based decision on when to stop preventive chemotherapy programs;
- The POC-CCA rapid test detecting this parasite antigen in urine can be used for mapping as well as for monitoring and evaluation in areas with intestinal, African schistosomiasis;
- Ultra-sensitive techniques, such as LAMP and UCP-LF CAA tests, are necessary in areas characterized by low prevalence and low-intensity infection, i.e. where elimination of the disease is considered a realistic goal.

- Antibody surveillance in children is the approach of choice in areas from where schistosomiasis has recently been eliminated with its value for older individuals increasing with time after transmission interruption;
- Ultrasonography is needed for subtle morbidity as well as for follow-up of chronic schistosomiasis; it will continue to be needed for control of post-transmission schistosomiasis (Giboda and Bergquist 2000), which still exists in Japan even if the disease was eliminated there 40 years ago.

With respect to the choice of approach, Table 21.1 below provides specific advice.

TABLE 21.1 General overview of available diagnostic approaches.

Diagnostic Approach/Method		Sensitivity	Specificity	Time	Cost
PARASITOLOGICAL DIAGNOSIS	Kato-Katz	Moderate	Very high	Moderate	Low
	Hatching test	Moderate	Very high	Long	Moderate
	Urine filtration	Moderate	Very high	Moderate	Low
	Flotac devices	Adequate	Very high	Moderate	Moderate
	Skin test	High	Adequate	Short[a]	Moderate
ANTIBODY DETECTION	CHR/CTF	High	Very high	Long	Moderate
	COPT	High	Very high	Long	Moderate
	IHA			Long	Moderate
	ELISA	High sensitivity and high specificity with fractionated antigens; crude antigens produce cross-reactions.		Long	Moderate
	RIA			Long	Moderate
	IFA			Long	Moderate
	Diagnostic kits[b]	Moderate[b]	Moderate[b]	Short	Low
ANTIGEN DETECTION	PCR	Very high	Very high	Long	High
	Real-time PCR	Very high	Very high	Long	High
	LAMP	Very high	Very high	Long	High
	CCA	High	High	Short	Low
	CAA	Very high	Very high	Moderate	Moderate
NON-SPECIFIC BIOMARKERS	Metabonomics	High	Low	Long	High
	Cytokines	High	Low	Long	High
	miRNA	High	Low	Long	High
GENERAL APPROACHES	Questionnaires	Adequate	Moderate	Moderate	Low
	Reagent strips	Moderate	Adequate	Short	Low
	ECP	Medium	Low	Long	High
	Calprotectin	Medium	Low	Long	Low

[a] Delayed-type may occur after 24 hours;
[b] Strongly dependent on the antigen preparation.

A framework of evidence-based control strategies that does not exclude people without apparent symptoms is required. With respect to schistosomiasis elimination, it is critical that the diagnostic techniques applied are sensitive enough to find also the lightest of infections. As control activities operate today, the long-term benefit of current population-based MDA programs remains in doubt because ongoing low-level re-infections associated with symptoms remain below the DALY radar, such as subtle but persistent anemia, malnutrition, diminished development and abridged performance status. Therefore, as more obvious morbidities are disappearing, accurate diagnostics become more and more imperative in effective schistosomiasis reduction and even elimination.

21.10   LITERATURE CITED

Abbasi I, Hamburger J, Kariuki C, Mungai PL, Muchiri EM, King CH. 2012. Differentiating *Schistosoma haematobium* from related animal schistosomes by PCR amplifying inter-repeat sequences flanking newly selected repeated sequences. American Journal of Tropical Medicine and Hygiene 87(6): 1059–64. doi: 10.4269/ajtmh.2012.12-0243.

Abdel-Fattah M, Al-Sherbiny M, Osman A, Charmy R, Tsang V. 2011. Improving the detection limit of quantitative diagnosis of anti-*S. haematobium* antibodies using falcon assay screening test (FAST) ELISA by developing a new standard curve. Parasitology Research 108: 1457–1463. doi: org/10.1007/s00436-010-2198-y.

Adriko M, Standley CJ, Tinkitina B, Tukahebwa EM, Fenwick A, Fleming FM, Sousa-Figueiredo JC, Stothard JR, Kabatereine NB. 2014. Evaluation of circulating cathodic antigen (CCA) urine-cassette assay as a survey tool for *Schistosoma mansoni* in different transmission settings within Bugiri District, Uganda. Acta Tropica 136: 50–57. doi: 10.1016/j.actatropica.2014.04.001.

Agnew A, Fulford AJ, De Jonge N, Krijger FW, Rodriguez-Chacon M, Gutsmann V, Deelder AM. 1995. The relationship between worm burden and levels of a circulating antigen (CAA) of five species of *Schistosoma* in mice. Parasitology 111(Pt 1): 67–76.

Ahmed MM, Hussein HM, el-Hady HM. 1993. Evaluation of cercarien hullen reaction (CHR) as a diagnostic test in chronic schistosomiasis and as a parameter for reinfection in acute cases. Journal of the Egyptian Society of Parasitology 23(2): 365–371.

Akpata R, Neumayr A, Holtfreter MC, Krantz I, Singh DD, Mota R, Walter S, Hatz Ch, Richter J. 2015. The WHO ultrasonography protocol for assessing morbidity due to *Schistosoma haematobium*. Acceptance and evolution over 14 years. Systematic review. Parasitology Research 114(4): 1279–1289. doi: 10.1007/s00436-015-4389-z. Erratum in: Parasitology Research 114(5): 2045–2046. doi: 10.1007/s00436-015-4430-2.

Amarir F, Sebti F, Abbasi I, Sadak A, Fellah H, Nhammi H, Ameur B, El Idrissi AL, Rhajaoui M. 2014. *Schistosoma haematobium* detection in snails by DraI PCR and Sh110/Sm-Sl PCR: further evidence of the interruption of schistosomiasis transmission in Morocco. Parasites and Vectors 7: 288. doi: org/10.1186/1756-3305-7-288.

Antwi S, Aboah KE, Sarpong CK. 2014. The unacknowledged impact of urinary schistosomiasis in children: 5 cases from Kumasi, Ghana. Ghana Medical Journal 48(4): 228–233.

Aoki Y, Sato K, Muhoho ND, Noda S, Kimura E. 2003. Cercariometry for detection of transmission sites for schistosomiasis. Parasitology International 52: 403–408. doi: org/10.1016/S1383-5769(03)00057-6.

Ayele B, Erko B, Legesse M, Hailu A, Medhin G. 2008. Evaluation of circulating cathodic antigen (CCA) strip for diagnosis of urinary schistosomiasis in Hassoba school children, Afar, Ethiopia. Parasite 15(1): 69–75.

Bala S, Marcos M, Szabo G. 2009. Emerging role of microRNAs in liver diseases. World Journal of Gastroenterology 6: 5633–5640. doi: 10.3748/wjg.15.5633.

Balog CI, Alexandrov T, Derks RJ, Hensbergen PJ, van Dam GJ, Tukahebwa EM, Kabatereine NB, Thiele H, Vennervald BJ, Mayboroda OA, Deelder AM. 2010. The feasibility of MS and advanced data processing for monitoring *Schistosoma mansoni* infection. Proteomics-Clinical Applications 4(5): 499–510. doi: 10.1002/prca.200900158.

Balog CI, Meissner A, Göraler S, Bladergroen MR, Vennervald BJ, Mayboroda OA, Deelder AM. 2011. Metabonomic investigation of human *Schistosoma mansoni* infection. Molecular Biosystematics 7(5): 1473–80. doi: 10.1039/c0mb00262c.

Barda B, Ianniello D, Zepheryne H, Rinaldi L, Cringoli G, Burioni R, Albonico M. 2014. Parasitic infections on the shore of Lake Victoria (East Africa) detected by Mini-FLOTAC and standard techniques. Acta Tropica 137: 140–146. doi: 10.1016/j.actatropica.2014.05.012.

Bartel DP. 2014. MicroRNAs: genomics, biogenesis, mechanism and function. Cell 116: 281–297. doi: org/10.1016/S0092-8674(04)00045-5.

Bekana T, Mekonnen Z, Zeynudin A, Ayana M, Getachew M, Vercruysse J, Levecke B. 2015. Comparison of Kato-Katz thick-smear and McMaster egg counting method for the assessment of drug efficacy against soil-transmitted helminthiasis in school children in Jimma Town, Ethiopia. Transactions of the Royal Society of Tropical Medicine and Hygiene 109(10): 669–671. doi: 10.1093/trstmh/trv073.

Berggren WL, Weller TH, 1967. Immunoelectrophoretic demonstration of specific circulating antigen in animals infected with *Schistosoma mansoni*. American Journal of Tropical Medicine and Hygiene 16: 606–612.

Bergquist R, Johansen MV, Utzinger J. 2009. Diagnostic dilemmas in helminthology: what tools to use and when? Trends in Parasitology 25(4): 151–156. doi: 10.1016/j.pt.2009.01.004.

Bergquist R. 2013. Good things are worth waiting for. American Journal of Tropical Medicine and Hygiene 88(3): 409–410. doi: 10.4269/ajtmh.12-0741.

Bergwerff AA, van Dam GJ, Rotmans JP, Deelder AM, Kamerling JP, Vliegenthart JF. 1994. The immunologically reactive part of immunopurified circulating anodic antigen from *Schistosoma mansoni* is a threonine-linked polysaccharide consisting of --> 6)-(beta-D-GlcpA-(1 --> 3))-beta-D-GalpNAc-(1 --> repeating units. Journal of Biological Chemistry 269: 31510–31517.

Bligh J, Schramm G, Chiodini PL, Doenhoff MJ. 2010. Serological analysis of the outcome of treatment of *Schistosoma mansoni* infections with praziquantel. Annals of Tropical Medicine and Parasitology 104(6): 511–20. doi: 0.1179/136485910X12786389891245.

Bogoch II, Andrews JR, Dadzie Ephraim RK, Utzinger J. 2012. Simple questionnaire and urine reagent strips compared to microscopy for the diagnosis of *Schistosoma haematobium* in a community in northern Ghana. Tropical Medicine and International Health 17: 1217–1221. doi: org/10.1111/j.1365-3156.2012.03054.x.

Borges DS, de Souza JS, Romanzini J, Graeff-Teixeira C. 2013. Seeding experiments demonstrate poor performance of the hatching test for detecting small numbers of *Schistosoma mansoni* eggs in feces. Parasitology International 2013 Dec; 62(6): 543–7. doi: 10.1016/j.parint.2013.08.002.

Bosompem KM, Ayi I, Anyan WK, Arishima T, Nkrumah FK, Kojima S. 1997. A monoclonal antibody-based dipstick assay for diagnosis of urinary schistosomiasis. Transactions of the Royal Society of Tropical Medicine and Hygiene 91(5): 554–556.

Bustinduy AL, Sousa-Figueiredo JC, Adriko M, Betson M, Fenwick A, Kabatereine N, Stothard JR. 2013. Fecal occult blood and fecal calprotectin as point-of-care markers of intestinal morbidity in Ugandan children with *Schistosoma mansoni* infection. PLoS Neglected Tropical Disease 7(11): e2542. doi: 10.1371/journal.pntd.0002542. eCollection 2013 Nov.

Cai P, Gobert GN, You H, Duke M, McManus DP. 2015. Circulating miRNAs: Potential novel biomarkers for hepatopathology progression and diagnosis of schistosomiasis japonica in two murine models. PLoS Neglected Tropical Diseases 9(7): e0003965. doi: 10.1371/journal.pntd.0003965. eCollection 2015.

Caldeira RL, Jannotti-Passos LK, Lira PM, Carvalho OS. 2004. Diagnostic of *Biomphalaria* snails and *Schistosoma mansoni*: DNA obtained from traces of shell organic materials. Memórias do Instituto Oswaldo Cruz 99(5): 499–502. doi: org/10.1590/S0074-02762004000500007.

Candido RR, Favero V, Duke M, Karl S, Gutiérrez L, Woodward RC, Graeff-Teixeira C, Jones MK, St Pierre TG. 2015. The affinity of magnetic microspheres for *Schistosoma* eggs. International Journal for Parasitology 2015 45(1): 43–50. doi: 10.1016/j.ijpara.2014.08.011.

Carvalho GB, Pacífico LG, Pimenta DL, Siqueira LM, Teixeira-Carvalho A, Coelho PM, Pinheiro CS, Fujiwara RT, Oliveira SC, Fonseca CT. 2014. Evaluation of the use of C-terminal part of the *Schistosoma mansoni* 200 kDa tegumental protein in schistosomiasis diagnosis and vaccine formulation. Experimental Parasitology 139: 24–32. doi: org/10.1016/j.exppara.2014.02.003.

Chand MA, Chiodini PL, Doenhoff MJ. 2010. Development of a new assay for the diagnosis of schistosomiasis, using cercarial antigens. Transactions of the Royal Society of Tropical Medicine and Hygiene 104(4): 255–258.

Chen X, Liang H, Zhang J, Zen K, Zhang C-Y. 2012. Secreted microRNAs: a new form of intercellular communication. Trends in Cell Biology 22: 125–132. doi: org/10.1016/j.tcb.2011.12.001.

Chou YH, Chiou HJ, Chiou SY, Lee SD, Hung GS, Wu SC, Kuo BI, Lee RC, Chiang JH, Chiang T, Yu C. 2003. Duplex doppler ultrasound of hepatic Schistosomiasis japonica: a study of 47 patients. American Journal of Tropical Medicine and Hygiene 68(1): 18–23.

Clements AC, Brooker S, Nyandindi U, Fenwick A, Blair L. 2008. Bayesian spatial analysis of a national urinary schistosomiasis questionnaire to assist geographic targeting of schistosomiasis control in Tanzania, East Africa. International Journal of Parasitology 38(3-4): 401–415.

Coelho PM, Jurberg AD, Oliveira AA, Katz N. 2009. Use of a saline gradient for the diagnosis of schistosomiasis. Memórias do Instituto Oswaldo Cruz 104(5): 720–723.

Colley D, Secor W. 2014. Immunology of human schistosomiasis. Parasite Immunology 36: 347–357. doi: org/10.1111/pim.12087.

Colley DG, Binder S, Campbell C, King CH, Tchuem Tchuenté LA, N'Goran EK, Erko B, Karanja DM, Kabatereine NB, van Lieshout L, Rathbun S. 2013. A five-country evaluation of a point-of-care circulating cathodic antigen urine assay for the prevalence of *Schistosoma mansoni*. American Journal of Tropical Medicine and Hygiene 88(3): 426–432. doi: 10.4269/ajtmh.12-0639.

Coons AH, Creech HJ, Jones RN. 1941. Immunological properties of an antibody containing a fluorescent group. Proceedings of The Society for Experimental Biology and Medicine 47: 200–202.

Coons AH. 1959. The diagnostic application of fluorescent antibodies. Schweizerische Zeitschrift für Pathologie und Bakteriologie 22: 700–723.

Corstjens PL, Nyakundi RK, de Dood CJ, Kariuki TM, Ochola EA, Karanja DM, Mwinzi PN, van Dam GJ. 2015. Improved sensitivity of the urine CAA lateral-flow assay for diagnosing active *Schistosoma* infections by using larger sample volumes. Parasites and Vectors. 22: 8: 241. doi: 10.1186/s13071-015-0857-7.

Corstjens PL, DE Dood CJ, Kornelis D, Tjon Kon Fat EM, Wilson RA, Kariuki TM, Nyakundi RK, Loverde PT, Abrams WR, Tanke HJ, van Lieshout L, Deelder AM, van Dam GJ. 2014. Tools for diagnosis, monitoring and screening of *Schistosoma* infections utilizing lateral-flow based assays and upconverting phosphor labels. Parasitology 141: 1841–1855. doi:10.1017/S0031182014000626.

Corstjens PL, van Lieshout L, Zuiderwijk M, Kornelis D, Tanke HJ, Deelder AM, van Dam GJ. 2008. Up-converting Phosphor Technology based lateral flow assay (UPT-LF) for detection of *Schistosoma* circulating anodic antigen (CAA) in serum. Journal of Clinical Microbiology 46: 171–176.

Coulibaly JT, N'Goran EK, Utzinger J, Doenhoff MJ, Dawson EM. 2013a. A new rapid diagnostic test for detection of anti-*Schistosoma mansoni* and anti-*Schistosoma haematobium* antibodies. Parasites and Vectors 6: 29. doi: 10.1186/1756-3305-6-29.

Coulibaly JT, N'Gbesso YK, Knopp S, N'Guessan NA, Silué KD, van Dam GJ, N'Goran EK, Utzinger J. 2013b. Accuracy of urine circulating cathodic antigen test for the diagnosis of *Schistosoma mansoni* in preschool-aged children before and after treatment. PLoS Neglected Tropical Diseases 7: e2109.

Coulibaly JT, Knopp S, N'Guessan NA, Silué KD, Fürst T, Lohourignon LK, Brou JK, N'Gbesso YK, Vounatsou P, N'Goran EK, Utzinger J. 2011. Accuracy of urine circulating cathodic antigen (CCA) test for *Schistosoma mansoni* diagnosis in different settings of Côte d'Ivoire. PLoS Neglected Tropical Diseases 5(11): e1384. doi: 10.1371/journal.pntd.0001384.

Davis A, Wegner DH. 1979. Multicentre trials of praziquantel in human schistosomiasis: design and techniques. Bulletin of the World Health Organization 57(5): 767–771.

De Jonge N, Kremsner PG, Krijger FW, Schommer G, Fillié YE, Kornelis D, van Zeyl RJ, van Dam GJ, Feldmeier H, Deelder AM. 1990. Detection of the schistosome circulating cathodic antigen by enzyme immunoassay using biotinylated monoclonal antibodies. Transactions of the Royal Society of Tropical Medicine and Hygiene 84(6): 815–818.

De Jonge N, De Caluwé P, Hilberath GW, Krijger FW, Polderman AM, Deelder AM. 1989. Circulating anodic antigen levels in serum before and after chemotherapy with praziquantel in schistosomiasis mansoni. Transactions of the Royal Society of Tropical Medicine and Hygiene 83(3): 368–372.

De Noya BA, Spencer L, Noya O. 1992. Pre-and post-treatment immunodiagnostic evaluation in human schistosomiasis mansoni. Memórias do Instituto Oswaldo Cruz 87 Suppl 4: 271–276.

Deelder AM, van Dam GJ, van Lieshout L. 2012. 'Response to: Ashton RA, *et al*. 2011. Accuracy of circulating cathodic antigen tests for rapid mapping of *Schistosoma mansoni* and *S. haematobium* infections in Southern Sudan. Tropical Medicine and International Health 16: 1099–1103. Tropical Medicine and International Health 17(3): 402–403. doi: 10.1111/j.1365-3156.2011.02930.x.

Deelder AM, Van Zeyl RJM, Fillié YE, Rotmans JP, Duchenne W. 1989a. Recognition of gut-associated antigens by immunoglobulin M in the indirect fluorescent antibody test for schistosomiasis mansoni. Transactions of the Royal Society of Tropical Medicine and Hygiene 83: 364–367.

Deelder AM, de Jonge N, Boerman OC, Fillié YE, Hilberath GW, Rotmans JP, Gerritse MJ, Schut DW. 1989b. Sensitive determination of circulating anodic antigen in *Schistosoma mansoni* infected individuals by an enzyme-linked immunosorbent assay using monoclonal antibodies. American Journal of Tropical Medicine and Hygiene 40(3): 268–272.

Deelder AM, Kornelis D, Van Marck EA, Eveleigh PC, Van Egmond JG. 1980. *Schistosoma mansoni:* characterization of two circulating polysaccharide antigens and the immunological response to these antigens in mouse, hamster and human infections. Experimental Parasitology 50: 16–32.

Deelder AM, Klappe HT, van den Aardweg GJ, van Meerbeke EH. 1976. *Schistosoma mansoni:* demonstration of two circulating antigens in infected hamsters. Experimental Parasitology 40: 189–197.

Dimmette RM. 1955. Liver biopsy in clinical schistosomiasis: comparison of wedge and needle types. Gastroenterology 29(2): 219–234.

Ding JZ, Gan XX, Shen HY, Shen LY, Yu J. 1998. Establishment and application of dot immunogold filtaration assay for the detection of anti-schistosome antibodies. Chinese Journal of Parasitic Disease Control 11: 308–310 (in Chinese).

Doenhoff MJ, Dunne DW, Bain J, Lillywhite JE, McLaren ML. 1985. Serodiagnosis of mansonian schistosomiasis with CEF6, a cationic antigen fraction of *Schistosoma mansoni* eggs. Developments in Biological Standardization 62: 63–73.

Downs JA, Corstjens PL, Mngara J, Lutonja P, Isingo R, Urassa M, Kornelis D, Dam GJ. 2015. Correlation of serum and dried blood spot results for quantitation of *Schistosoma* circulating anodic antigen: a proof of principle. Acta Tropica 150: 59–63. doi: 10.1016/j.actatropica.2015.06.026.

Ebrahim A, El-Morshedy H, Omer E, El-Daly S, Barakat R. 1997. Evaluation of the Kato-Katz thick smear and formol ether sedimentation techniques for quantitative diagnosis of *Schistosoma mansoni* infection. American Journal of Tropical Medicine and Hygiene 57: 706–708.

El Scheich T, Holtfreter MC, Ekamp H, Singh DD, Mota R, Hatz C, Richter J. 2014. The WHO ultrasonography protocol for assessing hepatic morbidity due to *Schistosoma mansoni*. Acceptance and evolution over 12 years. Parasitology Research 113(11): 3915–3925. doi: 10.1007/s00436-014-4117-0. Erratum in: Parasitology Research. 2015. 114(1): 347. doi: 10.1007/s00436-014-4209-x.

Emukah E, Gutman J, Eguagie J, Miri ES, Yinkore P, Okocha N, Jibunor V, Nebe O, Nwoye AI, Richards FO. 2012. Urine heme dipsticks are useful in monitoring the impact of praziquantel treatment on *Schistosoma haematobium* in sentinel communities of Delta State, Nigeria. Acta Tropica 122(1): 126–131.

Engels D, Sinzinkayo E, Gryseels B. 1996. Day-to-day egg count fluctuation in *Schistosoma mansoni* infection and its operational implications. American Journal of Tropical Medicine and Hygiene 54: 319–324.

Engvall E, Perlmann P. 1971. Enzyme-linked immunosorbent assay (ELISA). Quantitative assay of immunoglobulin G. Immunochemistry 8(9): 871–874.

Erko B, Medhin G, Teklehaymanot T, Degarege A, Legesse M. 2013. Evaluation of urine-circulating cathodic antigen (Urine-CCA) cassette test for the detection of *Schistosoma mansoni* infection in areas of moderate prevalence in Ethiopia. Tropical Medicine and International Health 18(8): 1029–1035. doi: 10.1111/tmi.12117.

Espírito-Santo MC, Alvarado-Mora MV, Dias-Neto E, Botelho-Lima LS, Moreira JP, Amorim M, Pinto PL, Heath AR, Castilho VL, Gonçalves EM, *et al*. 2014. Evaluation of real-time PCR assay to detect *Schistosoma mansoni* infections in a low endemic setting. BMC Infectious Diseases 14: 558. doi: 10.1186/s12879-014-0558-4.

Fagraeus A. 1948. Antibody production in relation to the development of plasma cells. *In vivo* and *in vitro* Experiments. Acta Medica Scandinavica 130 Suppl 204, 122 pp.

Fagundes Teixeira C, Neuhauss E, Ben R, Romanzini J, Graeff-Teixeira C. 2007. Detection of *Schistosoma mansoni* eggs in feces through their interaction with paramagnetic beads in a magnetic field. PLoS Neglected Tropical Diseases 1(2): e73.

Fairley NH, Williams FE. 1927. A preliminary report on an intradermal reaction in schistosomiasis. Medical Journal of Australia 2(24): 811–818.

Fenton EM, Mascarenas MR, Lopez GP, Sibbett SS. 2009. Multiplex lateral-flow test strips fabricated by two-dimensional shaping. ACS Applied Materials and Interfaces 1(1): 124–129. doi: 10.1021/am800043z.

Fernández-Soto P, Gandasegui Arahuetes J, Sánchez Hernández A, López Abán J, Vicente Santiago B, Muro A. 2014. A loop-mediated isothermal amplification (LAMP) assay for early detection of *Schistosoma mansoni* in stool samples: a diagnostic approach in a murine model. PLoS Neglected Tropical Diseases 8: e3126. http://dx. doi: org/10.1371/journal.pntd.0003126.

Flowers JR, Hammerberg B, Wood SL, Malarkey DE, van Dam GJ, Levy MG, McLawhorn LD. 2002. *Heterobilharzia americana* infection in a dog. Journal of the American Veterinary Medical Association 220: 193–196.

Gabriël S, De Bont J, Phiri IK, Masuku M, Riveau G, Schacht AM, Deelder AM, van Dam GJ, Vercruysse J. 2002. Transplacental transfer of schistosomal circulating anodic antigens in cows. Parasite Immunology 24: 521–525.

Gan HR. 1926. The intradermal test for antigen of *Schistosoma japonicum*. Chinese Medical Journal 387 (in Chinese).

Gao H, Xiao D, Song L, Zhang W, Shen S, Yin X, Wang J, Ke X, Yu C, Zhang J. 2015. Assessment of the diagnostic efficacy of enolase as an indication of active infection of *Schistosoma japonicum*. Parasitology Research. Sep 30.

Gazzinelli G, Pinto FJ, Pellegrino J, Memória JM. 1965. The intradermal test in the diagnosis of schistosomiasis mansoni. IX. Skin response to a purified fraction isolated from cercarial extracts. Journal of Parasitology 51(5): 75375–6.

Giboda M, Bergquist NR. 2000. Post-transmission schistosomiasis: a new agenda. Acta Tropica 77(1): 3–7.

Gobert GN, Stenzel DJ, Jones MK, McManus DP. 1997. Immuno localization of the fatty acid-binding protein Sj-FABPc within adult *Schistosoma japonicum*. Parasitology 115: 33–39.

Gold R, Rosen FS, Weller TH, 1969. A specific circulating antigen in hamsters infected with *Schistosoma mansoni*. Detection of antigen in serum and urine and correlation between antigenic concentration and worm burden. American Journal of Tropical Medicine and Hygiene 18: 545–552.

Gordon CA, Acosta LP, Gobert GN, Olveda RM, Ross AG, Williams GM, Gray DJ, Harn D, Li Y, McManus DP. 2015. Real-time PCR demonstrates high prevalence of *Schistosoma japonicum* in the Philippines: implications for surveillance and control. PLoS Neglected Tropical Diseases 9(1): e0003483. doi: 10.1371/journal.pntd.0003483. eCollection 2015.

Guan X1, Shi Y. 1996. Collaborative study on evaluation of immunodiagnostic assays in schistosomiasis japonica by treatment efficacy assessment. Collaboration group. Chinese Medical Journal (Engl) 109(9): 659–664.

Guo JJ, Zheng HJ, Xu J, Zhu XQ, Wang SY, Xia CM. 2012. Sensitive and specific target sequences selected from retrotransposons of *Schistosoma japonicum* for the diagnosis of schistosomiasis. PLoS Neglected Tropical Diseases 6: e1579. doi: org/10.1371/journal.pntd.0001579.

Hamburger J, Xu YX, Ramzy RM, Jourdane J, Ruppel A. 1998. Development and laboratory evaluation of a polymerase chain reaction for monitoring *Schistosoma mansoni* infestation of water. American Journal of Tropical Medicine and Hygiene 59(3): 468–73.

Hamburger J, Abbasi I, Kariuki C, Wanjala A, Mzungu E, Mungai P, Muchiri E, King CH. 2013. Evaluation of loop-mediated isothermal amplification suitable for molecular monitoring of schistosome-infected snails in field laboratories. American Journal of Tropical Medicine and Hygiene 88(2): 344–51. doi: 10.4269/ajtmh.2012.12-0208.

Hancock K, Tsang VC. 1986. Development and optimization of the FAST-ELISA for detecting antibodies to *Schistosoma mansoni*. Journal of Immunological Methods 92(2): 167–176.

Härter G, Frickmann H, Zenk S, Wichmann D, Ammann B, Kern P, Fleischer B, Tannich E, Poppert S. 2014. Diagnosis of neuroschistosomiasis by antibody specificity index and semi-quantitative real-time PCR from cerebrospinal fluid and serum. Journal of Medical Microbiology 63: 309–312. doi: org/10.1099/jmm.0.066142-0.

Hayunga EG, Möllegård I, Duncan JF, Sumner MP, Stek M Jr, Hunter KW Jr. 1986. Development of circulating antigen assay for rapid detection of acute schistosomiasis. Lancet 2(8509): 716–718.

Hayunga EG, Möllegård I, Duncan JF Jr, Sumner MP, Stek M Jr, Hunter KW Jr. 1987. Early diagnosis of *Schistosoma mansoni* in mice using assays directed against cercarial antigens isolated by hydrophobic chromatography. Journal of Parasitoligy 73(2): 351–362.

He X, Sai X, Chen C, Zhang Y, Xu X, Zhang D, Pan W. 2013. Host serum miR-223 is a potential new biomarker for *Schistosoma japonicum* infection and the response to chemotherapy. Parasites and Vectors 6: 272. doi: org/10.1186/1756-3305-6-272.

He W, Zhu YC, Hua W, Liu Y. 2000. Development of a rapid immunodiagnosis assay for schistosomiasis-colloidal dye strip immunoassay. Clinical Journal of Schistosomiasis Control 12(1): 18–20. (In Chinese).

Higuchi R, Fockler C, Dollinger G, Watson R. 1993. Kinetic PCR analysis: real-time monitoring of DNA amplification reactions. Biotechnology (NY) 11: 1026–1030.

Hirst GK. 1942. The quantitative determination of Influenza virus and antibodies by means of red cell agglutination. Journal of Experimental Medicine 75: 49–64.

Hotez P, Alvarado M, Basáñez M, Bolliger I, Bourne R, Boussinesq M, Brooker S, Brown A, Buckle G, Budke C, *et al.* 2014. The global burden of disease study 2010: interpretation and implications for the neglected tropical diseases. PLoS Neglected Tropical Diseases 8(7): e2865. doi: 10.1371/journal.pntd.0002865. eCollection 2014.

Huang TW, Ti TF, Cha CL, Hu CC. 1966. Indirect hemagglutination test (slide method) in the diagnosis of Schistosomiasis japonica. Chinese Medical Journal 85(2): 96–106 (in Chinese).

Hung YW, Remais J. 2008. Quantitative detection of *Schistosoma japonicum* cercariae in water by real-time PCR. Eases 2(11): e337. doi: 10. 1371/journal.pntd.0000337. Epub 2008 Nov 18.

Hunter GW, Yokogawa M, Akusawa M, Sano M, Araki K, Kobayashi M. 1982. Control of schistosomiasis japonica in the Nagatoishi area of Kurume, Japan. American Journal of Tropical Medicine and Hygiene 31: 760–770.

Jiang SF, Qiu JW, Liu J, Zhang XP, He YY, Ma XJ, Zhang L, Zhang MM, Cai L. 2009. Development of dot immunogold filtration assay kit for rapid detection of antibody to schistosome in human sera. Chinese Journal of Schistosomiasis Control 21: 500–502 (in Chinese).

Kagan I, Pellegrino J. 1961. A critical review of immunological methods for the diagnosis of bilharziasis. Bulletin of the World Health Organization 25: 611–674.

Kahama AI, Kremsner PG, van Dam GJ, Deelder AM. 1998. The dynamics of a soluble egg antigen of *Schistosoma haematobium* in relation to egg counts, circulating anodic and cathodic antigens and pathology markers before and after chemotherapy. Transactions of the Royal Society of Tropical Medicine and Hygiene 92(6): 629–633.

Kariuki TM, van Dam GJ, Deelder AM, Farah IO, Yole DS, Wilson RA, Coulson PS. 2006. Previous or ongoing schistosome infections do not compromise the efficacy of the attenuated cercaria vaccine. Infection and Immunity 74: 3979–3986.

Kato T, Miura M. 1954. On the comparison of some stool examination methods. Japanese Journal of Parasitology 3: 35.

Katz N, Chaves A, Pellegrino J. 1972. A simple device for quantitative stool thick-smear technique in Schistosomiasis mansoni. Revista do Instituto de Medicina Tropical de São Paulo 14: 397–400.

Kemp WM, Greene ND, Damian RT. 1974. Sharing of cercarien hüllen reaktion antigens between *Schistosoma mansoni* cercariae and adults and uninfected *Biomphalaria pfeifferi*. American Journal of Tropical Medicine and Hygiene 23(2): 197–202.

Kemp WM. 1972. Serology of the Cercarien hüllen reaktion of *Schistosoma mansoni*. Journal of Parasitology 58(4): 686–692.

Kihara JH, Njagi EN, Kenya EU, Mwanje MT, Odek AE, van Dam G, Kahama AI, Ouma JH. 2009. Urinary soluble egg antigen levels in *Schistosoma haematobium* infection in relation to sex and age of Kenyan school children following praziquantel treatment. Transactions of the Royal Society of Tropical Medicine and Hygiene 103(10): 1024–1030. doi: 10.1016/j.trstmh.2009.04.021.

King CH, Magak P, Salam EA, Ouma JH, Kariuki HC, Blanton RE; World Health Organization. 2003. Measuring morbidity in schistosomiasis mansoni: relationship between image pattern, portal vein diameter and portal branch thickness in large-scale surveys using new WHO coding guidelines for ultrasound in schistosomiasis. Tropical Medicine and International Health 8(2): 109–117.

King CH, Bertsch D. 2013. Meta-analysis of urine heme dipstick diagnosis of *Schistosoma haematobium* infection, including low-prevalence and previously-treated populations. PLoS Neglected Tropical Diseases 7(9): e2431.

King CH. 2015a. Health metrics for helminth infections. Acta Tropica 141(Pt B): 150–160. doi: 10.1016/j. actatropica.2013.12.001.

King CH. 2015b. It's time to dispel the myth of "asymptomatic" schistosomiasis. PLoS Neglected Tropical Diseases 9(2): e0003504. doi: 10.1371/journal.pntd.0003504.

Kinkel HF, Dittrich S, Bäumer B, Weitzel T. 2012. Evaluation of eight serological tests for diagnosis of imported schistosomiasis. Clinical and Vaccine Immunology 19(6): 948–953. doi: 10.1128/CVI.05680-11.

Knopp S, Corstjens PL, Koukounari A, Cercamondi CI, Ame SM, Ali SM, de Dood CJ, Mohammed KA, Utzinger J, Rollinson D, van Dam GJ. 2015. Sensitivity and specificity of a urine circulating anodic antigen test for the diagnosis of *Schistosoma haematobium* in low endemic settings. PLoS Neglected Tropical Diseases 14; 9(5): e0003752. doi: 10.1371/journal.pntd.0003752.

Knopp S, Glinz D, Rinaldi L, Mohammed KA, N'Goran EK, Stothard JR, Marti H, Cringoli G, Rollinson D, Utzinger J. 2009. FLOTAC: a promising technique for detecting helminth eggs in human faeces. Transactions of the Royal Society of Tropical Medicine and Hygiene 103(12): 1190–1194.

Koukounari A, Donnelly CA, Moustaki I, Tukahebwa EM, Kabatereine NB, Wilson S, Webster JP, Deelder AM, Vennervald BJ, van Dam GJ. 2013. A latent markov modelling approach to the evaluation of circulating cathodic antigen strips for schistosomiasis diagnosis pre- and post-praziquantel treatment in Uganda. PLoS Computational Biology 9(12): e1003402. doi: 10.1371/journal.pcbi.1003402.

Koukounari A, Sacko M, Keita AD, Gabrielli AF, Landouré A, Dembelé AR, Clements AC, Whawell S, Donnelly CA, Fenwick A, Traoré M, Webster JP. 2006. Assessment of ultrasound morbidity indicators of schistosomiasis in the context of large-scale programs illustrated with experiences from Malian children. American Journal of Tropical Medicine and Hygiene 75(6): 1042–1052.

Köhler G, Milstein C. 1975. Continuous cultures of fused cells secreting antibody of predefined specificity. Nature 256(5517): 495–497.

Lamberton PH, Kabatereine NB, Oguttu DW, Fenwick A, Webster JP. 2014. Sensitivity and specificity of multiple Kato-Katz thick smears and a circulating cathodic antigen test for *Schistosoma mansoni* diagnosis pre- and post-repeated-praziquantel treatment. PLoS Neglected Tropical Diseases 8(9): e3139. doi: 10.1371/journal.pntd.0003139. eCollection 2014.

Lei ZL, Zhou XN. 2015. Eradication of schistosomiasis: a new target and a new task for the national schistosomiasis control program in the People's Republic of China. Chinese Journal of Schistomiasis Control 27: 1–4. (in Chinese).

Lengeler C, de Savigny D, Mshinda H, Mayombana C, Tayari S, Hatz C, Degrémont A, Tanner M. 1991. Community-based questionnaires and health statistics as tools for the cost-efficient identification of communities at risk of urinary schistosomiasis. International Journal of Epidemiology 20: 796–807.

Lengeler C, Utzinger J, Tanner M. 2002. Questionnaires for rapid screening of schistosomiasis in sub-Saharan Africa. Bulletin of the World Health Organization 80(3): 235–242.

Leutscher PD, van Dam GJ, Reimert CM, Ramarakoto CE, Deelder AM, Ørnbjerg N. 2008. Eosinophil cationic protein, soluble egg antigen, circulating anodic antigen and egg excretion in male urogenital schistosomiasis. American Journal of Tropical Medicine and Hygiene 79: 422–426.

Levecke B, Brooker SJ, Knopp S, Steinmann P, Sousa-Figueiredo JC, Stothard JR, Utzinger J, Vercruysse J. 2014. Effect of sampling and diagnostic effort on the assessment of schistosomiasis and soil-transmitted helminthiasis and drug efficacy: a meta-analysis of six drug efficacy trials and one epidemiological survey. Parasitology 141(14): 1826–1840.

Li YH. 1991. Advance of the study on immunodiagnosis of schistosomiasis, Nanjing, Jiansu Science Technology Publishing House (in Chinese).

Lier T, Johansen MV, Hjelmevoll SO, Vennervald BJ, Simonsen GS. 2008. Real-time PCR for detection of low intensity *Schistosoma japonicum* infections in a pig model. Acta Tropica 105: 74–80.

Lin DD, Liu JX, Liu YM, Hu F, Zhang YY, Xu JM, Li JY, Ji MJ, Bergquist R, Wu GL, Wu HW. 2008. Routine Kato-Katz technique underestimates the prevalence of *Schistosoma japonicum*: a case study in an endemic area of the People's Republic of China. Parasitology International 57(3): 281–6. doi: 10.1016/j.parint.2008.04.005.

Liu F, Cui S, Hu W, Feng Z, Wang Z, Han Z. 2009. Excretory/secretory proteome of the adult developmental stage of human blood fluke, *Schistosoma japonicum*. Molecular and Cellular Proteomics 8: 1236–1251. doi: org/10.1074/mcp.M800538-MCP200.

Liu S, Wu Y, Lu S. 1958. Diagnosis value of COPT for schistosomiasis: observation of 1626 cases and animal experiment. Chinese Medical Journal 44: 640–642 (in Chinese).

Lodh N, Mwansa JCL, Mutengo MM, Shiff CJ. 2013. Diagnosis of *Schistosoma mansoni* without the stool: comparison of three diagnostic tests to detect *Schistosoma mansoni* infection from filtered urine in Zambia. American Journal of Tropical Medicine and Hygiene 89: 46–50. doi: org/10.4269/ajtmh.13-0104.

Lotfy WM. 2009. Development and evaluation of an egg hatching technique for diagnosis of schistosomiasis Mansoni. Parasitologists United Journal 2: 127–132.

Lozano R, Naghavi M, Foreman K, Lim S, Shibuya K, Aboyans V, Abraham J, Adair T, Aggarwal R, Ahn SY, *et al.* 2012. Global and regional mortality from 235 causes of death for 20 age groups in 1990 and 2010: a systematic analysis for the global burden of disease study 2010. Lancet 380: 2095–2128.

Makarova JA, Shkurnikov MU, Turchinovich AA, Tonevitsky AG, Grigoriev AI. 2015. Circulating microRNAs. Biochemistry (Mosc) 80(9): 1117–26. doi: 10.1134/S0006297915090035.

Marrack JR. 1934. The chemistry of antigens and antibodies. Report no. 230 of the medical Research Council, His Majesty's Stationary Office, London (second edition 1938).

Meurs L, Brienen E, Mbow M, Ochola EA, Mboup S, Karanja DM, Secor WE, Polman K, van Lieshout L. 2015. Is PCR the next reference standard for the diagnosis of *Schistosoma* in stool? A comparison with microscopy in Senegal and Kenya. PLoS Neglected Tropical Diseases 9(7): e0003959. doi: 10.1371/journal.pntd.0003959. eCollection 2015 Jul.

Midzi N, Ndhlovu PD, Nyanga L, Kjetland EF, Reimert CM, Vennervald BJ, Gomo E, Mudenge G, Friis H, Gundersen SG, Mduluza T. 2003. Assessment of eosinophil cationic protein as a possible diagnostic marker for female genital schistosomiasis in women living in a *Schistosoma haematobium* endemic area. Parasite Immunology 25(11–12): 581–588.

Midzi N, Butterworth AE, Mduluza T, Munyati S, Deelder AM, van Dam GJ. 2008. Use of circulating cathodic antigen strips for the diagnosis of urinary schistosomiasis. Transactions of the Royal Society of Tropical Medicine and Hygiene 103: 45–51.

Mott KE, Baltes R, Bambagha J, Baldassini B. 1982. Field studies of a reusable polyamide filter for detection of *Schistosoma haematobium* eggs by urine filtration. Tropenmedizin und Parasitologie 33(4): 227–228.

Mott KE, Dixon H, Osei-Tutu E, England EC. 1983. Relation between intensity of *Schistosoma haematobium* infection and clinical haematuria and proteinuria. Lancet 1983 1(8332): 1005–1008.

Mott KE, Dixon H, Osei-Tutu E, England EC, Ekue K, Tekle A. 1985. Indirect screening for *Schistosoma haematobium* infection: a comparative study in Ghana and Zambia. Bulletin of the World Health Organization 63: 135–142.

Murray CJL, Vos T, Lozano R, Naghavi M, Flaxman AD, Michaud C, Ezzati M, Shibuya K, Salomon JA, Abdalla S, *et al.* 2012. Disability-adjusted life years (DALYs) for 291 diseases and injuries in 21 regions, 1990–2010: a systematic analysis for the global burden of disease study 2010. Lancet 380: 2197–2223.

Mwinzi PN, Kittur N, Ochola E, Cooper PJ, Campbell CH Jr, King CH, Colley DG. 2015. Additional evaluation of the point-of-contact circulating cathodic antigen assay for *Schistosoma mansoni* infection. Frontiers in Public Health 3: 48. doi: 10.3389/fpubh.2015.00048.

Nash TE. 1978. Antibody response to a polysaccharide antigen present in the schistosome gut. I. Sensitivity and specificity. American Journal of Tropical Medicine and Hygiene 27(5): 939–943.

Nash TE, Prescott B, Neva FA. 1974. The characteristics of a circulating antigen in schistosomiasis. Journal of Immunology 112: 1500–1507.

Nash TE, Nasir-Ud-Din, Jeanloz RW. 1977. Further purification and characterization of a circulating antigen in schistosomiasis. Journal of Immunology 119: 1627–1633.

Nausch N, Dawson EM, Midzi N, Mduluza T, Mutapi F, Doenhoff MJ. 2014. Field evaluation of a new antibody-based diagnostic for *Schistosoma haematobium* and *S. mansoni* at the point-of-care in northeast Zimbabwe. BMC Infectious Diseases 14: 165. doi: org/10.1186/1471-2334-14.

Navaratnam AM, Mutumba-Nakalembe MJ, Stothard JR, Kabatereine NB, Fenwick A, Sousa-Figueiredo JC. 2012. Notes on the use of urine-CCA dipsticks for detection of intestinal schistosomiasis in preschool children. Transactions of the Royal Society of Tropical Medicine and Hygiene 106(10): 619–622. doi: 10.1016/j.trstmh.2012.06.010.

Nibbeling HA, Kahama AI, Van Zeyl RJ, Deelder AM. 1998. Use of monoclonal antibodies prepared against *Schistosoma mansoni* hatching fluid antigens for demonstration of *Schistosoma haematobium* circulating egg antigens in urine. American Journal of Tropical Medicine and Hygiene 58(5): 543–550.

Nilsson LA, van Dam GJ, Deelder AM, Eriksson B, Gabone RM, Schmeisser S. 2001. The fibre-web blood sampling technique applied to serological diagnosis of schistosomiasis mansoni. Transactions of the Royal Society of Tropical Medicine and Hygiene 95: 33–35.

Notomi T, Okayama H, Masubuchi H, Yonekawa T, Watanabe K, Amino N, Hase T. 2000. Loop-mediated isothermal amplification of DNA. Nucleic Acids Research 28: e63. doi: org/10.1093/nar/28.12.e63.

Nourel Din MS, Nibbeling R, Rotmans JP, Polderman AM, Krijger FW, Deelder AM. 1994. Quantitative determination of circulating soluble egg antigen in urine and serum of *Schistosoma mansoni*-infected individuals using a combined two-site enzyme-linked immunosorbent assay. American Journal of Tropical Medicine and Hygiene 50(5): 585–94.

Obeng BB, Aryeetey YA, de Dood CJ, Amoah AS, Larbi IA, Deelder AM, Yazdanbakhsh M, Hartgers FC, Boakye DA, Verweij JJ, van Dam GJ, van Lieshout L. 2008. Application of a circulating-cathodic-antigen (CCA) strip test and real-time PCR, in comparison with microscopy, for the detection of *Schistosoma haematobium* in urine samples from Ghana. Annals of Tropical Medicine and Parasitology 102: 625–633.

Ochodo EA, Gopalakrishna G, Spek B, Reitsma JB, van Lieshout L, Polman K, Lamberton P, Bossuyt PM, Leeflang MM. 2015. Circulating antigen tests and urine reagent strips for diagnosis of active schistosomiasis in endemic areas. Cochrane Database of Systematic Reviews 3: CD009579. doi: 10.1002/14651858.CD009579.pub2.

Oliver-Gonzales J. 1954. Anti-egg precipitations of human infected with *Schistosoma mansoni*. Journal of Infectious Diseases 95: 86–91.

Olsvik O, Popovic T, Skjerve E, Cudjoe KS, Hornes E, Ugelstad J, Uhlén M. 1994. Magnetic separation techniques in diagnostic microbiology. Clinical Microbiology Reviews 7: 43–54.

Olveda DU, Olveda RM, McManus DP, Cai P, Chau TN, Lam AK, Li Y, Harn DA, Vinluan ML, Ross AG. 2014a. The chronic enteropathogenic disease schistosomiasis. International Journal of Infectious Diseases. 28: 193–203. doi: 10.1016/j.ijid.2014.07.009.

Olveda DU, Olveda RM, Lam AK, Chau TN, Li Y, Gisparil AD 2nd, Ross AG. 2014b. Utility of Diagnostic Imaging in the diagnosis and management of schistosomiasis. Clinical Microbiology 3(2): 142.

Ouchterlony O. 1948. In vitro method for testing the toxin-producing capacity of diphtheria bacteria. Acta Pathologica et Microbiologica Scandinavica 25(1–2): 186–191. doi:10.1111/j.1699-0463.1948.tb00655.x.

Papadopoulos MC, Abel PM, Agranoff D, Stich A, Tarelli E, Bell BA, Planche T, Loosemore A. Saadoun S, Wilkins P, Krishna S. 2004. A novel and accurate diagnostic test for human African trypanosomiasis. Lancet 363(9418): 1358–1363.

Pardo J, Arellano JL, López-Vélez R, Carranza C, Cordero M, Muro A. 2007. Application of an ELISA test using *Schistosoma bovis* adult worm antigens in travellers and immigrants from a schistosomiasis endemic area and its correlation with clinical findings. Scandinavian Journal of Infectious Diseases 39(5): 435–440.

Pauling L, Campbell DH, Pressman D. 1943. The nature of the forces between antigen and antibody and of the precipitation reaction. Physiological Reviews 23(3): 203–219.

Pellegrino J, 1958. The intradermal test in the diagnosis of bilharziasis. Bulletin of the World Health Organization 18(5–6): 945–961.

Peng H, Song K, Huang C, Ye S, Song H, Hu W, Han Z, McManus DP, Zhao G, Zhang Q. 2008. Expression, immunolocalization and serodiagnostic value of a myophilin-like protein from *Schistosoma japonicum*. Experimental Parasitology 119(1): 117–24. doi: 10.1016/j.exppara.2008.01.017.

Poggensee G, Reimert CM, Nilsson LA, Jamaly S, Sjastad A, Roald B, Kjetland EF, Helling-Giese G, Richter J, Chitsulo L, Kumwenda N, Gundersen SG, Krantz I, Feldmeier H. 1996. Diagnosis of female genital schistosomiasis by indirect disease markers: determination of eosinophil cationic protein, neopterin and IgA in vaginal fluid and swab eluates. Acta Tropica 62(4): 269–80.

Polderman AM, Stuiver PC, Krepel H, Smith SJ, Smelt AH, Deelder AM. 1989. The changed picture of schistosomiasis in the Netherlands. Nederlands Tijdschrift voor Geneeskunde 133: 167-171.

Polman K, Stelma FF, Gryseels B, van Dam GJ, Talla I, Niang M, van Lieshout L, Deelder AM. 1995. Epidemiologic application of circulating antigen detection in a recent *Schistosoma mansoni* focus in northern Senegal. American Journal of Tropical Medicine and Hygiene 53: 152–157.

Pontes LA, Dias-Neto E, Rabello A. 2002. Detection by polymerase chain reaction of *Schistosoma mansoni* DNA in human serum and feces. American Journal of Tropical Medicine and Hygiene 66(2): 157–62.

Pujol FH, Alarcón de Noya B, Cesari IM. 1989. Immunodiagnosis of Schistosomiasis mansoni with APIA (alkaline phosphatase immunoassay). Immunological Investigations 18(9–10): 1071–1080.

Qian ZL, Deelder AM. 1983. *Schistosoma japonicum*: immunological response to circulating polysaccharide antigens in rabbits with a light infection. Experimental Parasitology 55(3): 394–403.

Qiu LZ, Xue HC. 1990. Experimental diagnosis. pp. 448–527. *In*: Mao SP, editor. *Schistosome Biology and Control of Schistosomiasis*. Publishing House for People's Health, Beijing (in Chinese).

Rabello A, Pontes LA, Dias-Neto E. 2002. Recent advances in the diagnosis of *Schistosoma* infection: the detection of parasite DNA. Memórias do Instituto Oswaldo Cruz 97 Suppl 1: 171–2.

Ramarakoto CE, Leutscher PD, van Dam G, Christensen NO. 2008. Ultrasonographical findings in the urogenital organs in women and men infected with *Schistosoma haematobium* in northern Madagascar. Transactions of the Royal Society of Tropical Medicine and Hygiene 102(8): 767–773. doi: 10.1016/j.trstmh.2008.03.007.

Reimert CM, Tukahebwa EM, Kabatereine NB, Dunne DW, Vennervald BJ. 2008. Assessment of *Schistosoma mansoni* induced intestinal inflammation by means of eosinophil cationic protein, eosinophil protein X and myeloperoxidase before and after treatment with praziquantel. Acta Tropica 105(3): 253–9. doi: 10.1016/j.actatropica.2007.11.004. Epub 2007 Nov 29.

Reimert CM, Mshinda HM, Hatz CF, Kombe Y, Nkulila T, Poulsen LK, Christensen NO, Vennervald BJ. 2000. Quantitative assessment of eosinophiluria in *Schistosoma haematobium* infections: a new marker of infection and bladder morbidity. American Journal of Tropical Medicine and Hygiene 62(1): 19–28.

Richards FO Jr, Tsang VC, Brand JA, Hancock K. 1989. Modification of the FAST-ELISA for field diagnosis of schistosomiasis mansoni with serum or blood samples. Tropical Medicine and Parasitology 40(3): 332–334.

Richter J, Hatz C, Campagne G, Bergquist NR, Jerkins JM. 1996. Ultrasound in schistosomiasis: a practical guide to the standardized use of ultrasonography for the assessment of schistosomiasis-related morbidity. (http://www.who.int/tdr/publications/documents/ultrasound-schistosomiasis. pdf) accessed 14.11.2015.

Richter J, Poggensee G, Helling-Giese G, Kjetland E, Chitsulo L, Koumenda N, Gundersen SG, Krantz I, Feldmeier H. 1995. Transabdominal ultrasound for the diagnosis of *Schistosoma haematobium* infection of the upper female genital tract: a preliminary report. Transactions of the Royal Society of Tropical Medicine and Hygiene 89(5): 500–501.

Robijn ML, Planken J, Kornelis D, Hokke CH, Deelder AM. 2008. Mass spectrometric detection of urinary oligosaccharides as markers of *Schistosoma mansoni* infection. Transactions of the Royal Society of Tropical Medicine and Hygiene 102(1): 79–83. Epub 2007 Nov 9.

Robijn ML, Koeleman CA, Hokke CH, Deelder AM. 2007. *Schistosoma mansoni* eggs excrete specific free oligosaccharides that are detectable in the urine of the human host. Molecular and Biochemical Parasitology 151(2): 162–72. Epub 2006 Nov 27.

Robinson E, Picon D, Sturrock HJ, Sabasio A, Lado M, Kolaczinski J, Brooke S. 2009. The performance of haematuria reagent strips for the rapid mapping of urinary schistosomiasis: field experience from Southern Sudan. Tropical Medicine and International Health 14(12): 1484–1487.

Rodriguez-Molina R, Gonzalez JO, De Sala AR. 1962. The circumoval precipitin test in *Schistosoma mansoni*. A study of 300 patients. Journal of the American Medical Association 182: 1001–1004.

Ross AG, McManus DP, Farrar J, Hunstman RJ, Gray DJ, Li YS. 2012. Neuroschistosomiasis. Journal of Neurology 259: 22–32. doi: org/10.1007/s00415-011-6133-7.

Sarhan RM, Aminou HA, Saad GAR, Ahmed OA. 2014. Comparative analysis of the diagnostic performance of adult, cercarial and egg antigens assessed by ELISA, in the diagnosis of chronic human *Schistosoma mansoni* infection. Parasitology Research 113: 3467–3476. doi: org/10.1007/s00436-014-4017-3.

Scherrer AU, Sjöberg MK, Allangba A, Traoré M, Lohourignon LK, Tschannen AB, N'Goran EK, Utzinger J. 2009. Sequential analysis of helminth egg output in human stool samples following albendazole and praziquantel administration. Acta Tropica 109(3): 226–231. doi: 10.1016/j.actatropica.2008.11.015.

Shak KG, Legolvian PC, Salib, M, Sabour M, Nooman Z. 1959. Needle biopsy of the liver and spleen in schistosomiasis. A histopathologic study. American Journal of Clinical Pathology 31(1): 46–59.

Shane HL, Verani JR, Abudho B, Montgomery SP, Blackstock AJ, Mwinzi PN, Butler SE, Karanja DM, Secor WE. 2011. Evaluation of urine CCA assays for detection of *Schistosoma mansoni* infection in Western Kenya. PLoS Neglected Tropical Diseases 25: 5(1): e951. doi: 10.1371/journal.pntd.0000951.

Smit GJ. 1961. The "cercarien hullen reaktion" and the cercaricidal reaction. Tropical and Geographical Medicine 13: 374–377.

Smith H, Doenhoff M, Aitken C, Bailey W, Ji M, Dawson E, Gilis H, Spence G, Alexander C, van Gool T. 2012. Comparison of *Schistosoma mansoni* soluble cercarial antigens and soluble egg antigens for serodiagnosing schistosome infections. PLoS Neglected Tropical Diseases 6(9): e1815. doi: 10.1371/journal.pntd.0001815.

Sousa-Figueiredo JC, Betson M, Kabatereine NB, Stothard JR. 2013. The urine circulating cathodic antigen (CCA) dipstick: a valid substitute for microscopy for mapping and point-of-care diagnosis of intestinal schistosomiasis. PLoS Neglected Tropical Diseases 7(1): e2008. doi: 10.1371/journal.pntd.0002008.

Speich B, Knopp S, Mohammed KA, Khamis IS, Rinaldi L, Cringoli G, Rollinson D, Utzinger J. 2010. Comparative cost assessment of the Kato-Katz and FLOTAC techniques for soil-transmitted helminth diagnosis in epidemiological surveys. Parasites and Vectors 14; 3: 71. doi: 10.1186/1756-3305-3-71.

Standley CJ, Lwambo NJ, Lange CN, Kariuki HC, Adriko M, Stothard JR. 2010. Performance of circulating cathodic antigen (CCA) urine-dipsticks for rapid detection of intestinal schistosomiasis in school children from shoreline communities of Lake Victoria. Parasites and Vectors 5: 3(1): 7.

Stavitsky AB, Jarchow CC. 1954. Micromethods for the study of proteins and antibodies. I. Procedure and general applications of hemagglutination and hemagglutination-inhibition reactions with tannic acid and protein-treated red blood cells. Journal of Immunology 72(5): 360–367.

Stete K, Krauth SJ, Coulibaly JT, Knopp S, Hattendorf J, Müller I, Lohourignon LK, Kern WV, N'Goran EK, Utzinger J. 2012. Dynamics of *Schistosoma haematobium* egg output and associated infection parameters following treatment with praziquantel in school-aged children. Parasites and Vectors 5: 298. doi: 10.1186/1756-3305-5-298.

Stothard JR, Sousa-Figueiredo JC, Standley C, Van Dam GJ, Knopp S, Utzinger J, Ameri H, Khamis AN, Khamis IS, Deelder AM, Mohammed KA, Rollinson D. 2009a. An evaluation of urine-CCA strip test and fingerprick blood SEA-ELISA for detection of urinary schistosomiasis in school children in Zanzibar. Acta Tropica 111(1): 64–70. doi: 10.1016/j.actatropica.2009.02.009.

Stothard JR, Sousa-Figueiredo JC, Simba Khamis I, Garba A, Rollinson D. 2009b. Urinary schistosomiasis-associated morbidity in school children detected with urine albumin-to-creatinine ratio (UACR) reagent strips. Journal of Pediatric Urology 5(4): 287–291.

Sørensen E, Bøgh HO, Johansen MV, McManus DP. 1999. PCR-based identification of individuals of *Schistosoma japonicum* representing different subpopulations using a genetic marker in mitochondrial DNA. International Journal for Parasitology 29: 1121–1128. doi: org/10.1016/S0020-7519(99)00040-5.

Talaat M, Evans DB. 1996. Costs, benefits and operational implications of using quantitative techniques to screen for schistosomiasis haematobium in Egypt. Southeast Asian Journal of Tropical Medicine and Public Health 27(1): 29–35.

Tanaka H, Matsuda H, Blas BL, Noseñas JS. 1975. Evaluation of a technique of circumoval precipitin test using blood taken on filter paper and a microtiter technique of complement fixation test of *Schistosoma japonicum*. Japanese Journal of Experimental Medicine 45(2): 105–111.

Tanaka H, Tsuji M. 1997. From discovery to eradication of schistosomiasis in Japan: 1847–1996. International Journal for Parasitology 27(12): 1465–1480.

Tang Y, Wang Y, Shi XH, Xu WM, Gan XX. 2008. Rapid detection of specidic IgM against *Schistosoma japonicum* by dot immunogold filtration assay. International Journal of Epidemiology and Infectious Disease 35: 316–318 (in Chinese).

Tchuem Tchuenté LA, Kueté Fouodo CJ, Kamwa Ngassam RI, Sumo L, Dongmo Noumedem C, Kenfack CM, Gipwe NF, Nana ED, Stothard JR, Rollinson D. 2012. Evaluation of circulating cathodic antigen (CCA) urine-tests for diagnosis of *Schistosoma mansoni* infection in Cameroon. PLoS Neglected Tropical Diseases 6(7): e1758. doi: 10.1371/journal.pntd.0001758.

Teesdale CH, Fahringer K, Chitsulo L. 1985. Egg count variability and sensitivity of a thin smear technique for the diagnosis of *Schistosoma mansoni*. Transactions of the Royal Society of Tropical Medicine and Hygiene 79(3): 369–373.

Tibble J, Teahon K, Thjodleifsson B, Roseth A, Sigthorsson G, Bridger S, Foster R, Sherwood R, Fagerhol M, Bjarnason I. 2000. A simple method for assessing intestinal inflammation in Crohn's disease. Gut 47(4): 506–13. doi: 10.1136/gut.47.4.506.

Tomita N, Mori Y, Kanda H, Notomi T. 2008. Loop-mediated isothermal amplification (LAMP) of gene sequences and simple visual detection of products. Nature Protocols 3: 877–882. doi: org/10.1038/nprot.2008.57.

Tong Q, Chen R, Zhang Y, Yang GJ, Kumagai T, Furushima-Shimogawara R, Lou D, Yang K, Wen L, Lu S, Ohta N, Zhou X. 2015. A new surveillance and response tool: risk map of infected *Oncomelania hupensis* detected by loop-mediated isothermal amplification (LAMP) from pooled samples. Acta Tropica 141: 170–177. doi: org/10.1016/j.actatropica.2014.01.006.

Truant AL, Elliott SH, Kelly MT, Smith JH. 1981. Comparison of formalin-ethyl ether sedimentation, formalin-ethyl acetate sedimentation and zinc sulfate flotation techniques for detection of intestinal parasites. Journal of Clinical Microbiology 13: 882–884.

Tsang VC, Hancock K, Kelly MA, Wilson BC, Maddison SE. 1983. *Schistosoma mansoni* adult microsomal antigens, a serologic reagent. II. Specificity of antibody responses to the *S. mansoni* microsomal antigen (MAMA). Journal of Immunology. 130(3): 1366–1370.

Van Dam GJ, Odermatt P, Acosta L, Bergquist R, de Dood CJ, Kornelis D, Muth S, Utzinger J, Corstjens PL. 2015a. Evaluation of banked urine samples for the detection of circulating anodic and cathodic antigens in *Schistosoma mekongi* and *S. japonicum* infections: a proof-of-concept study. Acta Tropica 141: 198–203. doi: 10.1016/j.actatropica.2014.09.003.

Van Dam GJ, Xu J, Bergquist R, de Dood CJ, Utzinger J, Qin ZQ, Guan W, Feng T, Yu XL, Zhou J, Zheng M, Zhou XN, Corstjens PL. 2015b. An ultra-sensitive assay targeting the circulating anodic antigen for the diagnosis of *Schistosoma japonicum* in a low-endemic area, People's Republic of China. Acta Tropica 141: 190–197. doi: 10.1016/j.actatropica.2014.08.004.

Van Dam GJ, de Dood CJ, Lewis M, Deelder AM, van Lieshout L, Tanke HJ, van Rooyen LH, Corstjens PL. 2013. A robust dry reagent lateral flow assay for diagnosis of active schistosomiasis by detection of *Schistosoma* circulating anodic antigen. Experimental Parasitology 135: 274–282.

Van Dam GJ, Wichers JH, Falcao Ferreira TM, Ghati D, van Amerongen A, Deelder AM. 2004. Diagnosis of schistosomiasis by reagent strip test for detection of circulating cathodic antigen. Journal of Clinical Microbiology 42: 5458-5461.

Van Dam GJ, Bogitsh BJ, van Zeyl RJ, Rotmans JP, Deelder AM. 1996. *Schistosoma mansoni*: in vitro and in vivo excretion of CAA and CCA by developing schistosomula and adult worms. Journal of Parasitology 82: 557–564.

Van Dam GJ, Bergwerff AA, Thomas-Oates JE, Rotmans JP, Kamerling JP, Vliegenthart JF, Deelder AM. 1994. The immunologically reactive O-linked polysaccharide chains derived from circulating cathodic antigen isolated from the human blood fluke *Schistosoma mansoni* have Lewis x as repeating unit. European Journal of Biochemistry 225: 467–482.

Van Epps HL. 2006. Michael Heidelberger and the demystification of antibodies. Journal of Experimental Medicine 203(1): 5. doi: 10.1084/jem.2031fta.

Van der Werf MJ, de Vlas SJ, Brooker S, Looman CW, Nagelkerke NJ, Habbema JD, Engels D. 2003. Quantification of clinical morbidity associated with schistosome infection in sub-Saharan Africa. Acta Tropica 86(2–3): 125–139.

Van Gool T, Vetter H, Vervoort T, Doenhoff MJ, Wetsteyn J, Overbosch D. 2002. Serodiagnosis of imported schistosomiasis by a combination of a commercial indirect hemagglutination test with S*chistosoma mansoni* adult worm antigens and an enzyme-linked immunosorbent assay with *S. mansoni* egg antigens. Journal of Clinical Microbiology 40(9): 3432–3437.

Van Lieshout L, Roestenberg M. 2015. Clinical consequences of new diagnostic tools for intestinal parasites. Clinical Microbiology and Infection 21(6): 520–528.

Van Lieshout L, Polderman AM, Deelder AM. 2000. Immunodiagnosis of schistosomiasis by determination of the circulating antigens CAA and CCA, in particular in individuals with recent or light infections. Acta Tropica 77(1): 69–80.

Van Lieshout L, de Jonge N, el-Masry N, Mansour MM, Bassily S, Krijger FW, Deelder AM. 1994. Monitoring the efficacy of different doses of praziquantel by quantification of circulating antigens in serum and urine of schistosomiasis patients. Parasitology 108(Pt 5): 519–526.

Van Lieshout L, Panday UG, De Jonge N, Krijger FW, Oostburg BF, Polderman AM, Deelder AM. 1995. Immunodiagnosis of schistosomiasis mansoni in a low endemic area in Surinam by determination of the circulating antigens CAA and CCA. Acta Tropica 59(1): 19–29.

Van 't Wout AB, De Jonge N, Wood SM, Van Lieshout L, Mitchell GF, Deelder AM. 1995. Serum levels of circulating anodic antigen and circulating cathodic antigen detected in mice infected with *Schistosoma japonicum* or *S. mansoni*. Parasitology Research 81(5): 434–437.

Van 't Wout AB, De Jonge N, Tiu WU, Garcia EE, Mitchell GF, Deelder AM. 1992. Schistosome circulating anodic antigen in serum of individuals infected with *Schistosoma japonicum* from the Philippines before and after chemotherapy with praziquantel. Transactions of the Royal Society of Tropical Medicine and Hygiene 86(4): 410–413.

Vidal CH, Gurgel FV, Ferreira ML, Coutinho AL, Azevedo Filho HR. 2010. Portal doppler ultrasound evaluation in myelitis by *Schistosoma mansoni*. Arquivos de Neuro-Psiquiatria 68(1): 67–71.

Vogel H, Minning W. 1949. Weitere beobachtungen uber die cercarien hullen reaktion. Zeitschrift fuer Tropenmedizin und Parasitologie 1: 378–386.

Wang S, Hu W. 2014. Development of "-omics"research in *Schistosoma* spp. and-omics-based new diagnostic tools for schistosomiasis. Frontiers in Microbiology 5: 313. doi: 10.3389/fmicb.2014.00313.

Wang Y, Holmes E, Nicholson JK, Cloarec O, Chollet J, Tanner M, Singer BH, Utzinger J. 2004. Metabonomic investigations in mice infected with *Schistosoma mansoni*: an approach for biomarker identification. Proceedings of the National Academy of Sciences USA 101(34): 12676–12681.

Wang C, Chen L, Yin X, Hua W, Hou M, Ji M, Yu C, Wu G. 2011. Application of DNA-based diagnostics in detection of schistosomal DNA in early infection and after drug treatment. Parasites and Vectors 4: 164. doi: org/10.1186/1756-3305-4-164.

Weerakoon KG, Gobert GN, Cai P, McManus DP. 2015. Advances in the diagnosis of human schistosomiasis. Clinical Microbiology Reviews 28(4): 939–67. doi: 10.1128/CMR.00137-14.

Wen LY, Chen JH, Ding JZ, Zhang JF, Lu SH, Yu LL, Shen LY, Wu GL, Zhou XN, Zheng J. 2005. Evaluation on the applied value of the dot immunogold filtration assay (DIGFA) for rapid detection of anti-*Schistosoma japonicum* antibody. Acta Tropica 96(2–3): 142–7.

WHO. 1993. The Control of Schistosomiasis. Second report of the WHO Expert Committee WHO Technical Report Series 830. Geneva. Switzerland.

WHO. 2000. Ultrasound in schistosomiasis: a practical guide to the standardized use of ultrasonography for the assessment of schistosomiasis-related morbidity. (http://apps.who.int/iris/bitstream/10665/66535/1/TDR_STR_SCH_00.1.pdf) accessed 15.11.2015.

WHO. 2012a. Elimination of schistosomiasis. (http://www.who.int/neglected_diseases/mediacentre/WHA_65.21_Eng.pdf) accessed on 08.11.2015.

WHO. 2012b. Accelerating work to overcome the global impact of neglected tropical diseases: a roadmap for implementation (http://www.who.int/neglected_diseases/NTD_RoadMap_2012_Fullversion.pdf) accessed on 08.11.2015.

WHO. 2015. Schistosomiasis fact sheet, updated May 2015. (http://www.who.int/mediacentre/factsheets/fs115/en/) accessed 06.11.2015.

Wilson MS, Mentink-Kane MM, Pesce JT, Ramalingam TR, Thompson R, Wynn TA. 2012. Immunopathology of schistosomiasis. Immunology and Cell Biology 85: 148–154.

Wilson AR, van Dam GJ, Kariuki TM, Farah IO, Deelder AM, Coulson PS. 2006. The detection limits for estimates of infection intensity in schistosomiasis mansoni established by a study in non-human primates. International Journal of Parasitology 36: 1241–1244.

Worrell CM, Bartoces M, Karanja DM, Ochola EA, Matete DO, Mwinzi PN, Montgomery SP, Secor WE. 2015. Cost analysis of tests for the detection of *Schistosoma mansoni* infection in children in western Kenya. American Journal of Tropical Medicine and Hygiene 92(6): 1233–9. doi: 10.4269/ajtmh.14-0644.

Wu L, Wu M, Tian D, Chen S, Liu B, Chen Q, Wang J, Cai Q, Ji B, Wang L, Zhang S, Ruan D, Zhu X, Guo Z. 2012. Clinical and imaging characteristics of cerebral schistosomiasis. Cell Biochemistry and Biophysics 62: 289–295. doi: org/10.1007/s12013-011-9294-1.

Xia CM, Rong R, Lu ZX, Shi CJ, Xu J, Zhang HQ, Gong W, Luo W. 2009. *Schistosoma japonicum*: a PCR assay for the early detection and evaluation of treatment in a rabbit model. Experimental Parasitology 121: 175–179. doi: org/10.1016/j.exppara.2008.10.017.

Xu J, Rong R, Zhang HQ, Shi CJ, Zhu XQ, Xia CM. 2010. Sensitive and rapid detection of *Schistosoma japonicum* DNA by loop-mediated isothermal amplification (LAMP). International Journal for Parasitology 40: 327–331. doi: org/10.1016/j.ijpara.2009.08.010.

Xu J, Feng T, Lin DD, Wang QZ, Tang L, Wu XH, Guo JG, Peeling RW, Zhou XN. 2011. Performance of a dipstick dye immunoassay for rapid screening of *Schistosoma japonicum* infection in areas of low endemicity. Parasites and Vectors 4: 87.

Xu X, Zhang Y, Lin D, Zhang J, Xu J, Liu YM, Hu F, Qing X, Xia C, Pan W. 2014. Serodiagnosis of *Schistosoma japonicum* infection: genome-wide identification of a protein marker and assessment of its diagnostic validity in a field study in China. Lancet Infectious Diseases 14: 489–497.

Yalow RS, Berson SA. 1960. Immunoassay of endogenous plasma insulin in man. Journal of Clinical Investigation 39: 1157–1175.

Yang F, Tan XD, Liu B, Yang C, Ni ZL, Gao XD, Wang Y. 2015. Meta-analysis of the diagnostic efficiency of the questionnaires screening for schistosomiasis. Parasitology Research 114(9): 3509–3519. doi: 10.1007/s00436-015-4579-8.

Yang K, Sun LP, Huang YX, Yang GJ, Wu F, Hang DR, Li W, Zhang JF, Liang YS, Zhou XN. 2012. A real-time platform for monitoring schistosomiasis transmission supported by Google earth and a web-based geographical information system. Geospatial Health 6(2): 195–203.

Yang K, Li W, Sun LP, Huang YX, Zhang JF, Wu F, Hang DR, Steinmann P, Liang YS. 2013. Spatio-temporal analysis to identify determinants of *Oncomelania hupensis* infection with *Schistosoma japonicum* in Jiangsu province, China. Parasites and Vectors 6: 138. doi: 10.1186/1756-3305-6-138.

Yu JM, de Vlas SJ, Yuan, HC, Gryseels B. 1998. Variations in fecal *Schistosoma japonicum* egg counts. American Journal of Tropical Medicine and Hygiene 59(3): 370–375.

Yu LL, Ding JZ, Wen LY, Lou D, Yan XL, Lin LJ, Lu SH, Lin DD, Zhou XN. 2011. Development of a rapid dipstick with latex immunochromatographic assay (DLIA) for diagnosis of schistosomiasis japonica. Parasites and Vectors 4: 157. doi: 10.1186/1756-3305-4-157.

Yu Q, Yang H, Feng Y, Zhu Y, Yang X. 2012a. Magnetic affinity enzyme-linked immunoassay for diagnosis of schistosomiasis japonicum in persons with low-intensity infection. American Journal of Tropical Medicine and Hygiene 87: 689–693. doi: org/10.4269/ajtmh.2012.11-0716.

Yu Q, Yang H, Feng Y, Yang X, Zhu Y. 2012b. Magnetic affinity enzyme-linked immunoassay based on recombinant 26 kDa glutathione-S-transferase for serological diagnosis of schistosomiasis japonica. Acta Tropica 124: 199–202. doi: org/10.1016/j.actatropica.2012.08.006.

Yu Q, Yang H, Guan F, Feng Y, Yang X, Zhu Y. 2014. Detection of IgG in sera of patients with schistosomiasis japonica by developing magnetic affinity enzyme-linked immunoassay based on recombinant 14-3-3 protein. Transactions of the Royal Society of Tropical Medicine and Hygiene 108(1): 37–41. doi: 10.1093/trstmh/trt097.

Zhang Y, Taylor MG, McCrossan MV, Bickle QD. 1999. Molecular cloning and characterization of a novel *Schistosoma japonicum* "irradiated vaccine-specific" antigen, Sj14-3-3. Molecular and Biochemical Parasitology 103(1): 25–34.

Zhang Y, Zhao J, Wang X, Xu X, Pan W. 2015. Evaluation of six novel antigens as potential biomarkers for the early immunodiagnosis of schistosomiasis. Parasites and Vectors 8: 447. doi: 10.1186/s13071-015-1048-2.

Zhang JF, Xu J, Bergquist R, Yu LL, Yan XL, Zhu HQ, Zhou XN, Wen LY. 2016. Development and application of diagnostics in the national schistosomiasis control programme in the People's Republic of China. Acta Tropica (in press).

Zhong ZR, Zhou HB, Li XY, Luo QL, Song XR, Wang W, Wen HQ, Yu L, Wei W, Shen JL. 2010. Serological proteome-oriented screening and application of antigens for the diagnosis of Schistosomiasis japonica. Acta Tropica 116(1): 1–8. doi: 10.1016/j.actatropica.2010.04.014.

Zhou YB, Yang MX, Tao P, Jiang QL, Zhao GM, Wei JG, Jiang QW. 2008. A longitudinal study of comparison of the Kato-Katz technique and indirect hemagglutination assay (IHA) for the detection of schistosomiasis japonica in China, 2001–2006. Acta Tropica 107(3): 251–254. doi: 10.1016/j.actatropica.2008.06.009.

Zhou XN, Xu J, Chen HG, Wang TP, Huang XB, Lin DD, Wang QZ, Tang L, Guo JG, Wu XH, Feng T, Chen JX, Guo J, Chen SH, Li H, Wu ZD, Peeling RW. 2011. Tools to support policy decisions related to treatment strategies and surveillance of Schistosomiasis Japonica towards elimination. PLoS Neglected Tropical Diseases 5(12): e1408. doi: 10.1371/journal.pntd.0001408.

Zhu Y, He W, Liang Y, Xu M, Yu C, Hua W, Chao G. 2002. Development of a rapid, simple dipstick dye immunoassay for schistosomiasis diagnosis. Journal of Immunological Methods. 2002 266(1–2): 1–5.

Zhu YC. 2005. Immunodiagnosis and its role in schistosomiasis control in China: a review. Acta Tropica 96(2–3): 130–136.

Zhu R, Dang H, Zhang LJ, Li HZ, Zheng CJ, Wu XH, Guo JG. 2009. National surveillance of schistosomiasis in China, 2005-2008. Chinese Journal of Schistosomiasis Control 21: 358–362.

Zhu Y, Hua W, Xu M, He W, Wang X, Dai Y, Zhao S, Tang J, Wang S, Lu S. 2012. A novel immunodiagnostic assay to detect serum antibody response against selected soluble egg antigen fractions from *Schistosoma japonicum*. PLoS One 7: e44032. doi: org/10.1371/journal.pone.0044032.

Zhu H, Xu J, Zhu R, Cao C, Bao Z, Yu Q. 2014. Comparison of the miracidium hatching test and modified Kato-Katz method for detecting *Schistosoma japonicum* in low prevalence areas of China. Southeast Asian Journal of Tropical Medicine and Public Health 45: 20–25.

CHAPTER

Control of Schistosomiasis

Alan Fenwick OBE[1],*, *Fiona M Fleming*[2] *and Lynsey Blair*[3]

22.1 INTRODUCTION

From the days of early Egyptian civilization (see Bergquist *et al.*, Chapter 2) through to the 1850's the symptoms of schistosomiasis (including blood in urine and liver enlargement) were known but the cause was not and no treatment existed. From the discovery of the adult worms by Bilharz in 1851 until 1918, despite the complete life cycle description by Pirajá da Silva in 1908, there was still no treatment identified for the disease (Bilharz 1853; Pirajá da Silva 1908). Subsequently, from 1918 until 1970 various control measures were attempted including treatments with less than effective drugs and snail control using ineffective molluscicides (Christopherson 1918; El-Nagar 1958; Dawood *et al.* 1966; Amin 1972). With improvements in formulations during the 1970s, better tools for treatment and snail control became available (Fenwick *et al.* 2006; King and Bertsch 2015). However they were too expensive for the poorest developing countries. At the turn of the century there were globally an estimated 200 million people infected with the disease, schistosomiasis remained a Neglected Tropical Disease (NTD) (Chitsulo 2000). From 2000 onwards, the price of praziquantel dropped by over 90% to less than 10 US cents per tablet making large scale treatment affordable for the first time (Fenwick *et al.* 2009). During the period 2003 through to 2015 control programs proliferated in sub-Saharan Africa as schistosomiasis control attracted more publicity, funding and visibility, underpinned by a World Health Assembly resolution to achieve control of schistosomiasis in endemic areas (WHA 54.19). By 2012, the momentum behind schistosomiasis control had grown to the extent that a further resolution was passed to intensify schistosomiasis control programs and eliminate morbidity by 2020 followed by elimination of transmission by 2030 (WHA 65.21).

Despite centuries of existence, we are only 160 years on from the discovery of schistosome worms and 100 years on from the discovery of the life cycle yet schistosomiasis has been eliminated from Japan and Puerto Rico, controlled to an acceptable level in Egypt and reduced to only small foci in the Caribbean Islands and most of the Middle East (Chitsulo 2000).

Improvements in access to clean water and sanitation with a simultaneous increase in socio-economic status inevitably plays an important role in reducing schistosomiasis in the population. However, for the millions living across sub-Saharan Africa without this access, reduction in serious morbidity and intensity of infection over the last decade has relied on treatment alone.

---

[1] Tropical Parasitology, SCI – Imperial College, Department of Infectious Disease Epidemiology, St Mary's Campus, Norfolk Place, London W2 1PG, UK.

[2] Monitoring, Evaluation and Research, Schistosomiasis Control Initiative, Imperial College, London.

[3] Implementation, Schistosomiasis Control Initiative, Imperial College, London.

* Corresponding author

22.2   METHODS OF CONTROL

22.2.1   Treatment

Treatment of schistosomiasis was first attempted in 1918 by Christopherson who instigated the use of antimony drugs using the regime of 14 daily inter-peritoneal injections with "Astiban" (Christopherson 1918). At that time, schistosomiasis was perceived as a serious disease in Egypt and Sudan warranting such radical treatment despite the drug's side effects. Individuals were diagnosed by microscopical examination of stool and urine samples to identify schistosome eggs.

During the 20th century a series of drugs were developed, tested and implemented but each proved unacceptable. Hycanthone seemed to be an effective injectable drug. However, large scale use at the population level resulted in a small percentage of patients suffering severe liver failure, ultimately leading to its withdrawal (Farid *et al.* 1972; Oostburg 1972). The pharmaceutical company CIBA produced Ambilhar specifically for *Schistosoma haematobium* but increased frequency of use led to an increase in reported neurological side effects (Shekhar 1991). Another anti-schistosomal compound which was effective only against *S. haematobium* was Metrifonate which was very inexpensive (Davis and Bailey 1969). The drawback was that to be effective metrifonate needed to be taken as three doses one week apart, which was very difficult logistically to administer in rural areas (Mgeni *et al.* 1990). Against *S. mansoni* the drug Oxamniquine produced by Pfizer at last appeared to be an excellent treatment. It was used extensively in Brazil, however resistance soon developed (Gentile and Oliveira 2008), while in Africa, the effective dose was 50% higher than in South America (Foster 1987) which made the treatment more expensive.

Praziquantel, the current drug of choice, is dealt with by Remigio Olveda and Ross (Chapter 23) but its arrival on the scene and the testing through the World Health Organization in the 1970's as a well-tolerated single dose treatment changed the potential for the control of schistosomiasis forever. For a further discussion of treatment (chemotherapy, see Remigio Olveda and Ross, Chapter 23).

22.2.2   Snail Control

In the 1920's, the belief was that the snail was in fact the weakest link in the life cycle – after all if there are no snails there is no transmission. The first recognized molluscicide was copper sulphate, used extensively in Egypt to treat the water bodies of the Nile Delta during the 1920's and 1930's (El-Gindy 1957). Laboratory testing assumed that a solution of copper sulphate drip fed or sprayed into running water, ponds and lakes would kill snails effectively. Sadly, this assumption was incorrect because most waters in Africa which support snails are silt laden and alkaline. The result was that copper sulphate was almost immediately precipitated as copper carbonate, rendering it ineffective with little chance of killing snails (WHO 1965).

In the late 1960's Shell Chemicals developed a molluscicide Frescon (N-trityl morpholine) specifically to target the snail hosts of schistosomiasis. This chemical in the form of an emulsifiable concentrate was, like copper sulphate, very efficient in the laboratory at killing fresh water snails. It was very extensively tested in Sudan in the Gezira Scheme from 1971 through 1976 but it was found that it did not kill snail eggs which meant that snail repopulation was rapid (Amin 1972). Also, N-trityl morpholine was a petrochemical by product and when the cost of fuel sky rocketed in the mid 1970's so did the price of Frescon (Duke and Moore 1976).

Around the same time, in Ethiopia Dr Akliku Lemma published the results of his work on the properties of a plant called *Phytolacca dodecandra* (known as Endod) as a molluscicide. Dr Lemma's suggestion was that this shrub be planted along the banks of irrigation canals and streams so that the fruit of the shrub would fall into the water and kill snails (Lemma *et al.* 1972; Lemma and Yau 1974a; Lemma and Yau 1974b; Goll *et al.* 1983; Erko *et al.* 2002). After some hype, the Lemma suggestion was never proved to work and never fully adopted. A number of other plant molluscicides have been tried by several investigators (McCullough *et al.* 1980), but what should be an inexpensive and environmentally friendly snail control tool has never fulfilled its promise.

The final and surviving molluscicide is Niclosamide. Marketed by Bayer, Niclosamide is safe for humans and is actually prescribed as an anthelmintic for certain worm infections, killing tapeworms on contact. As a 70% wettable powder or a 30% emulsifiable concentrate niclosamide kills molluscs at approximately 1 ppm (part per million). However the compound and the solvent of the emulsifiable concentrate also kill other aquatic creatures including fish. It is therefore not an acceptable product for all water bodies, especially if the local people served by that water body are dependent on fish from it for food or income. Like Frescon, Niclosamide has increased significantly in price since it was first used in snail control (Jobin 1979). However as treatment alone is unlikely to result in more than morbidity control at the population level, the move towards elimination will need to focus on other aspects of the life cycle, including snail control, so after a prolonged period in which it has not been used, it is likely that molluscicides may soon be back in fashion.

22.2.3 Other Methods of Snail Control

Snail predators have also been considered for use in control programs including competitor snails, lungfish, crevettes and ducks, with some limited success (Madsen 1990; Sokolow *et al.* 2015). In Sudan, an attempt was made to introduce competitive snails (*Marisa* species) in the hope that they would either consume or compete with the *Biomphalaria* or *Bulinus* intermediate host snails (Haridi and Jobin 1985; Mahmoud 2001). Competitor snails do seem to have replaced *Biomphalaria* snails in the Caribbean Island of St. Lucia where schistosomiasis has mostly disappeared (Pointier and Jourdane 2000).

The most effective method of snail transmission control would be through environmental change. The aim of the environmental change would be to make the existing snail habitat uninhabitable or ensure separation between hosts, i.e. to prevent snails becoming infected and/or prevent people becoming infected. This may involve draining water bodies such as village ponds, the creation of concrete platforms at water contact points or using pipes to deliver irrigation water instead of open canals (Xianyi *et al.* 2005; Balen *et al.* 2007).

22.3 CONTROL IN THE 20ᵀᴴ CENTURY

22.3.1 Africa

One of the first efforts to control schistosomiasis took place on the Egyptian-Sudanese border in the early 1920's during the building of the Gezira irrigation scheme (a 2 million acre canal network) to the south of Khartoum (El-Nagar 1958). Schistosomiasis was recognized as a problem which would be magnified if infected Egyptian laborers imported into Sudan to dig the canals brought the infection with them. Egyptians were examined at the border and treated if infected (Stephenson 1947).

During the 1950's, a control project commenced called "Egypt 49", involving the drugs and molluscicides available at the time, Ambilhar and copper sulphate (Dawood and Dazo 1966; Dawood *et al.* 1966; Farooq 1966). A follow up project focusing at the Fayoum Oasis south-west of Cairo was funded by the German Technical Cooperation Agency (GTZ) (Miller and Love 1989). They imported a fleet of "Unimog" vehicles which sprayed niclosamide into the many kilometers of canals on a regular basis. Eventually Egypt's control program became a success based on praziquantel (Salem *et al.* 2011). Having been proven to be an excellent drug to treat human infections, in 1988 praziquantel became available for purchase (Davis *et al.* 1979; Davis and Wegner 1979; Kardaman *et al.* 1983). The World Bank agreed to support the Egyptian control program by purchasing praziquantel and supporting delivery through the many doctors in Egypt to school-aged children and adults at risk (Reich *et al.* 1998). A series of TV advertisements were developed by the Ministry of Health and there was blanket coverage on TV urging people to go to the nearest health center for testing and treatment if found to be infected. Two key critical messages were delivered; 1) for the first time, treatment was with tablets, not an injection and 2) health education around the life cycle including water contact (Michelson *et al.* 1993). By 2000, prevalence of intestinal disease had dropped from 32% (in 1935) to 4.2% and prevalence of urogenital disease from 48% (in 1935) to only 3% (Barakat *et al.* 2014) (Fig. 22.1).

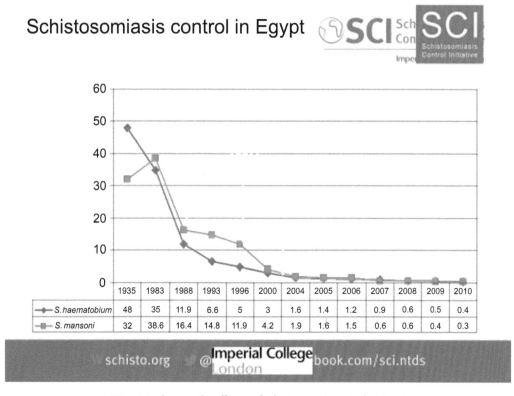

FIG. 22.1 Showing the efficacy of schistosomiasis control in Egypt.

In 1971 in Sudan, the National Research Council initiated a research and control program headed by the newly graduated Ph.D. from London, Dr Mutamad Ahmed El Amin. In 1978, the Blue Nile Health Project (BNHP) supported by WHO, the United States Agency for International Development (USAID) and the UKs Department for International Development (DFID), was launched with a 10 year remit to control schistosomiasis and malaria in the Gezira Irrigation Scheme between the Blue and White Niles. This area is almost completely flat, with almost 2 million people residing and working in agriculture there. The BNHP provided treatment for both schistosomiasis and malaria, provided indoor spraying against mosquitoes and provided families with concrete latrine slabs to those who constructed pit latrines in their home compounds, improving sanitation alongside treatment (el Gaddal 1985).

22.3.2   Caribbean and South America

Schistosomiasis was carried to the Caribbean Islands and to South America from Africa with slaves in the 19th century (Malek 1980). With the lack of sanitation, eggs from the adult worms reached fresh water where the larvae sought a suitable host snail. Miracidia from *S. haematobium* worms did not find a compatible host but *S. mansoni* miracidia found a *Biomphalaria* snail species (*B. glabrata*) which was able to support the parasite and lead to transmission, particularly in Brazil, Surinam, Venezuela and in the Caribbean Islands (Malek 1980). In the 1960's the British Medical Research Council and the Rockefeller Foundation combined to carry out the definitive Randomly Controlled Trial (RCT) to determine which control intervention from snail control, chemotherapy or improved access to water and hygiene would be the most effective. This RCT was set in parallel valleys on the Island of St Lucia and the trial ran for several years (Jordan *et al.* 1993). The conclusion was that each could affect transmission. However, the trial did not try changing the crops or introducing competitor snails. Eventually schistosomiasis was eliminated from St Lucia when the banana crops were changed and a *Thiaria* (*Thiara granifera*) snail was introduced which eliminated the *B. glabrata* host snail (Prentice 1983; Webbe 1987).

In Brazil the control of schistosomiasis was funded from 1975 through the government's nationwide program and was based on chemotherapy using, initially, oxamniquine and later praziquantel (Machado 1982). Between 1976 and 1987 mass treatment was carried out intensively, with more selective treatment of only those infected from 1987 to 2003 through a coordinated approach between states and municipalities and support from the federal government. During this time more that 12 million cases were treated which led to a shrinking in the distribution of schistosomiasis (Amaral *et al.* 2006). However, there are still pockets of high-risk of infection in northern and eastern areas of the country despite regular treatment interventions, highlighting that further interventions are likely required (Scholte *et al.* 2014).

22.3.3 Asia

In China, schistosomiasis was an extensive problem at the beginning of the 20th century with a widespread distribution and reported severe health consequences for millions of infected people (Chen 1999; Minggang and Zheng 1999). Indeed *Schistosoma japonicum* is a virulent parasite and is still a major public health problem within the country (Ross *et al.* 2001). The species also differs from the African species in that it is zoonotic and so there are many animal reservoirs of infection. Changes to agricultural practices alongside the treatment of millions of individuals since 1984, have significantly reduced the prevalence of infection across the country (Ross *et al.* 2001).

In The Philippines, the World Bank supported a schistosomiasis control program for *S. japonicum* during the late 20th century but foci of infection remain and the use of large scale treatment has been questioned (Inobaya *et al.* 2015).

22.4 CONTROL IN THE 21ST CENTURY

22.4.1 Setting the Scene: The World Health Organization (WHO)

The WHO became more closely involved with schistosomiasis control than ever before in the 1970's when Dr Andrew (Rikk) Davis masterminded the Phase III trials of praziquantel and the trials to compare a single dose at 40 mg/kg with a split dose of 20 mg/kg given on the same day several hours apart (Davis *et al.* 1979; Davis and Wegner 1979). It was found that the single dose was not significantly different from the split dose in either efficacy or acceptability, thus praziquantel has since been recommended to be administered as a single dose given after food.

The World Health Assembly speeded up implementation of schistosomiasis and soil-transmitted helminth (STH) control by passing a resolution in May 2001 urging member states to implement mass drug administration to reach at least 75% of school-aged children living in endemic areas by 2010 (WHO Expert Committee on the Control of Schistosomiasis (2001: Geneva Switzerland) and World Health Organization 2002). With the continuing success of ongoing programs across Africa and the start-up of new programs across many additional countries, the WHA then strengthened their resolve in 2012 by passing a resolution calling for progress to be made towards the elimination of schistosomiasis where feasible (WHO 2012). Alongside this, the African Regional Office of WHO initiated the development of national strategic plans for neglected tropical disease programs and assisted the completion of mapping schistosomiasis distribution, with a view to developing a strategy for schistosomiasis control and elimination in each country. The estimates of the number of individuals in need of treatment outlined the scale of the task ahead (see Table 22.1).

TABLE 22.1 Top ten highest burden countries ranked by estimated number of individuals requiring treatment for schistosomiasis across the African Region of WHO (WHO 2013a).

Country	Individuals requiring treatment for schistosomiasis
Nigeria	57, 814, 083
Ethiopia	21, 106, 522
Democratic Republic of Congo	17, 080, 905
Mozambique	12, 843, 508
Kenya	11, 125, 882
United Republic of Tanzania	9, 529, 480
Cameroon	9, 484, 894
Uganda	8, 079, 707
Malawi	6, 382, 717
Ghana	6, 356, 380

The WHO also produced a series of guidelines on how to treat at scale for schistosomiasis and STH, using the early principles of egg detection within communities to determine the overall prevalence in the population, see Table 22.2 (Crompton and WHO 2006; WHO 2011a).

TABLE 22.2 Strategy for the control of schistosomiasis (adapted from Preventive Chemotherapy in Human Helminthiasis (Crompton and WHO 2006).

Category	Prevalence among school-aged children	Action to be taken	
High-risk community	−50% by parasitological methods (intestinal and urinary schistosomiasis) or −30% by questionnaire for visible haematuria (urinary schistosomiasis)	Treat all school-age children (enrolled and not enrolled) once a year	Also treat adults considered to be at risk (from special groups to entire communities living in endemic areas)
Moderate-risk community	−10% but <50% by parasitological methods (intestinal and urinary schistosomiasis) or <30% by questionnaire for visible haematuria (urinary schistosomiasis)	Treat all school-age children (enrolled and not enrolled) once every 2 years	Also treat adults considered to be at risk (special risk groups only)
Low-risk community	<10% by parasitological methods (intestinal and urinary schistosomiasis)	Treat all school-age children (enrolled and not enrolled) twice during their primary schooling age (e.g., once on entry and once on exit)	Praziquantel should be available in dispensaries and clinics for treatment of suspected cases

22.4.2 Increasing Access: Praziquantel at the Ready

The availability of praziquantel to governments, the availability of the Active Pharmaceutical Ingredient (API), the availability of donated drug and the funds to purchase the tablets from generic suppliers have all changed significantly since the product was first developed in the 1970's.

FIG. 22.2 Praziquantel tablets. Photo P. Jourdan.

The original praziquantel formulation was developed by Bayer, but in the 1990's the South Korean Company Shin Poong developed a new synthesis technique and started marketing their product at a much lower price. By the time SCI was established and opened a tender in 2002 for purchasing praziquantel, there were several manufacturers who expressed interest at a price range of $0.07-$0.10 US cents per tablet. By 2007, the market was still small with USAID purchasing up to 80 million tablets a year and SCI also purchasing the same amount. In 2007, Merck KGaA began the Merck Praziquantel Donation Program committing to 200 million tablets over a 10 year period. However, in 2012, this was revised to scale up to 250 million tablets donated annually from 2016. In addition, praziquantel will continue to be purchased using USAID and DFID funding to maintain the market, with World Vision also donating to certain African countries.

The Pediatric Praziquantel Consortium (including Astellas Pharma, Merck KGaA and Farmanguinhos) is also actively developing a new child-appropriate formulation which is considered essential for treating children under 6 years old. Compared to the praziquantel formulation commercially available today, this formulation will use small, orally-dispersible tablets with an improved and acceptable taste.

The availability of effective drug has therefore increased dramatically in recent years, with the emphasis now on fundraising to finance their delivery.

22.4.3 Political Will: The Millenium Development Goals (MDGs)

When the MDGs (Fig. 22.3) were published at the turn of the century no country in sub-Saharan Africa could boast a national program for the treatment of schistosomiasis and STH and the MDGs at first sight were not designed with the control of schistosomiasis as a priority (Fenwick *et al.*

2009). However, a closer look gave those interested in schistosomiasis an advocacy tool with which to fundraise:

The 8 Millennium Development Goals

1 ERADICATE EXTREME POVERTY AND HUNGER

2 ACHIEVE UNIVERSAL PRIMARY EDUCATION

3 PROMOTE GENDER EQUALITY AND EMPOWER WOMEN

4 REDUCE CHILD MORTALITY

5 IMPROVE MATERNAL HEALTH

6 COMBAT HIV/AIDS, MALARIA AND OTHER DISEASES

7 ENSURE ENVIRONMENTAL SUSTAINABILITY

8 GLOBAL PARTNERSHIP FOR DEPLOYMENT

FIG. 22.3 The eight Millenium development goals.

Knowing the effect of schistosomiasis and STH on nutritional levels and cognitive capability, advocating for schistosomiasis and STH control could focus on the MDGs and highlight that four of them (MDGS 1, 2, 4 and 6) would never be achieved without widespread treatment of school-aged children (Fig. 22.2) and women of child bearing age against schistosomiasis and STH on a regular basis.

22.4.4   Igniting Momentum: Mobilizing Funds for Control

Armed with the MDG table, the list of treatments needed and a ready supply of praziquantel, safe in the knowledge that treatments were safe, effective and reasonably inexpensive, in 2001 the Bill and Melinda Gates Foundation (BMGF) was approached to support Neglected Tropical Disease Programs. The WHO assisted BMGF in the evaluation of an application from Professor Alan Fenwick in 2001 for $34 million over 5 years to initiate control activities in six sub-Saharan African countries and provide proof of principle that treatment at scale could be successful. This funding played a formative role in the creation of the Schistosomiasis Control Initiative (SCI) based at Imperial College London and the establishment of active schistosomiasis control programs within Burkina Faso, Mali, Niger, Tanzania, Uganda and Zambia (Fenwick *et al.* 2009). Five years later all the programs had proven the principle that, if given the opportunity and the resources, the countries could achieve success (Kabatereine *et al.* 2007; Toure *et al.* 2008; Clements *et al.* 2008; Fenwick *et al.* 2009; Fleming *et al.* 2009).

As momentum simultaneously grew in advocating for the previously independent preventive chemotherapy NTDs to be implemented as integrated programs, the BMGF added additional funding on the condition that the various implementing groups would work together to increase efficiency by coordinating their efforts. This was on the basis that disease distribution overlapped to such an extent that often the same individuals within the same demographic within the same geographical area were receiving treatment for each disease and often from the same health workers. To coordinate control efforts and treatment would reduce the burden on the limited health resources within the countries and improve acceptability by those receiving treatment (Molyneux *et al.* 2005).

The principal investigators of the various programs (Alan Fenwick at the SCI, Peter Hotez at George Washington University, Eric Ottesen and David Molyneux at the Global Alliance for the Elimination of Lymphatic Filariasis and Jacob Kumaresan at the International Trachoma Initiative) began to work more closely together and they approached both USAID and DFID for funding of integrated NTD programs (Fenwick *et al.* 2005; Molyneux *et al.* 2005; Hotez *et al.* 2007a, b). The positive response from both bilateral donors led to a significant increase in funding (in NTD terms) with USAID awarding $100 million in 2006 for NTD treatments over a 5 years period and DFID announcing £50 million for NTD control in 2008 over the next 5 years. As success continued, USAID awarded a further $450 million and DFID added £200 million. Primary implementing organizations for schistosomiasis control included the Schistosomiasis Control Initiative, Research Triangle International (RTI) and the Carter Centre. Meanwhile the World Bank responded to a request from the government of the Yemen and allocated $25 million for the control of schistosomiasis there.

In 2006 a new source of funding opened up more by chance than by design. The journalist Andrew Jack wrote a piece in the *Financial Times* about Bill Gates and mentioned the inexpensive treatments of NTDs, which interested a wealthy entrepreneur. With funding from Legatum and program management from an organization called Geneva Global, a five-year control program in Burundi and Rwanda implemented by SCI and the local Ministries of Health began (Ndayishimiye *et al.* 2014). This in turn led to the establishment of the END Fund, a private philanthropic initiative, which from 2012 aimed to raise funds for NTD control and elimination. By the end of 2015 the END Fund expects to have disbursed over $22 million to partners since its inception.

Within a decade, funding for NTDs including schistosomiasis had rapidly expanded but still showed no signs of slowing. The increased interest in effective altruism as a methodology for making charitable donations and focus on the low cost/high return on investment which deworming presents led to interest in schistosomiasis control first by Givingwhatwecan (www. givingwhatwecan.org), then by Givewell (www.givewell.org) and then by the Children's Investment Fund Foundation (CIFF). As a result of the positive recommendations from these organizations, additional funding was attracted which meant that countries with no funding for the control of schistosomiasis or deworming were suddenly able to access donated praziquantel and apply for funding for delivery within the context of their national strategic plans. Thus from 2010 through to 2015 additional control programs were instigated.

22.4.5 Expansion of Coverage

While the WHO recommendation defines what a control strategy for schistosomiasis using preventive chemotherapy might be, the reality on the ground is often complicated. Schistosomiasis is a focal disease, therefore defining the boundaries for treatment at large scale can be challenging and requires identifying the 'implementation unit', such as a village, a district or even a province/region. Similarly, the strategy may be limited by the availability of praziquantel or by the funds available for delivery. For these reasons, countries often adopt variable strategies dependent on what is achievable.

Some countries, mainly those with a long standing history of support have reached national coverage, have reduced the morbidity due to schistosomiasis and have reduced the treatment program. These countries include Egypt and Morocco. Other countries with treatment programs dating back to 2003 have reached national coverage and reduced prevalence and intensity of infection but still have areas where annual treatment is still warranted such as Uganda and Niger. However, there are countries where treatment could be reduced, such as Burundi and Mali. More recent countries to start treating at scale include Ghana, Cote d'Ivoire and Malawi and are making progress towards the elimination of morbidity due to regular treatment campaigns. However, at the other end of the spectrum, there are still countries within Africa which have only commenced treating since 2014, including Ethiopia and DRC and will not reach national coverage for some years to come due to both their geographical and population size.

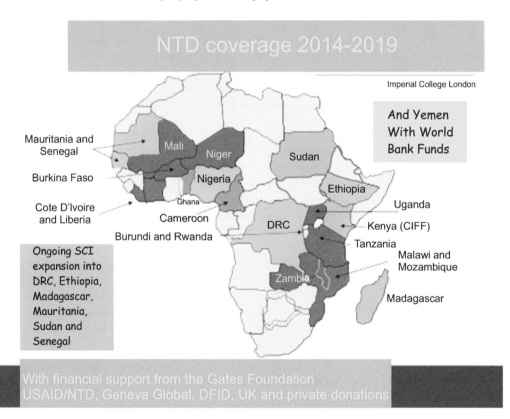

FIG. 22.4 NTD coverage in Africa.

Nigeria represents a special case. In two states, Nasarawa and Plateau, the Carter Center has been supporting treatment for many years with funding from the Izumi Foundation however there are other states which have a limited treatment history which have not yet embarked on control (Gutman *et al.* 2008; Njepuome *et al.* 2009). Nigeria alone will represent a challenge if schistosomiasis is to be controlled in sub-Saharan Africa with the WHO estimating up to 60 million people in need of treatment. In recognition of the scale of the problem, both USAID (through the Carter Centre and

RTI/Envision) and DFID (through Sightsavers, MITOSATH, Helen Keller International and CBM) are investing in the control of NTDs, including schistosomiasis, across the country.

By 2015, programs across Africa funded by DFID, USAID, END Fund and CIFF are active in Angola, Benin, Burkina Faso, Burundi, Cameroon, Central African Republic, Cote d'Ivoire, Democratic Republic of Congo, Ethiopia, Ghana, Guinea, Kenya, Liberia, Madagascar, Malawi, Mali, Mauritania, Mozambique, Namibia, Niger, Nigeria, Rwanda, Senegal, Sierra Leone, Sudan, Tanzania, Togo, Uganda, Zambia and Zimbabwe; extending the reach of schistosomiasis control beyond anything previously achieved (see also Fig. 22.4).

22.5 ELIMINATION VS CONTROL

As countries continue to scale-up and provide regular treatment for schistosomiasis, prevalence and intensity of infection is falling, thus controlling morbidity due to disease by preventing people developing serious consequences later in life. The question which then arises for those countries is 'what next'? Is the strategy for schistosomiasis to continue treatment indefinitely and ensure prevalence levels remain suppressed or can they proceed further and aim for elimination of transmission? One of the major limiting factors in switching to such an elimination strategy is the lack of sensitive diagnostic techniques to determine accurate disease prevalence in the population. For over 20 years, the gold standard diagnostic tests have been the Kato-Katz (KK) and urine filtration methods which measures egg output within a fixed volume of urine or stool sample (Katz *et al.* 1972; World Health Organization. 1994). These tests are known to become less sensitive at diagnosing low intensity infections which becomes a critical weakness in elimination if the tests are unable to identify accurately if a person is infected or not (Bergquist *et al.* 2009; Gomes *et al.* 2014).

The recently-developed Circulating Cathodic Antigen (CCA) urine dipstick test for *S. mansoni* has proven a more sensitive diagnostic for intestinal disease (see also Bergquist *et al.*, Chapter 21). In disease mapping of Rwanda and Burundi during 2014 using the CCA test, prevalence of 2–5% as detected by KK have been measured as over 40% by the CCA test, when including trace results as positive for infection. This influences the cut-offs currently recommended by WHO at which to treat within the population. As more evidence is generated on the use of CCA, it is hoped that new thresholds for treatment will be recommended by WHO to allow countries to develop strategies aligned to elimination (Kittur *et al.* 2016).

In addition to improved diagnostics, it is recognized that other interventions will be required to push schistosomiasis beyond the low prevalence which treatment alone can achieve. Within the islands of Zanzibar, the Zanzibar Elimination of Schistosomiasis Transmission (ZEST) project has been running since 2012. In addition to twice yearly treatment with praziquantel, snail control and behavioral change interventions are being trialed across the islands in an attempt to reduce prevalence of heavy infections to less than 1% (Knopp *et al.* 2012).

22.6 THE FUTURE OF CONTROL

The impressive forward movement in NTD control using mass drug administration has led to the publication of a regular report by the World Health Organization to monitor progress towards control and elimination goals (WHO 2011b, 2013b, 2015). Similarly, the BMGF also support the 'Uniting to Combat NTDs' project which produces an annual scorecard on progress (Uniting to Combat NTDs 2015).

Despite the significant progress on schistosomiasis control over the last decade in terms of increased advocacy, funds, drugs and political support, the 2015 scorecard clearly shows that schistosomiasis is lagging behind the other NTDs in terms of progress (Fig. 22.5). According to WHO figures in 2013, less than 20% of the target population received treatment against schistosomiasis.

ANNUAL SCORECARD

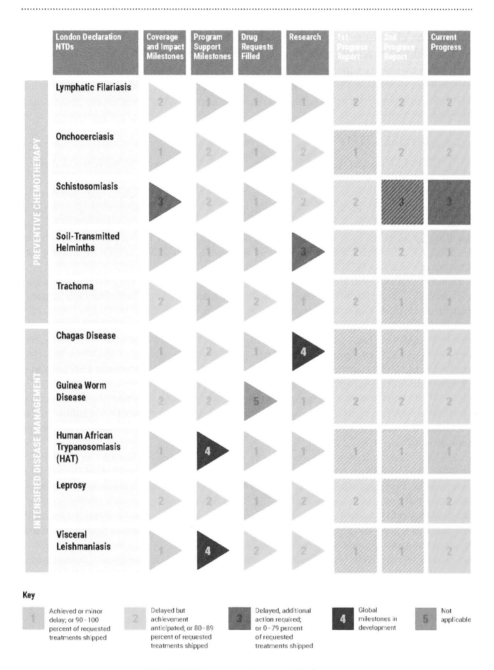

	London Declaration NTDs	Coverage and Impact Milestones	Program Support Milestones	Drug Requests Filled	Research	1st Progress Report	2nd Progress Report	Current Progress
PREVENTIVE CHEMOTHERAPY	Lymphatic Filariasis	2	1	1	1	2	2	2
	Onchocerciasis	1	2	1	2	1	2	2
	Schistosomiasis	3	2	1	2	2	3	3
	Soil-Transmitted Helminths	1	1	1	3	2	2	1
	Trachoma	2	1	2	1	2	1	1
INTENSIFIED DISEASE MANAGEMENT	Chagas Disease	1	2	1	4	1	1	2
	Guinea Worm Disease	2	2	5	1	2	2	2
	Human African Trypanosomiasis (HAT)	1	4	1	1	1	1	1
	Leprosy	2	2	1	2	2	1	2
	Visceral Leishmaniasis	1	4	2	2	1	1	2

Key

1 Achieved or minor delay; or 90–100 percent of requested treatments shipped	2 Delayed but achievement anticipated; or 80–89 percent of requested treatments shipped	3 Delayed, additional action required; or 0–79 percent of requested treatments shipped	4 Global milestones in development	5 Not applicable

FIG. 22.5 Progress against parasitic diseases.

Professor Sir Roy Anderson in a recent paper turned his attention to schistosomiasis. He stated that although schistosomiasis is endemic in 54 countries, it has one of the lowest geographical coverages by mass drug administration of all helminth diseases (Anderson *et al.* 2015). However, with increasing drug availability through donation, the goal is increasing coverage to 75% of at-risk children in endemic countries and elimination in some regions.

Using an age-structured deterministic model of schistosome transmission in a human community the effect of mass drug administration was examined. The model was fitted to baseline data from a longitudinal re-infection study in Kenya and validated against the subsequent re-infection data. The work suggests that the current WHO treatment goals should be successful in bringing about a major reduction in schistosome infection in treated communities and if continued over a 15 year period, they are likely to result in elimination, at least in areas with lower transmission. This very promising analysis suggests that by 2020 there will be a significant effect provided that countries and their supporters ensure that the available donated and purchased praziquantel are all used wisely and efficiently and not allowed to expire on the shelves of Medical Stores in Africa.

22.7   LITERATURE CITED

Amaral RS, Tauil PL, Lima DD, Engels D. 2006. An analysis of the impact of the schistosomiasis control programme in Brazil. Memórias do Instituto Oswaldo Cruz 101 Suppl 1: 79–85.

Amin MA. 1972. Large-scale assessment of the molluscicides copper sulphate and N-tritlymorpholine (Frescon) in the North group of the Gezira irrigated area, Sudan. Journal of Tropical Medicine and Hygiene 75(9): 169–175.

Anderson RM, Turner HC, Farrell SH, Yang J, Truscott JE. 2015. What is required in terms of mass drug administration to interrupt the transmission of schistosome parasites in regions of endemic infection? Parasites & Vectors 8: 553.

Balen J, Zhao ZY, Williams GM, McManus DP, Raso G, Utzinger J, Zhou J, Li YS. 2007. Prevalence, intensity and associated morbidity of *Schistosoma japonicum* infection in the Dongting Lake region, China. Bulletin of the World Health Organization 85(7): 519–526.

Barakat R, El Morshedy H, Farghaly A. 2014. Human schistosomiasis in the Middle East and North Africa Region. *In*: Rafati MAMaS, editor. *Neglected Tropical Diseases-Middle East and North Africa*, Neglected Tropical Diseases: Wien. Springer-Verlag. http://link.springer.com/book/10.1007/978-3-7091-1613-5.

Bergquist R, Johansen MV, Utzinger J. 2009. Diagnostic dilemmas in helminthology: what tools to use and when? Trends in Parasitology 25(4): 151–156.

Bilharz TM. 1853. Fernere beobachtungen über das die pfortader des menschen bewohnende distomum haematobium und sein verhältnis zu gewissen pathologischen bildungen, von Dr. Th. Bilharz in Cairo (aus brieflichen Mittheilungen an Professor v. Siebold vom 29 März 1852). Zeitschrift für Wissenschaftliche Zoologie 4: 72–76.

Chen M. 1999. Progress in schistosomiasis control in China. Chinese Medical Journal 112(10): 930–933.

Chitsulo. 2000. The global status of schistosomiasis and its control. Acta Tropica 77: 41–51.

Christopherson JB. 1918. Intravenous injections of antimonium tartaratum in bilharziosis. British Medical Journal 2(3024): 652–653.

Clements AC, Brooker S, Nyandindi U, Fenwick A, Blair L. 2008. Bayesian spatial analysis of a national urinary schistosomiasis questionnaire to assist geographic targeting of schistosomiasis control in Tanzania, East Africa. International Journal for Parasitology 38(3–4): 401–415.

Crompton DWT, WHO. 2006. *Preventive chemotherapy in human helminthiasis: coordinated use of anthelminthic drugs in control interventions: a manual for health professionals and programme managers.* Geneva: World Health Organization. 62 p.

Davis A, Bailey DR. 1969. Metrifonate in urinary schistosomiasis. Bulletin of the World Health Organization 41(2): 209–224.

Davis A, Biles JE, Ulrich AM. 1979. Initial experiences with praziquantel in the treatment of human infections due to *Schistosoma haematobium*. Bulletin of the World Health Organization 57(5): 773–779.

Davis A, Wegner DH. 1979. Multicentre trials of praziquantel in human schistosomiasis: design and techniques. Bulletin of the World Health Organization 57(5): 767–771.

Dawood IK, Dazo BC. 1966. Field tests on two new molluscicides (Molucid and WL 8008) in the Egypt-49 project area. Bulletin of the World Health Organization 35(6): 913–920.

Dawood IK, Dazo BC, Farooq M. 1966. Large-scale application of bayluscide and sodium pentachlorophenate in the Egypt-49 project area. Evaluation of relative efficacy and comparative costs. Bulletin of the World Health Organization 35(3): 357–367.

Duke BO, Moore PJ. 1976. The use of a molluscicide in conjunction with chemotherapy to control *Schistosoma haematobium* at the Barombi Lake foci in Cameroon. III. Conclusions and costs. Tropenmedizin und Parasitologie (Stuttgart) 27(4): 505–508.

El-Gindy MS. 1957. Field studies on the action of copper sulphate as a molluscocide for control of the snail vectors of schistosomiasis in Egypt. Journal of the Egyptian Medical Association 40(2): 111–121.

El-Nagar H. 1958. Control of schistosomiasis in the Gezira, Sudan. Journal of Tropical Medicine and Hygiene 61(9): 231–235.

el Gaddal AA. 1985. The Blue Nile Health Project: a comprehensive approach to the prevention and control of water-associated diseases in irrigated schemes of the Sudan. Journal of Tropical Medicine and Hygiene 88(2): 47–56.

Erko B, Abebe F, Berhe N, Medhin G, Gebre-Michael T, Gemetchu T, Gundersen SG. 2002. Control of *Schistosoma mansoni* by the soapberry Endod (Phytolacca dodecandra) in Wollo, northeastern Ethiopia: post-intervention prevalence. East African Medical Journal 79(4): 198–201.

Farid Z, Smith JH, Bassily S, Sparks HA. 1972. Hepatotoxicity after treatment of schistosomiasis with hycanthone. British Medical Journal 2(5805): 88–89.

Farooq M. 1966. Importance of determining transmisson sites in planning bilharziasis control. Field observations from the Egypt-49 project area. American Journal of Epidemiology 83(3): 603–612.

Fenwick A, Molyneux D, Nantulya V. 2005. Achieving the millennium development goals. Lancet 365(9464): 1029–1030.

Fenwick A, Rollinson D, Southgate V. 2006. Implementation of human schistosomiasis control: challenges and prospects. Advances in Parasitology 61: 567–622.

Fenwick A, Webster JP, Bosque-Oliva E, Blair L, Fleming FM, Zhang Y, Garba A, Stothard JR, Gabrelli AF, Clements ACA, *et al.* 2009. The schistosomiasis control initiative (SCI): rationale, development and implementation from 2002–2008. Parasitology 136(Special Issue 13): 1719–1730.

Fleming FM, Fenwick A, Tukahebwa EM, Lubanga RG, Namwangye H, Zaramba S, Kabatereine NB. 2009. Process evaluation of schistosomiasis control in Uganda, 2003 to 2006: perceptions, attitudes and constraints of a national programme. Parasitology 136(13): 1759–1769.

Foster R. 1987. A review of clinical experience with oxamniquine. Transactions of the Royal Society of Tropical Medicine and Hygiene 81(1): 55–59.

Gentile R, Oliveira G. 2008. Brazilian studies on the genetics of *Schistosoma mansoni*. Acta Tropica 108(2–3): 175–178.

Goll PH, Lemma A, Duncan J, Mazengia B. 1983. Control of schistosomiasis in Adwa, Ethiopia, using the plant molluscicide endod (*Phytolacca dodecandra*). Tropenmedizin Und Parasitologie 34(3): 177–183.

Gomes LI, Enk MJ, Rabello A. 2014. Diagnosing schistosomiasis: where are we? Revista da Sociedade Brasileira de Medicina Tropical 47(1): 3–11.

Gutman J, Fagbemi A, Alphonsus K, Eigege A, Miri ES, Richards FO, Jr. 2008. Missed treatment opportunities for schistosomiasis mansoni, in an active programme for the treatment of urinary schistosomiasis in Plateau and Nasarawa states, Nigeria. Annals of Tropical Medicine and Parasitology 102(4): 335–346.

Haridi AA, Jobin WR. 1985. Estimated risks and benefits from introducing *Marisa cornuarietis* into the Sudan. Journal of Tropical Medicine and Hygiene 88(2): 145–151.

Hotez P, Raff S, Fenwick A, Richards F, Jr, Molyneux DH. 2007a. Recent progress in integrated neglected tropical disease control. Trends in Parasitology 23(11): 511–514.

Hotez PJ, Molyneux DH, Fenwick A, Kumaresan J, Sachs SE, Sachs JD, Savioli L. 2007b. Control of neglected tropical diseases. The New England Journal of Medicine 357(10): 1018–1027.

Inobaya MT, Olveda RM, Tallo V, McManus DP, Williams GM, Harn DA, Li Y, Chau TN, Olveda DU, Ross AG. 2015. Schistosomiasis mass drug administration in the Philippines: lessons learnt and the global implications. Microbes and infection. Institut Pasteur 17(1): 6–15.

Jobin WR. 1979. Cost of snail control. American Journal of Tropical Medicine and Hygiene 28(1): 142–154.

Jordan PMD, Jordan PMDS, Webbe G, Sturrock RF. 1993. *Human Schistosomiasis*. Wallingford: CAB International. 480 p.

Kabatereine NB, Brooker S, Koukounari A, Kazibwe F, Tukahebwa EM, Fleming FM, Zhang Y, Webster JP, Stothard JR, Fenwick A. 2007. Impact of a national helminth control programme on infection and morbidity in Ugandan school children. Bulletin of the World Health Organization 85(2): 91–99.

Kardaman MW, Amin MA, Fenwick A, Cheesmond AK, Dixon HG. 1983. A field trial using praziquantel (BiltricideR) to treat *Schistosoma mansoni* and *Schistosoma haematobium* infection in Gezira, Sudan. Annals of Tropical Medicine and Parasitology 77(3): 297–304.

Katz N, Chaves A, Pellegrino J. 1972. A simple device for quantitative stool thick-smear technique in Schistosomiasis mansoni. Revista do Instituto de Medicina Tropical de Sao Paulo 14(6): 397–400.

King CH, Bertsch D. 2015. Historical perspective: snail control to prevent schistosomiasis. PLoS Neglected Tropical Diseases 9(4): e0003657.

Kittur N, Castleman JD, Campbell CH, Jr, King CH, Colley DG. 2016. Comparison of *Schistosoma mansoni* prevalence and intensity of infection, as determined by the circulating cathodic antigen urine assay or by the Kato-Katz fecal assay: a systematic review. American Journal of Tropical Medicine and Hygiene. pii: 15–0725. [Epub ahead of print].

Knopp S, Mohammed KA, Ali SM, Khamis IS, Ame SM, Albonico M, Gouvras A, Fenwick A, Savioli L, Colley DG, *et al.* 2012. Study and implementation of urogenital schistosomiasis elimination in Zanzibar (Unguja and Pemba islands) using an integrated multidisciplinary approach. BMC Public Health 12: 930.

Lemma A, Brody G, Newell GW, Parkhurst RM, Skinner WA. 1972. Studies on the molluscicidal properties of endod (*Phytolacca dodecandra*). I. Increased potency with butanol extraction. Journal of Parasitology 58(1): 104–107.

Lemma A, Yau P. 1974a. Studies on the molluscicidal properties of endod (*Phytolacca dodecandra*). III. Stability and potency under different environmental conditions. Ethiopian Medical Journal 12(3): 115–124.

Lemma A, Yau P. 1974b. Studies on the molluscicidal properties of endod (*Phytolacca dodecandra*). II. Comparative toxicity of various molluscicides to fish and snails. Ethiopian Medical Journal 12(3): 109–114.

Machado PA. 1982. The Brazilian program for schistosomiasis control, 1975–1979. American Journal of Tropical Medicine and Hygiene 31(1): 76–86.

Madsen H. 1990. Biological methods for the control of freshwater snails. Parasitology Today 6(7): 237–241.

Mahmoud AAF. 2001. *Schistosomiasis.* London: Imperial College Press. 510 p.

Malek EA. 1980. *Snail-transmitted Parasitic Diseases.* Volumes 1 and 2, Boca Raton, Fla: CRC Press. p. 334, 324.

McCullough FS, Gayral P, Duncan J, Christie JD. 1980. Molluscicides in schistosomiasis control. Bulletin of the World Health Organization 58(5): 681–689.

Mgeni AF, Kisumku UM, McCullough FS, Dixon H, Yoon SS, Mott KE. 1990. Metrifonate in the control of urinary schistosomiasis in Zanzibar. Bulletin of the World Health Organization 68(6): 721–730.

Michelson MK, Azziz FA, Gamil FM, Wahid AA, Richards FO, Juranek DD, Habib MA, Spencer HC. 1993. Recent trends in the prevalence and distribution of schistosomiasis in the Nile delta region. American Journal of Tropical Medicine and Hygiene 49(1): 76–87.

Miller MJ, Love EJ. 1989. *Parasitic Diseases: Treatment and Control.* Boca Raton, Fla: CRC Press. 360 p.

Minggang C, Zheng F. 1999. Schistosomiasis control in China. Parasitology International 48(1): 11–19.

Molyneux DH, Hotez PJ, Fenwick A. 2005. Rapid-impact interventions: how a policy of integrated control for Africa's neglected tropical diseases could benefit the poor. PLoS Medicine 2(11): e336.

Ndayishimiye O, Ortu G, Soares Magalhaes RJ, Clements A, Willems J, Whitton J, Lancaster W, Hopkins A, Fenwick A. 2014. Control of neglected tropical diseases in Burundi: partnerships, achievements, challenges and lessons learned after four years of programme implementation. PLoS Neglected Tropical Diseases 8(5): e2684.

Njepuome NA, Hopkins DR, Richards FO, Jr, Anagbogu IN, Pearce PO, Jibril MM, Okoronkwo C, Sofola OT, Withers PC, Jr, Ruiz-Tiben E, *et al.* 2009. Nigeria's war on terror: fighting dracunculiasis, onchocerciasis, lymphatic filariasis and schistosomiasis at the grassroots. American Journal of Tropical Medicine and Hygiene 80(5): 691–698.

Oostburg BF. 1972. Clinical trial with hycanthone in Schistosomiasis mansoni in Surinam. Tropical and Geographical Medicine 24(2): 148–151.

Pirajá da Silva MA. 1908. Contribuição para o estudo da Schistosomíase na Bahia. Brazilian Journal of Medical and Biological Research 22: 281–282.

Pointier JP, Jourdane J. 2000. Biological control of the snail hosts of schistosomiasis in areas of low transmission: the example of the Caribbean area. Acta Tropica 77(1): 53–60.

Prentice MA. 1983. Displacement of *Biomphalaria glabrata* by the snail *Thiara granifera* in field habitats in St. Lucia, West Indies. Annals of Tropical Medicine and Parasitology 77(1): 51–59.

Reich MR, Govindaraj R, Dumbaugh K, Yang B-m, Brinkmann A, El-Saharty S. 1998. *WHO Action Programme on Essential Drugs. World Health Organization. Division of Control of Tropical Diseases. International strategies for tropical disease treatments: experiences with praziquantel.* Geneva: World Health Organization. 94 p.

Ross AG, Sleigh AC, Li Y, Davis GM, Williams GM, Jiang Z, Feng Z, McManus DP. 2001. Schistosomiasis in the People's Republic of China: prospects and challenges for the 21st century. Clinical Microbiology Reviews 14(2): 270–295.

Salem S, Mitchell RE, El-Alim El-Dorey A, Smith JA, Barocas DA. 2011. Successful control of schistosomiasis and the changing epidemiology of bladder cancer in Egypt. BJU International 107(2): 206–211.

Scholte RG, Gosoniu L, Malone JB, Chammartin F, Utzinger J, Vounatsou P. 2014. Predictive risk mapping of schistosomiasis in Brazil using Bayesian geostatistical models. Acta Tropica 132: 57–63.

Shekhar KC. 1991. Schistosomiasis drug therapy and treatment considerations. Drugs 42(3): 379–405.

Sokolow SH, Huttinger E, Jouanard N, Hsieh MH, Lafferty KD, Kuris AM, Riveau G, Senghor S, Thiam C, N'Diaye A, *et al.* 2015. Reduced transmission of human schistosomiasis after restoration of a native river prawn that preys on the snail intermediate host. Proceedings of the National Academy of Science USA 112(31): 9650–9655.

Stephenson RW. 1947. Bilharziasis in the Gezira irrigated area of the Sudan. Transactions of the Royal Society of Tropical Medicine and Hygiene 40(4): 479–494.

Toure S, Zhang Y, Bosque-Oliva E, Ky C, Ouedraogo A, Koukounari A, Gabrielli AF, Bertrand S, Webster JP, Fenwick A. 2008. Two-year impact of single praziquantel treatment on infection in the national control programme on schistosomiasis in Burkina Faso. Bulletin of the World Health Organization 86(10): 780–787, A.

Uniting to Combat NTDs. 2015. *Delivering on Promises and Driving Progress.* 46 p.

Webbe G. 1987. The St Lucia project and control of schistosomiasis. Endeavour 11(3): 122–126.

WHO. 1965. Molluscicide screening and evaluation. Bulletin of the World Health Organization 33(4): 567–581.

WHO. 2011a. *Helminth control in school-age children: a guide for managers of control programmes.* Geneva: World Health Organization. 75 p. + annexes.

WHO. 2011b. *Working to overcome the global impact of neglected tropical diseases: first WHO report on neglected tropical diseases: update 2011.* Geneva: World Health Organization. viii, 14 p.

WHO. 2012. WHA65.21. *Elimination of schistosomiasis. Sixty-fifth World Health Assembly Geneva 21–26 May 2012 Resolutions, decisions and annexes.* Geneva: World Health Organization. pp. 36–37.

WHO. 2013a. *Schistosomiasis: progress report 2001–2011, strategic plan 2012–2020 (in IRIS).* Geneva: World Health Organization.

WHO. 2013b. *Sustaining the drive to overcome the global impact of neglected tropical diseases: second WHO report on neglected tropical diseases. (In IRIS).* Geneva: World Health Organization. xii, 138 p.

WHO. 2015. *Investing to overcome the global impact of neglected tropical diseases: third WHO report on neglected tropical diseases.* Geneva: World Health Organization. 191 p.

WHO Expert Committee on the Control of Schistosomiasis (2001: Geneva Switzerland), World Health Organization. 2002. *Prevention and control of schistosomiasis and soil-transmitted helminthiasis: report of a WHO expert committee.* Geneva: World Health Organization. 57 p.

World Health Organization. 1994. *Bench aids for the diagnosis of intestinal parasites.* Geneva: World Health Organization. 10 pl. in 1 folder.

Xianyi C, Liying W, Jiming C, Xiaonong Z, Jiang Z, Jiagang G, Xiaohua W, Engels D, Minggang C. 2005. Schistosomiasis control in China: the impact of a 10-year world bank loan project (1992–2001). Bulletin of the World Health Organization 83(1): 43–48.

Chemotherapy Against Schistosomiasis

Remigio M Olveda[1] and *Allen GP Ross*[2,*]

23.1 INTRODUCTION

In the past, antischistosomal pharmacologic agents against *Schistosoma hematobium*, *S. mansoni* (hyacanthone, metrifonate, oxamiquin and others) and *S. japonicum* (niridazole, antimonials) have been used but were found less effective due to the development of toxicities before therapeutic levels were reached. The situation dramatically changed when the drug praziquantel was released in 1979. Praziquantel was proven to be well tolerated and effective against the five species of schistosomes affecting humans. The drug Artemether was introduced in the early 1990s and has shown to be successful in the prevention of schistosomiasis. Corticosteroids are used as an adjuvant therapy for the treatment of Katayama syndrome within two months of freshwater contact and for the treatment of schistosomal encephalopathy during the oviposition stage. The treatment of bovines in China and The Philippines has now been introduced into integrated control programs given the significant role bovines play in human transmission. We now discuss the role of these chemotherapeutic agents in the treatment and control of schistosomiasis.

23.2 PRAZIQUANTEL

Praziquantel (PZQ), a pyrazinoisoquinoline derivative, is a safe and highly effective oral drug that is active against all schistosome species (Fig. 23.1A) (Ross *et al.* 2001, 2002; Xiao 2005; Zhou *et al.* 2005; Doenhoff and Pica-Mattoccia 2006; Utzinger *et al.* 2007; Doenhoff *et al.* 2008). Since its discovery in the mid-1970s, its safety and efficacy have ensured its widespread use. It is absorbed well but undergoes extensive first-pass hepatic clearance. PZQ is secreted in breast milk and is metabolized by the liver and its (inactive) metabolites are excreted in the urine. Side effects are mild and it can be used to treat young children and pregnant women. PZQ acts within one hour of ingestion, although its precise action on adult worms is unknown. It appears to cause tetanic contractions and tegumental vacuoles, causing worms to detach from the wall of the vein and die. Schistosome calcium ion channels have been suggested as the molecular target of PZQ, but

[1] Research Institute for Tropical Medicine, Corporate Ave, Muntinlupa, 1781 Metro Manila, The Philippines.
[2] Tropical Medicine and Global Health, Griffith University, Logan Campus, University Drive, Meadowbrook QLD 4131, Australia.
* Corresponding author

the evidence remains indirect and its precise mechanism of action remains unknown (Aragon *et al.* 2009). A recent study implicated myosin light chain as an alternative target of PZQ. Standard treatment of chronic schistosomiasis is 60 mg/kg of PZQ in divided doses; for mass chemotherapy a single dose (40 mg/kg) is used (Table 23.1) (Rabello 1997; Ross *et al.* 2002). A recent meta-analyses of 52 clinical trials showed that a dosage of 30–60 mg/kg praziquantel compared with placebo, produced a cure rate of around 76% (range from 67–83%) for human schistosomiasis (Liu *et al.* 2011). No significant differences in cure rates were found among the species *S. haematobium*, *S. japonicum* or *S. mansoni*. Cure rate of the drug at 40 mg/kg (the current dosage recommended by WHO) was 52% (range from 49–55%) compared to 91% (range from 88%–92%) when divided dosages were increased to 60, 80, 100 mg/kg (Liu *et al.* 2011).

Treatment failures (e.g., 60 mg/kg) with this dose have been reported, particularly in areas where schistosomiasis has been recently introduced (Ross *et al.* 2002). Whether this is because the drug works in concert with the host immune response, which has yet to develop or is due to migrating larvae not yet susceptible to PZQ is unknown. Drug susceptibility usually takes three weeks to develop. There is also some evidence, albeit controversial, that resistance to PZQ may be emerging in Africa, where there has been heavy exposure to the drug and where, worryingly, there are reports of *S. mansoni* and *S. haematobium* infections that are not responsive (Gryseels *et al.* 2006). Other drugs that have been used in the treatment of schistosomiasis are oxamniquine (Vansil) for *S. mansoni* and metrifonate (trichlorfon) for *S. haematobium*. Potential new targets for drug development against schistosomes include peroxiredoxin-1 (Aragon *et al.* 2009), thioredoxin glutathione reductase (Kumagai *et al.* 2009), tyrosine kinases (Shrivastava *et al.* 2005), tyrosine kinase receptors (Ahier *et al.* 2008; Ting-An *et al.* 2009), phosphatidylinositol kinases (Ahier *et al.* 2009) and neuropeptides (Bahia *et al.* 2009).

23.3 ARTEMETHER

PZQ cannot be used for chemoprophylaxis because of its short half-life (1–1.5 hours) and due to the fact it cannot kill schistosomula (the migrating larvae) that are 3–21 days old. Artemether (ART) (and other artemisinin derivatives), which comes from the leaves of the Chinese medicinal plant *Artemisia annua* (Fig. 23.1B) and is used for the treatment of malaria, is effective against juvenile schistosomes during the first 21 days of infection in animals and humans (Doenhoff and Pica-Mattoccia 2006) and it should kill all immature schistosomula if it is given every two weeks. Accordingly, it has been used as a chemoprophylactic in high-risk groups, such as flood relief workers and fishermen, in areas of China where schistosomiasis is endemic (Xiao 2005; Utzinger *et al.* 2007). The dosage required is lower than that needed for the treatment of malaria, but it is unlikely that ART would be used in other regions, such as Africa, where malaria is endemic because such use might lead to the selection of ART-resistant *Plasmodium falciparum* (Gray *et al.* 2010). In animals, combination therapy with PZQ plus ART is safe and results in higher worm reduction rates than therapy with PZQ alone (Xiao 2005; Utzinger *et al.* 2007).

Multiple doses of artemether over weekly or biweekly intervals resulted in protection rates from 65% to 97% due to its killing of juvenile schistosomula (Table 23.1) (Liu *et al.* 2011). Increasing doses and shortening medication intervals improved the efficacies of the drug. Praziquantel and artemisinin derivatives when used in combination resulted in a higher protection rate of 84% (range 64%–91%) than praziquantel used alone (Liu *et al.* 2011). The combination of praziquantel and artesunate greatly increased the protection rate to 96% (range 78%–99%) for preventing schistosome infection (Liu *et al.* 2011). Side effects of abdominal discomfort, fever, sweating and giddiness, were minimal and transient (Liu *et al.* 2011).

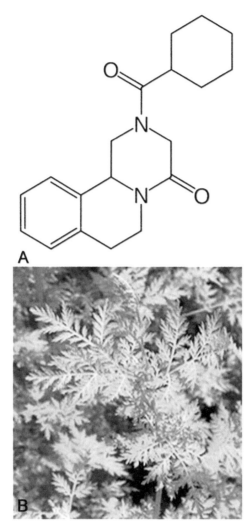

FIG. 23.1 Chemotherapy. **A**. Praziquantel. **B**. *Artemisia annua,* used in the preparation of artemether and other artemisinin derivatives.

23.4   ADJUVENT TREATMENT

Corticosteroids and anticonvulsants maybe needed as adjuvants to praziquantel in neuroschistoso-miasis (Ross *et al.* 2012). Corticosteroids (e.g., prednisone 1.5–2.0 mg/kg per day for three weeks) are used to treat Katayama syndrome within two months of freshwater contact (Table 23.1) and for the treatment of schistosomal encephalopathy during the oviposition (egg laying) stage (Ross *et al.* 2007, 2012). Corticosteroids help to alleviate acute allergic reactions and prevent mass effects caused by excessive granulomatous inflammation in the central nervous system. Anticonvulsants are used to treat seizures associated with cerebral schistosomiasis but lifelong use is rarely indicated (Ross *et al.* 2012). Phenobarbital and phenytoin are probably the most readily available anticonvul-sants in regions where schistosomiasis is endemic, but they are both known to markedly increase the metabolism and clearance of PZQ (Ross *et al.* 2012). If long-term anticonvulsant therapy is required in conjunction with PZQ, consideration should be given to the use of a non-hepatically metabolized anticonvulsant such as levetiracetam (Keppra) or agents that have less hepatic interaction with PZQ such as sodium valproate or carbamazepine (Ross *et al.* 2012).

TABLE 23.1 Treatment and prophylaxis for schistosomiasis.

Drug	Dosage regimen	Use in special groups			Main contraindication	Comments
		Pregnancy	Breast feeding	Children		
Praziquantel	*S. haematobium, S. intercalatum, S. mansoni, S. japonicum*, 40 mg/kg per day, two doses a day by mouth; *S. japonicum, S. mekongi* 60 mg/kg per day, three doses a day by mouth	Category B: usually safe, but benefits must outweigh risk	Discontinue breast feeding on day of treatment for up to 72 hours	Safe for age >4; not established for age <4	Hypersensitivity, ocular cysticercosis	Used to treat all schistosomes. Usually well tolerated. Caution while driving or performing other tasks requiring alertness on the day of treatment or the day after
Artesunate	*S. haematobium, S. mansoni, S. intercalatum, S. japonicum, S. mekongi*; prophylaxis: 6 mg/kg every 2–4 weeks by mouth	No data, not recommended	No data, not recommended	Safe	Hypersensitivity	Prophylactic against *S. haematobium, S. mansoni, S. japonicum*. Antischistosomula properties
Corticosteroids (prednisone)	Adult: 1.5–2.0 mg/kg per day by mouth for 3 weeks. Paediatric: 0.05–2.0 mg/kg per day, three doses a day by mouth	Category C: unknown effect, not recommended	Unknown effect, not recommended	Safe	Immunosuppressant	Used for treatment of Katayama syndrome (within 2 months of water contact) and treatment for sequelae of neuroschistosomiasis
Anticonvulsants	Consult neurologist	Category D: not recommended	Benefits must outweigh risk	Safe	Hypersensitivity	Treatment of seizures associated with neuroschistosomiasis

23.5 ANIMAL TREATMENT

Periodic treatment of bovines with PZQ, managed by local veterinarians, is often included in control programs for *S. japonicum* in China and is now being introduced in The Philippines. This is because they are large animals (water buffaloes often exceed 500 kg in weight), they deposit considerable amounts of excreta near or in surface water, they live for 10 to 12 years and they can carry large numbers of schistosomes and develop hepato-intestinal disease. In rice fields, where *S. japonicum* is endemic in The Philippines, up to 90% of water buffaloes and cattle are infected (Gordon *et al.* 2015a, b). PZQ is curative, but the animals become reinfected. In most areas of endemicity, a large proportion (80%) of the environmental contamination may be traced back to bovine defecation and a high proportion of schistosome eggs (Ross *et al.* 2001). The treatment doses (e.g., 25 mg/kg for water buffaloes; and 30 mg/kg for cattle) are lower than that used for humans because the collateral effects may be fatal, which is thought to result from portal occlusion by dead worms in heavily infected animals, as noted in other forms of cattle schistosomiasis (Ross *et al.* 2001). Older buffaloes tend to excrete fewer viable eggs than those less than 18 months old; this could reflect self-cure, decreased egg viability or decreased egg production by female worms and all have been described for other forms of bovine schistosomiasis (Ross *et al.* 2001).

23.6 CHAPTER SUMMARY

Praziquantel is the foremost-prescribed drug and is quite effective against the adult worms of all *Schistosoma* species if given at the correct dosage (e.g., 60 mg/kg). A derivative of the antimalarial drug, artemisinin, has recently been studied for the control of acute schistosomal infections. The addition of this drug to the treatment of schistosomiasis holds the potential for increasing the cure rate for acute schistosomiasis and possible preventing infection in travelers. However, to date no clinical trial has examined the role of ART alone or in combination with corticosteroids for the treatment of neurological symptoms complicating Katayama syndrome. The treatment of bovines within integrated control programs in Asia has commenced and it is hoped that this will lead to the elimination of the disease in China and The Philippines.

Preventive chemotherapy with 40 mg/kg of praziquantel has been endorsed and advocated by WHO for the global control of schistosomiasis, yet the drug does not prevent reinfection. While the global use of praziquantel is being scaled up, concerns are also growing about inadequate drug coverage, low cure rates, poor drug compliance and the potential for development of drug resistance. Despite the commitment to mass drug administration programs, it is becoming increasingly clear that the sustainable control of schistosomiasis will require an integrated and intersectorial approach that goes beyond deworming.

23.7 LITERATURE CITED

Ahier A, Khayath N, Vicogne J, Dissous C. 2008. Insulin receptors and glucose uptake in the human parasite *Schistosoma mansoni*. Parasite 15: 573–579.

Ahier A, Rondard P, Gouignard N, Khayath N, Huang S, Trolet J, Donoghue DJ, Gauthier M, Pin JP, Dissous C. 2009. A new family of receptor tyrosine kinases with a venus flytrap binding domain in insects, other invertebrates activated by aminoacids. PLoS One 4: e5651.

Aragon AD, Imani RA, Blackburn VR, Cupit PM, Melman SD, Goronga T, Webb T, Loker ES, Cunningham C. 2009. Towards an understanding of the mechanism of action of praziquantel. Molecular Biochemistry Parasitology 164: 57–65.

Bahia D, Oliveira LM, Mortara RA. 2009. Phosphatidylinositol and related-kinases: a genome-wide survey of classes and subtypes in the *Schistosoma mansoni* genome for designing subtype-specific inhibitors. Biochemical and Biophysical Research Communications 380: 525–530.

Doenhoff MJ, Pica-Mattoccia L. 2006. Praziquantel for the treatment of schistosomiasis: its use for control in areas with endemic disease and prospects for drug resistance. Expert Review of Anti-infective Therapy 4: 199–210.

Doenhoff MJ, Cioli D, Utzinger J. 2008. Praziquantel: mechanisms of action, resistance and new derivatives for schistosomiasis. Current Opinion Infectious Diseases 21: 659–667.

Gordon CA, Acosta LP, Gobert GN, Jiz M, Olveda RM, Ross AG, Gray DJ, Williams GM, Harn D, Li Y, McManus DP. 2015a. High prevalence of *Schistosoma japonicum* and *Fasciola gigantica* using molecular diagnostics in bovines from Northern Samar, the Philippines. PLoS Neglected Tropical Diseases 9(2): e0003108. doi: 10.1371/journal.pntd.0003108. eCollection 2015 Feb.

Gordon CA, Acosta LP, Gobert GN, Olveda RM, Ross AG, Williams GM, Gray DJ, Harn D, Li Y, McManus DP. 2015b. Real-time PCR demonstrates high human prevalence of *Schistosoma japonicum* in the Philippines: implications for surveillance and control. PLoS Neglected Tropical Diseases 9(1): e0003483. doi: 10.1371/journal.pntd.0003483. eCollection 2015 Jan.

Gray DJ, McManus DP, Li Y, Williams GM, Bergquist R, Ross AG. 2010. Schistosomiasis elimination: lessons from the past guide the future. Lancet Infectious Diseases 10: 733–736.

Gryseels B, Polman K, Clerinx J, Kesten L. 2006. Human schistosomiasis. Lancet 368: 1106–1118.

Kumagai T, Osada Y, Ohta N, Kanazawa T. 2009. Peroxiredoxin-1 from *Schistosoma japonicum* functions as a scavenger against hydrogen peroxide but not nitric oxide. Molecular Biochemistry Parasitology 164: 26–31.

Liu R, Dong HF, Guo Y, Zhao QP, Jiang MS. 2011. Efficacy of praziquantel and artemisinin derivatives for the treatment and prevention of human schistosomiasis: a systematic review and meta-analysis. Parasites and Vectors 4: 201.

Rabello A. 1997. Diagnosing schistosomiasis. Memórias do Instituto Oswaldo Cruz 92: 669–676.

Ross AG, Bartley PB, Sleigh AC, Olds GR, Yuesheng Li, Williams G, McManus DP. 2002. Schistosomiasis. New England Journal Medicine 346: 1212–1220.

Ross AG, Sleigh AC, Li Y, Davis GM, Williams GM, Jiang Z, Feng Z, McManus DP. 2001. Schistosomiasis in the People's Republic of China: prospects and challenges for the 21st century. Clinical Microbiology Reviews 14: 270–295.

Ross AG, Vickers D, Olds GR, Shah S, McManus DP. 2007. Katayama syndrome. Lancet Infectious Diseases 7(3): 218–24.

Ross AG, Hou X, Farrar J, Huntsman R, Gray DJ, McManus DP, Li YS. 2012. Neuroshistosomiasis. Journal of Neurology 259(1): 22–32. Epub 2011 Jun 15.

Shrivastava J, Qian BZ, Mcvean G, Webster JP. 2005. An insight into the genetic variation of *Schistosoma japonicum* in mainland China using DNA microsatellite markers. Molecular Ecology 4: 839–849.

Ting-An W, Hong-Xiang Z. 2009. PTK-pathways and TGF-beta signaling pathways in schistosomes. Journal Basic Microbiology 49: 25–31.

Utzinger J, Xiao SH, Tanner M, Keiser J. 2007. Artemisinins for schistosomiasis and beyond. Current Opinion in Investigational Drugs 8: 105–116.

Xiao SH. 2005. Development of antischistosomal drugs in China, with particular consideration to praziquantel and the artemesinins. Acta Tropica 96: 153–167.

Zhou XN, Wang YL, Chen MG, Wu XH, Jiang QW, Chen XY, Zheng J, Utzinger J. 2005. The public health significance and control of schistosomiasis in China – then and now. Acta Tropica 96: 97–105.

CHAPTER

Schistosomiasis Vaccine Development: The Missing Link

Robert Bergquist[1,] * and *Donald P McManus*[2]

24.1 INTRODUCTION

Schistosomiasis vaccine development represents an uphill battle, first due to the complicated immunology involved when the human host balances between resisting new infections and avoiding serious pathology when reacting against eggs (from previous infections) trapped in the tissues; and currently because of limited funding resulting from not reaching the goal with sufficient speed. It is indeed fortunate that most of the research groups involved in schistosomiasis vaccine development hold a steady tack and resist abandoning ship in midcourse, a tactic now supported by the Roadmap advanced by the World Health Organization (WHO 2012a) outlining how, when and where the neglected tropical diseases (NTDs) shall be conquered (WHO 2012). The historic meeting on 3 February 2012 in London, UK provided new hope by full endorsement of this Roadmap to eradicate, eliminate and intensify the control of schistosomiasis and 16 other NTDs by major governments and representatives of large pharmaceutical companies (WHO 2012b).

The worldwide reduction of schistosomiasis morbidity achieved through sustained praziquantel treatment has led to a strong emphasis on chemotherapy as the sole agent to achieve elimination of the disease. However, this progress notwithstanding, schistosomiasis prevalence remains largely unaffected as rapid reinfection quickly restores the prevailing levels of infection after each scheduled mass drug administration (MDA) intervention. Only in a minority of endemic areas, notably in the People's Republic of China, has a clear reduction of prevalence been achieved and this outcome must be attributed to effective snail control and other measures, as much as to chemotherapy, which emphasizes the need for complementary approaches. However, while snail control is the obvious adjunct in China, the snail species that play the role of intermediate host outside Southeast Asia are non-amphibious and therefore considerably more difficult to control. Another tentative supplement would be a vaccine making up for the short-term effect of chemotherapy. The development and positioning of an effective vaccine within the spectrum of disease control approaches is an option that would add a much needed long-term attribute to be applied after

[1] UNICEF/UNDP/World Bank/WHO Special Programme for Research and Training in Tropical Diseases (TDR), World Health Organization, Geneva, Switzerland.
[2] QIMR Berghofer Medical Research Institute, Brisbane, QLD, Australia.
* Corresponding author

MDA has cleared the deck. Current progress is promising, not only due to novel adjuvants that can selectively manipulate the immune responses, but also because immunological research makes it possible to assess the specific responses each antigen elicits (and what it needs to elicit) through the study of cell signaling. Supported by the introduction of molecular cloning (Cohen *et al.* 1973) and based on immunological studies of schistosomiasis infection, both in the field and the laboratory (Butterworth and Hagan 1987; Wilson and Coulson 1998; Bergquist and Colley 1998), the quest for a schistosomiasis vaccine became a priority in the final decades of the last century. Although persistent efforts to develop such a vaccine continue, the initially adequate allocation of funding was not sustained, which has limited the possibilities to rapidly reach the goal.

In order to successfully relocate between various, completely different environments with the aim of ending up in the human host to live there for years (sometimes decades), schistosomes have evolved intricate survival strategies (Jenkins *et al.* 2005). This makes the development of an effective schistosomiasis vaccine difficult, but the fact the parasite cannot replicate in its definitive host provides some respite. Thus, in contrast to infections due to viruses and bacteria, even a partial reduction in parasite burden should be useful.

It has been argued that immune responses directed against the adult worm might not be the major mechanism involved in parasite obliteration (Pearce *et al.* 1990). If this presumption is correct, the short interval between cercarial skin penetration and the presence of schistosomula in the lungs about 72 hours later would be the time when the parasite should be the most vulnerable for immune attack. This susceptibility is likely to at least partly spill over into the juvenile schistosome in the liver sinusoids by this which time the parasite is already resistant to antibody-dependent cellular cytotoxicity (ADCC) controlled by complement (El Ridi and Tallima 2009).

Animal experiments carried out over the last 50 years have provided a good understanding of parasite-host interaction. However, although the protective human immune responses to schistosome infection might be similar to what goes on in the experimental animal, important defensive mechanisms might be different, while population studies in endemic areas cannot provide all the data needed. It might therefore be useful to also carry out testing in non-human primates, even if only to strengthen the safety aspect of vaccination strategies. Vaccines based on studies performed only in the mouse model could even have undesirable effects if taken to human clinical trials prematurely (Lebens *et al.* 2004; Siddiqui *et al.* 2011). Furthermore, it has been speculated that schistosome vaccination trials in mice may be flawed, resulting in inaccurate protective efficacy data (Wilson 2016). In addition, there are obstacles to exploiting gathered information directly. For example, even if irradiated cercariae or schistosomula treated with artemether induce unprecedented protection experimentally (Mangold and Dean 1984; Bergquist *et al.* 2004; Fukushige *et al.* 2015), it cannot be realized for human use since such an attenuated vaccine would likely carry too high a risk for side effects. On the other hand, these findings motivate the search for the key antigens simulating these responses. A study on mice vaccinated with *Schistosoma mansoni* irradiated cercariae, has already identified a group of five such antigens, i.e. paramyosin, heat shock protein 70 (HSP-70), a 23 kDa integral membrane protein, triosephosphateisomerase (TPI) and glutathione S-transferase (GST) (Richter *et al.* 1993). The search for defined schistosomiasis vaccine candidates of this kind has intensified, resulting eventually in a large number of putative, protective antigens. This work followed a step-by-step approach that included identification of protective antigens, selection of the most promising ones, cloning them and finally making sure that they could be reproduced at large scale.

The rationale for schistosomiasis vaccine development is summarized in Table 24.1. However, two overarching themes missing from the table need to be borne in mind: 1). No vaccine has yet been accepted for public use; and 2). Chemotherapy followed by vaccination would integrate short-term effect with long-term protection.

TABLE 24.1 Key points in the ongoing discussion of the possibilities for elimination of schistosomiasis.

| Chemotherapy | | Supporting evidence for the realization of an effective schistosomiasis vaccine | Vaccination | |
Advantages	Drawbacks		Advantages	Drawbacks
Chemotherapy with praziquantel safe, effective and affordable	Transmission unaffected with rapid reinfection calling for swift re-treatment	Partial immunity against schistosome infection acquired naturally in endemic areas (Butterworth *et al.* 1992) High-level protection consistently realized with irradiated cercariae (Mangold and Dean 1984), which should be possible to emulate	Long-term effect	Cold-chain delivery needed. Possible adverse side effects
Infrastructure for routine diagnostics and repeated MDA interventions in place in many areas Current control programs successful	Most African endemic areas still outside regular treatment Drug compliance often suboptimal Drug-resistance a risk due to expanded drug delivery	Vaccines repeatedly proved cost-effective for long-term control of infectious diseases (Pickering *et al.* 2009) Veterinary vaccines against *Taeniasolium* (Gauci *et al.* 2013) and ovine cystic echinococcosis (Larrieu *et al.* 2013) shown high efficacy in the field	Absence of schistosome replication in the definitive (human) host makes partial effect acceptable Inhibition of the risk for rebound morbidity (Olveda *et al.* 1996)	Low-level infection not acceptable as due to increased adverse effects in the long term

Pre clinical work in experimental animal models and human population studies have focused on the investigation of immune mechanisms. The role of macrophages and T helper cell sub-populations in human schistosomiasis has been intensively studied and show that T_H1 predominantly activates the cellular arm of the immune system through interferon-gamma (IFNγ) and IL-2, whilst T_H2 stimulation is characterized by antibody production in association with IL-4 and IL-5 production, including immediate hypersensitivity (mediated by IgE) and eosinophilia typical of helminth infections. Indeed, high levels of specific IgE antibodies correlate strongly with post-treatment resistance (Hagan *et al.* 1991; Dunne *et al.* 1995; Walter *et al.* 2006). Lately, however, it has been argued that this mechanism should be avoided due to the risk of anaphylactic shock.

The immune response towards schistosomes can also be separated differently: 1) down-regulation of acute responses against released antigens from eggs trapped in tissues, which are the cause of granuloma formation, collagen deposition and in intestinal schistosomiasis, hepatic periportal fibrosis, chronic inflammation and anemia; and 2) age-dependent, concomitant immunity against re-infection resulting over time from repeated, natural parasite death leading to partial, natural immunity. This partial protection has been associated with increased eosinophilia, $CD23^+$ B cells, interleukin (IL)-5 and IgE antibodies directed against adult worm antigens together with low levels of IgG_4 antibodies against these worm components (Mo *et al.* 2014). These two types of immune responses are, however, not completely unrelated as some egg antigens also appear in adult worms and schistosomula, so the result is a balance between regulation and protection against new infections.

24.2 CYTOKINE ACTIVATION AND INTERPLAY

The immune responses during the first months post-infection are largely of the $CD4^+$ T_H1 type associated with macrophages activated to produce IL-6, IL-12, tumor necrosis factor-alpha (TNFα) and nitric oxide (NO) but shift later to a T_H2 response, which is firmly in place 3 months post-infection (Hesse *et al.* 2000; Pearce *et al.* 2004). The CD14 receptor has an important role in this connection as it recognizes a wide range of pathogen products and can initiate dependent and independent signaling involving toll-like receptors (TLR), which suggests that it may function as a regulator of macrophage responses as well as of the T_H1/T_H2 balance (Tundup *et al.* 2014). It seems that IL-10 plays a major role in the overall regulation of the immune response to schistosome infection (Corrêa-Oliveira *et al.* 1998; Bustinduy *et al.* 2015; Sanin *et al.* 2015), but more studies regarding this and the many other interleukins involved are needed, in particular with reference to the interaction between them, before a more complete understanding can emerge.

Apart from having been shown to be without adverse health effects, the outcome of the two vaccine candidates now in clinical trials, i.e. the recombinant 14 kDa fatty acid-binding protein (FABP) from *S. mansoni* (rSm14) and the recombinant 28 kDa glutathione S-transferase of *S. haematobium* (rSh28GST), underline the importance of the T_H1/T_H2 balance discussed above. The rSh28-GST protein, given together with alum, induced mainly a T_H2 response. Six months after the start of the trial, a significant increase in the specific antibody response including all IgG isotypes, IgE excepted, was recorded. Although there was an increase in the mean values of IL-2, IL-5, IL 13 and INFγ together with slight decreases of IL-10 and the transforming growth factor β (TGF) $_β$, 7 weeks after the start of the trial, the overall cytokine measurements did not provide a clear lead except for the significant increase of IL-5 and IL-13 as an indication of the antibody response seen (Riveau *et al.* 2012). The rSm14-FABP protein, given together with the adjuvant GLA-SE, a T_H1-promotor consisting of glucopyranosyl lipid A in a squalene-induced oil-in-water emulsion, resulted in a mixed T_H1/T_H2 response with a build-up of TNFα, IL-2 and IL-10 and a strong, specific IgG_{1-3} response without any increase in IgE (Santini-Oliveira *et al.* 2015).

24.3   ADJUVANTS

New types of adjuvants have become available for use in laboratory animals in recent years with some accepted also for in human Phase 1 trials (Lousada-Dietrich *et al.* 2011; Treanor *et al.* 2013; Santini-Oliveira *et al.* 2015). Adjuvants constitute an important issue with respect to parasite vaccine development as the ultrapure antigens of the types discussed here are generally weak immunogens by themselves. Small antigens, polypeptides in particular, need to be conjugated to an immunogenic carrier protein to produce acceptable immune responses. It must also be borne in mind that too much or too small amounts of antigen may induce tolerance rather than an active response, so there is a 'window of immunogenicity' that must be targeted (Hanly *et al.* 1995). This is not the place to discuss this area in detail and the reader is referred to an excellent general review of the field by Petrovsky and Aguilar (2004) and one especially concerned with schistosomiasis vaccines (Stephenson *et al.* 2014). Suffice it to say that aluminum salts (alum), the only widely used adjuvant in human vaccines, induces a good antibody (T_H2) response. This adjuvant is useful for vaccines against viral and some bacterial infections, but shows little cellular immune activation (T_H1), which may be the most important defense against schistosome infection.

To sustain an antibody response, a continual or intermittent supply of antigen is needed. This can be achieved either by sustained antigen release from an antigen depot at the injection site or through booster doses, where smaller booster doses, typically around half the priming dose, are favored because they preferably stimulate higher affinity clones of B cells (Hanly *et al.* 1994). Another approach is to formulate the antigen in emulsified micro-droplets with a stabilizing surfactant, which are transported to the draining lymph nodes by macrophages where the antigen is trapped by follicular dendritic cells where most of the necessary cell to cell interactions take place (Howard *et al.* 2004).

Many bacteria contain substances that activate macrophages as well as T and B cells. Thus, bacterial derivatives have been modified to maximize their desirable potential, while minimizing inflammatory reactions, e.g., some of the new generation adjuvants incorporate a chemical variant of endotoxin called monophosphoryl lipid A (MPL), a family of endotoxic TLR or bacterial cell wall constituents that can be formulated with liposomes, oil emulsions or aluminum salts (Baker *et al.* 1988; Stephenson *et al.* 2014). Even if combinations of antigens and adjuvants with particulate lipid or oil components can have properties which produce different responses in different animals, this approach has been successful in the production of new delivery techniques that include combinations of strategies and delivery mechanisms for uniquely formulating antigens and adjuvants (Alving *et al.* 2012). Following this thinking, a synthetic MPL has been developed and used in attempts to develop other anti-parasite vaccines, i.e. against malaria (Lousada-Dietrich *et al.* 2011) and *Leishmania* (Bertholet *et al.* 2009). Glucopyranosyl lipid A (GLA), shown to promote T_H1 responses leading to the differentiation of IFNγ-secreting T_H1 CD4$^+$ T cells (Coler *et al.* 2010), was formulated together with squalene to produce an oil-in-water emulsion (SE) that was approved for human use in a Phase I trial of an influenza vaccine (Treanor *et al.* 2013). This same formulation was eventually used for the formulation of the Sm14-FABP schistosomiasis vaccine (Santini-Oliveira *et al.* 2015).

The recent, increased use of recombinant, subunit and synthetic vaccines has made the need for better adjuvants acute. Major barriers in the way of new adjuvants are not just due to lack of knowledge how to produce them, but to a great part the possibility of potential toxicity and other unacceptable, adverse side-effects. Further, it does not seem possible for adjuvants to be approved as products in their own right (Petrovsky and Aguilar 2004), but only as part of a vaccine combination, which makes the whole field of adjuvant research *per se* precarious. Finally, the cost of developing a new adjuvant is becoming almost prohibitive, particularly as each antigen can only be registered as part of a specific combination. Against this background, it is hoped that the new products referred to above represent breakthroughs to new and useful perspectives. Above all, the acceptance for Phase I human trials of several vaccine products that include synthetic MPL provides reassurance that new multi-active adjuvants with specific, desired characteristics will appear.

24.4 CONCERTED ACTION

In the early 1990s, the UNICEF/UNDP/World Bank/WHO Special Programme for Research and Training in Tropical Diseases (TDR) instituted a program aimed at bridging the gap between pure research laboratories and the pharmacological industry relying on:

1. Independent testing of vaccine candidates in standard mouse model;
2. Study of correlate human immune reactions in endemic areas;
3. Exploration of adjuvants and antigen stability;
4. Industrial vaccine production; and eventually;
5. Clinical trials.

24.4.1 Independent Laboratory Testing

Two laboratories with long-term experience in experimental schistosomiasis were contracted to carry out parallel independent tests in C57BL6 and BALB/c mice for each vaccine candidate. The mice were immunized according to the specifications of the investigators providing the antigens, while a standardized protocol using the NMRI Puerto Rican strain of *S. mansoni* was followed for the challenges. The mice received 150 cercariae percutaneously by a one-hour exposure of the tail or abdomen with perfusions carried out six weeks later.

The original panel of antigens subjected to independent testing included six *S. mansoni* antigens (Table 24.2) that were reported to consistently deliver high levels of protection in experimental animals. The researchers representing these antigens were invited to send their antigen preparations to TDR to be forwarded to the two independent laboratories for parallel testing in the standardized mouse model agreed on. Reflecting the schistosome species, where from they had originated and the molecular weights, they were referred to Sm14-FABP, Sm23, Sm28-TPI, Sm28-GST, Sm62-IrV5 and Sm97-paramyosin.

TABLE 24.2 Overview of the original panel of vaccine candidates.

Antigen	Identity	Expression	Protection	Reference*
Sm14-FABP	Recombinant 14 kDa fatty acid-binding protein	From schistosomula to adult	65%	Moser *et al.* 1991; Tendler *et al.* 1995; Tendler *et al.* 1996
Sm23	23 kDa integral membrane protein tested as MAP**	All stages	40–50%	Reynolds *et al.* 1992; Richter *et al.* 1993; Köster *et al.* 1993
Sm28-GST	Recombinant glutathione S-transferase	From schistosomula to egg	30–60%	Boulanger *et al.* 1991; Trottein *et al.* 1992; Capron *et al.* 1994
Sm28-TPI	Triose phosphatase isomerase tested as MAP**	From schistosomula to egg	30–60%	Harn *et al.* 1992; Richter *et al.* 1993; Reynolds *et al.* 1994
Sm62-IrV5	Recombinant 62 kDa myosin from γ-irradiated cercariae	From schistosomula to egg	50–95%	Soisson *et al.* 1992; Hawn *et al.* 1993
Sm97-paramyosin	Native 97 kDa myofibrillar protein	From schistosomula to adult	30%	Lanar *et al.* 1986; Pearce *et al.* 1988; Flanigan *et al.* 1989

*Available at the time of the independent testing project;
**Synthetic loops of multiple antigen peptides.

Unexpectedly, none of the antigen preparations investigated reached the stated goal of consistent induction of 40% protection or better. In spite of this disappointing result, it was decided to continue the program as planned including investigation of the antigens for association with resistance versus susceptibility to re-infection (the human correlate study) in Brazil, Egypt and The Philippines. It was felt that instability of the vaccine candidate formulations, possibly accentuated by shipping under less than ideal conditions, had contributed to the unsatisfactory results and it was concluded that antigen presentation plays a more important role than previously appreciated (Bergquist 1998; Bergquist and Colley 1998). At the Egyptian site, a cohort of 225 long-term residents of an area endemic for *S. mansoni*, for whom historical data collected over the previous five years were available, was followed for an additional 6 months after treatment with praziquantel. The frequency of re-infection, cytokine spectrum and immune responses was recorded for each individual and antigen, resulting in precise correlations, both prospectively and retrospectively, with detailed epidemiological data covering a 66-month period (Al-Sherbiny *et al.* 2003). Markers for resistance and susceptibility to re-infection were identified in the search for the types of response to aim for and which ones to avoid. The results were much better than could have been assumed after the prior independent testing. Still, however, few antigens produced clear-cut results; e.g., a single antigen might have a significant positive correlation with INFγ production but also show a significant negative correlation with respect to the specific IgG_1 titer induced. The other two correlate studies (Ribeiro de Jesus *et al.* 2000; Acosta *et al.* 2002) showed similar results. These observations suggest that there were desirable, specific immune responses that were generated by the vaccine candidates, while indicating that there were also responses that must be avoided. Even if conclusive results were not achieved, the insights gained must be deemed useful for future antigen selection as well as for vaccine formulation.

With improved reagents and new antigens becoming available in the two decades since the first independent testing, a new human correlate study for the current frontrunners is urgently needed. A very recent study, based on a protein microarray containing schistosome proteins, moved in this direction (Pearson *et al.* 2015). Attempting to identify targets of protective IgG$_1$ immune responses in *S. haematobium*-exposed individuals that acquired praziquantel-induced resistance to schistosomiasis long-term treatment found numerous antigens with known vaccine potential, such as Smp80, tetraspanins and GSTs. Interestingly, corresponding IgG$_1$ responses were not elevated in exposed individuals who did not acquire praziquantel-induced resistance.

24.5 NEW CARDS ON THE TABLE

A majority of the 'old' antigens are still 'in the running', others are less 'hot' now, while a few 'newcomers' have been added to the list of promising antigens (Tables 24.2 and 24.3). To date, three vaccines candidates have made it into clinical trials: Sm14 and SmTSP-2 (Merrifield *et al.* 2016) against S. *mansoni* and Sh28GST against S. *haematobium* (although the GST antigen in the original panel was a *S. mansoni* construct, the final formulation is a *S. haematobium* vaccine candidate – Bilhvax). Two new candidates have appeared, i.e. SmP-80 (calpain) and SmTSP-2. The former is a calcium-dependent cysteine protease, found in the tegument and underlying musculature of *S. mansoni*, while the latter is a tetraspanin. SmP-80 supports schistosome immune evasion by recycling membrane components and the vaccine elicits a predominantly T_H1 response with strong anti-fecundity action (Siddiqui *et al.* 1993; Ahmad *et al.* 2010; Zhang *et al.* 2014). The tetraspanins (TSPs) belong to a super family of transmembrane proteins found in all multicellular eukaryotes (Hemler 2005). They are essential for tegument formation and consist of cell-surface proteins spanning the membrane four times forming two extracellular loops in the process (Sotillo *et al.* 2015).

Many tetraspanins have been found to induce protective immunity in vaccinated mice and constitute the target in naturally resistant human subjects, the most promising being SmTSP-2 (Tran *et al.* 2006, 2010; Sotillo *et al.* 2015). The target of the TSP-2 molecule comprises a 9 kDa extracellular domain corresponding to the part of an extracellular loop localized on the apical membrane of the tegument (Curti *et al.* 2015), a TSP-2 component that appears on the schistosomulum surface from day 5 as part of their development into the adult worm (Tran *et al.* 2010).

Paramyosin, a 97 kDa protein, one of the original *S. mansoni* vaccine candidates (James *et al.* 1996), has a corresponding *S. japonicum* counterpart (Sj97) that has recently been shown to induce 62–86% protection in mice without adjuvant (Jiz *et al.* 2015). These data suggest that this antigen represents a most promising candidate vaccine. It is a constituent of the parasite tegument and has also been found in the acetabular glands of schistosomula where an exogenous part of the antigen inhibits activation of the terminal pathway of complement, suggesting a key involvement in host immune evasion (Gobert and McManus 2005). Other important vaccine candidates are Sj26-GST and Sj28-GST, which have shown encouraging protective efficacy in different mammalian hosts (You and McManus 2015).

The current emphasis on genomic, transcriptomic and proteomic information has given important insights into the highly adapted relationship between schistosomes and their mammalian hosts. For example, the knowledge that the schistosome cannot synthesize glucose (Clemens and Basch 1989) or insulin (Hu *et al.* 2003), promoted the search for a schistosome insulin receptor, which was eventually identified in *S. mansoni* by Ahier *et al.* (2008) and in *S. japonicum* by Zhou *et al.* (2009). This receptor is now a most promising vaccine target, particularly in the *S. japonicum* field, where two types of insulin receptors (SjIR-1 and SjIR-2) that can bind human insulin have been isolated (You *et al.* 2010). Antibodies to the recombined protein of a ligand domain (LD) of *S. japonicum* SjIR-2 have been shown to confer stunting of adult worms and a highly significant reduction of the number of excreted, mature intestinal eggs in a murine vaccine/challenge model as well as lower density of liver granulomas in the definitive host (You *et al.* 2010, 2012, 2015).

TABLE 24.3 Overview of some current priority vaccine candidates for human application.

Vaccine identity	Location and role in parasite	Mechanisms involved	Protection level	Vaccination approach	Human immune responses
Sm14-FABP, a 14 kDa recombinant protein (Moser *et al.* 1991; Tendler *et al.* 1996; Tendler and Simpson 2008; Santini-Oliveira *et al.* 2015)	Throughout cytoplasmic matrix in gut and tegument (Brito *et al.* 2002) Needed for scavenging fatty acids from the host (Brito *et al.* 2002; Angelucci *et al.* 2004) May have a role in immune evasion	Attrition due to blockage of uptake of arachidonic, oleic and other fatty acids (Angelucci *et al.* 2004) Detailed immune mechanism unclear	37–50% reduction of worm burdens (Ribeiro *et al.* 2002)	Intramuscular injection of antigen with GLA-SE, a T_H1-promotor consisting of glucopyranosyl lipid A in a squalene-induced oil-in-water emulsion (Behzad *et al.* 2012)	Phase I trial in Brazil in adults, showed a mixed T_H1/T_H2 response with build-up of TNFα, IL-2 and IL-10. Strong specific IgG_{1-3} response without any IgE. No serious adverse events (Santini-Oliveira *et al.* 2015)
Sh28-GST (Bilhvax), a 28 kDa recombinant protein (Capron *et al.* 2002; Riveau *et al.* 2012)	Parenchyma, genital organs and tegument. Detoxifying enzyme (Smith *et al.* 1986)	Various but importantly antibody-dependent cellular cytotoxicity (ADCC) and anti-fecundity involving IgG, IgA and IgE antibodies (Grezel *et al.* 1993; Riveau *et al.* 1998; Capron *et al.* 2002)	About 40% reduction of worm burden in baboons, 40–50% in mice and 50–70% in rats (Riveau *et al.* 1998)	Intramuscular injection of antigen with alum as adjuvant	Phase I trials in European adults and Senegalese children showed mainly T_H2 response with specific IgG and IgA + IFN-γ, IL-5 and IL-13. No serious adverse events (Riveau *et al.* 2012). Phase II in Senegal finalized and Phase III initiated.
SmP-80 78 kDa protein subunit of calpain (Hota-Mitchell *et al.* 1997; Karmakar *et al.* 2014)	Calcium-dependent cysteine protease of musculature and surface, syncytial epithelium (Siddiqui *et al.* 1993)	Inhibition of the recycling of membrane components (Siddiqui *et al.* 1993). Mainly T_H1-dependent attrition (Ahmad *et al.* 2010). Strong anti-fecundity action (Mo *et al.* 2014)	About 50% reduction of worm burdens in mice and slightly better in baboons (Ahmad *et al.* 2010; 2011). 100% egg elimination, both in mice and baboons (Mo *et al.* 2014)	Various approaches including recombinant protein and plasmids. As adjuvant resiquimod (R848), alum and GLA-SE have been tried (Ahmad *et al.* 2010; 2011; Zhang *et al.* 2014)	In correlate studies, at-risk children and hyper-exposed adults show varying levels of IgG specific for SmP-80 with no or negligible IgE immunoreactivity (Ahmad *et al.* 2011)
SmTSP-2 9 kDa recombinant part of the extracellular schistosome tetraspanin-2 domain (Curti *et al.* 2013; Sotillo *et al.* 2015)	Crucial role in tegument formation when schistosomula changes into adult worm where it upholds tegument integrity (Tran *et al.* 2010; Sotillo *et al.* 2015)	Antibody-dependent attrition (Tran *et al.* 2006)	57–64 % reduction of worm burdens in mice (Tran *et al.* 2006)	Various approaches including plasmids	TSP-2 is strongly recognized by IgG_1 and IgG_3 but not IgE from naturally resistant individuals but not in chronically infected or 'naive' individuals (Tran *et al.* 2006)

Table 24.3 Contd.

Vaccine identity	Location and role in parasite	Mechanisms involved	Protection level	Vaccination approach	Human immune responses
SjLD-2 Recombinant protein of the ligand domain of *S. japonicum* insulin receptor no. 2 (SjIR-2). (You *et al.* 2010, 2012, 2015)	Schistosome insulin receptors (IR) can utilize host insulin for development and growth. SjIR-2 is located in the male parenchyma and the vitelline tissue of females (You *et al.* 2010)	Antibody-dependent blockage of the function of SjIR-2	Stunting of adult worms, significant reduction in fecal egg excretion and reduction in liver granuloma density in the host (You *et al.* 2012, 2015)	Intraperitoneal injection of mice with rSjLD-2 adjuvanted with quil A	Not investigated
Sj97-paramyosin (Pmy) A 97 kDa myofibrillar protein (Ramirez *et al.* 1996; Jiz *et al.* 2015)	Tegument of schistosomula and adult worm (core protein of myosin filaments in all invertebrate muscle fibers)	T_H2 biased response with a role for IgE (Leenstra *et al.* 2006; Jiz *et al.* 2009) and eosinophill-mediated killing of schistosomula (Kojima *et al.* 1987)	62–86% protection in mice without adjuvant (Jiz *et al.* 2015) 34% in water buffaloes (Mc-Manus *et al.* 2001)	Various – without or with quil A	Not done (primarily aimed for use as transmission-blocking vaccine in cattle and buffaloes)

24.6   COCKTAIL VACCINES

The idea of mixing vaccine antigens is not new. Indeed, it was already attempted at the time of the testing of the first antigen panel 25 years ago (Table 24.1), but the few combinations tried then did not improve protection levels and the results were never published. Similar results were seen when Zhang *et al.* (2001) used a cocktail of DNA plasmids encoding four different *S. japonicum* antigens: Sj14, Sj23, Sj28 and Sj62, respectively. However, when Yuan *et al.* (2007) constructed a DNA vaccine based on Sj14 and Sj23, the bivalent DNA expressed well and the protective immunity was higher than that of a univalent DNA version. In a different approach aimed at manipulating the immune response, a cocktail peptide-DNA dual vaccine candidate based on four *S. japonicum* T_H1-type epitope peptides induced a dominant T_H1-type response in mice, which was shown to down-regulate the development of hepatic fibrosis (Tao *et al.* 2009).

The main thrust of the work by Sotillo *et al.* (2015), referred to above, was a proteomic study of the apical membrane of *S. mansoni* schistosomula. This approach, one of the first of its kind, resulted in more than 450 proteins, 200 of which were shown to be involved in expression profiles at different developmental stages of the parasite. Interestingly, some of the identified molecules were already known vaccine targets, such as Sm-TSP-1, Sm-TSP-2, Sm29 and calpain (P-80). The proteomic approach is bound to provide us with more targets that we can simultaneously handle, reviving the idea of combining them, indeed an approach that is already taking place. Recent work by Pinheiro *et al.* (2014) and Gonçalves de Assis *et al.* (2015) reports partial protection in mice by a multivalent, chimeric vaccine composed of *S. mansoni* SmTSP-2 and Sm29 that produced a wide spectrum of specific immune response. Another recombinant fusion consisting of Sm14 and Sm29 showed 45–48% protection in mice and a significant reduction of tissue egg burdens compared with groups of mice immunized with the individual antigens (Mossallam *et al.* 2015). Da'Dara *et al.* (2008) investigated two DNA vaccines based on the heat shock protein SjHSP70; one fused with a Sj23 plasmid and the other with a Sj28-TPI plasmid. The authors reported a significant immunoprotective effect with both these cocktail vaccines when co-administered together with an IL-12 plasmid in water buffaloes. Cluster randomized intervention trials are currently underway around schistosomiasis-endemic areas in China and The Philippines aimed at determining the impact on schistosome transmission of a multi-component integrated control strategy, including bovine vaccination using a heterologous "prime-boost" delivery platform based on the Sj28-TPI vaccine (Gray *et al.* 2014).

24.7   TRANSMISSION-BLOCKING VACCINES

Since *S. japonicum* has a large range of definitive hosts besides humans, domestic animals play an important part in the continuing endemicity of schistosomiasis in China and The Philippines. For this reason, a transmission-blocking veterinary vaccine is being considered for this part of the world (McManus and Dalton 2006; You *et al.* 2010, 2012, 2015). Studies of re-infection in some groups of animals after treatment with praziquantel suggest that age-related resistance is likely to occur in water buffaloes but not in cattle (Wang *et al.* 2006). Immunological analysis support this view as the number of CD4$^+$ T cellsin cattle, in contrast to water buffaloes, correlate with worm development and a shift from T_H1 to T_H2 type of immunity (Yang *et al.* 2012). Based on the fact that these bovines are major reservoirs involved in the transmission of *S. japonicum*, further studies on their immunology might facilitate the selection of further transmission-blocking vaccine candidates suitable as complement to ongoing chemotherapy and snail control campaigns in China.

Although the immunology of cattle has not been studied as well as that in experimental animals, e.g., mice, previous research has shown that the worm burdens of Chinese can be reduced by 65–76% in bovines by immunization with irradiated schistosomula (Hsu *et al.* 1984). In addition, it has been shown that water buffaloes vaccinated with irradiated *S. japonicum* cercariae gain weight and also develop high resistance to reinfection early in life (Shi *et al.* 1990). However, a recent study has provided new insights on immunity against schistosomiasis in a natural host, such as the water buffalo, e.g., an intense T_H2 type immune response in at the site of cercarial penetration, which differs considerably from that seen in mice (McWilliam *et al.* 2013).

It is now clear that several *S. japonicum* vaccine candidates can induce 50–70% protection in vaccination/challenge experiments using mice as well as larger mammalian species (McManus and Loukas 2008). As further immunizations are likely to boost these levels, further development of these vaccine candidates for veterinary use is a realistic proposition. In an up-to-date review, You and McManus (2015) discuss currently known aspects of the host immune response in schistosomiasis japonica and the current status of vaccine development with respect to *S. japonicum* infections in mammalian hosts, where they give detailed information on more than 20 candidate vaccines. Additional studies on the immunology of buffaloes and cattle represent an important research area that will be vital in the process of selecting *S. japonicum* vaccine antigens and in defining optimum routes of immunization.

24.8   A NEW SYNTHESIS

It has been said that the road towards a successful schistosomiasis vaccine will be long and winding (Bergquist and Colley 1998). It was and it is, but the fact that two vaccine candidates have been accepted as safe for human use SH28-GST (Riveau *et al.* 2012) and Sm14-FABP (Santini-Oliveira *et al.* 2015) marks 'the end of the beginning'. However, it is thanks to WHO's Roadmap (WHO 2012a) and the London declaration of 2012 (WHO 2012b) on the one hand and the impressive progress achieved in China on the other, that the elimination concept has been floated. Two meetings on the theme of schistosomiasis elimination strategy and potential role of a vaccine, co-sponsored by the National Institute of Allergy and Infectious Diseases (NIAID) of the National Institutes of Health (NIH) in the United States and the Bill and Melinda Gates Foundation (BMGF), have recently been held (Mo *et al.* 2014; Mo and Colley 2016). The conclusions of these meetings are that although schistosomiasis elimination in some areas may be achievable through current MDA programs, global control and elimination would require an approach integrated with additional approaches, including vaccines. It was further stressed that clinical evaluation of a schistosomiasis vaccine in the field should be feasible. Some basic product characteristics of two types of vaccine: one for human use and a transmission-blocking vaccine for bovines were proposed to guide early-stage product development.

Together with existing tools, a schistosomiasis vaccine tailored to fit into the overall control program needs would help to achieve and sustain elimination. However, it would need to complement MDA without adding separate implementation costs. Defining a clear product development plan that reflects a vaccine strategy as complementary to the existing control programs to combat different forms of schistosomiasis would be important to develop an effective vaccine. The development risk is decreasing along with the growth of data. Thus, existing information and tools should allow partaking in a clinical efficacy field evaluation of the future role of a schistosomiasis vaccine in existing control programs guided by modeling. Against this background, it should possible to define a target product profile (TPP) that includes both target populations and vaccine characteristics. Indeed, with two vaccine candidates in clinical trials and increasing numbers of other candidates entering into preclinical development, disease elimination should be pursued with a vaccine in the baggage. To achieve this goal, it is time to strengthen collaborative research, continue unraveling the role of cytokines and identify opportunities that would facilitate driving the most promising candidates to the finishing line.

24.9   LITERATURE CITED

Acosta LP, Aligui GD, Tiu WU, McManus DP, Olveda RM. 2002. Immune correlate study on human *Schistosoma japonicum* in a well-defined population in Leyte, Philippines: I. Assessment of 'resistance' versus 'susceptibility' to *S. japonicum* infection. Acta Tropica 84(2): 127–136.

Ahier A, Khayath N, Vicogne J, Dissous C. 2008. Insulin receptors and glucose uptake in the human parasite *Schistosoma mansoni.* Parasite 15(4): 573–579.

Ahmad G, Zhang W, Torben W, Ahrorov A, Damian RT, Wolf RF, White GL, Carey DW, Mwinzi PN, Ganley-Leal L, Kennedy RC, Siddiqui AA. 2011. Preclinical prophylactic efficacy testing of Sm-p80-based vaccine in a nonhuman primate model of *Schistosoma mansoni* infection and immunoglobulin G and E responses to Sm-p80 in human serum samples from an area where schistosomiasis is endemic. Journal of Infectious Diseases 204(9): 1437–1449. doi: 10.1093/infdis/jir545.

Ahmad G, Zhang W, Torben W, Noor Z, Siddiqui AA. 2010. Protective effects of Sm-p80 in the presence of resiquimod as an adjuvant against challenge infection with *Schistosoma mansoni* in mice. International Journal of Infectious Diseases 14: 781–787. doi: 10.1016/j.ijid.2010.02.2266.

Alving CR, Peachman KK, Rao M, Reed SG. 2012. Adjuvants for human vaccines. Current Opinion in Immunology 24(3): 310–315. doi: 10.1016/j.coi.2012.03.008.

Al-Sherbiny M, Osman A, Barakat R, El Morshedy H, Bergquist R, Olds R. 2003. *In vitro* cellular and humoral responses to *Schistosoma mansoni* vaccine candidate antigens. Acta Tropica 88(2): 117–130.

Angelucci F, Johnson KA, Baiocco P, Miele AE, Brunori M, Valle C, Vigorosi F, Troiani AR, Liberti P, Cioli D, Klinkert MQ, Bellelli A. 2004. *Schistosoma mansoni* fatty acid binding protein: specificity and functional control as revealed by crystallographic structure. Biochemistry 43(41): 13000–13011.

Baker PJ, Hiernaux JR, Fauntleroy MB, Stashak PW, Prescott B, Cantrell JL, Rudbach JA. 1988. Ability of monophosphoryl lipid A to augment the antibody response of young mice. Infection and Immunity 56(12): 3064–3066.

Behzad H, Huckriede AL, Haynes L, Gentleman B, Coyle K, Wilschut JC, Kollmann TR, Reed SG, McElhaney JE. 2012. GLA-SE, a synthetic toll-like receptor 4 agonist, enhances T-cell responses to influenza vaccine in older adults. Journal of Infectious Diseases 205(3): 466–473. doi: 10.1093/infdis/jir769.

Bergquist R, Utzinger J, Chollet J, Shu-Hua X, Weiss NA, Tanner M. 2004. Triggering of high-level resistance against *Schistosoma mansoni* reinfection by artemether in the mouse model. American Journal of Tropical Medicine and Hygiene 71(6): 774–777.

Bergquist NR. 1998. Schistosomiasis vaccine development: progress and prospects. Memórias do Instituto Oswaldo Cruz 93 Suppl 1: 95–101.

Bergquist NR, Colley DG. 1998. Schistosomiasis vaccine: research to development. Parasitology Today 14(3): 99–104.

Bertholet S, Goto Y, Carter L, Bhatia A, Howard RF, Carter D, Coler RN, Vedvick TS, Reed SG. 2009. Optimized subunit vaccine protects against experimental leishmaniasis. Vaccine 27: 7036–7040.

Boulanger D, Reid GD, Sturrock RF, Wolowczuk I, Balloul JM, Grezel D, Pierce RJ, Otieno MF, Guerret S, Grimaud JA, et al. 1991. Immunization of mice and baboons with the recombinant Sm28GST affects both worm viability and fecundity after experimental infection with Schistosoma mansoni. Parasite Immunology 13: 473–490. doi: 10.1111/j.1365-3024.1991.tb00545.x.

Brito CF, Oliveira GC, Oliveira SC, Street M, Riengrojpitak S, Wilson RA, Simpson AJ, Correa-Oliveira R. 2002. Sm14 gene expression in different stages of the Schistosoma mansoni life cycle and immunolocalization of the Sm14 protein within the adult worm. Brazilian Journal of Medical and Biological Research 35: 377–381. doi: 10.1590/S0100-879X2002000300014.

Bustinduy AL, Sutherland LJ, Chang-Cojulun A, Malhotra I, DuVall AS, Fairley JK, Mungai PL, Muchiri EM, Mutuku FM, Kitron U, King CH. 2015. Age-stratified profiles of Serum IL-6, IL-10 and TNF-α cytokines among Kenyan Children with Schistosoma haematobium, Plasmodium falciparum and other chronic parasitic co-infections. 2015. American Journal of Tropical Medicine and Hygiene 92(5): 945–951. doi: 10.4269/ajtmh.14-0444.

Butterworth AE, Dunne DW, Fulford AJ, Thorne KJ, Gachuhi K, Ouma JH, Sturrock RF. 1992. Human immunity to Schistosoma mansoni: observations on mechanisms and implications for control. Immunological investigations 21: 391–407.

Butterworth AE, Hagan P. 1987. Immunity in human schistosomiasis. Parasitology Today 3(1): 11–16.

Capron A, Capron M, Riveau G. 2002. Vaccine development against schistosomiasis from concepts to clinical trials. British Medical Bulletin 62: 139–148.

Capron A, Riveau G, Grzych JM, Boulanger D, Capron M, Pierce R. 1994. Development of a vaccine strategy against human and bovine schistosomiasis. Background and update. Tropical and Geographical Medicine 46(4 Spec. No.): 242–246.

Clemens LE, Basch PF. 1989. Schistosoma mansoni: insulin independence. Experimental Parasitology 68(2): 223–229.

Coler RN, Baldwin SL, Shaverdian N, Bertholet S, Reed SJ, Raman VS, et al. 2010. A synthetic adjuvant to enhance and expand immune responses to influenza vaccines. PLoS One 5: e13677.

Cohen SN, Chang ACY, Boyer HW, Helling RB. 1973. Construction of biologically functional bacterial plasmids in vitro. Proceedings of the National Academy of Sciences USA 70(11): 3240–3244.

Corrêa-Oliveira R, Malaquias LC, Falcão PL, Viana IR, Bahia-Oliveira LM, Silveira AM, Fraga LA, Prata A, Coffman RL, Lambertucci JR, et al. 1998. Cytokines as determinants of resistance and pathology in human Schistosoma mansoni infection. Brazilian Journal of Medical and Biological Research 31(1): 171–177.

Curti E, Kwityn C, Zhan B, Gillespie P, Brelsford J, Deumic V, Plieskatt J, Rezende WC, Tsao E, Kalampanayil B, Hotez PJ, Bottazzi ME. 2013. Expression at a 20 L scale and purification of the extracellular domain of the Schistosoma mansoni TSP-2 recombinant protein: a vaccine candidate for human intestinal schistosomiasis. Human Vaccines & Immunotherapeutics 9(11): 2342–2350.

Da'Dara AA, Li YS, Xiong T, Zhou J, Williams GA, McManus DP, Feng Z, Yu XL, Gray DJ, Harn DA. 2008. DNA-based vaccine protects against zoonotic schistosomiasis in water buffalo. Vaccine 26: 3617–3625.

Dunne DW, Hagan P, Abath FG. 1995. Prospects for immunological control of schistosomiasis. Lancet 345: 1488–1491.

El Ridi R, Tallima H. 2009. Schistosoma mansoni ex vivo lung-stage larvae excretory-secretory antigens as vaccine candidates against schistosomiasis. Vaccine 27: 666–673. doi: 10.1016/j.vaccine.2008.11.039.

Flanigan TP, King CH, Lett RR, Nanduri J, Mahmoud AA. 1989. Induction of resistance to Schistosoma mansoni infection in mice by purified parasite paramyosin. Journal of Clinical Investigation 83(3): 1010–1014.

Fukushige M, Mitchell KM, Bourke CD, Woolhouse ME, Mutapi F. 2015. A meta-analysis of experimental studies of attenuated Schistosoma mansoni vaccines in the mouse model. Frontiers in Immunology 6: 85. doi: 10.3389/fimmu.2015.00085. eCollection 2015.

Gauci C, Jayashi C, Lightowlers MW. 2013. Vaccine development against the *Taenia solium* parasite: the role of recombinant protein expression in *Escherichia coli*. Bioengineered 4(5): 343–347. doi: 10.4161/bioe.23003.

Gobert GN, McManus DP. 2005. Update on paramyosin in parasitic worms. Parasitology International 54: 101–107.

Gonçalves de Assis NR, Batistoni de Morais S, Figueiredo BC, Ricci ND, de Almeida LA, da Silva Pinheiro C, Martins Vde P, Oliveira SC. 2015. DNA vaccine encoding the chimeric form of *Schistosoma mansoni* Sm-TSP2 and Sm29 confers partial protection against challenge infection. PLoS One 10(5): e0125075. doi: 10.1371/journal.pone.0125075.

Gray DJ, Li YS, Williams GM, Zhao ZY, Harn DA, Li SM, Ren MY, Feng Z, Guo FY, Guo JG, *et al.* 2014. Multi-component integrated approach for the elimination of schistosomiasis in the Peoples' Republic of China: design and baseline results of a 4-year cluster-randomised intervention trial. International Journal for Parasitology 44: 659–668.

Grezel D, Capron M, Grzych JM, Fontaine J, Lecocq JP, Capron A. 1993. Protective immunity induced in rat schistosomiasis by a single dose of the Sm28GST recombinant antigen: effector mechanisms involving IgE and IgA antibodies. European Journal of Immunology 23: 454–460. doi: 10.1002/eji.1830230223.

Hagan P, Blumenthal UJ, Dunn D, Simpson AJ, Wilkins HA. 1991. Human IgE, IgG4 and resistance to reinfection with *Schistosoma haematobium*. Nature 349: 243–245.

Harn DA, Gu W, Oligino LD, Mitsuyama M, Gebremichael A, Richter D. 1992. A protective monoclonal antibody specifically recognizes and alters the catalytic activity of schistosome triose-phosphate isomerase. Journal of Immunology 148(2): 562–567.

Hanly WC, Bennett BT, Artwohl JE. 1994. Abstract from a series of short articles on adjuvants published in the biologic resources laboratory bulletin (http://www.whale.to/a/adjuvants1.html – accessed 04 January 2016).

Hanly WC, Artwohl JE, Bennett BT. 1995. Review of polyclonal antibody production procedures in mammals and poultry. Journal of the Institute for Laboratory Animal Research (ILAR) 37(3): 93–118.

Hawn TR, Tom TD, Strand M. 1993. Molecular cloning and expression of SmIrV1, a *Schistosoma mansoni* antigen with similarity to calnexin, calreticulin and OvRal1. Journal of Biological Chemistry 268(11): 7692–7698.

Hemler ME. 2005. Tetraspanin functions and associated microdomains. Nature Reviews Molecular Cell Biology 6(10): 801–811. doi: 10.1038/nrm1736.

Hesse M, Cheever AW, Jankovic D, Wynn TA. 2000. NOS-2 mediates the protective anti-inflammatory and antifibrotic effects of the Th1-inducing adjuvant, IL-12, in a Th2 model of granulomatous disease. American Journal of Pathology 157(3): 945–955.

Hota-Mitchell S, Siddiqui AA, Dekaban GA, Smith J, Tognon C, Podesta RB. 1997. Protection against *Schistosoma mansoni* infection with a recombinant baculovirus-expressed subunit of calpain. Vaccine 15(15): 1631–1640.

Howard CJ, Charleston B, Stephens SA, Sopp P, Hope JC. 2004. The role of dendritic cells in shaping the immune response. Animal Health Research Reviews 5(2): 191–195.

Hsu SY, Xu ST, He YX, Shi FH, Shen W, Hsu HF, Osborne JW, Clarke WR. 1984. Vaccination of bovines against schistosomiasis japonica with highly irradiated schistosomula in China. American Journal of Tropical Medicine and Hygiene 33: 891–898.

Hu W, Yan Q, Shen DK, Liu F, Zhu ZD, Song HD, Xu XR, Wang ZJ, Rong YP, Zeng LC, *et al.* 2003. Evolutionary and biomedical implications of a *Schistosoma japonicum* complementary DNA resource. Nature Genetics 35: 139–147.

James SL, Pearce EJ, Sher A. 1987. Prospects for a nonliving vaccine against schistosomiasis based on cell-mediated immune resistance mechanisms. Memórias do Instituto Oswaldo Cruz 82 Suppl 4: 121–123.

Jenkins SJ, Hewitson JP, Jenkins GR, Mountford AP. 2005. Modulation of the host's immune response by schistosome larvae. Parasite Immunology 27: 385–393. doi: 10.1111/j.1365-3024.2005.00789.x.

Jiz M, Friedman JF, Leenstra T, Jarilla B, Pablo A, Langdon G, *et al.* 2009. Immunoglobulin E (IgE) responses to paramyosin predict resistance to reinfection with *Schistosoma japonicum* and are attenuated by IgG4. Infection and Immunity 77: 2051–2058. doi: 10.1128/IAI.00012-09.

Jiz MA, Wu H, Olveda R, Kurtis JD. 2015. Development of paramyosin as a vaccine candidate for schistosomiasis. Frontiers in Immunology 6: 347. doi: 10.3389/fimmu.2015.00347. eCollection 2015.

Karmakar S, Zhang W, Ahmad G, Torben W, Alam MU, Le L, Damian RT, Wolf RF, White GL, Carey DW, Carter D, Reed SG, Siddiqui AA. 2014. Use of an Sm-p80-based therapeutic vaccine to kill established adult schistosome parasites in chronically infected baboons. Journal of Infectious Diseases 209(12): 1929–40. doi: 10.1093/infdis/jiu031.

Kojima S, Niimura M, Kanazawa T. 1987. Production and properties of a mouse monoclonal IgE antibody to *Schistosoma japonicum*. Journal of Immunology 139: 2044-2049.

Köster B, Hall MRT, Strand M. 1993. *Schistosoma mansoni*: immunoreactivity of human sera with the surface antigen Sm23. Experimental Parasitology 77: 282–294.

Lanar DE, Pearce EJ, James SL, Sher A. 1986. Identification of paramyosin as schistosome antigen recognized by intradermally vaccinated mice. Science 31; 234(4776): 593–596.

Larrieu E, Herrero E, Mujica G, Labanchi JL, Araya D, Grizmado C, Calabro A, Talmon G, Ruesta G, Perez A, *et al.* 2013. Pilot field trial of the EG95 vaccine against ovine cystic echinococcosis in Rio Negro, Argentina: early impact and preliminary data. Acta Tropica 127(2): 143–151. doi: 10.1016/j.actatropica. 2013.04.009.

Lebens M, Sun JB, Czerkinsky C, Holmgren J. 2004. Current status and future prospects for a vaccine against schistosomiasis. Expert Review of Vaccines 3(3): 315–328.

Leenstra T, Acosta LP, Wu HW, Langdon GC, Solomon JS, Manalo DL, Su L, Jiz M, Jarilla B, Pablo AO, *et al.* 2006. T-Helper-2 cytokine responses to Sj97 predict resistance to reinfection with *Schistosoma japonicum*. Infection and Immunity 74: 370–381.

Lousada-Dietrich S, Jogdand PS, Jepsen S, Pinto VV, Ditlev SB, Christiansen M, Larsen SO, Fox CB, Raman VS, Howard RF, *et al.* 2011. A synthetic TLR4 agonist formulated in an emulsion enhances humoral and Type 1 cellular responses against GMZ2 – a GLURP-MSP3 fusion protein malaria vaccine candidate. Vaccine 29: 3284–3290.

Mangold BL, Dean DA. 1984. The migration and survival of gamma-irradiated *Schistosoma mansoni* larvae and the duration of host-parasite contact in relation to the induction of resistance in mice. Parasitology 88(Pt 2): 249–265.

McManus DP, Wong JY, Zhou J, Cai C, Zeng Q, Smyth D, Li Y, Kalinna BH, Duke MJ, Yi X. 2001. Recombinant paramyosin (rec-Sj-97) tested for immunogenicity and vaccine efficacy against *Schistosoma japonicum* in mice and water buffaloes. Vaccine 20(5–6): 870–878.

McManus DP, Loukas A. 2008. Current status of vaccines for schistosomiasis. Clinical Microbiology Reviews 21: 225–242.

McManus DP, Dalton JP. 2006. Vaccines against the zoonotic trematodes *Schistosoma japonicum*, *Fasciola hepatica* and *Fasciola gigantica*. Parasitology 133(Suppl): S43–61.

McWilliam HE, Piedrafita D, Li Y, Zheng M, He Y, Yu X, McManus DP, Meeusen EN. 2013. Local immune responses of the Chinese water buffalo, *Bubalus bubalis*, against *Schistosoma japonicum* larvae: crucial insights for vaccine design. PLoS Neglected Tropical Diseases 7(9): e2460. doi: 10.1371/journal.pntd. 0002460. eCollection 2013.

Merrifield M, Hotez PJ, Beaumier CM, Gillespie P, Strych U, Hayward T, Bottazzi ME. 2016. Advancing a vaccine to prevent human schistosomiasis. Vaccine pii: S0264-410X(16)30076-7. doi: 10.1016/j. vaccine.2016.03.o79.

Mo AX, Colley DG. 2016. Workshop report: Schistosomiasis vaccine clinical development and product characteristics. Vaccine 34(8): 995–1001.

Mo AX, Agosti JM, Walson JL, Hall BF, Gordon L. 2014. Schistosomiasis elimination strategies and potential role of a vaccine in achieving global health goals. American Journal of Tropical Medicine and Hygiene 90(1): 54–60. doi: 10.4269/ajtmh.13-0467.

Moser D, Tendler M, Griffiths G, Klinkert MQ. 1991. A 14 kDa *Schistosoma mansoni* polypeptide is homologous to a gene family of fatty acid binding proteins. Journal of Biological Chemistry 266: 8447–8454.

Mossallam SF, Amer EI, Ewaisha RE, Khalil AM, Aboushleib HM, Bahey-El-Din M. 2015. Fusion protein comprised of the two schistosomal antigens, Sm14 and Sm29, provides significant protection against *Schistosoma mansoni* in murine infection model. BMC Infectious Diseases 15: 147. doi: 10.1186/ s12879-015-0906-z.

Olveda RM, Daniel BL, Ramirez BD, Aligui GD, Acosta LP, Fevidal P, Tiu E, de Veyra F, Peters PA, Romulo R, *et al.* 1996. Schistosomiasis japonica in the Philippines: the long-term impact of population-based chemotherapy on infection, transmission and morbidity. Journal of Infectious Diseases 174: 163–172. 10.1093/infdis/174.1.163.

Pearce EJ, Kane CM, Sun J, Taylor JJ, McKee AS, Cervi L. 2004. Th2 response polarization during infection with the helminth parasite *Schistosoma mansoni.* Immunological Reviews 201: 117–126.

Pearce EJ, Hall BF, Sher A. 1990. Host-specific evasion of the alternative complement pathway by schistosomes correlates with the presence of a phospholipase C-sensitive surface molecule resembling human decay accelerating factor. Journal of Immunology 144: 2751–2756.

Pearce EJ, James SL, Hieny S, Lanar DE, Sher A. 1988. Induction of protective immunity against *Schistosoma mansoni* by vaccination with schistosome paramyosin (Sm97), a nonsurface parasite antigen. Proceedings of the National Academy of Sciences USA 85(15): 5678–5682.

Pearson MS, Becker L, Driguez P, Young ND, Gaze S, Mendes T, Li XH, Doolan DL, Midzi N, Mduluza T, *et al.* 2015. Of monkeys and men: immunomic profiling of sera from humans and non-human primates resistant to schistosomiasis reveals novel potential vaccine candidates. Frontiers in Immunology 6: 213. doi: 10.3389/fimmu.2015.00213. eCollection 2015.

Petrovsky N, Aguilar JC. 2004. Vaccine adjuvants: Current state and future trends. Immunology and Cell Biology 82: 488–496; doi: 10.1111/j.0818-9641.2004.01272.x.

Pickering LK, Baker CJ, Freed GL, Gall SA, Grogg SE, Poland GA, Rodewald LE, Schaffner W, Stinchfield P, Tan L, Zimmerman RK, Orenstein WA. 2009. Immunization programs for infants, children, adolescents and adults: clinical practice guidelines by the infectious diseases society of America. Clinical Infectious Diseases 49(6): 817–840. doi: 10.1086/605430.

Pinheiro CS, Ribeiro AP, Cardoso FC, Martins VP, Figueiredo BC, Assis NR, Morais SB, Caliari MV, Loukas A, Oliveira SC. 2014. A multivalent chimeric vaccine composed of *Schistosoma mansoni* SmTSP-2 and Sm29 was able to induce protection against infection in mice. Parasite Immunology 36(7): 303–12. doi: 10.1111/pim.12118.

Ramirez BL, Kurtis JD, Wiest PM, Arias P, Aligui F, Acosta L, Peters P, Olds GR. 1996. Paramyosin: a candidate vaccine antigen against *Schistosoma japonicum.* Parasite Immunology 18: 49–52. doi: 10.1046/j.1365-3024.1996.d01-4.x

Ribeiro F, Vieira Cdos S, Fernandes A, Araujo N, Katz N. 2002. The effects of immunization with recombinant Sm14 (rSm14) in reducing worm burden and mortality of mice infected with *Schistosoma mansoni.* Revista da Sociedade Brasileiradade Medicina Tropical 35(1): 11–17.

Ribeiro de Jesus A, Araújo I, Bacellar O, Magalhães A, Pearce E, Harn D, Strand M, Carvalho EM. 2000. Human immune responses to *Schistosoma mansoni* vaccine candidate antigens. Infection and Immunity 68(5): 2797–2803.

Richter D, Reynolds SR, Harn DA. 1993. Candidate vaccine antigens that stimulate the cellular immune response of mice vaccinated with irradiated cercariae of *Schistosoma mansoni.* Journal of Immunology 151(1): 256–265.

Riveau G, Deplanque D, Remoué F, Schacht AM, Vodougnon H, Capron M, Thiry M, Martial J, Libersa C, Capron A. 2012. Safety and immunogenicity of rSh28GST antigen in humans: phase 1 randomized clinical study of a vaccine candidate against urinary schistosomiasis. PLoS Neglected Tropical Diseases 6(7): e1704. doi: 10.1371/journal.pntd.0001704.

Riveau G, Poulain-Godefroy O, Dupré L, Remoué F, Mielcarek N, Locht C, Capron A. 1998. Glutathione S-Transferases of 28 kDa as major vaccine candidates against schistosomiasis. Memórias do Instituto Oswaldo Cruz 93, Suppl I: 87–94. http://dx. doi: org/10.1590/S0074-02761998000700012.

Reynolds SR, Dahl CE, Harn DA. 1994. T and B epitope determination and analysis of multiple antigenic peptides for the *Schistosoma mansoni* experimental vaccine triose-phosphate isomerase. Journal of Immunology 152(1): 193–200.

Reynolds SR, Shoemaker CB, Harn DA. 1992. T and B cell epitope mapping of Sm23, an integral membrane protein *S. mansoni.* Journal of Immunology 149: 3995–4001.

Sanin DE, Prendergast CT, Mountford AP. 2015. IL-10 Production in macrophages is regulated by a TLR-driven CREB-mediated mechanism that is linked to genes involved in cell metabolism. Journal of Immunology 195(3): 1218–32. doi: 10.4049/jimmunol.1500146.

Santini-Oliveira M, Coler RN, Parra J, Veloso V, Jayashankar L, Pinto PM, Ciol MA, Bergquist R, Reed SG, Tendler M. 2015. Schistosomiasis vaccine candidate Sm14/GLA-SE: Phase 1 safety and immunogenicity clinical trial in healthy, male adults. Vaccine 34(4): 586–594. doi: 10.1016/j.vaccine.2015.10.027.

Shi YE, Jiang CF, Han JJ, Li YL, Ruppel A. 1990. *Schistosoma japonicum:* an ultraviolet-attenuated cercarial vaccine applicable in the field for water buffaloes. Experimental Parasitology 71: 100–106.

Siddiqui AA, Siddiqui BA, Ganley-Leal L. 2011. Schistosomiasis vaccines. 7(11): 1192–1197. doi: 10.4161/hv.7.11.17017.

Siddiqui AA, Zhou Y, Podesta RB, Karcz SR, Tognon CE, Strejan GH, Dekaban GA, Clarke MW. 1993. Characterization of Ca (2+)-dependent neutral protease (calpain) from human blood flukes, *Schistosoma mansoni*. Biochimicaet Biophysica Acta 1181: 37–44. doi: 10.1016/0925-4439(93)90087-H.

Soisson LM, Masterson CP, Tom TD, McNally MT, Lowell GH, Strand M. 1992. Induction of protective immunity in mice using a 62-kDa recombinant fragment of a *Schistosoma mansoni* surface antigen. Journal of Immunology 149(11): 3612–3620.

Sotillo J, Pearson M, Becker L, Mulvenna J, Loukas A. 2015. A quantitative proteomic analysis of the tegumental proteins from *Schistosoma mansoni* schistosomula reveals novel potential therapeutic targets. International Journal of Parasitology 45(8): 505–516. doi: 10.1016/j.ijpara.2015.03.004.

Smith DB, Davern KM, Board PG, Tiu WU, Garcia EG, Mitchell GF. 1986. Mr 26,000 antigen of *Schistosoma japonicum* recognized by resistant WEHI 129/J mice is a parasite glutathione S-transferase. Proceedings of the National Academy of Sciences USA 83: 8703–8707.

Stephenson R, You H, McManus DP, Toth I. 2014. Schistosome vaccine adjuvants in preclinical and clinical research. Vaccines (Basel) 2(3): 654–85. doi: 10.3390/vaccines2030654.

Tao FF, Yang YF, Wang H, Sun XJ, Luo J, Zhu X, Liu F, Wang Y, Su C, Wu HW, Zhang ZS. 2009. Th1-type epitopes-based cocktail PDDV attenuates hepatic fibrosis in C57BL/6 mice with chronic *Schistosoma japonicum* infection. Vaccine 27(31): 4110–7. doi: 10.1016/j.vaccine.2009.04.073.

Tendler M, Simpson AJ. 2008. The biotechnology-value chain: development of Sm14 as a schistosomiasis vaccine. Acta Tropica 108(2–3): 263–266. doi: 10.1016/j.actatropica.2008.09.002.

Tendler M, Brito CA, Vilar MM, Serra-Freire N, Diogo CM, Almeida MS, Delbem AC, Da Silva JF, Savino W, Garratt RC, Katz N, Simpson AS. 1996. A *Schistosoma mansoni* fatty acid-binding protein, Sm14, is the potential basis of a dual-purpose anti-helminth vaccine. Proceedings of the National Academy of Sciences USA 93(1): 269–273.

Tran MH, Freitas TC, Cooper L, Gaze S, Gatton ML, Jones MK, Lovas E, Pearce EJ, Loukas A. 2010. Suppression of mRNAs encoding tegument tetraspanins from *Schistosoma mansoni* results in impaired tegument turnover. PLoS Pathogens 6: e1000840. doi: 10.1371/journal.ppat.1000840.

Tran MH, Pearson MS, Bethony JM, Smyth DJ, Jones MK, Duke M, Don TA, McManus DP, Correa-Oliveira R, Loukas A. 2006. Tetraspanins on the surface of *Schistosoma mansoni* are protective antigens against schistosomiasis. Nature Medicine 12: 835–840. doi: 10.1038/nm1430.

Treanor JJ, Essink B, Hull S, Reed S, Izikson R, Patriarca P, Goldenthal KL, Kohberger R, Dunkle LM. 2013. Evaluation of safety and immunogenicity of recombinant influenza hemagglutinin (H5/Indonesia/05/2005) formulated with and without a stable oil-in-water emulsion containing glucopyranosyl-lipid A (SE + GLA) adjuvant. Vaccine 31(48): 5760–5765.

Trottein F, Vaney MC, Bachet B, Pierce RJ, Colloc'h N, Lecocq JP, Capron A, Mornon JP. 1992. Crystallization and preliminary X-ray diffraction studies of a protective cloned 28 kDa glutathione S-transferase from *Schistosoma mansoni*. Journal of Molecular Biology 224(2): 515–518.

Tundup S, Srivastava L, Nagy T, Harn D. 2014. CD14 influences host immune responses and alternative activation of macrophages during *Schistosoma mansoni* infection. Infection and Immunity 82(8): 3240–3251. doi: 10.1128/IAI.01780-14.

Walter K, Fulford AJ, McBeath R, Joseph S, Jones FM, Kariuki HC, Mwatha JK, Kimani G, Kabatereine NB, Vennervald BJ, *et al.* 2006. Increased human IgE induced by killing *Schistosoma mansoni in vivo* is associated with pretreatment Th2 cytokine responsiveness to worm antigens. Journal of Immunology 177(8): 5490–5498.

Wang T, Zhang S, Wu W, Zhang G, Lu D, Ornbjerg N, Johansen MV. 2006. Treatment and reinfection of water buffaloes and cattle infected with *Schistosoma japonicum* in Yangtze River Valley, Anhui Province, China. Journal of Parasitology 92: 1088–1091.

Wilson RA, Coulson PS. 1998. Why don't we have a schistosomiasis vaccine? Parasitology Today 14(3): 97–99.

Wilson RA, Li XH, Castro-Borges W. 2016. Do schistosome vaccine trials in mice have an intrinsic flaw that generates spurious protection data? Parasites & Vectors 9(1): 89.

WHO. 2012a. http://apps.who.int/iris/bitstream/10665/70809/1/WHO_HTM_NTD_2012.1_eng. pdf (accessed 30 December 2015).

WHO. 2012b. http://www.who.int/neglected_diseases/London_meeting_follow_up/en/ (accessed 30 December 2015).

Yang J, Fu Z, Feng X, Shi Y, Yuan C, Liu J, Hong Y, Li H, Lu K, Lin J. 2012. Comparison of worm development and host immune responses in natural hosts of *Schistosoma japonicum*, yellow cattle and water buffalo. BMC Veterinary Research 8: 25. doi: 10.1186/1746-6148-8-25.

You H, McManus DP. 2015. Vaccines and diagnostics for zoonotic schistosomiasis japonica. Parasitology 142(2): 271–289. doi: 10.1017/S0031182014001310. Epub 2014 Oct 31.

You H, Gobert GN, Cai P, Mou R, Nawaratna S, Fang G, Villinger F, McManus DP. 2015. Suppression of the insulin receptors in adult *Schistosoma japonicum* impacts on parasite growth and development: further evidence of vaccine potential. 2015. PLoS Neglected Tropical Diseases 9(5): e0003730. doi: 10.1371/journal.pntd.0003730. eCollection 2015.

You H, Gobert GN, Duke MG, Zhang W, Li Y, Jones MK, McManus DP. 2012. The insulin receptor is a transmission blocking veterinary vaccine target for zoonotic *Schistosoma japonicum*. International Journal for Parasitology 42: 801–807.

You H, Zhang W, Jones MK, Gobert GN, Mulvenna J, Rees G, Spanevello M, Blair D, Duke M, Brehm K, McManus DP. 2010. Cloning and characterisation of *Schistosoma japonicum* insulin receptors. PLoS One. 2010 Mar 24; 5(3): e9868. doi: 10.1371/journal.pone.0009868.

Yuan H, You-En KS, Long-Jiang Y, Xiao-Hua Z, Liu-Zhe L, Cash M, Lu Z, Zhi L, Deng-Xin S. 2007. Studies on the protective immunity of *Schistosoma japonicum* bivalent DNA vaccine encoding Sj23 and Sj14. Experimental Parasitololgy 115(4): 379–386.

Zhang W, Ahmad G, Le L, Rojo JU, Karmakar S, Tillery KA, Torben W, Damian RT, Wolf RF, White GL, *et al.* 2014. Longevity of Sm-p80-specific antibody responses following vaccination with Sm-p80 vaccine in mice and baboons and transplacental transfer of Sm-p80-specific antibodies in a baboon. Parasitology Research 113(6): 2239–2250. doi: 10.1007/s00436-014-3879-8.

Zhang Y, Taylor MG, Johansen MV, Bickle QD. 2001. Vaccination of mice with a cocktail DNA vaccine induces a Th1-type immune response and partial protection against *Schistosoma japonicum* infection. Vaccine 20(5–6): 724–730.

Zhou Y, Zheng H, Chen Y, Zhang L, Wang K, Guo J, Huang Z, Zhang B, Huang W, Jin K, *et al.* 2009. The *Schistosoma japonicum* genome reveals features of host–parasite interplay. Nature 460: 345–351.

CHAPTER

Geospatial Surveillance and Response Systems for Schistosomiasis

John Malone[1,*], *Robert Bergquist*[2] and *Laura Rinaldi*[3]

25.1 INTRODUCTION

All Earth-based spatial and temporal references should ideally be 'relatable' to one another and ultimately to a real physical location. The understanding of this concept has opened new avenues of scientific inquiry, the essence of which is the geographic information system (GIS), links unrelated information by using location as the key index variable. Visualization is an integral part of GIS, which captures, integrates, stores, manages, analyzes, shares and displays the data. Locations and extents in space and time in this structure may be recorded by coordinates representing longitude, latitude, elevation and time. GIS allows interpretation and visualizing of data in ways that reveal relationships and patterns by presenting them in a cartographic environment. Although GIS is a broad term that can refer to a number of different technologies, processes and methods, we deal here only with medical and veterinary epidemiological applications. Data collected from various sources, such as cartographic records, archives and field research can be bundled together with remotely sensed (RS) data from Earth-observing satellites and this information in the form of electronic data layers (overlays) capable of displaying this information can be shown visually (Fig. 25.1).

Satellite-based RS permits the visualization in the form of different environmental datalayers represented by raster and vector formats. The former uses a net of adjacent polygons (or cells) to provide a virtual cover of a given part of a territory. These pixels will eventually contain the attributed values of the objects they are assigned to represent. This type of visualization is typically utilized to represent continuous phenomena, such as land cover, land use, topography and climate data. The vectors are basically different as they consist of storing tables containing the coordinates together with instructions how to find points that they represent and whether or not these points are independent. In principle, all lines are represented by vector chains and all areas by polygons, while attributes can be entered into tables using alphanumeric characters to label specific classes and/or property categories.

[1] Pathobiological Sciences, School of Veterinary Medicine, Louisiana State University, Baton Rouge, LA, USA.

[2] UNICEF/UNDP/World Bank/WHO Special Programme for Research and Training in Tropical Diseases (TDR), World Health Organization, Geneva, Switzerland.

[3] Section of Parasitology and Parasitic Diseases, Department of Veterinary Medicine and Animal Productions, University of Naples Federico II, Naples, Italy.

* Corresponding author

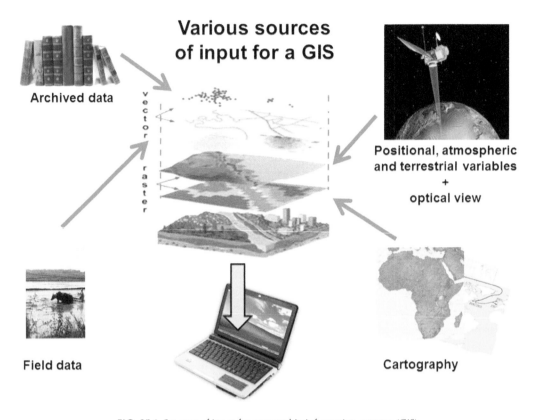

FIG. 25.1 Sources of input for geographic information systems (GIS).

Improved computers and widespread use of the Internet for exchange of information have created the prerequisite for a broad acceptance of GIS and RS approaches within the epidemiological sciences. While classical epidemiology largely consists of tables and general geographical maps, GIS facilitates access to a wider spectrum of information through collection and processing data from the field and Earth-observing satellites. With respect to diseases restricted by environmental requirements, the epidemiologists of today have access to a multitude of ecological and climatic data never before available in such amounts and with such ease. Spurred by faster and cheaper computers, better network processing, electronic data publishing and data capture techniques, the field has experienced a period of extremely rapid growth in the last few decades.

Disease mapping and environmental risk assessment using digital geospatial data resources are now established analytical tools in both human and veterinary public health. Spatial epidemiology has become defined as a sub-discipline of epidemiology, whose primary purpose is to describe and explain the spatial aspects of disease. The development of near real-time surveillance systems, based on GIS, global positioning systems (GPS) and RS, facilitate the establishment of accurate, up-to-date early-warning systems (EWS). Cartographic representations of the burden of disease are becoming increasingly helpful in a world where computer-generated map applications represent a straightforward way of visualizing large numbers of epidemiological datasets in a geographical context. For example, risk assessment frameworks linking spatial data from virtual globes together with GIS software packages can assemble state-of-the-art modules resulting in user-friendly platforms for evaluation, prediction and risk profiling (Yang *et al.* 2012). In this way, the epidemiological status can be shared in real time not only with decision-makers in the Health Ministries, but also with the individual researchers who continually extend the evidence base. The progress so far is encouraging and will no doubt contribute to the new field of surveillance-response systems (Tambo *et al.* 2014; Bergquist *et al.* 2015).

The leading supplier of GIS software, web GIS and geodatabase management tools is the Environmental Systems Research Institute (ESRI) based in California, USA. This enterprise started by providing map-based analyses of topographic maps from the United States Geological Survey (USGS) as a base layer manually overlaid with different transparent, gridded, plastic sheets representing separate geographic layers such as geology, soils, topography, etc. Later, ESRI developed a computerized, polygon information overlay system that was used as a framework for data integration allowing people to visualize problems, which contributed to finding better solutions more rapidly. This concept, introduced in the mid-1970s, eventually led to the release of ArcInfo, the first commercial GIS software, in 1982. ESRI's first desktop solution, ArcView, had a major impact on the industry, opening up the possibilities of GIS to more users. During the late 1990s, ArcView was replaced by ArcGIS, a modular and scalable GIS platform.

The growth of the Internet helped distribute GIS more widely and other desktop GIS platforms, including web map servers, spatial database management systems software development frameworks and libraries, have come into play. There is now a large field of various commercial software applications besides ArcView as well as many open-source GIS software packages (see Table 25.1). The field has thus become multi-participatory allowing the advantage of cloud computing opportunities that facilitate GIS access for anyone, anywhere.

TABLE 25.1 Alternative GIS software applications (off-the-shelf GIS packages).

Name	Provider	Access	Operating system
ArcGIS	ESRI[a]	Licence	Windows
Car Bayes	CRAN R project[b]	Open source	Windows, Mac OSX, Linux
GeoBUGS*	BUGS project[c]	Open source	Windows
Python	PSF[d]	Open source	Windows, Mac OSX, Linux
Quantum GIS	OSGeo[e]	Open source	Windows, Mac OSX, Linux
R	R project	Open source	Windows, Mac OSX, Linux
SaT Scan	Martin Kulldorff[f]	Open source	Windows
SIGEpi	PAHO/WHO[g]	Cost-recovery	Windows
WinBUGS	BUGS project[c]	Open source	Windows

* GeoBUGS is an add-on to WinBUGS, developed by Department of Epidemiology and Public Health, Imperial College, St Mary's Hospital, London, UK.
[a] Environmental Systems Research Institute, Redlands, CA, US.
[b] Comprehensive R Archive Network, authored and maintained by Duncan Lee, Glasgow University, UK;
[c] Bayesian inference Using Gibbs Sampling; (Geman and Geman 1984).
[d] Python Software Foundation; (https://www.python.org/).
[e] Open Source Geospatial Foundation; (http://www.osgeo.org/)
[f] with Information Management Services Inc.
[g] Pan American Health Organization (Area of Health Analysis and Information Systems) and the World Health Organization.

Presentation of geography and environmental variables associated with health data produced the concept of disease ecology. RS information provides useful insights on geo-climatic, ecological and anthropogenic factors related to transmission levels and patterns of many communicable diseases. Representation of data, such as prevalence and intensity of infection in the form of maps facilitates interpretation, synthesis and recognition of outbreak frequencies as well as cluster phenomena (Rinaldi *et al.* 2006). The facility of producing maps continue to be one of the most useful functions of GIS in epidemiology as it permits pattern recognition (not immediately obvious when seen in a table) and thereby facilitates decision-making (Paolino *et al.* 2005). The recognition of the map as provider of 'first glance' information is obvious nowadays. Still, changing the mentality of 'traditional' epidemiologists has taken time as it was difficult to convince people about the utility in visualizing health data. It is in this connection important to take into account the difference between vector and raster maps, in particular to decide when it is more appropriate using the latter than relying primarily on the former. Several types of analysis are better dealt with in a GIS environment than in any other way and the GIS approach is indeed routinely used for the description of neighborhoods, buffers, overlay analysis, network analysis, network modeling and calculation of spatial parameters (Ward and Carpenter 2000). However, it is important to understand that the use of GIS does by no means facilitate overcoming the two major concerns of any empirical research: data availability and data quality (Kistemann *et al.* 2002; Rinaldi *et al.* 2015).

GIS can be constructed with several datalayers, thus permitting the simultaneous visualization of health data with climatic and environmental data. This holds promise for the understanding of environmental-health linkages and the generation of new hypotheses about the distribution patterns of diseases. The visualization of epidemiological data is particularly required in the current era of climate change, global transports as well as movements of humans and animals when the establishment of EWS requires present and potential future geographical scenarios of many different epidemics. For example, areas into which certain diseases can be expected to expand in the near future have been shown for schistosomiasis in northern China (Yang *et al.* 2006; Zhou *et al.* 2008). Other important drivers for the (re-) emergence and spread of vector-borne parasites include vector habitat changes (Bhunia *et al.* 2011), pollution, resistance to pesticides and drugs and the general fall-out from globalization of commerce and travel (Harrus and Baneth 2005). The changing distribution of previously strictly localized endemic veterinary and human infections already include a large number of infections (Rinaldi *et al.* 2015).

EWS can be defined as the sum of all initiatives leading to improved awareness and knowledge of the distribution of diseases or infections that might permit forecasting the further evolution of an outbreak. Disease surveillance systems and EWS are prime examples of approaches with immediate practical application and the construction of virtual globes for epidemiological research are becoming increasingly important for displaying results in an easily understandable manner. Early identification is an important first step towards implementing effective interventions to control epidemics and reduce their impact (Estrada-Peña *et al.* 2007). The capability to produce frequently updated data on environmental variables pertinent to vector-borne disease transmission makes GIS and RS useful resources for the development of EWS (Ceccato *et al.* 2005). Surveillance must feature characteristics independent from the techniques used for case collection, an aspect that is evident when the geographical dimension is considered and where discontinuities in the spatial rate of a disease is of interest. Surveillance must also include diagnostics with high sensitivity and specificity for the identification of possible new cases of disease or any variation of the natural rate (Johansen *et al.* 2010; Rinaldi and Cringoli 2014). The difficulty in interpreting clusters of cases of disease lies in the appreciation whether to attribute them to an epidemic or to a minor alteration in the occurrence of the disease in question (Elliott *et al.* 2004; Elliott 2011).

Schistosomiasis, like many other parasitic diseases, not only depends on interaction between the human definitive host and the snail intermediate hosts, but environmental variables play an important part, in particular with respect to the latter. Climate conditions restrict snail distributions through variation of a number of parameters, e.g., rainfall, humidity, shade and temperature, the latter also governing the speed of maturation of the infectious agent inside the snail. The capabilities in terms of spatial, temporal and spectral resolution of the sensors of the satellite-based instruments and epidemiological, computer-based models are leading to a much improved understanding of which geographical areas can support the disease. In addition, risk-mapping, new methods of surveillance and access to large databases promise to improve our ability to perceive the complex relationship between environment and infection.

The wider use of GIS and RS became possible by the convergence of computer technology, earth sciences and spatial statistics in the last 20–30 years. The Landsat program (http://landsat.gsfc.nasa.gov/) became the symbol of a new vision in the 1970s, with launch of its first MSS sensor (80 m ground resolution in four spectral bands) a tantalizing harbinger of great things to come. Although access to these images was originally limited by high cost (USD 4,000/scene), one young student realized the possibilities on offer. Dr Barnett Cline published his ideas over 45 years ago (Cline 1970) followed up with an invited retrospect for the first issue of *Geospatial Health* (Cline 2006). However, his peers were initially slow to follow his lead and reliance on satellite-based datasets did not begin to pick up until 10–15 years later, when the first research papers in this area started to appear more regularly, partly explained by the reduced cost of RS datasets. Today GIS, RS and GPS are well-known tools of the trade and few scientists working in the epidemiological field can manage without them. However, training is important and there are inherent problems with some of the new technologies that should be appreciated before they are taken aboard (Herbreteau *et al.* 2007). Still, it cannot be denied that GIS provides a new way to address classic concepts of 'landscape epidemiology' and the essential nidality of disease (Pavlovskii 1945) by virtue of its potential to match the relative suitability of various environments to the parasite life cycle and transmission dynamics of host-parasite systems.

25.2   THE PIONEER YEARS – 1980 TO 2000

Schistosomiasis was the topic of pioneer studies done by Cross *et al.* (1984) using the Landsat MSS (https://lta.cr.usgs.gov/MSS) satellite data and rainfall-temperature weather variables for geospatial risk assessment in The Philippines. Other early applications of geospatial methods to human and animal disease agents before 1995 includes Rift Valley fever (RVF) (Linthicum *et al.* 1987), East Coast Fever (Lessard *et al.* 1990), African trypanosomiasis (Rogers and Randolph 1993) and fascioliasis (Malone *et al.* 1992; Zukowski *et al.* 1992, 1993).

It was not until the early 1990's that Malone and others reported successful use of land surface day/night temperature difference (dT) data from the Advanced Very High Resolution Radiometer (AVHRR) satellite imagery (1-km spatial resolution) emanating from the polar-orbiting operational environmental satellites of the National Oceanic and Atmospheric Administration (NOAA) to predict the environmental risk of schistosomiasis in the Nile Delta of Egypt. Annual and seasonal composite maps of dT patterns were proposed to reflect stable landforms, soils and climate-irrigation-water table factors that influence *S. mansoni* propagation and transmission in the Nile Delta (Malone *et al.* 1994).

Subsequent studies by the Schistosomiasis Research Project (SRP), a 10-year effort aimed at reducing schistosomiasis in Egypt funded by the Egyptian Ministry of Health and the United States Agency for International Development (USAID), confirmed and extended these results using GIS and higher-resolution (30 m) Landsat thematic mapper data (https://lta.cr.usgs.gov/TM) to develop and validate a village-scale control program model for schistosomiasis in the Kafr El Sheikh Governorate in the Nile Delta of Egypt (Malone *et al.* 1997). Soil type, water table and water quality at 79 of 154 rural health unit sites were related to the environmental suitability for *S. mansoni*. Model validation was done using data collected on snail population bionomics-infection rates, water quality, underground water table and cercariometry (measurement of the infectious parasite stage–cercariae–shed by snail host vectors) at 13 hydrologically representative sites (Abdel-Rahman *et al.* 2001). Next, *S. mansoni* prevalence values from a 5-year prevalence database of rural health units (RHU) in Kafr El Sheikh were compared within a GIS to class values extracted from a tasseled cap transformation (Tcap) of Landsat TM imagery and from AVHRR dT maps. Use of 4 Tcap classes with dT values extracted from 5×5 km areas centered on RHUs explained 74% of the variation in RHU prevalence. Taken together, these data provided biology-based evidence that RS-detected environmental features could be used to predict *S. mansoni* prevalence in irrigated agricultural areas (Abdel-Rahman *et al.* 2001).

Concurrent with SRP studies in Egypt, broad new interest was developing on the use of geospatial methods for schistosomiasis elsewhere, including Madagascar (Jeanne 1996), Brazil (Bavia *et al.* 1999), Zimbabwe (Mukaratirwa *et al.* 1999), Ethiopia-East Africa (Kristensen *et al.* 2001; Malone *et al.* 2001) and China (Zhou *et al.* 2001). In South America, Bavia *et al.* (1999) constructed a GIS based on maps of regional environmental features, to study *S. mansoni* prevalence in 30 representative municipalities and snail host distribution in Bahia State, Brazil and the spatial and temporal dynamics of infection identifying environmental factors that influence the distribution of schistosomiasis. In Ethiopia and East Africa, a biology-based climate suitability threshold map for *S. mansoni* and *S. haematobium* was constructed in a GIS covering the Intergovernmental Authority on Development (IGAD) region of East Africa based on a compilation of prevalence data from literature reports (Malone *et al.* 2001). In related studies, Kristensen *et al.* (2001) developed a GIS model on geospatial distribution of the snail vectors of schistosomiasis in Ethiopia based on a number of historical malacological surveys. In Zimbabwe, extensive data on schistosomiasis and snail distribution from earlier collaboration with the Danish Bilharziasis Laboratory (DBL) were geolocated and entered into a GIS to map risk of urinary and intestinal schistosomiasis and other snail-borne diseases (Mukaratirwa *et al.* 1999). In China, prompted by the need to assess health effects of environmental change associated with the impending construction of the Three Gorges Dam on the Yangtze River, Zhou and others began work on a very productive series of publications, continuing to the present, of GIS/RS studies on risk assessment and control of the *S. japonicum-Oncomelania* biological system (Zhou *et al.* 1999). Coupled with these events, early reviews, conferences and editorial commentaries advocated wider use of RS/GIS citing the untapped potential of this new field of epidemiologic endeavor (Hugh-Jones 1989; Malone 1995; Bergquist 2001). In a special volume of Advances in Parasitology (2000) devoted to progress in use of GIS and RS, Brooker and Michael (2000) promoted their view of the potential of adopting this approach to the study of the epidemiology and control of schistosomiasis and other helminths in Africa.

In 2000, the Rockefeller Foundation's Bellagio Center sponsored a team residency workshop that resulted in formation of the 'GIS Network On Schistosomiasis Information Systems' (GnosisGIS), a scientific group dedicated to promoting health applications in the geospatial sciences (www. GnosisGIS.org) that has since been expanded to include various other diseases as well as social sciences. The *International Journal of Health Geographics*, first published in 2002 and the journal *Geospatial Health* (http://geospatialhealth.net), inaugurated in 2006, functioned as a forum for the growing number of papers addressing use of GIS/RS in the health sciences. Previously, new applications touching on health were published in more general geography journals or technology-based publications such as *Remote Sensing of Environment* or the *Journal of Photogrammetry and Remote Sensing*. Recognition of the potential health applications for schistosomiasis, along with malaria and other diseases, was catching on. Classic studies on schistosomiasis in Africa from the colonial and post-colonial era could now be re-examined in a new light where space and time played the decisive role.

25.3 AFRICA REVISITED

In 2002, at the World Summit on Sustainable Development held in South Africa, a report entitled *Down to Earth. Geographical Information for Sustainable Development in Africa* was presented. This report, highlighted recent research progress on GIS and schistosomiasis in Egypt and East Africa (National Academy of Sciences, 2002). It had been 10 years since the United Nations adopted Agenda 21 in Rio de Janeiro, a watershed event that marked a new emphasis on sustainability of disease control by international development agencies.

Armed by the success of schistosomiasis control programs in Egypt, China and Brazil, international political consensus shifted to implementing national schistosomiasis control programs in sub-Saharan Africa based on the rationale of preventive chemotherapy to reduce intensity of infection and thus morbidity. In 2002, development and implementation of the 'Schistosomiasis Control Initiative' (SCI) (http://www.sci-ntds.org) began with major funding from the Bill and Melinda Gates Foundation to initiate partnerships with Ministry of Health control programs in six

African countries. Based on affordable new drugs costing less than USD 0.50 per person per year and the recognition that co-infection by soil transmitted helminths (STHs) that often accompanies schistosomiasis, a policy of addressing integrated worm control for both was adopted (Kabatereine *et al.* 2007; Fenwick *et al.* 2009). A key component of control programs in each SCI-supported country was the implementation of a GIS/RS mapping and modeling system to record results of large-scale, cross-sectional national surveys to be used with existing data to provide regional-scale baseline data on the distribution and intensity of schistosomiasis and STHs. These maps were used to guide the development of optimal treatment strategies in each country (Clements *et al.* 2006, 2009, 2010) and to monitor and evaluate success of control program interventions. Thus, project-generated GIS map data and satellite RS data became linked to spatial information on climate, elevation, proximity to lakes and other hydrological features. In addition, these activities initiated innovative Bayesian geostatistical prediction models that facilitated planning of real world, large-scale schistosomiasis control programs. SCI project results were turned over to the Ministry of Health in each country for future use and integration with other neglected tropical disease (NTD) control programs.

In 2005, a new initiative was conceived – the *Contrast* project – a multi-disciplinary alliance to optimize schistosomiasis control and transmission surveillance in sub-Saharan Africa proposed by a consortium of partners from seven African and four European countries. Research focused on four molecular, biological, spatial and social themes (Bergquist 2013; Utzinger *et al.* 2013), out of which the spatial one centered on extensive use of GIS/RS mapping and modeling to identify risk factors that govern the frequency and transmission dynamics of schistosomiasis. Sponsored by the European Commission's 6th Framework (FP6), *Contrast* built upon SCI implementation of geospatial methods for schistosomiasis by developing climate suitability risk maps and empirical statistical maps based on Bayesian statistical models. The overall approach emphasized detailed knowledge of the distribution and abundance of snail hosts (Chammartin 2013; Standley *et al.* 2012; Schur *et al.* 2013; Stensgaard *et al.* 2013). It also brought together existing electronic information (published and unpublished) into a single database in an open-source Google Earth platform accessible from the web (Stensgaard *et al.* 2009), which made it possible to share information and manage data well beyond the 4-year project timeline. Importantly, *Contrast*'s accumulated experience on transmission control, facilitated by geospatial methods, provided timely evidence to justify the shift from exclusive focus on morbidity control to adoption of the schistosomiasis elimination agenda in low-transmission countries (Rollinson *et al.* 2013; WHO 2012). In May 2012, the World Health Assembly passed resolution WHA 65.21, calling upon member states to intensify schistosomiasis control and wherever possible, to attempt transmission interruption and initiate interventions towards local elimination (WHO 2012). This marked a new era in schistosomiasis research and control aimed at the much more ambitious goal of elimination of schistosomiasis as a public health problem.

The combined SCI and *Contrast* experience was part of the stimulus that led to the NTD program launched by the USAID (Linehan *et al.* 2011; Hanson *et al.* 2012). In 2006, the US congress authorized funding for a 3-stage 'roll-out package' to integrate and scale up control of the NTDs to national level in four 'fast-track' countries in Africa with planned expansion to additional countries. The program, in 2015 deemed a major success, has since expanded with new partnerships in 25 low-income countries in Africa, Asia and the Americas (Haiti) (http://www.ntd-ngdonetwork.org/). Administered by RTI International (http://www.rti.org), partners from non-governmental organizations (NGOs), including SCI, were contracted to administer national programs with Ministries of Health in each country. In seven years, more than 1 billion treatments for NTDs were delivered using $ 6.7 billion in donated medicines in integrated control programs.

Envision, a sub-project of USAID's NTD program operating since 2011 (http://www.ntdenvision.org/), provides technical and financial assistance to national NTD programs to manage and implement strategic control and elimination activities, including GIS/RS mapping, aiming to control and eliminate seven NTDs by the year 2020. USAID and the *Department for*

International Development (DfID) in the United Kingdom have a newly established agreement to work with the World Health Organization (WHO) in support of a coordinated global initiative to map seven NTDs (schistosomiasis, lymphatic filariasis, onchocerciasis, trachoma and STH in each administrative district in Africa for each disease.

Geospatial mapping using GIS/RS for surveillance and response systems had unquestionably emerged as mainstream epidemiologic practice and had become an integral component of global efforts to integrate control of schistosomiasis with other NTDs. However, these investments pose important unsettled questions: does mass treatment below the level needed to interrupt transmission of schistosomiasis and other NTDs sentence us to the expense of an eternal cycle of re-infection? Or will it deliver us to the first step of elimination? (King 2011; Rollinson 2013).

Concerns about the potential effects of impending climate change on vector-borne diseases was the focus of one additional major project funded in 2010–2014 by the European Commission's 7th Framework (FP7)-*Healthy Futures*. The aim of his project was to contribute to reducing the future burden of three, water-related high-impact vector-borne diseases (VBD) in Africa–malaria, schistosomiasis and RVF. The project consortium was comprised of an inter-disciplinary group of climatologists, disease modelers and experts in the environmental, health and socio-economics sciences, together with staff in government health ministries in the East African Community (EAC). A total of fifteen institutions made up the consortium, located in ten different countries, five African (Rwanda, Kenya, Uganda, Tanzania, South Africa) and five European (Ireland, Sweden, Austria, Italy, UK). VBD's were expected to be sensitive to changes in environmental conditions such as increased ambient temperature or changes in the timing and levels of rainfall associated with climate change. Dynamical simulation models were developed for each of the three targeted diseases based on data generated by the moderate resolution imaging spectroradiometer (MODIS) and the tropical rainfall measuring mission (TRMM) satellite instruments. The climate surrogate data gathered covered the EAC at 1-km^2 resolution at Earth surface. Model output risk maps were produced using ArcGIS software (http://www.esri.com/software/arcgis) based on current climate data and long term climate change scenario projections as proposed by the Intergovernmental Panel on Climate Change – IPCC (2013). Predicted changes in the distribution and transmission patterns for malaria (Tompkins *et al.* 2013; Hagenlocher and Castro 2015), schistosomiasis (McCreech *et al.* 2015) and RVF (Taylor *et al.* 2016) were represented as maps covering the EAC and decision support frameworks were developed for use by the scientific community and stakeholders in the EAC.

An integrated, open-source Atlas presenting the key results of the *Healthy Futures* project along the lines of the paper by Hagenlocher and Castro (2015) is underway. This online resource will provide information on past, present and future conditions of risk for malaria, schistosomiasis and RVF and allow the exploration and visualization of results through web-based interactive tools. The Atlas embodies a guided access to information on climate change, potentiality of disease occurrence and population vulnerability with respect to these three diseases in the EAC region through direct access to downloadable datasets and metadata integrated in the *Healthy Futures* Metadata Portal (*Healthy Futures* 2014). Current available information can be directly accessed through the Healthy Futures website (http://www.healthyfutures.eu).

The objective of the Atlas is to communicate and visualize complex information in a guided and simple, yet interactive, manner. Information can be queried based on three prime selection criteria: (i) infectious diseases targeted by the *Healthy Futures* project; (ii) time, allowing comparisons of current conditions with a range of future projections, while allowing access to information on historic outbreaks and (iii) different components of risk. Future climate change projections based on two representative concentration pathways (RCPs) emission scenarios RCP 4.5 (mid-level change) and RCP 8.5 (high-level change) for each decade to 2100 throughout the EAC study area will thus be made available. Relative values of social vulnerability are mapped based on a range of indicators, such as susceptibility to disease (e.g., immunity, malnutrition, poverty, conflict, remoteness) and lack of resilience (e.g., education level, health facilities, number of dependents), while social and susceptibility indicators are weighted and combined in the form of a composite map indicator of geospatial risk (Healthy Futures 2014). The portal also organizes and documents datasets in a standardized manner allowing the data to be searchable via keywords, titles and abstracts and with time frames and spatial definitions. Datasets can be queried based on keywords and spatial search functionalities where access and/or contact information is provided. The original Metadata Portal was installed and hosted by the International Livestock Research Institute in Nairobi (ILRI) and was configured and customized through University of Salzburg partners (Z_GIS, PLUS). The metadata platform software used is freely available from ESRI (http://www.esri.com/software/arcgis/geoportal). The Portal uses the CSW (Catalogue Service for the Web) standard of the Open Geospatial Consortium (OGC) (http://www.opengeospatial.org), which makes it interoperable with other metadata portals and programs.

If successfully adopted and further developed, the Atlas will be among the first of its kind in geospatial health research to offer public health practitioners, scientists and stakeholders a tool to enable identification of likely VBD hotspots under different climate change scenarios at policy-relevant time-intervals over the coming century. Twelve articles emanating from the *Healthy Futures* 'Remote Sensing of Environment project' has been published in a special issue of *Geospatial Health* (Geospatial health, issue 1s, Volume 11, 146 p. http://www.geospatialhealth.net) in 2016.

The emergence of GIS and satellite generated data in schistosomiasis research and integration into real world control strategies are well chronicled by major projects dominated by research teams working in Africa and have been the focus of this report. The reader is referred to an excellent review by Mayangadze *et al.* (2015) to round out the list of other publications and reports on more individual efforts from the scientific literature.

25.4   FAREWELL TO THE GOD OF PLAGUES

This heading refers to a famous poem by Chairman Mao of the People's Republic of China indicating the start of the first serious effort to stop schistosomiasis in the country. In the early 1950s, 100 million people lived in endemic areas and more than 10 million were actually infected (Utzinger *et al.* 2005). There were also about 1.2 million infected cattle and an estimated 14,300 km^2 of habitats of the intermediate host snail *Oncomelania hupensis* (Chen and Feng 1999). The most recent published data give reason to believe that there are now less than 250,000 infected people (Li *et al.* 2013). The fight against schistosomiasis in China is not exclusively centered on chemotherapy as in the rest of the world. The fact that *O. hupensis* is amphibious makes its elimination somewhat easier than the snail control in Africa and South America. The success of the Chinese schistosomiasis control program (Utzinger *et al.* 2005) relies partly on large-scale environmental management, burying snails by retrenching waterways and impeding their survival by cleaning riverbanks and cementing canal sides. The additional approach of fencing in areas and beaches along the major rivers in an effort to prevent contamination has also paid off.

As early as 1984, Cross *et al.* used meteorological and Landsat satellite images to identify schistosomiasis endemic areas in The Philippines. The growth of spatial analysis based on RS data and GIS tools in Southeast Asia was at first slow and sporadic but work on spatial distribution and temporal variation picked up with time. The 'International Symposium on Schistosomiasis', held in Shanghai, China (Xiaonong *et al.* 2001) marked the beginning in earnest of geospatial tools for schistosomiasis control in China. The symposium had as theme 'Schistosomiasis Control in the 21st Century' and was held to draw attention to the new challenges in control of schistosomiasis.

Even if the snails can survive dry periods for long periods, water is necessary for survival and reproduction, while the temperature determines the rate of growth and development and vegetation affects shade as well as providing food. The strong focus on snail habitats, made GIS and RS important tools that were used for identification of land cover, land use, water bodies, vegetation, temperature, humidity (wetness), land surface temperatures (LST) and indices such as the normalized difference vegetation index (NDVI) and the modified normalized difference water index (MNDWI) (Wang *et al.* 2012; Wang and Zhuang 2015). These new approaches, including spatial statistics, are exceptionally useful for extracting and handling environmental data and they were applied for this purpose as soon as reliable data became available on (Xiaonong *et al.* 2001; Zhou *et al.* 2001; Weng 2002; Yang *et al.* 2005; Guo *et al.* 2005).

Wu *et al.* (2002) combined RS information with ground-based ecological data to identify snail habitats and monitor the changes in one of the major lakes in central China. Based on the understanding that snail habitats are associated with grasslands in and around water bodies, they reported that RS had a sensitivity of 90–96% and a specificity of 61–69% for correctly classifying snail habitats. A much later study in the same area (Wang *et al.* 2012) reported that a simple combination of two indexes, MNDWI and the NDVI, made it possible to directly estimate the snail habitats quantitatively. Early on, Guo *et al.* (2002, 2005; Yang *et al.* 2006) emphasized the possibility of rapidly providing detailed updated information with wide geographical coverage, thereby highlighting further advantages of RS technology over manual snail documentation. Bayesian statistical analysis and graphical display can be applied for the demonstration of the distribution and variation of snail habitats, both at large and very small scales (Yang *et al.* 2005). Risk mapping is well exemplified by a framework for rapid assessment developed by Yang *et al.* (2012). They bundled spatial data from Google Earth® with a GIS package, creating a WebGIS platform making real-time data on schistosomiasis risk accessible via the Internet. This platform has the potential to improve support systems and strengthen control activities, as it can rapidly locate high-risk areas, retrieve the data needed as well as provide detailed, up-to-date information on the performance of the control program.

To predict the potential impact on transmission of schistosomiasis japonica in China caused by climate change, Zhou *et al.* (2008) collected data from 193 weather stations from 1951 to 2000 and extrapolated the trend for the next 40 years. Comparing a tentative average, temperature rise of 2.2°C with known accumulated temperature models for *Oncomelania* snails and *S. japonicum*, taking into account both temperatures for snail survival and parasite growth inside the snail, a climate-transmission model for schistosomiasis was established that pointed towards a clear risk of a considerable expansion northward by schistosomiasis by the year 2050.

An earlier review by Malone *et al.* (2010), focusing on the implementation of a geospatial health infrastructure in Southeast Asia for control of schistosomiasis and other helminth infections, points out that in spite of the growth of the geospatial technologies, health workers have not rapidly taken advantage of the widely available, low-cost spatial data resources for epidemiological modeling. The review proposes a model training program and suggests the establishment of regional working groups to facilitate development and use of geospatial health infrastructure resources. The situation has improved during the last few years and GIS and RS are now well established in China, while their use in The Philippines is still in the build-up stage (Leonardo *et al.* 2007). One explanation could be the different latitude and geography of the endemic areas in the two countries. While China has transmission seasons, The Philippines has year-round transmission with the endemic areas spread between many islands with different geography.

Schistosomiasis japonica has a long history in The Philippines with the disease ensconced in more than half of the country's 78 provinces. The main stay of control is MDA with praziquantel, in contrast to China largely without backup of snail control. The prevalence ranges from 1% to 50% within different endemic zones with strong evidence that water buffaloes and cattle contribute significantly to disease transmission (Olveda *et al.* 2014). Apart from the paper by Cross *et al.* (1984), referred to above, relatively few papers on the use of geospatial technology have appeared. However, Leonardo *et al.* (2005) have effectively used such approaches to show how environmental factors affect the spatial distribution of disease. Topography and slope showed up as the main physical factors influence the vegetation cover, land use and soil type in the study area, while irrigation networks affected the magnitude and distribution of schistosomiasis more than any other environmental factor.

While schistosomiasis is retreating in China, it is still strongly entrenched in The Philippines (Bergquist and Tanner 2010). Two new schistosomiasis foci have been reported in the northernmost and central parts of the country as late as in 2002 and 2005 (Leonardo *et al.* 2015). GIS-generated maps showing the spatial distribution of snails in and around the foci indicated proximity of human habitation to snail habitats as an important epidemiological factor, while the onset was due to migration of people from endemic provinces.

25.5 DIASPORA – THE NEW WORLD

In Brazil, a national program for the control of schistosomiasis (Super intendência da Campanha de Saúde Pública–SUCAM) (http://www.jusbrasil.com.br/topicos/2706867/superintendencia-da-campanha-de-saude-publica) was launched in 1975, which directed the activities for large-scale chemotherapy. Although the disease is still far from eliminated, the program was deemed successful after 20–25 years of consistent, strong activities in the endemic areas, where control measures were repeatedly implemented resulting in a decrease in both infection prevalence and incidence of hepatosplenic cases with associated death (Katz 1998; Sarvel *et al.* 2011).

The adoption of geospatial approaches to schistosomiasis control in the Americas emerged in a similar time frame and pattern as in Africa and Asia. Early work by Bavia *et al.* (1999, 2001) in Bahia State in northeast Brazil led the way on GIS risk assessment for schistosomiasis. Analysis of environmental factors and prevalence in 30 representative municipalities in a GIS showed that population density and the duration of the annual dry period were the most significant determinants in the area studied (Bavia *et al.* 1999). A follow-up study showed addition of data on temperature difference and NDVI collected by the satellite-borne AVHRR sensor that could be used for development of a GIS environmental risk assessment model for schistosomiasis in Brazil (Bavia *et al.* 2001).

In 2003, the Oswaldo Cruz Foundation (FIOCRUZ), the US State Department and the Inter-American Institute for Global Change Research (http://www.iai.int) sponsored a workshop in Rio de Janeiro, Brazil (Confalonieri and Marinho 2005) that brought together representatives from Ministries of Health, academic researchers and RS specialists to review progress on infectious disease control with emphasis on diseases endemic in the Americas and to build a consensus on developing geospatial health capacity. Two of the workshop papers addressed GIS with respect to schistosomiasis, one on use of the fine-resolution IKONOS satellite imagery (http://www.geoimage.com.au/satellite/ikonosin) for a community study at the household scale in coastal Kenya (Kitron *et al.* 2005); the second on climate-based national-scale mapping of the potential distribution and of schistosomiasis and its three snail vector species in Brazil (Malone 2005). The latter study generated a climate suitability map of probability surfaces using Floromap software (http://www.dapa.ciat.cgiar.org/floramap/) and principal components analysis and represented in ESRI's ArcView GIS software.

Joining the consensus in Brazil on the potential value of geospatial methods, Gazzinelli and Kloos 2007 further promoted use of spatial tools in a brief review of schistosomiasis and its intermediate host snails, while Guimaraes *et al.* (2006, 2008) reported successful use of social, meteorological and RS-derived digital elevation and NDVI data in a GIS to delimit risk for schistosomiasis at the municipality level in Minas Gerais. In a subsequent study in 2010, they used regression modeling of environmental, social, biological and RS-derived variables to develop regional risk maps for both schistosomiasis and *Biomphalaria* spp., the snail intermediate host (Guimaraes *et al.* 2010) GIS was further used for the identification of priority areas for interventions in tourism areas endemic for schistosomiasis (Carvalho *et al.* 2010), while other studies on the north coast of Pernambuco State using spatial analysis at the village household scale related water contact patterns to infection risk (Paredes *et al.* 2010).

Again, the concept of geospatial health applications was catching on. A special issue of *Geospatial Health*, published in 2012, was devoted to geospatial applications for NTDs, including schistosomiasis, in South America and the Caribbean (Malone and Bergquist 2012). Two papers addressed schistosomiasis, with emphasis on the snail intermediate host in Brazil (Barboza *et al.* 2012; Scholte *et al.* 2012). Of particular interest for the Brazilian distribution of the disease is the presence of two compatible species: *Biomphalaria glabrata* and *B. straminea* and that the competitive selection makes *B. glabrata* dominate in irrigation systems, while *B. straminea* is more common in natural water sources (Barboza *et al.* 2012).

By this time there was an increasing global political commitment to reduce or eliminate the combined burden of multiple NTDs by development of cost-effective control strategies (Hotez 2008). In response, WHO created a Department for the control of the NTDs formulating a 'Global Plan to Combat Neglected Tropical Diseases 2008–2015' (Ault *et al.* 2012). In the Americas, this political commitment was expressed through the Council Resolution CD49.R19 in 2009 of the Pan American Health Organization (PAHO) entitled 'Elimination of Neglected Diseases and other Infections Related to Poverty' (Ault *et al.* 2012), while a strategic plan was adopted for the control and elimination of NTDs in Latin America and the Caribbean (PAHO, 2010). The Inter-American Development Bank (IDB) joined in a partnership with PAHO and the 'Global Network for NTD' (http://www.globalnetwork.org/) to provide technical and financial support for the control and elimination of the NTDs in Latin America (Colston 2012). Targeted at national governments and NGOs, the partnership promoted the integration of NTD prevention, control and elimination activities into existing systems and efforts, such as primary care, clean water, improved sanitation and housing initiatives. Given the importance of socioeconomic and environmental risk factors in the persistence of transmission of NTDs, geospatial mapping and modeling was recognized early on to be useful for the prediction of the distribution and prevalence of these diseases and to identify areas where hotspots or disease overlap occur. Significantly, all NTD projects carried out by the IDB included a component on mapping and modeling technical assistance to strengthen national and local health surveillance.

25.6 THE FUTURE FOR GEOSPATIAL SCIENCES

In the published literature on health applications of the geospatial sciences since the 1980s, malaria and schistosomiasis were the focus of the first and second most numerous articles, respectively. It is likely schistosomiasis will continue to be a barometer of progress in geospatial health sciences when the attention of researchers is drawn to emerging issues such as the efficiencies of integrated control.

Various major scientific groups are interested in public health applications of the geospatial sciences. For example, NASA's Earth Science area has moved towards a strategic goal that includes the study of climate and environmental change and the potential impact on public health issues, such as infectious diseases, emergency preparedness and response (https://www.earthobserva-tions.org/documents/cop/he_henv/20110329_Geneva/16_Haynes.pdf). NASA's Public Health Program has been chronicled by Luvall (2013). The Group on Earth Observations (GEO) (https://www.earthobservations.org/geoss.php), an international agency with support from over 100 governmental departments, NGOs and scientific organizations has an interest in health and so has the International Society for Photogrammetry and Remote Sensing (ISPRS). Yet another group, the International Medical Geology Association (MEDGEO) (http://rock.geosociety.org) has similar goals in its stated mission in medical geology, the science dealing with the relationships between geological factors and health.

In light of this convergence of interests, it is of great potential value that the International Society of Geospatial Health – GnosisGIS (http://www.GnosisGIS.org) has made efforts to cross-fertilize and reinforce linkages of diverse interest groups working on health applications. One idea envisioning this unifying future was proposed in a recent special issue of *Geospatial Health*, the official journal of GnosisGIS, e.g. the creation of a 'Global Health Observation System of Systems' (Malone *et al.* 2014). This idea was put forward to enable a 'virtual globe' vehicle for scientists and organizations working on seemingly very different health issues to work in an over arching collaboration to further public health using a standardized, interoperable, open source global resource data portal within the concept of a 'GeoHealth network' (Box 25.1).

Box 25.1

Mandate of the Group on Earth Observations Health Network (GeoHealth), an open data portal on health mapping based on remote sensing.

GeoHealth will collaborate on activities relating to the GEO societal benefit area (SBA) on health of the Global Earth Observation System of Systems (GEOSS).

GeoHealth will encourage governmental departments, inter-governmental bodies and NGOs to organize and improve mapping and predictive modeling of the distribution of infectious, vector-borne diseases globally and make these data and forecasts accessible to health experts, decision makers and managers.

GeoHealth aims to be a community of practice and a task in the GEO workplan. Partnership should be guided by a Steering Committee comprising the key stakeholders, initially the ISPRS VIII/2 Working Group on Health and GnosisGIS.

GeoHealth draws on GEO's work on data-sharing principles to promote full and open exchange of data and on the GEOSS common infrastructure to enable interoperability through adoption of consistent standards.

The virtual globe concept is not new, but the project is now coming into its own. Many and various efforts in this direction have been made over the last 10–15 years. However, the field did not take off until a user-friendly applications started to appear (Boulos 2005). Intuitive technologies, such as Google Earth enable scientists around the world to share data and in a readily understandable fashion without the need for much technical assistance. In 2008, Elvidge and Tuttle felt that three-dimensional software modeling of the Earth leading to virtual globes would revolutionize Earth observation, data access and integration, while Stensgaard *et al.* (2009) and Yang *et al.* (2012) used Google Earth for the management and control of vector-borne diseases, e.g., schistosomiasis. The authors of this paper felt that the use of this approach led to a better understanding of the epidemiology of infectious diseases and the multidimensional environment in which they occur.

Is it possible to develop a dynamical 3-dimensional (3-D) or even 4-D (adding the temporal dimension) models of disease, e.g., in the form of a bi-weekly global schistosomiasis report? We are on the cusp of succeeding in this endeavor, facilitated by global Earth-observation satellite systems, big data, climatology advances and new satellite-borne sensors, as is evident from the 40 years of Landsat legacy data, vividly presented as a slide show available at http://www.oosa.unvienna.org/pdf/pres/copuos2012/panel-01. pdf. Perhaps most crucially, however, is the sub-meter resolution environmental data now available from Worldview-2 and Worldview-3 satellites (De Roeck *et al.* 2014). Add to this the elective, value-added potential of low-altitude sensors on drone airborne vehicles as a source of very high-resolution data collection within a user-set agenda (Capolupo *et al.* 2014). Future NASA satellite and international space station missions such as HyspIRI (http://hyspiri.jpl.nasa.gov/) will provide further enhanced capability to map vector-borne and other environmentally sensitive diseases based on global hyperspectral visible and multispectral thermal data products ($60m^2$ resolution, 5-day thermal and 19-day hyperspectral repeat intervals) that will enable structural and functional classification of ecosystems and the measurement of key environmental parameters (temperature, soil moisture). The planned ECOSTRESS instrument on the international space station (http://www.nasa.gov/jpl/nasas-ecostress) will monitor plant health using surface temperature measurements (and derived evapotranspiration values) with minimum daily repeat coverage at varying times during the day and at least one measurement at night at approximately 38×57 m spatial resolution (Luvall 2013). Will these advances finally allow seamless mapping and modelling of diseases, not only at continental scales ($1 km^2$) and local community-agricultural field scales (30 m²), but for the first time also at the habitat-household scale (< 1 m²) within individual communities? If so, this is a technology whose time has come.

25.7 LITERATURE CITED

Abdel-Rahman MS, El-Bahy MM, Malone JB, Thompson NM, El-Bahy NM. 2001. Geographical information systems as a tool for control program management for schistosomiasis in Egypt. Acta Tropica 79: 49–57.

Ault SK, Nicholls RS, Saboya MI. 2012. The pan American health organization's role and perspectives on the mapping and modeling of the neglected tropical diseases in Latin America and the Caribbean: an overview. Geospatial Health 6(3): S7–S9.

Barboza DM, Zhang C, Santos NC, Silva MM, Rollemberg CV, de Amorim FJ, Ueta MT, de Melo CM, de Almeida JA, Jeraldo Vde L, de Jesus AR. 2012. *Biomphalaria* species distribution and its effect on human *Schistosoma mansoni* infection in an irrigated area used for rice cultivation in northeast Brazil. Geospatial Health 6(3): S103–109.

Bavia ME, Malone JB, Hale L, Dantes A, Marroni L, Reis R. 2001. Use of thermal and vegetation index data from earth observing satellites to evaluate the risk of schistosomiasis in Bahia, Brazil. Acta Tropica 79: 79–85.

Bavia ME, Hale LT, Malone JB, Braud DH, Shane S. 1999. Geographic information systems and the environmental risk of schistosomiasis in Bahia, Brazil. American Journal of Tropical Medicine and Hygiene 60(4): 566–572.

Bergquist NR. 2001. Vector-borne parasitic diseases: new trends in data collection and risk assessment. Acta Tropica 27; 79(1): 13–20.

Bergquist R, Yang GJ, Knopp S, Utzinger J, Tanner M. 2015. Surveillance and response: Tools and approaches for the elimination stage of neglected tropical diseases. Acta Tropica 141(Pt B): 229–234. doi: 10.1016/j.actatropica.2014.09.017.

Bergquist R. 2013. Closing in on 'perhaps the most dreadful of the remaining plagues': an independent view of the multidisciplinary alliance to optimize schistosomiasis control in Africa. Acta Tropica 128(2): 179–81. doi: 10.1016/j.actatropica.2013.08.016.

Bergquist R, Tanner M. 2010. Controlling schistosomiasis in Southeast Asia: a tale of two countries. Advances in Parasitology 72: 109–44. doi: 10.1016/S0065-308X(10)72005-4.

Bhunia GS, Kesari S, Chatterjee N, Pal DK, Kumar V, Ranjan A, Das P. 2011. Incidence of visceral leishmaniasis in the Vaishali district of Bihar, India: spatial patterns and role of inland water bodies. Geospatial Health 5(2): 205–215.

Boulos MN. 2005. Web GIS in practice III: creating a simple interactive map of England's strategic health authorities using Google maps API, Google earth KML and MSN virtual earth map control. International Journal of Health Geography 21(4): 22. doi: 10.1186/1476-072X-4-22.

Brooker S, Michael E. 2000. The potential of geographical information systems and remote sensing in the epidemiology and control of human helminth infections. Advances in Parasitology 47: 245–288.

Capolupo A, Pindozzi S, Okello C, Boccia L. 2014. Indirect field technology for detecting areas object of illegal spills harmful to human health: application of drones, photogrammetry and hydrological models. Geospatial Health 8(3): S699–707. doi: 10.4081/gh.2014.298.

Carvalho OS, Scholte RG, Guimarães RJ, Freitas CC, Drummond SC, Amaral RS, Dutra LV, Oliveira G, Massara CL, Enk MJ. 2010. The Estrada real project and endemic diseases: the case of schistosomiasis, geoprocessing and tourism. Memorias Institute Oswaldo Cruz 105(4): 532–536.

Ceccato P, Connor SJ, Jeanne I, Thomson MC. 2005. Application of geographical information systems and remote sensing technologies for assessing and monitoring malaria risk. Parassitologia 47(1): 81–96.

Chammartin F, Hürlimann E, Raso G, N'Goran EK, Utzinger J, Vounatsou P. 2013. Statistical methodological issues in mapping historical schistosomiasis survey data. Acta Tropica 128: 345–352.

Chen MG, Feng Z. 1999. Schistosomiasis control in China. Parasitology International 48(1): 11–19.

Clements AC, Deville MA, Ndayishimiye O, Brooker S, Fenwick A. 2010. Spatial codistribution of neglected tropical diseases in the East African Great Lakes region: revisiting the justification for integrated control. Tropical Medicine and International Health 15: 198–207. doi: 10.1111/j.1365-3156.2009.02440.x.

Clements AC, Firth S, Dembelé R, Garba A, Touré S, Sacko M, Landouré A, Bosqué-Oliva E, Barnett AG, Brooker S, Fenwick A. 2009. Use of Bayesian geostatistical prediction to estimate local variations in *Schistosoma haematobium* infection in Western Africa. Bulletin of the World Health Organization 87: 921–929.

Clements AC, Moyeed R, Brooker S. 2006. Bayesian geostatistical prediction of the intensity of infection with *Schistosoma mansoni* in East Africa. Parasitology 133: 711–719.

Cline BL. 2006. Invited editorial for the inaugural issue of geospatial health. Geospatial Health 1(1): 3–5.

Cline BL. 1970. New eyes for epidemiologists: aerial photography and other remote sensing techniques. American Journal of Epidemiology 92(2): 85–89.

Colston J. 2012. The neglected tropical diseases (NTD) initative for Latin America and the Caribbean of the inter-American development bank and the role of geospatial analysis in health programmes. Geospatial Health 6(3): S11–S14.

Confalonieri UEC, Marinho DP. 2005. Remote sensing and the control of infectious diseases: proceedings from an inter-American workshop. Rio de Janeiro: ENSP/FIOCRUZ, 104 pp.

Cross ER, Perrine R, Sheffield C. 1984. Predicting areas endemic for schistosomiasis using weather variables and a Landsat database. Military Medicine 149: 542–544.

De Roeck E, Van Coillie F, De Wulf R, Soenen K, Charlier J, Vercruysse J, Hantson W, Ducheyne E, Hendrickx G. 2014. Fine-scale mapping of vector habitats using very high resolution satellite imagery: a liver fluke case-study. Geospatial Health 8(3): S671–683. doi: 10.4081/gh.2014.296.

Elliott MR. 2011. A simple method to generate equal-sized homogenous strata or clusters for population-based sampling. Annals of Epidemiology 21(4): 290–6. doi: 10.1016/j.annepidem.2010.11.016.

Elliott P, Wartenberg D. 2004. Spatial epidemiology: current approaches and future challenges. Environmental Health Perspectives 112(9): 998–1006.

Elvidge CD, Tuttle BT. 2008. How virtual globes are revolutionizing Earth observation data access and integration. The International Archives of the Photogrammetry, Remote Sensing and Spatial Information Sciences B6a: 137–139.

Estrada-Peña A, Vatansever Z, Gargili A, Buzgan T. 2007. An early warning system for Crimean-Congo haemorrhagic fever seasonality in Turkey based on remote sensing technology. Geospatial Health 2(1): 127–135.

Fenwick A, Webster JP, Bosque-Oliva E, Blair L, Fleming FM, Zhang Y, Garba A, Stothard JR, Gabrielli AF, Clements ACA, Katbatereine NB, Toure S, Dembele R, Nyandindi U, Mwansa J, Koukounari A. 2009. The schistosomiasis research initiative (SCI): rationale, development and implementation from 2002–2008. Parasitology 136(13): 1707–1718.

Gazzinelli A, Kloos H. 2007. The use of spatial tools in the study of *Schistosoma mansoni* and its intermediate host snails in Brazil: a brief review. Geospatial Health 2(1): 51–58.

Guimarães RJ, Freitas CC, Dutra LV, Moura AC, Amaral RS, Drummond SC, Scholte RG, Carvalho OS. 2008. Schistosomiasis risk estimation in Minas Gerais State, Brazil, using environmental data and GIS techniques. Acta Tropica 108(2–3): 234–41. doi: 10.1016/j.actatropica.2008.07.001. Epub 2008 Jul 22.

Guimarães RJ, Freitas CC, Dutra LV, Scholte RG, Martins-Bedé FT, Fonseca FR, Amaral RS, Drummond SC, Felgueiras CA, Oliveira GC, Carvalho OS. 2010. A geoprocessing approach for studying and controlling schistosomiasis in the state of Minas Gerais, Brazil. Memorias Institute Oswaldo Cruz 105(4): 524–31.

Guo JG, Vounatsou P, Cao CL, Utzinger J, Zhu HQ, Anderegg D, Zhu R, He ZY, Li D, Hu F. 2005. A geographic information and remote sensing based model for prediction of *Oncomelania hupensis* habitats in the Poyang Lake area, China. Acta Tropica 96: 213–222. doi: 10.1016/j.actatropica.2005.07.029.

Guo J, Lin D, Hu G, Ning A, Liu H, Lu S, Li D, Wu X, Wang R, Chen M, Zheng J, Tanner M. 2002. Rapid identification of *Oncomelania hupensis* snail habitat in the Poyang Lake region by geographic information system (GIS) and remote sensing (RS). Zhonghua Liu Xing Bing Xue Za Zhi 23(2): 99–101 (In Chinese).

Hagenlocher M, Castro MC. 2015. Mapping malaria risk and vulnerability in the United Republic of Tanzania: a spatial explicit model. Population Health Metrics 3; 13(1): 2. doi: 10.1186/s12963-015-0036-2. eCollection 2015.

Hanson C, Weaver A, Zoerhoff KL, Kabore A, Linehan M, Doherty A, Engels D, Savioli L, Ottesen EA. 2012. Integrated implementation of programs targeting neglected tropical diseases through preventive chemotherapy: identifying best practices to roll out programs at national scale. American Journal of Tropical Medicine and Hygiene 86(3): 508–13. doi: 10.4269/ajtmh.2012.11-1589.

Harrus S, Baneth G. 2005. Drivers for the emergence and re-emergence of vector-borne protozoal and bacterial diseases. International Journal of Parasitology 35(11–12): 1309–1318.

Healthy Futures. 2014. Latest posting: Project News, Issue 7, December, 2014. (http://www. healthyfutures.eu/index.php?option=com_k2&view=item&layout=item&id=171&Itemid=275&lang=en).

Herbreteau V, Salem G, Souris M, Hugot JP, Gonzalez JP. 2007. Thirty years of use and improvement of remote sensing, applied to epidemiology: from early promises to lasting frustration. Health Place 13(2): 400–403.

Hotez PJ, Bottazzi ME, Franco-Paredes C, Ault SK, Periago MR. 2008. The neglected tropical diseases of Latin America and the Caribbean: a review of disease burden and distribution and a roadmap for control and elimination. PLoS Neglected Tropical Diseases 2: e300.

Hugh-Jones ME. 1989. Applications of remote sensing to the identification of habitats of parasites and disease vectors. Parasitology Today 5: 244–251.

IPCC. 2013. Climate change 2013. The physical science basis. Working group 1 contribution to the fifth assessment report of the intergovernmental panel on climate change. Summary for policymakers. Eds Stocker TF, Qin D, Plattner GK, Tirngor MMB, Allen S, Boschung J, Nauels A, Xia Y, Bex V, Midgely PM. 27pp. (https://www.ipcc.ch/pdf/assessment-report/ar5/wg1/WGIAR5_SPM_brochure_en.pdf).

Jeanne I. 2000. Malaria and schistosomiasis: 2 examples using systems of geographical information and teledetection in Madagascar. Bulletin de la Societe de Pathologie Exotique 93(3): 208–214.

Johansen MV, Sithithaworn P, Bergquist R, Utzinger J. 2010. Towards improved diagnosis of zoonotic trematode infections in Southeast Asia. Advances in Parasitology 73: 171–195. doi: 10.1016/S0065-308X(10)73007-4.

Kabatereine NB, Brooker S, Koukounari A, Kazibwe F, Tukahebwa EM, Fleming FM, Zhang Y, Webster JP, Stothard JR, Fenwick A. 2007. Impact of a national helminth control programme on infection and morbidity in Ugandan school children. Bulletin of the World Health Organization 85: 19–99.

Katz N. 1998. Schistosomiasis control in Brazil. Memorias Institute Oswaldo Cruz, Rio de Janeiro 93, Suppl 1: 33–35.

King CH, Olbrych SK, Soon M, Singer ME, Carter J, Colley DG. 2011. Utility of repeated praziquantel dosing in the treatment of schistosomiasis in high-risk communities in Africa: a systematic review. PLoS Neglected Tropical Diseases 5(9): e1321. doi: 10.1371/journal.pntd.0001321.

Kistemann T, Dangendorf F, Schweikart J. 2002. New perspectives on the use of geographical information systems (GIS) in environmental health sciences. International Journal of Hygiene and Environmental Health 205(3): 169–181.

Kristensen TK, Malone JB, McCarroll JC. 2001. Use of satellite remote sensing and geographic information systems to model the distribution and abundance of snail intermediate hosts in Africa: a preliminary model for *Biomphalaria pfeifferi* in Ethiopia. Acta Tropica 2001; 79: 73–78.

Leonardo L, Rivera P, Saniel O, Antonio Solon J, Chigusa Y, Villacorte E, Christoper Chua J, Moendeg K, Manalo D, Crisostomo B, Sunico L, Boldero N, Payne L, Hernandez L, Velayudhan R. 2015. New endemic foci of schistosomiasis infections in the Philippines. Acta Tropica 141(Pt B): 354–360. doi: 10.1016/j.actatropica.2013.03.015.

Leonardo LR, Crisostomo BA, Solon JA, Rivera PT, Marcelo AB, Villasper JM. 2007. Geographical information systems in health research and services delivery in the Philippines. Geospatial Health 1(2): 147–155.

Leonardo LR, Rivera PT, Crisostomo BA, Sarol JN, Bantayan NC, Tiu WU, Bergquist NR. 2005. A study of the environmental determinants of malaria and schistosomiasis in the Philippines using remote sensing and geographic information systems. Parassitologia 47(1): 105–114.

Lessard P, L'Eplattenier R, Norval RA, Kundert K, Dolan TT, Croze H, Walker JB, Irvin AD, Perry BD. 1990. Geographical information systems for studying the epidemiology of cattle diseases caused by *Theileria parva*. Veterinary Record 17; 126(11): 255–262.

Li S, Zheng H, Gao Q, Zhang L-J, Zhu R, Xu J, Guo J, Xiao N, Zhou XN. 2013. Endemic status of schistosomiasis in People's Republic of China in 2012. Chinese Journal of Schistosomiasis Control 25(6): 557–563 (in Chinese).

Linehan M, Hanson C, Weaver A, Baker M, Kabore A, Zoerhoff KL, Sankara D, Torres S, Ottesen EA. 2011. Integrated implementation of programs targeting neglected tropical diseases through preventive chemotherapy: proving the feasibility at national scale. American Journal of Tropical Medicine and Hygiene 84(1): 5–14. doi: 10.4269/ajtmh.2011.10-0411.

Linthicum KJ, Bailey CL, Davies FG, Tucker CJ. 1987. Detection of Rift Valley fever viral activity in Kenya by satellite remote sensing imagery. Science 27; 235(4796): 1656–1659.

Luvall JC. 2013. NASA's public health program: how we can use NASA satellite data to study global public health issues. Nasa technical reports server. paperback, June 27, 2013. http://www.amazon.com/NASAs-Public-Health-Program-Satellite/dp/1289044635.

Malone JB, Tourre YM, Faruque F, Luvall JC, Bergquist NR. 2014. Towards establishment of GeoHealth, an open-data portal for health mapping and modelling based on earth observations by remote sensing. Geospatial Health 8(3): S599–S602.

Malone JB, Bergquist NR. 2012. Mapping and modelling neglected tropical diseases and poverty in Latin America and the Caribbean. Geospatial Health 6(3): S1–5.

Malone JB, Yang GJ, Leonardo L, Zhou XN. 2010. Implementing a geospatial health data infrastructure for control of Asian schistosomiasis in the People's Republic of China and the Philippines. Advances in Parasitology 73: 71–100. doi: 10.1016/S0065-308X(10)73004-9.

Malone JB, Bavia ME, Amaral R, McNally K, Nieto P, Carniero D, Neves A. 2005. Climate-based prediction of the potential distribution of schistosomiasis in Brazil. 2005. *In*: Confaloneri UEC, Marinho DP, editors. Remote Sensing and the Control of Infectious Diseases: Proceedings from an Inter-American Workshop. Rio de Janeiro: ENSP/FIOCRUZ, 104 pp.

Malone JB. 2005. Biology-based mapping of vector-borne parasites by geographic information systems and remote sensing. Parasitologia 47: 27–50.

Malone JB, Yilma JM, McCarroll JC, Erko B, Mukaratirwa S, Zhou XY. 2001. Satellite climatology and environmental risk of *Schistosoma mansoni* in Ethiopia and East Africa. Acta Tropica 79: 59–72.

Malone JB, Abdel Rahman MS, ElBahy MM, Huh OK, Shafik M, Bavia ME. 1997. Geographic information systems and the distribution of *Schistosoma mansoni* in the Nile Delta. Parasitology Today 13(3): 112–119.

Malone JB. 1995. The geographic understanding of snail-borne disease in endemic areas using satellite surveillance Memorias Institute Oswaldo Cruz, Rio de Janeiro 90(2): 205–209.

Malone JB, Huh OK, Fehler DP, Wilson PA, Wilensky DE, Holmes RA, Elmagdoub AI. 1994. Temperature data from satellite imagery and the distribution of schistosomiasis in Egypt. American Journal of Tropical Medicine and Hygiene 50(6): 714–722.

Malone JB, Fehler DP, Loyacano AF, Zukowski SH. 1992. Use of Landsat MSS imagery and soil type in a geographic information system to assess site specific risk of fascioliasis on Red River basin farms in Louisiana. Annals of the New York Academy of Science 653: 389–397.

Mayangadze T, Chimbari MJ, Gebreslasie M, Mukaratirwa S. 2015. Application of geospatial technology in schistosomiasis modeling in Africa: a review. Geospatial Health 10(2): 326. doi: 10.4081/gh.2015.32.

McCreesh N, Nikulin G, Booth M. 2015. Predicting the effects of climate change on *Schistosoma mansoni* transmission in eastern Africa. Parasites and Vectors 6; 8: 4. doi: 10.1186/s13071-014-0617-0.

National Academy of Sciences. 2002. Down to earth. Geographic information for sustainable development in Africa. Committee on the geographic foundation for Agenda 21 committee on geography. Mapping Science Committee. National Research Council: 172 p.

Olveda DU, Li Y, Olveda RM, Lam AK, McManus DP, Chau TN, Harn DA, Williams GM, Gray DJ, Ross AG. 2014. Bilharzia in the Philippines: past, present and future. International Journal of Infectious Diseases 18: 52–56. doi: 10.1016/j.ijid.2013.09.011.

PAHO. 2010. Control and elimination of five neglected diseases in Latin America and the Caribbean, 2010-2015. Analysis of progress, priorities and lines of action for lymphatic filariasis, schistosomiasis, onchocerciasis, trachoma and soil transmitted helminthiasis PAHO, 93 pp. Available at http://new.paho.org/q/dmdocuments/2010/Prioritization_NTD_PAHO_Dec_17_2010_En.pdf.

Paolino L, Sebillo M, Cringoli G. 2005. Geographical information systems and on-line GI Services for health data sharing and management. Parassitologia 47(1): 171–175.

Paredes H, Santos RS, Resendes APC, Albuquerque J, Bocanegra S, Gomes EC, Baarbosa CS. 2010. Spatial pattern, water use and risk levels associated with the transmission of schistosomiasis on the north coast of Pernambuco, Brazil. Cad. Saude Publica, Rio de Janeiro 26(5): 1013–1023.

Pavlovskii EN. 1945. The ecological parasitology. Journal of General Biology 6: 65–92.

Rinaldi L, Hendrickx G, Cringoli G, Biggeri A, Ducheyne E, Catelan D, Morgan E, Williams D, Charlier J, Von Samson-Himmelstjerna G, Vercruysse J. 2015. Mapping and modelling helminth infections in ruminants in Europe: experience from GLOWORM. Geospatial Health 19: 257–259. doi: 10.4081/gh.2015.347.

Rinaldi L, Cringoli G. 2014. Exploring the interface between diagnostics and maps of neglected parasitic diseases. Parasitology 141: 1803–1810. doi: 10.1017/S0031182013002230.

Rinaldi L, Musella V, Biggeri A, Cringoli G. 2006. New insights into the application of geographical information systems and remote sensing in veterinary parasitology. Geospatial Health 1: 33–47.

Rogers DJ, Randolph SE. 1993. Detection of tsetse and ticks in Africa: past, present and future. Parasitology Today 9: 266–271.

Rollinson D, Knopp S, Levitz S, Stothard JR, Tchuente LA, Garba A, Mohammed KA, Schur N, Person B, Colley DG, Utzinger J. 2013. Time to set the agenda for schistosomiasis elimination. Acta Tropica 128: 423–440.

Sarvel AK, Oliveira AA, Silva AR, Lima AC, Katz N. 2011. Evaluation of a 25-year-program for the control of schistosomiasis mansoni in an endemic area in Brazil. PLoS Neglected Tropical Diseases 5(3): e990. doi: 10.1371/journal.pntd.0000990.

Scholte RGC, Carvlho OS, Malone JB, Utzinger J, Vounatsou P. 2012. Spatial distribution of *Biomphalaria* spp., the intermediate host snails of *Schistosoma mansoni*, in Brazil. Geospatial Health 6(3): S95–S101.

Schur N, Hürlimann E, Stensgaard AS, Chimfwembe K, Mushinge G, Simoonga C, Kabatereineh NB, Kristensene TK, Utzinger J, Vounatsou P. 2013. Spatially explicit *Schistosoma* infection risk in eastern Africa using Bayesian geostatistical modelling. Acta Tropica 128: 365–377.

Standley CJ, Vounatsou P, Gosoniu L, Jørgensen A, Adriko M, Lwambo NJ, Lange CN, Kabatereine NB, Stothard JR. 2012. The distribution of *Biomphalaria* (Gastropoda: Planorbidae) in Lake Victoria with ecological and spatial predictions using Bayesian modelling. Hydrobiologia 683: 249–264.

Stensgaard AS, Saarnak CFL, Utzinger J, Vounatsou P, Simoonga C, Mushinge G, Rahbek C, Møhlenberg F, Kristensen TK. 2009. Virtual globes and geospatial health: the potential of new tools in the management and control of vector-borne diseases. Geospatial Health 3: 127–141.

Stensgaard A, Utzinger J, Vounatsou P, Hürlimann E, Schur N, Saarnak CF, Simoonga C, Mubita P, Kabatereine NB, Tchuenté LAT, Rahbek C, Kristensen TK. 2013. Large-scale determinants of intestinal schistosomiasis and intermediate host snail distribution across Africa: does climate matter? Acta Tropica 128: 378–90.

Tambo E, Ai L, Zhou X, Chen JH, Hu W, Bergquist R, Guo JG, Utzinger J, Tanner M, Zhou XN. 2014. Surveillance-response systems: the key to elimination of tropical diseases. Infectious Diseases of Poverty 27(3): 17. doi: 10.1186/2049-9957-3-17. eCollection 2014.

Taylor D, Hagenlocher M, Jones AE, Kienberger S, Leedale J, Morse AP. 2016. Environmental change and Rift Valley fever in eastern Africa: projecting beyond HEALTHY FUTURES. Geospatial Health (in press).

Tompkins AM, Ermert V. 2013. A regional-scale, high resolution dynamical malaria model that accounts for population density, climate and surface hydrology. Malaria Journal 12: 65. doi: 10.1186/1475-2875-12-65.

Utzinger J, Brattig NW, Kristensen TK. 2013. Schistosomiasis research in Africa: how the CONTRAST alliance made it happen. Acta Tropica 128: 182–195.

Utzinger J, Zhou XN, Chen MG, Bergquist R. 2005. Conquering schistosomiasis in China: the long march. Acta Tropica 96(2–3): 69–96.

Wang Y, Zhuang D. 2015. A rapid monitoring and evaluation method of schistosomiasis based on spatial information technology. International Journal of Environmental Research and Public Health 12(12): 15843–15859. doi: 10.3390/ijerph121215025.

Wang ZL, Zhu R, Zhang ZJ, Yao BD, Zhang LJ, Gao J, Jiang QW. 2012. Identification of snail habitats in the Poyang Lake region, based on the application of indices on joint normalized difference vegetation and water. Zhonghua Liu Xing Bing Xue Za Zhi. 2012 Aug; 33(8): 823–7 (in Chinese).

Ward MP, Carpenter TE. 2000. Techniques for analysis of disease clustering in space and in time in veterinary epidemiology. Preventive Veterinary Medicine 45: 257–284.

Weng QH. 2002. Land use change analysis in the Zhujiang Delta of China using satellite remote sensing, GIS and stochastic modelling. Journal of Environmental Management 64: 273–284. doi: 10.1006/jema.2001.0509.

WHO. 2012. Accelerating work to overcome the global impact of neglected tropical diseases: a roadmap for implementation. http://www.who.int/neglected_diseases/NTD_RoadMap_2012_Fullversion.pdf.

Wu WP, Davis G, Liu HY, Seto E, Lu SB, Zhang J, Hua ZH, Guo JG, Lin DD, Chen HG, Peng G, Feng Z. 2002. Application of remote sensing for surveillance of snail habitats in Poyang Lake, China. Zhongguo Ji Sheng Chong Xue Yu Ji Sheng Chong Bing Za Zhi 20(4): 205–208 (In Chinese).

Xiaonong Z, Minggang C, McManus D, Bergquist R. 2001. Schistosomiasis control in the 21st century. 2001. Proceedings of the international symposium on schistosomiasis, Shanghai, July 4-6, 2001. Acta Tropica 79(1): 97–106.

Yang K, Sun LP, Huang YX, Yang GJ, Wu F, Hang DR, Li W, Zhang JF, Liang YS, Zhou XN. 2012. A real-time platform for monitoring schistosomiasis transmission supported by Google earth and a web-based geographical information system. Geospatial Health 6(2): 195–203.

Yang GJ, Vounatsou P, Tanner M, Zhou XN, Utzinger J. 2006. Remote sensing for predicting potential habitats of *Oncomelania hupensis* in Hongze, Baima and Gaoyou lakes in Jiangsu province, China. Geospatial Health 1(1): 85–92.

Yang GJ, Vounatsou P, Zhou XN, Tanner M, Utzinger J. 2005. A Bayesian-based approach for spatio-temporal modeling of county level prevalence of *Schistosoma japonicum* infection in Jiangsu province, China. International Journal of Parasitology 35: 155–162. doi: 10.1016/j.ijpara.2004.11.002.

Zhou XN, Yang GJ, Yang K, Wang XH, Hong QB, Sun LP, Malone JB, Kristensen TK, Bergquist NR, Utzinger J. 2008. Potential impact of climate change on schistosomiasis transmission in China. American Journal of Tropical Medicine and Hygiene 78(2): 188–94.

Zhou XN, Malone JB, Kristensen TK, Bergquist NR. 2001. Application of geographic information systems and remote sensing to schistosomiasis control in China. Acta Tropica 79(1): 97–106.

Zhou X, Kristensen TK, Hong Q, Sun L. 1999. Analysis for spatial distribution of Oncomelania snail in mainland China by geographic information system (GIS) database. Zhonghua Yu Fang Yi Xue Za Zhi 33 (6): 343–5 (in Chinese).

Zukowski SH, Hill JM, Jones FW, Malone JB. 1992. Development and validation of a soil-based geographic information system model of habitat of *Fossaria bulimoides*, a snail intermediate host of *Fasciola hepatica*. Preventive Veterinary Medicine 11: 221–227.

Zukowski SH, Jones FW, Malone JB. 1993. Fascioliasis in cattle in Louisiana II. Development of a system to use soil maps in a geographic information system to estimate disease risk on Louisiana coastal marsh rangeland. Veterinary Parasitology 47: 51–65.

CHAPTER

Future Directions: The Road to Elimination

Allen GP Ross[1],*, Remigio M Olveda[2] and Li Yuesheng[3,4]

26.1 INTRODUCTION

Preventive chemotherapy with 40 mg/kg of praziquantel has been endorsed and advocated by the World Health Organization (WHO) for the global control of schistosomiasis globally yet the drug does not prevent reinfection (WHO 2006). We discuss the many issues related to this control strategy which is now implemented in many schistosomiasis-endemic countries.

26.2 MASS DRUG ADMINISTRATION

Considerable optimism surrounds Mass Drug Administration (MDA) for the control of schistosomiasis globally, for which praziquantel (PZQ) has served as the cornerstone since its inception in 1979. Numerous studies have claimed that preventive chemotherapy (i.e. 40 mg/kg of praziquantel), given annually or biannually, can significantly reduce the prevalence and intensity of infection and control morbidity in the long term (Molyneux 2004; Molyneux and Hotez 2005; Brady *et al.* 2006; Engels *et al.* 2006; Fenwick 2006; Kabatereine *et al.* 2007; Hotez 2008; Hotez *et al.* 2009; Bockarie *et al.* 2013). In the last decade, close to one billion US dollars has been raised for MDA campaigns against neglected tropical diseases (NTDs) largely from international donors (e.g., Merck KgaA, World Bank, Unites States Agency for International Development (USAID), British Department for International Development (DIFD), Geneva Global and the Bill & Melinda Gates Foundation) and delivered vertically to local endemic communities through national health care services largely using unpaid volunteers (Brady *et al.* 2006; Merck 2013). In total, 28 million children have been treated for schistosomiasis to date through the Merck Praziquantel Donation Program (Merc 2013). Since 2007, Merck has been providing WHO annually (free of charge) with up to 25 million tablets (Hotez *et al.* 2009). In the medium term, the company will increase that number tenfold to 250 million tablets per year (Rollinson *et al.* 2013). The donation commitment for the entire continent of Africa amounts to € 17.2 million Euros (Rollinson *et al.* 2013).

---

[1] Tropical Medicine and Global Health, Griffith University, Logan Campus, University Drive, Meadowbrook QLD 4131, Australia.

[2] Research Institute for Tropical Medicine, Corporate Ave, Muntinlupa, 1781 Metro Manila, The Philippines.

[3] Hunan Institute of Parasitic Diseases, Yueyang, Hunan Province, World Health Organisation Collaborating Centre for Research and Control on Schistosomiasis in Lake Region, Yueyang, Hunan, 414000, People's Republic of China.

[4] QIMR Berghofer Medical Research Institute, 300 Herston Rd, Brisbane QLD 4006, Australia.

* Corresponding author

Despite the commitment to MDA programs, it is becoming increasingly clear that the sustainable control of schistosomiasis will require an integrated, inter-sectorial approach that goes beyond deworming (Parker and Tim 2011). While the global use of PZQ is being scaled up, there are also growing concerns about inadequate drug coverage, low cure rates, poor drug compliance, lack of baseline information prior to the commencement of MDA programs, their inadequate monitoring and evaluation once commenced and the potential for the development of drug resistance (Parker and Tim 2011). PZQ is effective against all the human schistosome species and is currently the drug of choice to treat infection (Liu *et al.* 2011; Olliaro *et al.* 2011). However, parasitological cure depends on the treatment dose. A recent systematic review and meta-analysis of 52 clinical trials showed that, compared with placebo, a dosage of 30–60 mg/kg PZQ produced a cure rate of around 76% (range from 67–83%) for human schistosomiasis (Liu *et al.* 2011). No significant differences in cure rates were found among subjects infected with *S. haematobium, S. japonicum* or *S. mansoni*. The cure rate of the drug at a 40 mg/kg dosage (which is the current dose recommended by the WHO) was 52% (range from 49–55%) compared with 91% (range from 88%–92%) when dosages were increased to 60, 80, 100 mg/kg, divided into two or more doses (Liu *et al.* 2011. In the early 1980s and again in 2011, WHO, in an attempt to optimize PZQ use for the treatment of schistosomiasis, launched a series of multi-country trials, comparing the efficacy and safety of 40 mg/kg and 60 mg/kg in schistosome-infected patients in Asia, Africa and the Americas (Olliaro *et al.* 2011). In these clinical trials the 40 mg/kg dose was found to be effective (92% cure rate) and better tolerated than the higher 60 mg/kg dose (Olliaro *et al.* 2011). However, given the small sample size in each country (approximately 200 per site) the generalizability of these findings are questionable.

Since 2000, close to one billion US dollars has been raised for MDA programs of NTDs from various international donors in order to 'rescue' the bottom billion from neglected tropical diseases (NTDs) (Hotez *et al.* 2009). The drugs provided are reported to cure infection, control morbidity, prevent reinfection and possibly lead to elimination (WHO 2006; Bockarie *et al.* 2013). Moreover, an 'audacious goal' has been put forward to try to eliminate several of these diseases in Africa by 2020 (Hotez 2011). Unfortunately, what has been advocated by the proponents of MDA for schistosomiasis will simply not work for a number of fundamental reasons. Firstly, 'preventive chemotherapy' does not prevent reinfection or alter the life cycle of the disease-causing parasites (Gray *et al.* 2010). The patient is typically weak, malnourished, immunosuppressed and living in a rudimentary home with no running water or sanitation thus they are highly susceptible to immediate reinfection following treatment. Moreover, they must contact schistosome cercariae contaminated waters in order to bathe, wash clothes and fish. Secondly, the drugs are not getting to the people who need them the most. Many countries are reporting less than < 50% drug coverage among the highest risk groups (Parker and Tim 2011). The medication is typically delivered by unpaid volunteers to remote villages on foot or by boat through fragmented or non-existing health services. Once there, the same unpaid staff are expected to educate and treat the local inhabitants and to return months or years later to conduct monitoring and evaluation. Thirdly, drug compliance is very low with many countries reporting < 50% compliance (Muhumuza *et al.* 2013). Patients are often asked to take empirical treatment for a disease they know very little about even if they are not infected or show no symptoms. It has been reported that up to 80% of those who ultimately take the drug suffer from transient side-effects such as dizziness, syncope, vomiting and diarrhea (Muhumuza *et al.* 2013). Once observed and reported by others in the community, compliance quickly drops. Fourthly, the efficacy of the drug itself is in question. As indicated earlier, the WHO claims that PZQ has an efficacy of 92% when given at a dosage of 40 mg/kg, yet the recent meta-analysis also referred to earlier compared data from 52 clinical trials and found that the cure rate was only 52% for the same dosage (Liu *et al.* 2011; Olliaro *et al.* 2011). In sum, 'preventive chemotherapy' is not the silver bullet for schistosomiasis control programs that it is purported to be.

26.3 INTEGRATED CONTROL AND THE ROAD TO ELIMINATION

Given the current fragile state of the global economy it is uncertain how long donor funding for MDA programs will last. Many of these current programs are ill conceived, are run in parallel with current National Health Services and have no exit strategy (Gray *et al.* 2010; Parker and Tim 2011). Once donor funds dry up, how will the control work be maintained? China is often illustrated as the global success story for schistosomiasis control and elimination but no other country in the world will be able replicate what they have done. In the last 50 years alone China has spent close to one billion US dollars on controlling the disease through integrated control measures (Ross *et al.* 2013). China currently employs 20,000 full-time staff to work on the disease with an annual operating budget of $120 million US dollars (Ross *et al.* 2013). What other country in the world has the political commitment, human resources and the financial capital to duplicate this? So how should donor funding be spent if the MDA strategy is inherently flawed? The funds should be spent on poverty reduction and focus on breaking the parasite life cycle, thereby preventing transmission. Proven integrated control strategies that could assist in breaking the life cycle of the disease are: targeted mollusciciding of the snail intermediate host, sanitation, vaccination, irrigation and health education (Fig. 26.1) (Gray *et al.* 2010). A recently trialled health education video against soil transmitted helminths showed a significant reduction in transmission as result of increase awareness and improved hand washing after defecation (Bieri *et al.* 2013). In the communities with the highest prevalence (>50%), wells should be dug, toilets installed and agricultural practises enhanced in order to improve the nutrition and immune status of the people. A healthy well-nourished population once treated (i.e. with 60 mg/kg of PZQ) has a better chance to ward off infection and to plan its own future (Ross *et al.* 2014). Ultimately these communities will be responsible for their own future wellbeing. Although this inter-sectoral approach will require considerable effort and careful monitoring it is ultimately the only strategy that will lead to sustainability (Gray *et al.* 2010; Parker and Tim 2011).

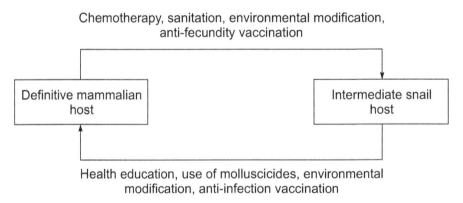

FIG. 26.1 Interventions that can target transmission pathways in the schistosome life cycle.

26.4 CHAPTER SUMMARY

Mass drug administration with 40 mg/kg of praziquantel is presently advocated for the control and elimination of schistosomiasis globally. However, targeting only one aspect of the parasite's life cycle will not lead to elimination. Sustainable control of schistosomiasis will require an integrated, inter-sectorial approach that goes beyond deworming. The road to elimination will take time and considerable human and financial resources but ultimately will be achieved.

26.5 LITERATURE CITED

Bieri FA, Gray DJ, Williams GM, Raso G, Li YS, Yuan L, He Y, Li RS, Guo FY, Li SM, McManus DP. 2013. Health-education package to prevent worm infections in Chinese school children. New England Journal Medicine 368(17): 1603–12. doi: 10.1056/NEJMoa1204885.

Bockarie MJ, Rebollo M, Molyneuz DH. 2013. Preventive chemotherapy as a strategy of elimination of neglected tropical parasitic diseases: endgame challenges. Philosophical Transactions of the Royal Society B: Biological Sciences 368: 1–11.

Brady MA, Hooper PJ, Ottesen EA. 2006. Projected benefits from integrating NTD programs in sub-Saharan Africa. Trends in Parasitology 22(7): 285–291.

Engels D, Savioli L. 2006. Reconsidering the underestimated burden caused by neglected tropical diseases. Trends in Parasitology 22(8): 363–366.

Fenwick A, Molyneux D, Nantulya V. 2005. Achieving the millennium development goals. Lancet 365: 1029–1030.

Fenwick A. 2006. New initiatives against Africa's worms. Transactions of the Royal Society of Tropical Medicine and Hygiene 100: 200–207.

Gray DJ, McManus DP, Li YS, Williams GM, Bergquist R, Ross AG. 2010. Schistosomiasis elimination: Lessons from the past guide at future. Lancet Infectious Diseases 10: 733–6.

Hotez PJ. 2008. *Forgotten People, Forgotten Diseases: The Neglected Tropical Diseases and their Impact on Global Health and Development.* Washington: ASM Press. 254 p.

Hotez PJ, Fenwick A, Savioli L, Molyneux DH. 2009. Rescuing the bottom billion through the control of neglected tropical diseases. Lancet 373: 1570–75.

Hotez P. 2011. Enlarging the "audacious goal": Elimination of the world's high prevalence neglected tropical diseases. Vaccine 292: D104–110.

Kabatereine NB, Brooker S, Koukounari A, Kazibwe F, Tukahebwa EM, Fleming FM, Zhang Y, Webster JP, Stothard JR, Fenwick A. 2007. Impact of a national helminth control programme on infection and morbidity in Ugandan school children. Bulletin of the World Health Organization 85: 91–99.

Liu R, Dong HF, Guo Y, Zhao QP, Jiang MS. 2011. Efficacy of praziquantel and artemisinin derivatives for the treatment and prevention of human schistosomiasis: a systematic review and meta-analysis. Parasites and Vectors 4: 201. doi: 10.1186/1756-3305.

Merck KGaA. News Release. 2013. Merck praziquantel donation program to treat schistosomiasis starts in Sudan. August 25, 2013: 1–4.

Molyneux DH. 2004. 'Neglected diseases' but unrecognized successes – challenges and opportunities for infectious disease control. Lancet 364: 380–383.

Molyneux DH, Hotez PJ, Fenwick A. 2005. Rapid-impact interventions: how a policy of integrated control for Africa's neglected tropical diseases could benefit the poor. PLoS Medicine 2(11): e 336.

Muhumuza S, Olsen A, Katahoire A, Nuwaha F. 2013. Uptake of preventive treatment for intestinal schistosomiasis among school children in Jinja district, Uganda: a cross sectional study. PLoS One 7; 8(5): e63438. doi: 10.1371/journal.pone.0063438.

Olliaro PL, Vaillant MT, Belizario VJ, Lwambo NJ, Ouldabdallahi M, Pieri OS, Amarillo ML, Kaatano GM, Diaw M, Domingues AC, Favre TC, Lapujade O, Alves F, Chitsulo L. 2011. A multicentre randomized controlled trial of the efficacy and safety of single-dose praziquantel at 40 mg/kg vs. 60 mg/kg for treating intestinal schistosomiasis in the Philippines, Mauritania, Tanzania and Brazil. PLoS Neglected Tropical Diseases. 5(6): e1165. doi: 10.1371/journal.pntd.0001165.

Parker M, Tim A. 2011. Does mass drug administration for the integrated treatment of neglected tropical diseases really work? Assessing evidence for the control of schistosomiasis and soil-transmitted helminths. Health Research Policy and Systems 9: 3.

Rollinson D, Knopp S, Levitz S, Stothard R, Tchenete LT, Garba A, Mohammed KA, Schur N, Person B, Colley DG, Utzinger J. 2013. Time to set the agenda for schistosomiasis elimination. Acta Tropica 128(2): 423–40.

Ross AG, Olveda RM, Acosta L, Harn DA, Chy D, Li Y, Gray DJ, Gordon CA, McManus DP, Williams GM. 2013. Road to the elimination of schistosomiasis from Asia: the journey is far from over. Microbes Infection 15(13): 858–865.

Ross AG, Olveda RM, McManus DP, Harn DA, Li YS, Gray DJ, Williams GM. 2014. Can mass drug administration lead to the sustainable control schistosomiasis in the Philippines? Journal of Infectious Diseases. 2014 Jul 28. pii: jiu 416.

World Health Organization. 2006. Preventive chemotherapy in human helminths. Coordinated use of anthelminthic drugs in control interventions: a manual for health professionals and programme managers. Geneva, Switzerland: World Health Organization.

Index

Printed and bound by CPI Group (UK) Ltd, Croydon, CR0 4YY

01/11/2024

01782604-0010